AF903787

TURBULENCE IN COASTAL AND CIVIL ENGINEERING

ADVANCED SERIES ON OCEAN ENGINEERING

Editor-in-Chief
Philip L-F Liu (*Cornell University and National University of Singapore*)

Published

Vol. 39 Liquefaction Around Marine Structures
by Mutiu Sumer

Vol. 40 Design and Construction of Berm Breakwaters
by Jentsje Van der Meer and Sigurdur Sigurdarson

Vol. 41 Dynamics of Coastal Systems (Second Edition)
by Job Dronkers

Vol. 42 Theory and Applications of Ocean Surface Waves (Third Edition)
by Chiang C. Mei, Michael Aharon Stiassnie and Dick K.-P. Yue

Vol. 43 Japan's Beach Erosion: Reality and Future Measures (Second Edition)
by Takaaki Uda

Vol. 44 Satellite SAR Detection of Sub-Mesoscale Ocean Dynamic Processes
by Quanan Zheng

Vol. 45 Ocean Surface Waves: Their Physics and Prediction (Third Edition)
by Stanisław Ryszard Massel

Vol. 46 Tsunami: To Survive from Tsunami (Second Edition)
by Tomotsuka Takayama, Kazumasa Katoh, Fumihiko Imamura, Yoshiaki Kawata, Susumu Murata and Shigeo Takahashi

Vol. 47 Coastal Engineering: Theory and Practice
by Vallam Sundar and Sannasi A. Sannasiraj

Vol. 51 Turbulence in Coastal and Civil Engineering
by B. Mutlu Sumer and David R. Fuhrman

Forthcoming

Vol. 48 Introduction to Coastal Engineering and Management (Third Edition)
by J. William Kamphuis

Vol. 49 Design of Coastal Hazard Mitigation Alternatives for Rising Seas
by David Basco

Vol. 50 Tsunami: Engineering Perspective for Mitigation, Protection and Modeling
by V. Sundar, S. A. Sannasiraj, K. Murali and V. Sriram

*For the complete list of titles in this series, go to https://www.worldscientific.com/series/asoe

Advanced Series on Ocean Engineering — Volume 51

TURBULENCE IN COASTAL AND CIVIL ENGINEERING

B. Mutlu Sumer
BM SUMER Consultancy & Research, Turkey
Previously at Technical University of Denmark, Denmark

David R. Fuhrman
Technical University of Denmark, Denmark

World Scientific

NEW JERSEY • LONDON • SINGAPORE • BEIJING • SHANGHAI • HONG KONG • TAIPEI • CHENNAI • TOKYO

Published by

World Scientific Publishing Co. Pte. Ltd.
5 Toh Tuck Link, Singapore 596224
USA office: 27 Warren Street, Suite 401-402, Hackensack, NJ 07601
UK office: 57 Shelton Street, Covent Garden, London WC2H 9HE

British Library Cataloguing-in-Publication Data
A catalogue record for this book is available from the British Library.

Advanced Series on Ocean Engineering — Vol. 51
TURBULENCE IN COASTAL AND CIVIL ENGINEERING

Copyright © 2020 by World Scientific Publishing Co. Pte. Ltd.

All rights reserved. This book, or parts thereof, may not be reproduced in any form or by any means, electronic or mechanical, including photocopying, recording or any information storage and retrieval system now known or to be invented, without written permission from the publisher.

For photocopying of material in this volume, please pay a copying fee through the Copyright Clearance Center, Inc., 222 Rosewood Drive, Danvers, MA 01923, USA. In this case permission to photocopy is not required from the publisher.

ISBN 978-981-3234-30-7 (hardcover)
ISBN 978-981-3234-31-4 (ebook for institutions)
ISBN 978-981-3234-32-1 (ebook for individuals)

For any available supplementary material, please visit
https://www.worldscientific.com/worldscibooks/10.1142/10829#t=suppl

Printed in Singapore

Preface

The 1980s and 1990s have witnessed the publication of three excellent books on coastal hydrodynamics and sediment transport, Sea Bed Mechanics (1984) by J.F.A. Sleath; Mechanics of Coastal Sediment Transport (1992) by J. Fredsøe and R. Deigaard; and Coastal Bottom Boundary Layers and Sediment Transport (1992) by P. Nielsen, all three of which discussed turbulence but with limited coverage of the subject, as the main focus of these books was not turbulence but rather sediment transport. There are two equally excellent books on turbulence in the Civil Engineering literature, both of which came out in 1993, Turbulence in Open-Channel Flows by I. Nezu and H. Nakagawa; and Turbulence Models and their Application in Hydraulics: A State of the Art Review, by W. Rodi.

Much progress has taken place since the publications of the aforementioned books, particularly in the measurements of turbulence in both waves and currents, using cutting-edge technology (such as Laser Doppler Anemometry, Particle Image Velocimetry, and hot-film technique for measurements of wall shear stress) as well as in the Computational Fluid Dynamics (CFD) simulations of turbulent flows in these flow environments, using advanced modelling of turbulence. As a result, substantial knowledge has accumulated on the subject over the past 20–30 years. The authors have been actively involved in the development of this new research, at various platforms, as leaders of and participants in many national and international research programs, and as supervisors of more than thirty PhD and Postdoctoral projects undertaken at the Technical University of Denmark (DTU).

The topic of turbulence and turbulent flows has been taught as a Masters course at DTU since the mid 1980s, first by B. Mutlu Sumer (BMS), and later by BMS and David R. Fuhrman (DRF) jointly, starting from 2006.

BMS prepared lecture notes for the course in 2001, entitled "Lecture Notes on Turbulence". The success of the latter over the years encouraged the authors to turn it into a text book. The present book is actually built on the aforementioned lecture notes. However, quite an extensive amount of new knowledge originating from the new research has been incorporated into the existing material.

The principal aim of the present book is to describe in detail turbulent flow processes in coastal and civil engineering. With a clear understanding of the fundamental processes, it is relatively easier for a practicing engineer to assess complex problems involving turbulent flows in coastal and civil engineering, in order to make engineering predictions and recommendations on how to cope with potential risks and optimize designs. The book is essentially intended for professionals and researchers in the area of Coastal and Civil Engineering, and as a text book for graduate/post graduate students. The contents of the book will, however, additionally provide sufficient background in the study of turbulent flows relevant to many other disciplines, such as Wind Engineering, Mechanical Engineering, and Environmental Engineering.

We note that, not long ago, the hydraulic engineering community has welcomed a recent book on turbulence modelling, Large Eddy Simulation in Hydraulics (2013) by W. Rodi, G. Constantinescu and T. Stoesser. We believe that the present book, and the existing body of literature including the previously mentioned book by Rodi et al., and those cited in the preceding paragraphs, form a complementary source of information on turbulence in the fields of Coastal and Civil Engineering.

An important part of BMS's research on turbulence during the past 30 years has been supported by several research programs and grants, both national and international (mainly European Union), notably, 1987–1990 "Turbulence Around Offshore Structures" of the Danish Research Council (STVF); 1989–1996 "Marine Technique I" and subsequently "Marine Technique II" of the Danish Research Council (STVF); 1989–1995 MAST projects 0035-C and MAS2 CT 92-0027 "Coastal Morphodynamics", funded by European Communities; 1997–2000 "Surf and Swash Zone Mechanics" SASME, a European Union MAS3 research program, funded by the European Union; 2005–2008 "Exploitation and Protection of Coastal Zones", EPCOAST, a Danish Research Council Research Frame Program, funded by Danish Research Council (FTP); 2008–2012 "Seabed Wind Farm Interaction", a research frame program, funded by Danish Council for Strategic Research (DSF)/Energy and Environment; 2013–2017 "ASTARTE: Assessment, STrategy And Risk reduction for Tsunamis in Europe", a research program under the European Union FP-7 European Community for Research, Technological Development and Demonstration Activities under the funding scheme of Collaborative Project; and many PhD and Postdoctoral grants, for the most part, from DTU.

BMS has interacted in his research on turbulence with his many colleagues, former students and former research associates. In this context, Prof. Jørgen Fredsøe, his colleague at DTU, has a special place in the joint effort to build content in turbulence research within topics such as wave boundary layers, implications for sediment transport, and mathematical modelling, amongst others. Prof. Rolf Deigaard, also a colleague at DTU, took part in the previously mentioned effort at DTU with invaluable contributions, until his move to Danish Hydraulic Institute (DHI) in 2006. BMS would also like to mention the names of his former students and research associates he has interacted with in his research on turbulence. These include former PhD students, in reverse chronological order: Dr. Karsten L. Jensen, Dr. Bjarne Jensen, Dr. Stefan Carstensen, Dr. Iris P. Buxbom, Dr. Carsten L. Lodahl, Mr. Michael Arnskov, Dr. Thorkild S. Laursen, and Dr. Bjørn L. Jensen; and former research associates, again, in reverse chronological order, Dr. Cuneyt Baykal, Dr. Anil Guner, Dr. Ozgur Kirca, Dr. Figen (Hatipoglu) Dixen, Dr. Lloyd H. C. Chua, Dr. Sevket Cokgor, Dr. Andrzej Kozakiewicz, Dr. Nian-Sheng Cheng, and Prof. Hitoshi Tanaka. Finally, BMS thanks his wife Tijen and sons Kaan and Eren U. for their unconditional love.

DRF's research in turbulence and related subjects has likewise been supported by several research programs and grants over the past decade. Among these are the above-mentioned projects: 2008–2012 "Seabed Wind Farm Interaction" and 2013–2017 "ASTARTE" projects, both of which involved collaboration with BMS. Additionally, DRF gratefully acknowledges the support of the following projects: 2013–2017 "Management of Sea Bed and Wind Farm Interaction", supported by the ForskEL program; as well as 2018–2021 "SWASH: Simulating WAve Surf zone Hydrodynamics and sea bed morphology", an ongoing project at the time of publication supported by Independent Research Fund Denmark (Grant no. 8022-00137B).

Similarly, DRF is also grateful for interaction and collaboration with numerous colleagues, students, and former research associates in turbulence related teaching and research at DTU. His colleague at DTU, Dr. Stefan Carstensen, deserves special acknowledgment, with whom DRF has jointly taught the Turbulence Theory / Turbulent Flows course at DTU since 2015. DRF would also like to thank his colleagues Dr. Knud Erik Meyer and Dr. Clara M. Velte for their support using the air jet facilities at the DTU Fluid Mechanics Laboratory, and especially Mr. Benny Edelsten for help collecting the experimental data set provided as part of Exercise 3 (Chapter 10). DRF likewise acknowledges many years of support and collaboration with the other coastal engineering faculty at DTU, namely Prof. Erik Damgaard Christensen and Prof. Per A. Madsen.

DRF would likewise like to mention the names of several PhD and Postdoctoral students he has had the pleasure of collaborating with in many years of turbulence related research, in reverse chronological order: Dr. Bjarke Eltard Larsen, Dr. Titi Sui, Dr. Jonatan Margalit, Dr. Jinzhao Li, Dr. Uğur Çalişkan, Dr. Cuneyt Baykal and Dr. Niels Gjøl Jacobsen. He also gratefully acknowledges two of his former MSc students: Dr. Signe Schløer and Ms. Johanna Sterner, who were the original developers of the MatRANS model, featured prominently in Chapters 5, 6, and 10. Finally, DRF thanks his wife Hülya and daughter Aylin Isabella for their continuous love and support.

The authors would like to express their sincere appreciation for the continuous support and excellent working environment provided by the Administration at DTU Mechanical Engineering. The authors would also like to express their gratitude to the staff of the Hydraulic Engineering Laboratory at DTU Mechanical Engineering, who have, over the years, skillfully provided assistance to the authors, their students and their research associates. BMS would also like to thank Mr. Hans Jørn Poulsen, who prepared a substantial portion of the figures, and also BM SUMER Consultancy & Research for providing financial assistance in: (1) having a significant portion of the figures drawn, by Ms. Selma Karatepe, and also (2) obtaining the Clark Little Photography License for the image reproduced from `http://www.clarklittlephotography.com` in Fig. 7.25.

The authors are grateful for helpful comments received from several colleagues who have proof-read and provided valuable feedback on excerpts or full chapters, including: Dr. V. S. Ozgur Kirca, Dr. Stefan Carstensen, Prof. Per A. Madsen, Dr. Bjarke Eltard Larsen and Dr. Titi Sui.

The textbook which follows contains references to several supplemental digital files, especially in Chapters 4, 5 and 10. All supplemental digital files are available online in the DTU Data Collection located at:

`https://doi.org/10.11583/DTU.c.4508648`

A corrigenda document will be added and maintained at the collection above, should any errors become known to the authors.

Contents

Preface v

List of symbols and abbreviations xxi

1. Introduction and the origin of turbulence 1

1.1 Introduction . . . 1
1.2 Flow over a flat plate . . . 3
1.3 Flow in wave boundary layers . . . 5
1.4 Flow in wave boundary layers in solitary motion . . . 6
1.5 Steady flow in pipes . . . 7
1.6 Oscillatory flow in pipes . . . 8
1.7 Flow around a circular cylinder in a steady current . . . 9
1.8 Flow induced by breaking waves . . . 11
1.9 References . . . 12

Appendix to Chapter 1: Transition to turbulence over a flat plate 15

1.A Description of transition process . . . 15
1.A.1 Problem statement . . . 15
1.A.2 Emmons' (1951) observations . . . 15
1.A.3 Detailed description of a single turbulent spot . . 17
1.A.4 Critical Reynolds number for transition . . . 23
1.A.5 Transition and turbulent spots over rough boundaries . . . 26
1.B Hydrodynamic stability theory for transition . . . 27
1.B.1 The Orr-Sommerfeld equation . . . 27
1.B.2 The eigenvalue problem . . . 31

1.B.3 Experimental verification of the hydrodynamic stability theory 34
1.C Direct numerical simulation (DNS) of transition 35
1.D References 39

2. Basic equations for turbulent flows 43

2.1 Method of averaging and Reynolds decomposition 43
2.2 Continuity equation 45
2.3 Equations of motion 45
2.4 Energy equations 47
2.4.1 Energy equation for laminar flows 47
2.4.2 Energy equation for the mean flow 48
2.4.3 Energy equation for the fluctuating flow 49
2.5 Reference 52

Appendix to Chapter 2: Derivation of the continuity and Navier-Stokes equations 53

2.A Continuity equation 53
2.B Navier-Stokes equation 55
2.C References 63

3. Steady turbulent boundary layer flows 65

3.1 Flow close to a wall 65
3.1.1 Idealized flow in the half space $y > 0$ 65
3.1.2 Mean and turbulence characteristics of flow close to a smooth wall 71
3.1.3 Mean and turbulence characteristics of flow close to a rough wall 79
3.2 Flow across the entire section 92
3.3 Turbulence modelling approach 96
3.3.1 Turbulence modelling of flow close to a wall . . . 97
3.3.2 Turbulence modelling of flow across the entire section 106
3.4 Flow resistance 107
3.4.1 Smooth circular pipe 107
3.4.2 Rough circular pipe 108
3.4.3 Transitional circular pipe 109
3.4.4 Non-circular pipe and open channel 109

3.5 Bursting process 109
3.5.1 Description of the bursting process 112
3.5.2 Flow visualization of the bursting process 114
3.5.3 Contribution of ejection and sweep events to the Reynolds stress 118
3.5.4 Bursting process over rough walls 120
3.5.5 Important characteristics of the bursting process . 123
3.6 Implications of bursting process for sediment transport . . 126
3.6.1 Particle motions near the bottom in an open channel: Mechanism of particle suspension 126
3.6.2 Lift forces on moving particles near boundaries . . 136
3.6.3 Initiation of suspension of particles from the bottom 140
3.6.4 Suspension of particles released away from the bottom 146
3.7 References 150

Appendix to Chapter 3: Dimensional analysis 155

3.A Buckingham's Pi theorem 155
3.B Example: Law of the wall 156

4. Statistical, correlation and spectral analysis 159

4.1 Statistical analysis 160
4.1.1 Probability density function 160
4.1.2 Statistical moments 161
4.1.3 Matlab example: Basic statistics and creating a p.d.f. 165
4.2 Correlation analysis 171
4.2.1 Space correlations 171
4.2.2 Time correlations 176
4.2.3 Taylor's frozen turbulence approximation 178
4.2.4 Matlab example: Time correlation and scales . . . 181
4.3 Spectral analysis 184
4.3.1 General considerations 184
4.3.2 Energy balance in wave number space 187
4.3.3 Kolmogorov's theory: Universal equilibrium range and inertial subrange 191
4.3.4 Vortex stretching 195

4.3.5 One-dimensional spectra 200
4.3.6 Aliasing in one-dimensional spectra 206
4.3.7 Matlab example: The discrete Fourier transform . 209
4.3.8 Matlab example: Creating a one-dimensional energy density spectrum 214
4.4 References 219

Appendix to Chapter 4: Proofs of some common isotropic turbulence relationships 223

4.A Proof of the isotropic relationship between the correlation functions $f(r)$ and $g(r)$, Eq. 4.24 223
4.B Proof of Eqs. 4.30 and 4.31, relating λ_f and λ_g to the variance of fluctuating velocity gradients 225
4.C Proof of isotropic relation Eq. 4.66, the governing equation for the correlation function $Q_{ij}(\mathbf{r})$ 226
4.D Proof that the integral of $T(k)$ over all wave numbers is zero, Eq. 4.77 230
4.E Proof of the isotropic relationship between the one-dimensional spectra $F(k)$ and $G(k)$, Eq. 4.118 230
4.F Proof of the isotropic relationship between $E(k)$ and $F(k)$, Eq. 4.119 231
4.G Proof of the isotropic dissipation rate ε, Eq. 4.83 232
4.H References 235

5. Wave boundary layers 237

5.1 Laminar wave boundary layers 239
5.1.1 Velocity distribution across the boundary layer depth 239
5.1.2 Bed shear stress 244
5.2 Laminar-to-turbulent transition 245
5.2.1 General description: Turbulent spots 245
5.2.2 Critical Reynolds number for transition 254
5.2.3 Imprint of turbulent spot in bed shear stress signal 255
5.2.4 Laminar-to-turbulent transition in terms of friction coefficient 257
5.2.5 Turbulent spots over a rough bed 259
5.2.6 Transition in terms of flow resistance and phase . 261
5.3 Turbulent wave boundary layers: Smooth bed 264

5.3.1 Oscillating tunnel in Jensen et al. (1989) study . . 265
5.3.2 Ensemble averaging 267
5.3.3 General description of turbulent wave boundary layer 268
5.3.4 Mean flow velocity 269
5.3.5 Turbulence quantities 273
5.3.6 Effect of Reynolds number 275
5.3.7 Wall-normal pressure gradient 276
5.3.8 Boundary-layer thickness 281
5.4 Turbulent wave boundary layers: Rough bed 282
5.4.1 Experimental setup in Jensen et al. (1989) rough-bed experiments 283
5.4.2 Comparison of rough-bed and smooth-bed results 284
5.4.3 Boundary-layer thickness 287
5.5 Flow resistance in wave boundary layers 288
5.5.1 Wave friction coefficient for smooth-bed wave boundary layer 289
5.5.2 Wave friction coefficient for rough-bed wave boundary layer 290
5.5.3 Wave friction coefficient for transitional-bed wave boundary layer 295
5.6 Combined wave and current boundary layers 296
5.6.1 Experimental setup in Lodahl et al. (1998) study 298
5.6.2 Laminar-to-turbulent transition in combined oscillatory flow and current 300
5.6.3 Mean wall shear stress 307
5.6.4 Friction coefficient for combined flow 312
5.6.5 Wave-induced apparent roughness 314
5.6.6 Oscillating component of the wall shear stress in combined flow 315
5.6.7 Flow resistance in combined wave and current boundary layers 316
5.7 Boundary layers in wave flumes 319
5.7.1 Waves alone 320
5.7.2 Combined waves and current 322
5.8 Bursting process in wave boundary layers 325
5.9 Miscellaneous wave boundary layer examples 328
5.10 Solitary wave boundary layers 331
5.10.1 Laminar solitary wave boundary layers 331

5.10.2 Transitional and turbulent solitary wave boundary layers . . . 334
5.10.3 Experimental facility in Sumer et al. (2010) study 335
5.10.4 Flow regimes in solitary wave boundary layers . . 336
5.10.5 Flow resistance in solitary wave boundary layers . 347
5.10.6 Velocity profiles in solitary wave boundary layers 351
5.10.7 Remarks on practical applications . . . 354
5.11 Tsunami wave boundary layers . . . 356
5.11.1 Idealized tsunami signals . . . 357
5.11.2 Signals based on tsunami observations . . . 362
5.11.3 Time variation of the boundary layer thickness and bed shear stress . . . 367
5.11.4 Comparison with tsunami field observations in the presence of energetic wind waves . . . 370
5.12 Mathematical modelling of turbulent wave boundary layers . . . 373
5.12.1 One-dimensional vertical (1DV) RANS equations 373
5.12.2 Treatment of convective acceleration terms 375
5.12.3 Pressure gradient . . . 377
5.12.4 Two-equation k-ω turbulence closure . . . 378
5.12.5 Boundary conditions . . . 379
5.12.6 Numerical solution . . . 379
5.12.7 Standard Wilcox (2006) k-ω closure model results 380
5.12.8 Low Reynolds number Wilcox (2006) k-ω closure model results . . . 383
5.13 References . . . 388

Appendix to Chapter 5: Some essentials of linear potential flow wave theory 401

5.A Governing equations . . . 401
5.B Surface elevation and velocity kinematics beneath a progressive regular wave . . . 403
5.C Linear dispersion relation . . . 404
5.D Wave drift . . . 404
5.E Energy flux . . . 406
5.F Radiation stress . . . 406
5.G Shallow water approximations . . . 407
5.H Deep water approximations . . . 408
5.I Shoaling . . . 409

5.J Link to wave boundary layer quantities 410
5.K References 410

6. Streaming in wave boundary layers 413

6.1 Streaming beneath sinusoidal progressive waves 414
6.1.1 Laminar theory 414
6.1.2 Turbulent streaming beneath progressive waves . 420
6.2 Streaming in converging-diverging oscillatory flow 424
6.2.1 Laminar theory 424
6.2.2 Turbulent streaming in oscillatory converging-diverging flow 428
6.3 Streaming due to changing bottom roughness 434
6.3.1 Anisotropic turbulence model of Fuhrman et al. (2011) 435
6.3.2 Normally-directed oscillatory flow 438
6.3.3 Obliquely-directed oscillatory flow 443
6.4 Streaming beneath non-sinusoidal waves 445
6.4.1 Streaming due to free-stream velocity skewness . . 446
6.4.2 Streaming due to free-stream acceleration skewness 454
6.5 Examples of streaming beneath real waves 460
6.6 Importance of streaming and wave shape in coastal sediment transport 463
6.7 References 465

Appendix to Chapter 6: The vorticity equation 473

6.A Derivation of the vorticity equation 473
6.B Vorticity generation due to anisotropic normal Reynolds stresses 474

7. Flow and turbulence in breaking waves 477

7.1 Wave breaking and breaker types 479
7.2 Flow induced by spilling waves 481
7.2.1 The surface roller model 482
7.2.2 A simple model of undertow 485
7.2.3 Comparison with measured undertow profile of Ting and Kirby (1994) 493
7.3 Flow induced by plunging waves 495

7.3.1 Experimental setup in Sumer et al. (2013) study . 498
7.3.2 Description of plunging breaking waves in Sumer et al. (2013) study 500
7.3.3 Bed shear stress in Sumer et al. (2013) study . . . 508
7.3.4 Mechanism of sediment suspension by plunging waves 518
7.4 Flow induced by surging waves 524
7.4.1 Experiments of Jensen et al. (2014) 524
7.4.2 Description of plunging and surging breaking waves in Jensen et al. (2014) study 526
7.5 Flow induced by plunging solitary waves 530
7.5.1 Experimental setup in Sumer et al. (2011) study . 532
7.5.2 Description of plunging solitary wave in Sumer et al. (2011) study 534
7.5.3 Bed shear stress in Sumer et al. (2011) study . . . 539
7.5.4 Sediment-bed experiments of Sumer et al. (2011) study 547
7.6 Numerical simulation of breaking waves 548
7.6.1 Background 548
7.6.2 Stability analysis of two-equation closure models . 550
7.6.3 Model description for breaking wave simulations . 558
7.6.4 Simulation of spilling breaking waves of Ting and Kirby (1994) 559
7.6.5 Simulation of plunging breaking waves of Ting and Kirby (1994) 567
7.6.6 Summary and concluding remarks 573
7.7 References 573

Appendix to Chapter 7: The depth-integrated momentum equation 587

8. Diffusion and dispersion 591

8.1 One-particle analysis 592
8.1.1 Diffusion for small times 595
8.1.2 Diffusion for large times 596
8.2 Conservation of mass: Eulerian analysis 602
8.3 Longitudinal dispersion 604
8.3.1 Mechanism of longitudinal dispersion 604
8.3.2 Application of one-particle analysis 606

8.4 Longitudinal dispersion in an open channel 609
8.4.1 Formulation 610
8.4.2 Zeroth moment of concentration 612
8.4.3 Mean particle velocity 613
8.4.4 Longitudinal dispersion coefficient 615
8.5 Longitudinal dispersion in rivers 623
8.6 Longitudinal dispersion in an oscillating tunnel 625
8.7 Dispersion in the surf zone 636
8.8 References 637

Appendix to Chapter 8: Calculation of the settling velocity 641

8.A General formulation 641
8.B Small Reynolds number: Stokes law 642
8.C Larger Reynolds numbers 642
8.D References 643

9. Mathematical modelling of turbulence 645

9.1 The closure problem 646
9.2 Types of turbulence models 647
9.3 Mixing length model of Prandtl (1925) 648
9.3.1 The eddy viscosity 648
9.3.2 Prandtl's mixing length hypothesis 650
9.3.3 Prandtl's second hypothesis 650
9.3.4 The mixing length and solution for flow above a smooth wall 651
9.3.5 The mixing length and solution for flow above transitional and rough walls 654
9.3.6 Turbulence quantities 655
9.4 The k-ω model of Wilcox (2006) 655
9.4.1 Model equations 656
9.4.2 Wall boundary conditions 662
9.4.3 Numerical computation 668
9.4.4 An application example 670
9.5 Large eddy simulation (LES) 675
9.5.1 Model equations 675
9.5.2 An application example 677
9.6 Scaling of computational costs in the modelling of turbulent flows 683

9.6.1 Direct numerical simulation (DNS) 683
9.6.2 Large eddy simulation (LES) 684
9.6.3 Turbulence energy equation models 685
9.7 References 687

Appendix to Chapter 9: The exact k equation 693
9.A Derivation of the exact k equation 693
9.B Reference 694

10. Hands-on exercises 695
10.1 Introduction 695
10.2 Data sets 696
10.3 MatRANS k-ω turbulence model in Matlab 696
10.4 Exercise 1: Analysis of a turbulent boundary layer in an open channel 698
10.5 Exercise 2: A simple numerical model of dispersion in a turbulent boundary layer flow 701
10.6 Exercise 3: Statistical, correlation, and spectral analysis of turbulent air jet flow 706
10.7 Exercise 4: Turbulence modelling of oscillatory wave boundary layer flows 708
10.8 References 714

Author index 715

Subject index 725

List of symbols and abbreviations

The main symbols used in the book are listed below. In some cases, the same symbol is used for more than one quantity. This is to maintain generally-accepted conventions in different fields. In most cases, however, their use is restricted to a single chapter. When relevant, this is marked in the following list by chapter numbers in parentheses. Abbreviations used in the book are also listed at the end.

Latin symbols

a	amplitude of oscillatory motion in free stream in waves, or half of stroke of water particle displacement in free stream in solitary waves
A	inverse of von Karman constant, $A = 1/\kappa$
c	wave celerity
	concentration (Ch. 8)
c_g	group velocity
$< c >$	cross-sectional average concentration
C_D	drag coefficient
c_f	total (dimensionless) skin friction for flow over a flat plate
C_L	lift coefficient
C_M	inertia coefficient
C_p	pth moment of concentration
D	pipe diameter of flow in a pipe
	cylinder diameter for flow around a cylinder (Ch. 1)
	turbulent diffusion coefficient (Ch. 8)
D_x, D_y, D_z	turbulent diffusion coefficient in x, y, z directions
D_u	Velocity defect function

D_1	longitudinal dispersion coefficient
d	particle/grain size
$E(k)$	energy spectrum (3D)
E	energy density
E_f	energy flux
F	lift force on a particle
$F(k)$, $G(k)$	energy spectra (1D)
Fr	Froude number
$f(r)$, $g(r)$	correlation coefficients
f_w, f_w^*	wave friction coefficient
g	acceleration due to gravity
g_x, g_y, g_z	volume force in $x-$, $y-$ and $z-$directions, respectively
h	water depth in the sea, flow depth in open channel
H	wave height
H_0	deep-water wave height
K	kinetic energy per unit volume of fluid, turbulent energy
k	turbulent kinetic energy per unit mass (Ch. 5, 6, 9)
	roughness height
	wave number of turbulence in spectral representation (Ch. 4)
	wave number (Ch. 7)
k_s	Nikuradse's equivalent sand roughness
L	wave length of surface waves
L_0	deep-water wave length
ℓ	lenth scale (turbulence modelling)
ℓ_m	mixing length
$O(\cdot)$	order of magnitude
p	pressure
p'	fluctuating pressure, or
	pressure associated with disturbance in stability analysis
P	pressure of the mean flow in stability analysis
$p(\cdot)$	probability density function of the argument
r	separation distance between two measurement points (Ch. 4)
r_h	hydraulic radius
R	radius of pipe in pipe flow, or
	runup (Ch. 7)
Re	Reynolds number
Re_c	current Reynolds number

Re_τ	friction Reynolds number
Re_w	wave boundary layer Reynolds number
Re_x	Reynolds number for flow over a flat plate, based on x, distance from the leading edge
Re_δ	Reynolds number for flow over a flat plate, based on δ, the boundary layer thickness
Re_θ	Reynolds number based on momentum thickness
R_E	Eulerian time correlation function
R_L	Lagrangian time correlation function
S_{xx}	radiation stress
t	time
T	wave period, or period of oscillatory flow
T_E	Eulerian macro time scale
T_L	Lagrangian macro time scale
u, v, w	velocity components in the x, y, z directions
u', v', w'	fluctuating velocities in the x, y, z directions
u_i	velocity in the x_i direction
u'_i	fluctuating velocity in the x_i direction
$\mathbf{u}$	velocity vector
U_0	free-stream velocity, or the velocity at free surface in open channel, or that at the centreline of pipe
U_∞	approach velocity
U_c	current velocity
U_f	friction velocity
U_m	cross-sectional or depth averaged velocity
U_{0m}	maximum value of undisturbed orbital velocity, or maximum value of free-stream velocity in oscillatory flow
V	cross-sectional or depth averaged velocity, or Lagrangian velocity (Ch. 8)
w_s	fall velocity of sediment particle
x	particle position (Ch. 8)
x, y, z	Cartesian coordinates
x_i	coordinate axes in a Cartesian system
y	particle position relative to the mean particle position (Ch. 8)
y_0	roughness length

Greek symbols

α	wave number in hydrodynamic stability analysis
β	beach slope, or fall-velocity (Rouse) parameter, $\beta = w_s/(\kappa U_f)$ (Ch. 8)
γ	specific weight of water/fluid
Γ	flux due to turbulent diffusion
γ_p	specific weight of particle/sediment particle
δ	boundary layer thickness
$\delta(\cdot)$	Dirac delta function
δ_{ij}	Kronecker delta
δ_v	viscous sublayer thickness
δ_1	Stokes length
ε	dissipation rate of turbulent kinetic energy, or turbulent diffusion coefficient (Ch. 8)
η	surface elevation
η_K	Kolmogorov (smallest) length scale of turbulence (Ch. 4)
θ	boundary layer momentum thickness, or Shields parameter (Ch. 3), or phase function, e.g. $\theta = \omega t - kx$ (Ch. 6)
θ_s	critical Shields parameter for initiation of suspension
κ	von Karman constant
λ	spacing between low speed wall streaks, or wave length of disturbance in hydrodynamic stability analysis (Ch. 1), or proportionality factor in stress – rate-of-strain relation (Ch. 2), or mean rate at which sediment settle out of suspension (Ch. 8)
λ_f, λ_g	micro length scales of turbulence
λ_1, λ_2	stress limiter coefficients (turbulence modelling)
Λ_f, Λ_g	macro length scales of turbulence
μ	dynamic viscosity, or non-dimensional mean velocity (Ch. 8)
μ_s	non-dimensional mean sediment/particle velocity
μ_t	dynamic turbulence (eddy) viscosity
ν	kinematic viscosity
ν_t	kinematic turbulence (eddy) viscosity
ξ, η, ς	components of half of vorticity vector
ξ	normalized vertical coordinate $(= y/\delta_1)$ (Ch. 6)
ρ	density of water/fluid

ρ_p	density of particle/sediment particle
σ_{ij}	elements of stress tensor
σ_{xx}, σ_{yy}, σ_{zz}	diagonal elements of stress tensor (normal stresses)
τ	shear stress, or separation time between two measurements (Ch. 4)
τ_K	Kolmogorov (smallest) time scale of turbulence (Ch. 4)
$\overline{\tau}$	period-averaged mean shear stress
τ_0	wall/bed shear stress
τ_c	mean wall shear stress corresponding to the current component of combined wave and current
τ_m	period-averaged mean bed shear stress in combined wave and current (Section 5.6.7)
τ_{max}	maximum bed shear stress in combined wave and current (Section 5.6.7)
τ_w	maximum value of bed shear stress corresponding to the wave component of combined wave and current (Section 5.6.7)
τ_E	Eulerian micro time scale
τ_{0m}	maximum value of wall/bed shear stress in wave boundary layer
τ_{xy}, τ_{xz}, τ_{yz}	off-diagonal components of stress tensor (shear stresses)
υ_K	Kolmogorov velocity scale ($= \eta_K/\tau_K$)
ϕ	phase lead of the wall/bed shear stress over the free stream velocity in wave boundary layers
Φ	energy dissipation per unit volume of fluid
ψ	stream function
ω	angular frequency of waves, $\omega = 2\pi/T$, or specific dissipation rate (primarily Ch. 9, and turbulence modelling)
$\boldsymbol{\omega}$	vorticity vector
ω_1, ω_2, ω_3	components of vorticity vector
ωt	phase in wave boundary layers

Miscellaneous

∇	gradient operator, $(\frac{\partial}{\partial x}, \frac{\partial}{\partial y}, \frac{\partial}{\partial z})$
∇^2	Laplacian operator, $\frac{\partial^2}{\partial x^2} + \frac{\partial^2}{\partial y^2} + \frac{\partial^2}{\partial z^2}$
$\bar{\cdot}$ (overbar)	time average, or ensemble average
$\langle\cdot\rangle$	period average (Ch. 6 and 7), or cross-sectional average (Ch. 8)
Superscript +	for example, $y^+ = \frac{yU_f}{\nu}$

Abbreviations

CFD	computational fluid dynamics
CTA	constant temperature anenometry
DES	detached eddy simulation
DNS	direct numerical simulation
DTU	Technical University of Denmark
FFT	fast Fourier transform
LDA	laser Doppler anemometry
LES	large eddy simulation
MAC	marker-and-cell (numerical method)
MWL	mean water level
N.-S.	Navier-Stokes
ODE	ordinary differential equation
PDE	partial differential equation
p.d.f.	probability density function
PIV	particle image velocimetry
RANS	Reynolds-averaged Navier-Stokes
r.m.s.	root mean square
RWA	reverse wrap-around (storage order from FFT, Ch. 4)
SPH	smooth particle hydrodynamics
s.r.f.	stationary random function
SWL	still water level
VOF	volume of fluid (numerical method)

Chapter 1

Introduction and the origin of turbulence

1.1 Introduction

Turbulence could possibly be best described by reference to Reynolds' (1883) experiment in a tube to investigate, in his words, "... the circumstances which determine whether the motion of water shall be direct or sinuous ... in parallel channels". By "direct", Reynolds (1883) means laminar, and by "sinuous" he means turbulent.

Reynolds (1883) describes his experimental setup as follows: "The water was drawn through the tubes out of a large glass tank, in which the tubes were immersed, arrangements being made so that a streak or streaks of highly colored water entered the tubes with the clear water." As for the results of his experiments, he writes: "The general results were as follows:

(1) When the velocities were sufficiently low, the streak of color extended in a beautiful straight line through the tube" (see Fig. 1.1a, reproduced from Reynolds' (1883) original publication, along with Fig. 1.1b).

(2) If the water in the tank had not quite settled to rest, at sufficiently low velocities, the streak would shift about the tube, but there was no appearance of sinuosity.

(3) As the velocity was increased by small stages, at some point in the tube, always at a considerable distance from the trumpet or intake, the color band would all at once mix up with the surrounding water, and fill the rest of the tube with a mass of colored water ...", see Fig. 1.1b.

As Reynolds' (1883) experiment reveals, flows of fluid (water, air, etc.) occur in two distinctly different regimes: one where the flow is regular and smooth, called the laminar regime (Fig. 1.1a), and the second the opposite, called the turbulent regime (Fig. 1.1b). In the laminar regime, fluid particles follow regular, smooth paths (Fig. 1.2a), while in the turbulent regime,

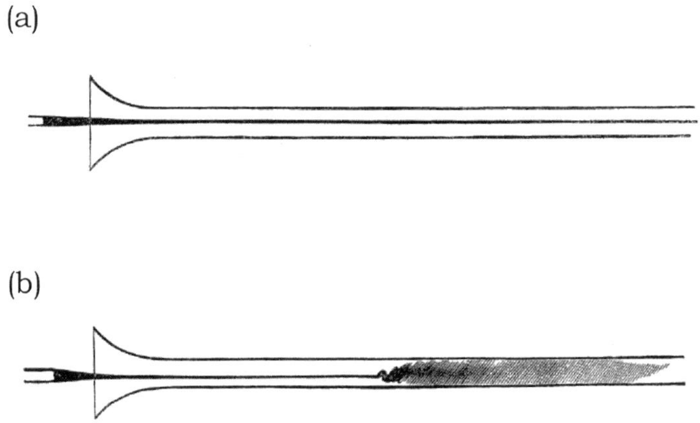

Fig. 1.1 Reynolds' (1883) tube experiment. Illustrations (reproduced from the original publication of Reynolds, 1883) describing the (a) laminar and (b) turbulent regimes of the flow.

they follow irregular, random paths (Fig. 1.2b).

This book is concerned with turbulent flows, especially those in coastal and civil engineering. The following is a partial list of turbulent flows encountered in these fields: Flow around marine (coastal, offshore) and hydraulic structures, flow around buildings, high rises, towers and bridges, flow in tubes and ducts, flow in open channels, canals, waterways and rivers, flow over the sea bed, flow in wave boundary layers in the ocean and in coastal areas, flow induced by breaking waves, flow (including the boundary layers) induced by tsunami waves, etc. While this book is directed towards coastal and civil engineers, the contents will likewise provide sufficient background in the study of turbulent flows to also be relevant to many other disciplines, such as those related to wind, mechanical, environmental, and chemical engineering.

One intriguing question in the general field of turbulence concerns the origin of turbulence. Much research has been done both experimentally and theoretically to address this question ever since the publication of Reynolds' (1883) historical experiment on transition. The queries in this context are: (1) What governs the laminar-to-turbulence transition? What are the governing parameters that control the transition? (2) What is the way in which this transition occurs? Does the transition occur gradually, or does it occur suddenly? (3) What is the mechanism behind the transition?

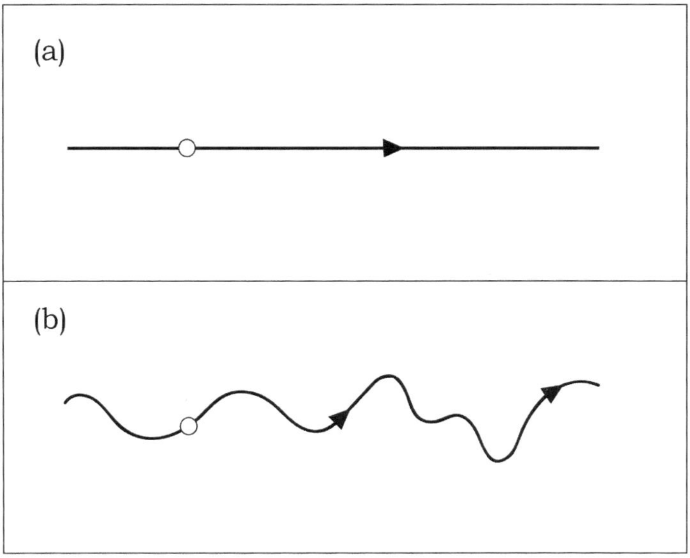

Fig. 1.2 Paths of fluid particles in a pipe flow. (a) Laminar regime. (b) Turbulent regime.

The remainder of this section will address these queries with reference to the following examples: Flow over a flat plate; Flow in wave boundary layers; Flow in wave boundary layers in solitary motion; Flow in pipes; Flow in oscillatory motion in pipes; Flow around a circular cylinder; and Flow induced by breaking waves.

Each case will now be considered individually in what follows. Along the way, references will be made to many forthcoming figures or sections, so as to provide the reader a first overview of, and introduction to, some of the contents contained within.

1.2 Flow over a flat plate

The zero-pressure-gradient boundary-layer flow over a flat plate (Figs. 1.3 and 1.4) is an idealized case of many flows encountered in practice, such as flow in the atmospheric boundary layer, flow over the sea bottom, and several flows in the aeronautical and maritime industry.

Laminar-to-turbulent transition *over a flat plate* is discussed in great

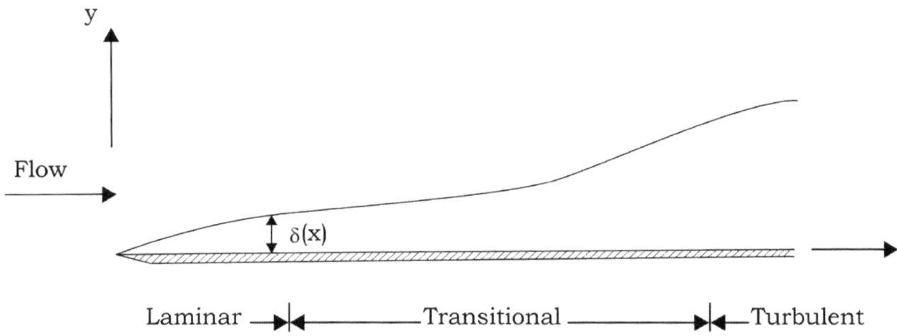

Fig. 1.3 Definition sketch. Flow over a flat plate at zero incidence.

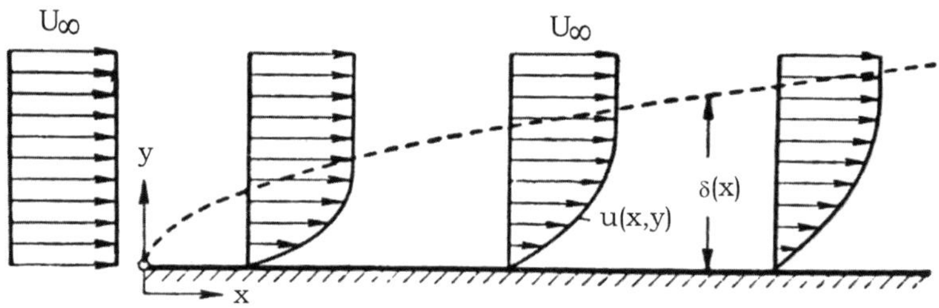

Fig. 1.4 Boundary layer on a flat plate. Close-up immediately downstream of the leading edge. Development of the boundary-layer velocity profile.

detail in the Appendix at the end of this chapter. The way in which the laminar-to-turbulent transition occurs in the other cases (which are to be discussed in the following paragraphs) is, for the most part, similar to this canonical flat-plate case. Therefore, it is hoped that including a detailed discussion of the process of transition in this special case will help the reader get a good feel of the transition process in the general study of turbulence.

As discussed in Appendix 1.A, the laminar-to-turbulent transition over a flat plate with a smooth surface is governed by the Reynolds number, $Re_x = \frac{xU_\infty}{\nu}$, in which x is the distance from the leading edge of the plate, U_∞ the approach velocity of the flow, and ν the kinematic fluid viscosity. Experiments show that the transition first occurs when $Re_{x\ cr} = 3 \times 10^5 - 5 \times 10^5$, with the emergence of tiny turbulent patches, called turbulent spots (Figs. 1.7–1.10). These individual spots grow in size and in population as they move downstream until they merge, and the whole boundary layer becomes completely turbulent, as sketched in Fig. 1.13.

The transition can be explained in terms of the theory of hydrodynamic stability according to which the transition occurs simply because the laminar boundary-layer flow over the flat plate becomes unstable against small disturbances for $Re_x > Re_{x\ cr}$, as discussed in great detail in Appendix 1.B. For such large Reynolds numbers, any small disturbance grows (theoretically unlimited), eventually leading to the inception of turbulence.

Appendix 1.A.5 also discusses the rough wall case, and its implications for the turbulent transition.

The reader is referred to the chapter Appendix for details of the transition process, including how the hydrodynamic stability theory handles the transition process (Appendix 1.B), and also some recent work on the direct numerical simulation (DNS) of the transition process (Appendix 1.C).

1.3 Flow in wave boundary layers

In shallow waters, the orbital motion of water particles under a progressive wave degenerates into a bottom-parallel, straight line, oscillatory motion at the seabed. For each half cycle of this oscillatory motion, a "new" time-dependent boundary layer develops over the seabed. This cyclic boundary layer is called the wave boundary layer, Fig. 5.1.

As in the previous example, the laminar-to-turbulent transition in wave boundary layers is also governed by the Reynolds number, the wave-boundary-layer Reynolds number in the present case, $Re = \frac{aU_m}{\nu}$, in which a is the amplitude of the orbital motion of water particles near the bed (formally in the free stream region i.e. just outside the boundary layer), and U_{0m} the maximum value of the horizontal component of the free stream orbital velocity. Observations show that the wave boundary layer remains laminar for Reynolds numbers smaller than $Re_{cr} = 1.5 \times 10^5$. The subject is discussed in great detail in Section 5.2 in Chapter 5.

Fig. 5.7 shows a time sequence of flow pictures near the bed in a horizontal plane at a distance from the bed of $y/\delta_1 = 0.24$ for the Reynolds number at $Re = 4.5 \times 10^5$ in which δ_1 is a scale of the thickness of the boundary layer, called the Stokes length (Eq. 5.13). The sequence illustrates how the transition to turbulence develops at various stages of the flow half cycle. It is seen that turbulence first emerges in the form of tiny turbulent patches, turbulent spots, just as in the flat-plate case in the previous example, the tiny yellow areas in Fig. 5.7. Fig. 5.7 indicates that the aforementioned turbulent spots grow in size, and increase in population, as the flow develops (Fig. 5.7b), and merge (Fig. 5.7c), which eventually form a fully developed turbulent boundary layer for further phases during the half cycle of the wave boundary layer. As seen clearly, what happens in space (in x) in the previous flat-plate example happens in phase (ωt) in the present wave boundary-layer case. Here, ω is the angular frequency of the wave, and t the time.

It is particularly interesting to note that the transition is also a function of the phase in the wave boundary layers. Even for very large Reynolds numbers, the flow may remain laminar for quite some time during the boundary layer development in the wave half cycle. For instance, the flow is still laminar for a time interval until $\omega t = 20^0 - 25^0$ even for $Re = 10^6$, a Reynolds number order of magnitude larger than Re_{cr}, Fig. 5.17.

The reader can consult Chapter 5, Section 5.2 for a very detailed analysis and discussions of laminar-to-turbulent transition in wave boundary layers, including topics such as the imprint of turbulent spots in the bed shear stress signal, the transition in terms of the friction coefficient, the transition in terms of flow resistance and phase, and the effect of a rough boundary on transition.

1.4 Flow in wave boundary layers in solitary motion

Although much research has been carried out on boundary layers under harmonic progressive waves, this is not so for boundary-layer flows under solitary waves (Fig. 5.64). Flow regimes in a (transient) solitary wave boundary layer on a smooth bed are discussed in Section 5.10.4 in Chapter 5. Observations, presented in Section 5.10.4, show that the transition to turbulence is governed, again, by the Reynolds number, $Re = \frac{aU_{0m}}{\nu}$, the Reynolds number of the solitary wave boundary layer, in which U_{0m} is the maximum value of the free-stream velocity, defined by Eq. 5.100, and a is the half of the stroke of the water particle displacement in the free-stream region, Eqs. 5.105 and 5.106.

It is seen from Section 5.10.4 that, for Re larger than 5×10^5, the bed shear stress experiences single, or multiple or, sometimes quite dense spikes, produced by the passage of turbulent spots (Fig. 5.73), first emerging toward the end of the flow period, and gradually spreading to earlier phases of the motion with increasing Re, in the same way as in harmonic progressive waves.

The reader is referred to Section 5.10.4 for details of the turbulent transition in solitary wave boundary layers.

1.5 Steady flow in pipes

Historically, Reynolds (1883) (see the discussion in the beginning of this chapter) was the first to recognize the non-dimensional combination of the four quantities, namely the pipe diameter (D), the mean flow velocity (V), the dynamic viscosity ($\mu = \nu\rho$), and the fluid density (ρ), $\rho\frac{DV}{\mu}$, as the governing parameter of the process of turbulent transition in pipes with smooth walls. This non-dimensional group is now known as the Reynolds number, following the proposal of Prandtl (1910) (see Launder, 2015). In a paper twelve years after the publication of his 1883 paper, Reynolds (1895, p. 124) established the critical value of this non-dimensional group for transition as "... steady direct motion in round tubes is stable or unstable according to

$$\rho\frac{DV}{\mu} < 1900: \ \text{stable}, \qquad \rho\frac{DV}{\mu} > 2000: \ \text{unstable} \tag{1.1}$$

the number being thus a criterion of the possible maintenance of sinuous or eddying motion." Later research showed that the critical value of the Reynolds number depends considerably on the degree of disturbance, the initial turbulence, at the intake of the pipe. Monin and Yaglom (1973, pp. 75-76) and Schlichting (1979, p. 451) report critical values of the Reynolds numbers like $O(10^4)$ that various investigators have achieved by arranging special inlet conditions, and thereby reducing the initial turbulence massively.

Now, recall that we have explained the turbulent transition for flow over a flat plate in terms of the theory of hydrodynamic stability according to which, we argued, the transition occurs simply because the laminar boundary-layer flow over the flat plate is unstable against small disturbances for Reynolds numbers larger than $Re_{x\ cr}$. For such large Reynolds numbers, any small disturbance grows, eventually leading to the inception of turbulence. The implementation of hydrodynamic stability theory for pipe flows shows, however, that the pipe flow is strictly stable for all Reynolds number, Schlichting (1979, pp. 542–543), and Monin and Yaglom (1973, pp. 142–147). This is despite the fact that: (1) Flows in pipes do undergo transition, first revealed by Reynolds' (1883) experiments, Fig. 1.1;

and (2) It is also difficult to envisage that parabolic velocity profiles in channels can be made unstable with very small disturbances whereas the same profiles in pipes cannot, Schlichting (1979, p. 542).

This problem has been the subject of a great many investigations over the past century, since the hydrodynamic stability theory had been introduced to study flow instabilities in the 1920s. Recent years have witnessed studies where the turbulent transition has been studied by direct numerical simulation (DNS). Most recently, Wu et al. (2015) utilized DNS, investigating laminar through gradual transition to fully developed turbulence in pipes. They anticipated that weakly *finite* and localized disturbances at the laminar inlet would induce gradual transition, eventually leading to a state of fully developed turbulence. They remarked that finite amplitude is needed because pipe flow is linearly stable with respect to infinitesimal disturbances. Wu et al. (2015) tested several scenarios for the finite initial disturbances at the inlet. They demonstrated that the "mystery" attributed to the breakdown process of the Reynolds pipe transition can be unveiled, partially resolving with a direct, spatially evolving simulation that carries weakly but finitely perturbed laminar inflow through gradual rather than abrupt transition arriving at the fully developed turbulent state. Likewise, the DNS of the Reynolds' (1883) color band visualization, when resolved in very fine spatial dimensions, indicates that mixing-up of color with the surrounding fluid in Reynolds' experiment actually contains hairpin packets long before breakdown (Wu et al., 2015, Fig. 2 E and F).

1.6 Oscillatory flow in pipes

In the oscillatory flow in a (smooth) pipe, the laminar-to-turbulent transition is governed by two parameters, namely the Reynolds number, $Re = \frac{aU_m}{\nu}$, and the pipe-radius-to-Stokes-length ratio, $\frac{R}{\delta_1}$ in which δ_1 is the Stokes length, Eq. 5.13. Laminar-to-turbulent transition is studied in great detail in pipes in the case of an oscillating flow alone, and also in the case of combined oscillating flow and current in Section 5.6.2 in Chapter 5. Fig. 5.45 displays the curve of neutral stability in the plane $(Re, \frac{R}{\delta_1})$. As seen from the figure, for $\frac{R}{\delta_1} \to \infty$, the curve of neutral stability approaches asymptotically to the line $Re_{cr} = 1.5 \times 10^5$, the critical value of the Reynolds number for transition in the case of the wave boundary layer (see the discussion above under flow in wave boundary layers). The figure shows that we need larger and larger Reynolds number for turbulent transition with decreasing pipe-radius parameter.

In the interesting case of combined oscillating flow and current (representing the combined waves and current case in coastal-engineering practice), a third, obvious parameter, is the current Reynolds number. Therefore, there are three parameters governing the transition, Re_w, Re_c, and $\frac{R}{\delta_1}$ in which we use the sub-indices w and c, to differentiate the wave-boundary-layer, and current Reynolds numbers, $Re_w = \frac{aU_m}{\nu}$ and $Re_c = \frac{DV}{\nu}$, respectively. Depending on the values of these three parameters, laminar-to-turbulent transition, or turbulent-to-laminar transition (called re-laminarization) can be determined. Fig. 5.46 gives an example for the curve of neutral stability in the combined flow case where both laminar-to-turbulent transition and turbulent-to-laminar transition (re-laminarization) are observed clearly.

Oscillating flow (including the combined oscillating flow and current) in a circular pipe has been investigated by Lodahl et al. (1998) in an extensive experimental study. No flow visualization has been carried out in Lodahl et al.'s (1998) study. Nevertheless, very extensive measurements have been made of the wall shear stress, using hot-film anemometry. The wall shear stress signals obtained in Lodahl et al.'s experiments suggest that the initiation of turbulence is associated with the emergence of turbulent spots in the way as it happens in wave boundary layers. (See also Section 5.2.3, Chapter 5, and also Carstensen et al. (2010) for an extensive discussion on the imprint of turbulent spots on the wall shear stress signal.)

The reader is referred to Section 5.6.2 in Chapter 5 for a detailed discussion of transition processes, taken from Lodahl et al. (1998). The mechanism behind the most interesting case of the turbulent-to laminar transition (re-laminarization) in combined oscillating flow and current is discussed in Section 5.6.2.3, Chapter 5, again, from Lodahl et al. (1998).

1.7 Flow around a circular cylinder in a steady current

Flow around a circular cylinder in a steady current exhibits the process of turbulent transition twice as the Reynolds number (the governing parameter of flow around a smooth circular cylinder) is increased. Fig. 1.5 summarizes the regimes of flow around a cylinder as a function of the Reynolds number, the cylinder Reynolds number defined by $Re = \frac{DU}{\nu}$, in which D is the cylinder diameter, and U the approach velocity.

The first transition occurs in the cylinder wake when the Reynolds num-

a)	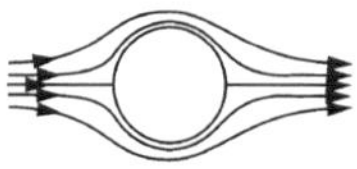	No separation. Creeping flow	Re < 5
b)	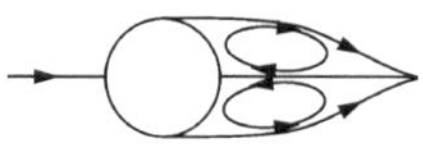	A fixed pair of symmetric vortices	5 < Re < 40
c)	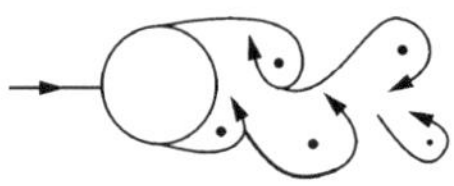	Laminar vortex street	40 < Re < 200
d)	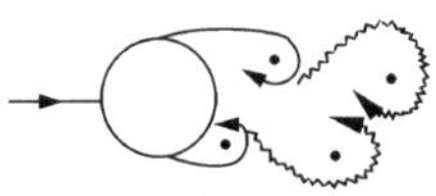	Transition to turbulence in the wake	200 < Re < 300
e)	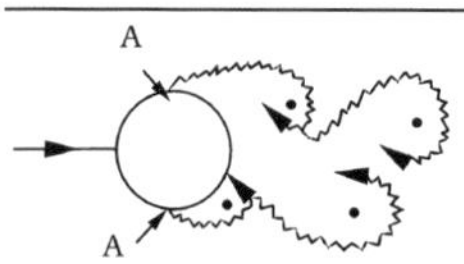	Wake completely turbulent. A:Laminar boundary layer separation	$300 < Re < 3 \times 10^5$ Subcritical
f)	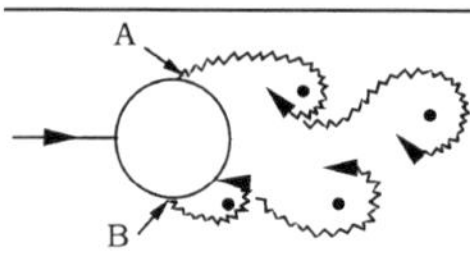	A:Laminar boundary layer separation B:Turbulent boundary layer separation;but boundary layer laminar	$3 \times 10^5 < Re < 3.5 \times 10^5$ Critical (Lower transition)
g)	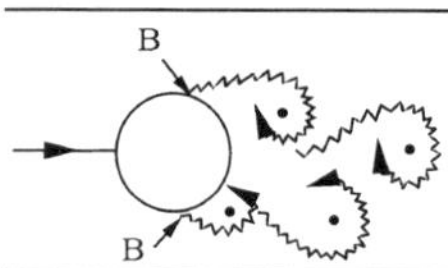	B: Turbulent boundary layer separation;the boundary layer partly laminar partly turbulent	$3.5 \times 10^5 < Re < 1.5 \times 10^6$ Supercritical
h)	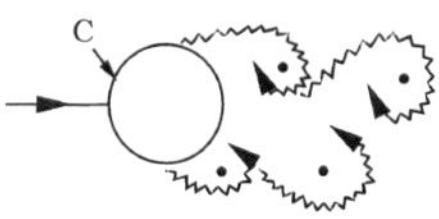	C: Boundary layer completely turbulent at one side	$1.5 \times 10^6 < Re < 4 \times 10^6$ Upper transition
i)	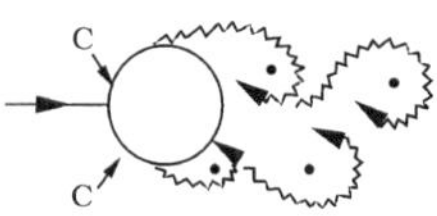	C: Boundary layer completely turbulent at two sides	$4 \times 10^6 < Re$ Transcritical

Fig. 1.5 Regimes of flow around a smooth, circular cylinder in steady current. Taken from Sumer and Fredsøe (2006).

ber is increased to 200, Fig. 1.5d. The region of transition to turbulence moves towards the cylinder, as Re is increased in the range $200 < Re < 300$. The vortex shedding becomes three dimensional, namely the shedding occurs in cells in the spanwise direction, in this range. For $Re > 300$, the wake is completely turbulent. The boundary layer over the cylinder surface remains laminar, however, for increasing Re over a very wide range of Re, namely $300 < Re < 3 \times 10^5$, Fig. 1.5e.

With a further increase in Re, transition to turbulence occurs this time in the boundary layer, the second transition. This transition first occurs at the point of separation, and on one side of the cylinder only, and therefore the boundary layer separation is turbulent at one side of the cylinder, and laminar at the other side. This occurs over the narrow range $3 \times 10^5 < Re < 3.5 \times 10^5$, Fig. 1.5f.

In the range $3 \times 10^5 < Re < 1.5 \times 10^6$, the region of transition to turbulence in the boundary layer moves upstream of the cylinder surface towards the stagnation point as the Reynolds number is increased. In this range, the boundary layer over the cylinder surface is partly laminar and partly turbulent, Fig. 1.5g.

The boundary layer on one side becomes fully turbulent when Re reaches the value of about 1.5×10^6, Fig. 1.5h. So, in this flow regime, the boundary layer is completely turbulent on one side of the cylinder, and partly laminar and partly turbulent on the other side.

Finally, when Re is increased further such that $Re > 4.5 \times 10^6$, the boundary layer over the cylinder surface is virtually turbulent everywhere, Fig. 1.5i.

The above description is for steady current flow around the cylinder. In the case when the cylinder is subject to waves, the laminar-to-turbulent transition is governed not only by the Reynolds number but also by the Keulegan-Carpenter number, among other parameters. The reader can consult Sumer and Fredsøe (2006) for various scenarios, and the implications of the latter for forces on coastal and offshore structures.

1.8 Flow induced by breaking waves

Waves approaching shoreline undergo shoaling where the wave height increases continuously. At some depth, the wave undergoing shoaling becomes unstable and breaks. With this, the wave dissipates energy in the form of turbulence. The turbulence generated by breaking waves is a different category of turbulence, different from those described in the previous examples.

Flow and turbulence induced by breaking waves is discussed extensively in Chapter 7. The way in which waves break in shallow waters depends on the slope of the beach and the characteristics of the wave. There are three types of breaking, namely spilling breaker, plunging breaker, and surging breaker. Fig. 7.1 illustrates these breakers, with the beach slope marked in the illustrations.

The types of wave breaker, and therefore turbulence generated, are governed by the so-called surf similarity parameter, ξ_0

$$\xi_0 = \frac{\tan(\beta)}{\sqrt{H_0/L_0}} \tag{1.2}$$

in which $\tan(\beta)$ is the beach slope, H_0 the deep-water wave height, and L_0 the deep-water wave length. Based on the surf similarity parameter, the previously mentioned three types of wave breaker are classified according to Eqs. 7.6–7.8. When ξ_0 is smaller than 0.5, no breaking will occur, and therefore no (breaking-induced) turbulence will be generated. Hence, in this case, the only turbulence that will exist is the turbulence generated in the wave boundary layer, if the wave-boundary-layer Reynolds number is sufficiently large.

The reader is referred to Chapter 7 for an extensive account of flow and turbulence in breaking waves.

1.9 References

(1) Carstensen, S., Sumer, B.M. and Fredsøe, J. (2010): Coherent structures in wave boundary layers. Part 1. Oscillatory motion. J. Fluid Mech., vol. 646, 169–206.

(2) Launder, B.E. (2015): First steps in modelling turbulence and its origins: A commentary on Reynolds (1895) "On the dynamical theory of incompressible viscous fluids and the determination of the criterion". Phil. Trans. R. Soc. A, 373, article no. 20140231.

(3) Lodahl, C.R., Sumer, B.M. and Fredsøe, J. (1998): Turbulent combined oscillatory flow and current in a pipe. J. Fluid Mech., vol. 373, 313–348.

(4) Monin, A.S. and Yaglom, A.M. (1973): Statistical Fluid Mechanics: Mechanics of Turbulence. MIT Press, Cambridge, Mass.

(5) Prandtl, L. (1910): Eine Beziehung zwischen Warmeaustasch und Stromungswiderstand der Flussigkeiten. Phys. Z., vol. 11, 1072–1078.

(6) Reynolds, O. (1883): An experimental investigation of the circumstances which determine whether the motion of water shall be direct or sinuous, and of the law of resistance in parallel channels. Phil. Trans. Roy. Soc. Lond., vol. 174, 935–982.
(7) Reynolds, O. (1895): On the dynamical theory of incompressible viscous fluids and the determination of the criterion. Phil. Trans. Roy. Soc. A, vol. 186, 123–164.
(8) Schlichting, H. (1979): Boundary Layer Theory. McGraw-Hill.
(9) Sumer, B.M. and Fredsøe, J. (2006): Hydrodynamics Around Cylindrical Structures. Advanced Series on Ocean Engineering. World Scientific.
(10) Wu, X., Moin, P., Adrian, R.J. and Baltzer, J.R. (2015): Osborne Reynolds pipe flow: Direct simulation from laminar through gradual transition to fully developed turbulence. Proc. Nat. Academ. Sci. U.S.A., vol. 112, 7920–7924.

Appendix to Chapter 1: Transition to turbulence over a flat plate

1.A Description of transition process

1.A.1 *Problem statement*

Fig. 1.3 displays the definition sketch concerning transition to turbulence over a flat plate. We have a uniform flow with zero incidence and with almost zero pressure gradient over a flat plate. The "undisturbed" flow velocity is U_∞, Fig. 1.4. The flow over the plate consists of two regions over the half space $y > 0$: boundary layer with a boundary-layer thickness of $\delta(x)$, and free-stream (potential flow) region above the boundary layer. The boundary layer begins to develop from the leading edge of the plate, Fig. 1.4. This boundary layer, being laminar in the initial stage, is subject to transition to turbulence further downstream, as sketched in Fig. 1.3, which will eventually turn fully turbulent for further downstream portions of the flow. The inception of turbulence in the transitional regime occurs in the form of tiny turbulent patches near the wall, called turbulent spots, as will be detailed in the following sections.

1.A.2 *Emmons' (1951) observations*

Emmons (1951) was the first to observe turbulent spots (incidentally, accidentally) when experimenting with a water-table analogy to supersonic flow. (The latter analogy had been prepared at Harvard as a student demonstration.) On this table, water flows in a 1/8- to 1/4-in. (3.2-6.4 mm) deep layer, over a nearly horizontal piece of plate glass about 3 by 6 ft. (91.4 cm by 183 cm). Fig. 1.6 shows a plan view of the glass plate with water flowing over it, reproduced from Emmons' (1951) original publication. Emmons (1951) describes the turbulent spots as follows:

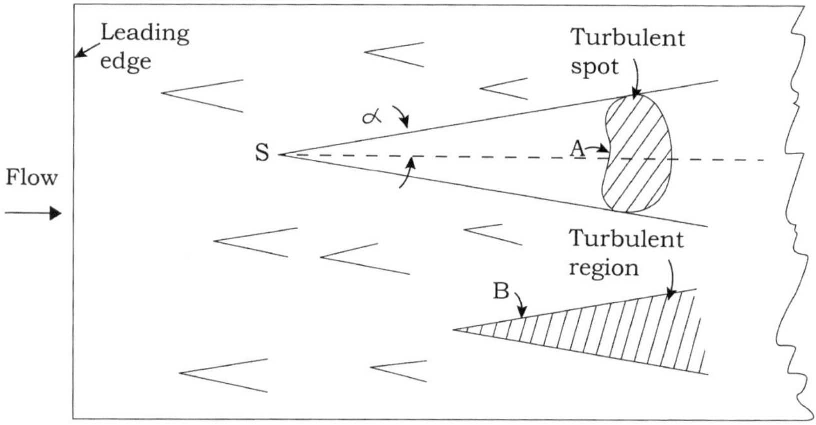

Fig. 1.6 Transition on a flat plate. Emmons' (1951) experiment. Plan view. Adapted from Emmons (1951).

"... At some instant and at some point (both random), a tiny spot of turbulence appeared as at S. The spot was small and of irregular shape. This spot grew as it was washed along with the boundary-layer fluid. The approximate shape and size of the spot after a period of growth is shown at A, as is also the locus of successive spot positions. Each turbulent spot was washed off the end of the plate. At random times, spots appeared at random locations over the plate surface. ... (Turbulent spots) could be increased in density and frequency at a given location on the plate either by increasing the water velocity or adding disturbances by placing an object (finger) in the deep (low velocity) water well ahead of the plate. A disturbance (as a vertical wire) placed directly into the water on the plate acted as a continuous source, as at B of Fig. 1.6, and thus filled a sector of the plate surface with turbulent water.

The growing spot approximately preserved its shape as it grew, and the locus of successive spot positions appeared straight and at the same angle as the continuous turbulence from a wake.

The turbulence in transition is seen to be essentially intermittent. Each point of the plate is covered by a laminar boundary layer, except during those periods of time during which a turbulent spot moves over it. Since spots grow as they move, the downstream points of the plate are covered for greater fractions of the time by a turbulent layer. In fact, the spots

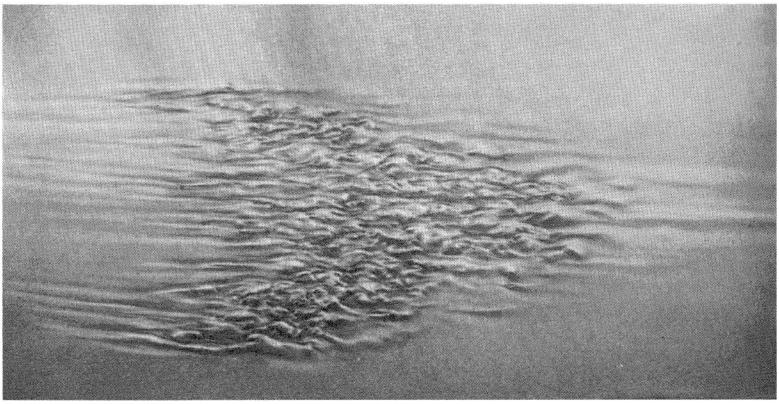

Fig. 1.7 Aluminium visualization of sublayer streaks for turbulent spots in water. Flow from left to right. $U_\infty = 12$ cm/s. Cantwell et al. (1978).

soon overlap and cover downstream portions continuously."

We note that Emmons' (1951) sketch of a turbulent spot, A in Fig. 1.6, slightly misrepresents the actual shape of the spot, the arrowhead shape, see for example Figs. 1.7 and 1.8.

1.A.3 *Detailed description of a single turbulent spot*

Quite a substantial amount of information has accumulated on turbulent spots from the beginning of the 1970s, following Emmons' (1951) observations and a few follow-up publications that appeared thereafter (Schubauer and Klebanoff, 1955; and Elder, 1960). We will, in this subsection, describe the flow processes that take place in a single turbulent spot, based on, notably, Cantwell et al. (1978) and Gad-el-Hak et al. (1981).

1.A.3.1 *Features of a single turbulent spot*

Fig. 1.7 displays a plan-view image of a turbulent spot obtained by Cantwell et al. (1978). The authors employed an extremely dense suspension of aluminium flakes in the experimental water channel for their flow visualization. The "flat plate" boundary-layer flow was established on the glass side walls of the channel. Spots were generated by touching the wall momentarily with a small rod. The authors report that, even under strong illumination, it was not possible to see more than 2 or 3 mm into the fluid because of the extremely dense suspension of aluminium flakes. Therefore, their flow

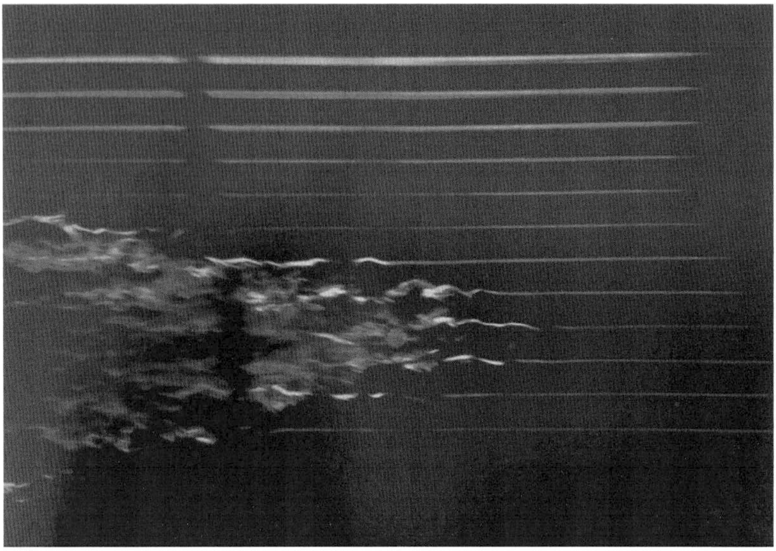

Fig. 1.8 Plan view of the spot at $y = 0.1$ cm. The dye lines undergo large-amplitude oscillations before being overtaken by the spot. Gad-el-Hak et al. (1981).

visualization is strictly confined to the region very near the wall. (Note that Cantwell et al., 1978, used a second method of flow visualization as well in their study.)

Fig. 1.7 shows that the turbulent spot is an arrowhead-shaped isolated area of turbulence in an otherwise laminar boundary layer. The flow-visualization image of Gad-el-Hak et al. (1981) of a turbulent spot reproduced in Fig. 1.8 also clearly shows that the flow is turbulent in the spot structure while it is laminar outside, which was made visible by streaks of colour introduced into the flow upstream. Observations by Cantwell et al. (1978) and Gad-el-Hak et al. (1981) and many others show that the flow in this isolated area of turbulence actually "bursts" with strong oscillations. Furthermore, from Figs. 1.7 and 1.8, it is clear that the spot has a distinct streaky feature. These streaks can be followed for long distances, and their appearance is evidence of the presence of longitudinal vortices. Moreover, Cantwell et al. (1978, pp. 646–647) report that they measured the spacing between the streaks, using an optical correlation technique, and obtained a mean value of $\lambda^+ = \lambda U_f/\nu = 86$, a value close enough to the generally accepted value $\lambda^+ = 100$ in fully-developed turbulent boundary layers (see Chapter 3, Section 3.5.2) in which $U_f = \sqrt{\tau_0/\rho}$ is the wall friction velocity,

Fig. 1.9 Side view of the turbulent spot obtained, using aluminium technique. Flow from left to right. Adapted from Cantwell et al. (1978).

τ_0 is the wall shear stress, ρ the fluid density, and ν the kinematic viscosity. This is an important result in the sense that it indicates that a turbulent spot "mimics" a fully-developed turbulent boundary layer.

Fig. 1.9 illustrates the side view of a turbulent spot obtained under conditions similar to Fig. 1.7 in Cantwell et al.'s (1978) study. In order to photograph the side view of turbulent spots, the following procedure was utilized. A half spot was generated by touching the wall at the free surface, photographing the side wall as usual. A mirror was mounted above the channel at 45^0 to show also a longitudinal section through the spots seen at the free surface, and this view was also photographed simultaneously (Cantwell et al., 1978, Fig. 7a). For comparison, Cantwell et al. (1978, Fig. 7b) present a similar side-view image, obtained in their study, of a fully-developed turbulent boundary layer at a slightly higher velocity, with transition induced by a trip wire near the entrance to the channel. This comparison shows that the two side-view pictures, i.e. the side view picture of a single turbulent spot and that of a fully-developed turbulent boundary layer, are evidently quite similar. This is also an important result, providing further evidence that a turbulent spot can be viewed as a building block of a fully-developed turbulent boundary layer. We shall return to this point later.

Fig. 1.10, reproduced from Gad-el-Hak et al. (1981), illustrates the time evolution of an individual spot in plan view. These experiments were made in a towing tank, and the turbulent spots (Figs. 1.8 and 1.10) were made visible by a technique which utilized fluorescent dye, i.e., dye which is visible only when excited by a strong light source of the appropriate wave length. A 5 watt Argon Laser with a cylindrical lens was used to produce a sheet of light, positioned parallel to the flat plate at $y = 1.3$ cm (measured from the plate), at almost twice the undisturbed laminar boundary-layer thickness at this location.

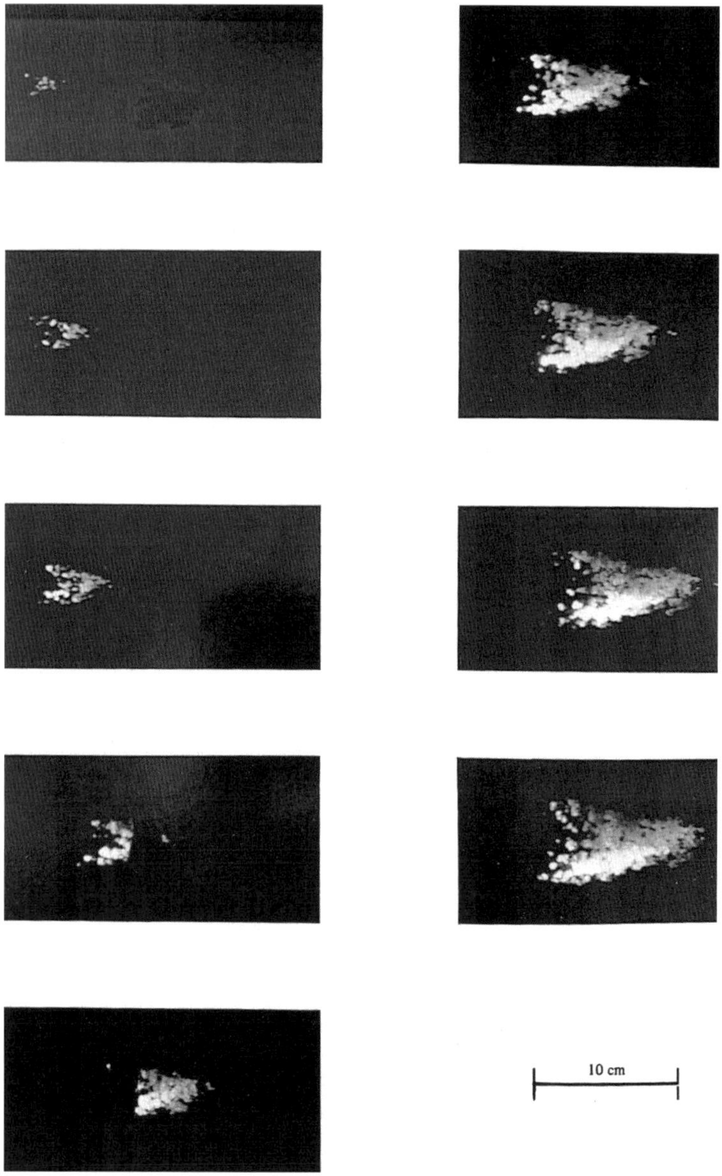

Fig. 1.10 Sequence of plan views at $y = 1.3$ cm, illustrating the time evolution of a turbulent spot. Flow from left to right. The picture frames are 0.2 s apart, except the very last two, which are 0.1 s apart. The first image (the top image on the left column) corresponds to 1.7 s after the generation of the spot. Gad-el-Hak et al. (1981).

1.A.3.2 *Imprint of turbulent spot in the velocity signal*

Fig. 1.11 (reproduced from Cantwell et al., 1978) illustrates the imprint of a turbulent spot in the velocity signal very near the wall, at $y = 0.038$ cm (Fig. 1.11, the bottom trace), and that at a distance $y = 0.953$ cm (Fig. 1.11, the top trace) at the measuring station $x = 119$ cm, along with the time variations of the ensemble means at the same y locations. As seen, the velocity near the wall is increased tremendously, by a factor of 5, during the passage of the spot. (This suggests that the wall shear stress would also undergo the same kind of increase during the passage of a spot.) The decrease in the velocity at $y = 0.953$ cm, on the other hand, is a manifestation that this point is no longer in the free-stream region during the passage of the spot.

The fluctuations in the velocity signal, particularly those near the wall (Fig. 1.11), reveal that the flow in this isolated area of turbulence is rather "violent", in line with our description in the preceding paragraphs from flow visualization studies.

Fig. 1.12, from the same study (Cantwell et al., 1978), presents the time variations of the ensemble-mean velocity at various y locations across the boundary layer thickness, corresponding to the passage of turbulent spots at the same measuring station at $x = 119$ cm.

1.A.3.3 *Natural and artificial generation of turbulent spots*

Inception of turbulence (and therefore the generation of turbulent spots) occurs naturally, depending on various factors including the intensity of turbulence in the free-stream flow. Spots are also generated artificially. The latter is the preferred choice in the experimental investigation of turbulent spots, which obviously enables one to conduct controlled experiments. While, in Emmons' (1951) experiments, spots were generated naturally, Cantwell et al. (1978) in their aluminium-powder experiments generated spots artificially by touching the wall momentarily with a small rod; Gad-el-Hak et al. (1981) generated turbulent spots by a solenoid valve that ejected a small amount of fluid from a hole of 0.5 mm; Chambers and Thomas (1983) generated them by solenoid actuated pins, or air jets, or compressed air operated diaphragms; and Browand and Plocher (1985) by periodically puffing a fluid jet from a small hole, to give but a few examples.

Disturbances referred to in the preceding paragraph to generate turbulent spots artificially are all controlled disturbances. There may be, however, unintentional, and therefore uncontrolled, artificial disturbances that

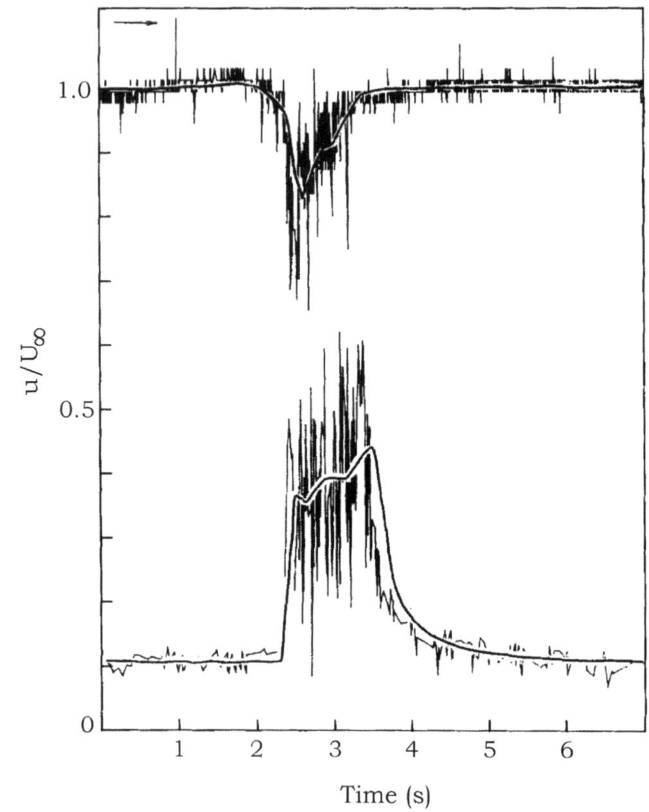

Fig. 1.11 Imprint of turbulent spot in the velocity signal. Velocity traces and time variations of ensemble mean velocity. Bottom: $y = 0.038$ cm. Top: $y = 0.953$ cm. At measuring station $x = 119$ cm. Cantwell et al. (1978).

affect the laminar-to-turbulent transition, such as the uncontrolled free-stream turbulence whereby turbulent motions are carried in by the fluid from the free stream; or those carried in through the fluid as sound waves; those produced in the boundary layer by surface irregularities; and those produced by the vibrations of the plate, as noted by Emmons (1951, p. 492).

Regardless of the way in which spots are generated, artificially (controlled, or uncontrolled) or naturally, individual spots grow in size and in population as they convect downstream until they merge and the whole boundary layer becomes completely turbulent, as sketched in Fig. 1.13.

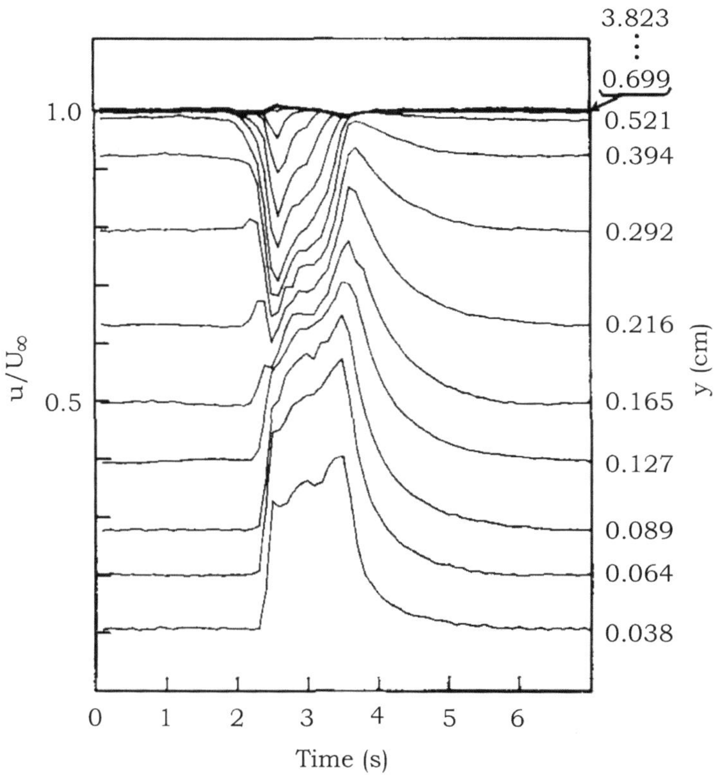

Fig. 1.12 Time variations of the ensemble-mean velocity at various y locations across the boundary layer thickness, corresponding to the passage of turbulent spots at measuring station $x = 119$ cm. Cantwell et al. (1978).

1.A.4 *Critical Reynolds number for transition*

Experiments show that no turbulent spots are observed for the Reynolds numbers Re_x smaller than 3×10^5–5×10^5 (Schlichting 1979, p. 474, and Monin and Yaglom, 1973, p. 77), marking the critical value of the Reynolds number for transition to turbulence in which

$$Re_x = \frac{xU_\infty}{\nu} \tag{1.3}$$

the length x being the distance from the leading edge (Fig. 1.3). Hence

$$Re_{x\ cr} = 3 \times 10^5 - 5 \times 10^5 \tag{1.4}$$

As is known, the Reynolds number is actually (proportional to) the ratio of the inertial forces and the viscous forces. The larger the value of the

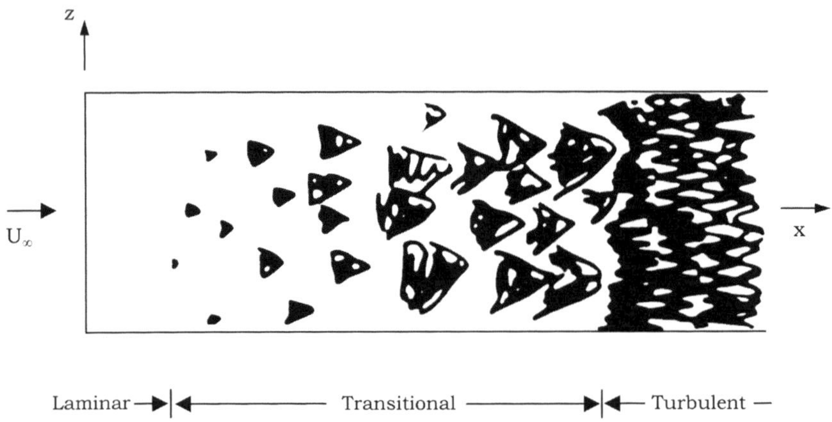

Fig. 1.13 With the inception of individual spots, the spots grow in size and in population as they convect downstream to eventually merge, and the whole boundary layer becomes fully turbulent. Plan view. A schematic description.

Reynolds number, the larger the inertia forces, and the smaller the viscous forces. Therefore, any disturbance introduced into the flow will grow potentially unlimited, leading to turbulence (emergence of turbulent spots) for larger values of the Reynolds number as the viscous forces will not be able to damp out the disturbances when the Reynolds number is sufficiently large, larger than the critical value of the Reynolds number.

The Reynolds number can also be defined in terms of the boundary-layer thickness δ

$$Re_\delta = \frac{\delta U_\infty}{\nu} \tag{1.5}$$

in which δ is (for the present purposes) defined as the value of y for which the velocity becomes 99% of U_∞:

$$y = \delta : \quad u(y) = 0.99 U_\infty \tag{1.6}$$

From the laminar theory, δ is (Schlichting, 1979, p. 47, and Monin and Yaglom, 1973, p. 140):

$$\delta = 5\sqrt{\frac{\nu x}{U_\infty}} \tag{1.7}$$

and from Eqs. 1.3, 1.5 and 1.7:

$$Re_\delta = 5\sqrt{Re_x} \tag{1.8}$$

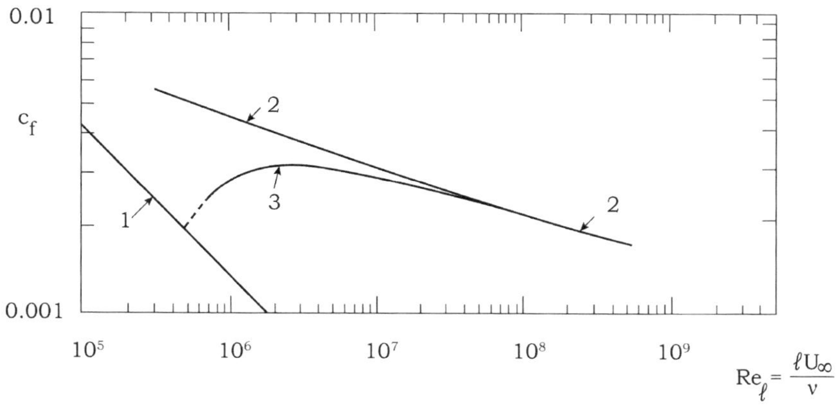

Fig. 1.14 Flow resistance for flow over a flat plate, Curve 1 representing the laminar solution, and Curves 2 and 3 representing the experimental data compiled by Schlichting (1979). The diagram reveals the flow regimes, laminar, turbulent and transitional, represented by Curves 1, 2 and 3, respectively. Adapted from Schlichting (1979, Fig. 21.2).

Therefore, the critical value of the Reynolds number for transition for the boundary-layer-thickness Reynolds number will be from Eqs. 1.4 and 1.8, rounding the numbers (Monin and Yaglom, 1973, p. 77):

$$Re_{\delta\ cr} = 2750 - 3500 \tag{1.9}$$

Laminar-to-turbulent transition can also be illustrated by plotting the flow resistance as a function of the Reynolds number. Fig. 1.14 illustrates such plotting in which c_f, the total (dimensionless) skin friction, is defined by

$$c_f = \frac{F_D}{\frac{1}{2}\rho U_\infty^2 bl} \tag{1.10}$$

in which F_D is the skin-friction drag of a plate (on one side) of length l, namely

$$F_D = b\int_0^l \tau_0(x')\, dx' \tag{1.11}$$

with b being the width of the plate, and τ_0 the wall shear stress.

In Fig. 1.14, Curve 1 represents the laminar solution; and Curve 2 and Curve 3 represent experimental data, compiled by Schlichting (1979, Fig. 21.2).

First of all, the three flow regimes discussed in the preceding paragraphs, namely, laminar, turbulent and transitional, are all revealed by Fig. 1.3,

Curve 1 representing the laminar regime, Curve 2 the turbulent regime, and Curve 3 the transitional regime.

Secondly, the diagram implies that, as the Reynolds number is increased, the laminar-to-turbulent transition may occur in two ways, either through a laminar-transitional-turbulent route, the natural transition, or through a laminar-turbulent route, bypassing the transitional regime, the so-called "bypass" transition.

Thirdly, the critical Reynolds number for the natural transition from the diagram appears to be $Re_{x\ cr} = 5 \times 10^5$, and that for the bypass transition $Re_{x\ cr} = 3 \times 10^5$. Note that this range, namely $Re_{x\ cr} = 3 \times 10^5$–$5 \times 10^5$, is the same as that indicated in Eq. 1.4.

1.A.5 *Transition and turbulent spots over rough boundaries*

To the authors' knowledge, no study is yet available investigating transition (and generation of turbulent spots) over a flat plate with a rough wall. Carstensen et al. (2012) studied the transition to turbulence in wave boundary layers over rough boundaries. We will hereby refer to Carstensen et al.'s (2012) work, as it is believed that their case closely resembles the present one, and therefore their results could give us a first indication of how the transition over a rough-wall flat plate would develop.

A detailed description of Carstensen et al.'s (2012) work can be found in Section 5.2.5 in Chapter 5. This work was actually an extension to rough-boundary situations of their previous work on smooth walls (Carstensen et al., 2010). In Carstensen et al.'s (2012) experiment, a fixed sand-grain bed in an oscillating tunnel was achieved by gluing sand of fairly uniform size one-layer deep on the bottom wall of their oscillating tunnel. Their observations showed that turbulent spots (Fig. 5.18) occur over rough boundaries as well. They note that the roughness Reynolds number, $k_s^+ (= k_s U_f / \nu)$ in their experiments was always $k_s^+ > O(60)$ when turbulent spots first emerged, implying that the boundary in their experiments behaved as a nearly fully rough wall. Here, k_s is Nikuradse's equivalent sand roughness, U_f the friction velocity, and ν the kinematic viscosity. The reader is referred to the video accompanying Carstensen et al.'s (2012) paper, the URL address `http://dx.doi.org/10.1063/1.4767536`, to get a full view of turbulent spots over a rough boundary. From Carstensen et al.'s (2012) study, it appears that turbulent spots develop in the same way as in the smooth-wall case. One interesting result, not unexpected, however, is that the critical Reynolds number for transition in the rough-wall case is much smaller than that of the smooth-wall situation (Fig. 5.19).

1.B Hydrodynamic stability theory for transition

In the previous section, we have discussed the laminar-to-turbulent transition over a flat plate, based on experimental observations. We have now detailed knowledge of the way in which the transition develops, from the inception of turbulence to the fully developed stage of the turbulent boundary layer. We have, however, not looked into why this transition occurs. What are the physics behind the observed transition process? What are the governing mechanisms? In the present section, we will partly address these questions by way of resorting to hydrodynamic stability theory.

We will, in the following paragraphs, summarize the hydrodynamic stability theory of laminar-to-turbulent transition over a flat plate where the transition develops naturally with no artificial disturbance (controlled or uncontrolled) introduced. It is important to note that involving an artificial disturbance leading to the so-called bypass transition, discussed in the previous section, essentially "violates" the conditions under which the current stability theory is developed. Therefore, the analysis will not cover bypass-transition situations.

Our presentation is mainly based on the extensive review of the subject in the book of Schlichting (1979, pp. 456–483).

1.B.1 *The Orr-Sommerfeld equation*

We will develop the hydrodynamic stability theory for transition over a flat plate in six steps.

First step. We start off with the Navier-Stokes (N.-S.) equations (Eq. 2.91). We decompose the hydrodynamic quantities in the N.-S. equations into (1) a mean flow (whose stability is the subject of this study), and (2) a disturbance superimposed on it:

$$u = U + u', \qquad v = V + v', \qquad w = W + w' \tag{1.12}$$

and

$$p = P + p' \tag{1.13}$$

in which U, V and W are the velocity components of the mean flow in a Cartesian coordinate system, in x, y and z directions, respectively, P is the pressure of the mean flow, and the quantities with prime superscript are the corresponding quantities associated with the disturbance.

Second step. Next, we consider that the flow over the flat plate is a two-dimensional flow, and likewise we assume that the disturbance is also a two-dimensional disturbance. (We will return to the latter in a short while.) Regarding the mean flow, we further assume that

$$U = U(y), \qquad V, W \equiv 0 \tag{1.14}$$

considering that the dependence of U on the x coordinate is much smaller than that on y; and likewise V and W can, to a first approximation, be set equal to nil. The pressure, on the other hand, is like

$$P = P(x, y) \tag{1.15}$$

keeping the x dependence, as the pressure gradient $\partial P/\partial x$ maintains the flow. Regarding the two-dimensional disturbance, we write formally

$$u' = u'(x, y, t), \qquad v' = v'(x, y, t), \qquad p' = p'(x, y, t) \tag{1.16}$$

Therefore, the resultant motion, with the disturbance superimposed, will be

$$u = U + u', \qquad v = 0 + v' = v', \qquad w = 0 + 0 = 0, \qquad p = P + p' \tag{1.17}$$

Third step. We assume that the mean flow, Eqs. 1.14 and 1.15, is a solution of the N.-S. equations, and further we require that the resultant motion, Eq. 1.17, must also satisfy the N.-S. equations. Now, substituting Eq. 1.17, the resultant motion, into the N.-S. equations (Eq. 2.91) for a two-dimensional, non-steady flow, and neglecting quadratic terms in the disturbance velocity components, we get

$$\frac{\partial u'}{\partial t} + U\frac{\partial u'}{\partial x} + v'\frac{dU}{dy} + \frac{1}{\rho}\frac{\partial P}{\partial x} + \frac{1}{\rho}\frac{\partial p'}{\partial x} = \nu\left(\frac{d^2U}{dy^2} + \nabla^2 u'\right) \tag{1.18}$$

$$\frac{\partial v'}{\partial t} + U\frac{\partial v'}{\partial x} + \frac{1}{\rho}\frac{\partial P}{\partial y} + \frac{1}{\rho}\frac{\partial p'}{\partial y} = \nu\nabla^2 v' \tag{1.19}$$

$$\frac{\partial u'}{\partial x} + \frac{\partial v'}{\partial y} = 0 \tag{1.20}$$

in which ∇^2 is the Laplacian operator, which in two-dimensions is $\nabla^2 = \partial^2/\partial x^2 + \partial^2/\partial y^2$. Since the mean flow satisfies the N.-S. equations, the above equations will reduce to the following three equations

$$\frac{\partial u'}{\partial t} + U\frac{\partial u'}{\partial x} + v'\frac{dU}{dy} + \frac{1}{\rho}\frac{\partial p'}{\partial x} = \nu\nabla^2 u' \tag{1.21}$$

$$\frac{\partial v'}{\partial t} + U\frac{\partial v'}{\partial x} + \frac{1}{\rho}\frac{\partial p'}{\partial y} = \nu\nabla^2 v' \tag{1.22}$$

$$\frac{\partial u'}{\partial x} + \frac{\partial v'}{\partial y} = 0 \tag{1.23}$$

We have three equations, and three unknowns $u' = u'(x, y, t)$, $v' = v'(x, y, t)$, and $p' = p'(x, y, t)$. We will, in the next steps, further eliminate p' from Eqs. 1.21 and 1.22, and thus, together with Eq. 1.23, will have two equations, and two unknowns $u' = u'(x, y, t)$ and $v' = v'(x, y, t)$.

Fourth step. Now, we introduce a stream function $\psi(x, y, t)$ associated with our disturbance:

$$u' = \frac{\partial \psi}{\partial y} \tag{1.24}$$

$$v' = -\frac{\partial \psi}{\partial x} \tag{1.25}$$

(ψ as defined above satisfies the continuity equation, Eq. 1.23.) We assume that the disturbance is in the form of a single oscillation in x and in t, and given as

$$\psi(x, y, t) = \phi(y)e^{i(\alpha x - \beta t)} \tag{1.26}$$

The quantity α in the above equation is a real quantity, and represents the wave number of the oscillating disturbance (wave disturbance), with the associated wave length

$$\lambda = \frac{2\pi}{\alpha} \tag{1.27}$$

(λ should not be confused with the streak spacing discussed in the previous section. We want to keep Schlichting's, 1979, notation unchanged in the analysis here.) The quantity β, on the other hand, is complex,

$$\beta = \beta_r + i\beta_i \tag{1.28}$$

in which i is the imaginary unit $i = \sqrt{-1}$, β_r is the real part, and β_i the imaginary part of β.

The quantity β_r represents the angular frequency of the disturbance wave while β_i represents the amplification factor, and determines the degree of amplification or damping. The disturbance is damped (i.e. exponential decay) if $\beta_i < 0$, and therefore our laminar flow ($U = U(y)$, $P(x, y)$) is stable, while if $\beta_i > 0$, the instability sets in (i.e. exponential growth of the disturbance).

We can also introduce a new quantity c, defined as the ratio of β and α:

$$c = \frac{\beta}{\alpha} = c_r + c_i \tag{1.29}$$

with c_r representing the velocity of propagation of the disturbance wave in the x-direction, and c_i representing, again, the damping or amplification of the disturbance, depending on the sign of c_i.

Fifth step. We now substitute Eqs. 1.24 and 1.25 in Eqs. 1.21 and 1.22, and further eliminate p' from these equations, we get the following ordinary, fourth order differential equation for $\phi(y)$

$$(U - c)(\phi'' - \alpha^2\phi) - U''\phi = -\frac{i}{\alpha Re}(\phi'''' - 2\alpha^2\phi'' + \alpha^4\phi) \tag{1.30}$$

in which the primes represent differentiation with respect to y, and Re is the Reynolds number, defined by (Eq. 1.5)

$$Re = \frac{\delta U_m}{\nu} \tag{1.31}$$

and Eq. 1.30 has been normalized such that all lengths are divided by the boundary layer thickness δ, and velocities by the maximum velocity U_m of the main flow. Eq. 1.30 is known as the Orr-Sommerfeld equation.

The solution to the Orr-Sommerfeld equation is to be sought with the boundary conditions for the disturbance

$$y = 0: \qquad u' \text{ and } v' = 0: \qquad \phi = 0 \text{ and } \phi' = 0 \tag{1.32}$$

and

$$y = \infty: \qquad u' \text{ and } v' = 0: \qquad \phi = 0 \text{ and } \phi' = 0 \tag{1.33}$$

The first set of the boundary conditions above is the no-slip boundary condition for the disturbance, and the second one states that the disturbance vanishes as $y \to \infty$.

Sixth step. We assumed in the second step above that the disturbance is two-dimensional (Eq. 1.16) whereas, in reality, a disturbance should be three dimensional. Therefore, objection may be raised to this assumption. Schlichting (1979, p. 460) notes, however, that this objection was removed by Squire (1933) who proved that a two-dimensional flow becomes unstable at a higher Reynolds number when the disturbance is assumed three-dimensional than when it is assumed two-dimensional. This means that two-dimensional disturbances constitute the "worst-case scenario" for two dimensional flows. Therefore, the value of the critical Reynolds number, or of the lowest limit of stability, is obtained by considering two-dimensional disturbances, as has been done in the present analysis summarized in the preceding paragraphs.

1.B.2 *The eigenvalue problem*

With the differential equation, Eq. 1.30, and the boundary conditions Eqs. 1.32 and 1.33, the problem of stability has been reduced to that known as the eigenvalue problem.

Given the mean flow $U(y)$ (the Blasius profile, Schlichting, 1979, pp. 135-138, and Table 7.1), the Orr-Sommerfeld equation, Eq. 1.30, contains four parameters, Re, α, c_r and c_i. For a given value of Re, and likewise, for a given value of α (or the wave length, $\lambda = \frac{2\pi}{\alpha}$), the Orr-Sommerfeld equation (Eq. 1.30) together with the boundary conditions (Eqs. 1.32 and 1.33) provide one eigenfunction $\phi(y)$ and one eigenvalue c $(= c_r + ic_i)$.

The sign of c_i, the key parameter in the hydrodynamic stability analysis, determines whether the disturbance is amplified ($c_i > 0$), indicating instability, or damped ($c_i < 0$), indicating stability.

The result of this exercise is usually plotted in a (Re, α) plane, called the stability diagram with two regions in the diagram, namely Stable and Unstable areas. The line separating these two regions is called the curve of neutral stability ($c_i = 0$).

The way in which this diagram is constructed may be formally summarized as follows: (1) For a given Reynolds number, solve the aforementioned eigenvalue problem, namely find the eigenfunctions $\phi(y)$ and the eigenvalues c $(= c_r + ic_i)$ for a given set of wave numbers α. (2) Plot the quantity c_i as function of the wave number α for which the eigenvalues c $(= c_r + ic_i)$ are found. (This is for a given value of Re.) (3) Subsequently, from the latter plot, find the broadest α range where $c_i > 0$, the flow being unstable in this range of the wave number. (4) Plot the latter on the plane (Re, α), one α range for one value of Re. (5) Repeat Steps 1-4 for many Re numbers of interest. (6) This will lead to a stability diagram with Stable and Unstable areas, separated with the curve of neutral stability.

Tollmien (1929) was the first to develop the hydrodynamic stability theory and produce the stability diagram for the boundary-layer flow over a flat plate. Fig. 1.15 displays this diagram, reproduced from Schlichting (1979, Fig. 16.11), in which δ_1 is the displacement thickness of the boundary layer, defined by:

$$U_\infty \delta_1 = \int_{y=0}^{\infty} (U_\infty - u) dy \qquad (1.34)$$

This boundary-layer thickness is, from the laminar solution (Blasius pro-

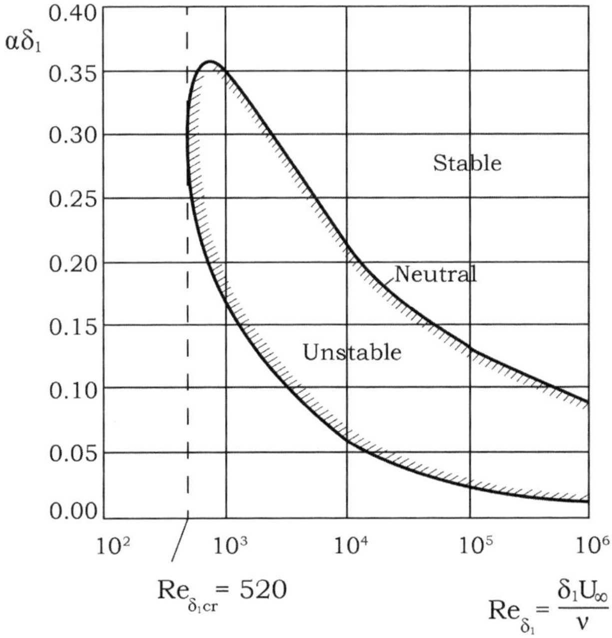

Fig. 1.15 Curve of neutral stability for the disturbance wave length $\alpha\delta_1$ as a function of the Reynolds number for the boundary layer on a flat plate at zero incidence. Theory is due to Tollmien (1929), and the numerical calculations are due to Jordinson (1970). Adapted from Schlichting (1979).

file), calculated as (Schlichting, 1979, p. 141)

$$\delta_1 = 1.7208\sqrt{\frac{\nu x}{U_\infty}} \tag{1.35}$$

and the Reynolds number in the diagram is based on the displacement thickness of the boundary layer:

$$Re_{\delta_1} = \frac{\delta_1 U_\infty}{\nu} \tag{1.36}$$

As seen clearly, the flow is unstable against the introduced disturbance waves only inside the area marked Unstable in Fig. 1.15. Outside this area, the flow is stable, i.e., the introduced disturbance waves will be damped. Incidentally, the disturbance waves in the hydrodynamic stability theory are traditionally called Tollmien-Schlichting waves. Tollmien-Schlichting waves were photographically captured, e.g., Fig. 1.16 (van Dyke, 1982).

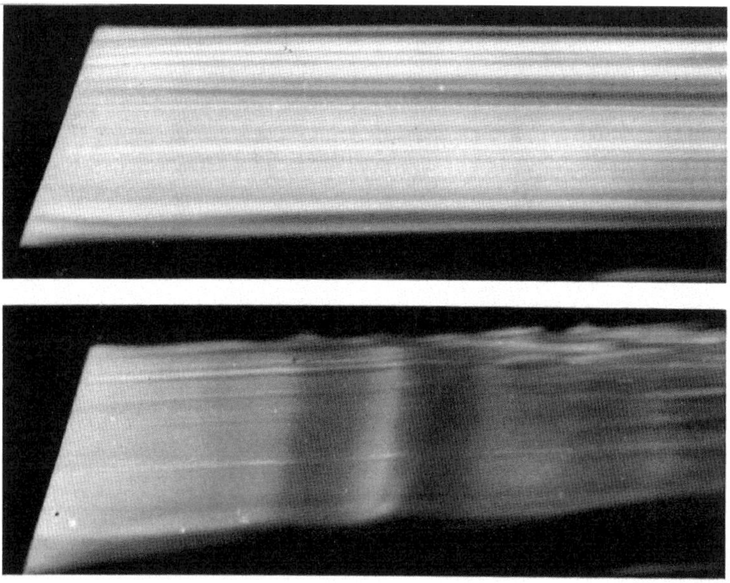

Fig. 1.16 Top panel: Laminar boundary layer, $Re_x = 2 \times 10^4$. Bottom panel: Two-dimensional Tollmien-Schlichting waves, $Re_x = 1 \times 10^5$. Adapted from van Dyke (1982). Original source: ONERA photographs, Werle (1980).

Returning to Fig. 1.15, as seen from the curve of neutral stability in the figure, the main flow is stable against the small disturbances for Re_{δ_1} smaller than 520 regardless of the disturbance-wave number, marking the value of the critical Reynolds number:

$$Re_{\delta_1\ cr} = 520 \tag{1.37}$$

Note that the theory behind the stability calculations presented in Fig. 1.15 (reproduced from Schlichting, Fig. 16.11) is due to Tollmien (1929), and the numerical calculations are due to Jordinson (1970). As for the numerical value of the critical Reynolds number, Schlichting (1979, p. 469) notes: "In his first paper of 1929, W. Tollmien indicated a value of 420; a subsequent calculation by Schlichting (1933) yielded the value of 575. The numerical value of 520 quoted was taken from a contemporary investigation carried out by Jordinson (1970)." From the latter figure, it is seen that the largest wave number for unstable disturbance waves is

$$\alpha\delta_1 = 0.36 \tag{1.38}$$

which corresponds to the smallest unstable wave length

$$\lambda_{\min} = \frac{2\pi}{0.36}\delta_1 = 17.5\delta_1 \simeq 6\delta \tag{1.39}$$

From Eqs. 1.3, 1.36 and 1.35:

$$Re_{\delta_1} = 1.7208\sqrt{Re_x} \tag{1.40}$$

or

$$Re_x = 0.3377 Re_{\delta_1}^2 \tag{1.41}$$

Therefore, the critical value of the Reynolds number $Re_{\delta_1\ cr} = 520$, when converted to Re_x from Eq. 1.41 is:

$$Re_{x\ cr} = 0.9 \times 10^5 \tag{1.42}$$

This value of the critical Reynolds number is markedly smaller than that for the transition obtained from experiments, discussed in the previous section (Eq. 1.4), $Re_{x\ cr} = 3\times10^5$–5×10^5 for natural transition, corresponding to the first sign of inception of turbulence. Schlichting (1979, pp. 470–471) explains this "mismatch" as follows. The position where the boundary layer becomes first unstable according to theory (Schlichting calls this position the point of instability) must always be expected to lie upstream of the experimentally observed point of transition since actual turbulence is created along the route from the point of instability to the point of transition owing to the amplification of the unstable disturbances. The reader is referred to Schlichting (1979) for a detailed discussion of the difference between the critical Reynolds number for the point of instability (obtained from the stability theory), and that for the laminar-to-turbulent transition (observed experimentally).

1.B.3 *Experimental verification of the hydrodynamic stability theory*

Schubauer and Skramstad (1943) verified the hydrodynamic stability theory by experiments conducted in a specially designed wind tunnel with an extremely low level of turbulence. The idea was to *simulate* precisely the same conditions in a carefully designed experiment as those stipulated by the theory, so that comparison between theory and experiment can be made on the same basis.

To this end, a thin metal strip extending over a width of about 30 cm, 0.05 mm thick and 2.5 mm deep was placed at a distance of 0.15 mm from the wall of the flat plate. This metal strip was excited by a magnetic field induced by an alternating current. This enabled two dimensional disturbances to be generated, as required by the theory. The experiments were conducted with disturbance waves with desired ranges of frequency, wave number, and amplitudes. In the experiments, the disturbance waves generated amplified, damped and neutral oscillations. The latter were measured with the aid of a hot-wire anemometer.

The results of the measurements are plotted in Schlichting (1979, Fig. 16.18) in a similar format to Fig. 1.15 and compared with the curve of neutral stability due to Tollmien (1929) shown in Fig. 1.15, and a very good agreement has been observed. The reader is referred to Schlichting (1979, pp. 477–479, Fig. 16.18) for details of this comparison exercise.

1.B.3.1 *Concluding remarks*

The theory of hydrodynamic stability offers us an explanation for the question of why the transition to turbulence occurs. According to the theory, the transition occurs simply because the laminar boundary-layer flow over a flat plate is unstable against small disturbances, and this is for Reynolds numbers larger than $Re_{x\ cr} = 3 \times 10^5$–$5 \times 10^5$. Therefore, for such large Reynolds numbers, any small disturbance grows (theoretically unlimited), leading to transition to turbulence in the boundary layer.

Now, we have seen in the previous section, Section 1.A, that the transition to turbulence occurs with the emergence of tiny turbulent patches, called turbulent spots, described in greater detail in Section 1.A.

The hydrodynamic stability theory does not, however, give explanations for how an "infant" spot comes into existence, an important, if not the most important, stage in the laminar-to-turbulent transition process. It does not answer questions like: What exactly happens? What are the physics behind the emergence of a turbulent spot? What is (are) the mechanism(s) governing this inception?

1.C Direct numerical simulation (DNS) of transition

Recently, Wu et al. (2017) conducted a Direct Numerical Simulation (DNS) of a zero-pressure-gradient, smooth, flat-plate boundary layer flow. Two simulations were conducted, one for a range of the Reynolds number Re_θ of $80 < Re_\theta < 450$ (Wu et al., 2017, Fig. 1; This Reynolds number is defined below) where the laminar-to-turbulent transition takes place, and the other for a range of the Reynolds number of $2400 < Re_\theta < 2650$ (Wu et al., 2017, Fig. 2) where the flow is in the fully-developed turbulent state.

The former range,

$$80 < Re_\theta < 450 \tag{1.43}$$

is, in terms of the current-length Reynolds number,

$$1.5 \times 10^4 < Re_x < 4.6 \times 10^5 \tag{1.44}$$

Here Re_θ is the momentum-thickness Reynolds number (used by the authors),

$$Re_\theta = \frac{\theta U_\infty}{\nu} \tag{1.45}$$

and θ is the momentum thickness, defined by

$$\theta = \int_0^\infty \frac{u(y)}{U_\infty}\left(1 - \frac{u(y)}{U_\infty}\right) dy \tag{1.46}$$

and given, for laminar boundary layers, as (Schlichting, 1979, p. 141)

$$\theta = 0.664\sqrt{\frac{\nu x}{U_\infty}} \tag{1.47}$$

From Eqs. 1.3, 1.45 and 1.47, the relationship between the above two Reynolds numbers is

$$Re_\theta = 0.664\sqrt{Re_x} \tag{1.48}$$

Wu et al.'s (2017) momentum-thickness Reynolds number range in Eq. 1.43 was converted to that of the current-length Reynolds in Eq. 1.44, using Eq. 1.48. We note that, in the conversion, we assumed that the laminar solution in Eq. 1.47 (and therefore Eq. 1.48) holds true also for the upper end of the range $80 < Re_\theta < 450$.

For the laminar-to-turbulent transition simulation, Wu et al. (2017) experimented with what they called "bypass transition in the narrow sense". They defined this transition as the superposition on a Blasius boundary layer by a continuous free stream flow of homogeneous isotropic turbulence at an initial turbulence intensity of 1–4%. They emphasized that the flow so defined is a subclass of bypass transition in the general sense, which refers to laminar boundary-layer transition caused by finite amplitude perturbations arising from unrestricted sources, such as passing wakes, roughness, or blowing and suction (Wu et al., 2014). Wu et al. (2017) further state that the idea put forward in their study is that turbulent spot inception mechanism in bypass transition in the narrow sense is analogous to the secondary instability that occurs in boundary-layer natural transition.

To achieve the stated objective, Wu et al. (2017) designed (and eventually executed) a DNS of the spatially developing zero-pressure-gradient, smooth plate boundary layer from an initial laminar state through a well-controlled process of bypass transition in the narrow sense to the fully turbulent state over the previously mentioned Reynolds number range, $80 < Re_\theta < 450$.

Vortical structures inside turbulent spots in Wu et al. (2017) were made visible, using the Swirling-Strength Criterion. This method uses the imaginary part of the complex eigenvalues of the velocity gradient tensor, λ_{ci}, to make vortices visible. Without resorting to this kind of method (there are similar methods such as Q-criterion, λ_2-criterion, etc.), no clear images of vortex structures inside the body of a turbulent spot can be achieved. These vortical structures are normally all invisible to the naked eye in a potential flow visualization study, in the laboratory, or in a numerical simulation. However, these vortical structures become fully visible when visualized with the help of one of the aforementioned vortex-identification methods. (This is analogous to creating images of blood vessels in the human body, using a computed tomography angiography (CTA) where marker (dye) can be placed with a simple injection in the arm.)

Wu et al. (2017) report that, through their DNS simulations, two previously unobserved phenomena were discovered:

The first is associated with the laminar-to-turbulent transition. Their simulation results showed that the free-stream turbulence perturbs the *spanwise* vortex lines inside the laminar boundary layer, producing strong, localized, *spanwise vortices*. These spanwise vortex lines quickly kink and are subsequently lifted up and stretched downstream into so-called Λ vortices, so named after their shape. The Λ vortices then evolve into hairpin packets of vortices, to form an infant turbulent spot. (Wu et al., 2017, call the latter transitional turbulent spot, to differentiate it from a second kind of turbulent spot, what they call turbulent-turbulent spot, which will be discussed shortly.) The turbulent spots further evolve and mature, as they move downstream, where they "infect" surrounding low-speed streaks and cause them to meander and breakdown, resulting in a sudden spreading and growth of the existing turbulent spots. Wu et al. (2017, Supporting Information, Movie S5) presented evidence with backtracking in time of the mature spots that the mature spots have evolved from small hairpin packets.

The second phenomenon discovered is the previously mentioned turbulent-turbulent spot. Downstream in the fully turbulent region of the simulation, spatially intermittent concentrations of small-scale vortices occur in the buffer layer region and extend down to the wall. They call such inner-layer concentrations of vortices with high values of swirling strength λ_{ci} turbulent-turbulent spots. They argue that these latter spots originate from hairpin packets in the turbulent boundary layer just as turbulent spots (transitional turbulent spots) do in the laminar to turbulent transition.

Wu et al. (2017) address the question how their results reconcile with the results from earlier studies of turbulent boundary layers what they call the streak-centered theories whereby streaks are considered to actively participate in a self-sustaining bursting cycle, namely, streak generation, lift up, oscillation, breakdown, and streak regeneration (see Section 3.5 in Chapter 3). In this regard, Wu et al. (2017) make the following points.

1. Wu et al. remark (2017, p. E5298) that, at this time, they are unable to entirely rule out the possibility that the streak-vortex self-sustaining cycle mechanism could also be a potential factor at play, co-existing with Wu et al.'s (2017) turbulent-turbulent spot mechanism.

2. The observations regarding the passive nature of the streaks versus the more dominating active role of turbulent-turbulent spots seem to be at odds with the observations of Kline and co-workers (Kline et al., 1967; Offen and Kline, 1974 and 1975), who reported that dye/smoke was scooped up, presumably by a vortex, and then went through a violent breakup process. Wu et al. (2017) in this context remark that their view is that smoke lift up, oscillation and breakup may be a by-product of the turbulent-turbulent spot dynamics.

3. Wu et al. (2017) also remark that, at this time, it is unclear whether the inception of turbulent-turbulent spots is aided by perturbations arising from the deep penetration of outer free stream flow into the inner layer.

4. They emphasize the following: One hallmark of the turbulent zero-pressure-gradient smooth flat-plate boundary layer is the deep penetration of free stream irrotational flow or free stream turbulence into the inner layer. This feature is absent in a channel or Couette flow (flow between two plates in which one is moving tangentially relative to the other). This distinguishing characteristic perhaps holds the key in reconciling their observations with the existing streak-vortex-centered theories developed using the channel and Couette flow, although, they say, this is purely a conjecture at this time.

Finally, we mention the following results of a study from the same group

(Wallace et al., 2010, and Park et al., 2012) regarding comparison of turbulent spots and fully-developed turbulent boundary layers for the flow over a flat plate. As mentioned earlier, Cantwell et al.'s (1978) measurement of the spacing between the low-speed wall streaks inside turbulent spots is in good accord with that in fully-developed turbulent boundary layers (Appendix 1.A.3.1). Likewise, the same authors showed that the side view of a single turbulent spot and that of a fully-developed turbulent boundary layer are quite similar (Appendix 1.A.3.1). Wallace et al. (2010) (see also Park et al., 2012 as well as Wu et al., 2017) from their DNS simulations showed unequivocally that statistics of turbulence (distributions of the mean velocity, r.m.s. velocity and vorticity component fluctuations, kinetic energy production and dissipation rates and enstrophy) are quite similar for the transition and for the fully-developed stage of turbulence, provided that only time instances when the flow is turbulent are used to determine the statistics for the transitional case. All this evidence strongly suggests that there is little difference between the structure and transport processes of a developed turbulent boundary layer and of turbulent spots that emerge in transition.

1.D References

(1) Browand, F. and Plocher, D.A. (1985): Image processing for sediment transport. In 21st IAHR Congress, pp. 19–23, Melbourne, Australia.

(2) Cantwell, B., Coles, D. and Dimotakis, P. (1978): Structure and entrainment in the plane of symmetry of a turbulent spot. J. Fluid Mech., vol. 87, 641–672.

(3) Carstensen, S., Sumer, B.M. and Fredsøe, J. (2010): Coherent structures in wave boundary layers. Part 1. Oscillatory motion. J. Fluid Mech., vol. 646, 169–206.

(4) Carstensen S., Sumer, B.M. and Fredsøe, J. (2012): A note on turbulent spots over a rough bed in wave boundary layers. Phys. Fluids, vol. 24, article no. 115104.

(5) Chambers, F.W. and Thomas, A.S.W. (1983): Turbulent spots, wave packets, and growth. Phys. Fluids, vol. 26, 1160–1162.

(6) Elder, J.W. (1960): An experimental investigation of turbulent spots and breakdown to turbulence. J. Fluid Mech., vol. 9, 235–246.

(7) Emmons, H.W. (1951): The laminar-turbulent transition in a boundary layer. Part I. J. Aero. Sci., vol. 18, 490–498.

(8) Gad-el-Hak, M., Blackwelder, R.F. and Riley, J.J. (1981): On the growth of turbulent regions in laminar boundary layers. J. Fluid Mech., vol. 110, 73–95.
(9) Jordinson, R. (1970): The flat plate boundary layer. Part 1. Numerical integration of the Orr-Sommerfeld equation. J. Fluid Mech., vol. 43, 801–811.
(10) Kline, S.J., Reynolds, W.C., Schraub, F.A. and Runstadler, P.W. (1967): The structure of turbulent boundary layers. J. Fluid Mech., vol. 30, 741–773.
(11) Monin, A.S. and Yaglom, A.M. (1973): Statistical Fluid Mechanics: Mechanics of Turbulence. MIT Press, Cambridge, Mass.
(12) Park, G.I., Wallace, J.M., Wu, X. and Moin, P. (2012): Boundary layer turbulence in transitional and developed states. Phys. Fluids, vol. 24, article no. 035105.
(13) Offen, G.R. and Kline, S.J. (1974): Combined dye-streak and hydrogen bubble visual observations of a turbulent boundary layer. J. Fluid Mech., vol. 62, p. 223.
(14) Offen, G.R. and Kline, S.J. (1975): A proposed model of the bursting process in turbulent boundary layers. J. Fluid Mech., vol. 70, 209–228.
(15) Schubauer, G.B. and Klebanoff, P.S. (1955): Contributions on the mechanics of boundary-layer transition. N.A.C.A. Tech. Note No. 3489.
(16) Schubauer, G.B. and Skramstad, H.K. (1943): Laminar boundary layer oscillations and stability of laminar flow. National Bureau of Standards Research Paper 1772. Reprint of confidential NACA Rep. dated April 1943.
(17) Schlichting, H. (1933): Zur Entstehung der Turbulenz bei der Plattenstromung. Nachr. Ges. Wiss. Gottingen, Math. Phys. Klasse 182–208. See also ZAMM 13, 171–174. 1933.
(18) Schlichting, H. (1979): Boundary Layer Theory. McGraw-Hill.
(19) Squire, H.B. (1933): On the stability of three-dimensional distribution of viscous fluid between parallel walls. Proc. Royal Soc. Lond. A, vol. 142, 621–628.
(20) Tollmien, W. (1929): Uber die Entstehung der Turbulenz. Mitt. Nachr. Ges. Wiss. Gottingen, Math. Phys. Klasse 21–44. English translation in NACA TM 609 (1931).
(21) van Dyke, M. (1982): An Album of Fluid Motion. The Parabolic Press, Stanford, California.

(22) Wallace, J.M., Park, G.I., Wu, X. and Moin, P. (2010): Boundary layer turbulence in transitional and developed states. Center for Turbulence Research. Proceedings of the Summer Program 2010, 77–86.
(23) Werle, H. (1980): Rech. Aerosp. 1980-5, 35–49.
(24) Wu, X., Moin, P. and Hickey, J.-P. (2014): Boundary layer bypass transition. Phys. Fluids, vol. 26, article no. 091104.
(25) Wu, X., Moin, P., Wallace, J.M., Skarda, J., Lozano-Duran, A. and Hickey, J.-P. (2017): Transitional-turbulent spots and turbulent-turbulent spots in boundary layers. Proc. Nat. Academ. Sci. U.S.A., vol. 114, E5292–E5299.

Chapter 2

Basic equations for turbulent flows

In this chapter, we will derive the basic equations for turbulent flows. These are the continuity equation, the equations of motion, and the energy equation. In a turbulent flow, any hydrodynamic quantity (such as any component of the velocity, or the pressure, or the temperature, etc.) fluctuates with time. Fig. 2.1 schematically illustrates a typical time series of any quantity (the i-component of the velocity in the example in the figure). Traditionally, this kind of behaviour is handled by decomposing the quantity in consideration into two components, namely a mean value and a fluctuating value around the mean. The next section will focus on the method of averaging to achieve mean values. Subsequently, the fundamental equations of hydrodynamics (the continuity equation and the equations of motion) will, by time averaging, be "transformed" to those for turbulent flows, leading to three sets of equations, namely the continuity equation, the Reynolds equations, and the energy equations, the fundamental equations for turbulent flows.

2.1 Method of averaging and Reynolds decomposition

The *mean value* of a hydrodynamic quantity, for example, the i-component of the velocity, u_i, is defined by

$$\overline{u}_i = \frac{1}{T} \int_{t_0}^{t_0+T} u_i \, dt \tag{2.1}$$

in which t_0 is any arbitrary time, and T is the time over which the mean is taken. Clearly T should be sufficiently large to give a reliable mean value.

This is called the time averaging. There are also other kinds of averaging, such as ensemble averaging, or moving averaging. We shall adopt the

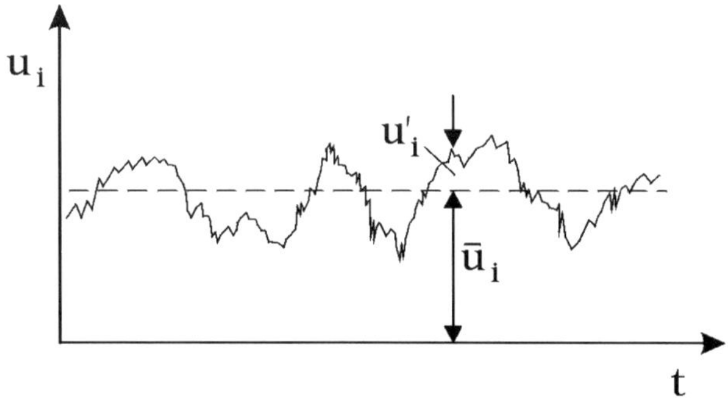

Fig. 2.1 Time series of u_i.

ensemble averaging in the case of wave boundary layers, as will be detailed in Chapter 5.

The instantaneous value of the velocity u_i may be written (see Fig. 2.1)

$$u_i = \overline{u}_i + u'_i \tag{2.2}$$

in which u'_i is called the *fluctuating part* (or *fluctuation*) of u_i, and $\overline{u}_i$ is its mean. This is called the *Reynolds decomposition*, after Reynolds (1895).

The Reynolds decomposition can be applied to any hydrodynamic quantity. Let f and g be any such quantities (any velocity component, pressure, etc.). The following relations are known as the Reynolds conditions (here, the overbar denotes the time averaging, Eq. 2.1):

$$\begin{aligned}
\overline{f+g} &= \overline{f} + \overline{g} \\
\overline{a\, f} &= a\, \overline{f} \\
\overline{a} &= a \\
\overline{\frac{\partial f}{\partial s}} &= \frac{\partial \overline{f}}{\partial s} \\
\overline{\overline{f}\, g} &= \overline{f}\, \overline{g} \\
\overline{\overline{f}} &= \overline{f} \\
\overline{f'} &= 0 \\
\overline{\overline{f}\, \overline{g}} &= \overline{f}\, \overline{g} \\
\overline{\overline{f}\, h'} &= \overline{f}\, \overline{h'} = 0
\end{aligned} \tag{2.3}$$

in which $a =$ constant, and $s = x_1, x_2, x_3, t$.

2.2 Continuity equation

For an incompressible fluid (such as water, to a very close approximation in most circumstances), the continuity equation (from the conservation of mass, see Appendix 2.A) is

$$\frac{\partial u_i}{\partial x_i} = 0 \tag{2.4}$$

where Einstein's summation convention is used, namely there is summation over the repeated indices:

$$\frac{\partial u_1}{\partial x_1} + \frac{\partial u_2}{\partial x_2} + \frac{\partial u_3}{\partial x_3} = 0 \tag{2.5}$$

in which u_i is the i-component of the velocity.

Averaging (Eq. 2.1) gives:

$$\overline{\frac{\partial u_i}{\partial x_i}} = \frac{\partial \overline{u}_i}{\partial x_i} = 0 \tag{2.6}$$

Subtracting the preceding equation from Eq. 2.4 gives

$$\frac{\partial u_i'}{\partial x_i} = 0 \tag{2.7}$$

Eq. 2.6 is the continuity equation for the mean velocity and Eq. 2.7 is the continuity equation for its fluctuating part.

2.3 Equations of motion

For an incompressible Newtonian fluid (such as water), the *Navier-Stokes (N.-S.) equation* is, from Appendix 2.B:

$$\rho \left(\frac{\partial u_i}{\partial t} + u_j \frac{\partial u_i}{\partial x_j} \right) = \rho g_i + \frac{\partial \sigma_{ij}}{\partial x_j} \tag{2.8}$$

in which t is the time, g_i is the volume force (such as the gravity force), x_i are the Cartesian coordinates, and σ_{ij} is the stress on the fluid, given as

$$\sigma_{ij} = -p\delta_{ij} + \mu \left(\frac{\partial u_i}{\partial x_j} + \frac{\partial u_j}{\partial x_i} \right) \tag{2.9}$$

in which p is the pressure, μ is the dynamic viscosity, and δ_{ij} is the Kronecker delta defined as

$$\begin{aligned} \delta_{ij} &= 1 \text{ for } i = j, \\ \delta_{ij} &= 0 \text{ for } i \neq j \end{aligned} \tag{2.10}$$

Eq. 2.9 is the constitutive equation for an incompressible Newtonian fluid. For a detailed derivation of the Navier-Stokes equation, refer to Appendix 2.B.

From the continuity equation (Eq. 2.4), the N.-S. equation becomes

$$\rho \frac{\partial u_i}{\partial t} + \frac{\partial}{\partial x_j}(\rho u_i u_j) = \rho g_i + \frac{\partial \sigma_{ij}}{\partial x_j} \tag{2.11}$$

and averaging each term results in

$$\rho \frac{\partial \overline{u}_i}{\partial t} + \frac{\partial}{\partial x_j}(\rho \overline{u_i u_j}) = \rho \overline{g}_i + \frac{\partial \overline{\sigma}_{ij}}{\partial x_j} \tag{2.12}$$

Now, the second term on the left hand side of the preceding equation (using the Reynolds decomposition) yields

$$\frac{\partial}{\partial x_j}(\rho \overline{u_i u_j}) = \rho \overline{u}_j \frac{\partial \overline{u}_i}{\partial x_j} + \frac{\partial}{\partial x_j}\left(\rho \overline{u_i' u_j'}\right) \tag{2.13}$$

Inserting the preceding equation in Eq. 2.12

$$\rho \left(\frac{\partial \overline{u}_i}{\partial t} + \overline{u}_j \frac{\partial \overline{u}_i}{\partial x_j} \right) = \rho \overline{g}_i + \frac{\partial}{\partial x_j}\left(\overline{\sigma}_{ij} - \rho \overline{u_i' u_j'} \right) \tag{2.14}$$

This equation is known as the *Reynolds equation*. (It is also commonly referred to as the *Reynolds-averaged Navier-Stokes (RANS) equation*, especially in numerical modelling circles; This terminology will be utilized interchangeably in this book.)

As seen, the equation of motion for the mean flow (Eq. 2.14) is quite similar to that of a laminar flow (Eq. 2.8). Comparison of the two equations indicates that, in the case of the turbulent flow, there is an additional stress, namely $-\rho \overline{u_i' u_j'}$. This additional stress is called the *Reynolds stress*. There are nine components of the Reynolds stress:

$$-\rho \overline{u_i' u_j'} = \begin{pmatrix} -\rho \overline{u_1' u_1'} & -\rho \overline{u_1' u_2'} & -\rho \overline{u_1' u_3'} \\ -\rho \overline{u_2' u_1'} & -\rho \overline{u_2' u_2'} & -\rho \overline{u_2' u_3'} \\ -\rho \overline{u_3' u_1'} & -\rho \overline{u_3' u_2'} & -\rho \overline{u_3' u_3'} \end{pmatrix} \tag{2.15}$$

As seen, the Reynolds stress forms a second order tensor, called the Reynolds stress tensor. It can readily be seen that the Reynolds stress tensor is symmetric, $-\rho \overline{u_i' u_j'} = -\rho \overline{u_j' u_i'}$.

To summarize at this point, there are four equations for the mean flow, the continuity equation (Eq. 2.6) and the three equations of motion (Eq. 2.14) while there are ten unknowns, namely, the three components of the velocity, $\overline{u}_i$, the pressure, $\overline{p}$, and the six components of the Reynolds stress, $-\rho \overline{u_i' u_j'}$. Therefore, the system is not closed. This problem is known as the *closure problem of turbulence.* We shall return to the closure problem later. See also Section 9.1.

In the case of laminar flows, there are four unknowns (namely, the three components of the velocity, u_i, and the pressure, p) and four equations, namely the continuity equation (Eq. 2.4) and the three N.-S. equations (Eqs. 2.8 with 2.9). Therefore, the system is closed.

2.4 Energy equations

2.4.1 *Energy equation for laminar flows*

The N.-S. equation (Eqs. 2.8 and 2.9; see also Appendix 2.B, Eq. 2.91) may be written as:

$$\rho\left(\frac{\partial u_i}{\partial t}+u_\alpha\frac{\partial u_i}{\partial x_\alpha}\right)=\rho g_i+\frac{\partial}{\partial x_\alpha}\left(-p\delta_{i\alpha}+\mu\left(\frac{\partial u_i}{\partial x_\alpha}+\frac{\partial u_\alpha}{\partial x_i}\right)\right) \tag{2.16}$$

or, dividing both sides by ρ, and using the continuity equation

$$\frac{\partial u_i}{\partial t}+u_\alpha\frac{\partial u_i}{\partial x_\alpha}=g_i-\frac{1}{\rho}\frac{\partial p}{\partial x_i}+\nu\frac{\partial^2 u_i}{\partial x_\alpha\partial x_\alpha} \tag{2.17}$$

where $\nu=\mu/\rho$ is the kinematic viscosity. Now, multiplying both sides of the preceding equation by u_j yields

$$u_j\frac{\partial u_i}{\partial t}+u_ju_\alpha\frac{\partial u_i}{\partial x_\alpha}=u_jg_i-u_j\frac{1}{\rho}\frac{\partial p}{\partial x_i}+\nu u_j\frac{\partial^2 u_i}{\partial x_\alpha\partial x_\alpha} \tag{2.18}$$

The first term on the left hand side is

$$u_j\frac{\partial u_i}{\partial t}=\frac{\partial}{\partial t}(u_iu_j)-u_i\frac{\partial u_j}{\partial t} \tag{2.19}$$

Solving for $\frac{\partial u_j}{\partial t}$ from Eq. 2.17, and inserting this into the preceding equation gives

$$u_j\frac{\partial u_i}{\partial t}=\frac{\partial}{\partial t}(u_iu_j)-u_i\left(g_j-\frac{1}{\rho}\frac{\partial p}{\partial x_j}+\nu\frac{\partial^2 u_j}{\partial x_\alpha\partial x_\alpha}-u_\alpha\frac{\partial u_j}{\partial x_\alpha}\right) \tag{2.20}$$

Substituting the latter equation in Eq. 2.18, and setting $i=j$,

$$\frac{\partial}{\partial t}\left(\frac{1}{2}\rho u_iu_i\right)+u_i\frac{\partial p}{\partial x_i}+\rho u_iu_\alpha\frac{\partial u_i}{\partial x_\alpha}=\rho u_ig_i+\mu u_i\frac{\partial^2 u_i}{\partial x_\alpha\partial x_\alpha} \tag{2.21}$$

Adding the term

$$-\mu\left(\frac{\partial u_i}{\partial x_\alpha}+\frac{\partial u_\alpha}{\partial x_i}\right)\frac{\partial u_i}{\partial x_\alpha}$$

on both sides of the equation and also the term

$$\mu u_i \frac{\partial^2 u_\alpha}{\partial x_i \partial x_\alpha}$$

on the right hand side of the equation (this term is actually zero due to continuity), and after some algebra, the following equation is obtained

$$\frac{\partial K}{\partial t} + \frac{\partial}{\partial x_\alpha}\left[K u_\alpha + p u_\alpha - u_i \mu\left(\frac{\partial u_i}{\partial x_\alpha} + \frac{\partial u_\alpha}{\partial x_i}\right)\right] = \rho u_\alpha g_\alpha - \Phi \tag{2.22}$$

in which K is the kinetic energy per unit volume of fluid

$$K = \frac{1}{2}\rho u_i u_i = \frac{1}{2}\rho\left(u^2 + v^2 + w^2\right) \tag{2.23}$$

and Φ is the energy dissipation per unit volume of fluid and per unit time

$$\Phi = \mu\left(\frac{\partial u_i}{\partial x_\alpha} + \frac{\partial u_\alpha}{\partial x_i}\right)\frac{\partial u_i}{\partial x_\alpha} \tag{2.24}$$

2.4.2 *Energy equation for the mean flow*

The equation of motion for the mean flow (the Reynolds equation, Eq. 2.14) reads:

$$\rho\frac{\partial \overline{u}_i}{\partial t} + \rho u_\alpha \frac{\partial \overline{u}_i}{\partial x_\alpha} = \rho \overline{g}_i - \frac{\partial \overline{p}}{\partial x_j} + \mu \frac{\partial^2 \overline{u}_i}{\partial x_\alpha \partial x_\alpha} + \frac{\partial}{\partial x_\alpha}\left(-\rho\overline{u'_i u'_\alpha}\right) \tag{2.25}$$

Now, in the preceding subsection, the energy equation, Eq. 2.22, is obtained from the equation of motion, Eq. 2.17. In a similar manner, the energy equation for the mean flow can be obtained from the equation of motion for the mean flow (i.e., from Eq. 2.25). The result is

$$\begin{aligned}\frac{\partial K_0}{\partial t} + \frac{\partial}{\partial x_\alpha}\left[K_0 \overline{u}_\alpha + \rho\overline{u'_i u'_\alpha}\,\overline{u}_i + \overline{p}\,\overline{u}_\alpha - \overline{u}_i \mu\left(\frac{\partial \overline{u}_i}{\partial x_\alpha} + \frac{\partial \overline{u}_\alpha}{\partial x_i}\right)\right] = \\ \rho \overline{u}_\alpha \overline{g}_\alpha - \Phi_0 - \left(-\rho\overline{u'_i u'_\alpha}\right)\frac{\partial \overline{u}_i}{\partial x_\alpha}\end{aligned} \tag{2.26}$$

in which K_0 is the kinetic energy per unit volume of fluid for the mean flow

$$K_0 = \frac{1}{2}\rho \overline{u}_i \overline{u}_i = \frac{1}{2}\rho\left(\overline{u}^2 + \overline{v}^2 + \overline{w}^2\right) \tag{2.27}$$

and Φ_0 is the energy dissipation per unit volume of fluid and per unit time, again, for the mean flow

$$\Phi_0 = \mu\left(\frac{\partial \overline{u}_i}{\partial x_\alpha} + \frac{\partial \overline{u}_\alpha}{\partial x_i}\right)\frac{\partial \overline{u}_i}{\partial x_\alpha} \tag{2.28}$$

Now, note the last term on the right hand side of Eq. 2.26. We shall return to this term in the next subsection.

2.4.3 *Energy equation for the fluctuating flow*

Averaging Eq. 2.22 gives

$$\frac{\partial}{\partial t}\left(\frac{1}{2}\rho\overline{u_i u_i}\right)+\frac{\partial}{\partial x_\alpha}\left[\frac{1}{2}\rho\overline{u_i u_i u_\alpha}+\overline{p u_\alpha}-\overline{u_i\mu\left(\frac{\partial u_i}{\partial x_\alpha}+\frac{\partial u_\alpha}{\partial x_i}\right)}\right]=\rho\overline{u_\alpha g_\alpha}-\overline{\Phi} \tag{2.29}$$

Furthermore, Eq. 2.26 may be written as:

$$\frac{\partial}{\partial t}\left(\frac{1}{2}\rho\overline{u}_i\overline{u}_i\right)+\frac{\partial}{\partial x_\alpha}\left[\frac{1}{2}\rho\overline{u}_i\overline{u}_\alpha\overline{u}_i+\rho\overline{u_i'u_\alpha'}\,\overline{u}_i+\overline{p}\,\overline{u}_\alpha-\overline{u}_i\mu\left(\frac{\partial\overline{u}_i}{\partial x_\alpha}+\frac{\partial\overline{u}_\alpha}{\partial x_i}\right)\right]=$$

$$=\rho\overline{u}_\alpha\overline{g}_\alpha-\Phi_0-\left(-\rho\overline{u_i'u_\alpha'}\right)\frac{\partial\overline{u}_i}{\partial x_\alpha} \tag{2.30}$$

Subtracting Eq. 2.30 from Eq. 2.29 gives

$$\frac{\partial K_t}{\partial t}+\frac{\partial}{\partial x_\alpha}\left[K_t\overline{u}_\alpha+\frac{1}{2}\rho\overline{u_i'u_i'u_\alpha'}+\overline{p'u_\alpha'}-\overline{u_i'\mu\left(\frac{\partial u_i'}{\partial x_\alpha}+\frac{\partial u_\alpha'}{\partial x_i}\right)}\right] \tag{2.31}$$
$$=\rho\overline{u_\alpha'g_\alpha'}-\Phi_t+\left(-\rho\overline{u_i'u_\alpha'}\right)\frac{\partial\overline{u}_i}{\partial x_\alpha}$$

in which K_t is the kinetic energy per unit volume of fluid for the fluctuating flow (i.e., the kinetic energy of turbulence)

$$K_t=\frac{1}{2}\rho\overline{u_i'u_i'}=\frac{1}{2}\rho\left(\overline{u'^2}+\overline{v'^2}+\overline{w'^2}\right) \tag{2.32}$$

and Φ_t is the viscous dissipation of the turbulent energy per unit volume of fluid and per unit time

$$\Phi_t=\mu\overline{\left(\frac{\partial u_i'}{\partial x_\alpha}+\frac{\partial u_\alpha'}{\partial x_i}\right)\frac{\partial u_i'}{\partial x_\alpha}} \tag{2.33}$$

Eq. 2.26 is the energy equation for the mean flow, and Eq. 2.31 is that for turbulence. As seen, the term $\left(-\rho\overline{u_i'u_\alpha'}\right)\frac{\partial\overline{u}_i}{\partial x_\alpha}$ is common in both equations, but with opposite signs. This implies that if this term is an energy gain for turbulence, then it will be an energy loss for the mean flow, or vice versa. Now, we shall show that this term is an energy gain for turbulence.

Integrating Eq. 2.31 over a volume V gives

$$\frac{\partial}{\partial t}\int_V K_t\,dV+\int_S F_\alpha n_\alpha\,dS=\int_V\rho\overline{u_\alpha'g_\alpha'}\,dV-\int_V\Phi_t\,dV$$
$$+\int_V\left(-\rho\overline{u_i'u_\alpha'}\right)\frac{\partial\overline{u}_i}{\partial x_\alpha}\,dV \tag{2.34}$$

in which

$$F_\alpha = K_t \overline{u}_\alpha + \frac{1}{2}\rho\overline{u_i' u_i' u_\alpha'} + \overline{p' u_\alpha'} - \overline{u_i' \mu\left(\frac{\partial u_i'}{\partial x_\alpha} + \frac{\partial u_\alpha'}{\partial x_i}\right)} \tag{2.35}$$

and S is the surface encircling the volume V. Here, the volume integral in the second term on the left hand side of Eq. 2.34 is converted to a surface integral, using the Green-Gauss theorem.

The first term on the right hand side of Eq. 2.34 drops when the volume force is taken as the gravity force, since $g_\alpha' = 0$. The second term on the left hand side of Eq. 2.34, on the other hand, is the energy flux at the surface S. Therefore, it is not a source term.

Now, consider a situation where there is no influx of turbulent energy. In this case, the maintenance of turbulence (i.e., $\frac{\partial}{\partial t}\int_V K_t \, dV \geq 0$ in Eq. 2.34) is possible only when $\int_V \left(-\rho\overline{u_i' u_\alpha'}\right) \frac{\partial \overline{u}_i}{\partial x_\alpha} \, dV$ in Eq. 2.34 is positive, meaning that the term $\left(-\rho\overline{u_i' u_\alpha'}\right) \frac{\partial \overline{u}_i}{\partial x_\alpha}$ in the turbulence energy equation (Eq. 2.31) should be positive. In other words, this term is an energy gain for turbulence, and therefore it is an energy loss for the mean flow. This is an important result, because it implies that turbulence extracts its energy from the mean flow, and the rate at which the energy is extracted from the mean flow is $\left(-\rho\overline{u_i' u_\alpha'}\right) \frac{\partial \overline{u}_i}{\partial x_\alpha}$.

An important implication of the above result is that turbulence is generated only when there exists a velocity gradient in the mean flow $\frac{\partial \overline{u}_i}{\partial x_\alpha}$. When there is no velocity gradient, no turbulence will be generated. In this case, any field of turbulence introduced into the flow will be dissipated (the decay of turbulence).

Example 2.1. Discuss the turbulence energy generation and the total energy dissipation (per unit volume of fluid and per unit time) for a turbulent boundary layer flow in an open channel (Fig. 2.2).

The energy extracted from the mean flow is $\left(-\rho\overline{u_i' u_\alpha'}\right) \frac{\partial \overline{u}_i}{\partial x_\alpha}$. For the present flow, this will be

$$\left(-\rho\overline{u'v'}\right) \frac{\partial \overline{u}}{\partial y}$$

This is the energy extracted from the mean flow per unit time and per unit volume of fluid (Fig. 2.2c, full line). As seen from the figure, the energy "generation" for (or "production" of) turbulence is zero at the free surface while it increases tremendously near the wall (though it decays rapidly to zero at the wall, since $\overline{u'v'} = 0$ at the wall).

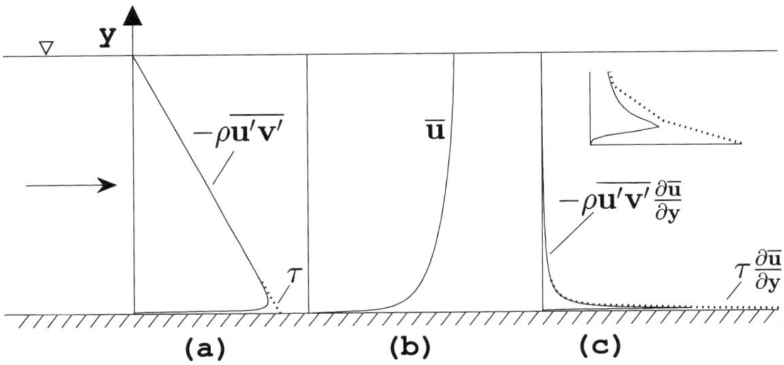

Fig. 2.2 Conceptual sketch for a turbulent boundary layer in an open channel, depicting profiles of (a) the Reynolds stress $-\rho\overline{u'v'}$ (full line) and the total shear stress $\tau = \mu\partial\overline{u}/\partial y - \rho\overline{u'v'}$ (dotted line), (b) the mean velocity $\overline{u}$, and (c) turbulent energy extracted from the mean flow $-\rho\overline{u'v'}\partial\overline{u}/\partial y$ (full line) as well as the total dissipation of the mean flow $\tau\partial\overline{u}/\partial y$ (dotted line). The inset in (c) depicts a zoomed in portion near the bed.

Note that while it is proved more formally above that the production of turbulent kinetic energy will generally be positive, this can also be seen from simple physical considerations in the present example i.e. a steady boundary layer flow with $\partial\overline{u}/\partial y > 0$ (Fig. 2.2). At any fixed y position in this boundary layer, when the instantaneous vertical fluctuation is $v' > 0$ it will tend to coincide with $u' < 0$, since the fluid will have come from the region of reduced mean $\overline{u}$ velocity below. Similarly, instants when $v' < 0$ will tend to coincide with $u' > 0$, since the fluid will then have come from the higher mean $\overline{u}$ velocity region above. Hence, the gradient in the mean flow in this example will cause a tendency for u' and v' to have opposite sign i.e. they will be negatively correlated such that $\overline{u'v'} \leq 0$. Since the mean velocity gradient in this case is again positive, this means that the sign of the turbulence energy production term identified in the preceding paragraph will indeed be positive i.e. $-\rho\overline{u'v'}\partial\overline{u}/\partial y \geq 0$.

From Eq. 2.26, the total energy dissipation from the mean flow, on the other hand, is

$$\Phi_0 + \left(-\rho\overline{u'_i u'_\alpha}\right)\frac{\partial\overline{u}_i}{\partial x_\alpha}$$

in which Φ_0 is given in Eq. 2.28. The total energy dissipation will therefore in this special case simplify to

$$\tau\frac{\partial\overline{u}}{\partial y} = \left(\mu\frac{\partial\overline{u}}{\partial y}\right)\frac{\partial\overline{u}}{\partial y} + \left(-\rho\overline{u'v'}\right)\frac{\partial\overline{u}}{\partial y}$$

From the preceding relation, it is seen that the energy dissipation is very large near the wall, and it decreases with the distance from the wall, similar to the variation of the turbulent energy generation (Fig. 2.2c, dotted line). Unlike the turbulent production term, however, the total energy dissipation is large and finite right at the wall, where it is dominated by the viscous contribution $\mu\left(\partial\overline{u}/\partial y\right)^2$. The near wall behavior in these quantities is emphasized in the inset in Fig. 2.2c, which depicts a zoomed in portion near the bed.

2.5 Reference

(1) Reynolds, O. (1895): On the Dynamical Theory of Incompressible Viscous Fluids and the Determination of the Criterion. Phil. Trans. Roy. Soc. A, vol. 186, 123–164.

Appendix to Chapter 2: Derivation of the continuity and Navier-Stokes equations

2.A Continuity equation

The continuity equation is derived from the principle of the conservation of mass. Consider a small control volume the size dx, dy, dz (Fig. 2.3). The influx of mass in the x-direction is (Fig. 2.3)

$$\rho u \, dy \, dz \tag{2.36}$$

in which u is the x-component of the velocity, and ρ is the density of the fluid. The efflux of mass in the same direction, on the other hand, is (Fig. 2.3)

$$\left(\rho u + \frac{\partial \rho u}{\partial x} dx \right) \, dy \, dz \tag{2.37}$$

Thus the net influx of mass in the x-direction is

$$-\frac{\partial \rho u}{\partial x} dx \; dy \, dz \tag{2.38}$$

and similar expressions for the other two directions:

$$-\frac{\partial \rho v}{\partial y} dy \; dx \, dz \tag{2.39}$$

$$-\frac{\partial \rho w}{\partial z} dz \; dx \, dy \tag{2.40}$$

in which v and w are the velocity components in the y- and z-directions, respectively.

The rate of change of mass in the volume dx, dy, dz is

$$\frac{\partial}{\partial t}(\rho \, dx \, dy \, dz) \tag{2.41}$$

in which t is the time.

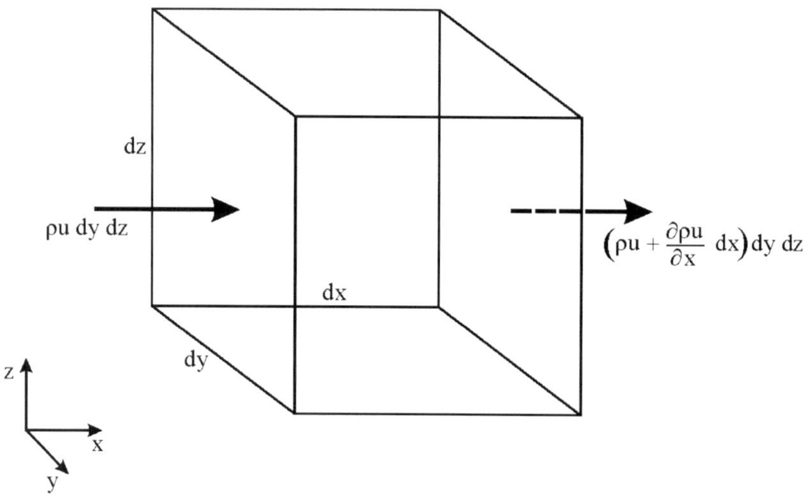

Fig. 2.3 Definition sketch for conservation of mass.

The conservation of mass requires the rate of change of mass in the volume dx, dy, dz should be equal to the total net influx of mass:

$$\frac{\partial}{\partial t}(\rho\, dx\, dy\, dz) = -\frac{\partial \rho u}{\partial x} dx\; dy\, dz - \frac{\partial \rho v}{\partial y} dy\; dx\, dz - \frac{\partial \rho w}{\partial z} dz\; dx\, dy$$

or

$$\frac{\partial \rho}{\partial t} + \frac{\partial \rho u}{\partial x} + \frac{\partial \rho v}{\partial y} + \frac{\partial \rho w}{\partial z} = 0 \tag{2.42}$$

This is the continuity equation for a compressible fluid. If the fluid is incompressible, ρ will be constant, and therefore Eq. 2.42 simplifies to:

$$\frac{\partial u}{\partial x} + \frac{\partial v}{\partial y} + \frac{\partial w}{\partial z} = 0 \tag{2.43}$$

or in vector form

$$\nabla \cdot \mathbf{u} = 0 \tag{2.44}$$

where $\mathbf{u} = (u, v, w)$ is the velocity vector and $\nabla = \left(\frac{\partial}{\partial x}, \frac{\partial}{\partial y}, \frac{\partial}{\partial z}\right)$ is the gradient operator. Using Einstein's summation notation, Eq. 2.43 can be written more compactly:

$$\frac{\partial u_i}{\partial x_i} = 0 \tag{2.45}$$

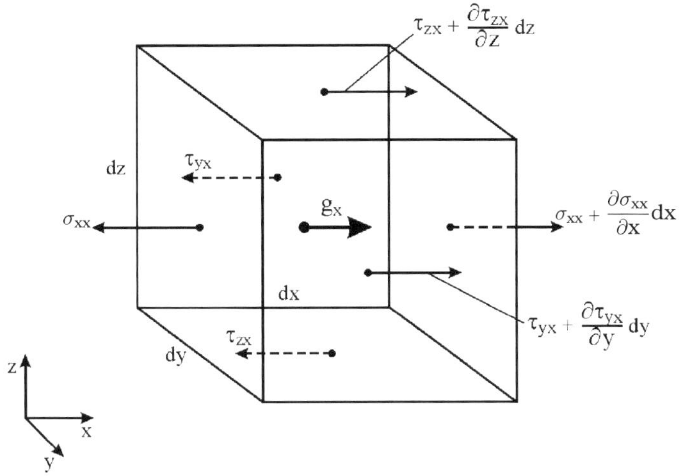

Fig. 2.4 Forces on a small element of fluid.

2.B Navier-Stokes equation

In this section, we will derive the Navier-Stokes equations. We will achieve this in six steps.

1) Equation of motion

Consider a small element of fluid, dx, dy, dz (Fig. 2.4). There are two kinds of forces acting on this fluid element: (1) surface forces (normal stresses and shear stresses); and (2) volume or body forces, such as the gravity force. These forces are indicated in Fig. 2.4 where only the components in the x-direction are shown, to keep the figure relatively simple. Here, σ_{xx} is the normal stress in the x-direction, τ_{yx} and τ_{zx} are the shear stresses in the x-direction (all surface forces); and g_x is the x-component of the volume force per unit volume.

From Newton's second law, the equation of motion in the x-direction for the considered fluid element is:

$$\rho \frac{du}{dt} = \rho g_x + \frac{\partial \sigma_{xx}}{\partial x} + \frac{\partial \tau_{yx}}{\partial y} + \frac{\partial \tau_{zx}}{\partial z} \tag{2.46}$$

Similarly, for the other two directions:

$$\rho \frac{dv}{dt} = \rho g_y + \frac{\partial \tau_{xy}}{\partial x} + \frac{\partial \sigma_{yy}}{\partial y} + \frac{\partial \tau_{zy}}{\partial z} \tag{2.47}$$

$$\rho\frac{dw}{dt} = \rho g_z + \frac{\partial\tau_{xz}}{\partial x} + \frac{\partial\tau_{yz}}{\partial y} + \frac{\partial\sigma_{zz}}{\partial z} \tag{2.48}$$

2) The stress tensor

The stress components on the right hand sides of the above equations form a tensor, called the stress tensor:

$$\begin{pmatrix} \sigma_{xx} & \tau_{yx} & \tau_{zx} \\ \tau_{xy} & \sigma_{yy} & \tau_{zy} \\ \tau_{xz} & \tau_{yz} & \sigma_{zz} \end{pmatrix} \tag{2.49}$$

The stress tensor is symmetric, meaning that $\tau_{yx} = \tau_{xy}$, etc. This can be demonstrated with reference to the equation of motion of the fluid element in Fig. 2.4 for the instantaneous rotational motion, Schlichting (1979). Using the symmetry conditions, the equations of motion (Eqs. 2.46–2.48) can, for convenience, be written as

$$\rho\frac{du}{dt} = \rho g_x + \frac{\partial\sigma_{xx}}{\partial x} + \frac{\partial\tau_{xy}}{\partial y} + \frac{\partial\tau_{xz}}{\partial z} \tag{2.50}$$

$$\rho\frac{dv}{dt} = \rho g_y + \frac{\partial\tau_{yx}}{\partial x} + \frac{\partial\sigma_{yy}}{\partial y} + \frac{\partial\tau_{yz}}{\partial z} \tag{2.51}$$

$$\rho\frac{dw}{dt} = \rho g_z + \frac{\partial\tau_{zx}}{\partial x} + \frac{\partial\tau_{zy}}{\partial y} + \frac{\partial\sigma_{zz}}{\partial z} \tag{2.52}$$

with the stress tensor

$$\begin{pmatrix} \sigma_{xx} & \tau_{xy} & \tau_{xz} \\ \tau_{yx} & \sigma_{yy} & \tau_{yz} \\ \tau_{zx} & \tau_{zy} & \sigma_{zz} \end{pmatrix} \tag{2.53}$$

We shall return to the above equations shortly. Next, we will discuss the diagonal components of the stress tensor.

If the fluid is at rest (the static situation), the shear stresses will vanish, and normal stresses will be equal in all directions (isotropy)

$$\sigma_{xx} = \sigma_{yy} = \sigma_{zz} = -p \tag{2.54}$$

in which the quantity p is the static fluid pressure, being identical with the thermodynamic pressure in the equation of state (Schlichting, 1979, p. 51).

If the fluid is in motion, the normal stresses are not equal in all directions. For convenience, however, we define the pressure to be the mean normal stress with minus sign, and denote it by the same symbol as in the static situation, p:

$$p = -\frac{1}{3}(\sigma_{xx} + \sigma_{yy} + \sigma_{zz}) \tag{2.55}$$

This quantity reduces to the static fluid pressure when the fluid is at rest. Calling this quantity pressure is an appropriate generalization of the elementary notion of pressure. It is important to note that this quantity is analogous to the static-fluid pressure in the sense that it will appear to be a measure of the local intensity of the "squeezing" of the fluid, Batchelor (1967, p. 141). It may be noted that, from tensor theory, this quantity, $p = -\frac{1}{3}(\sigma_{xx} + \sigma_{yy} + \sigma_{zz})$, is an invariant under the rotation of the axes of reference, Batchelor (1967, p. 141).

We now return to the stress tensor. It is convenient to write the stress tensor as the sum of two parts: (1) an isotropic part, and (2) a remaining non-isotropic part:

$$\begin{pmatrix} \sigma_{xx} & \tau_{xy} & \tau_{xz} \\ \tau_{yx} & \sigma_{yy} & \tau_{yz} \\ \tau_{zx} & \tau_{zy} & \sigma_{zz} \end{pmatrix} = \underbrace{\begin{pmatrix} -p & 0 & 0 \\ 0 & -p & 0 \\ 0 & 0 & -p \end{pmatrix}}_{\text{First Part}} + \underbrace{\begin{pmatrix} \sigma_{xx}+p & \tau_{xy} & \tau_{xz} \\ \tau_{yx} & \sigma_{yy}+p & \tau_{yz} \\ \tau_{zx} & \tau_{zy} & \sigma_{zz}+p \end{pmatrix}}_{\text{Second Part}} \tag{2.56}$$

with p defined by Eq. 2.55. The First Part is actually the stress tensor having the same form as that in a fluid at rest although the value of p for a moving fluid is not necessarily the same as in the same fluid at rest. The Second Part, termed as the deviatoric stress tensor, can be interpreted as the stress tensor in excess of the static situation, being due entirely to the motion of the fluid.

3) The rate-of-strain tensor

Let us return to the fluid element in Fig. 2.4, reproduced in Fig. 2.5. We consider Points A and B in Fig. 2.5, diagonally located at the bottom and top corners of the fluid element (dx, dy, dz). Owing to the presence of the velocity field, Point A will be displaced to Point A', and Point B to Point B' in a small time interval dt. The rate at which the fluid element is strained is actually determined by the relative motion of the two points, Point A and Point B.

Now, let the velocity at Point A be (u, v, w), and that at Point B $(u + du, v+dv, w+dw)$. The relative motion of Point A and Point B is described by (du, dv, dw):

$$du = \frac{\partial u}{\partial x}dx + \frac{\partial u}{\partial y}dy + \frac{\partial u}{\partial z}dz \tag{2.57}$$

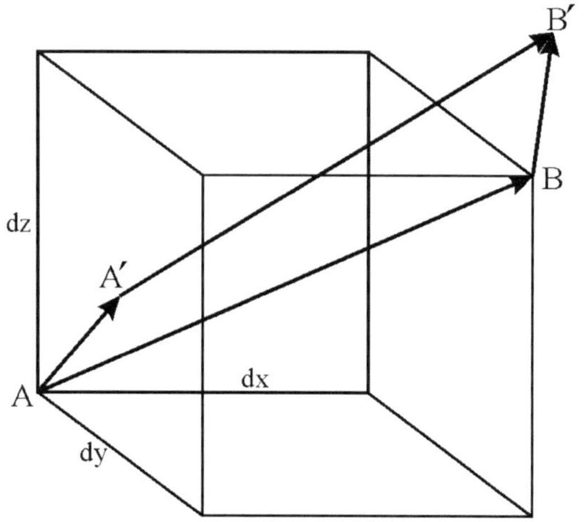

Fig. 2.5 Deformation of small fluid element. Point A is displaced to A', and Point B to B' in small time interval dt.

$$dv = \frac{\partial v}{\partial x}dx + \frac{\partial v}{\partial y}dy + \frac{\partial v}{\partial z}dz \tag{2.58}$$

$$dw = \frac{\partial w}{\partial x}dx + \frac{\partial w}{\partial y}dy + \frac{\partial w}{\partial z}dz \tag{2.59}$$

or by the following nine-element matrix

$$\begin{pmatrix} \partial u/\partial x & \partial u/\partial y & \partial u/\partial z \\ \partial v/\partial x & \partial v/\partial y & \partial v/\partial z \\ \partial w/\partial x & \partial w/\partial y & \partial w/\partial z \end{pmatrix} \tag{2.60}$$

It is convenient to re-arrange the expressions for the relative motion (du, dv, dw) in the following form:

$$du = (s_{xx}\, dx + s_{xy}\, dy + s_{xz}\, dz) + (\eta\, dz - \zeta\, dy) \tag{2.61}$$

$$dv = (s_{yx}\, dx + s_{yy}\, dy + s_{yz}\, dz) + (\zeta\, dx - \xi\, dz) \tag{2.62}$$

$$dw = (s_{zx}\, dx + s_{zy}\, dy + s_{zz}\, dz) + (\xi\, dy - \eta\, dx) \tag{2.63}$$

First Term Second Term

with

$$s_{ij} = \begin{pmatrix} s_{xx} & s_{xy} & s_{xz} \\ s_{yx} & s_{yy} & s_{yz} \\ s_{zx} & s_{zy} & s_{zz} \end{pmatrix} = \begin{pmatrix} \frac{\partial u}{\partial x} & \frac{1}{2}\left(\frac{\partial v}{\partial x} + \frac{\partial u}{\partial y}\right) & \frac{1}{2}\left(\frac{\partial w}{\partial x} + \frac{\partial u}{\partial z}\right) \\ \frac{1}{2}\left(\frac{\partial u}{\partial y} + \frac{\partial v}{\partial x}\right) & \frac{\partial v}{\partial y} & \frac{1}{2}\left(\frac{\partial w}{\partial y} + \frac{\partial v}{\partial z}\right) \\ \frac{1}{2}\left(\frac{\partial u}{\partial z} + \frac{\partial w}{\partial x}\right) & \frac{1}{2}\left(\frac{\partial v}{\partial z} + \frac{\partial w}{\partial y}\right) & \frac{\partial w}{\partial z} \end{pmatrix} \tag{2.64}$$

and

$$\xi = \frac{1}{2}\left(\frac{\partial w}{\partial y} - \frac{\partial v}{\partial z}\right), \qquad \eta = \frac{1}{2}\left(\frac{\partial u}{\partial z} - \frac{\partial w}{\partial x}\right), \qquad \zeta = \frac{1}{2}\left(\frac{\partial v}{\partial x} - \frac{\partial u}{\partial y}\right) \tag{2.65}$$

The expressions for the relative motion (du, dv, dw) consist of two terms: A First Term and a Second Term, as indicated in Eqs. 2.61–2.63.

We first discuss the Second Term. This term actually gives rise to the components of the vorticity vector $\boldsymbol{\omega}$, which is formally defined as

$$\boldsymbol{\omega} = (\omega_1, \omega_2, \omega_3) = 2(\xi, \eta, \zeta) \tag{2.66}$$

with the component terms given in Eq. 2.65 (For derivation of the so-called vorticity vector equation, see Appendix 6.A). The vorticity represents a rigid-body rotation of the fluid element, and therefore cannot be responsible for the strain experienced by the fluid element.

From the above considerations, the remaining part of the expressions for the relative motion (du, dv, dw), namely the First Term, should presumably be associated with the strain that the fluid element experiences. This term is represented by the tensor s_{ij} displayed in Eq. 2.64. This tensor is called the rate-of-strain (or strain rate) tensor. From Eq. 2.64, it is seen that s_{ij} is a symmetric tensor, $s_{ij} = s_{ji}$.

There are two kinds of elements of the rate-of-strain tensor: the diagonal components, $s_{xx}(= \frac{\partial u}{\partial x})$, etc.; and the off-diagonal components $s_{xy}(= \frac{1}{2}(\frac{\partial v}{\partial x} + \frac{\partial u}{\partial y}))$, etc. The diagonal components represent the rate of elongation; e.g., $s_{xx}(= \frac{\partial u}{\partial x})$ represents the rate of elongation in the x-direction experienced by the fluid element. The off-diagonal components, on the other hand, represent the rate of angular strain or deformation; e.g., $s_{xy}(= s_{yx})$ represents the rate of angular deformation in the plane normal to the z-axis, as illustrated in Fig. 2.6.

4) The stress and rate-of-strain relationship

We now return to the stress tensor in Eq. 2.56. *The components of the deviatoric stress tensor, the Second Part of the stress tensor in* Eq. 2.56, *are essentially created by the components of the rate-of-strain tensor,* Eq. 2.64. (This is because the remaining other effects, namely (1) the translational motion of the fluid element characterized by the velocity (u, v, w), and (2) the rigid-body rotation of the fluid element characterized by the vorticity components (ξ, η, ζ), cannot generate surface forces on the fluid element.)

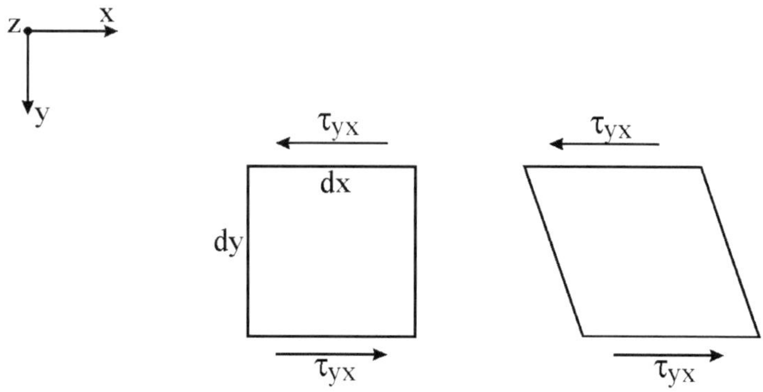

Fig. 2.6 Angular deformation of small fluid element.

On this premise, it may be concluded that the expressions for the components of the deviatoric stress tensor,

$$\begin{pmatrix} \sigma_{xx}+p & \tau_{xy} & \tau_{xz} \\ \tau_{yx} & \sigma_{yy}+p & \tau_{yz} \\ \tau_{zx} & \tau_{zy} & \sigma_{zz}+p \end{pmatrix}$$

can contain in them only the velocity gradients $\frac{\partial u}{\partial x}$, etc. in appropriate combinations implied by Eq. 2.64. For an *isotropic Newtonian fluid*, these expressions are

$$\sigma_{xx} + p = \lambda\, \nabla \cdot \mathbf{u} + 2\mu \frac{\partial u}{\partial x} \tag{2.67}$$

$$\sigma_{yy} + p = \lambda\, \nabla \cdot \mathbf{u} + 2\mu \frac{\partial v}{\partial y} \tag{2.68}$$

$$\sigma_{zz} + p = \lambda\, \nabla \cdot \mathbf{u} + 2\mu \frac{\partial w}{\partial z} \tag{2.69}$$

$$\tau_{xy} = \tau_{yx} = \mu \left(\frac{\partial v}{\partial x} + \frac{\partial u}{\partial y} \right) \tag{2.70}$$

$$\tau_{yz} = \tau_{zy} = \mu \left(\frac{\partial w}{\partial y} + \frac{\partial v}{\partial z} \right) \tag{2.71}$$

$$\tau_{zx} = \tau_{xz} = \mu \left(\frac{\partial u}{\partial z} + \frac{\partial w}{\partial x} \right) \tag{2.72}$$

in which $\nabla \cdot \mathbf{u} = \frac{\partial u}{\partial x} + \frac{\partial v}{\partial y} + \frac{\partial w}{\partial z}$, where $\mathbf{u} = (u, v, w)$ is the velocity vector and the quantities λ and μ are the factors of proportionality. Eqs. 2.67–2.72 are the constitutive equations for an isotropic Newtonian fluid. As seen, these relations are postulated to be linear; they must remain unchanged by a rotation of the coordinate system, or by an interchange of axes to ensure isotropy. Consult Schlichting (1979, pp. 58–60) for a detailed account of the analysis leading to Eqs. 2.67–2.72.

As seen from Eqs. 2.67–2.69, there are two contributions to the normal stresses. For example, for the x-component of the normal stress (Eq. 2.67): One contribution is from the elongation of the fluid element in the x-direction, i.e., $2\mu\frac{\partial u}{\partial x}$; and the second contribution is from the volumetric dilatation, the rate of change in volume, represented by $\lambda\ \nabla \cdot \mathbf{u}$. Clearly, the latter will be nil from the continuity equation (Eq. 2.44), if the fluid is incompressible.

Of the proportionality factors λ and μ, the quantity μ turns out to be the dynamic viscosity of the fluid. Regarding the first factor, λ, the physical interpretation is not clear. This is discussed in the next sub-section.

5) Stokes' hypothesis, and the constitutive equation a for Newtonian fluid

Schlichting (1979, p. 60) notes that the physical interpretation of the factor λ is still being disputed, even though the value which should be given to it is not. This numerical value is determined with the help of a hypothesis, known as Stokes' hypothesis. Stokes (1845) advanced this hypothesis according to which it is necessary to assume

$$3\lambda + 2\mu = 0, \qquad \text{or} \qquad \lambda = -\frac{2}{3}\mu \tag{2.73}$$

(for a detailed account of the latter, refer to Schlichting, 1979, p. 60). This relates the value of λ to the viscosity μ for a compressible fluid, thus reducing the number of fluid properties to one, as in the case of an incompressible fluid.

With this, the constitutive equations for a Newtonian fluid will read

$$\sigma_{xx} = -p - \frac{2}{3}\mu\ \nabla \cdot \mathbf{u} + 2\mu\frac{\partial u}{\partial x} \tag{2.74}$$

$$\sigma_{yy} = -p - \frac{2}{3}\mu\ \nabla \cdot \mathbf{u} + 2\mu\frac{\partial v}{\partial y} \tag{2.75}$$

$$\sigma_{zz} = -p - \frac{2}{3}\mu\ \nabla \cdot \mathbf{u} + 2\mu\frac{\partial w}{\partial z} \tag{2.76}$$

$$\tau_{xy} = \tau_{yx} = \mu\left(\frac{\partial v}{\partial x} + \frac{\partial u}{\partial y}\right) \tag{2.77}$$

$$\tau_{yz} = \tau_{zy} = \mu\left(\frac{\partial w}{\partial y} + \frac{\partial v}{\partial z}\right) \tag{2.78}$$

$$\tau_{zx} = \tau_{xz} = \mu\left(\frac{\partial u}{\partial z} + \frac{\partial w}{\partial x}\right) \tag{2.79}$$

For an incompressible Newtonian fluid, the constitutive equations will reduce to

$$\sigma_{xx} = -p + 2\mu\frac{\partial u}{\partial x} \tag{2.80}$$

$$\sigma_{yy} = -p + 2\mu\frac{\partial v}{\partial y} \tag{2.81}$$

$$\sigma_{zz} = -p + 2\mu\frac{\partial w}{\partial z} \tag{2.82}$$

$$\tau_{xy} = \tau_{yx} = \mu\left(\frac{\partial v}{\partial x} + \frac{\partial u}{\partial y}\right) \tag{2.83}$$

$$\tau_{yz} = \tau_{zy} = \mu\left(\frac{\partial w}{\partial y} + \frac{\partial v}{\partial z}\right) \tag{2.84}$$

$$\tau_{zx} = \tau_{xz} = \mu\left(\frac{\partial u}{\partial z} + \frac{\partial w}{\partial x}\right) \tag{2.85}$$

or, using Einstein's summation convention

$$\sigma_{ij} = -p\delta_{ij} + \mu\left(\frac{\partial u_i}{\partial x_j} + \frac{\partial u_j}{\partial x_i}\right) \tag{2.86}$$

in which σ_{ij} is the elements of the stress tensor, i.e., $\sigma_{11} = \sigma_{xx}$, $\sigma_{12} = \tau_{xy}$, etc., and δ_{ij} is the Kronecker delta defined by

$$\delta_{ij} = 1 \qquad \text{for } i = j, \tag{2.87}$$

$$\delta_{ij} = 0 \qquad \text{for } i \neq j \tag{2.88}$$

6) Navier-Stokes equation

The equations of motion from Eqs. 2.46–2.48, using Einstein's summation convention for convenience, can be written as:

$$\rho \frac{du_i}{dt} = \rho g_i + \frac{\partial \sigma_{ij}}{\partial x_j} \tag{2.89}$$

or, with the acceleration term fully written,

$$\rho \left(\frac{\partial u_i}{\partial t} + u_j \frac{\partial u_i}{\partial x_j} \right) = \rho g_i + \frac{\partial \sigma_{ij}}{\partial x_j} \tag{2.90}$$

in which u_i is the i component of the velocity, and g_i that of the volume force. Inserting the constitutive equations for an incompressible Newtonian fluid, Eq. 2.86, into the equation of motion, Eq. 2.90, yields:

$$\rho \left(\frac{\partial u_i}{\partial t} + u_j \frac{\partial u_i}{\partial x_j} \right) = \rho g_i + \frac{\partial}{\partial x_j} \left(-p\delta_{ij} + \mu \left(\frac{\partial u_i}{\partial x_j} + \frac{\partial u_j}{\partial x_i} \right) \right) \tag{2.91}$$

This equation is known as the Navier-Stokes equation.

This equation was first derived by Navier (1822), and later independently by Stokes (1845) in a manner that is consistent with the modern understanding presented above. For a detailed historical account of the early incarnations of the Navier-Stokes equation, see Darrigol (2002).

2.C References

(1) Batchelor, G.K. (1967): An Introduction to Fluid Dynamics. Cambridge University Press.

(2) Darrigol, O. (2002): Between hydrodynamics and elasticity theory: The first five births of the Navier-Stokes equation. Archive for History of Exact Sciences, vol. 56, 95–150.

(3) Navier, C.L.M.H. (1822): Memoire sur les lois du mouvement des fluides. Mem. Acad. Sci. Inst. France, vol. 6, 389–440.

(4) Schlichting, H. (1979): Boundary Layer Theory. McGraw-Hill.

(5) Stokes, G.G. (1845): On the theories of internal friction of fluids in motion. Trans. Cambridge Phil. Soc., vol. 8, 287–305.

Chapter 3

Steady turbulent boundary layer flows

In this chapter, we will study steady, turbulent boundary layers over a wall, or in a channel, or in a pipe. Examples of turbulent boundary layers over a wall may be: flow over a flat plate, flow over the seabed, flow in wide rivers, flow over the earth surface (the atmospheric boundary layer), etc.

We will first concentrate on the flow close to a wall (a general analysis), and then we will study the flow close to a smooth wall, and subsequently that close to a rough wall. The latter will include analyses of flow near the bed of a channel, and that near the wall of a pipe, etc. Finally, we will turn our attention to flows in channels/pipes, considering the entire channel/pipe section.

Of particular interest in our study is the bursting process, essentially the flow near the wall viewed as a quasi-cyclic process, recognized only relatively recently, as opposed to the classic approach where the (turbulent) flow is perceived as a random process. We will give an extensive account of the bursting process, and its implications for sediment transport. The latter will be with reference to four examples, namely: (1) Particle motions near the bottom in an open channel and mechanism of particle suspension; (2) Lift forces on moving particles near boundaries; (3) Initiation of suspension of particles from the bottom; and (4) Suspension of particles released away from the bottom.

3.1 Flow close to a wall

3.1.1 *Idealized flow in the half space $y > 0$*

We will first consider an idealized case where the flow takes place in the half space $y > 0$ (Fig. 3.1). We will further consider that there is no pressure

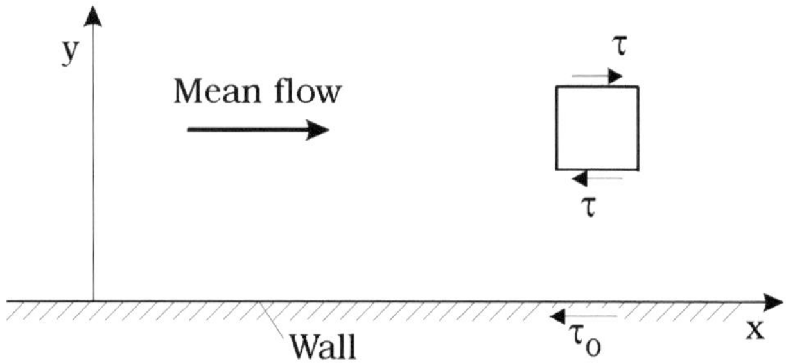

Fig. 3.1 Shear stress in the idealized half space $y > 0$.

gradient in the horizontal, x-direction. Although we consider this "hypothetical" flow for convenience, this kind of flow can actually be generated by the shear created by a drum-belt system. This corresponds to a classical "Couette flow" i.e. the flow of a viscous fluid in the space between two surfaces, one of which is moving tangentially relative to the other.

The analysis of this subsection will help develop a complete investigation of the mean flow and turbulence close to a smooth wall (Section 3.1.2) and a rough wall (Section 3.1.3) in real-life situations.

Shear stress

The basic equations are the continuity equation and the two components of the Reynolds equation. The continuity equation is automatically satisfied. The y-component of the Reynolds equation gives an equation for the Reynolds stress $-\rho\overline{v'^2}$. The x-component of the Reynolds equation (Eq. 2.14), on the other hand, reads

$$\rho\left(\frac{\partial\overline{u}}{\partial t}+\overline{u}\frac{\partial\overline{u}}{\partial x}+\overline{v}\frac{\partial\overline{u}}{\partial y}\right)=\rho\overline{g}_x-\frac{\partial\overline{p}}{\partial x}+\mu\left(\frac{\partial^2\overline{u}}{\partial x^2}+\frac{\partial^2\overline{u}}{\partial y^2}\right)+\frac{\partial}{\partial x}(-\rho\overline{u'^2})+\frac{\partial}{\partial y}(-\rho\overline{u'v'}) \tag{3.1}$$

Since $\frac{\partial\overline{u}}{\partial t}=0$, $\overline{v}=0$, $\overline{g}_x=0$, $\frac{\partial\overline{p}}{\partial x}=0$, $\overline{u}$ is a function of only y, and the flow is uniform in the x-direction (i.e., $\frac{\partial\overline{u}}{\partial x}=0$), then the Reynolds equation will read

$$\mu\frac{d^2\overline{u}}{dy^2}+\frac{d}{dy}(-\rho\overline{u'v'})=0 \tag{3.2}$$

or

$$\frac{d}{dy}\left(\mu\frac{d\overline{u}}{dy}+(-\rho\overline{u'v'})\right)=\frac{d\tau}{dy}=0 \tag{3.3}$$

in which τ is the total shear stress:

$$\tau = \mu \frac{d\overline{u}}{dy} + (-\rho\overline{u'v'}) \tag{3.4}$$

which is composed of two parts, namely the viscous part (the first term), and the part induced by turbulence, the Reynolds stress (the second term).

Integrating Eq. 3.3 gives

$$\tau = c, \qquad c = \text{ constant} \tag{3.5}$$

The boundary condition at the wall is:

$$y = 0: \qquad \tau = \tau_0 \tag{3.6}$$

in which τ_0 is the wall shear stress. From Eqs. 3.4 and 3.6, one obtains

$$\tau\left(= \mu \frac{d\overline{u}}{dy} + (-\rho\overline{u'v'})\right) = \tau_0 \tag{3.7}$$

This is an important result. It shows that the shear stress is constant over the entire depth in this idealized flow, and equal to the wall shear stress, τ_0.

Mean velocity

Our objective is to determine the velocity $\overline{u}$ as a function of y. To this end, we resort to dimensional analysis. A brief account of the latter is given in the Appendix at the end of this chapter.

Now, the mean properties of the flow (such as $\overline{u}$) depend on the following quantities

$$\tau_0,\ y,\ \mu,\ \rho \tag{3.8}$$

Here we consider a smooth wall, therefore no parameter representing the wall roughness is included. (We will develop a similar analysis in Section 3.1.3 for the case of a rough wall.)

The flow depends on τ_0 because the shear stress τ, which is equal to τ_0, governs the flow, and therefore the flow velocity; the larger the shear stress, the larger the flow velocity.

Likewise, the flow depends on the distance from the wall, y, because the closer to the wall, the smaller the velocity (just at the wall, obviously the velocity will be zero due to the no-slip condition).

Clearly, the flow should also depend on the fluid properties, namely μ, and ρ.

Hence

$$\overline{u} = F(\tau_0,\ y,\ \mu,\ \rho) \tag{3.9}$$

or

$$\Phi(\overline{u},\ \tau_0,\ y,\ \mu,\ \rho) = 0 \tag{3.10}$$

From dimensional analysis (Appendix 3.A), the preceding dimensional function Φ can be converted to a non-dimensional function ϕ (Appendix 3.B)

$$\phi\left(\frac{\overline{u}}{U_f},\ \frac{yU_f}{\nu}\right) = 0 \tag{3.11}$$

in which $\frac{\overline{u}}{U_f}$, and $\frac{yU_f}{\nu}$ are two non-dimensional quantities, U_f is the friction velocity defined by

$$U_f = \sqrt{\frac{\tau_0}{\rho}} \tag{3.12}$$

and ν is the kinematic viscosity

$$\nu = \frac{\mu}{\rho} \tag{3.13}$$

Solving for $\frac{\overline{u}}{U_f}$ from Eq. 3.11 gives

$$\frac{\overline{u}}{U_f} = f(y^+) \tag{3.14}$$

in which

$$y^+ = \frac{yU_f}{\nu} \tag{3.15}$$

Eq. 3.14 is known as *the law of the wall.* The quantities U_f and ν are called the *inner flow parameters,* or *wall parameters.*

Example 3.1. Boundary layer in an open channel: Flow close to the bed of the channel.

Now, consider a fully developed turbulent boundary layer flow in an open channel (Fig. 3.2). The flow depth is h. The flow is driven by gravity, with the volume force $g_x = gS$ in which g is the acceleration due to gravity, and S is the slope of the channel. The x-component of the Reynolds equation, Eq. 3.3, will, in the present case, read

$$\frac{d}{dy}\left(\mu\frac{d\overline{u}}{dy} + (-\rho\overline{u'v'})\right) = \frac{d\tau}{dy} = -\rho g_x = -\rho g S = -\gamma S \tag{3.16}$$

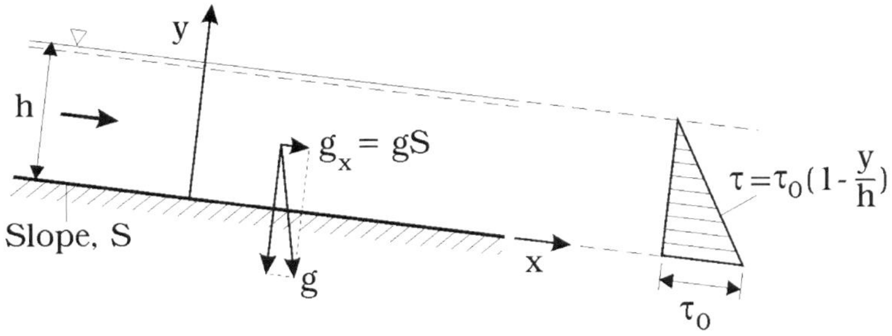

Fig. 3.2 Flow in an open channel.

in which $\gamma = \rho g$ is the specific weight of the fluid. Integrating gives

$$\tau = -\gamma S y + c; \qquad c = \text{constant} \tag{3.17}$$

The boundary condition at the free surface is:

$$y = h: \qquad \tau = 0 \tag{3.18}$$

From Eqs. 3.17 and 3.18, c is found to be $c = \gamma Sh$. Then the shear stress distribution is

$$\tau = \gamma Sh \left(1 - \frac{y}{h}\right) \tag{3.19}$$

The bed shear stress is obtained by inserting $y = 0$ in the above equation:

$$\tau_0 = \gamma Sh \tag{3.20}$$

Hence, the shear stress distribution may be written as

$$\tau = \tau_0 \left(1 - \frac{y}{h}\right) \tag{3.21}$$

(see the sketch in Fig. 3.2).

Our main concern in this example is the flow close to the bed. We will study the flow across the entire flow depth later, in Section 3.2. Now, close to the bed, the distance y is small compared with h, i.e., $y \ll h$, and therefore the shear stress can be written as

$$\tau = \tau_0 \left(1 - \frac{y}{h}\right) \simeq \tau_0 \tag{3.22}$$

i.e., close to the bed, the shear stress can be considered to remain unchanged with respect to y, and equal to the bed shear stress, $\tau \simeq \tau_0$. This thin layer of fluid is called the *constant stress layer*.

As seen, the flow in the constant stress layer is similar to the idealized flow considered in the preceding paragraphs; in both flows, the shear stress

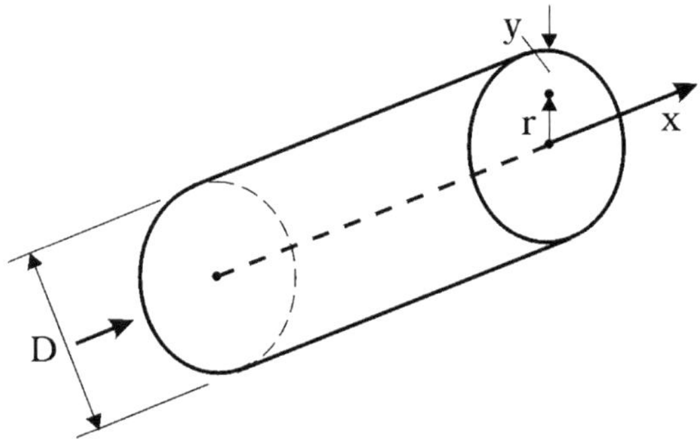

Fig. 3.3 Flow in a pipe.

is constant across the depth. The immediate implication of this result is that, the velocity distribution close to the bed (i.e., in the constant stress layer) is given by the law of the wall (Eq. 3.14):

$$\frac{\overline{u}}{U_f} = f(y^+) \tag{3.23}$$

Example 3.2. Boundary layer in a pipe: Flow close to the wall of the pipe.

Now, consider a fully developed turbulent boundary layer flow in a pipe (Fig. 3.3).

Similar to the previous example, the x-component of the Reynolds equation in the present case is:

$$\frac{1}{r}\frac{\partial}{\partial r}(r\tau) = \frac{\partial \overline{p}}{\partial x} \tag{3.24}$$

Here, the gravity term is included in the pressure gradient term for convenience. The flow is driven by the pressure gradient $\frac{\partial \overline{p}}{\partial x}$.

Integrating Eq. 3.24 gives

$$r\tau = \frac{\partial \overline{p}}{\partial x}\frac{r^2}{2} + c \tag{3.25}$$

in which c is a constant.

The boundary condition at the center line of the pipe is

$$r = 0: \qquad \tau = 0 \tag{3.26}$$

which gives $c = 0$. The boundary condition at the wall of the pipe is

$$r = \frac{D}{2}: \qquad \tau = \tau_0 \tag{3.27}$$

From Eqs. 3.25 and 3.27, one gets

$$\tau = 2\tau_0 \frac{r}{D} \quad \text{or} \quad \tau = \tau_0 \left(1 - \frac{2y}{D}\right) \tag{3.28}$$

in which $y = \frac{D}{2} - r$ is the distance from the pipe wall.

Again, similar to the previous example, for $y \ll \frac{D}{2}$, i.e., close to the wall, the shear stress can be considered to be constant, and approximately equal to the wall shear stress $\tau \simeq \tau_0$, the constant stress layer. Similar to the boundary layer flow in an open channel, in this thin layer of fluid near the wall of the pipe, the velocity is given by the law of the wall (Eq. 3.14)

$$\frac{\overline{u}}{U_f} = f(y^+) \tag{3.29}$$

3.1.2 *Mean and turbulence characteristics of flow close to a smooth wall*

In this subsection, mean and turbulence characteristics of a turbulent boundary layer flow close to a smooth wall will be studied in greater detail. As seen in the previous examples, the shear stress near the wall is approximately constant, and can be taken as equal to the wall shear stress (the constant stress layer). This subsection is concerned with the flow in the constant stress layer, with $y \ll h$. The analysis regarding the mean velocity will be advanced for two extreme cases: namely, for small values of the non-dimensional distance from the wall (Eq. 3.15) y^+, and for large values of y^+.

Mean velocity close to a smooth wall for small values of y^+: The viscous sublayer

As seen in the preceding paragraphs, the total shear stress is composed of two parts, the viscous part and the turbulent part (Eq. 3.4). Fig. 3.4 shows the contribution of each part to the total shear stress as a function of the distance from the wall. The figure indicates that, very near the wall, the turbulent part is small compared with the viscous part, $(-\rho\overline{u'v'}) \ll \mu\frac{d\overline{u}}{dy}$. Hence

$$\tau = \tau_0 = \mu \frac{d\overline{u}}{dy} + (-\rho\overline{u'v'}) \simeq \mu \frac{d\overline{u}}{dy} \tag{3.30}$$

or

$$\tau_0 \simeq \mu \frac{d\overline{u}}{dy} \tag{3.31}$$

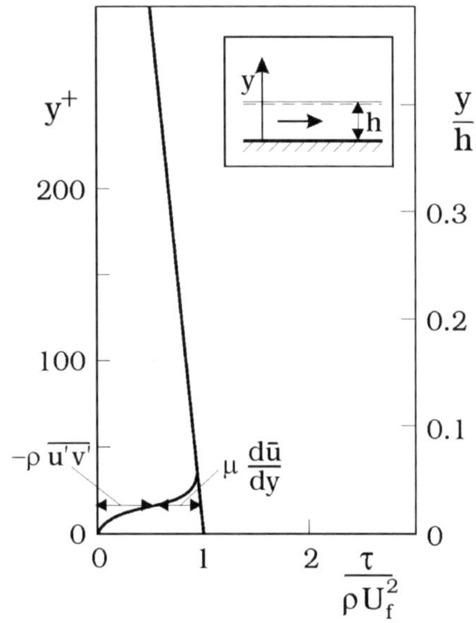

Fig. 3.4 Shear stress distribution in open channel. Adapted from Grass (1971).

Integrating gives

$$\overline{u} = \frac{U_f^2}{\nu} y + c \tag{3.32}$$

The no-slip condition at the wall requires:

$$y = 0: \qquad \overline{u} = 0 \tag{3.33}$$

This gives $c = 0$. Therefore, the mean velocity distribution will be

$$\overline{u} = \frac{U_f^2}{\nu} y \tag{3.34}$$

or, in the form corresponding to the law of the wall (Eq. 3.14 and Eq. 3.15):

$$\frac{\overline{u}}{U_f} = y^+ \tag{3.35}$$

The velocity varies linearly with the distance from the wall. This analytical expression agrees well with the experiments for $y^+ < 5$. Beyond $y^+ = 5$, the data begins to deviate from the preceding expression (Fig. 3.5). The layer in which the analytical expression in Eq. 3.35 agrees with the experimental data is called the *viscous sublayer.* The thickness of this layer is, from Fig. 3.5:

$$\delta_v = 5\frac{\nu}{U_f} \quad \text{or alternatively} \quad \delta_v^+ = 5 \tag{3.36}$$

It must be mentioned that, although this very thin layer of fluid is called the viscous sublayer, turbulence from the main body of the flow does penetrate into this layer. This can be seen in the velocity signal and also in the wall shear stress signal in the form of strong turbulent fluctuations.

To get a sense of how thick this layer is, for example, for $U_f = O(1$ cm/s), $\delta_v = 5\frac{\nu}{U_f} = 5 \times 10^{-2} \times O(1) = O(0.05 \text{ cm}) = O(0.5 \text{ mm})$, a very small thickness.

Mean velocity close to a smooth wall for large values of y^+: The logarithmic layer

Let us now consider larger y^+. First of all, it is important to note that, although the non-dimensional distance from the wall y^+ is large, the actual value of y is still small compared with h so that $\tau \simeq \tau_0$, namely y is still assumed to be in the constant stress layer.

We begin with Eq. 3.9, and differentiate both sides of Eq. 3.9 with respect to y:

$$\frac{d\overline{u}}{dy} = \frac{d}{dy}F(\tau_0,\ y,\ \mu,\ \rho) \tag{3.37}$$

We shall return the above equation shortly.

We now return to Fig. 3.4; the figure shows that, for large values of y^+, the viscous part of the total shear stress is small compared with the turbulent part, $\mu\frac{d\overline{u}}{dy} \ll (-\rho\overline{u'v'})$. Since the viscous part of the shear stress is negligible for large values of y^+, i.e., the viscosity is not important for the shear stress, it should not be important for the velocity variation $\frac{d\overline{u}}{dy}$ either, as the latter variation is directly associated with the shear stress. Therefore, dropping the viscosity μ in Eq. 3.37 yields:

$$\frac{d\overline{u}}{dy} = \frac{d}{dy}F(\tau_0,\ y,\ \rho) \tag{3.38}$$

or

$$\Psi\left(\frac{d\overline{u}}{dy},\ \tau_0,\ y,\ \rho\right) = 0 \tag{3.39}$$

From dimensional analysis (Appendix 3.A), this dimensional equation is converted to the following non-dimensional equation:

$$\psi\left(\frac{d\overline{u}}{dy}\frac{1}{\frac{U_f}{y}}\right) = 0 \tag{3.40}$$

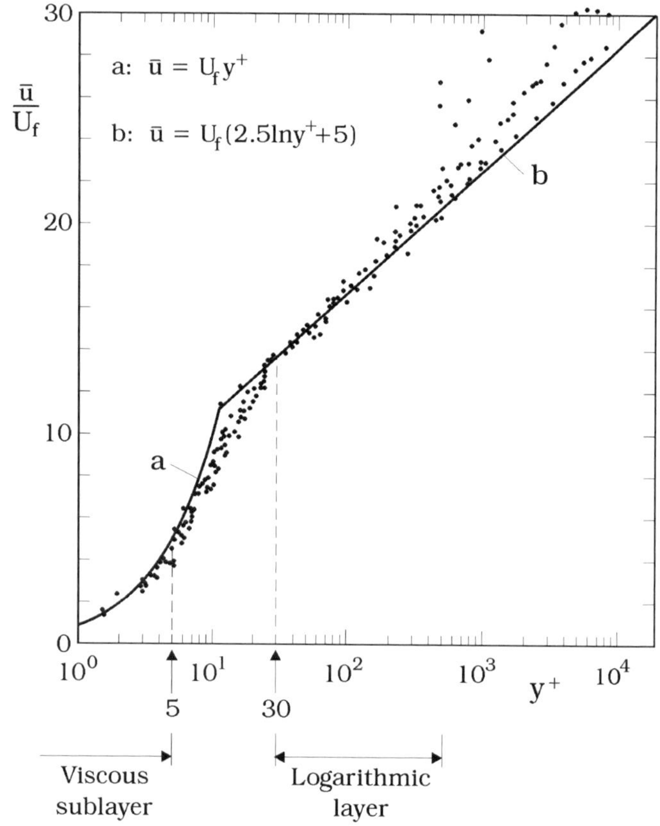

Fig. 3.5 Velocity distribution. Adapted from Monin and Yaglom (1973).

This equation has a single variable, namely

$$\frac{d\overline{u}}{dy}\frac{1}{\frac{U_f}{y}}$$

Solving this variable from Eq. 3.40 will give

$$\frac{d\overline{u}}{dy}\frac{1}{\frac{U_f}{y}} = A \tag{3.41}$$

in which A is the solution to Eq. 3.40, a constant.

Integrating Eq. 3.41 gives

$$\overline{u} = AU_f \ \ln y \ + \ B_1 \tag{3.42}$$

or in the form of the law of the wall (Eq. 3.14 and Eq. 3.15):

$$\overline{u} = U_f \; (A \ln y^+ \; + \; B) \tag{3.43}$$

in which B is another constant. This expression is known as the *logarithmic law.*

Experiments conducted in pipe flows, in channel flows, and in boundary layer flows have all confirmed the logarithmic law with

$$A = 2.5, \quad B = 5.1 \tag{3.44}$$

The constant A is also written in the literature in the form of $1/\kappa$, in which κ is called the von Karman constant, $\kappa = 0.4$. It is to be noted that A is a universal constant; namely, it always assumes the same value, $A = 2.5$, irrespective of the wall roughness (smooth, rough, or transitional), as will be demonstrated later.

Fig. 3.5 compares the logarithmic law with the experimental data. It is seen that the logarithmic law agrees quite well with the data in the range $30 < y^+ < 500$. It may be noted, however, that the upper bound of the range $30 < y^+ < 500$ is actually related to the condition $y \ll h$, or equivalently to the condition $y^+ \ll Re_\tau$ in which $Re_\tau = \frac{hU_f}{\nu}$ is the so-called friction Reynolds number. If the Reynolds number is small, then the upper bound of the range will be inevitably small. For example, if the friction Reynolds number is, say, $Re_\tau = 1000$, then the range in which the log law is satisfied will arguably be $30 < y^+ < 100$ so that the condition $y^+ \ll Re_\tau$ is presumably satisfied. The layer in which the logarithmic law is satisfied is called the *logarithmic layer.*

The lower bound of the logarithmic layer is often taken as $y^+ = 70$ in the literature. The layer of fluid which lies between the viscous sublayer and the logarithmic layer is called the *buffer layer,* $5 < y^+ < 70$. Clearly, in this layer, both the viscous effects and the turbulence effects are equally important.

To sum up, for the velocity distribution to satisfy the logarithmic law, the following two conditions have to be met:

1. The distance from the wall y should be in the constant stress layer, $\tau \simeq \tau_0$, and for this, the following inequality needs to be satisfied:

$$y \ll h \qquad \text{or} \qquad y^+ \ll Re_\tau = \frac{U_f h}{\nu} \tag{3.45}$$

in which h is the flow depth in the case of the open channel flow, or the radius in the case of the pipe flow, or the boundary layer thickness in the case of the turbulent boundary layer over a wall. (The above condition can be taken as $y < 0.1h$ (equivalently, $y^+ < 0.1Re_\tau$ where $Re_\tau = hU_f/\nu$ is again the friction Reynolds number). However, the data given in Monin and Yaglom (1973, p. 288–289) implies that it can be relaxed even to $y < (0.2$–$0.3)h$ (equivalently, $y^+ < (0.2$–$0.3)Re_\tau$).

2. At the same time, the distance y should be sufficiently large so that the variation of the velocity, $\frac{d\bar{u}}{dy}$, is independent of the viscosity. For this, y^+ should be

$$y^+ > 30 - 70 \tag{3.46}$$

These conditions (Eqs. 3.45 and 3.46) are to be strictly observed when fitting the logarithmic law (Eq. 3.43) to a measured set of velocity data.

To summarize the above, the logarithmic law, Eq. 3.43, generally holds for say 30–$70 < y^+ \ll Re_\tau = hU_f/\nu$.

Turbulence quantities close to a smooth wall

By "turbulence quantities", we mean the Reynolds stresses, namely

$$-\rho\overline{u_i'u_j'} = \begin{pmatrix} -\rho\overline{u'^2} & -\rho\overline{u'v'} & -\rho\overline{u'w'} \\ -\rho\overline{v'u'} & -\rho\overline{v'^2} & -\rho\overline{v'w'} \\ -\rho\overline{w'u'} & -\rho\overline{w'v'} & -\rho\overline{w'^2} \end{pmatrix} \tag{3.47}$$

For the present flow (steady and streamwise uniform flow close to a smooth wall), $-\rho\overline{u'w'}$ should be zero because there is no variation of the velocity in the z-direction (see Fig. 3.6). Also, $-\rho\overline{v'w'}$ should be zero because there is no mean flow in the y-z plane (no secondary currents, for example). Also recall that the Reynolds stress tensor is symmetric. Therefore, there are only four independent Reynolds stress components.

These are the three diagonal components, and one off-diagonal component, namely, $-\rho\overline{u'v'}$, or, for convenience, dropping $-\rho$ in the diagonal

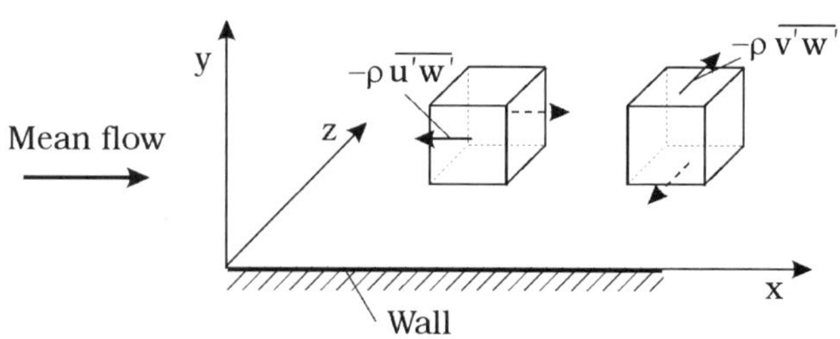

Fig. 3.6 Zero components of the Reynolds stresses.

components, and ρ in the off-diagonal components:

$$\overline{u'^2},\ \overline{v'^2},\ \overline{w'^2},\ \text{and}\ -\overline{u'v'} \tag{3.48}$$

These are the turbulence quantities.

Now, close to the wall, these quantities should depend on the same independent quantities as in the case of the mean velocity (Eq. 3.9), namely

$$\tau_0,\ y,\ \mu,\ \rho \tag{3.49}$$

and, from dimensional analysis (Appendix 3.A), we obtain the following non-dimensional relations

$$\sqrt{\overline{u'^2}} = U_f\ f_1(y^+), \qquad \sqrt{\overline{v'^2}} = U_f\ f_2(y^+),$$
$$\sqrt{\overline{w'^2}} = U_f\ f_3(y^+), \qquad -\overline{u'v'} = U_f^2\ f_4(y^+) \tag{3.50}$$

The functions $f_1, ..., f_4$ are universal functions, and their explicit forms are to be determined from experiments, or from turbulence modelling.

For large values of y^+, we know that the variation of the velocity with y, i.e. $\frac{d\bar{u}}{dy}$, is independent of the viscosity (see the argument in conjunction with Eq. 3.38). Since the turbulence is generated by the velocity gradient $\frac{d\bar{u}}{dy}$, and since $\frac{d\bar{u}}{dy}$ is independent of the viscosity for such large values of y^+, then the turbulence should also be independent of the viscosity. Therefore, for large values of y^+, the viscosity should drop out in the expressions in Eq. 3.50, meaning that the turbulence quantities in Eq. 3.50 should tend to constant values:

$$\frac{\sqrt{\overline{u'^2}}}{U_f} = f_1(y^+) \to A_1, \tag{3.51}$$

$$\frac{\sqrt{\overline{v'^2}}}{U_f} = f_2(y^+) \to A_2, \tag{3.52}$$

$$\frac{\sqrt{\overline{w'^2}}}{U_f} = f_3(y^+) \to A_3, \tag{3.53}$$

$$\frac{-\overline{u'v'}}{U_f^2} = f_4(y^+) \to A_4, \tag{3.54}$$

Fig. 3.7 shows the experimental results. As expected, the universal functions $f_1, ..., f_4$ approach zero, as y^+ goes to zero, and also, in conformity with Eqs. 3.51–3.54, they tend to constant values as y^+ takes very large

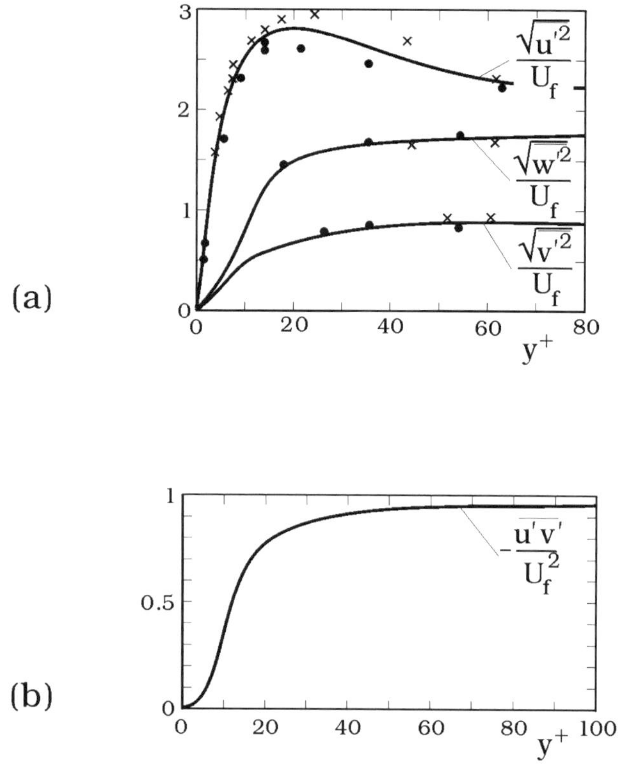

Fig. 3.7 Turbulence quantities. Circles: Laufer (1954), crosses: Klebanoff (1955), and solid line in (b): According to Laufer's measurements. Adapted from Monin and Yaglom (1973).

values. From the figure, the latter constant values are found to be $A_1 \simeq 2.3$, $A_2 \simeq 0.9$, $A_3 \simeq 1.7$, and $A_4 \simeq 1$.

Turbulent energy budget close to a smooth wall

From Eq. 2.31, the turbulent energy equation for a steady boundary layer flow is

$$\frac{\partial F_\alpha}{\partial x_\alpha} - (-\rho\overline{u_i' u_\alpha'})\frac{\partial \overline{u}_i}{\partial x_\alpha} + \Phi_t = 0 \tag{3.55}$$

in which Φ_t is given in Eq. 2.33, and F_α

$$F_\alpha = K_t \overline{u}_\alpha + \frac{1}{2}\rho\overline{u_i' u_i' u_\alpha'} + \overline{p' u_\alpha'} - \overline{u_i'\mu\left(\frac{\partial u_i'}{\partial x_\alpha} + \frac{\partial u_\alpha'}{\partial x_i}\right)} \tag{3.56}$$

The second term in Eq. 3.55 represents the turbulence production, and the third term the viscous dissipation of turbulent energy.

The first term, on the other hand, represents the flux of turbulent energy; it consists of four parts:

(1) $K_t\overline{u}_\alpha$ represents the transport of turbulent energy by convection;
(2) $\frac{1}{2}\rho\overline{u'_i u'_i u'_\alpha}$ represents the transport due to turbulent diffusion;
(3) $\overline{p'u'_\alpha}$ represents the transport due to pressure fluctuations; and
(4) $-\overline{u'_i\mu\left(\frac{\partial u'_i}{\partial x_\alpha}+\frac{\partial u'_\alpha}{\partial x_i}\right)}$ represents the transport due to viscous dissipation.

For the present boundary layer flow, the "turbulent energy budget", Eq. 3.55, reduces to

$$\frac{\partial F_\alpha}{\partial x_\alpha} - (-\rho\overline{u'v'})\frac{\partial \overline{u}}{\partial y} + \Phi_t = 0 \tag{3.57}$$

It can easily be seen that F_α in the present case consists of only items 2, 3 and 4 above.

Fig. 3.8 shows the contributions to the turbulent energy budget of the previously mentioned effects plotted versus the distance from the wall y^+. It is seen that most of the energy production and energy dissipation takes place near the wall in the buffer layer, namely $5 < y^+ < 70$.

3.1.3 *Mean and turbulence characteristics of flow close to a rough wall*

Consider that the wall is now covered with roughness elements (Fig. 3.9). Let the height of the roughness elements be k. Assume that k is sufficiently large (larger than the thickness of the viscous sublayer) so that the flow will be affected by the presence of the roughness elements.

Mean velocity close to a rough wall

In this case, there is one additional parameter to describe the flow, namely the roughness height, k. Therefore, Eq. 3.9 will read

$$\overline{u} = F(\tau_0,\ y,\ \mu,\ \rho,\ k) \tag{3.58}$$

Now, the total shear stress is given as in the case of the smooth wall (Eq. 3.4), namely

$$\tau = \mu\frac{d\overline{u}}{dy} + (-\rho\overline{u'v'}) \tag{3.59}$$

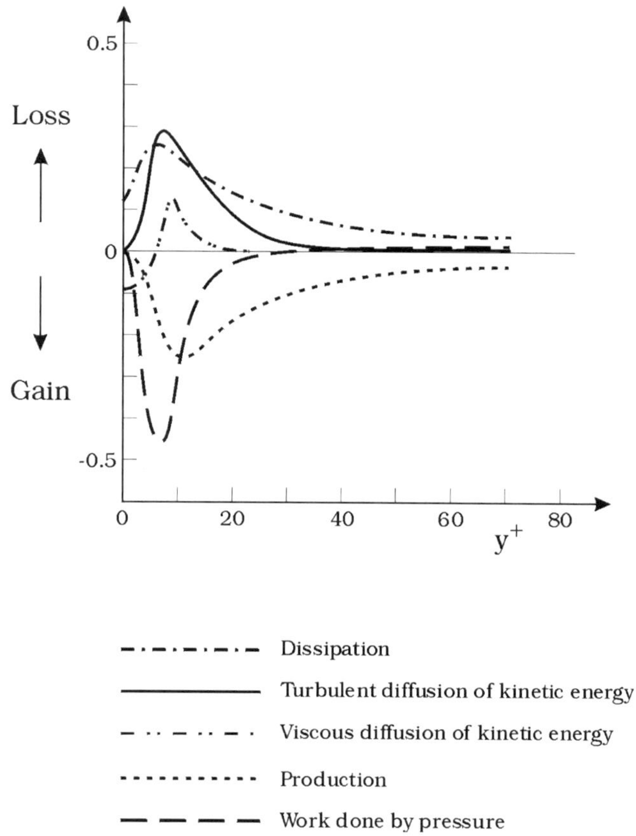

Fig. 3.8 Turbulent energy budget. Adapted from Laufer (1954).

However, the turbulent part of the total shear stress, $-\rho\overline{u'v'}$, will, in the present case, consist of two parts:

1. The first part is associated with the mean velocity gradient, $\frac{d\bar{u}}{dy}$, in the same way as in the case of the smooth wall; and

2. The second part is associated with the vortex shedding from the roughness elements. (Observations show that lee-wake vortices are constantly shed from the roughness elements into the main body of the flow, Sumer et al., 2001.)

Clearly, for large distances from the wall, (1) the viscous part $\mu\frac{d\bar{u}}{dy}$ in the total shear stress (Eq. 3.59) is negligible, as discussed in the previous section, and (2) the turbulent part associated with the vortex shedding is

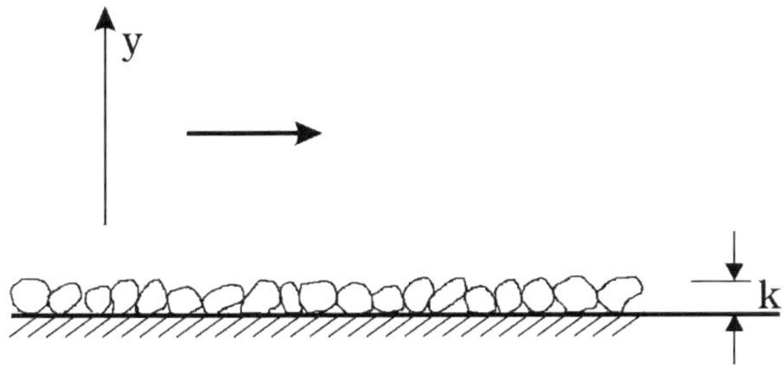

Fig. 3.9 Rough wall.

also negligible, because the vortices shed into the flow cannot penetrate to large distances due to their limited life time. This means that, for large distances from the wall, the shear stress is practically independent of the viscosity and the wall roughness.

Since the shear stress for large distances from the wall is independent of the viscosity and the wall roughness, then the velocity variation $\frac{d\overline{u}}{dy}$ should also be independent of these effects. Hence, dropping out these quantities (i.e., μ and k) in Eq. 3.58, yields:

$$\frac{d\overline{u}}{dy} = \frac{d}{dy}F(\tau_0,\ y,\ \rho) \tag{3.60}$$

This is precisely the same equation as in the case of the smooth wall (Eq. 3.38). Therefore, the result obtained in conjunction with Eq. 3.38 is equally valid for the present rough-wall case as well (Eq. 3.42), i.e.,

$$\overline{u} = AU_f\ \ln y\ +\ B_1 \tag{3.61}$$

Here, A should be the same constant as that in the case of the smooth wall, i.e., $A = 2.5$, meaning that A must be a *universal constant,* irrespective of the category of the wall (smooth, or rough, or transitional).

In the present context, the preceding expression may be written as

$$\overline{u} = AU_f\ \ln\left(\frac{y}{y_0}\right) \tag{3.62}$$

in which y_0 is called the *roughness length* (not to be confused with the roughness height k). Eq. 3.58 suggests that this new quantity should be a function of μ, k, τ_0 and ρ,

$$y_0 = y_0(\mu,\ k,\ \tau_0,\ \rho) \tag{3.63}$$

and from dimensional analysis (Appendix 3.A),

$$y_0 = k\ g\left(\frac{kU_f}{\nu}\right) \tag{3.64}$$

The explicit form of the function $g(\frac{kU_f}{\nu})$ can be found from experiments.

Nikuradse (1933) carried out such experiments (see Schlichting, 1979, p. 616); he coated the walls of circular pipes with sand grains of given size. The sand grains were glued on the wall in a densely packed manner. To differentiate Nikuradse's roughness from others, we will use the symbol k_s for the height of sand roughness in Nikuradse's experiments. Now, changing k by k_s, Eq. 3.64 will be

$$y_0 = k_s\ g\left(\frac{k_sU_f}{\nu}\right) \tag{3.65}$$

or

$$y_0 = k_s\ g(k_s^+) \tag{3.66}$$

with

$$k_s^+ = \frac{k_sU_f}{\nu} \tag{3.67}$$

According to Nikuradse's experiments, there are three categories of wall:

1. Hydraulically smooth wall when $k_s^+ < 5$. This is when the roughness height is smaller than the thickness of the viscous sublayer, Eq. 3.36. Clearly, in this case, the roughness elements are covered by the viscous sublayer, and therefore not directly exposed to the main body of the flow. Hence, the wall, although it contains roughness elements, acts as a smooth wall.

2. Completely rough wall (i.e. hydraulically rough) when $k_s^+ > 70$. In this case, the viscous sublayer is completely destructed, and the roughness elements are completely exposed to the main body of the flow.

3. Transitional wall when $5 < k_s^+ < 70$.

Hydraulically smooth wall. In this case, the function g takes the following form

$$g(k_s^+) = \frac{a}{k_s^+} \tag{3.68}$$

This is because k_s will drop out only when g is given in the form depicted in the preceding equation. Here a is a constant. Inserting the preceding expression into Eq. 3.65 gives:

$$y_0 = k_s \frac{a}{\frac{k_sU_f}{\nu}} = \frac{a\nu}{U_f}$$

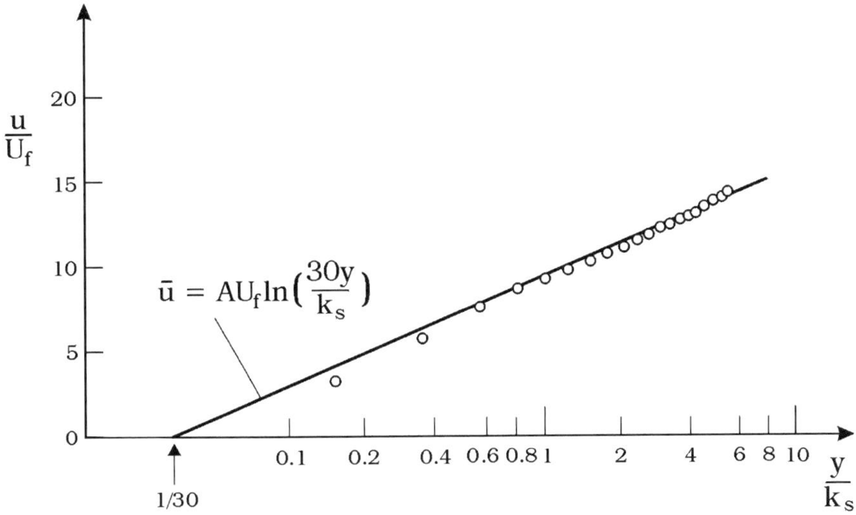

Fig. 3.10 Velocity distribution. Rough wall. Adapted from Grass (1971).

and from Eq. 3.62

$$\overline{u} = AU_f \ \ln\left(\frac{y}{y_0}\right) = AU_f \ \ln\left(\frac{yU_f}{a\nu}\right)$$

Comparing the preceding equation with Eqs. 3.43 and 3.44 gives the constant a as 1/9. With this value, the function $g(k_s^+)$ becomes:

$$g(k_s^+) = \frac{1/9}{k_s^+} \tag{3.69}$$

Completely rough wall. The function g is, in this case, given as

$$g(k_s^+) = b \tag{3.70}$$

This is because the viscosity does not play any role, and therefore should drop out; this is possible only when g is given as in the preceding form. Here, b is another constant. Nikuradse's experiments show that this constant is

$$g(k_s^+) = \frac{1}{30} \tag{3.71}$$

Substituting this into Eq. 3.65, the roughness length is found to be

$$y_0 = \frac{k_s}{30} \tag{3.72}$$

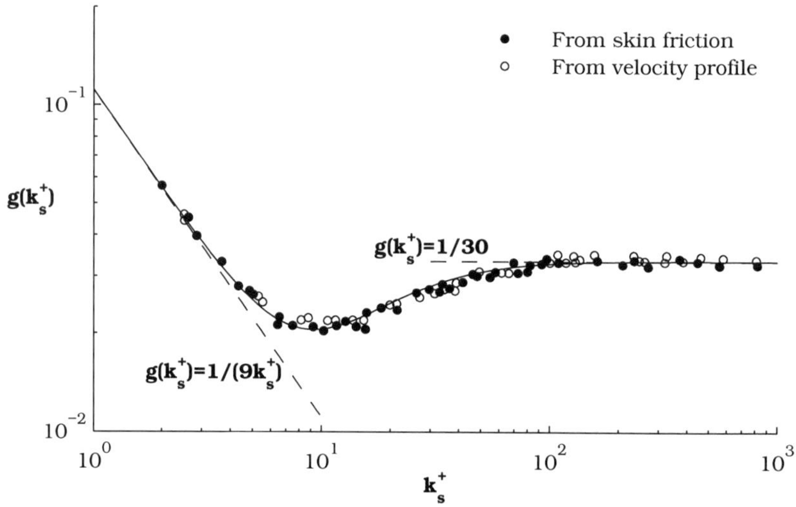

Fig. 3.11 Variation of function $g(k_s^+)$. Data: Nikuradse (1933), reproduced from Monin and Yaglom (1973). Solid curve: Eq. 3.74.

and from Eq. 3.62

$$\overline{u} = AU_f \ \ln\left(\frac{30y}{k_s}\right) \tag{3.73}$$

This is the *logarithmic law for flow over a completely rough wall.* Fig. 3.10 compares the above expression with the experimental data (Grass, 1971, Fig. 4). As seen, the logarithmic law begins to deviate from the measured velocity profile for $y \lesssim 0.2k_s$. This is not unexpected, because we have seen that, for the velocity profile to satisfy the logarithmic law, y should be sufficiently large, so large that the variation of the velocity $\frac{d\overline{u}}{dy}$ is independent of the wall roughness; otherwise, the velocity distribution will clearly not satisfy the logarithmic law, as revealed by Fig. 3.10.

Transitional wall. In this transitional wall case, the function g is given in Fig. 3.11. The data can be represented by the following empirical expression:

$$g(k_s^+) = \frac{1/9}{k_s^+} + \frac{1}{30}\exp[-140(k_s^+ + 6)^{-1.7}] \tag{3.74}$$

As seen from Fig. 3.11, the above expression approaches to

$$g(k_s^+) = \frac{1/9}{k_s^+} \tag{3.75}$$

as $k_s^+ \to 0$ (the smooth wall), c.f., Eq. 3.69, and it approaches to

$$g(k_s^+) = \frac{1}{30} \tag{3.76}$$

for $k_s^+ > 70$ (the rough wall), c.f., Eq. 3.71.

Nikuradse's equivalent sand roughness

Wall roughness is governed by various factors such as the height of the roughness elements, the shape of the roughness elements, and the packing pattern. To be able to compare the results obtained with different wall roughnesses, we need a common "base line". The concept of "Nikuradse's equivalent sand roughness" facilitates this.

The idea is to find Nikuradse's equivalent sand roughness for any kind of wall roughness. Nikuradse's equivalent sand roughness, k_s, is defined such that the velocity distribution across the depth is given by Eq. 3.73. In practice, it is easy to find k_s; the measured velocity distribution is plotted on a semi-logarithmic diagram, and the y-intercept of the straight line which represents the logarithmic law (Eq. 3.73) will be $\frac{k_s}{30}$.

Values of Nikuradse's equivalent sand roughness are reported in the literature for various wall roughnesses, and they are generally in the range $k_s \cong (2-4)k$ (see Kamphuis, 1974, and the review in the paper by Bayazit, 1983). It may be noted that the upper bound of the previously mentioned range can sometimes reach values as much as $5k$ and even more, depending on the shape of the roughness elements, the packing pattern as well as the roughness-height-to-flow-depth ratio (the latter is for large values of this ratio such as $> O(0.3)$) (Bayazit, 1983). In particular, $k_s \cong 2k$ for walls covered by sand, pebbles, or stones the size in the range 0.54-46 mm (Kamphuis, 1974); and $k_s \cong (3-6)k$ for the ground covered with ordinary grass or agricultural crops (Monin and Yaglom, 1973). Bayazit (1983) reports the following values for Nikuradse's equivalent sand roughness for various boundaries, Table 3.1.

Finally, $k_s \cong 2.5k$ for a river bed where the sediment grains are in motion (k being the grain size). This is when the bed remains plane (i.e., no bed forms are developed) (Engelund and Hansen, 1967). This corresponds to a weak sediment transport, and before bed ripples emerge. When the bed is covered with ripples, k_s is found to be $k_s \cong (2\text{–}3)k$ with 2-D ripples (Fredsøe et al., 1999), k now being the ripple height. In the case of 3-D ripples, the previous relation can, to a first approximation, be used. For very high velocities, ripples are washed away and the bed becomes plane again; this

Table 3.1 Nikuradse's equivalent sand roughness. Here variable d refers to various sediment grain diameters, with the subscripted numbers corresponding to the percentage finer by weight.

Author (See Bayazit, 1983, for references)	Roughness	k_s
Leopold et al. (1964)	Gravel	$3.5d_{84}$
Limerinos (1970)	Gravel	$3.5d_{84}$
Bayazit (1976)	Closely packed spheres	$2.5d$
Charlton et al. (1978)	Gravel	$3.5d_{90}$
Hey (1979)	Gravel	$3.5d_{84}$
Thompson and Campbell (1979)	Gravel	$4.5d_{90}$
Gladki (1979)	Gravel	$2.5d_{80}$
Denker (1980)	Closely-packed cylinders	$2d$
Bray (1980)	Gravel	$3.5d_{84}$
Griffths (1981)	Gravel	$5d_{50}$

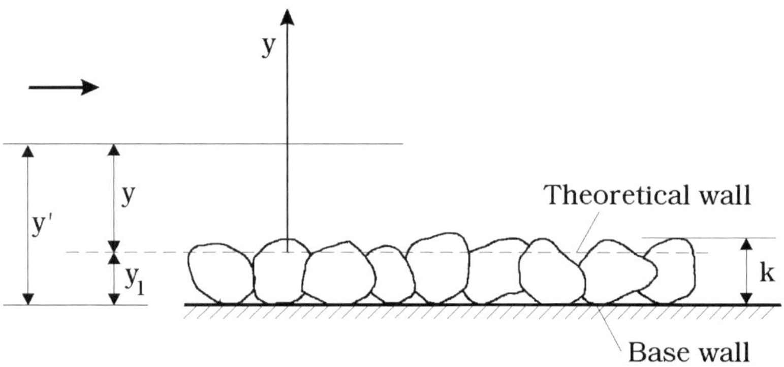

Fig. 3.12 Theoretical wall.

sediment-transport regime is called the sheet-flow regime (Sumer et al., 1996). In this case, the k_s value can be calculated from the following empirical relations (Sumer et al., 1996):

1. No-suspension sediment transport ($w_s/U_f > 0.8 - 1$):

$$\frac{k_s}{d} = 2 + 0.6\theta^{2.5}; \qquad 0 < \theta \lesssim 2 \tag{3.77}$$

2. Suspension sediment transport ($w_s/U_f < 0.8 - 1$):

$$\frac{k_s}{d} = 4.5 + \frac{1}{8}\exp\left[0.6\left(\frac{w_s}{U_f}\right)^4 \theta^2\right]\theta^{2.5} \tag{3.78}$$

in which here $k = d$ is the sediment grain size, w_s the fall (or settling) velocity of sediment grains (see the Appendix to Chapter 8 for methods of estimating w_s in practice) and θ the Shields parameter defined by

$$\theta = \frac{U_f^2}{g(s-1)d} \tag{3.79}$$

in which s is the specific gravity of sediment grains and g the acceleration due to gravity. For quartz based sediment $s \approx 2.65$.

Theoretical wall

The exact location of the wall is not a well defined quantity when the wall is rough. For this reason, we introduce the notion of a so-called theoretical wall. Formally, *the theoretical wall is defined as the location from which the y distances in* Eq. 3.73 *are measured* (Fig. 3.12).

Now, for convenience, change the coordinate to y', the distance from the base wall. Now suppose that the theoretical wall lies at the distance y_1 from the wall (Fig. 3.12), then the relationship between y and y' is:

$$y = y' - y_1 \tag{3.80}$$

Hence, the logarithmic law (Eq. 3.73) will read, in terms of y_1 and y':

$$\overline{u} = AU_f \ \ln\left(\frac{30(y'-y_1)}{k_s}\right) \tag{3.81}$$

For a given measured velocity profile $\overline{u}(y')$, and taking $A = 2.5$, the quantities (1) the friction velocity, U_f, (2) Nikuradse's equivalent sand roughness, k_s and (3) the location of the theoretical wall, y_1, can be determined from Eq. 3.81.

The procedure may possibly be best described by reference to the following example:

In the laboratory, time-averaged velocity profiles are measured with a Laser Doppler Anemometer over a stone-covered bed in an open channel (the stones the size of $k = 3.8$ cm, and the flow depth $h = 40$ cm from the top of the stones). The measurements are made at four, equally spaced, vertical sections between the crests of two neighbouring stones. Then, the space-averaged velocity profile is obtained from the measured, four velocity distributions. Subsequently, the following procedure is adopted to get the location of the theoretical wall (namely, y_1, or alternatively Δy, see the definition sketch in the inset in Fig. 3.13a), U_f and k_s.

1. Plot the time- and space-averaged velocity, $\overline{u}$, in a semi-log graph for various values of y_1, or alternatively for various values of Δy. The measured velocity profiles are plotted for eight values of Δy, namely $\Delta y/k = 0$; 0.1;

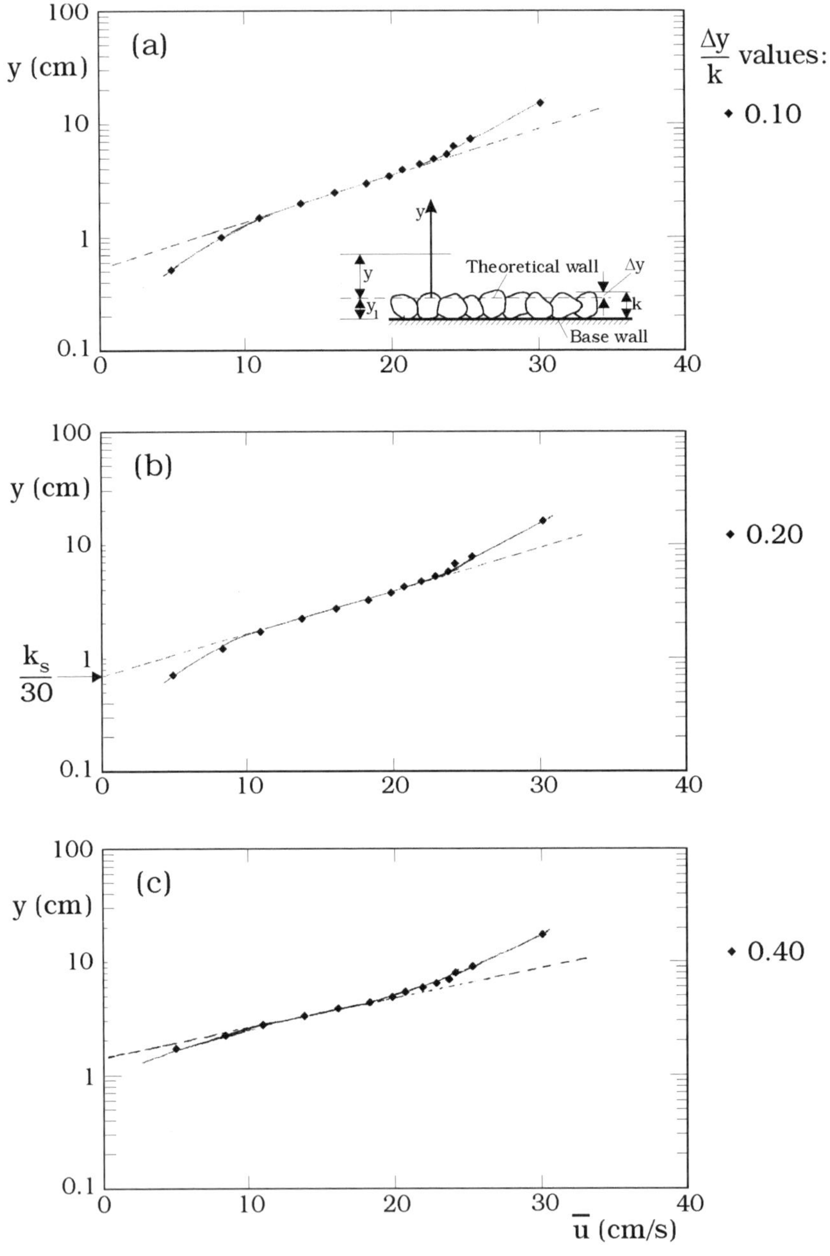

Fig. 3.13 Semilog plots of the velocity distribution for three different theoretical-wall locations.

Table 3.2 Thickness of the logarithmic layer for various values of the location of the theoretical wall in the example.

$\Delta y/k$	Logarithmic layer	Thickness of the logarithmic layer
0	$1.5 < y(\text{cm}) < 5$	3.5 cm
0.1	$1.5 < y(\text{cm}) < 5$	3.5 cm
0.15	$1.5 < y(\text{cm}) < 5$	3.5 cm
0.2	$1.5 < y(\text{cm}) < 6$	4.5 cm
0.25	$1.5 < y(\text{cm}) < 5$	3.5 cm
0.3	$2.5 < y(\text{cm}) < 5.5$	3.0 cm
0.35	$2.5 < y(\text{cm}) < 5$	2.5 cm
0.4	$2.5 < y(\text{cm}) < 5$	2.5 cm

0.15; 0.2; 0.25; 0.3; 0.35; 0.4. Only three profiles are displayed here, in Fig. 3.13, to keep the figure relatively simple.

2. Identify the straight line portion of each curve in Fig. 3.13. For this, look at the interval $0.2k_s \lesssim y \lesssim 0.1h$ to start with. (The latter interval can be taken as $0.2k_s \lesssim y \lesssim (0.2-0.3)h$ with the upper bound relaxed if necessary.) This is the interval where the *logarithmic layer* is supposed to lie. Here the upper bound $0.1h$ (or $(0.2-0.3)h$ in the relaxed case), ensures that the y levels lie in the constant stress layer, while the lower bound, $0.2k_s$, ensures that the variation of $\overline{u}$ with respect to y is not influenced by the boundary roughness, two conditions necessary for the velocity distribution to satisfy the logarithmic law.

3. Identify the case where the thickness of the logarithmic layer (where the velocity is represented with a straight line) is largest. This is to include all potential measurement points which satisfy the condition $0.2k_s \lesssim y \lesssim 0.1h$.

Now, the previously mentioned eight velocity profiles, when analyzed, reveal the results tabulated in Table 3.2.

4. As seen from the preceding table, the case where $\Delta y/k = 0.2$ gives the thickest logarithmic layer. Therefore, adopt this location as the location of the theoretical wall (Fig. 3.13b).

5. The straight line portion of this velocity profile (corresponding to $\Delta y/k = 0.2$, Fig. 3.13b) has a slope equal to AU_f, Eq. 3.81. From this information, find U_f.

6. Extend the straight line portion of the velocity profile (corresponding to $\Delta y/k = 0.2$) to find its y-intercept; this is equal to $\frac{k_s}{30}$ (Fig. 3.13b). From this information, find k_s.

Returning to the theoretical wall, research shows that the origin of the distance $y(= y' - y_1)$, namely the level of the theoretical wall, lies (0.15-0.35)k below the top of the roughness elements (Bayazit 1976, 1983), also revealed by the example in the preceding paragraphs.

Turbulence quantities close to a rough wall

The turbulence quantities given in Eq. 3.48 in the case of the rough wall should depend on the same independent quantities as in the case of the smooth wall (Eq. 3.49) except that (1) μ will drop out (no influence of viscosity) and (2) k_s will be added to the list of the independent quantities:

$$\tau_0,\ y,\ \rho,\ k_s \tag{3.82}$$

From dimensional analysis (Appendix 3.A), we get the following non-dimensional relations

$$\begin{aligned}\sqrt{\overline{u'^2}} &= U_f\ g_1\left(\frac{y}{k_s}\right), & \sqrt{\overline{v'^2}} &= U_f\ g_2\left(\frac{y}{k_s}\right),\\ \sqrt{\overline{w'^2}} &= U_f\ g_3\left(\frac{y}{k_s}\right), & -\overline{u'v'} &= U_f^2\ g_4\left(\frac{y}{k_s}\right)\end{aligned} \tag{3.83}$$

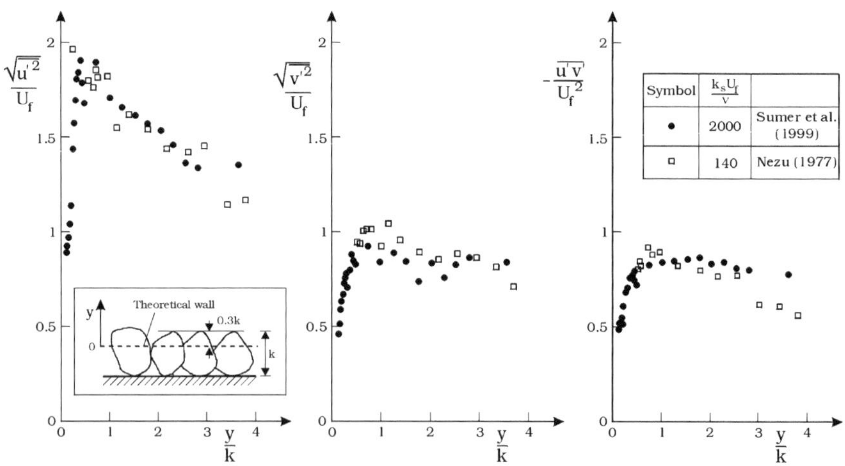

Fig. 3.14 Turbulence quantities. Rough wall. Averaging over both time and space. Sumer et al. (2001).

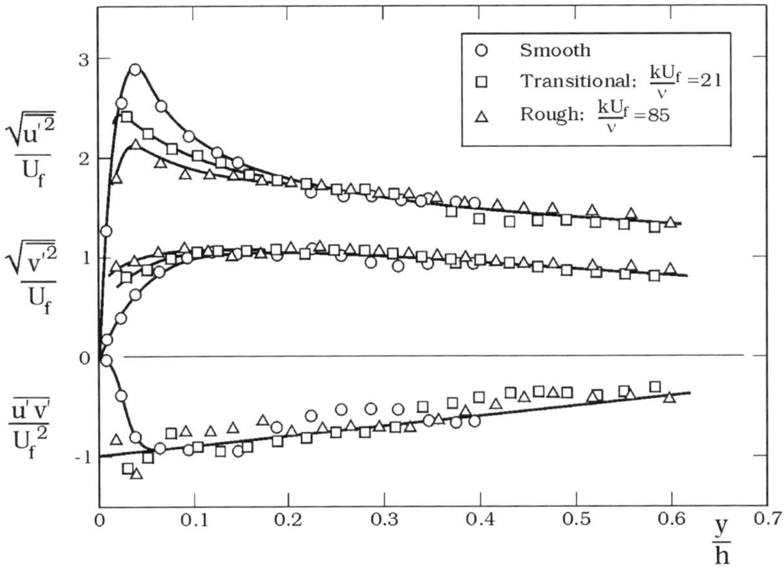

Fig. 3.15 Turbulence quantities. Grass (1971).

Fig. 3.14 shows data regarding $\sqrt{\overline{u'^2}}$, $\sqrt{\overline{v'^2}}$, and $-\overline{u'v'}$. The data are plotted in a format slightly different from that in the above equation, namely, the actual roughness height is used to normalize y, rather than Nikuradse's equivalent sand roughness. Notice the large difference in the roughness Reynolds numbers of the two sets of data. Despite this, the good agreement between these two data may imply that the functions $g_1, .., g_4$ are universal.

Fig. 3.15 compares the turbulence quantities obtained in another study (Grass, 1971); in these tests, everything else is maintained unchanged, but the wall roughness is changed, to see the influence of the wall roughness on the end results. From the figure, the following two points are worth emphasizing:

1. The influence of the wall roughness disappears with the distance from the wall, as expected.

2. The distribution of turbulence across the depth is more uniform in the case of the rough wall than in the case of the smooth wall.

3.2 Flow across the entire section

In the preceding section, we have been concerned with the flow close to a wall (a wall of an open channel, that of a pipe, etc.). Close to the wall (Fig. 3.16), the lowermost three layers, namely the viscous sublayer, the buffer layer, and the logarithmic layer, form the constant stress layer where the condition $y \ll h$ is satisfied, implying that the shear stress can be considered to remain practically unchanged with y, and is equal to the wall shear stress.

In the present section, we will turn our attention to the flow across the entire section of the channel/pipe. Clearly, in this case, there will be an additional parameter, namely the flow depth in the case of the channel-flow (h in Fig. 3.16), or the pipe radius in the case of the pipe flow, or the boundary-layer thickness in the case of the boundary-layer flow.

Mean flow across the entire section

In the analysis developed for flows close to a smooth wall, the mean flow velocity has been written as a function of four quantities, τ_0, y, μ, ρ (Eq. 3.9). When we consider the flow across the entire section, however, two changes will take place:

1. There will be one additional parameter, namely the flow depth (or the pipe radius, or the boundary-layer thickness) affecting the flow, as mentioned in the preceding paragraph (Fig. 3.2); and

2. The shear stress τ (Fig. 3.2),

$$\tau = \tau_0 \left(1 - \frac{y}{h}\right) \tag{3.84}$$

will replace the wall shear stress τ_0.

Hence, the velocity is now dependent of the following five parameters

$$\overline{u} = F(\tau,\ y,\ \mu,\ \rho,\ h) \tag{3.85}$$

However, from Eqs. 3.84 and 3.85, the preceding equation may be written as

$$\overline{u} = F(\tau_0,\ y,\ \mu,\ \rho,\ h) \tag{3.86}$$

From dimensional analysis (Appendix 3.A), the following non-dimensional function may be obtained

$$\overline{u} = U_f\ \varphi\left(\frac{U_f h}{\nu},\ \frac{y}{h}\right) \tag{3.87}$$

Now, we will follow the same line of thought as in the preceding sections; namely, the variation of the velocity $\frac{d\overline{u}}{dy}$ at large distances from the wall is independent of the viscosity. For large distances from the wall, dropping out the viscosity in Eq. 3.86,

$$\frac{d\overline{u}}{dy} = \frac{d}{dy}F(\tau_0,\ y,\ \rho,\ h)$$

or

$$\Psi_1\left(\frac{d\overline{u}}{dy},\ \tau_0,\ y,\ \rho,\ h\right) = 0 \tag{3.88}$$

and from dimensional analysis we obtain

$$\frac{d\overline{u}}{dy} = \frac{U_f}{h}\ \varphi_1\left(\frac{y}{h}\right) \tag{3.89}$$

Integrating gives

$$\overline{u} = U_f \int \varphi_1\left(\frac{y}{h}\right)\ d\left(\frac{y}{h}\right) + c$$

and using the boundary condition

$$y = h: \qquad \overline{u} = U_0$$

the velocity

$$\overline{u} - U_0 = U_f \int_{y=h}^{y} \varphi_1\left(\frac{y}{h}\right)\ d\left(\frac{y}{h}\right)$$

or denoting the integral by another function, $f_1(\frac{y}{h})$,

$$\overline{u} - U_0 = U_f\ f_1\left(\frac{y}{h}\right) \tag{3.90}$$

in which U_0 is the velocity at the free surface in the case of the open-channel flow, and the center line velocity in the case of the pipe flow. The preceding equation is known as the *velocity-defect law.* This velocity distribution is satisfied in the outer-flow region, Fig. 3.16.

The explicit form of the function $f_1(\frac{y}{h})$ can be found in the following way.

As sketched in Fig. 3.16, the outer region and the logarithmic layer is expected to overlap over a certain y interval. This is because the condition that y is sufficiently large so that the viscosity μ drops out is common to both layers, and therefore the two layers should overlap over a y interval. Hence, in the overlapping region, the two velocity distributions, namely the

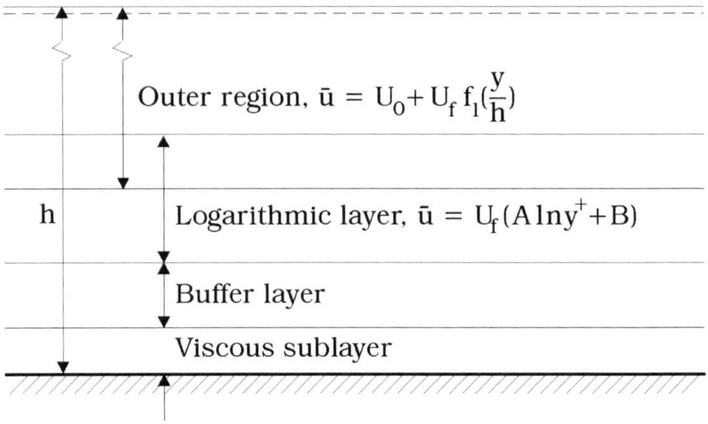

Fig. 3.16 Flow regions in a turbulent boundary layer.

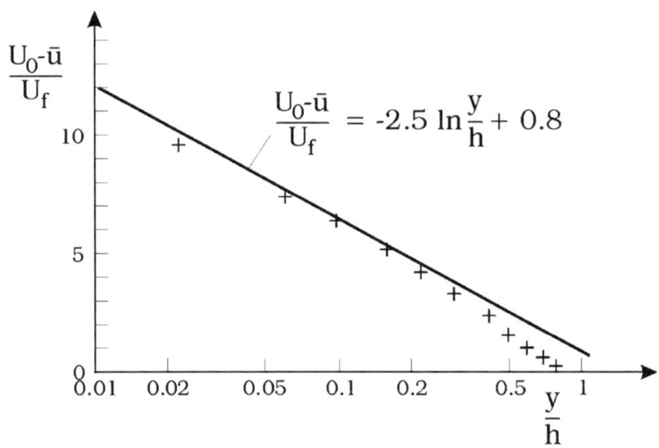

Fig. 3.17 Velocity distribution over the entire section of the pipe. Adapted from Monin and Yaglom (1973). Data from Laufer (1954)

logarithmic law (Eq. 3.43) and the velocity-defect law (Eq. 3.90), should be identical; therefore from Eqs. 3.43 and 3.90,

$$AU_f \ \ln\left(\frac{yU_f}{\nu}\right) \ + \ BU_f \equiv U_0 + U_f \ f_1\left(\frac{y}{h}\right) \tag{3.91}$$

and solving for the function f_1 gives

$$f_1\left(\frac{y}{h}\right) = A\ln\left(\frac{y}{h}\right) + B' \tag{3.92}$$

Therefore, the velocity-defect law

$$\overline{u} - U_0 = U_f \left(A \ln\left(\frac{y}{h}\right) + B' \right) \tag{3.93}$$

A is the same universal constant as in the analysis for the near-wall flow, i.e., $A = 2.5$. Constant B', on the other hand, is found from experiments as $B' = -0.8$ for pipe flows (Monin and Yaglom, 1973). However, if Eq. 3.93 is assumed to be applicable for the entire cross section, then B' has to be $B' = 0$.

Experiments show that the velocity-defect law begins to deviate from the measured velocity profiles (not very significantly, however) for $\frac{y}{h} > 0.2$–0.3 (Fig. 3.17).

When the general form of the equation for the velocity is considered (Eq. 3.87), it will be seen that U_0 is a function of the Reynolds number. From Eq. 3.91,

$$U_0 = AU_f \ \ln\left(\frac{hU_f}{\nu}\right) + \ U_f(B - B') \tag{3.94}$$

The velocity-defect law is also given in terms of the cross-sectional average velocity. In the case of the *open-channel flow*

$$V = \frac{1}{h} \int_{y=0}^{h} \overline{u}(y) \ dy$$

and

$$V = U_0 - U_f(A - B') \tag{3.95}$$

In this case, the velocity-defect law (from Eqs. 3.93 and 3.95) will read as

$$\overline{u} - V = AU_f \left(\ln\left(\frac{y}{h}\right) + 1 \right) \tag{3.96}$$

Note that, from the preceding equation, the velocity $\overline{u}$ is equal to the mean flow velocity, $\overline{u} = V$, when $y = 0.37h$.

Finally, it can readily be seen that the velocity-defect law in its form in Eq. 3.93 is equally valid for the case of the rough wall as well. However, the velocity U_0 in this case will be different from that in Eq. 3.94:

$$U_0 = AU_f \ln\left(\frac{30h}{k_s}\right) - U_f B' \tag{3.97}$$

Turbulence quantities across the entire section

The functional forms of the turbulence quantities depicted in Eq. 3.50 will include one additional parameter in the present case, namely $h^+ = \frac{hU_f}{\nu} (= Re_\tau)$ or alternatively the Reynolds number $Re = \frac{hV}{\nu}$.

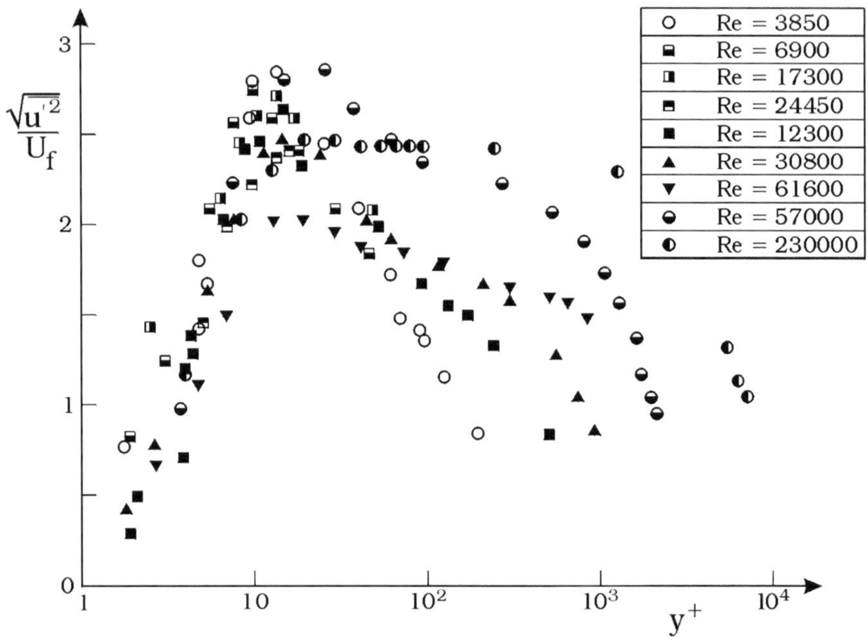

Fig. 3.18 Streamwise turbulence intensity distributions. Data compiled by Wei and Willmarth (1989).

Figs. 3.18 and 3.19 illustrates how this latter parameter influences the turbulence quantities $\sqrt{u'^2}$ and $\sqrt{v'^2}$ (Wei and Willmarth, 1989). As seen, near the wall, the effect of *Re* disappears, as expected (see Section 3.1.2). However, away from the wall, the Reynolds number significantly influences the turbulence quantities. As *Re* increases, $\frac{\sqrt{u'^2}}{U_f}$, and $\frac{\sqrt{v'^2}}{U_f}$ increase. This is basically due to the increase in the flow depth when *Re* increases; indeed, the flow depth in terms of the wall parameters is like $h^+ \sim Re^p$, $p > 0$ in which again $h^+ (= \frac{hU_f}{\nu})$. (As *Re* increases, the turbulence generated near the wall will be diffused to larger and larger h^+, as revealed by Figs. 3.18 and 3.19).

3.3 Turbulence modelling approach

We have, in the preceding sections, studied turbulent shear flows in an open channel, in a pipe, or in a boundary layer over a wall (smooth and rough).

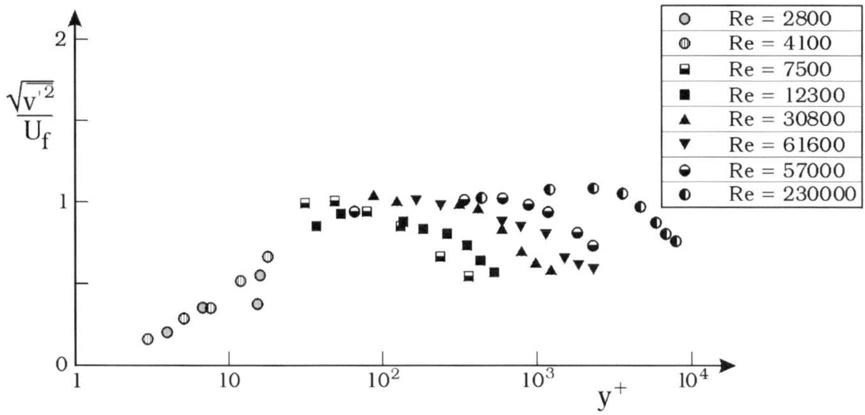

Fig. 3.19 Wall-normal turbulence intensity distributions. Data compiled by Wei and Willmarth (1989).

We have obtained valuable information on mean flow and turbulence. We have been able to describe the velocity distributions in the constant stress layer, and obtained analytical expressions for the mean flow velocity in the viscous sublayer and in the logarithmic layer in the smooth-wall case, and, again, in the logarithmic layer in the rough- and the transitional-wall cases. However, we were unable to obtain explicit expressions for the velocity distribution in the buffer layer (Fig. 3.16) in the smooth-wall case, and also in the layer just above the wall, adjacent to the logarithmic layer in the rough- and transitional-wall cases.

This can be achieved with the aid of *turbulence modelling*. A detailed account of the turbulence modelling is given in Chapter 9. Here, we restrict ourselves only with the so-called mixing length model (this is also covered in Section 9.3). We will use the latter model to describe the velocity distributions in the previously mentioned cases as well as across the entire section.

3.3.1 *Turbulence modelling of flow close to a wall*

Turbulence modelling of flow close to a smooth wall

Our objective is to obtain the velocity distribution, $\overline{u}$, near the wall (in the constant stress layer). The x-component of the Reynolds equation leads to the following equation (Eq. 3.7):

$$\tau\left(=\mu\frac{d\overline{u}}{dy}+(-\rho\overline{u'v'})\right)=\tau_0 \tag{3.98}$$

in which there are two unknowns, namely $\overline{u}$, and $-\rho\overline{u'v'}$. To close the system, we need a second equation. Drawing an analogy to Newton's friction law

$$\tau=\mu\frac{du}{dy} \tag{3.99}$$

the turbulence-induced shear stress can be written as

$$-\rho\overline{u'v'}=\mu_t\frac{d\overline{u}}{dy} \tag{3.100}$$

in which μ_t is called the dynamic turbulence (or eddy) viscosity. This is the simplest turbulence model.

The second step is to determine the turbulence viscosity. Making an analogy to the kinetic theory of gases, μ_t is written as

$$\mu_t=\rho\ell_m V_t \tag{3.101}$$

in which ℓ_m is called the *mixing length,* interpreted as the length of "penetration" of fluid particles in the cross-flow direction, or the length in the same direction over which fluid particles/parcels keep their coherence. The quantity V_t in the above equation is the turbulence velocity, a characteristic velocity of the turbulent motion, for example, $V_t\sim\sqrt{\overline{v'^2}}$. The hypothesis leading to Eq. 3.101 is known as Prandtl's mixing-length hypothesis. The preceding equation can also be found in Chapter 9 as Eq. 9.7.

There are two quantities to be ascribed in the preceding equation to determine the turbulence viscosity: ℓ_m and V_t. According to Prandtl's second hypothesis, these two quantities are related in the following way (see Chapter 9):

$$V_t=\ell_m\left|\frac{d\overline{u}}{dy}\right| \tag{3.102}$$

Clearly, it is expected that (1) the larger the mixing length, the larger the turbulence velocity; and also (2) the larger the velocity gradient $\left|\frac{d\overline{u}}{dy}\right|$, the larger the turbulence generation (Section 2.4.3), and therefore the larger the turbulence velocity, revealing Eq. 3.102.

From Eqs. 3.101 and 3.102, the turbulence viscosity is

$$\mu_t=\rho\ell_m^2\left|\frac{d\overline{u}}{dy}\right| \tag{3.103}$$

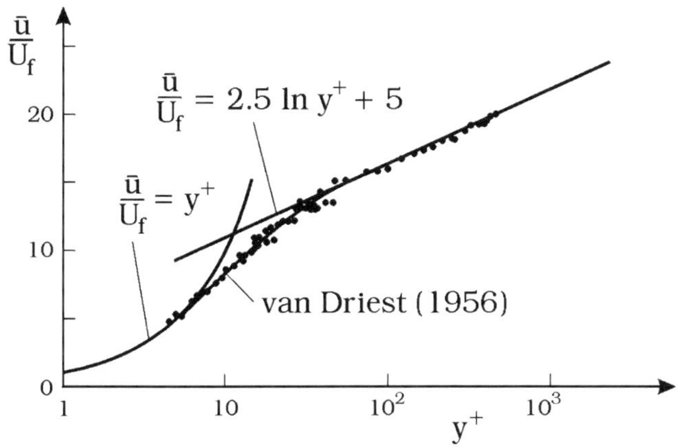

Fig. 3.20 Velocity distribution close to the wall, resolved also in the buffer layer. Adapted from Nezu and Rodi (1986).

Now, the third step is to determine the mixing length in the preceding equation. van Driest (1956) gives the mixing length as follows

$$\ell_m = \kappa y \left[1 - \exp\left(-\frac{y^+}{A_d}\right)\right] \tag{3.104}$$

in which κ is the von Karman constant (=0.4), and A_d is called the damping coefficient, taken as 25. As seen,

$$\ell_m \to \kappa y, \text{ for large values of } y^+ \tag{3.105}$$

while

$$\ell_m \to 0, \text{ as } y^+ \to 0 \tag{3.106}$$

The latter implies that there will be no turbulence mixing as the wall is approached.

Eqs. 3.100, 3.103 and 3.104 are the model equations (three equations); so, there are four equations altogether (along with the Reynolds equation, Eq. 3.98), and four unknowns, namely, $\overline{u}$, $-\rho\overline{u'v'}$, μ_t, and ℓ_m (the latter already given by Eq. 3.104). Hence, the system is closed. Therefore the unknowns can be determined. Let us focus on $\overline{u}$.

From Eqs. 3.98, 3.100 and 3.103, one gets

$$\rho\ell_m^2 \left(\frac{d\overline{u}}{dy}\right)^2 + \mu\frac{d\overline{u}}{dy} = \tau_0 \tag{3.107}$$

Solving for $\frac{d\overline{u}}{dy}$ from the preceding equation, and then substituting Eq. 3.104 into the solution, and subsequently integrating gives

$$\overline{u} = 2U_f \int_0^{y^+} \frac{dy^+}{1 + \left\{1 + 4\kappa^2 y^{+2}\left[1 - \exp\left(-\frac{y^+}{A_d}\right)\right]^2\right\}^{1/2}} \tag{3.108}$$

This is called the *van Driest velocity distribution,* Fig. 3.20.

1. As seen from the figure, for small values of y^+, the van Driest profile goes to the linear velocity distribution in the viscous sublayer (Eq. 3.35),

$$\frac{\overline{u}}{U_f} = y^+ \tag{3.109}$$

while, for large values of y^+, it goes to the logarithmic velocity distribution in the logarithmic layer (Eq. 3.43),

$$\overline{u} = U_f\ (A \ln y^+ + B) \tag{3.110}$$

reconciling with the previously obtained information.

2. Furthermore, the van Driest profile agrees remarkably well with the measured velocity profile including the buffer layer ($5 < y^+ < 70$), Fig. 3.20.

Regarding the other two unknowns, $-\rho\overline{u'v'}$ and μ_t, inserting Eqs. 3.104 and 3.108 in Eq. 3.103, μ_t is obtained, and from this and Eqs. 3.108 and 3.100, $-\rho\overline{u'v'}$ is obtained.

Turbulence modelling of flow close to transitional and rough walls

In this case, the mixing length may be given as (Cebeci and Chang, 1978) as

$$\ell_m = \kappa(y + \Delta y)\left[1 - \exp\left(-\frac{(y^+ + \Delta y^+)}{A_d}\right)\right] \tag{3.111}$$

1. y in the above equation is the distance from the wall; it is measured from the level where the velocity $\overline{u}$ is zero. This level lies almost at the top of the roughness elements. (y here should not be confused with the distance measured from the theoretical wall. As will be pointed out later (Item 3 below), apparently $y + \Delta y$ is the distance measured from the theoretical wall).

2. Δy is the so-called coordinate displacement, or the coordinate shift (see Example 3.3 below), and given by Cebeci and Chang (1978)

$$\Delta y^+ = 0.9\left[\sqrt{k_s^+} - k_s^+ \exp\left(-\frac{k_s^+}{6}\right)\right]; \qquad 5 < k_s^+ < 2000 \tag{3.112}$$

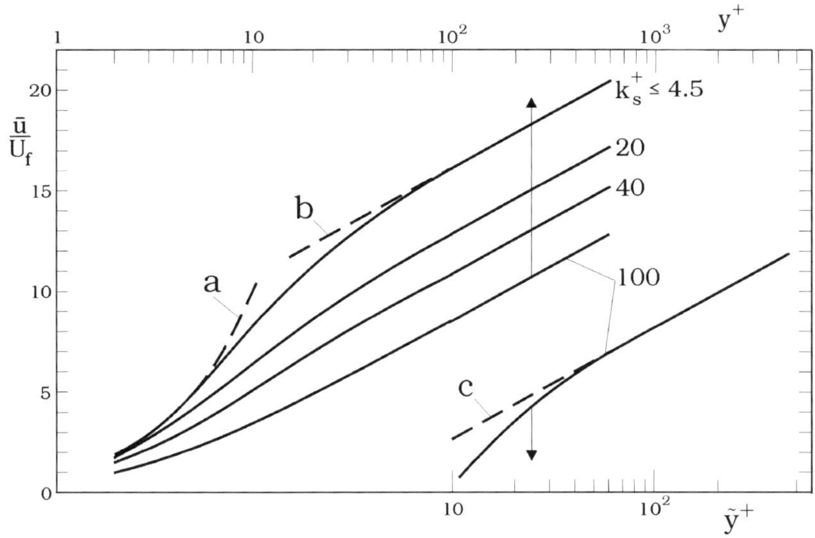

Fig. 3.21 van Driest velocity profiles for different values of roughness, where $\tilde{y}^+ = \tilde{y}U_f/\nu$, with $\tilde{y}$ being measured from the theoretical wall.

in which k_s^+ is the roughness Reynolds number (k_s being Nikuradse's equivalent sand roughness), Eq. 3.67, $k_s^+ = \frac{k_s U_f}{\nu}$.

3. Apparently, the distance $y + \Delta y$ is the distance measured from the theoretical wall, as will be shown in Example 3.3.

Similar to the case of the smooth wall, $\overline{u}$ can be obtained from the model equations, with ℓ_m given in Eq. 3.111:

$$\overline{u} = 2U_f \int_0^{y^+} \frac{dy^+}{1 + \left\{1 + 4\kappa^2 (y^+ + \Delta y^+)^2 \left[1 - \exp\left(-\frac{(y^+ + \Delta y^+)}{A_d}\right)\right]^2\right\}^{1/2}} \tag{3.113}$$

Fig. 3.21 displays the velocity profiles obtained from Eq. 3.113 including the smooth-wall profile. In Fig. 3.21, a represents the linear profile in the viscous sublayer (Eq. 3.109), b the logarithmic profile in the case of the hydraulically smooth wall (Eq. 3.110), and c the logarithmic profile in the case of the completely rough wall (Eq. 3.73), namely, by putting $\tilde{y} = y + \Delta y$,

$$\overline{u} = AU_f \ \ln\left(\frac{30\tilde{y}}{k_s}\right) \tag{3.114}$$

$\tilde{y}$ being measured from the theoretical wall.

Example 3.3. Rotta's (1962) theory about the coordinate shift Δy.

The velocity distribution *for large y^+ values* is written in the following form:

$$\overline{u} = U_f \left[\frac{1}{\kappa} \ln y^+ + c(k_s^+) \right] \tag{3.115}$$

in which y is measured from the level where the velocity $\overline{u} = 0$ (not to be confused with y measured from the theoretical wall). The function $c(k_s^+)$ in the above equation is determined from Nikuradse's experiments (Fig. 3.22). Note that $1 < k_s^+ < 1000$ (see the figure).

For the smooth-wall case, the preceding equation reduces to

$$\overline{u} = U_f \left[\frac{1}{\kappa} \ln y^+ + c(0) \right] \tag{3.116}$$

Now, in the case of the smooth wall, we know that there are three distinct regions, namely,

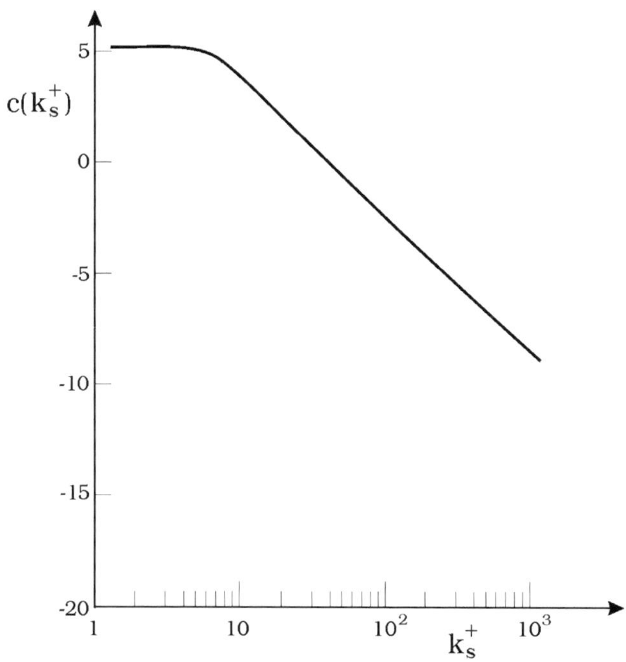

Fig. 3.22 Function $c(k_s^+)$ in Rotta's (1962) theory.

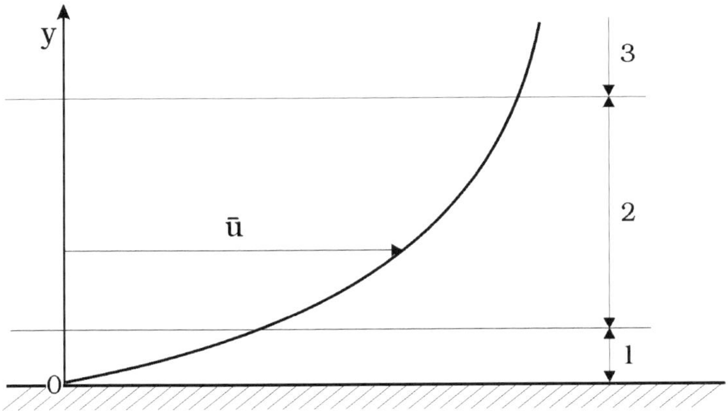

Fig. 3.23 Three regions in the flow over a smooth wall.

1. $y^+ < 5$ where the momentum transfer across the depth is due to viscosity (Region 1, corresponding to the viscous sublayer, Fig. 3.23);

2. $5 < y^+ < 70$ where this transfer is due to both viscosity and turbulence (Region 2, corresponding to the buffer layer, Fig. 3.23); and

3. $y^+ > 70$ where it is practically due to turbulence alone (Region 3, the logarithmic layer, Fig. 3.23).

In the case of the transitional/rough wall, Region 2 and Region 3 are still present (Fig. 3.24a). However, Region 1 is practically not existent. Now, we imagine a fictitious region (Region 1) just beneath Region 2, where a fictitious momentum transfer takes place. Let us assume that this momentum transfer is due to viscosity. Let the thickness of this layer be Δy. So, with the addition of this fictitious layer, the flow over the rough wall will be "identical" to that over the smooth wall. Therefore, we can implement the smooth-wall velocity distribution (Eq. 3.116) for the present case (with the wall coordinate $y + \Delta y$, Fig. 3.24b):

$$\overline{u} + U'(\Delta y) = U_f \left[\frac{1}{\kappa}\ln(y^+ + \Delta y^+) + c(0)\right] \tag{3.117}$$

in which $\overline{u}$ is the actual velocity and $U'(\Delta y)$ is the fictitious velocity that is to be added to $\overline{u}$, to get the fictitious smooth-wall velocity (see the figure). Considering large values of y^+, therefore $y^+ + \Delta y^+ \simeq y^+$, the above equation becomes

$$\overline{u} + U'(\Delta y) \simeq U_f \left[\frac{1}{\kappa}\ln(y^+) + c(0)\right] \tag{3.118}$$

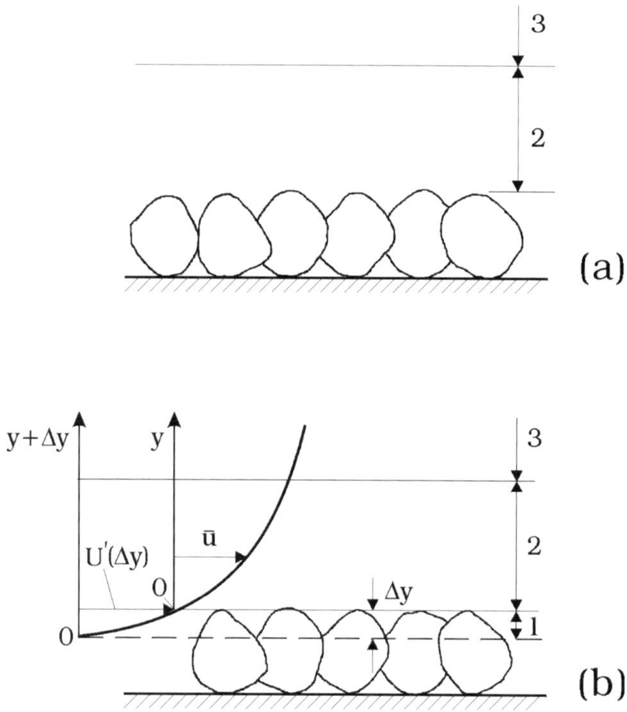

Fig. 3.24 Flow over a rough wall in Rotta's (1962) theory.

The fictitious velocity $U'(\Delta y)$ can be calculated from, for example, Eq. 3.108, replacing y^+ in Eq. 3.108 by Δy^+. (Δy^+ cannot be calculated from Eq. 3.116 because the latter equation is valid only for large values of y^+.) Therefore, from Eqs. 3.118 and 3.108, the actual velocity is obtained as

$$\overline{u} = U_f \left[\frac{1}{\kappa}\ln(y^+) + c(0)\right] - U'(\Delta y^+) \tag{3.119}$$

On the other hand, this velocity is given independently in Eq. 3.115. So, setting these two equations equal,

$$U_f \left[\frac{1}{\kappa}\ln(y^+) + c(0)\right] - U'(\Delta y^+) \equiv U_f \left[\frac{1}{\kappa}\ln y^+ + c(k_s^+)\right] \tag{3.120}$$

one gets

$$c(k_s^+) - c(0) + \frac{U'(\Delta y^+)}{U_f} = 0 \tag{3.121}$$

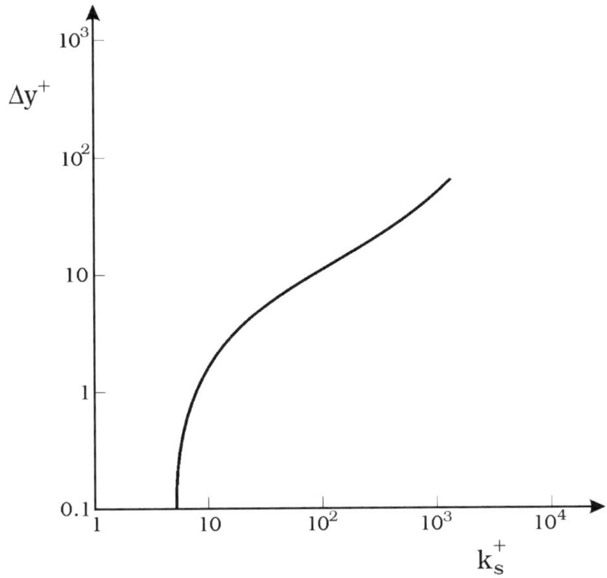

Fig. 3.25 Coordinate shift in Rotta's (1962) theory.

Given the function $c(k_s^+)$ (Fig. 3.22), and $U'(\Delta y^+)$ from Eq. 3.108,

$$U'(\Delta y^+) = 2U_f \int_0^{\Delta y^+} \frac{dy^+}{1 + \left\{1 + 4\kappa^2 \Delta y^{+2} \left[1 - \exp\left(-\frac{\Delta y^+}{A_d}\right)\right]^2\right\}^{1/2}}$$

the coordinate shift Δy^+ can be calculated from Eq. 3.121. Rotta (1962) carried out this calculation. Fig. 3.25 shows the result.

It may be noted that the variation depicted in Fig. 3.25 agrees fairly well with the expression given by Cebeci and Chang (1978) in Eq. 3.112.

As a final remark, the coordinate shift normalized by the roughness height, $\frac{\Delta y}{k}$, can be found from Eq. 3.112 (where k_s may be taken as $2k$). The result of this exercise is plotted in Fig. 3.26. As seen, the $\frac{\Delta y}{k}$ values exhibited in the figure agree rather well with the experimental observations that the level of the theoretical wall lies $(0.15\text{–}0.35)k$ below the top of the roughness elements, given in conjunction with the theoretical wall in Section 3.1.3. This implies that the coordinate shift Δy may be interpreted as the distance of the theoretical wall from the top of the roughness elements.

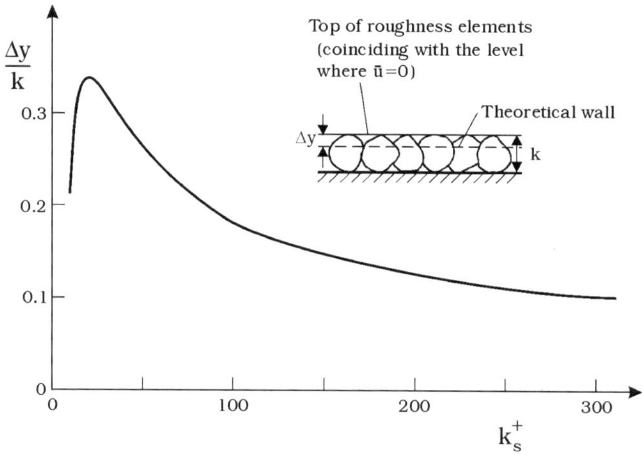

Fig. 3.26 Coordinate shift Δy from Rotta (1962) and Cebeci and Chang (1978). (k_s is taken as $2k$).

3.3.2 *Turbulence modelling of flow across the entire section*

The velocity distribution given in Eq. 3.113 is obviously valid near the wall. When applied for the entire depth, the velocities predicted from this equation are apparently smaller than the measured velocities. This difference is corrected by the so-called Coles' wake function (see Coleman and Alanso, 1983)

$$\frac{\Pi}{\kappa}\,\omega\left(\frac{y^+}{h^+}\right)=\left(\frac{\Pi}{\kappa}\right)\,2\,\sin^2\left(\frac{\pi}{2}\frac{y^+}{h^+}\right) \tag{3.122}$$

in which Π is a wake strength coefficient, h is the flow-depth/pipe-radius, or the boundary-layer thickness.

When this equation is used together with Eq. 3.113, it does not produce a profile with a derivative $\frac{d\bar{u}}{dy}$ equal to zero at the center line of a pipe, or at the free-stream edge of the boundary layer over a wall (clearly, in the case of pipe flows and boundary-layer flows, this derivative has to be zero). To avoid this drawback, Coles' wake function has been corrected. The final form of the velocity expression (including the corrected wake function) is as follows (Coleman and Alanso, 1983):

$$\frac{\overline{u}}{U_f} = \int_0^{y^+} \frac{2\ dy^+}{1+\left\{1+4\kappa^2(y^+ + \Delta y^+)^2\left[1-\exp\left(-\frac{(y^+ + \Delta y^+)}{A_d}\right)\right]^2\right\}^{\frac{1}{2}}}$$
$$+\left(\frac{y^+}{h^+}\right)^2\left(1-\frac{y^+}{h^+}\right)+\left(\frac{2\Pi}{\kappa}\right)\left(\frac{y^+}{h^+}\right)^2\left[3-2\frac{y^+}{h^+}\right] \tag{3.123}$$

Regarding the Π values, this parameter basically depends on two factors: (1) the driving pressure gradient, and (2) the degree of turbulence in the outer region. A convergent flow (favourable pressure gradient) tends to produce low values of Π, while a divergent flow (adverse pressure gradient) produces high values of Π. Likewise, if the turbulence in the outer region is intense, Π will have low values, while, the turbulence in the outer region is weak, Π will have large values (Coleman and Alanso, 1983). The latter authors report a range of Π from 0 to 0.55.

3.4 Flow resistance

In coastal and civil engineering applications, one of the frequently asked questions is how to predict the friction velocity U_f, i.e.,

$$U_f = \sqrt{\frac{\tau_0}{\rho}} \tag{3.124}$$

when the mean flow velocity, V, is given. This issue may be best described by reference to a flow in a circular pipe.

3.4.1 *Smooth circular pipe*

For a *smooth* circular pipe, the flow velocity is given by Eq. 3.43 with coefficients given in Eq. 3.44. Schlichting (1979, p. 603) takes it in the following, slightly different form:

$$\overline{u} = U_f\ (2.5 \ln y^+ \ + \ 5.5) \tag{3.125}$$

Assuming that this velocity distribution is, to a first approximation, valid across the entire section, and therefore putting $y = R$, the pipe radius, will give:

$$U_0 = U_f\left[2.5 \ln\left(\frac{RU_f}{\nu}\right)+5.5\right] \tag{3.126}$$

On the other hand, we have obtained in the preceding paragraphs a relationship between the mean flow velocity and the maximum velocity (Eq. 3.95), which reads

$$V = U_0 - (A - B')U_f \tag{3.127}$$

Now, Schlichting (1979, p. 609) takes this relation in the following form

$$V = U_0 - 3.75U_f \tag{3.128}$$

Combining Eq. 3.126 with Eq. 3.128, we obtain the following equation (Schlichting, 1979, p. 610)

$$\frac{1}{\sqrt{8}}\frac{V}{U_f} = 2.035 \log\left(\frac{VD}{\nu}\sqrt{8}\frac{U_f}{V}\right) - 0.91 \tag{3.129}$$

in which D is the pipe diameter, and "log" is the common logarithm (the logarithm to base 10). The coefficients can be adjusted slightly so that the preceding equation agrees with the experimental data. This latter exercise gives the following relation

$$\frac{1}{\sqrt{8}}\frac{V}{U_f} = 2.0 \log\left(\frac{VD}{\nu}\sqrt{8}\frac{U_f}{V}\right) - 0.8 \tag{3.130}$$

This is Prandtl's universal law of friction for smooth pipes (Schlichting, 1979, p. 611).

3.4.2 *Rough circular pipe*

In the case of a *rough* circular pipe, we can do the same exercise, but this time with the velocity distribution given by Eq. 3.73

$$\overline{u} = AU_f \ \ln\left(\frac{30y}{k_s}\right) \tag{3.131}$$

Putting $y = R$ in the above equation

$$U_0 = AU_f \ \ln\left(\frac{30R}{k_s}\right) \tag{3.132}$$

and combining Eqs. 3.132 and 3.128, we get (Schlichting, 1979, p. 621)

$$\frac{U_f}{V} = \frac{1}{\sqrt{8}(2\log\left(\frac{R}{k_s}\right) + 1.68)} \tag{3.133}$$

A comparison with Nikuradse's experimental results shows that closer agreement can be obtained, if the constant 1.68 is replaced by 1.74 (Schlichting, 1979, p. 621):

$$\frac{U_f}{V} = \frac{1}{\sqrt{8}(2\log\left(\frac{R}{k_s}\right) + 1.74)} \tag{3.134}$$

3.4.3 *Transitional circular pipe*

In the case of a *transitional-wall-category circular pipe*, Colebrook and White obtain the following resistance relation (Schlichting, 1979, p. 621)

$$\frac{1}{\sqrt{8}}\frac{V}{U_f} = 1.74 - 2\log\left(\frac{k_s}{R} + \frac{18.7}{(\frac{VD}{\nu})(\sqrt{8}\frac{U_f}{V})}\right) \tag{3.135}$$

Given the mean-flow velocity, the friction velocity can easily be calculated from Eqs. 3.130, 3.134 or 3.135.

3.4.4 *Non-circular pipe and open channel*

In the case of a non-circular pipe flow or an open channel flow, the preceding resistance relations can be used provided that the pipe diameter D should be replaced by $4r_h$ (and hence the radius $R = D/2 = 2r_h$), in which r_h is the hydraulic radius, equal to the cross-sectional flow area divided by the wetted perimeter (Schlichting, 1979, p. 622). For example, in the case of an open channel with a rough bed, the resistance relation can be found as

$$\frac{V}{U_f} = 5.75\log\left(\frac{14.8r_h}{k_s}\right) \tag{3.136}$$

or

$$\frac{V}{U_f} = 2.5\ln\left(\frac{14.8r_h}{k_s}\right) \tag{3.137}$$

It may be noted that the ratio V/U_f may be traditionally written in terms of the friction coefficient, f, defined by

$$U_f = \sqrt{\frac{f}{2}}V \qquad \text{or} \qquad \frac{V}{U_f} = \sqrt{\frac{2}{f}} \tag{3.138}$$

It is also noted that for a wide river or channel having rectangular cross-section, that the hydraulic radius, r_h, is well approximated by the channel flow depth h.

3.5 Bursting process

We will start off with a close look at velocity time series near the wall in a turbulent boundary layer flow. Fig. 3.27 shows two time series; the bottom time series is the streamwise velocity recorded at a distance of $y = 0.2\delta_v$ in which δ_v is the thickness of the viscous sublayer (Eq. 3.36), and the top time series is the wall shear stress, measured at the same section.

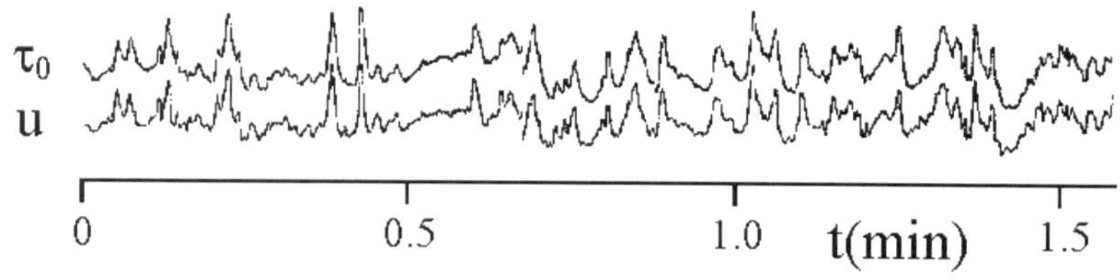

Fig. 3.27 Streamwise velocity and wall shear stress traces. Velocity is measured at $y = 0.2\delta_v$, where δ_v is the thickness of the viscous sublayer. Adapted from Eckelmann (1974).

These measurements were conducted by Eckelmann (1974) in an oil channel flow. Use of oil enabled Eckelmann (1974) to resolve the viscous sublayer sufficiently accurately, with the thickness of the viscous sublayer being like $\delta_v = O(1 \text{ cm})$. The Reynolds number of the flow was approximately 8000 in which the Reynolds number is based on the maximum centre-line velocity and the channel width.

When closely inspected, the velocity and wall shear stress traces in Fig. 3.27 do not look entirely random, but rather "quasi" periodic. This is particularly the case for the part of the signal from approximately $t = 0.6$ min until approximately $t = 1.4$ min. Fig. 3.28 presents three further sets of velocity and wall shear stress time series from the same study. The velocities were measured at $y = \delta_v$ in Fig. 3.28a; at $y = 1.7\delta_v$ in Fig. 3.28b; and at $y = 12\delta_v$ in Fig. 3.28c, respectively. The velocity time series in Figs. 3.28a–c show that the aforementioned quasi-periodicity in the velocity traces seems to disappear as one moves away from the wall (c.f., Fig. 3.27 and 3.28c). Incidentally, there is a phase lag between the wall shear stress signal and the velocity signal in Fig. 3.27 and Figs. 3.28a,b; the wall shear stress signal lags behind the velocity signal, as expected.

Observations similar to Eckelmann (1974) opened up a whole new research area in the 1960s and 1970s where the near-wall flow is viewed as a quasi-cyclic process, as opposed to the classic approach where the flow process was perceived as a random process. This quasi-cyclic process has become known as the *bursting process*. Reviews of the subject can be found in Laufer (1975), Hinze (1975, pp. 659–668), Cantwell (1981), Grass (1983), Grass et al. (1991) and Nezu and Nakagawa (1993) among others.

Most recently, Direct Numerical Simulation (DNS) has been conducted by Wu et al. (2017) of a zero-pressure-gradient, smooth, flat plate, boundary-layer flow, see the detailed description of this study provided in

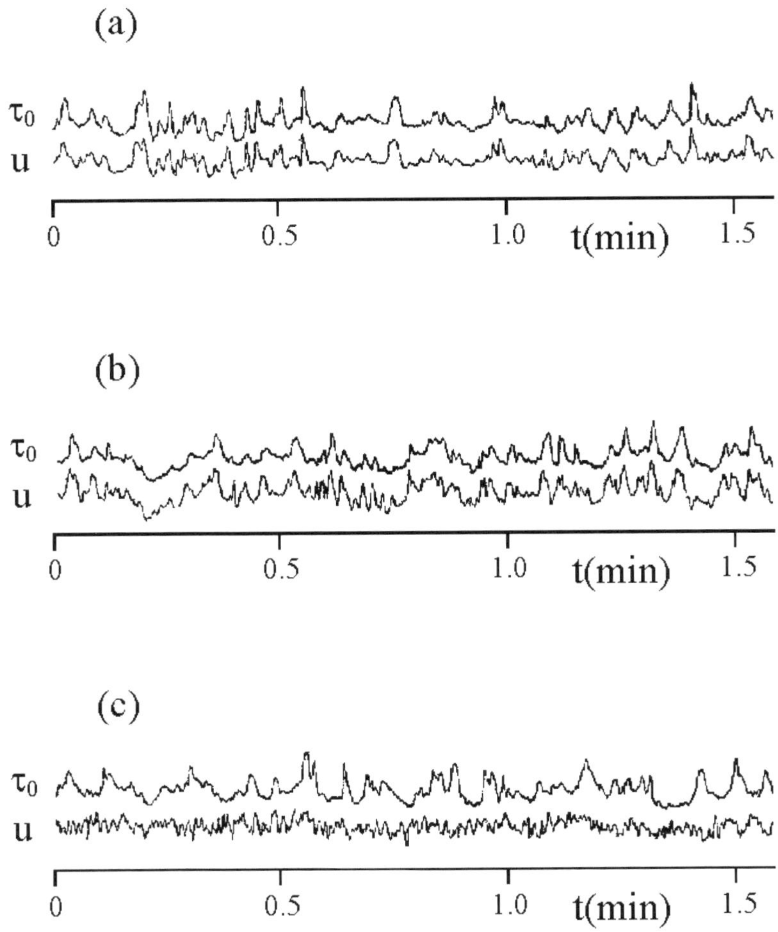

Fig. 3.28 Streamwise velocity and wall shear stress traces. The streamwise velocity is measured at (a) δ_v, (b) $1.7\delta_v$ and (c) $12\delta_v$, with δ_v being the thickness of the viscous sublayer. Adapted from Eckelmann (1974).

Appendix 1.C. Wu et al.'s (2017) simulations for the range of the Reynolds number $2400 < Re_\theta < 2650$ revealed coherent structures attributed by the authors to the bursting process. The reader is referred to Appendix 1.C, where the authors address the question of how their results reconcile with the earlier studies discussed in the reviews cited in the preceding paragraph.

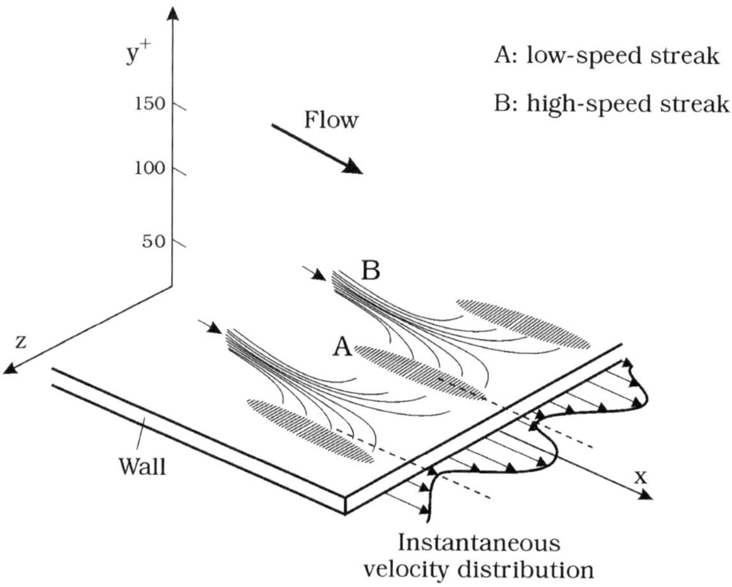

Fig. 3.29 Snapshot of flow near the wall.

3.5.1 *Description of the bursting process*

The description of the bursting process in the following paragraphs is mainly based on the work of Offen and Kline (1973, 1975).

Fig. 3.29 gives a snapshot (schematic) of the flow picture near the wall. As seen, the flow consists of a transverse alternation of low-speed (A) and high-speed (B) regions. These are called low-speed and high-speed wall streaks, respectively. The low-speed and high-speed streaks form locally and temporarily.

A low-speed wall streak is subject to an impulsive "suction". This causes this coherent, low-speed fluid to entrain vigorously into the main body of the flow. This is called the *ejection* event. See Fig. 3.30, Frames 1 and 2.

The fluid ejected into the flow grows in size, as it is convected downstream, while keeping its coherent entity. This continues for some period of time (Fig. 3.30, Frame 3), and finally, at some point ($t = T$), the previously mentioned coherent structure breaks up (Fig. 3.30, Frame 4). At this point, some fluid from the coherent structure returns back to the near-wall region (Fig. 3.30, Frame 5). This fluid impinges on the wall (in the form of a high-speed wall streak, called the *in-rush* or *sweep* event), and spreads out

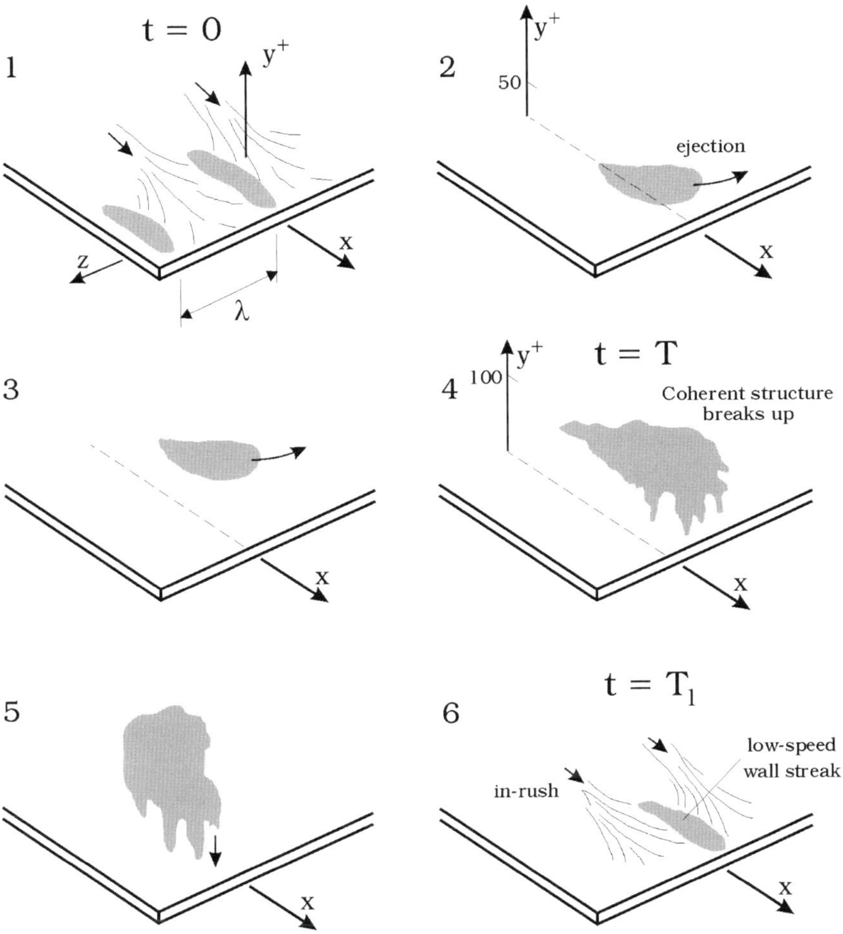

Fig. 3.30 Sequence of flow pictures over one bursting period.

sideways (Fig. 3.30, Frame 6). Two such neighbouring high-speed streaks eventually merge (Fig. 3.30, Frame 6), and retard the flow there, and, as a result, a new low-speed wall streak is formed. The time period spent from the formation of the low-speed streak at time $t = 0$ (Fig. 3.30, Frame 1) to the formation of the next one at time $t = T_1$ (Fig. 3.30, Frame 6) is called the *bursting period*. After the formation of this new low-speed streak, the same chain of events occurs, and the process continues in a repetitive manner.

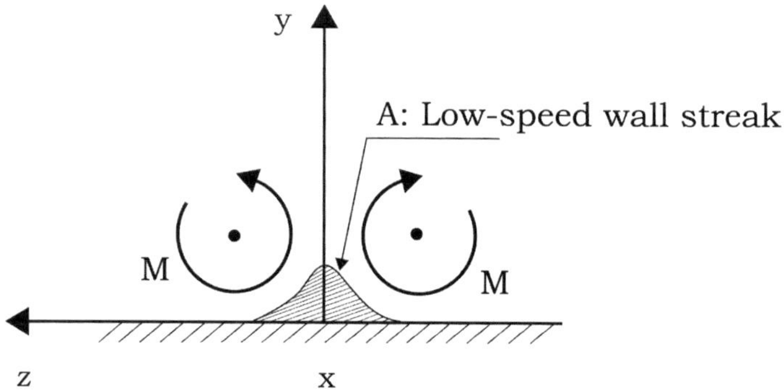

Fig. 3.31 Cross-section view of the flow in Frame 6 in the previous figure. M: Counter-rotating streamwise vortices generated by the two neighbouring high-speed fluids impinging on the wall. A: Low-speed wall streak between them, generated by the same process.

Fig. 3.31 depicts a cross section, passing through the alternating high- and low-speed wall streaks in Frame 6 in Fig. 3.30. It may be inferred from Frame 6 in Fig. 3.30 that two counter-rotating streamwise vortices (marked Vortices M in Fig. 3.31) are generated as the two neighbouring high-speed fluids impinge on the wall, with the low-speed wall streak lying between them (marked A in Fig. 3.31).

3.5.2 *Flow visualization of the bursting process*

The low-speed wall streaks can be made visible in a simple laboratory experiment: Sprinkle sand onto the bottom of an open water channel, and observe the pattern of the sediment on the bottom shortly after. Fig. 3.32 illustrates a sequence of flow pictures, 1/12 s apart, obtained in such an experiment with intensively illuminated fine sand, the size 0.1 mm, moving over the bottom (Grass, 1971). The vertical arrows indicate the development of an in-rush event where the fluid is swept sideways into the low-speed wall streaks, visualized by sand. The longitudinal streaks in this sequence represent the low-speed wall streaks, some being formed earlier, and some being in the process of development. These longitudinal streaks visualized by sand should be low-speed fluid areas as the sand in these areas do not appear blurry or streaky.

Low-speed wall streaks can also be made visible, using hydrogen-bubble technique in water flows, Fig. 3.33, reproduced from Kline et al. (1967).

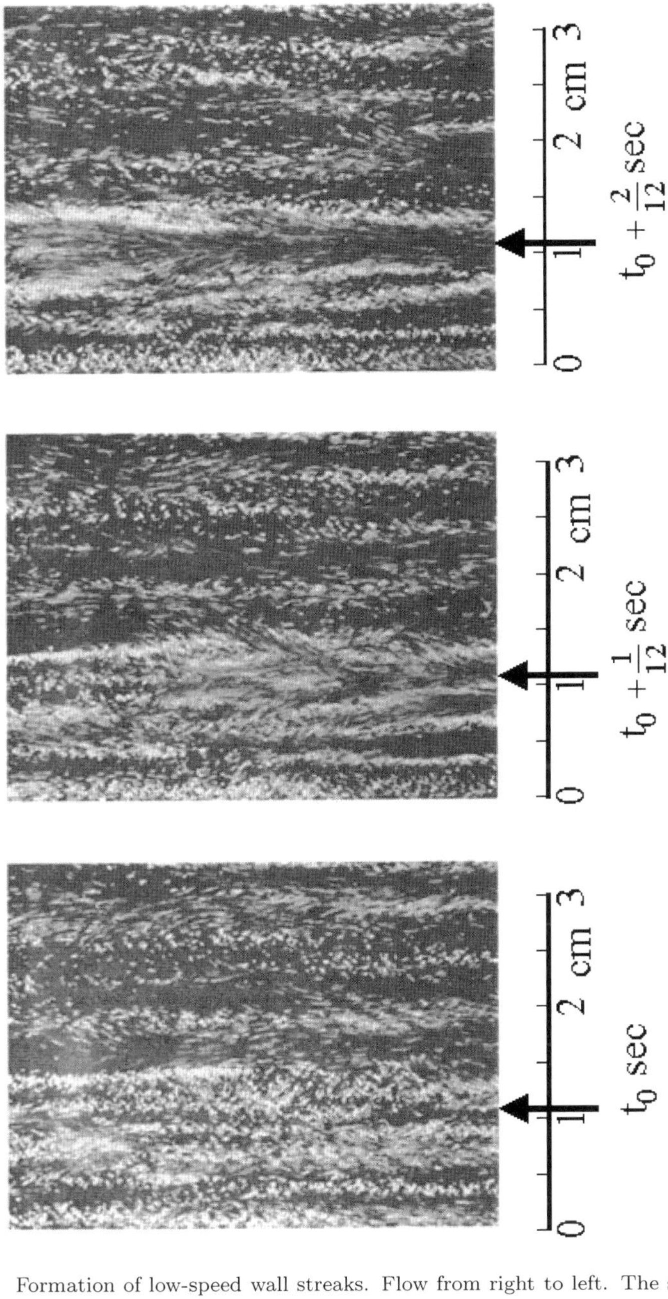

Fig. 3.32 Formation of low-speed wall streaks. Flow from right to left. The sequence illustrates the sweep (or in-rush) event where the fluid is swept sideways, visualized by sand. Grass (1971).

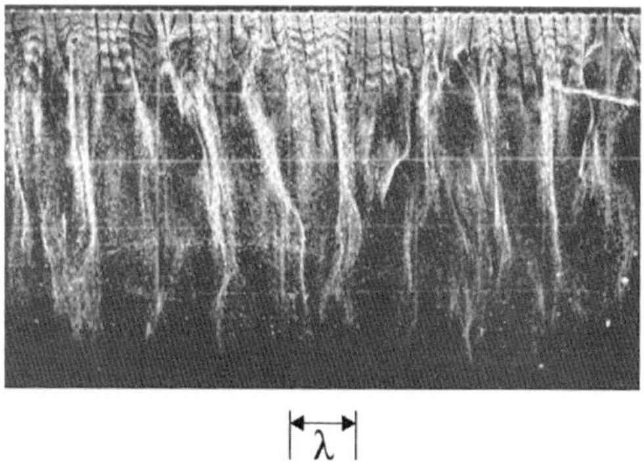

Fig. 3.33 Low-speed wall streaks, visualized by hydrogen-bubble technique. Plan view. Flow from top to bottom. Adapted from Kline et al. (1967).

Typically, in the hydrogen-bubble experiments, a very thin wire, insulated at constant intervals, carries an electric current in pulses. This produces tiny hydrogen bubble blocks that make the flow visible, with high degree of resolution in space and time.

The previously mentioned experiments, and many others, show that the spacing between the low-speed wall streaks, also called the cross-flow wave length, λ (see Fig. 3.30, Frame 1, and Fig. 3.33), satisfies

$$\lambda^+ = \frac{\lambda U_f}{\nu} \simeq 100 \tag{3.139}$$

We note that, as expected, λ scales with the inner flow parameters, the friction velocity, U_f, and the kinematic viscosity, ν.

Side views of the bursting process can also be made visible in a laboratory experiment, using various techniques. Fig. 3.34 illustrates the side view of an ejection event (Fig. 3.34a), and that of a sweep event (Fig. 3.34c) obtained in an open water channel experiment (Grass, 1971), using the hydrogen bubble technique. The hydrogen-bubble images in the figure are accompanied by the corresponding instantaneous u and v velocity profiles (Fig. 3.34b) in which square symbols are for the velocity profiles associated with the ejection event (photograph on left), and triangular symbols are for the velocity profiles associated with the sweep event (photograph on right). Circular symbols are for the mean velocity velocity profile, $\overline{u}$. It may be noted that Grass (1971) did his experiments with three kinds of bottom roughness, namely $k^+ = 0$ (smooth wall), $k^+ = 20.7$ (transitional wall), and $k^+ = 84.7$ (rough wall). The images in Fig. 3.34 are for the

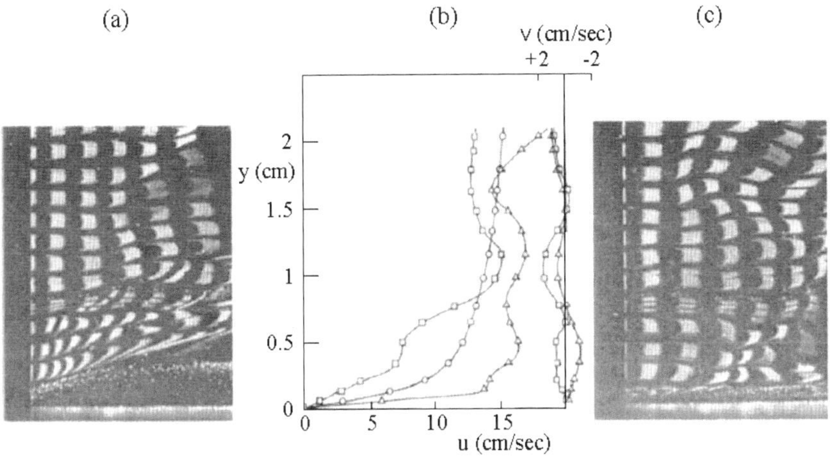

Fig. 3.34 Ejection (Panel (a)), and sweep (Panel (c)) in side view, visualized by hydrogen bubble technique. Panel (b): Velocity profiles associated with ejection (square symbols); those associated with sweep (triangular symbols); and mean velocity, $\overline{u}$, profile (circular symbols). Grass (1971).

smooth-wall experiments. The rough-wall case is studied in Section 3.5.4.

Incidentally, the hydrogen bubble images would enable one to obtain instantaneous velocities by measuring the distance that the corner of a particular bubble block traveled in a short time interval, 1/50 second in Grass' (1971) experiments. Grass (1971) obtained the longitudinal, u, and the vertical velocities, v, in this way. This produced an adequate sample of instantaneous point u, v velocities from which he was able to obtain mean and turbulence quantities. He then validated his measurements against the existing information: His mean velocity data (Grass, 1971, Fig. 4) compared very well with the logarithmic law (Eq. 3.43 for the smooth-wall case and Eq. 3.73 for the rough-wall case). Likewise, his turbulence data for the smooth-wall case (Grass, 1971, Fig. 6) compared, again, very well with Laufer's turbulence data (Fig. 3.7). Clearly, this gives confidence in the use of Grass' (1971) hydrogen-bubble data in the analysis of the bursting process.

We now return to Fig. 3.34. The figure clearly shows that the instantaneous velocity profile associated with the ejection event exhibits a considerable amount of deficit, and that associated with the sweep event, likewise, a considerable amount of excess with respect to the mean velocity profile, $\overline{u}(y)$, near the wall. From the figure, it is clear that:

(1) the ejection event can be represented by

$$u' < 0 \text{ and } v' > 0 \tag{3.140}$$

while:

(2) the sweep event can be represented by

$$u' > 0 \text{ and } v' < 0 \tag{3.141}$$

in which $u' = u - \overline{u}$, and $v' = v - \overline{v} = v - 0 = v$ are the fluctuating velocity components in the streamwise and vertical directions, respectively.

Note the correspondence in the above, with the fact that $\overline{u'v'} < 0$ in steady uniform channel flow, i.e. that the horizontal and vertical turbulent fluctuations are negatively correlated.

3.5.3 *Contribution of ejection and sweep events to the Reynolds stress*

Wallace et al. (1972) and Brodkey et al. (1974) defined four types of motion according to the sign of the fluctuating velocity components, depicted in Table 3.3. As seen from the table, the first two types of motion are the ejection and sweep motions (Eqs. 3.140 and 3.141). Owing to the background turbulence, there are two additional motions, called interaction motions. These interaction motions are called wallward interaction if $v' < 0$, and outward interaction if $v' > 0$.

On the basis of this categorization, Brodkey et al. (1974) determined the contributions of each motion to the Reynolds shear stress by conditional (time) averaging from the data they obtained in an oil channel. The Reynolds number based on the centre-line velocity and the channel width was 7700. Fig. 3.35 displays the results; the vertical axis $-\rho\overline{u'v'}_c/(-\rho\overline{u'v'})$ in the figure represents the contribution of each motion, with $-\rho\overline{u'v'}_c$ being the conditional averaged Reynolds shear stress (conditional averaging according to Table 3.3), and $-\rho\overline{u'v'}$ the Reynolds shear stress.

Table 3.3 Four types of motion near the wall.

Sign of u'	Sign of v'	Sign of $u'v'$	Type of motion
$-$	$+$	$-$	Ejection
$+$	$-$	$-$	Sweep
$-$	$-$	$+$	Interaction (wallward)
$+$	$+$	$+$	Interaction (outward)

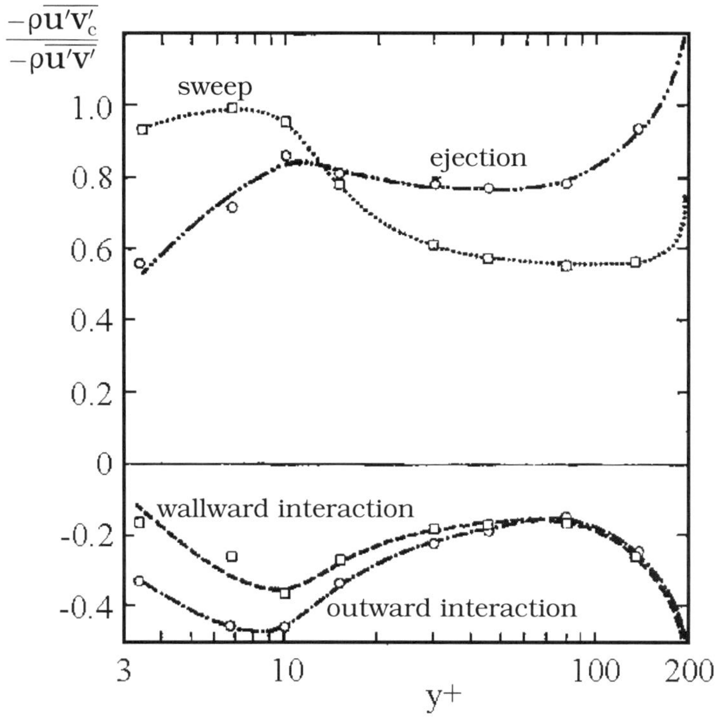

Fig. 3.35 Classified Reynolds stresses normalized with local average Reynolds stress. Top panel: Dots: Sweep, and Dashed-double-dots line: Ejection. Bottom panel: Dashed line: Wallward interaction, and Dashed-dot line: Outward interaction. Brodkey et al. (1974).

As seen from Fig. 3.35, the major contributions to the Reynolds shear stress come from the ejection and sweep motions although the interaction motions can also contribute to the Reynolds stress, but their contributions being a factor of 2 less, at best.

Fig. 3.36 shows the fraction of time during which the previously mentioned motions take place. Roughly, the ejection and sweep motions occur 30% of the time each while the interaction motions occur 20% of the time each.

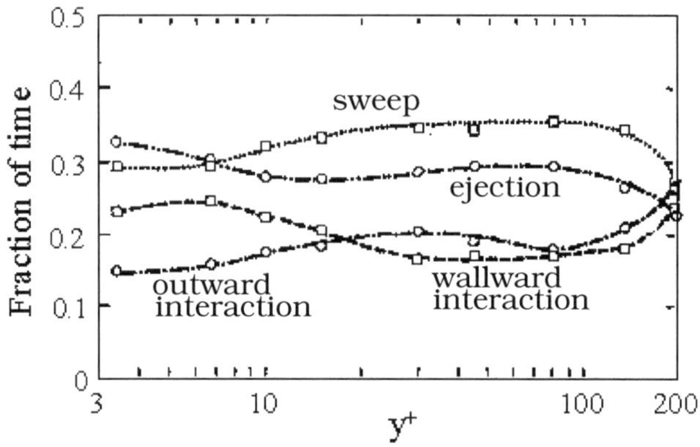

Fig. 3.36 Fraction of time the signals are in the respective categories. Dots: Sweep; Dashed-double-dots line: Ejection; Dashed line: Wallward interaction; and Dashed-dot line: Outward interaction. Brodkey et al. (1974).

3.5.4 *Bursting process over rough walls*

Fig. 3.37 shows a side view of ejection (Fig. 3.37a) and sweep (Fig. 3.37c) events in the case of the rough bottom with $k^+ = 84.7$ in Grass' (1971) experiments. Similar to Fig. 3.34, the hydrogen-bubble images are accompanied with the instantaneous velocity profiles associated with these events (Fig. 3.37b). As seen, the picture is much the same except that the vertical velocity profiles near the wall in the case of the rough wall are not as "clear-cut" as in the case of the smooth wall, where it was very distinct that $v' > 0$ for ejection and $v' < 0$ for sweep phases near the wall. Also, the longitudinal velocity profile associated with the ejection event, although there is a definite deficit with respect to the mean profile, does not exhibit the same degree of deficit as in the case of the smooth wall.

Grass (1971) notes that differences in the bursting process between the smooth- and rough-wall flows appear to be mainly associated with the detailed mechanics of low-momentum fluid entrainment at the wall surface. In this respect, he notes that it is envisaged that the smooth wall viscous sublayer fluid and the fluid trapped between the roughness elements form a "passive" reservoir of low-momentum fluid, which is drawn on during ejection phases. He also differentiates between the two cases with the following two observations: (1) The entrainment in the rough-wall case was extremely violent, with ejected fluid rising almost vertically from between the interstices of the roughness elements; and (2) The long twisting streamwise

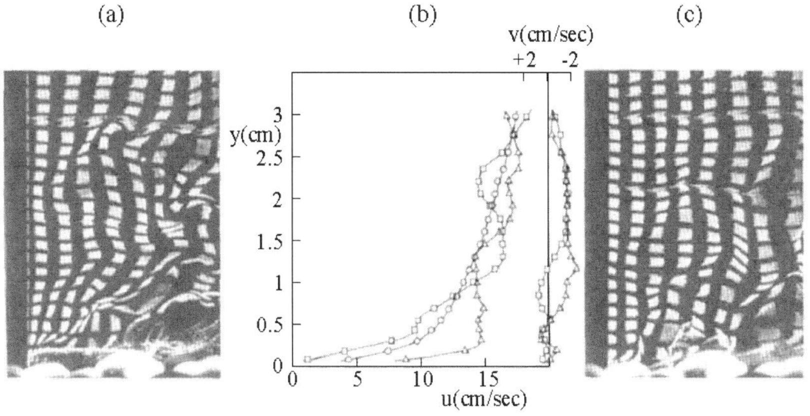

Fig. 3.37 Ejection (Panel (a)), and sweep (Panel (c)) in side view, visualized by hydrogen bubble technique. Panel (b): Velocity profiles associated with ejection (square symbols); those associated with sweep (triangular symbols); and mean velocity, $\overline{u}$, profile (circular symbols). Rough bed with $k^+ = 84.7$. Grass (1971).

vortices, very apparent near the smooth wall during inrush-ejection cycles, were much less conspicuous in the transitional- and rough-wall cases.

Grass et al. (1991, pp. 60–61), in a later study, systematically investigated the effect of the wall roughness on low speed wall streaks, conducting experiments in an open-channel flow, using the hydrogen bubble technique. They carried out four kinds of experiments, one experiment with a hydraulically smooth wall, and three experiments with rough walls with spherical roughness elements having size $k = 1.15$ mm, 6 mm and 12 mm, respectively. The results indicated that, just as in the smooth wall case, the spanwise flow structure adjacent to the rough boundary exhibits a well defined cross-flow wave length, λ (recall the definition of λ, Fig. 3.30, Frame 1). The experiments showed that this wave length is directly proportional to the size of the roughness elements, k, under fully rough conditions, with

$$\frac{\lambda}{k} \simeq 3 \tag{3.142}$$

Earlier, Aydin and Leutheusser (1985) studied coherent structures associated with the bursting process in a turbulent open-channel flow with the principle variable of the study being the texture of the channel bottom; they tested three kinds of bottom, namely smooth, longitudinally ribbed, and artificially roughened, using spherical beads in the latter case (the spherical beads are arranged in an open diamond configuration). The authors used the hydrogen-bubble technique with spanwise bubble wires placed at different elevations. Their observations clearly showed the presence of the low-speed wall streaks for all three bottom textures. Of particular interest is the spacing between the streaks, the cross-flow wave length, λ. Aydin and Leutheusser's (1985) result for the spherical-bead roughness appears to be quite similar to Grass et al.'s (1991) relation $\lambda/k \simeq 3$, Eq. 3.142. Note that the roughness height of the spherical beads in Aydin and Leutheusser's experiments (1985) was $k = 0.58$ cm, corresponding to a value of the roughness Reynolds number of $k_s^+ = 231$, revealing that the bottom behaved as a fully rough wall in the experiment.

Similar results have been reported by Defina (1996). The latter author experimented with an open water channel with the bottom roughness achieved with spherical particles, the size approximately 1 cm, arranged in the densest possible packing array. The roughness Reynolds number of the experiments was in the range $k_s^+ = 120$–340. The tracer used for the flow visualization of the low-speed wall streaks was dyed milk. The tracer was given through small holes drilled at the interstices between the particles across the spanwise direction, the spacing between the holes being about 0.5 cm. The low-speed wall streaks were photographed from above. The results showed that, near the bottom, the spacing between the streaks was $\lambda/k_s \simeq 4$, or in terms of k, $\lambda/k \simeq 2.7$, in agreement with the previously described data (Eq. 3.142).

It is also to be noted that Defina's (1996) experiments further showed that (1) the spacing between the streaks increased linearly with the distance from the bed; and more importantly (2) the streaks completely disappeared when $k_s > O(0.5h)$. We will return to this second issue later in the section.

Further data were obtained by Detert et al. (2010). These authors used three different bed particles: spheres; uniform gravel; and gravel from the river Rhine, which, in the tests, actually formed a 0.10 m thick sediment layer. The spheres were arranged in the densest packing arrangement while the gravel were arranged in a loose packing arrangement. The roughness Reynolds number was $k_s^+ = 377$ for the spherical particle roughness, 643–969 for the uniform-gravel roughness, and 2245 for the Rhine-gravel roughness. PIV measurements in horizontal planes revealed turbulent structures with regions of lower and higher velocities that were elongated in the streamwise direction, with the spacing between the low-speed regions being $\lambda/k_s \simeq 3$–4 near the bed, essentially confirming the results of the previous research. Detert et al.'s (2010) results also showed that the spacing between the low-speed regions increased linearly with the distance from the bed, in agreement with Defina (1996).

A further piece of evidence for the presence of the bursting process over rough walls comes from Carstensen et al. (2012). The latter authors studied turbulent spots over a rough boundary (obtained by gluing 0.25 cm coarse sand onto the base bottom of an oscillating tunnel) in wave boundary layers. A detailed account of Carstensen et al.'s (2012) study is given in Chapter 5. The flow-visualization video of Carstensen et al. (2012) clearly reveals the existence of low-speed wall streaks although they are slightly "contaminated" by the presence of the roughness elements on the boundary. The reader can consult the latter video, which is a supplementary material to the paper of Carstensen et al. (2012), with the URL address `http://dx.doi.org/10.1063/1.4767536`. In Carstensen et al.'s (2012) tests, the roughness Reynolds number was estimated to be $k_s^+ > O(60)$ when the turbulent spots first emerged (Carstensen et al., 2012, Section D), indicating that the wall was nearly completely rough. It is interesting to note that, when closely inspected, Carstensen et al.'s (2012) video confirms Grass et al.'s (1991) empirical relation $\lambda/k \simeq 3$, referred to in the preceding paragraphs, Eq. 3.142.

Finally, we will discuss briefly the question of whether or not the bursting process occurs over walls with very large roughnesses.

No study is yet available, investigating bursting process over such boundaries. As discussed in Section 3.1.3, turbulence over a rough wall is contributed by two effects:

1. The first effect is linked to the mean velocity gradient, $d\overline{u}/dy$, in the same way as in the case of the smooth wall. The bursting process is associated with this part of turbulence.

2. The second effect is linked to the vortex shedding from the roughness elements.

It is expected that, for very large roughnesses, the turbulence generated by the second effect will overshadow that generated by the first one. Therefore it is most likely that there will be no bursting process over such boundaries. The finding of Defina (1996) discussed in the preceding paragraphs that the low-speed wall streaks disappear when $k_s > O(0.5h)$ supports this argument. This issue, however, appears to be worth exploring further.

3.5.5 *Important characteristics of the bursting process*

The rest of this section is concerned with quantitative information on the bursting process.

1. The mean spacing of the low-speed wall streaks (Fig. 3.30, Frame 1, and Fig. 3.33) is, Eq. 3.139:

$$\lambda^+ = \frac{\lambda U_f}{\nu} \simeq 100 \tag{3.143}$$

for the smooth-wall case (see, for example, Lee et al., 1974). This quantity scales with the roughness height k for the rough-wall case (Grass et al., 1991), Eq. 3.142:

$$\frac{\lambda}{k} \simeq 3 \tag{3.144}$$

2. Fluid ejections originate in

$$5 < y^+ < 50 \tag{3.145}$$

for the smooth-wall case (Nychas et al., 1973). Notice that this is precisely the same range where e.g. turbulence production is the strongest, see Fig. 3.8. Corino and Brodkey (1969), on the other hand, report that lower half of the viscous sublayer $y^+ < 2.5$ is essentially passive and the rest ($2.5 < y^+ < 5$) active, being influenced by the quasi-cyclic, bursting events occurring in $5 < y^+ < 70$.

For the rough-wall case, Sumer and Deigaard's (1981) experiments indicate that fluid ejections originate from elevations like

$$0.5 < \frac{y}{k} < 2 \tag{3.146}$$

(y is measured from the theoretical wall). Note, however, that the latter is based on single particle experiments where the particles were slightly heavier than water with specific gravities in the range $s = 1.008$–1.009, and with a size of 3 mm (spherical particles). The lower bound of the range in Eq. 3.146 actually reflects the finite size of the particles used in Sumer and Deigaard's (1981) experiments. It is expected that the ejected fluid can even originate from between the interstices of the roughness elements (Grass, 1971); see the discussion in Section 3.5.4, related to Grass' (1971) study. The latter is also supported by the observations made in Sumer et al.'s study (2001) on suction removal of sediment from between armour blocks.

3. The x- and z-extents of **ejected** low-speed fluid structures (not the entire length, but the ejected portion of the low-speed wall streak) involve dimensions of the order of

$$20 \text{ to } 40 \text{ in } x^+, \text{and} \tag{3.147}$$

$$15 \text{ to } 20 \text{ in } z^+ \tag{3.148}$$

The length of the low-speed wall streak itself is of the order of

$$L^+ = \frac{LU_f}{\nu} = 1000 \tag{3.149}$$

Nino and Garcia (1996) report lengths on the order of 1000 to 2000 wall units.

4. Ejected fluid elements reach y^+ values of

$$y^+ = 80 - 100 \tag{3.150}$$

(Nychas et al., 1973). Praturi and Brodkey (1978) report that, in some rare cases, the ejected fluid elements travel up to 300 y^+.

For rough walls, the previously mentioned Sumer and Deigaard's (1981) slightly-heavier-than-water particle experiments suggest that ejected fluid element reach y distances like

$$y \simeq 3k \tag{3.151}$$

(y is measured from the theoretical wall).

5. The mean streamwise distance from the onset of lift-up of a low-speed wall streak to the break-up of any sign of coherency is about

$$1300 \text{ in } x^+ \tag{3.152}$$

(Offen and Kline, 1973).

6. The mean time between the onset of ejection and break-up (the life time of a burst) is

$$\begin{aligned} \frac{TU_0}{h} &= 2.3 \text{ for the smooth boundary data, and} \\ &= 1.3 \text{ for the rough boundary data} \end{aligned} \tag{3.153}$$

from the study by Jackson (1976). As seen, T scales with the outer flow parameters, namely U_0, the free-stream velocity (or the velocity at the free surface in the case of the open channel flow), and h is the boundary-layer thickness/the flow depth.

7. Finally, the mean bursting period is

$$\frac{T_1 U_0}{h} = 5 \tag{3.154}$$

(Jackson, 1976). Note that this, too, scales with the outer flow parameters.

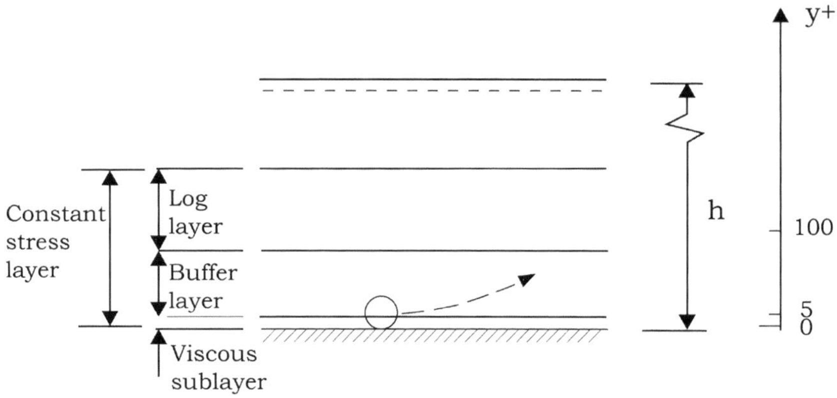

Fig. 3.38 Definition sketch. Particle suspension from (near) bottom.

3.6 Implications of bursting process for sediment transport

In this section, implications of the turbulent bursting process for sediment transport will be discussed with reference to four examples: (1) Particle motions near the bottom in an open channel and mechanism of particle suspension; (2) Lift forces on moving particles near boundaries; (3) Initiation of suspension of particles from the bottom; and (4) Suspension of particles released away from the bottom.

Each example is now considered individually.

3.6.1 *Particle motions near the bottom in an open channel: Mechanism of particle suspension*

Observations made in a laboratory channel show that a small heavy particle may be lifted from the bottom into the body of the flow and suspended (Fig. 3.38). Such a particle, in suspension, travels most of the time close to the bottom (owing to the gravity) but does not leave the main body of the flow; it is "resuspended" when it returns to the bottom, or to locations close to the bottom. Hence there must be some mechanism by which the heavy particle is brought up into suspension from the bottom (or from near the bottom). This mechanism should be closely associated with the turbulence structure near the bottom, namely the bursting process.

This problem has been studied experimentally in two consecutive works by Sumer and Oguz (1978) and Sumer and Deigaard (1981). The former study did this exercise in an open water channel with a smooth bottom, while the latter study used an open water channel with a smooth as well as a rough bottom. This section essentially summarizes the results of Sumer and Oguz's (1978) and Sumer and Deigaard's (1981) studies.

Experiments of Sumer and Oguz (1978) and Sumer and Deigaard (1981). Sumer and Oguz (1978) and Sumer and Deigaard (1981) used single spherical particles in their experiments. The particles in these experiments were designed, taking into account the following considerations.

First, (1) the particles should be big enough to be exposed to the action of turbulence (i.e., the bursting process), and yet (2) not too big for this action to be smeared out. For the latter, clearly, the particle size should be smaller than or, at least, in the same order of magnitude as the dimensions implied in Items 1–3 in the Section 3.5.5, with Eqs. 3.143, 3.145, and 3.147–3.149 for smooth bottoms, and those implied in Items 1 and 2 in Section 3.5.5, with Eqs. 3.144 and 3.146 for rough bottoms.

Second, the particles should be maintained in the flow, and not fall out of suspension due to gravity. This condition implies that the parameter w_s/U_f should not be much greater than unity, one of the criteria for particle suspension (see a detailed discussion in Section 3.6.4). Here, w_s is the fall velocity of the particle in still water, and U_f the bottom friction velocity.

The aforementioned conditions necessitated large size particles with extremely small fall velocities, and presumably with extremely small specific gravities. Such particles are not commercially available, and therefore Sumer and Oguz (1978) and Sumer and Deigaard (1981) produced the particles themselves. In the former study, the particles were made of adhesive wax, while, in the latter, they were made from plastic with tiny holes drilled in them which were then sealed with bees' wax. It may be noted that the particle size had to be selected slightly larger than the dimensions required by item 2 in the first condition above. This was to get a better resolution in the photographic recording of the particle trajectories.

The particle properties in the Sumer and Oguz (1978) and Sumer and Deigaard (1981) experiments are summarized in Tables 3.4 and 3.5, respectively.

In Tables 3.4 and 3.5, d is the particle size, s the specific gravity of the particle, d^+ the particle size in wall units, defined by

$$d^+ = \frac{dU_f}{\nu} \tag{3.155}$$

Table 3.4 Particle properties in Sumer and Oguz's (1978) experiments. Smooth bottom.

Particle	d (mm)	d^+	s	w_s (cm/s)	$\frac{w_s}{U_f}$	θ
A	3.1	34	1.0007	0.22	0.18	6.7
B	4.0	44	1.0028	0.85	0.70	1.33
C	2.8	28	1.0046	0.77	0.64	1.14

Table 3.5 Particle properties in Sumer and Deigaard's (1981) experiments. Smooth and rough bottoms.

Particle	d (mm)	d^+	s	w_s (cm/s)	$\frac{w_s}{U_f}$	θ
Smooth bottom:						
A	2.9	44	1.0029	0.64	0.44	2.5
B	3.0	46	1.0075	1.33	0.92	0.93
C	3.1	47	1.0258	3.06	2.1	0.29
Rough bottom with 3.6 mm pebbles, $k^+ = 81$:						
D1	3.0	67	1.0090	1.49	0.67	1.8
D2	3.0	67	1.0081	1.41	0.63	2.0

k^+ the bottom roughness height in wall units:

$$k^+ = \frac{kU_f}{\nu} \tag{3.156}$$

and θ the Shields parameter

$$\theta = \frac{U_f^2}{g(s-1)d} \tag{3.157}$$

in which g is the acceleration due to gravity. Although the definition of the Shields parameter breaks down in the case of a single particle (no sediment bed), this quantity is presented in the tables to illustrate the strength of the flow with reference to the particle that it is affecting.

Sumer and Oguz (1978) used a photographic technique in combination with a stroboscope to record the side view motion of single particles at small time increments 0.06–0.09 s. Sumer and Deigaard (1981) extended the former study, by utilizing a stereo-photogrammetric system coupled with a stroboscope to record the motion of single particles in three dimensions at similar time increments, the latter being 0.086 s.

Fig. 3.39 presents a photograph of the side view of a typical particle trajectory originating from the bottom in Sumer and Oguz's (1978) experiment. The particle in Fig. 3.39 is Particle C of the Sumer and Oguz (1978)

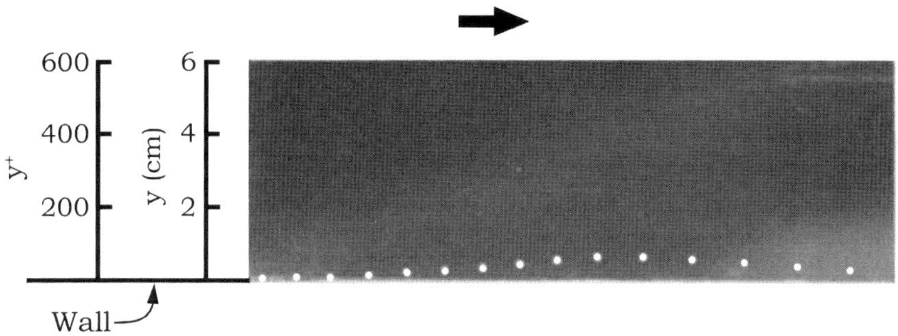

Fig. 3.39 Trajectory of a particle lifted up into the flow from the bottom and suspended. Particle C, Table 3.4. Particle images are 0.086 apart. Sumer and Oguz (1978).

experiments, Table 3.4. The flow is from left to right. The particle images are 0.086 s apart. As seen from the figure, the particle is lifted up from the bottom, and brought up into suspension. It is maintained in suspension until it returns back to the bottom.

Fig. 3.40 displays three trajectories from Sumer and Deigaard's (1981) study, Fig. 3.40a (Particle A), Fig. 3.40b (Particle C), and Fig. 3.40c (Particle D), Table 3.5. The top and middle panels in the figure correspond to the smooth-bottom case, and the bottom panel to the rough-bottom case. The vertical scales in Fig. 3.40 are considerably exaggerated for illustration purposes.

As seen from Fig. 3.40, the trajectory of an individual particle consists of an alternation between upward and downward trajectories. The common feature in all trajectory records is that most of the downward motions of particles coming close to the bottom terminated very near the bottom or at the bottom. Fig. 3.41 displays the frequency histogram of particle position y_T^+ at the termination of the downward paths of those particles which return back to the bottom region. This is for Particle B of Sumer and Deigaard (1981). With the termination of their downward trajectories, the particles are lifted up and brought up into suspension to continue their motion with upward trajectories.

Fig. 3.42 illustrates a typical frequency histogram of particle position in the vertical across the flow depth. This is for Particle C of the Sumer and Oguz (1978) study. The figure indicates that the histogram is strongly weighted towards the bottom. This is due to gravity. The particle spends most of the time near the bottom but does not fall out of suspension; it is

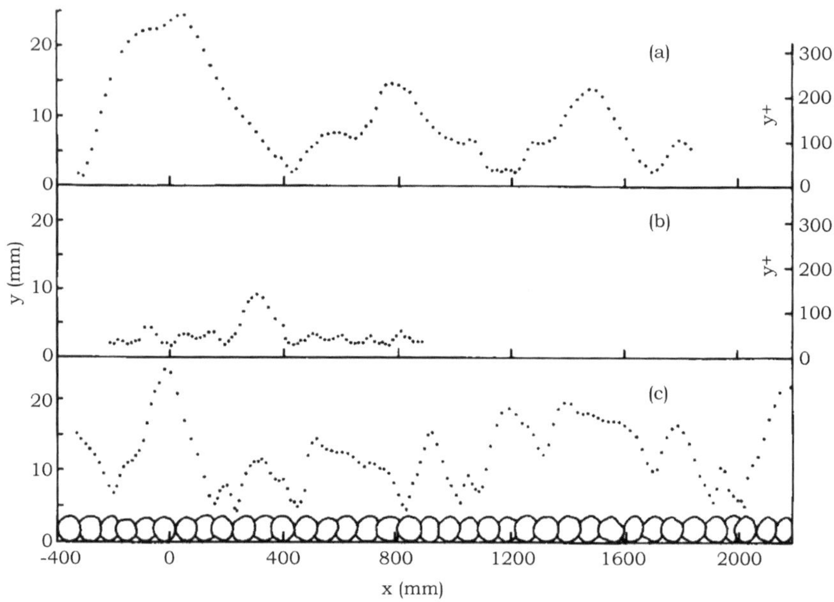

Fig. 3.40 Typical trajectories of individual particles. Flow from left to right. (a): Particle A, smooth bottom; (b) Particle C, smooth bottom; and (c) Particle D1, rough bottom. See Tables 3.4 and 3.5 for particle properties. Dots are spaced 0.086 s apart. The vertical scales are considerably exaggerated. Sumer and Deigaard (1981).

"resuspended" whenever it comes very near the bottom, or "lands" on the bottom. The question here is what brings the particle up into suspension, despite the effect of gravity. The next subsection will focus on this issue.

Mechanism of suspension from/near bottom. Sumer and Oguz (1978) and Sumer and Deigaard (1981), from their single particle experiments, obtained data, including frequency histograms of vertical positions of particles at the termination of upward trajectories of individual particles; those of downward trajectories of individual particles; those of vertical positions of particles at the origin of upward trajectories; their statistical properties; conditional statistics of particle velocities in the streamwise direction and in the vertical direction; conditional statistics of spanwise motion of individual particles from their three-dimensional trajectories; etc. Based on (1) this extensive set of data along with the existing information (which come from the work by Sutherland, 1967, Engelund, 1970, Engelund and Gravesen, 1972, Grass, 1971 and 1974, and Jackson, 1976), and (2) the

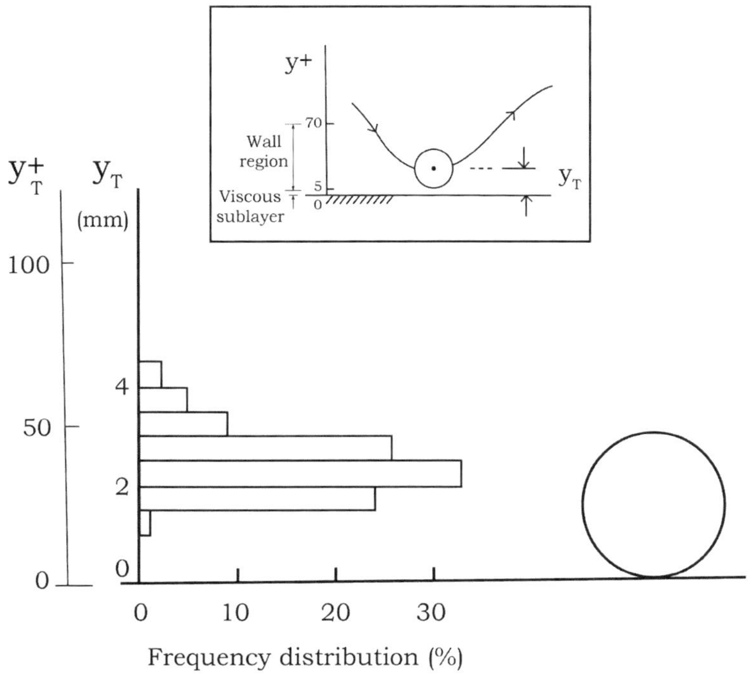

Fig. 3.41 Frequency histograms of y^+ at termination of the downward paths of those particles returning back to the wall region, after completing their their upward paths. Particle B, Smooth bottom, Table 3.5. Circle: Particle when it is on the bottom (to scale). Sumer and Deigaard (1981).

agreement between the latter data and the bursting process, Sumer and Oguz (1978) and Sumer and Deigaard (1981) proposed a mechanism responsible for suspension of heavy particles from/near the bottom. The following paragraphs summarize this mechanism.

We shall begin the description with a particle on the bottom. The flow pattern very close to the bottom consists of a spanwise alternation of low-speed streaks and high-speed zones (Fig. 3.29). When the high-speed fluid hits the bottom, it spreads out sideways. Spreading fluid pushes the particle (wandering in its immediate vicinity) to the adjacent low-speed wall streak whereby the particle finds itself swept into a low-speed wall streak. This process can also be viewed in terms of the streamwise vortices associated with the low-speed wall streaks, Vortices M in Fig. 3.31. The particle caught

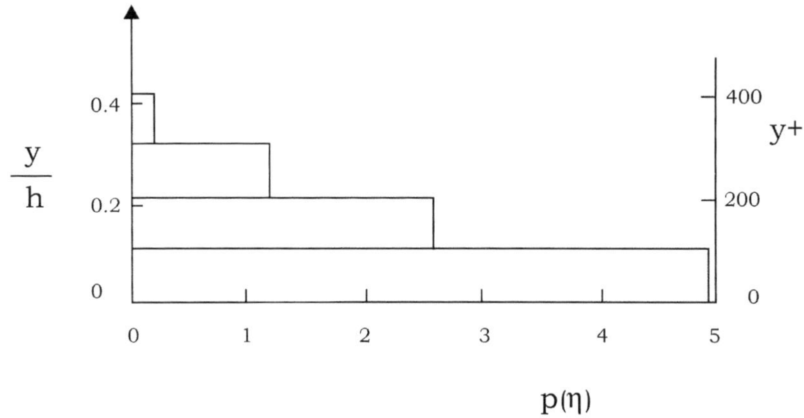

Fig. 3.42 Frequency histograms representing the probability density distribution of particle position in the vertical. Particle C, smooth bottom, Table 3.4. Sumer and Oguz (1978).

in a streamwise vortex is carried by this vortex in the spanwise direction (z-direction), and deposited momentarily in the low-speed wall streak which forms between Vortices M, marked A in Fig. 3.31.

The fact that (1) a particle on the bottom (or very near the bottom) finds itself swept into a low-speed wall streak, and (2) it is big enough to be exposed to the action of turbulence (i.e., the bursting process) makes it possible for the particle to be subjected to the action of the mechanism which brings about the lift-up of the wall streak. As mentioned in Section 3.5.1, a low-speed wall streak is subject to an impulsive "suction", or lift. This causes the low-speed fluid to entrain into the main body of the flow, the ejection event (Frame 2, Fig. 3.30), described in the previous section. With the ejection of this low-speed fluid, the particle (being in the body of this coherent fluid) will entrain into the main body of the flow, and travel upwards in the same fluid up until the coherence of this fluid (the burst) breaks up. Recall the sequence of events of the bursting process, Fig. 3.30. Fig. 3.43 illustrates schematically the aforementioned suspension process.

Now, if the particle is not too heavy, it will not fall out of the coherent fluid it is traveling with, up until the latter breaks up. With the breakup of this rising coherent fluid, the particle will fall towards the bottom due to gravity. The particle at this stage will presumably travel in a high-speed fluid zone. In the case if the termination of the downward trajectory of

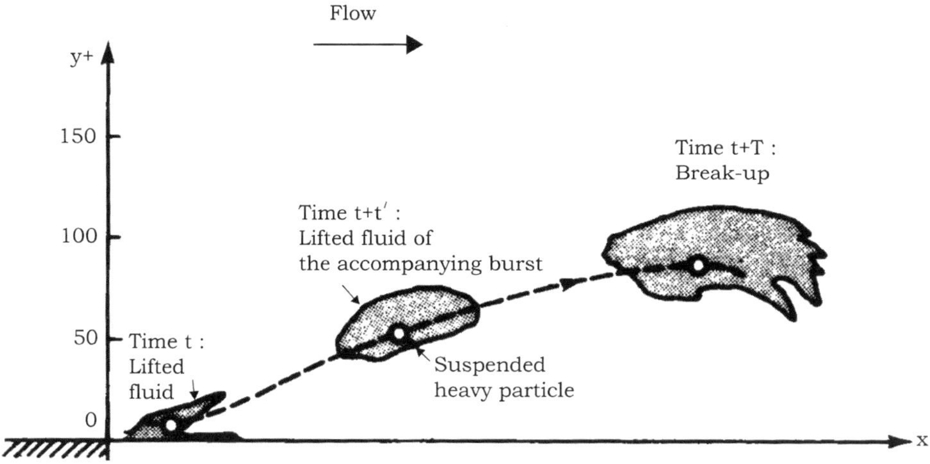

Fig. 3.43 Schematic view of the side-view motion of suspended heavy particle. Sumer and Deigaard (1981).

the particle is in the wall region, at $y_T^+ < 70$, the particle will inevitably be swept into a low-speed wall streak, and eventually get resuspended, in exactly same manner as that described in the preceding paragraph.

From the above considerations, it appears that a particle wandering near or on the bottom always ends up with a low-speed wall streak. This is the key component of the mechanism responsible for the particle suspension from/near the bottom. In this context, an interesting argument comes from Lumley (1978, p. 309): In the case of a boundary layer above a horizontal wall in which heavy "inertia-free" particles are suspended, the vertical velocity of the particle at any time is

$$v_p = v - w_s \tag{3.158}$$

in which v is the vertical velocity of fluid seen by the particle, and w_s the fall velocity of the particle in still fluid. Averaging the above equation over possible particle trajectories:

$$\overline{v}_p = \overline{v} - \overline{w}_s \tag{3.159}$$

As the particles are maintained in suspension all the time, the average vertical velocity of the particle will be nil, $\overline{v}_p = 0$, and therefore, from the preceding equation, and considering $\overline{w}_s = w_s =$ constant:

$$\overline{v} = w_s \tag{3.160}$$

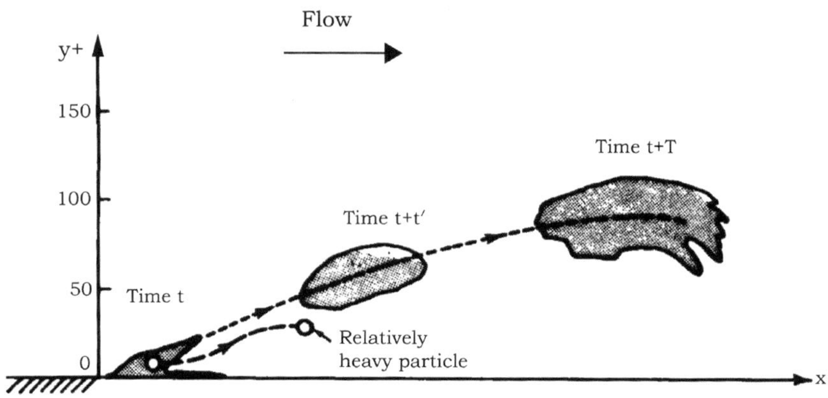

Fig. 3.44 Schematic view of the side-view motion of a relatively heavy particle. Sumer and Deigaard (1981).

that is, the vertical component of the fluid velocity seen by the particle (on the average) must be equal to the fall velocity, w_s. This means that the particle must seek out (on the average) rising currents, to offset the tendency to fall out. This simple argument shows, in an elegant way, that the particles near or at the bottom must move in the spanwise direction to locations where the rising currents originate. These rising currents are, according to the suspension mechanism described in the preceding paragraphs, presumably the ejection events of the bursting process.

In the case of *relatively heavy particles* (such as Particle C in the Sumer and Deigaard, 1981, experiments; see Fig. 3.40b), the particle ejected from/near the wall region cannot be maintained in the lifted fluid but leaves the main body of the lifted fluid due to gravity before the coherence of the lifted fluid (the burst) breaks up. See Fig. 3.44 for a schematic description. Note that the latter coincides with what is conventionally known as the concept of crossing-trajectories effect, i.e., the particle falls out of an eddy before the eddy loses its coherence. With the particle falling out of the rising fluid, it starts to return back near the bottom. The particle wandering very close to the bottom is finally swept into a low-speed wall streak from where it is ejected again into the flow. This process causes the particle to have a series of short hops, as is observed in Fig. 3.40b.

In the case of *rough bottoms*, Sumer and Deigaard's (1981) rough-bottom results show that the particle motions close to a rough bottom are very similar in character to those close to a smooth bottom (cf., Figs. 3.40a,c).

Their data showed that, although the lifting-up and rise of particles occur more intensely in the case of rough bottom, no significant variation in the measured values of various quantities in both cases was observed.

The above mentioned qualitative, even quantitative (partially), similarity suggests that the mechanism responsible for particle motions in both cases would be similar, i.e., the similar sequence of events occurs near a rough bottom, which would be the main agent responsible for particle suspension, as in the case of the smooth bottom.

Multiple particles and sediment bed. Grass' (1971) flow visualization in Fig. 3.32 where the flow was visualized using fine sand, the size 0.1 mm, clearly show that the multiple particles (that actually form a sediment bed, albeit a so-called "starving" sediment bed) are pushed into the low-speed wall streaks. With this, the sediment in the low-speed wall streaks will be subject to the same mechanism which brings about the lift-up of the wall streak, similar to the processes described for suspension of single particles described in the preceding paragraphs.

The motion of sediment particles lying over a smooth as well as a rough bed was investigated by Nino and Garcia (1996), using particle flow visualization and particle tracking techniques. In Nino and Garcia's (1996) study, natural sand particles were used in large quantities, in contrast to the single particles used in Sumer and Oguz's (1978) and Sumer and Deigaard's (1981) studies. The results of Nino and Garcia's sediment experiments (1996) confirm the findings of Sumer and Oguz's (1978) and Sumer and Deigaard's (1981) single particle experiments, both qualitatively and, for the most part, quantitatively, despite the very large differences in the particle size and the particle specific gravity; the sediment size ranging from d = 0.038–0.53 mm in the Nino and Garcia experiments in contrast to the particle size d = 2.8–3.1 mm in the Sumer and Oguz and Sumer and Deigaard experiments; and the specific gravity of the sediment in the Nino and Garcia experiments being $s = 2.65$ versus s = 1.0007–1.0258 in the Sumer and Oguz and Sumer and Deigaard experiments. It should be noted that, in view of the observed agreement between the two studies, Nino and Garcia (1996) adapted the mechanism of particle suspension of Sumer and Deigaard (1981) (Figs. 3.43 and 3.44) to illustrate the mechanism of suspension of multiple sediment particles from a sediment bed.

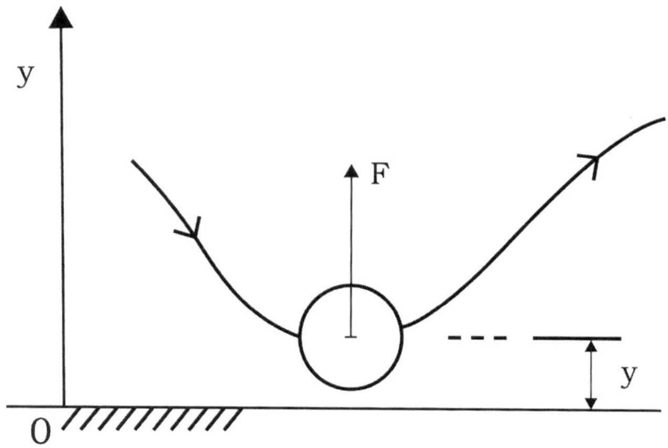

Fig. 3.45 Lift force on the particle at the time of suspension where the particle's downward motion terminates and its upward motion starts.

3.6.2 *Lift forces on moving particles near boundaries*

We have, in the previous section, described the mechanism of particle suspension from/near the bottom in an open channel with smooth and rough boundaries. Our focus in the present section will be the lift force on the particle at the time of suspension where the particle's downward motion terminates and its upward motion starts, Fig. 3.45. This problem has been studied by Sumer (1984). The present section summarizes Sumer's (1984) study.

The vertical component of the equation of motion of a spherical particle at the time of suspension from/near the bottom reads:

$$\frac{\pi d^3}{6}\rho_p\frac{dv_p}{dt} = -\frac{\pi d^3}{6}(\gamma_p - \gamma) + F_D - \frac{\pi d^3}{6}\frac{\partial p}{\partial y} + \frac{1}{2}\frac{\pi d^3}{6}\rho\frac{d}{dt}(v - v_p) \quad (3.161)$$

in which d is the particle size, ρ and ρ_p the densities of fluid and particle, γ and γ_p the specific weights of fluid and particle, v and v_p fluid and particle velocities in the vertical direction, p the pressure in excess of hydrostatic pressure, y the distance from the bottom, and t time.

The force terms on the right-hand-side of the above equation are as follows:

1. The first term is the submerged weight of the particle.
2. The second term, F_D, is the drag force on the particle in the vertical direction.

3. The third term is the force due to the pressure gradient in the vertical direction, the pressure gradient which evidently drives the flow of low-speed fluid outward at the time of suspension, corresponding to the ejection event, see Frame 2 in Fig. 3.30, and also see Time t, Lifted fluid in Fig. 3.43. The minus sign takes care of the direction of the pressure-gradient force; $-\frac{\pi d^3}{6}\frac{\partial p}{\partial y}$ is in the positive y direction, i.e., in the upward direction.

4. The fourth term is the apparent-mass force in the vertical direction, the factor $\frac{1}{2}$ being the apparent-mass coefficient for a sphere. The apparent-mass force is the force to accelerate the particle relative to the fluid in its immediate surrounding, with an acceleration $\frac{d}{dt}(v - v_p)$.

It may be noted that, although they are two entirely different physical processes, the third term above is analogous to the so-called Froude-Krylov force, and the fourth term to the apparent-mass force in the in-line direction on a marine object subjected to waves, Sumer and Fredsøe (2006, their Eqs. 4.26 and 4.30).

A slightly different version of Eq. 3.161 was first employed by Engelund (1970). Eq. 3.161 can be written as

$$F = \frac{\pi d^3}{6}\rho_p \frac{dv_p}{dt} + \frac{\pi d^3}{6}(\gamma_p - \gamma) \tag{3.162}$$

with F:

$$F = F_D - \frac{\pi d^3}{6}\frac{\partial p}{\partial y} + \frac{1}{2}\frac{\pi d^3}{6}\rho \frac{d}{dt}(v - v_p) \tag{3.163}$$

F representing the lift force exerted on the particle at the instant of suspension from/near the bottom, Fig. 3.45.

Sumer (1984) determined the lift force, F, employing the trajectory records obtained in Sumer and Deigaard's (1981) work, calculating the acceleration $\frac{dv_p}{dt}$ at the time of suspension, and subsequently F from Eq. 3.162. Sumer (1984) did this exercise for Particles A, B, D1 and D2 (see Table 3.5). Trajectories for Particle C were not used, as this particle, being considerably heavier than water, experienced the crossing-trajectories effect discussed in the previous section, and therefore proved inappropriate for this latter exercise.

Fig. 3.46 displays the lift force, F, obtained by Sumer (1984), corresponding to the times of suspension of particles from/near the bottom, plotted versus the distance from the bottom, y. (The distance y is measured from the bottom to the centroid of the particle.) In the case of the rough bottom, y is measured from the theoretical bottom to the centroid of the particle. The lowermost two data points in the case of the smooth

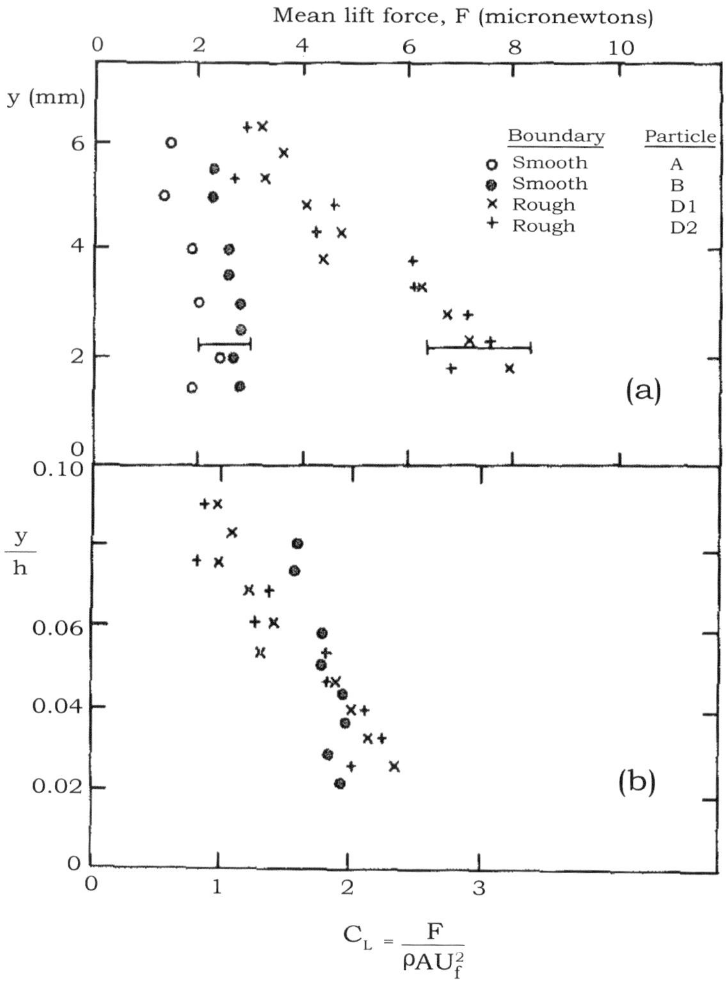

Fig. 3.46 Mean lift force vs. distance from the bottom. Error bar denotes mean standard deviations over the depth. Legend for (b) is the same as that for (a). Sumer (1984).

bottom in Fig. 3.46a correspond to the particle position when the particle is on the bottom at the time of suspension. Likewise, the lowermost four data points in the case of the rough bottom correspond to the particle position when the particle is on the bottom (either at the top of a roughness element or in the interstices of the roughness elements). The rest of the data points within distances $y = O(2\text{ mm})$ indicate the extremely close proximity of the particles to the bottom at the time of suspension. In the figure, the lift force data are plotted in normalized form in the bottom panel. Here, h is the flow depth, A is the area of the sphere in a horizontal projection, U_f is the bottom friction velocity, and C_L is a lift coefficient defined by

$$C_L = \frac{F}{\rho A U_f^2} = \frac{F}{\rho(\frac{\pi d^2}{4})U_f^2} \tag{3.164}$$

In the latter non-dimensional plot, only the data corresponding to Particles B, D1 and D2 are included, since the properties of these particles are almost the same (Table 3.5).

From Fig. 3.46, Sumer (1984) concluded the following:

1. The lift force in the case of the smooth bottom reaches a maximum value (albeit, a weak variation) at about 2.5 mm. This corresponds to the y position from which most ejections originate, which corresponds to $y^+ = 35$–40.

2. The lift force decreases appreciably with the distance from the bottom.

3. The lift force increases markedly with the bed roughness. Despite the fact that the flow Reynolds number was kept almost unchanged in these tests, the lift force increased by up to three times as the bottom condition changed from smooth to rough.

4. The extra effect of a rough wall tends to disappear with the distance from the wall.

5. It may be noted that Sumer (1984) also discussed other potential lift forces, which could arise from proximity effect of the bottom, e.g. the so-called Magnus effect, and elastic rebound on collision with the bottom, and concluded that these forces proved not relevant in the present context.

Finally, Sumer (1984) argued that the major contribution to the measured lift force comes from the pressure-gradient term in Eq. 3.163; i.e., the lift force essentially caused by the upward-directed pressure gradient, the pressure gradient that drives the flow of low-speed fluid outward (ejection) at the time of suspension. This is actually expected since the particles used in Sumer and Deigaard's (1981) experiments are only very slightly heavier than water (save Particle C, however, Table 3.5), and therefore the drag force, $F_D \sim (v - v_p)^2$, and the apparent-mass force, $\frac{1}{2}\frac{\pi d^3}{6}\rho\frac{d}{dt}(v - v_p)$, in Eq. 3.163 will be not very significant. This is because the particle and the fluid surrounding the particle at the time of suspension will have vertical velocities, v_p and v, fairly close to each other (so close that the boundary layer over the surface of the particle - due to the motion of the particle relative to the surrounding fluid - will not be fully developed), leading to small, negligible drag and apparent-mass forces.

This argument has been supported by a recent PhD study carried out at the Technical University of Denmark (for which BMS was the principal supervisor, and a paper from this study is in preparation at the time of this writing, Karsten L. Jensen and B. Mutlu Sumer, 2019). In this work, the upward-directed pressure gradient was measured, utilizing neutrally-buoyant particles in an experimental setting similar to that of Sumer and Deigaard (1981). The equation of motion (Eq. 3.161), with the terms $-\frac{\pi d^3}{6}(\gamma_p - \gamma)$, F_D, $\frac{1}{2}\frac{\pi d^3}{6}\rho\frac{d}{dt}(v - v_p)$ all being zero since the particles were neutrally buoyant, leads to

$$\frac{\pi d^3}{6}\rho_p\frac{dv_p}{dt} = -\frac{\pi d^3}{6}\frac{\partial p}{\partial y} \tag{3.165}$$

such that the upward-directed pressure gradient was calculated from the measured particle acceleration. This exercise was done for all trajectories with the upward-directed acceleration satisfying $\frac{dv_p}{dt} > 0$, and the conditional statistics (namely, the mean value and the standard deviation) were calculated. We note that this is actually like measuring the wall-normal pressure gradient with a "flying probe". The data obtained in the study is slightly different from that of Sumer (1984) in that, in Sumer's (1984) study, the pressure gradient was "sampled" at the instants of particle suspension, as discussed in the preceding paragraphs, whereas, in Karsten L. Jensen's PhD study, the data was sampled with the condition $\frac{dv_p}{dt} > 0$. Nevertheless, the two sets of data agree quite well, supporting the argument that the contributions to the lift force from the drag and the apparent-mass terms in Sumer's (1984) study are small and negligible, and the lift force is almost solely caused by the wall-normal pressure gradient, responsible for impulsive suction of the low-speed wall streak into the flow in the ejection phase of the bursting process, by which the particle is entrained into the flow.

3.6.3 *Initiation of suspension of particles from the bottom*

Consider a flow of fluid over a bed composed of particles having size d. We ask the following question: What is the criterion that describes the initiation of suspension of particles from the bottom. It is to be noted that this is different from the classic Shields criterion that describes the initiation of motion of particles at the bottom. Our concern in this section is not the initiation of motion but the initiation of suspension.

The initiation of suspension from the bottom is governed by the following parameters:

$$\tau_0,\ \gamma_p - \gamma,\ d,\ \rho,\ \text{and}\ \mu \tag{3.166}$$

in which τ_0 is the bottom shear stress, $\gamma_p - \gamma = \gamma(s-1)$ is the submerged specific gravity of particles (s being the specific gravity of particles, $s = \frac{\gamma_p}{\gamma}$), ρ is the fluid density, and μ is the fluid viscosity.

The quantities τ_0, ρ and μ are involved because the flow near the bottom in the constant stress layer ($y/h \ll 1$) is described solely by these quantities. This is for a bottom acting as a hydraulically smooth boundary (Section 3.1.2). In the case of a rough boundary (Section 3.1.3), there is an additional quantity, namely the bottom roughness, which is proportional to the sediment size d. So, the quantities τ_0, ρ, μ and d in Eq. 3.166 are involved because the flow near the bottom (in the constant stress layer, $y/h \ll 1$) are governed by these quantities. We stress here that these are the quantities that control the flow near the bottom in the constant stress layer, $y/h \ll 1$ (Sections 3.1.2 and 3.1.3). Outside the constant stress layer, one more parameter would kick in, namely the flow depth, h. However, the process we study, the initiation of suspension from the bottom, involves the flow near the bottom, and therefore, the latter quantity will not be included in the present dimensional considerations.

On the other hand, the quantities $\gamma_p - \gamma$, and d in Eq. 3.166 are involved because these quantities characterize the sediment particles. As noticed, the particle size d is involved in the formulation both (1) as a parameter associated with the flow near the bottom, and (2) as a quantity associated with the sediment.

On dimensional grounds, the above set of dimensional parameters in Eq. 3.166 can be written in the following non-dimensional form

$$\theta_s = \theta_s\left(\frac{dU_f}{\nu}\right) \tag{3.167}$$

with $\nu = \mu/\rho$, the kinematic viscosity; $U_f = (\tau_0/\rho)^{1/2}$, the bottom friction velocity; $\frac{dU_f}{\nu}$ the particle Reynolds number; θ is the Shields parameter (Eq. 3.157), and θ_s is the critical value of the Shields parameter corresponding to the initiation of suspension of sediment from the bottom. Suspension occurs when

$$\theta > \theta_s\left(\frac{dU_f}{\nu}\right) \tag{3.168}$$

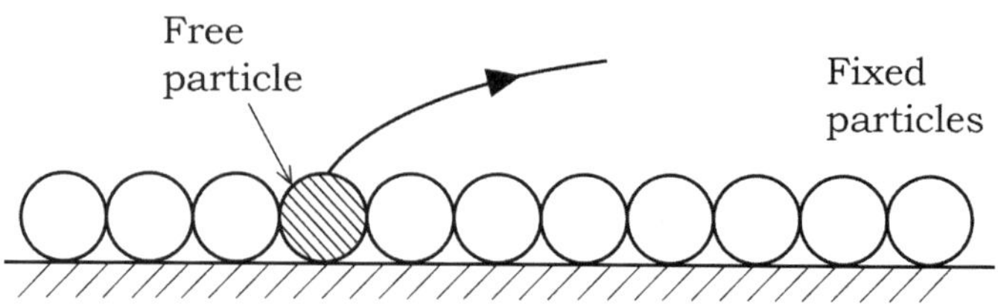

Fig. 3.47 Experimental arrangement to study the initiation of suspension from the bottom of a particle which is a member of a sediment bed. The instant where the free particle is lifted up into the flow, and suspended, is determined by visual observation. Sumer (1986).

Experimental determination of the function $\theta_s(\frac{dU_f}{\nu})$ is a challenging problem, as controlling of a test to investigate the conditions for the initiation of suspension of particles (of a sediment bed) proves extremely difficult. To circumvent this problem, a bed of particles can be arranged such that all the particles but one are glued (fixed) on the bottom of an open channel (Fig. 3.47), and the initiation of suspension can be studied with the free particle (Fig. 3.47). Sumer and Irvali (reported in Sumer, 1986) did this exercise. Fig. 3.48 is reproduced from Sumer (1986), illustrating trajectories of the "free" particle which was a member of the bed composed of 6 mm spherical particles glued in a densely packed manner one layer deep to the bottom of an open channel.

The procedure followed in a typical test of Sumer and Irvali's experiments was as follows: (1) Maintain the free particle in its place with the help of a blunt-tip point gauge; (2) Run the flow corresponding to a particular value of the particle Reynolds number, $\frac{dU_f}{\nu}$; (3) Remove the point gauge gently, to observe if the particle is lifted up into the flow ("sucked"); (4) If yes, photograph its trajectory from the side with a camera (with its shutter open) in combination with a stroboscope, similar to Sumer and Oguz (1978); and (5) Repeat the experiment for a wide range of particle specific gravity, $s = 1.0002$–1.01, using each time the same particle. The latter enabled Sumer and Irvali to test a different value of θ for a given value of the particle Reynolds number tested. In this way, Sumer and Irvali determined the initiation of suspension in an accurate manner. Three different values of the particle Reynolds number were tested, namely $\frac{dU_f}{\nu} = 19$, 47 and 70. While the particle Reynolds number $\frac{dU_f}{\nu} = 19$ falls into the transitional-

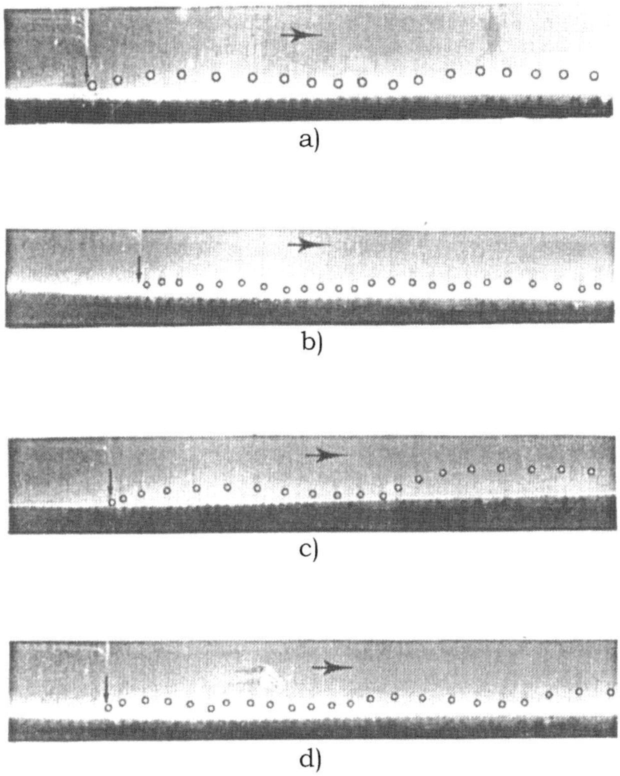

Fig. 3.48 Trajectory photographs of bed material particle which is put into suspension by flow. See the text for particle parameters for each case. Adapted from Sumer (1986).

wall category ($k_s^+ = 2.5\frac{dU_f}{\nu} = 2.5 \times 19 = 47.5$, see Section 3.1.3), the last two particle Reynolds numbers definitely correspond to the completely rough wall regime. Values of the particle Reynolds number smaller than $\frac{dU_f}{\nu} = 19$ was unattainable in the experiments due to experimental constraints. The particle parameters for each case in Fig. 3.48 are as follows: (a) $\frac{dU_f}{\nu} = 47$, $\theta = 0.53$, $\frac{w_s}{\kappa U_f} = 3.9$; (b) $\frac{dU_f}{\nu} = 47$, $\theta = 0.32$, $\frac{w_s}{\kappa U_f} = 5.4$; (c) $\frac{dU_f}{\nu} = 70$, $\theta = 1.5$, $\frac{w_s}{\kappa U_f} = 2.4$; (d) $\frac{dU_f}{\nu} = 70$, $\theta \approx 0.37$, $\frac{w_s}{\kappa U_f} = 5.5$ in which κ is the von Karman constant ($\simeq 0.4$).

We note that the particles (including the loose, free particle) in the above experiments were presumably subjected to the lift force caused by

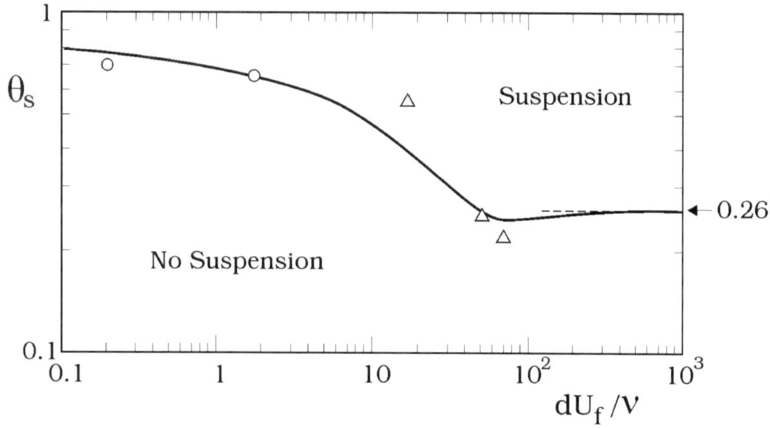

Fig. 3.49 Criterion for initiation of suspension from the bottom. Circles: Mantz (1978, Table 1 Run 2, Table 2, Run 2). Triangles: Sumer (1986, Fig. 8). Dashed line: Shields (1936). Solid line: Eq. 3.172. Reproduced from Sumer and Fredsøe (2002).

the near-bottom turbulent flow passing overhead, and as a result, the free particle was sucked from the bottom and brought up into suspension, with a mechanism similar to that described in the preceding paragraphs for single and multiple particles over smooth and rough bottoms.

The data obtained by Sumer and Irvali along with two other sets of data are plotted in terms of the critical Shields parameter for the initiation of suspension, θ_s, versus the particle Reynolds number by Sumer and Fredsøe (2002, p. 336). Fig. 3.49 displays this plot, reproduced from Sumer and Fredsøe (2002). The other two sets of data are from (1) Mantz (1978, Table 1 Run 2, and Table 2 Run 2), and (2) Shields (1936).

From Fig. 3.49, it is seen that the critical value $\theta_s(\frac{dU_f}{\nu})$ tends to a constant value for large values of the particle Reynolds number. This is not unexpected, as the viscosity will not play an important role for large values of the particle Reynolds number since the sediment bed will behave as a completely rough boundary in this case. Hence, $\theta_s(\frac{dU_f}{\nu})$ should asymptotically attain a constant value for large values of $\frac{dU_f}{\nu}$. From the data in Fig. 3.49, this constant value appears to be 0.26. This asymptotic value can also be worked out from lift-force considerations, as will be demonstrated in the following paragraphs.

For this, we will, to a first approximation, assume that a particle which is a member of a sediment bed, will experience the same kind of lift force,

F, as that measured on a single particle over a rough bottom, described in the previous section. (This approximation is arguably justifiable for the lift force on a single particle over a rough bottom, made of like particles. It is unjustifiable, however, for the lift force on a single particle over a smooth bottom.)

Now, suspension occurs when the lift force F (Eq. 3.164) is equal to the submerged weight of the particle G:

$$F = G \tag{3.169}$$

or

$$C_L \rho \left(\frac{\pi d^2}{4}\right) U_f^2 = \frac{\pi d^3}{6}(\gamma_p - \gamma) \tag{3.170}$$

which gives:

$$\theta_s = \frac{2}{3C_L} \tag{3.171}$$

θ_s being the critical Shields parameter for the initiation of suspension for a fully rough bottom. Taking the normalized lift force C_L for the particle on the bottom in the rough bottom case (the maximum value in Fig. 3.46b), $C_L = 2.4$, one obtains $\theta_s = 0.28$. This value is remarkably close to the previously discussed asymptotic value $\theta_s = 0.26$, revealed by the Shields (1936) data in Fig. 3.49.

Incidentally, the data obtained by Sumer and Irvali had been plotted earlier (Sumer, 1986, Fig. 8) on the Shields diagram together with the Shields (1936) data but without the data from the Mantz (1978) study, as the author (BMS) had apparently overlooked Mantz (1978) publication at the time. Therefore, in the absence of Mantz data, the left branch of the critical curve for suspension in this earlier plot (Sumer, 1986, Fig. 8) appeared to look differently from that in Fig. 3.49.

Finally, the trend in Fig. 3.49 can be represented by the following expression:

$$\theta_s = \left(\frac{dU_f}{\nu}\right)^{-0.05} \left[0.7\exp\left(-0.04\frac{dU_f}{\nu}\right)\right] + 0.26\left[1 - \exp\left(-0.025\frac{dU_f}{\nu}\right)\right] \tag{3.172}$$

The problem of initiation of suspension from the bottom has been addressed recently by Bose and Dey (2013); also, see the discussion by Sumer (2015) to the latter publication. Bose and Dey (2013) developed a theory describing the initiation of suspension. The curve that they called the threshold of suspension in their paper (Bose and Dey, 2013, Fig. 8) appears to compare well with the trend line fitted to the data plotted in Fig. 3.49, given in Eq. 3.172.

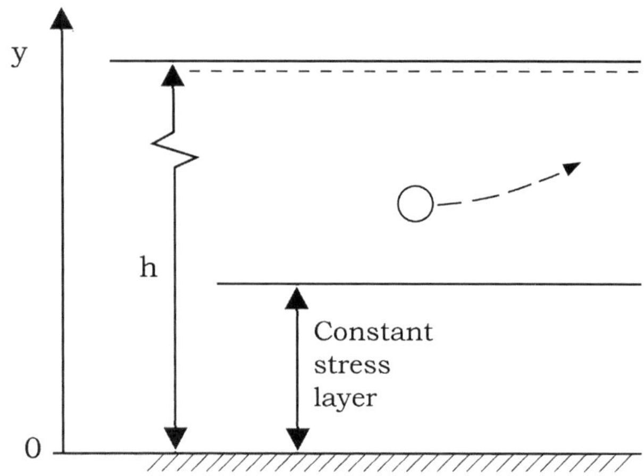

Fig. 3.50 Suspension of small heavy particles released into the flow not from the bottom but away from the bottom. We consider away-from-the-bottom locations to lie above the constant stress layer.

3.6.4 *Suspension of particles released away from the bottom*

In the previous sections, we have studied suspension of small heavy particles from the bottom. We have seen that the suspension of particles is governed by the turbulence activity near the bottom, namely the bursting process. The flow is controlled by the parameters associated with the constant stress layer ($y/h \ll 1$) in which y is the distance from the bottom, and h the flow depth.

In the present section, we will study suspension of small heavy particles released into the flow not from the bottom but away from the bottom. *We consider away-from-the-bottom locations to lie above the constant stress layer* (Fig. 3.50).

In this suspension process, the particles will also be supported against gravity by turbulence. However, this turbulence is not the near-wall turbulence, but rather the turbulence in the outer region, outside the constant stress layer.

A simple model describing suspension of heavy particles in this region can be worked out as follows. A particle will, at any time, find itself in a coherent parcel of fluid (an eddy), rising upwards with a velocity of V. If there is no crossing-trajectories effect, i.e., if the particle is not too heavy to fall out of the eddy, it will fall relative to the eddy with a velocity approximately equal to its fall velocity in still fluid, w_s. Hence, the particle will remain in suspension if

$$w_s < V \tag{3.173}$$

The velocity V is expected to be in the same order of magnitude as the r.m.s. value of the fluctuating component of the fluid velocity in the vertical direction, $\sqrt{\overline{v'^2}}$:

$$V \sim \sqrt{\overline{v'^2}} \tag{3.174}$$

The quantity $\sqrt{\overline{v'^2}}$ just above the constant stress layer is like $\sqrt{\overline{v'^2}} \sim U_f$, regardless of the wall category (whether it is smooth, transitional, or rough), see Fig. 3.15. Therefore, from Eq. 3.173, the particle is maintained in suspension in the main body of the flow if

$$\frac{w_s}{U_f} < 1 \tag{3.175}$$

Similar conclusions were obtained in the past by researchers such as Batchelor (1965) and Bagnold (1966) among others.

The above suspension criterion in the form of $\frac{w_s}{U_f} = 1$ is plotted in the Shields plane in terms of the Shields parameter, θ, and the particle Reynolds number $\frac{dU_f}{\nu}$ in Fig. 3.51. To achieve this, the following relations are used

$$\theta_{w_s/U_f=1} = \frac{U_f^2}{g(s-1)d} = \frac{w_s^2}{g(s-1)d} \tag{3.176}$$

$$w_s = \left(\frac{4}{3}\frac{(s-1)gd}{C_D}\right)^{\frac{1}{2}} \tag{3.177}$$

For the derivation of Eq. 3.177, see Appendix 8.A. From the above

$$\theta_{w_s/U_f=1} = \frac{4}{3}\frac{1}{C_D} \tag{3.178}$$

in which C_D is the drag coefficient for a sphere, and taken as (from Cheng, 2009):

$$C_D = \frac{24}{\mathrm{Re}}(1 + 0.27\,\mathrm{Re})^{0.43} + 0.47\left[1 - \exp(-0.04(\mathrm{Re})^{0.38})\right] \tag{3.179}$$

with

$$\mathrm{Re} = \frac{w_s d}{\nu} = \frac{U_f d}{\nu} \tag{3.180}$$

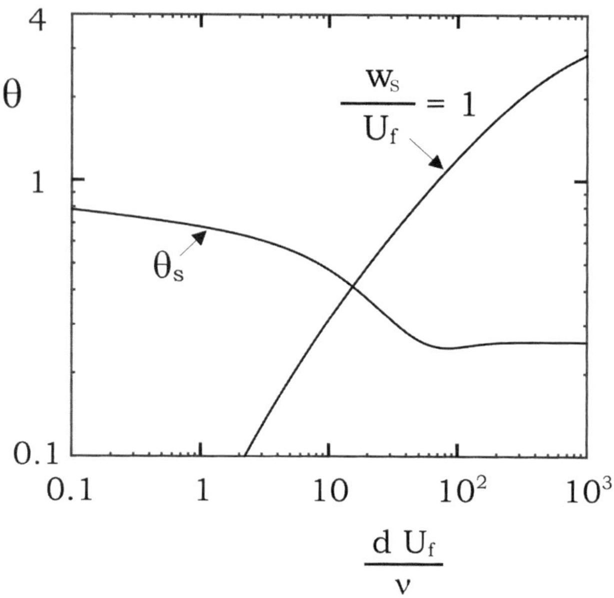

Fig. 3.51 Suspension criterion $\frac{w_s}{U_f} = 1$ plotted in the Shields plane in terms of the Shields parameter, θ, and the particle Reynolds number $\frac{dU_f}{\nu}$. The area above the curve $\frac{w_s}{U_f} = 1$ represents the Suspension regime ($\frac{w_s}{U_f} < 1$) where the particles released in the flow away from the bottom are maintained in suspension in the main body of the flow. The critical curve, θ_s, for the initiation of suspension from the bottom is reproduced from Fig. 3.49.

The area above the curve $\frac{w_s}{U_f} = 1$ represents the Suspension regime ($\frac{w_s}{U_f} < 1$) where the particles released in the flow away from the bottom are maintained in suspension in the main body of the flow. The area below the curve $\frac{w_s}{U_f} = 1$, on the other hand, represents the No-suspension regime ($\frac{w_s}{U_f} > 1$) where the particles released in the flow away from the bottom are not maintained in suspension in the main body of the flow. Fig. 3.51 also displays the critical curve, θ_s, for the initiation of suspension from the bottom, reproduced from Fig. 3.49.

Fig. 3.52 displays the same critical curves as in the previous figure, but with a "pattern" code, to make the figure relatively easy to read.

As seen from the figure, for very small particles (small values of $\frac{dU_f}{\nu}$), while very large values of the Shields parameter are required to bring the particle into suspension from the bottom, small values of the Shields parameter are sufficient to maintain the particles in suspension if they are

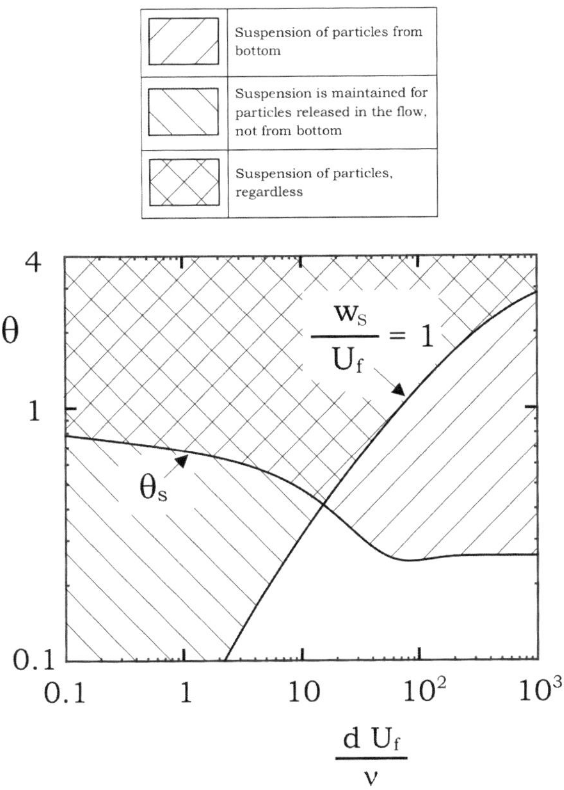

Fig. 3.52 Suspension criteria for heavy particles. See the legend at the top for the pattern code. The critical curves are reproduced from the previous figure.

released in the flow above the constant stress layer.

For very large particles (large values of $\frac{dU_f}{\nu}$), on the other hand, very large values of the Shields parameter are required to maintain the particles in suspension in the main body of the flow. In contrast, not very large values of the Shields parameter are required to bring them into suspension from the bottom. In the latter case, the particles will be suspended from the bottom but they will never make it to the main body of the flow above the constant stress layer unless θ is increased so much that $\frac{w_s}{U_f} < 1$, the area above the critical curve $\frac{w_s}{U_f} = 1$.

3.7 References

(1) Aydin, M. and Leutheusser, H.J. (1985): Flow visualization of coherent structures in turbulent open-channel flow. Proceedings of International Symposium on Refined Flow Modelling and Turbulence Measurements, The University of Iowa, Iowa City, Iowa, USA, September 16-18, vol. 2, p. E22-1–E22-10.

(2) Bagnold, R.A. (1966): An Approach to the Sediment Transport Problem for General Physics. Geological Survey Professional Paper 422-I. Geological Survey, Washington, DC.

(3) Batchelor, G.K. (1965): The motion of small particles in turbulent flow. Proc. 2nd Austr. Conf. on Hydraul. and Fluid Mech., 19, p. 019–041, Univ. of Auckland.

(4) Bayazit, M. (1976): Free surface flow in a channel of large relative roughness. J. Hydraul. Res., vol. 14, 115–126.

(5) Bayazit, M. (1983): Flow structure and sediment transport mechanics in steep channels. In: Mechanics of Sediment Transport, Proc. of Euromech 156, Istanbul, 12-14. July 1982, A.A. Balkema (Rotterdam), p. 197–206.

(6) Bose, S.K. and Dey, S. (2013): Sediment entrainment probability and threshold of sediment suspension: Exponential-base approach. J. Hydraul. Eng.-ASCE, vol. 139, 1099–1106.

(7) Brodkey, R.S., Wallace, J.M. and Eckelmann, H. (1974): Some properties of truncated signals in bounded shear flows. J. Fluid Mech., vol. 63, 209–224.

(8) Cantwell, B.J. (1981): Organized motion in turbulent flow. Ann. Rev. Fluid Mech., vol 13, 457–515.

(9) Carstensen, S., Sumer, B.M. and Fredsøe, J. (2012): A note on turbulent spots over a rough bed in wave boundary layers. Phys. Fluids, vol. 24, article no. 115104, doi: 10.1063/1.4767536.

(10) Cebeci, T. and Chang, K.C. (1978): Calculation of incompressible rough-wall boundary-layer flows. AIAA J., vol. 16, 730–735.

(11) Cheng, N.-S. (2009): Comparison of formulas for drag coefficient and settling velocity of spherical particles. Powder Technology, vol. 189, 395–398.

(12) Coleman, N.L. and Alanso, C.V. (1983): Two-dimensional channel flows over rough surfaces. J. Hydraul. Eng.-ASCE, vol. 109, 175–188.

(13) Corino, E.R. and Brodkey, R.S. (1969): A visual investigation of the wall region in turbulent flow. J. Fluid Mech., vol. 37, 1–30.

(14) Defina, A. (1996): Transverse spacing of low-speed streaks in a channel over a rough bed. In: Coherent Flow Structures in Open Channels. Eds.: P.J. Ashworth, S.J. Bennett, J.L. Best and S.J. McLelland. Wiley, p. 87–99.
(15) Detert, D., Nikora, V. and Jirka, G.H. (2010): Synoptic velocity and pressure fields at the water-sediment interface of streambeds. J. Fluid Mech., vol. 660, 55–86.
(16) Eckelmann, H. (1974): The structure of the viscous sublayer and the adjacent wall region in a turbulent channel flow. J. Fluid Mech., vol. 65, 439–459.
(17) Engelund, F. (1970): A note on the mechanics of sediment suspension. Basic Res. Prog. Rep., Hydraulic Lab., Tech. Univ. Denmark, vol. 21, p. 7–14.
(18) Engelund, F. and Gravesen, H. (1972): Settling of suspended sediment in a divergent channel. Prog. Rep., Hydraulic Lab., Tech. Univ. Denmark, vol. 27, p. 33–36.
(19) Engelund, F. and Hansen, E. (1967): A Monograph on Sediment Transport in Alluvial Streams. Teknisk Forlag, Copenhagen.
(20) Fredsøe, J. Andersen, K.H. and Sumer, B.M. (1999): Wave plus current over a ripple-covered bed. Coast. Eng., vol. 38, 177–221.
(21) Grass, A.J. (1971): Structural features of turbulent flow over smooth and rough boundaries. J. Fluid Mech., vol. 50, 233–255.
(22) Grass, A.J. (1974): Transport of fine sand on a flat bed: turbulence and suspension mechanics. Euromech 48, Inst. Hydrodyn. Hydraul Engng., Tech. Univ. Denmark, p. 33.
(23) Grass, A.J. (1983): The influence of boundary layer turbulence on the mechanics of sediment transport. In: Mechanics of Sediment Transport, Proc. of Euromech 156, Istanbul, 12-14. July 1982, A.A. Balkema (Rotterdam), p. 3–17.
(24) Grass, A.J., Stuart, R.J. and Mansour-Tehrani, M. (1991): Vortical structures and coherent motion in turbulent flow over smooth and rough boundaries. Phil. Trans. R. Soc. Lond. A, vol. 336, 35–65.
(25) Hinze, J.O. (1975): Turbulence. McGraw-Hill.
(26) Jackson, R.G. (1976): Sedimentological and fluid-dynamic implications of the turbulent bursting phenomenon in geophysical flows. J. Fluid Mech., vol. 77, 531–560.
(27) Jensen, K.L. and Sumer, B.M. (2019): Experimental investigation of wall-normal pressure gradient in a turbulent boundary layer in an open channel. Paper in preparation. From the PhD study of Karsten

L. Jensen supervised by B.M. Sumer at Technical University of Denmark. DTU Mechanical Engineering, "Vertical pressure gradient and particle motions in wave boundary layers", October 2014.
(28) Kamphuis, J.W. (1974): Determination of sand roughness for fixed beds. J. Hydraulic Res., vol. 12, 193–203.
(29) Klebanoff, P.S. (1955): Characteristics of turbulence in a boundary layer with zero pressure gradient. NACA, Report No. 1247.
(30) Kline, S.J., Reynolds, W.C., Schraub, F.A. and Runstadler, P.W. (1967): The structure of turbulent boundary layers. J. Fluid Mech., vol. 30, 741–773.
(31) Laufer, J. (1954): The Structure of Turbulence in Fully Developed Pipe Flow. NACA, Report No. 1174.
(32) Laufer, J. (1975): New trends in experimental turbulence research. Ann. Rev. Fluid Mech., vol. 7, 307–326.
(33) Lee, M.K., Eckelman, L.D., Hanratty, T.J. (1974): Identification of turbulent wall eddies through the phase relation of the components of the fluctuating velocity gradient. J. Fluid Mech., vol. 66, 17–33.
(34) Lumley, J.L. (1978): Two-phase and non-Newtonian flows. In: Turbulence, 2nd edn. (ed. P. Bradshaw), Topics in Applied Physics, vol. 12, 289, Springer.
(35) Mantz, P.A. (1978): Bedforms produced by fine, cohesionless, granular and flakey sediments under subcritical water flows. Sedimentology, vol. 25, 83–103.
(36) Monin, A.S. and Yaglom, A.M. (1973): Statistical Fluid Mechanics: Mechanics of Turbulence. MIT Press, Cambridge, Mass.
(37) Nezu, I. and Nakagawa, H. (1993): Turbulence in Open-Channel Flow. IAHR Monograph, A.A. Balkema (Rotterdam).
(38) Nezu, I. and Rodi, W. (1986): Open-channel flow measurements with a laser Doppler anemometer. J. Hydraul. Eng.-ASCE, vol. 112, 335–355.
(39) Nikuradse, J. (1933): Stromungsgesetze in rauhen Rohren. Forsch. Arb. Ing.-Wes. No. 361.
(40) Nino, Y. and Garcia, M.H. (1996): Experiments on particle-turbulent interactions in the near wall region of an open channel flow: implications for sediment transport. J. Fluid Mech., vol. 326, 285–319.
(41) Nychas, S.G., Hershey H.C. and Brodkey, R.S. (1973): A visual study of turbulent shear flow. J. Fluid Mech., vol. 61, 513–540.
(42) Offen, G.R. and Kline, S.J. (1973): Experiments on the velocity characteristics of "Bursts" and on the Interactions Between the Inner and Outer Regions of a Turbulent Boundary Layer. Dept. of Mech. Engrg.,

Stanford University, Report MD-31.
(43) Offen, G.R. and Kline, S.J. (1975): A proposed model of the bursting process in turbulent boundary layers. J. Fluid Mech., vol. 70, 209–228.
(44) Praturi, A.K. and Brodkey, R.S. (1978): A stereoscopic visual study of coherent structures in turbulent shear flow. J. Fluid Mech., vol. 89, 251–272.
(45) Rotta, J.C. (1962). Turbulent boundary layers in incompressible flow. Progress in Aerospace Science, vol. 2, p. 1–219.
(46) Schlichting, H. (1979). Boundary-Layer Theory. McGraw-Hill.
(47) Shields, A. (1936): Application of Similarity Principles and Turbulence Research to Bedload Movement. W.P. Ott and J.C. van Uchelen, eds., Soil Conservation Service Cooperative Laboratory, California Institute of Technology, Pasadena, CA. (Translation of A. Shields' Thesis Work).
(48) Sumer, B.M. (1984): Lift forces on moving particles near boundaries. J. Hydraul. Eng.-ASCE, vol. 110, 1272–1278.
(49) Sumer, B.M. (1986): Recent developments on the mechanics of sediment suspension. General-Lecture Paper, in the book: Euromech 192: Transport of Suspended Solids in Open Channels, A.A. Balkema Publishers, Rotterdam.
(50) Sumer, B.M. (2015): Discussion of Sediment entrainment probability and threshold of sediment suspension: Exponential-base approach. J. Hydraul. Eng.-ASCE, vol. 141, article no. 07014023.
(51) Sumer, B.M., Cokgor, S. and Fredsøe, J. (2001): Suction removal of sediment from between armour blocks. J. Hydraul. Eng.-ASCE, vol. 127, 293–306.
(52) Sumer, B.M. and Deigaard, R. (1981): Particle motions near the bottom in turbulent flow in an open channel - Part 2. J. Fluid Mech., vol. 109, 311–337.
(53) Sumer, B.M. and Fredsøe, J. (2002): The Mechanics of Scour in the Marine Environment. World Scientific.
(54) Sumer, B.M. and Fredsøe, J. (2006): Hydrodynamics Around Cylindrical Structures. World Scientific.
(55) Sumer, B.M., Kozakiewicz, A., Fredsøe, J. and Deigaard, R. (1996): Velocity and concentration profiles in the sheet flow layer of movable bed. J. Hydraul. Eng.-ASCE, vol. 122, 549–558.
(56) Sumer, B.M. and Oguz, B. (1978): Particle motions near the bottom in turbulent flow in an open channel. J. Fluid Mech., vol. 86, 109–127.
(57) Sutherland, A.J. (1967): Proposed mechanism for sediment entrainment by turbulent flow. J. Geophys. Res., vol. 72, 6183–6194.

(58) van Driest, E.R. (1956): On turbulent flow near a wall. J. Aeronautical Sciences, vol. 23, 1007–1011.
(59) Wallace, J.M., Brodkey, R.S. and Eckelmann, H. (1972): The wall region in turbulent shear flow. J. Fluid Mech., vol. 54, 39–48.
(60) Wei, T. and Willmarth, W.W. (1989): Reynolds-number effects on the structure of a turbulent channel flow. J. Fluid Mech., vol. 204, 57–95.
(61) Wu, X., Moin, P., Wallace, J.M., Skarda, J., Lozano-Duran, A. and Hickey, J.-P. (2017): Transitional-turbulent spots and turbulent-turbulent spots in boundary layers. Proc. Nat. Academy Sci. U.S.A., vol. 114, E5292–E5299.

Appendix to Chapter 3: Dimensional analysis

Dimensional analysis is a powerful tool in fluid mechanics to transform a dimensional function into a non-dimensional one. In the present book, we resort to the dimensional analysis on various occasions, particularly in Chapter 3, but also elsewhere. This appendix highlights the theory behind the dimensional analysis. The way in which the dimensional analysis is applied is described with reference to an example.

3.A Buckingham's Pi theorem

We may write the unit of any hydrodynamic quantity in the following form

$$[A] = L^{\alpha} T^{\beta} K^{\gamma} \tag{3.181}$$

in which $[A]$ is the unit of A, and L, T, and K are the fundamental units, namely, L is the length, T is the time, and K is the force. α, β and γ are positive or negative integers, or they may be zero. (Note that mass could alternatively be utilized instead of the force, this is a free choice.)

Buckingham's Pi theorem states the following: A dimensional function

$$f(A_1,\ A_2,\ ...,\ A_n) = 0 \tag{3.182}$$

can always be converted to the following non-dimensional function

$$F(\pi_1, \pi_2, ..., \pi_{n-r}) = 0 \tag{3.183}$$

in which A_i $(i = 1, ..., n)$ are the dimensional quantities, and r is the number of the fundamental units (normally $r = 3$, i.e., the length, the time and the force). The quantities π_i $(i = 1, ..., n-r)$ are the non-dimensional quantities and calculated from

$$\begin{aligned}\pi_1 &= A_1^{x_1} A_2^{y_1} A_3^{z_1} A_4 \\ \pi_1 &= A_1^{x_2} A_2^{y_2} A_3^{z_2} A_4 \\ &\cdots \\ \pi_{n-r} &= A_1^{x_{n-r}} A_2^{y_{n-r}} A_3^{z_{n-r}} A_4\end{aligned} \tag{3.184}$$

Note that the above combinations of A_i (for each π_i) should be selected so that

(1) The unit of any one of A_i should not be a combination of the units of the others; and
(2) Each and every fundamental unit should be included in at least one of the A_i's.

The advantage of the dimensional analysis is that, first of all, it "translates" the dimensional function into a non-dimensional function in a systematic way. Secondly and more importantly, in doing so, the number of the quantities included in the dimensional function is reduced by r (normally $r = 3$).

3.B Example: Law of the wall

In Section 3.1.1, the flow velocity is written as a function of the following non-dimensional quantities (Eq. 3.9)

$$\overline{u} = F(\tau_0,\ y,\ \mu,\ \rho) \tag{3.185}$$

or

$$f(\tau_0,\ \mu,\ \rho,\ \overline{u},\ y) = 0 \tag{3.186}$$

As seen, $n = 5$, and therefore the number of non-dimensional quantities will be $n - r = 5 - 3 = 2$. Hence

$$F(\pi_1, \pi_2) = 0 \tag{3.187}$$

in which the non-dimensional quantities π_1 and π_2 are

$$\pi_1 = \tau_0^{x_1} \mu^{y_1}\ \rho^{z_1}\ \overline{u} \tag{3.188}$$

and

$$\pi_2 = \tau_0^{x_2} \mu^{y_2}\ \rho^{z_2} y \tag{3.189}$$

The left and right hand sides of Eq. 3.188 should have the same units. Writing the units of the quantities on both sides of the equation according to Eq. 3.181,

$$L^0T^0K^0 = (KL^{-2})^{x_1}(KL^{-2}T)^{y_1}(KL^{-4}T^2)^{z_1}(LT^{-1}) \tag{3.190}$$

or

$$L^0T^0K^0 = L^{-2x_1-2y_1-4z_1+1}T^{y_1+2z_1-1}K^{x_1+y_1+z_1} \tag{3.191}$$

which requires

$$\begin{aligned} -2x_1 - 2y_1 - 4z_1 + 1 &= 0, \\ y_1 + 2z_1 - 1 &= 0, \\ x_1 + y_1 + z_1 &= 0 \end{aligned} \tag{3.192}$$

This gives

$$x_1 = -\frac{1}{2}, \quad y_1 = 0, \text{ and } \quad z_1 = \frac{1}{2} \tag{3.193}$$

Hence the first non-dimensional quantity

$$\pi_1 = \tau_0^{-1/2}\mu^0\rho^{1/2}\overline{u} = \frac{\overline{u}}{(\frac{\tau_0}{\rho})^{1/2}} = \frac{\overline{u}}{U_f} \tag{3.194}$$

Regarding the second non-dimensional quantity, from Eq. 3.189

$$L^0T^0K^0 = (KL^{-2})^{x_2}(KL^{-2}T)^{y_2}(KL^{-4}T^2)^{z_2}L \tag{3.195}$$

and

$$\begin{aligned} -2x_2 - 2y_2 - 4z_2 + 1 &= 0, \\ y_2 + 2z_2 &= 0, \\ x_2 + y_2 + z_2 &= 0 \end{aligned} \tag{3.196}$$

and

$$x_2 = \frac{1}{2}, \quad y_2 = -1, \text{ and } \quad z_2 = \frac{1}{2} \tag{3.197}$$

Then, the second non-dimensional quantity is

$$\pi_2 = \tau_0^{1/2}\mu^{-1}\rho^{1/2}y = \frac{y(\frac{\tau_0}{\rho})^{1/2}}{\frac{\mu}{\rho}} = \frac{yU_f}{\nu}\ (= y^+) \tag{3.198}$$

Therefore, the non-dimensional function will be

$$F\left(\frac{\overline{u}}{U_f}, \frac{yU_f}{\nu}\right) = 0 \tag{3.199}$$

or solving for the velocity

$$\frac{\overline{u}}{U_f} = f\left(\frac{yU_f}{\nu}\right) \qquad \text{or} \qquad \frac{\overline{u}}{U_f} = f(y^+) \tag{3.200}$$

This relation is known as the law of the wall (Section 3.1.1).

As seen, utilizing the dimensional analysis, the non-dimensional function in Eq. 3.186 with five dimensional quantities has been transformed to a non-dimensional function in Eq. 3.199 with two non-dimensional quantities.

Chapter 4

Statistical, correlation and spectral analysis

As has been seen in the previous chapters, hydrodynamic quantities such as the velocity components in a turbulent flow may appear to be random variables. Although more recent research has shown that the near-wall turbulent-boundary-layer process is quasi-repetitive in nature (Chapter 3, bursting process), unless special pattern recognition techniques are used to detect these repetitive, quasi-cyclic events, the turbulence signals even near a wall appear to be largely random. Therefore, it is imperative to use statistical and other analysis techniques to study turbulence.

Towards this end, three topics will be discussed in the present chapter. First we will briefly study the probability density function of the turbulent velocity, and how this relates to basic statistical quantities. We will then turn attention to correlation analysis, with focus on relating the correlation function to characteristic scales of turbulence. Finally, we will discuss spectral analysis of turbulence, with simultaneous focus on all scales as seen either from the wave number of frequency domain.

The present chapter is intended as merely an introduction to the main classical topics presented, giving the interested reader the necessary background to explore the topic and more recent developments further. In the much of the theoretical analysis that follows, it will be assumed that the turbulence is *isotropic*, and we will attempt to make clear when this assumption is made explicitly. The main text is likewise supplemented by the chapter Appendix, which contains several formal proofs of some commonly seen isotropic turbulence relationships, such that all of the material presented in this chapter is self contained.

With the present chapter's focus on *analysis*, we additionally supplement the text with five practical examples coded in Matlab. Much effort has been put into making these examples compact, such that they

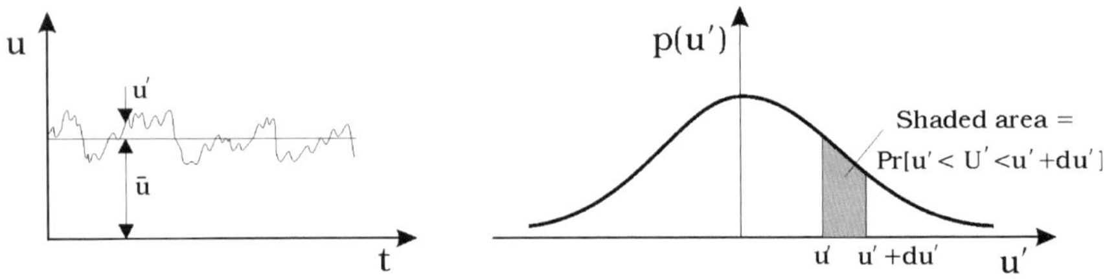

Fig. 4.1 Example (left) time series and (right) probability density function for a fluctuating velocity.

each fit neatly on a single page. The scripts are provided with the intention and hope that the interested reader may easily modify them to fit their own analysis needs. The scripts can be downloaded as a single file `MatlabExamples.zip` at:

`https://doi.org/10.11583/DTU.c.4508648`

4.1 Statistical analysis

4.1.1 *Probability density function*

Consider e.g. the x-component of the velocity, u, such as is depicted in Fig. 4.1 (left), which is assumed to arise from a stationary and ergodic process (i.e. that its statistical properties will not vary with time, and that they may be deduced from a single, sufficiently long sample). The velocity u, or its fluctuating component u' for that matter, may be considered as a random variable. Thus, u (or u') must have a *probability density function* (p.d.f.) such that

$$p(u')\, du' = \Pr[u' < U' < u' + du'] \tag{4.1}$$

and

$$\int_{-\infty}^{\infty} p(u')\, du' = 1 \tag{4.2}$$

in which $p(u')$ is the p.d.f. of the fluctuating velocity u', and $\Pr[u' < U' < u' + du']$ is the probability of occurrence of the fluctuating velocity between u' and $u' + du'$.

Experiments show that in many turbulent flows $p(u')$ can be reasonably represented, at least to a first approximation, by the Gaussian (normal) distribution

$$p(u') = \frac{1}{\sigma_u\sqrt{2\pi}} \exp\left(-\frac{u'^2}{2\sigma_u^2}\right) \tag{4.3}$$

in which σ_u is the standard deviation of the velocity of u:

$$\sigma_u = \sqrt{\overline{(u-\overline{u})^2}} = \sqrt{\overline{u'^2}} \tag{4.4}$$

The significance of σ_u is that the probability of u' lying within a given range corresponds to e.g.

$$Pr[-\sigma_u < U' < \sigma_u] \approx 0.6827 \tag{4.5}$$

$$Pr[-2\sigma_u < U' < 2\sigma_u] \approx 0.9545 \tag{4.6}$$

$$Pr[-3\sigma_u < U' < 3\sigma_u] \approx 0.9973 \tag{4.7}$$

Or in other words, the probability of u' falling outside the range $-\sigma_u < u' < \sigma_u$ is approximately 31.7%, outside the range $-2\sigma_u < u' < +2\sigma_u$ is only 4.6%, while that outside the range $-3\sigma_u < u' < +3\sigma_u$ is much smaller, namely 0.3%. It is emphasized that these values (Eqs. 4.5–4.7 and in the preceding sentence) assume that the p.d.f. follows the Gaussian distribution. Hence, provided a distribution is Gaussian (or reasonably so), the vast majority of measurements can be expected to lie within ±3 standard deviations of the mean.

4.1.2 *Statistical moments*

The statistical moments (or central moments) also give some information on the statistical properties of the velocities and other quantities:

(1) The first statistical moment (about zero) is the mean

$$\overline{u} = \int_{-\infty}^{\infty} u\, p(u)\, du \tag{4.8}$$

(2) The second central moment (about the mean) gives the variance

$$\sigma_u^2 = \int_{-\infty}^{\infty} (u-\overline{u})^2\, p(u)\, du = \int_{-\infty}^{\infty} u'^2\, p(u')\, du' \tag{4.9}$$

(3) The third central moment,

$$\overline{u'^3} = \int_{-\infty}^{\infty} (u-\overline{u})^3\, p(u)\, du = \int_{-\infty}^{\infty} u'^3\, p(u')\, du' \tag{4.10}$$

gives the skewness:

$$S_u = \frac{\overline{u'^3}}{(\sigma_u^2)^{\frac{3}{2}}} \tag{4.11}$$

(4) The fourth central moment,

$$\overline{u'^4} = \int_{-\infty}^{\infty} (u - \overline{u})^4 \, p(u) \, du = \int_{-\infty}^{\infty} u'^4 \, p(u') \, du' \tag{4.12}$$

gives the kurtosis:

$$F_u = \frac{\overline{u'^4}}{(\sigma_u^2)^2}$$

Notice that while the mean and variance are dimensional quantities, the skewness and kurtosis are normalized by the standard deviation raised to the appropriate power, such that they are dimensionless. As reference values, it can be noted that normally distributed signals have zero skewness, whereas their kurtosis is equal to 3.

As illustrations, intended to help provide a more physical interpretation to some of the (perhaps less familiar) higher statistical moments above, several segments coming from signals having various statistical properties are presented in Fig. 4.2. Fig. 4.2a depicts a time series segment (left) of a Gaussian signal, with associated p.d.f. depicted to the right. A Gaussian signal by definition has a symmetric p.d.f., and hence has zero skewness. In contrast, Fig. 4.2b depicts a similar, but positively skewed signal. Skewness is a measure of the asymmetry of the p.d.f. about the mean; This example demonstrates positive skewness, characterized by the signal having raised peaks and troughs, and correspondingly a left-leaning p.d.f. with long rightward tail. An example of a negatively skewed signal is not shown, but can easily be envisioned simply by turning this signal upside down, and mirroring the associated p.d.f. such that it is instead rightward leaning. Figs. 4.2c and 4.2d, respectively, depict similar examples having high and low kurtosis. Kurtosis is a measure of the "tailedness" (rather than the often-claimed "peakedness", see Westfall 2014) of a p.d.f. The signal having high kurtosis, Fig. 4.2c, indeed demonstrates exaggerated peaks in both the positive and negative directions, thus giving rise to more pronounced tails in the associated p.d.f. (In the case shown, the signal tends to remain near the single peak in the p.d.f., here corresponding to the mean, the majority of the time). In contrast, the signal having low kurtosis, Fig. 4.2d, wanders much more uniformly about its mean, and does therefore not exhibit pronounced tails in the corresponding p.d.f. It is again noted that as the p.d.f.s in Figs. 4.2c and 4.2d are symmetric, both of these signals have zero skewness.

As noted above, a Gaussian (normal) distribution may be taken as a reasonable first approximation for real-life turbulent fluctuations. In reality,

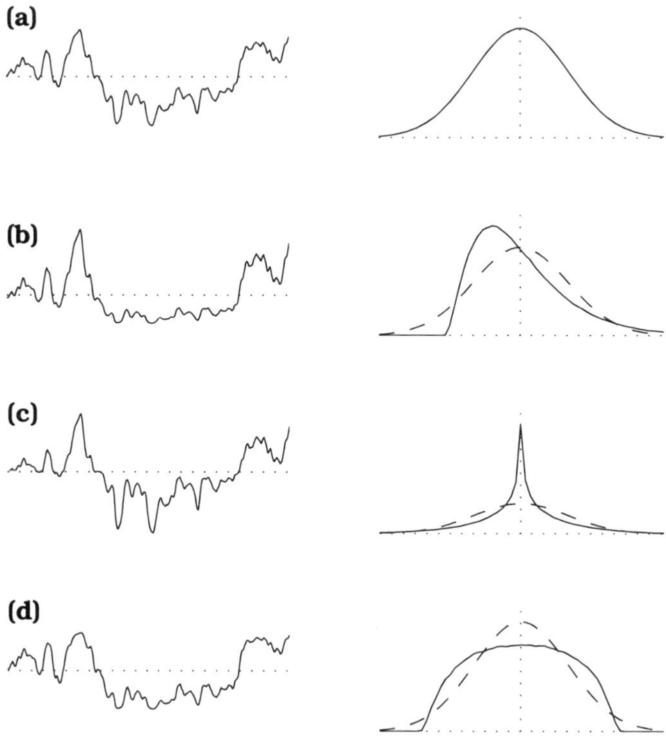

Fig. 4.2 Example (left) time series segments and corresponding (right) probability density functions for a signal that (a) is normally distributed, (b) is positively skewed, (c) has high kurtosis, and (d) has low kurtosis. On sub-plots (b)–(d) the Gaussian distribution from (a) is also shown (dashed line) for comparison.

however, experiments show that the p.d.f.s of turbulent fluctuating velocity components may differ somewhat from the Gaussian distribution, especially near a wall. As examples Fig. 4.3 shows p.d.f.s (in the form of frequency histograms) for the two components of the turbulent velocities, u and v, near a wall for three different boundaries, namely the smooth (left), transitional (middle) and rough walls (right), taken from Grass (1971). Note the asymmetry, hence non-zero skewness, in all distributions, which is typical of such observations particularly near the wall.

To further elucidate the near-wall region Eckelmann (1974) systematically investigated the p.d.f. of the streamwise fluctuating velocity in a smooth oil filled channel, such that a thick viscous sublayer region was present. The resulting p.d.f.s for various y^+ values are presented in Fig. 4.4.

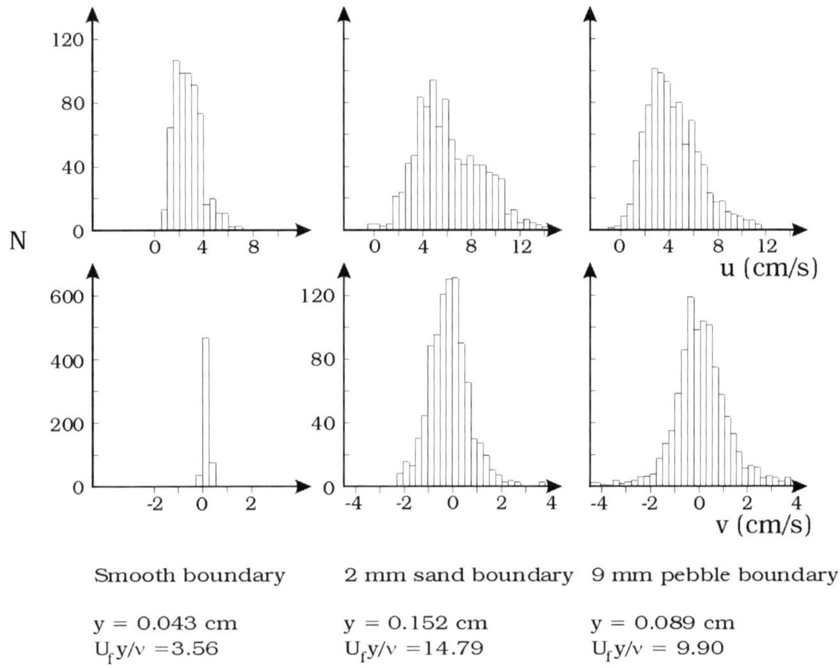

Fig. 4.3 Frequency histograms of u and v velocity fluctuations. Adapted from Grass (1971).

These results demonstrate that very near the wall ($y^+ < 13$) the streamwise fluctuations exhibit positive skewness, whereas above this value the turbulent signals have negative skewness. Similar results were likewise found by Kreplin (1973), which are presented in Fig. 4.5, where both the distribution of streamwise skewness (Fig. 4.5a) and kurtosis (Fig. 4.5b) versus y^+ are depicted.

Following Eckelmann (1974), the non-zero skewness exhibited in boundary layer near a wall may be linked to the bursting process (discussed in Section 3.5) in the following way: Positive skewness indicates that the most probable (i.e. median) velocity is less than the local mean (this corresponds e.g. to the peak in the left-leaning p.d.f. shown in Fig. 4.2b). Since measured negative turbulent fluctuations (i.e. velocities below the mean) at a point can be interpreted as the momentary passing of a low speed streak, this means that very near the wall ($y^+ < 13$), low speed streaks must be relatively wide. Conversely, very near the wall high speed streaks must be

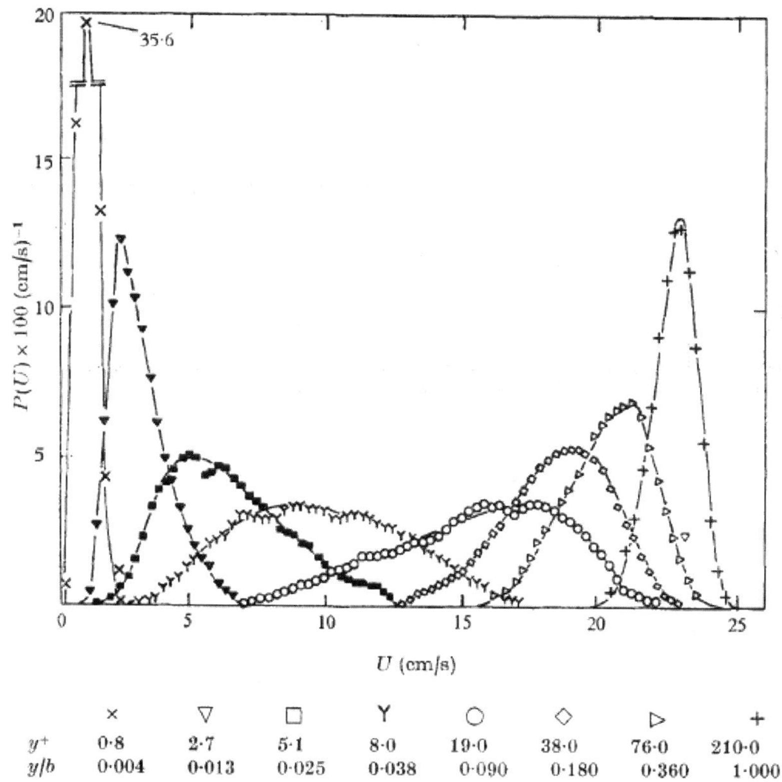

Fig. 4.4 Probability density functions of the instantaneous streamwise velocity u in a steady boundary layer flow at various y^+ positions. The Reynolds number is $Re_U = U_c b/\nu = 8200$, where $U_c = 22.5$ cm/s is the maximum center-line velocity in the channel, $b = 22$ cm is the channel width, and $\nu = 0.06$ cm^2/s (oil). This figure is reprinted from Eckelmann (1974), with permission from Cambridge University Press.

relatively narrow. For $y^+ > 13$, where the streamwise turbulent velocity exhibits negative skewness (Fig. 4.4), the opposite would hold.

4.1.3 *Matlab example: Basic statistics and creating a p.d.f.*

Let us now begin the first of our five Matlab examples, by presenting a function `turbgen.m` (Program 1) which generates a discrete normally distributed (synthetic) turbulent velocity signal of length `N`, with discrete time step (sampling interval) `dt`, based on a specified variance `sigma2` and macro time scale `T_E` (this quantity will be presented later in Section 4.2). In

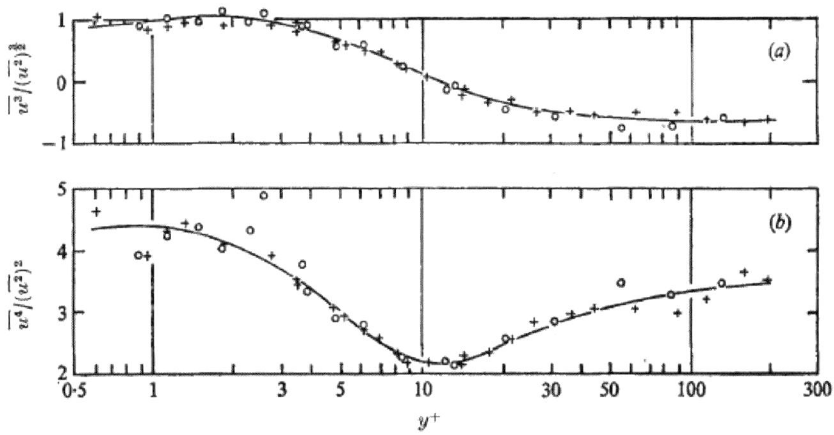

Fig. 4.5 Distribution of (a) the skewness and (b) the kurtosis of the streamwise velocity fluctuations u over the channel half-width, from Kreplin (1973), where $Re_U = 4800$ (○) and $Re_U = 7100$ (+). This figure is reprinted from Eckelmann (1974), with permission from Cambridge University Press.

short, this function works by first specifying an analytic one-dimensional energy density spectrum (covered in Section 4.3.5) in the so-called Fourier (frequency) domain, and then inverse transforming to obtain the desired discrete signal. For readers unfamiliar with these notions, they will be covered later in the present chapter; For the time being, this function may be considered simply as a "black box", convenient for generating a "turbulent" signal (time series) with certain known characteristics. These may then be independently detected via analysis.

Note that, as scientists/engineers most commonly work with time (rather than space) series, we have adopted this convention for the Matlab examples in this chapter, though they may be easily modified to work for space series. The `turbgen.m` function is utilized in Program 2, as well as in many of the other forthcoming Matlab examples. For educational purposes, as hinted above, these examples will generally follow the strategy of: (1) generating a signal with known characteristics and then (2) using analysis to independently detect the desired characteristics from the signal alone. It is hence intended and hoped that the methods demonstrated may be readily applied by students and practicing engineers to other real life turbulent signals, where such characteristics would likely not be known *a priori*.

Program 1: turbgen.m

```
1 function [u,t] = turbgen(N,dt,sigma2,T_E,spec,ns,seed)
2 % Generates a discrete turbulent velocity time series
3 % based on an analytical spectrum with random phasing.
4 % Input: N=series length, dt=time step (s), sigma2=
5 % variance (m^2/s^2), T_E=Macro time scale (s), ns
6 % determines the noise level (>0=on, 0=off), seed
7 % initializes the randon number generator ([]=off),
8 % and spec is the type of spectrum (see line 17).
9
10 % Reset random number generator, if necessary
11 if not(isempty(seed)); rng(seed); end;
12
13 % Determine discrete positive frequencies
14 f_N = 1/(2*dt); df = f_N/(N/2); f = [df:df:f_N];
15
16 % Prepare 1D energy density spectrum
17 switch spec % 1: Dryden, 2: von Karman, 3: Kaimal
18   case 1, S = 4*f*T_E./(1 + (2*pi*f*T_E).^2);
19   case 2, S = 4*f*T_E./(1 + 70.78*(f*T_E).^2).^(5/6);
20   case 3, S = 4*f*T_E./(1 + 6*f*T_E).^(5/3);
21 end % The spectra above are dimensionless
22 S = S.*(1 + ns*randn(1,N/2)); % Add noise, if ns>0
23 S = sigma2*S./f; % Dimensionalize (m^2/s)
24
25 % Convert to Fourier amplitudes
26 phi = [2.*pi.*rand(1,N/2-1) 0]; % Random phase shifts
27 U = sqrt(2*S*df).*(cos(phi) + 1i*sin(phi)); % (m/s)
28 U = [0 U(1:N/2-1)/2 U(N/2) ...
29      real(U(end-1:-1:1))/2-1i*imag(U(end-1:-1:1))/2];
30      % Re-arrange to reverse-wrap-around (RWA) order
31
32 % Generate time series using inverse FFT
33 u = N*ifft(U); % Turbulent velocity time series (m/s)
34 t = [0:N-1]*dt; % Corresponding time (s)
```

Program 2: ExamplePDF.m is a subsequent Matlab script in which a discrete synthetic signal is first generated using turbgen.m. For the sake of repeatability the seed 1842 is used, corresponding to the birth year of Osborne Reynolds. Using built-in Matlab functions (for simplicity) basic sample statistics are first calculated in lines 6–10, corresponding to sample approximations of those presented in Eqs. 4.8–4.12. After execution, this script results in the following screen output:

```
ubar =

  -5.1159e-18

sigma2 =

    0.0040

S_u =

    0.0604

F_u =

    2.9771

area =

     1
```

As seen, the mean (variable ubar) and skewness (S_u) are very near zero (the latter will be even closer if the series length N is further increased), and the kurtosis (F_u) is very near 3, in accordance with normally distributed data. The sample variance (sigma2) is also effectively equal to that specified as input, hence largely confirming successful generation of the intended signal.

The graphical output of this script is depicted in Fig. 4.6. Here a segment of the time series is first depicted (top), which at a glance indeed resembles a turbulent velocity signal. The script then calculates a probability density function based on the discrete signal, which is compared with the Gaussian normal distribution. The computed sample p.d.f. is presented in Fig. 4.6 (bottom), which matches the expected Gaussian distribution

Program 2: ExamplePDF.m

```
% Generate time series using turbgen.m
N = 1e5; T_E = 0.2; dt = 0.01; sigma2 = 0.004;
[u,t] = turbgen(N,dt,sigma2,T_E,1,0,1842);

% Calculate basic sample statistics
ubar = mean(u) % Mean (m/s)
sigma = std(u); % Std. dev. (m/s)
sigma2 = var(u) % Replace w/ sample variance (m^2/s^2)
S_u = skewness(u) % Skewness
F_u = kurtosis(u) % Kurtosis

% Plot a segment of the time series
subplot(2,1,1); plot(t,u,'k-');
xlabel('t (s)'); ylabel('u^\prime (m/s)'); grid on;
xlim(85.5+[0 15*T_E]); set(gca,'xtick',[0:1:t(end)])
title('Segment of time series'); ylim([-3 3].*sigma)

% Create a dimensional p.d.f.
ubins = [-5:0.1:5].*sigma; % Vector of bins
n = hist(u,ubins); % Creates histogram
ahist = trapz(ubins,n); % Total area of histogram
n = n/ahist; % Normalize to yield unit area
subplot(2,1,2); plot(ubins,n,'k-'); % Plot p.d.f.
grid on; xlim([-5 5].*sigma); % Set axis limits
xlabel('u^\prime (m/s)')
ylabel('p(u^\prime) (m/s)^{-1}'), title('p.d.f.');
area = trapz(ubins,n) % CHECK: integral must = 1.0!

% Compare with normal (Gaussian) distribution
pn = 1/sigma/sqrt(2*pi)*exp(-ubins.^2/2/sigma2);
hold on; plot(ubins,pn,'r--'); hold off;
legend('Computed','Gaussian','location','NorthEast')
```

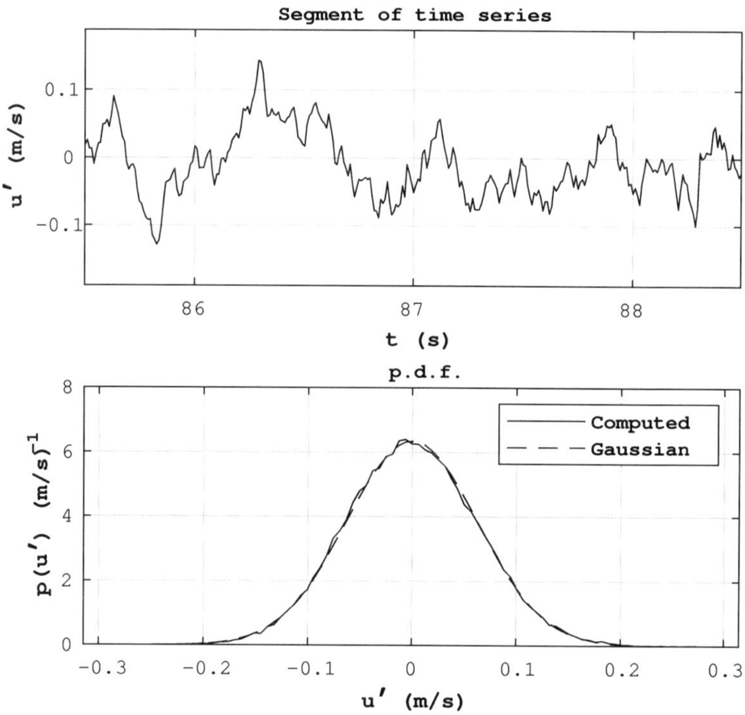

Fig. 4.6 Output from Matlab script `ExamplePDF.m`.

very well. The p.d.f. in this example is generated using the built-in Matlab function `hist`, which counts the number of values that fall within discrete bins, as specified in the `ubins` variable. The result is then normalized (lines 21 and 22), such that the area beneath the p.d.f. is equal to unity. This is subsequently checked using numerical integration (line 27, see also the end of the screen output above, which is easy and always a good idea in practice!).

This example demonstrates calculation of basic statistics, and creating a corresponding p.d.f. based on an example turbulent velocity signal. In present versions of Matlab, the built-in `histogram` function, first introduced in version R2014b, may alternatively be utilized for the latter. For educational purposes we prefer the provided example, however, as it gives a bit more detail as to what is done "behind the scenes" to produce a proper p.d.f. based on a discrete signal, with integral assured to be equal to unity.

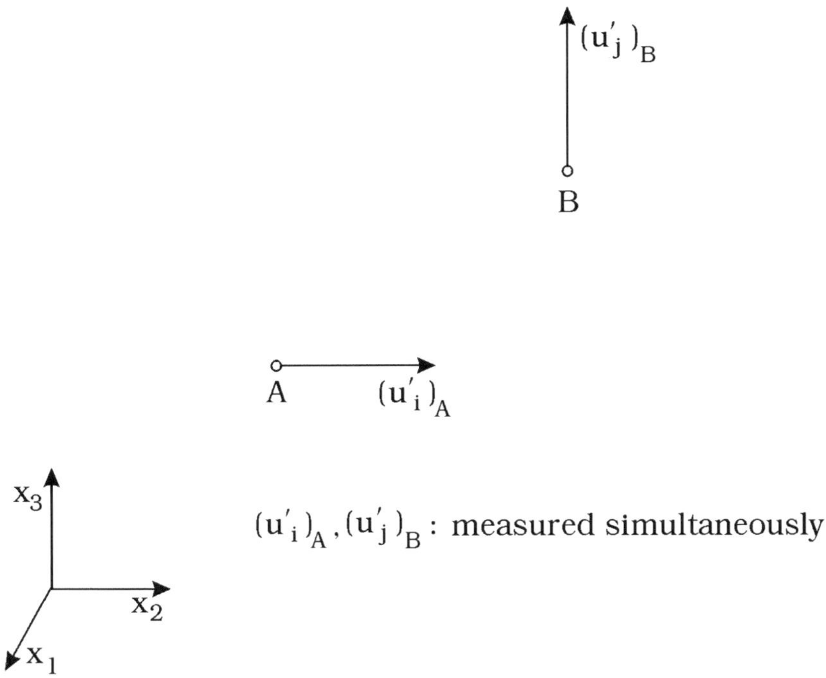

Fig. 4.7 Two-point double correlation measurement involving two velocity components measured simultaneously at points A and B.

4.2 Correlation analysis

Let us now turn our attention to correlation analysis. We shall specifically consider two kinds of correlations: (1) space correlations, and (2) time correlations. We will first focus on space correlations.

4.2.1 *Space correlations*

Definitions. Fig. 4.7 provides a definition sketch. Consider the i-component of the fluctuating velocity at Point A, $(u'_i)_A$, and the j-component of the fluctuating velocity at Point B, $(u'_j)_B$. Suppose that these velocities are measured simultaneously. Two-point double correlation is defined by

$$(Q_{ij})_{AB} = \overline{(u'_i)_A \ (u'_j)_B} \tag{4.13}$$

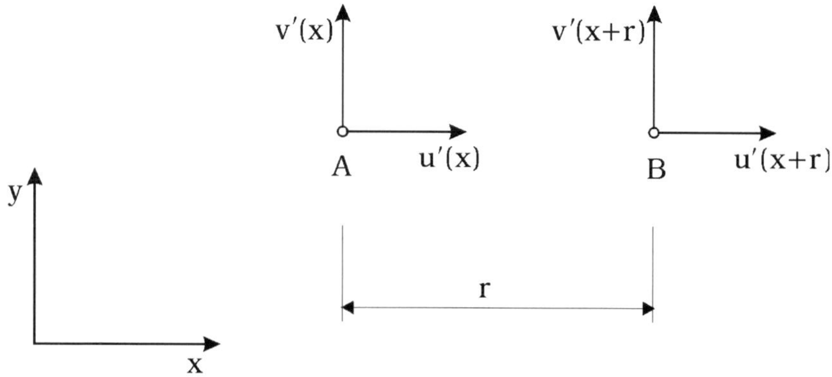

Fig. 4.8 Two-point double correlation measurement involving simultaneous velocity components separated by a distance r.

in which the overbar indicates time averaging (Eq. 2.1). There are other correlations as well: three-point, four-point,..., and triple, quadruple, etc., correlations.

The most important two-point double correlations used in turbulence (Fig. 4.8) are

$$\overline{u'(x)\ u'(x+r)} \tag{4.14}$$

and

$$\overline{v'(x)\ v'(x+r)} \tag{4.15}$$

in which r is the separation distance between the two points. For convenience, these correlations are usually normalized by $\overline{u'^2}$ and $\overline{v'^2}$, respectively:

$$f(r) = \frac{\overline{u'(x)\ u'(x+r)}}{\overline{u'^2}} \tag{4.16}$$

$$g(r) = \frac{\overline{v'(x)\ v'(x+r)}}{\overline{v'^2}} \tag{4.17}$$

where $f(r)$ and $g(r)$ are the so-called *correlation functions.*

The physical interpretation of the correlation functions is as follows. The larger the correlation, the motions at the two measurement points are statistically more alike, and therefore it is more likely that these two points belong to the same coherent structure. This would obviously give an idea about the length scale of the coherent structures, as will be detailed later in the section.

Properties of two-point double correlations. The following properties of the two-point double correlations hold for turbulent signals:

(1) The correlation functions are unity at $r = 0$:
$$f(0) = g(0) = 1 \tag{4.18}$$
(from Eqs. 4.16 and 4.17).

(2) There is symmetry between positive and negative separations:
$$f(r) = f(-r), \text{ and } g(r) = g(-r) \tag{4.19}$$
(stemming from assumed homogeneity).

(3) The correlation functions are smaller than or equal to unity:
$$f(r) \leq 1, \text{ and } g(r) \leq 1 \tag{4.20}$$
This can easily be seen from
$$\begin{aligned}\overline{(u'(x) - u'(x+r))^2} &= \overline{u'^2(x)} + \overline{u'^2(x+r)} - 2\overline{u'(x)u'(x+r)} \\ &= 2(\overline{u'^2} - \overline{u'(x)u'(x+r)})\end{aligned}$$
Since the left hand side of the equation ≥ 0, then $\overline{u'^2} - \overline{u'(x)u'(x+r)} \geq 0$, hence $f(r) \leq 1$.

(4) The correlation functions tend to zero as the separation distance goes to infinity
$$f(r) \to 0, \text{ and } g(r) \to 0, \text{ as } r \to \infty \tag{4.21}$$
This is because when r is very large, there will be as many $u'(x)u'(x+r)$ products with the negative sign as with the positive sign, therefore, when averaged, $\overline{u'(x)u'(x+r)}$ will be zero. Physically, r is so large that the two points will not lie in the same coherent structure, therefore, their motions will be uncorrelated, meaning that the correlation functions should be zero for such large r distances.

(5) While $f(r)$ is always positive, $g(r)$ can take negative values as well. To show this, we employ
$$rg(r) = rf(r) + \frac{1}{2}r^2\frac{\partial f}{\partial r} = \frac{1}{2}\frac{\partial}{\partial r}(r^2 f(r)) \tag{4.22}$$
This is the relationship between $f(r)$ and $g(r)$ in isotropic turbulence, as will be detailed in the following paragraphs. Integrating gives
$$\int_0^\infty rg(r)dr = \frac{1}{2}[r^2 f(r)]_0^\infty = 0 \tag{4.23}$$
provided that $r^2 f(r) = 0$ as $r \to \infty$, which is indeed the case since $f(r)$ is infinitely differentiable. For the integral on the left hand side of the equation to be zero, clearly $g(r)$ must have positive and negative values (this argument again assumes isotropic turbulence).

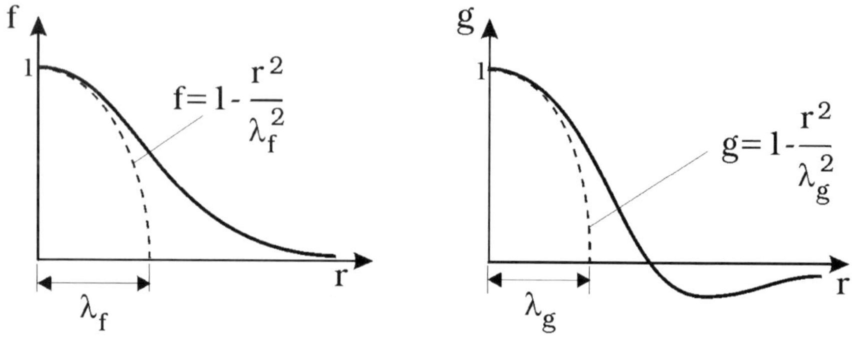

Fig. 4.9 Correlation functions f and g, and micro scales.

The general shapes of the functions of $f(r)$ and $g(r)$ are shown in Fig. 4.9.

As hinted above, there are two special classes of turbulence which are often assumed, in part because they make the mathematics of turbulence more tractable. The first, **homogeneous turbulence**, is such that the statistical properties are invariant to a *translation* of the coordinate axes. The second, **isotropic turbulence**, is a special class of turbulence where statistical properties are invariant to a *rotation* of the coordinate axes. This implies, among other things, that $\overline{u'^2} = \overline{v'^2} = \overline{w'^2}$. Isotropic turbulence implies that it is homogeneous, but not vice versa. Turbulence in the "core" region in a wind tunnel (where the wall effects are practically nonexistent) may be considered to be nearly isotropic. In such a turbulence field, the two correlation functions are related via the following equation

$$g(r) = f(r) + \frac{r}{2}\frac{\partial}{\partial r}f(r) \tag{4.24}$$

This is formally proved in Appendix 4.A (see also e.g. Hinze (1959), p. 150, or von Karman and Howarth (1938) who were the first to prove this relationship between $f(r)$ and $g(r)$). Eq. 4.24 was used above, leading to Eq. 4.22.

Scales of turbulence: Micro scales. Expanding $f(r)$ as a Taylor series about $r = 0$ gives

$$f(r) = 1 + \frac{1}{2}\left[\frac{\partial^2 f}{\partial r^2}\right]_{r=0} r^2 + ... \tag{4.25}$$

Note that the terms with r^n (n being an odd number, $n = 1, 3, 5, ..$) will drop because $f(r) = f(-r)$ i.e. because $f(r)$ is symmetric about $r = 0$.

Now, the formal definition of the Taylor (1935) micro length scale of turbulence, λ_f, is

$$\frac{1}{\lambda_f^2} = -\frac{1}{2}\left[\frac{\partial^2 f}{\partial r^2}\right]_{r=0} \tag{4.26}$$

Inserting the preceding equation in Eq. 4.25, one gets

$$f(r) = 1 - \frac{r^2}{\lambda_f^2} \tag{4.27}$$

As seen from Fig. 4.9, this is a parabolic fit to the correlation function for small values of r, and λ_f is the x-intercept of the fitted parabola. Likewise,

$$g(r) = 1 - \frac{r^2}{\lambda_g^2} \tag{4.28}$$

in which λ_g is another micro length scale, formally defined according to

$$\frac{1}{\lambda_g^2} = -\frac{1}{2}\left[\frac{\partial^2 g}{\partial r^2}\right]_{r=0} \tag{4.29}$$

These length scales can also be expressed in terms of the fluctuating-velocity gradients:

$$\frac{1}{\lambda_f^2} = \frac{1}{2\overline{u'^2}}\overline{\left(\frac{\partial u'}{\partial r}\right)^2} \tag{4.30}$$

and

$$\frac{1}{\lambda_g^2} = \frac{1}{2\overline{v'^2}}\overline{\left(\frac{\partial v'}{\partial r}\right)^2} \tag{4.31}$$

These equivalencies are proved in Appendix 4.B.

It turns out that the micro length scales λ_f and λ_g are not independent of each other. In *isotropic turbulence*, from Eq. 4.24 we have:

$$\left.\frac{\partial^2}{\partial r^2} g(r)\right|_{r=0} = 2\left[\frac{\partial^2}{\partial r^2} f(r)\right]_{r=0} \tag{4.32}$$

Inserting Eqs. 4.27 and 4.28 then leads to the relationship

$$\lambda_f = \sqrt{2}\lambda_g \tag{4.33}$$

Taylor (1935) states that these scales "may roughly be regarded as a measure of the diameters of the smallest eddies which are responsible for the dissipation of energy," though this should not be interpreted as corresponding to the smallest scales in a turbulent flow. Rather, as will be discussed

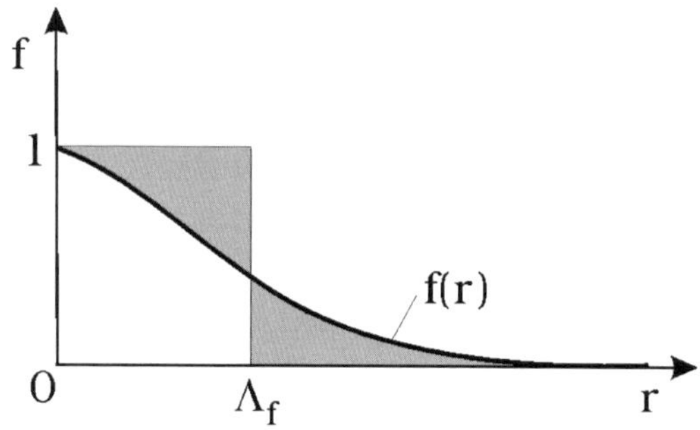

Fig. 4.10 Definition of the macro scale.

further in Section 4.3.3, these micro scales are often taken as characteristic of the so-called inertial subrange.

Scales of turbulence: Macro scales. The macro length scales of turbulence are defined by

$$\Lambda_f = \int_0^\infty f(r)dr \tag{4.34}$$

and

$$\Lambda_g = \int_0^\infty g(r)dr \tag{4.35}$$

This length scale is indicated graphically in Fig. 4.10 where the area of the rectangle $\Lambda_f \times 1$ is equal to the area under the correlation curve. These are also sometimes called the *integral scales* of turbulence. Λ_f or Λ_g can be interpreted as the characteristic length scales of the large coherent structures (or the so-called *eddies*, the rotating entities) of turbulence.

Similar to the micro scales, the macro scales are also related to each other. In *isotropic turbulence*, integrating Eq. 4.24 gives

$$\int_0^\infty g(r)dr = \int_0^\infty f(r)dr + \frac{1}{2}[rf(r) \mid_0^\infty - \frac{1}{2}\int_0^\infty f(r)dr \tag{4.36}$$

which leads simply to

$$\Lambda_f = 2\Lambda_g \tag{4.37}$$

4.2.2 *Time correlations*

There are two kinds of time correlations: Eulerian time correlations, and Lagrangian time correlations. The *Eulerian time correlation* is defined by

$$R_E(\tau) = \frac{\overline{u'(\tau)u'(\tau + t)}}{\overline{u'^2}} \tag{4.38}$$

where τ is the separation time between measurements. This is for the x-component of the velocity. Likewise, other Eulerian correlation function can also be defined for the y- and z- components of the velocity. Similar to space correlations, the time correlations satisfy

$$R_E(0) = 1 \tag{4.39}$$

$$R_E(\tau) \leq 1 \tag{4.40}$$

and

$$R_E(\tau) \to 0, \text{ as } \tau \to \infty \tag{4.41}$$

(Similar relations for the other components of the velocity likewise hold, should these be of interest.)

The micro and macro time scales can also be defined, similar to the space correlations. The Eulerian micro time scale is defined by

$$\frac{1}{\tau_E^2} = -\frac{1}{2}\left[\frac{\partial^2 R_E(\tau)}{\partial \tau^2}\right]_{\tau=0} \quad \Rightarrow \quad \tau_E = \left[-\frac{2}{(\partial^2 R_E(\tau)/\partial \tau^2)_{\tau=0}}\right]^{1/2} \tag{4.42}$$

(cf. Eq. 4.26), or

$$R_E(\tau) = 1 - \frac{\tau^2}{\tau_E^2} \tag{4.43}$$

(cf. Eq. 4.27). Analogous to Eq. 4.30, the micro time scale may equivalently be written as

$$\frac{1}{\tau_E^2} = \frac{1}{2\overline{u'^2}}\overline{\left(\frac{\partial u'}{\partial \tau}\right)^2} \quad \Rightarrow \quad \tau_E = \sqrt{\frac{2\overline{u'^2}}{\overline{(\partial u'/\partial \tau)^2}}} \tag{4.44}$$

The macro time scale, on the other hand, is defined by

$$T_E = \int_0^\infty R_E(\tau)d\tau \tag{4.45}$$

The macro time scale can be interpreted as the characteristic time it takes for a coherent structure (or an eddy) having characteristic size Λ_f to pass the measurement point.

The *Lagrangian time correlation* is defined in the same way as in the case of the Eulerian time correlation. However, the velocity is now measured by tracking the fluid particle, as the particle travels. Therefore, the Eulerian time series (which is the basis for the Eulerian time correlation) is to be replaced by the time series from the Lagrangian particle tracking. Then, all the equations given for the Eulerian correlation (Eqs. 4.38–4.45) are equally valid for the Lagrangian correlation.

So, for example, the Lagrangian correlation function (cf. Eq. 4.38) is defined as

$$R_L(\tau) = \frac{\overline{V'(\tau)V'(\tau + t)}}{\overline{V'^2}} \tag{4.46}$$

in which V' is the Lagrangian fluctuating velocity, and, for later use, the Lagrangian macro time scale is defined according to

$$T_L = \int_0^\infty R_L(\tau)\, d\tau \tag{4.47}$$

(cf. Eq. 4.45).

It may be noted that, in the laboratory, tracking of neutrally-buoyant particles can be achieved by means of an ordinary camera (or, two special cameras used in the field of photogrammetry for three-dimensional tracking), combined with a stroboscope (Sumer and Oguz, 1978, and Sumer and Deigaard, 1981, where tracking of small, but relatively heavy, particles was achieved), or using video techniques.

Of particular interest is the physical interpretation for the Lagrangian macro time scale. The Lagrangian macro time scale may be interpreted as the time during which a coherent structure maintains its "identity". Also, the length scale Λ_L, obtained by application of Eq. 4.56 below, may be interpreted as the characteristic distance traveled by coherent structures before breaking up.

4.2.3 *Taylor's frozen turbulence approximation*

We may expect that time and length scales discussed above are related. To find this relation, Taylor (1938) made the so-called *frozen-turbulence approximation*. To explain this notion, consider the coherent structure at time t in Fig. 4.11. The size of this coherent structure depicted is Λ_f, the macro length scale of turbulence. Consider the measurement point A. Now, under the frozen-turbulence approximation, the coherent structure is assumed to propagate with constant form (i.e. remain unchanged), as it is convected downstream at rate $\overline{u}$, the mean velocity at the measurement point. The time for the coherent structure to pass point A is T_E (by virtue of the physical interpretation in the previous section). Therefore, the distance that the coherent structure travels can, to a first approximation, be written as $\overline{u}T_E$. From the figure, this distance is equal to the size of the coherent structure, hence

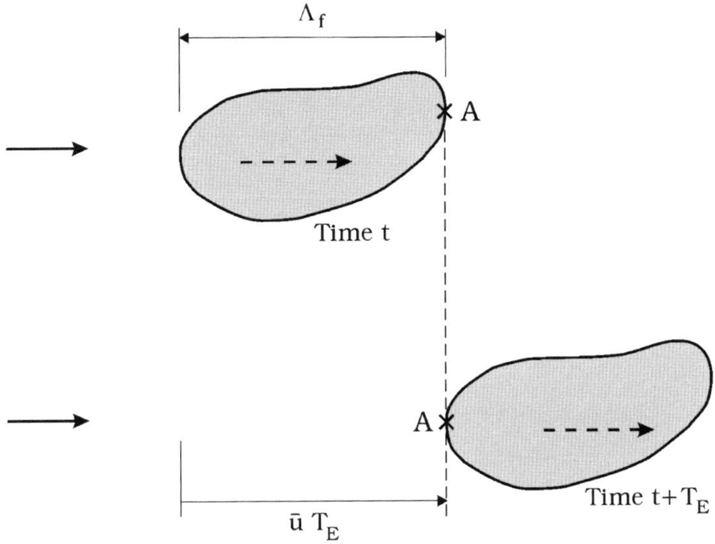

Fig. 4.11 Taylor's frozen turbulence approximation.

$$\Lambda_f = \overline{u}T_E \tag{4.48}$$

This relationship is important, as it enables the macro length scale to be calculated from knowledge of the corresponding time scale (the latter being much easier to measure than the former).

More generally, the frozen turbulence approximation can be mathematically stated as taking the material time derivative operator, d/dt, to be zero:

$$\frac{d}{dt} = \frac{\partial}{\partial t} + \overline{u}\frac{\partial}{\partial x} = 0 \qquad \text{or} \qquad \frac{\partial}{\partial t} = -\overline{u}\frac{\partial}{\partial x} \tag{4.49}$$

which is exact in circumstances of *pure advection* i.e. where coherent structures propagate without change in form (hence being "frozen"). Based on this relationship, the frozen turbulence approximation may similarly be applied to relate time derivatives to spatial derivatives.

A better understanding of the assumptions required to justify application of the frozen turbulence approximation, Eq. 4.49, may be obtained from inspecting the equations governing fluid motion. For the sake of simplicity, let us consider an idealized situation where a turbulent flow field is free of external pressure gradients, such that the mean flow $\overline{u}$ is both steady and spatially homogeneous. In this case, the streamwise component of the

Navier-Stokes equation becomes

$$\frac{\partial u'}{\partial t} + \overline{u}\frac{\partial u'}{\partial x} = -u_i'\frac{\partial u'}{\partial x_i} - \frac{1}{\rho}\frac{\partial p'}{\partial x} + \nu\frac{\partial^2 u'}{\partial x_k \partial x_k} \tag{4.50}$$

For the frozen turbulence approximation, Eq. 4.49, to be valid it is clear that the terms on the right hand side must be small relative to those on the left hand side. To formally investigate the relative importance of the individual terms, it is convenient to make the following replacements

$$u_i' \to \sigma_u u_i'^*, \qquad x_i \to \lambda x_i^*, \qquad t \to \frac{\lambda}{\overline{u}}t^*, \qquad p' \to \rho\sigma_u^2 p'^* \tag{4.51}$$

within Eq. 4.50, where an asterisk superscript denotes a (properly scaled) dimensionless variable, and λ is a generic characteristic length scale for an arbitrary coherent structure of interest. Simplifying the result then leads to the non-dimensionalized form of Eq. 4.50:

$$\frac{\partial u'^*}{\partial t^*} + \frac{\partial u'^*}{\partial x^*} = \frac{\sigma_u}{\overline{u}}\left(-u_i'^*\frac{\partial u'^*}{\partial x_i^*} - \frac{\partial p'^*}{\partial x^*}\right) + \frac{\nu}{\lambda\overline{u}}\frac{\partial^2 u'^*}{\partial x_k^* \partial x_k^*}, \tag{4.52}$$

From the above, it is now clear that for the frozen turbulence approximation to be valid, the *turbulence intensity*, $\sigma_u/\overline{u}$, must be small, i.e. that

$$\frac{\sigma_u}{\overline{u}} = \frac{\sqrt{\overline{u'^2}}}{\overline{u}} \ll 1 \tag{4.53}$$

Moreover, for the viscous terms to be negligible, the frozen turbulence approximation likewise requires that the Reynolds number for the scales in question be large, i.e. that

$$Re_\lambda = \frac{\lambda\overline{u}}{\nu} \gg 1 \tag{4.54}$$

For sufficiently large Reynolds numbers, the latter condition may be satisfied for all length scales in a turbulent flow. On this basis we may further generalize Eq. 4.48 to also include e.g. the micro scales, leading to the following approximations:

$$\overline{u} = \frac{\Lambda_f}{T_E} = \frac{\lambda_f}{\tau_E} = \frac{\omega}{k} = \frac{2\pi f}{k} = \frac{r}{\tau} \tag{4.55}$$

where ω is a characteristic angular frequency, f is the corresponding ordinary frequency, and $k = 2\pi/\lambda$ is a characteristic wave number (these quantities will be discussed at length in Section 4.3). Note that utilizing the relationship between r and τ above similarly enables e.g. approximation of the spatial correlation function $f(r)$ from the time correlation function $R_E(\tau)$. Similarly, the frozen turbulence approximation is likewise commonly utilized e.g. to convert frequency spectra to wave number spectra.

Such spectral analysis is covered in detail within Section 4.3. (For a more advanced method for the temporal-to-spatial mapping of turbulent spatial structures and spectra see e.g. Buchhave and Velte, 2017).

Analogously, the Lagrangian macro time and length scales can be related by

$$\overline{V} = \frac{\Lambda_L}{T_L} \qquad \text{or} \qquad \Lambda_L = \overline{V} T_L \tag{4.56}$$

where $\overline{V}$ is the mean Lagrangian particle velocity (in a turbulent open channel or pipe flow this would be equal to the cross-sectionally averaged velocity, see Chapter 8). The Lagrangian macro length scale, Λ_L, can be interpretted physically as discussed at the end of Section 4.2.2.

4.2.4 *Matlab example: Time correlation and scales*

We now present a Matlab example computing the Eulerian time correlation function, Program 3: `ExampleCorrelation.m`. The script again uses a synthetic turbulent signal generated using `turbgen.m` based on the same input as used in Program 2. The time correlation function R_E is computed discretely for each discrete time lag index (variable `n` in the script), and in this example the calculation is continued while $R_E > 0$ (small negative values will eventually be encountered when the function is near zero, due to the finite sample size), though this criterion could easily be modified. (Note also that the built in Matlab function `xcorr` may also be utilized to compute the correlation function in practice, though we prefer the present example, as it is more illustrative.) Based on the correlation function, both the macro and micro time scales are computed, resulting in the screen output (from lines 17, 30 and 31):

```
T_E =

    0.2019

tau_E =

    0.0514

tau_E =

    0.0514
```

All of the above would have units of s, provided that S.I. units are used, as presumed in the example. Note that the micro time scale calculated from both Eqs. 4.42 and 4.44 (lines 30 and 31), which are indeed seen to be equivalent from the screen output above.

The resulting Eulerian time correlation function R_E is depicted graphically in Fig. 4.12 (top) where it is seen to match precisely the function

$$R_E = \exp\left(-\frac{\tau}{T_E}\right) \tag{4.57}$$

This is the known correlation function corresponding to the synthetic signal analyzed in this example (corresponding to a Dryden spectrum, discussed later in Section 4.3.5). In more general (i.e. real) cases, it is emphasized that this should only be considered as a reference curve, and not a target.

Fig. 4.12 (bottom) depicts a zoomed in portion of the correlation function, where the calculated macro and micro time scales are presented graphically, in line with Figs. 4.9 and 4.10 (there shown for the analogous space

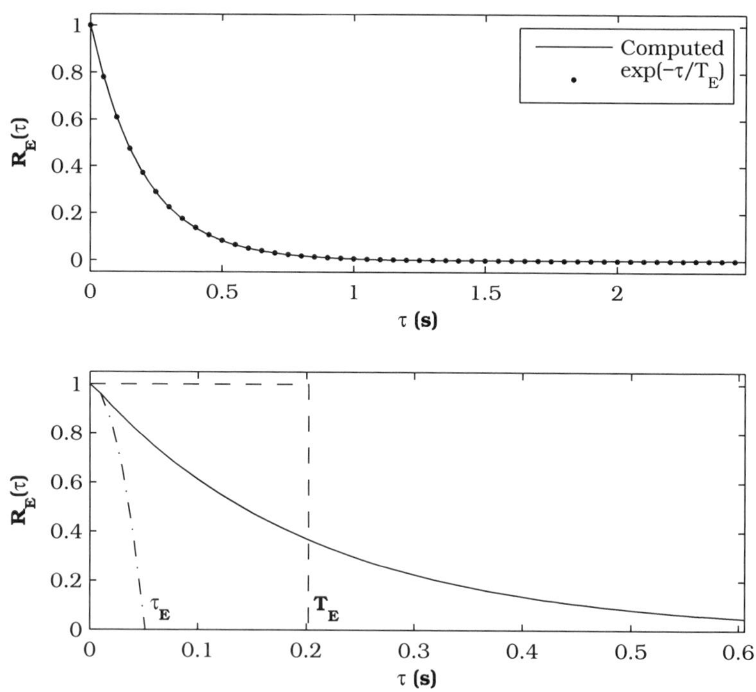

Fig. 4.12 Output from Matlab script `ExampleCorrelation.m`.

Program 3: ExampleCorrelation.m

```
% Generate time series using turbgen.m
N = 1e5; T_E = 0.2; dt = 0.01; sigma2 = 0.004;
[u,t] = turbgen(N,dt,sigma2,T_E,1,0,1842);
sigma2 = var(u); % Replace w/ sample var. (m^2/s^2)

% Compute / plot correlation coefficient
n = 1; R_E(n) = mean(u.^2)/sigma2; % Initialize
while R_E(n) > 0 % Compute while positive
    n = n + 1; % Update counter
    R_E(n) = mean(u(1:end-n+1).*u(n:end))/sigma2;
end
subplot(2,1,1); plot(t(1:n),R_E,'k-'); hold on;
xlabel('\tau (s)'); ylabel('R_E(\tau)');
xlim([0 t(n)]); ylim([-0.05 1.05])

% Calculate macro time scale & compare w/ target
T_E = trapz(t(1:n),R_E) % Compute macro time scale (s)
plot(t(1:5:end),exp(-t(1:5:end)/T_E),'k.'); hold off;
legend('Computed','exp(-\tau/T_E)');

% Zoom in & illustrate T_E graphically
subplot(2,1,2); plot(t(1:n),R_E,'k-');
xlim([0 3*T_E]); ylim([0 1.05])
xlabel('\tau (s)'); ylabel('R_E(\tau)');
hold on; plot([0 T_E T_E],[1 1 0],'b--');
text(T_E+0.005,0.07,'T_E')

% Calculate / illustrate micro time scale
R_tt = 2*(R_E(2)-R_E(1))/dt^2; % R_E''(tau) at tau=0
tau_E = 1/sqrt(-0.5*R_tt) % Micro time scale (s)
tau_E = sqrt(2*sigma2/mean((diff(u)/dt).^2))
  % The two estimations of tau_E above are equivalent
plot(t,1-t.^2/tau_E^2,'r-.'); hold off;
text(tau_E+0.005,0.07,'\tau_E')
```

scales). The computed macro time scale is expectedly very close to the input value $T_E = 0.2$ s, with the micro time scale being well below this value. Portions of this script can be easily modified to compute correlation functions and micro and macro scales for other discrete time or space series.

4.3 Spectral analysis

In turbulent flows, there is a continuous, broad range of length/time scales. For example, smoke coming out of a smoke stack has flow structures with varying sizes; Larger eddies (coherent structures) contain smaller eddies; these smaller eddies contain even smaller eddies, and so on. The correlation analysis described in the previous section cannot address the issue of the previously mentioned continuous, broad range of length/time scales. (It provides only certain characteristic length and/or time scales e.g. the micro and the macro scales, but nothing more). This continuous, broad range of scales can instead be studied by use of *spectral analysis*, the final main topic to be covered in this chapter. For convenience, much of the theoretical analysis will be developed specifically for isotropic turbulence.

As might by now be expected, there are two kinds of spectral analysis. One involves the space correlations, and the other involves time correlations. The spectrum function in the first case is resolved in the wave number domain, while that in the second case is resolved in the frequency domain. We will primarily consider in this section the former spectrum analysis. However, the analysis in the latter is analogous, provided that the wave number is replaced by the frequency. (In fact, these can be related to one another via the frozen turbulence approximation, as discussed in Section 4.2.3.)

4.3.1 *General considerations*

The Fourier transform of a two-point double correlation Q_{ij} (see Eq. 4.13 for the definition of Q_{ij}) is

$$\Phi_{ij}(\mathbf{k}) = \frac{1}{(2\pi)^3} \int_{-\infty}^{\infty} Q_{ij}\,(\mathbf{r})\, e^{-i\mathbf{k}\cdot\mathbf{r}} d\mathbf{r} \tag{4.58}$$

in which $\mathbf{r}$ is the separation vector (Fig. 4.13), $d\mathbf{r} = dr_1 dr_2 dr_3$, $\mathbf{k}$ is the wave number vector and $i = \sqrt{-1}$ is the imaginary unit. The inverse transform,

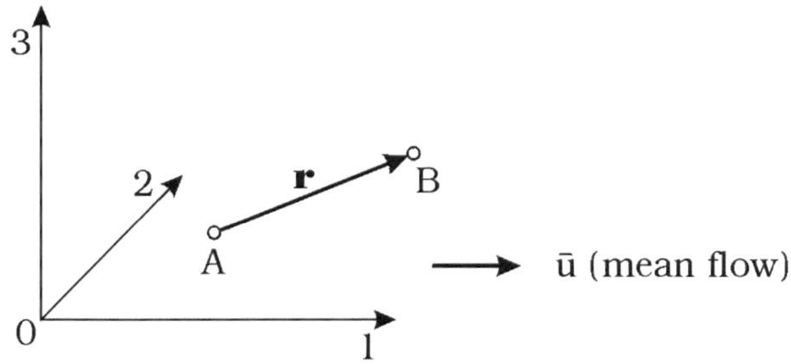

Fig. 4.13 Definition sketch.

on the other hand, is

$$Q_{ij}\ (\mathbf{r}) = \int_{-\infty}^{\infty} \Phi_{ij}(\mathbf{k})\ e^{i\mathbf{k}\cdot\mathbf{r}} d\mathbf{k} \tag{4.59}$$

in which $d\mathbf{k} = dk_1 dk_2 dk_3$ is an element of volume in wave number space.

Now, for $r = 0$, the preceding equation becomes

$$Q_{ij}\ (0) = \int_{-\infty}^{\infty} \Phi_{ij}(\mathbf{k})\ d\mathbf{k}$$

Setting $i = j$ then leads to

$$Q_{ii}(0) = \int_{-\infty}^{\infty} \Phi_{ii}(\mathbf{k})\ d\mathbf{k} \tag{4.60}$$

Here, $Q_{ii}(0)$ is two times the mean kinetic energy of turbulence per unit mass

$$Q_{ii}(0) = \overline{u_i' u_i'} = 2\ \left[\frac{1}{2}(\overline{u'^2} + \overline{v'^2} + \overline{w'^2})\right] \tag{4.61}$$

We write Eq. 4.60 in the following form

$$Q_{ii}(0) = \int_{-\infty}^{\infty}\int_{-\infty}^{\infty}\int_{-\infty}^{\infty} \Phi_{ii}(k_1, k_2, k_3)\ dk_1 dk_2 dk_3 \tag{4.62}$$

Since the left hand side of the above equation is proportional to the turbulent energy, then $\Phi_{ii}(k_1, k_2, k_3)$ will be like the energy density in the wave number space. We shall return to this in the following paragraphs.

In the case of *isotropic turbulence*, there will be spherical symmetry. Therefore, the spherical coordinates can be used in the wave number space. Eq. 4.62 in this case will read

$$\begin{aligned} Q_{ii}(0) &= \int_{k=0}^{\infty}\int_{A(k)} \Phi_{ii}(k)\ dk\ dA(k) \\ &= \int_{k=0}^{\infty} \Phi_{ii}(k)\ dk \int_{A(k)} dA(k) = \int_{k=0}^{\infty} (4\pi k^2)\Phi_{ii}(k)\ dk \end{aligned}$$

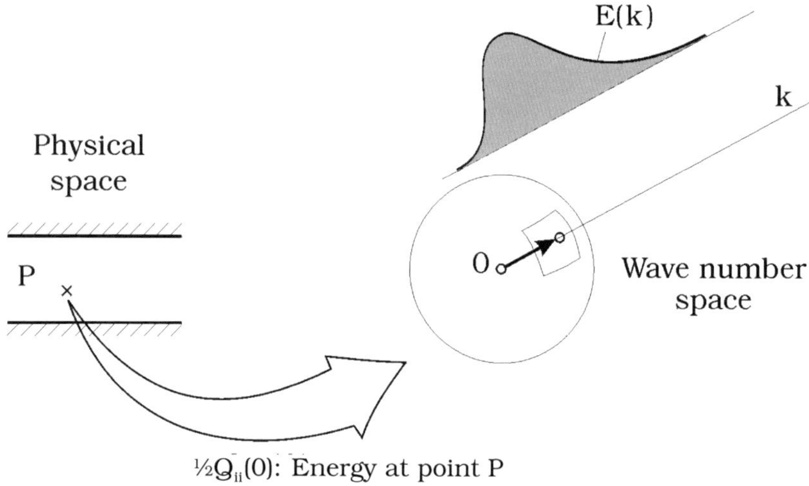

Fig. 4.14 Energy spectrum. Isotropic turbulence.

(where $4\pi k^2$ is the surface area of the sphere in k space) or

$$\frac{1}{2}Q_{ii}(0) = \frac{1}{2}\overline{u_i'u_i'} = \frac{K_t}{\rho} = \int_0^\infty E(k)\, dk \tag{4.63}$$

in which

$$E(k) = 2\pi k^2 \Phi_{ii}(k) \tag{4.64}$$

We emphasize that in the immediately above $k = \sqrt{k_1^2 + k_2^2 + k_3^2}$ should be interpreted as the modulus of the wave number vector $\mathbf{k}$. It is hence strictly positive, though any of the individual components (k_1, k_2, k_3) may take positive or negative values, as indicated in some of the integrals above.

(1) Eq. 4.63 is an especially important equation! As pointed out in the preceding paragraphs, the left hand side of the equation is the turbulent kinetic energy (per unit mass), therefore, $E(k)$ will be the energy density in the wave number space k. The function $E(k)$ is called the *energy spectrum function.*
(2) It may be noted that the spherical symmetry enables us to replace the three-dimensional spectrum function $\Phi_{ii}(k_1, k_2, k_3)$ (Eq. 4.62) with a one-dimensional spectrum function, $E(k)$.
(3) The relation given in Eq. 4.63 can be envisaged as follows (see the sketch in Fig. 4.14): The velocity time series of the three components of the velocity are recorded at the measurement point P. From these

time series, the turbulent energy per unit mass, $\frac{1}{2}Q_{ii}(0)$, is calculated, using Eq. 4.61. Then, according to Eq. 4.63, this energy is decomposed into infinite number of small energy components, i.e., $E(k)\,dk$, in the wave number space. Each of these small energies corresponds to a wave number k, which itself corresponds to a wave length (or to a turbulence "length scale")

$$\lambda = \frac{2\pi}{k} \tag{4.65}$$

As seen, the objective set at the beginning of this section, namely the study of the continuous, broad range of length scales of turbulence, apparently can be achieved by the application of the energy spectrum function $E(k)$. In the following section, we will have a closer look at this function.

4.3.2 *Energy balance in wave number space*

In *isotropic turbulence*, the previously mentioned correlation function $Q_{ij}(\mathbf{r})$ satisfies the following equation (see e.g. Hinze (1959), p. 159, and Raudkivi and Callander, 1975, p. 163)

$$\frac{\partial}{\partial t}Q_{ij}(\mathbf{r}) = S_{ij}(\mathbf{r}) + 2\nu\frac{\partial^2}{\partial r_k \partial r_k}Q_{ij}(\mathbf{r}) \tag{4.66}$$

in which

$$S_{ij}(\mathbf{r}) = \frac{\partial}{\partial r_k}(S_{ik,j} + S_{kj,i}) \tag{4.67}$$

in which

$$(S_{ik,j})_{AB} = \overline{(u_i')_A(u_k')_A(u_j')_B} \tag{4.68}$$

and

$$(S_{i,kj})_{AB} = \overline{(u_i')_A(u_k')_B(u_j')_B} \tag{4.69}$$

It should be noted that Eq. 4.66 stems from the N.-S. equation, and a detailed formal proof of the above is provided in Appendix 4.C.

The Fourier transform of Eq. 4.66 is

$$\frac{\partial}{\partial t}\Phi_{ij}(\mathbf{k}) = \Omega_{ij}(\mathbf{k}) - 2\nu k^2\Phi_{ij}(\mathbf{k}) \tag{4.70}$$

in which

$$\Omega_{ij}(\mathbf{k}) = \frac{1}{(2\pi)^3}\int_{-\infty}^{\infty} S_{ij}(\mathbf{r})\; e^{-i\mathbf{k}\cdot\mathbf{r}} d\mathbf{r} \tag{4.71}$$

and the inverse transform reads

$$S_{ij}(\mathbf{r}) = \int_{-\infty}^{\infty} \Omega_{ij}(\mathbf{k})\, e^{-i\mathbf{k}\cdot\mathbf{r}} d\mathbf{k} \tag{4.72}$$

Now, putting $i = j$ in Eq. 4.70 gives

$$\frac{\partial}{\partial t}\Phi_{ii}(\mathbf{k}) = \Omega_{ii}(\mathbf{k}) - 2\nu k^2 \Phi_{ii}(\mathbf{k}) \tag{4.73}$$

and multiplying both sides of the latter equation by $2\pi k^2$, and recalling that $E(k) = 2\pi k^2 \Phi_{ii}(k)$ (cf. Eq. 4.64), one obtains

$$\frac{\partial}{\partial t}E(k) = T(k) - 2\nu k^2 E(k) \tag{4.74}$$

in which

$$T(k) = 2\pi k^2 \Omega_{ii}(k) \tag{4.75}$$

This equation (Eq. 4.74) represents the "turbulent energy balance" in the wave number space.

To study the physical meaning of each term in the equation, we integrate Eq. 4.74 from $k = 0$ to ∞:

$$\frac{\partial}{\partial t}\int_0^{\infty} E(k)dk = \int_0^{\infty} T(k)dk - 2\nu \int_0^{\infty} k^2 E(k)dk \tag{4.76}$$

Now Hinze (1959), p. 178, has shown that:

$$\int_0^{\infty} T(k)dk = 0 \tag{4.77}$$

Our own formal proof of this is offered in Appendix 4.D. Hence Eq. 4.76 reduces to

$$\frac{\partial}{\partial t}\int_0^{\infty} E(k)dk = -2\nu \int_0^{\infty} k^2 E(k)dk \tag{4.78}$$

Recall that $\int_0^{\infty} E(k)dk$ is the turbulent energy per unit mass (Eq. 4.63); Hence, the left hand side of the above equation will represent the rate of change of the turbulent energy. Now, the integral on the right hand side of the equation, $\int_0^{\infty} k^2 E(k)dk$, is always positive. This means that

$$\frac{\partial}{\partial t}\int_0^{\infty} E(k)dk < 0 \tag{4.79}$$

From the above analysis, the following deductions can be made:

(1) Turbulent energy in isotropic turbulence decreases with time (Eq. 4.79). This is due to viscous dissipation, governed by the term on the right hand side of Eq. 4.78.

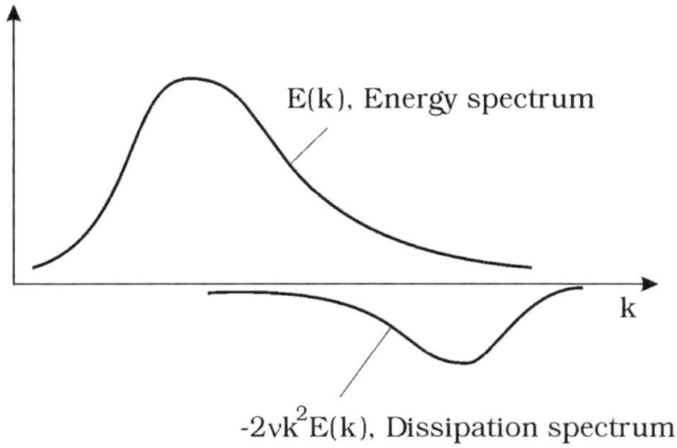

Fig. 4.15 Energy spectrum and dissipation spectrum. (Note that peaks are well apart).

(2) As seen from Eq. 4.78, the latter term is weighted by the factor k^2, and clearly is more effective at large wave numbers (i.e., at small size eddies) than at small wave numbers (i.e., at large size eddies), Fig. 4.15.
(3) The fact that the first term on the right hand side in Eq. 4.76 is zero implies that it is neither a source term nor a sink term. It can be shown that this term, namely $T(k)$, is responsible for the *transfer* of turbulent energy from small wave numbers to large wave numbers (see Example 4.1 below).

Summarizing at this point:

(1) Most of the turbulent energy is contained in small wave numbers (large eddies, or large scale motion, Fig. 4.15);
(2) Most of the dissipation occurs at large wave numbers (small eddies, or small scale motion, Fig. 4.15); and
(3) Between the small wave numbers and large wave numbers, there is a range of wave numbers responsible for transferring turbulent energy from small to large wave numbers.

Example 4.1. Transfer of turbulent energy in the wave number space.

Integrate Eq. 4.74 from $k = 0$ to k:

$$\frac{\partial}{\partial t}\int_0^k E(k)dk = \int_0^k T(k)dk - 2\nu\int_0^k k^2 E(k)dk \tag{4.80}$$

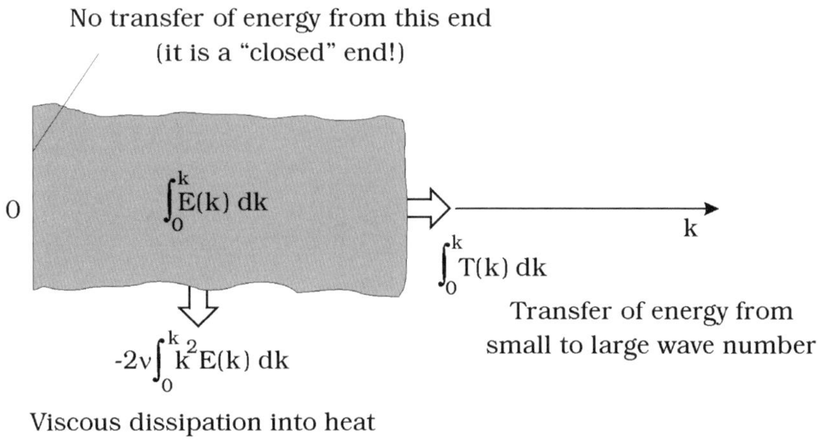

Fig. 4.16 Transfer of energy in wave number space.

Consider the interval in the wave number space from $k = 0$ to k (Fig. 4.16). The first term on the left hand side of the above equation represents the change in the turbulent energy in this interval per unit time. The third term represents the viscous dissipation in this interval per unit time (Fig. 4.16).

Note that as we are no longer integrating to $k = \infty$, the term involving $T(k)$ can no longer be taken as equal to zero. Rather, experiments show that the sign of the term $\int_0^k T(k)dk$ is the same as that of the viscous dissipation. Therefore, $\int_0^k T(k)dk$ should be an efflux, rather than an influx. This efflux cannot take place at the $k = 0$ end of the considered interval, because $k = 0$ forms a closed end (the wave number modulus cannot be negative!) Therefore, this efflux must take place at the other end of the interval, namely at the arbitrary upper wave number k (Fig. 4.16).

So, the picture is as follows: In the interval $(0, k)$ in the wave-number domain, some part of the turbulent energy is dissipated into heat (viscous dissipation), and some part is transferred to wave numbers larger than k. The latter means that *on average, there is a continuous flow of turbulent energy from small wave numbers to large wave numbers* i.e. a so-called *energy cascade*, as first described by Richardson (1926). The term responsible for this is $T(k)$. This term originates from the inertial (nonlinear convective) terms in the N.-S. equations. Therefore, the previously mentioned transfer of energy is sometimes called the *inertial transfer*.

4.3.3 *Kolmogorov's theory: Universal equilibrium range and inertial subrange*

In the process of turbulent energy transfer from small wave numbers (large-scale motion) to large wave numbers (small-scale motion), a point is reached beyond which the turbulence will "forget" the influence of the initial large-scale motion. From this point to the wave numbers where the viscous dissipation occurs, there is a range of wave numbers in which the character of turbulence will be the same in all flows (in streams, jets, atmospheric boundary-layer turbulence, etc.). This range is called the *universal equilibrium range.* It is called universal because the character of turbulence in this range is independent of the flow environment; It is called equilibrium because the energy input is in balance with the energy dissipation, as will be detailed later in this section. Kolmogorov's celebrated theory is concerned with this range.

Kolmogorov's first hypothesis states that, irrespective of the large-scale motion, the turbulence in the universal equilibrium range is isotropic. Now, on dimensional grounds, the energy spectrum function in this range is a function of the following quantities

$$E = f(\varepsilon,\ \nu,\ k) \tag{4.81}$$

in which ε is the dissipation rate of turbulent kinetic energy per unit mass (thus having units of turbulent kinetic energy per unit mass per time i.e. m^2s^{-3} in S.I. units):

$$\varepsilon = \frac{\Phi_t}{\rho} = \nu \overline{\left(\frac{\partial u_i'}{\partial x_\alpha} + \frac{\partial u_\alpha'}{\partial x_i}\right)\frac{\partial u_i'}{\partial x_\alpha}} \tag{4.82}$$

As shown above, the quantity ε is closely related to the quantity Φ_t defined previously in Eq. 2.33. Note that in isotropic turbulence, the dissipation from Eq. 4.82 can be expressed in terms of either Taylor micro scale according to:

$$\varepsilon = 30\nu\frac{\overline{u_1'^2}}{\lambda_f^2} = 15\nu\frac{\overline{u_1'^2}}{\lambda_g^2} \tag{4.83}$$

This is formally proved in Appendix 4.G, and can be utilized to obtain an estimate of ε in real flows once the micro scale has been determined from a correlation analysis. The interpretation of Eq. 4.83 will be discussed in more detail later in this section.

Representing the dissipation rate of turbulent energy, the quantity ε must obviously be involved. The kinematic viscosity ν should be involved because the energy is dissipated into heat through viscosity. The wave number, k, is involved because the turbulent energy process is primarily dependent on the wave number (Eq. 4.74). (k is taken instead of k_1, k_2 and k_3 because of the spherical symmetry allowed in isotropic turbulence!). Note that E (rather than Φ_{ii}) is adopted here to represent the energy spectrum because we assume the turbulence to be isotropic.

From dimensional analysis, one gets

$$\frac{E(k)}{\varepsilon^{1/4}\nu^{5/4}} = \frac{E(k)}{\upsilon_K^2 \eta_K} = \theta(\eta_K k) \qquad \text{or} \qquad E(k) = \upsilon_K^2 \eta_K \cdot \theta(\eta_K k) \quad (4.84)$$

in which θ is a non-dimensional function of $\eta_K k$, the quantity υ_K is

$$\upsilon_K = (\nu\varepsilon)^{1/4} \quad (4.85)$$

and η_K is

$$\eta_K = \left(\frac{\nu^3}{\varepsilon}\right)^{1/4} \quad (4.86)$$

Here the S.I. unit of υ_K is

$$[\upsilon_K] = \left(\frac{\mathrm{m}^2}{s}\frac{\mathrm{m}^2}{s}\left(\frac{\mathrm{m}}{\mathrm{s}\cdot\mathrm{m}}\right)^2\right)^{1/4} = \frac{\mathrm{m}}{\mathrm{s}}$$

and that of η_K is

$$[\eta_K] = \left(\left(\frac{\mathrm{m}^2}{\mathrm{s}}\right)^3 \frac{1}{\frac{\mathrm{m}^2}{\mathrm{s}}\left(\frac{\mathrm{m}}{\mathrm{s\cdot m}}\right)^2}\right)^{1/4} = \mathrm{m}$$

The quantity υ_K is hence a characteristic velocity (the Kolmogorov velocity scale), while η_K is a characteristic length (the Kolmogorov length scale). Based on the quantities υ_K and η_K, the following characteristic time scale may likewise be defined

$$\tau_K = \frac{\eta_K}{\upsilon_K} = \left(\frac{\nu}{\varepsilon}\right)^{1/2} \quad (4.87)$$

The above characteristic length η_K and time τ_K (the Kolmogorov time scale) represent the large wave number range (small scale motion) where viscous dissipation occurs i.e. turbulent kinetic energy is converted into heat. This can be most easily seen by defining a Reynolds number based on these characteristic scales

$$Re_K = \frac{\upsilon_K \eta_K}{\nu} = 1 \quad (4.88)$$

This is, essentially by definition, unity, clearly indicating the importance of viscosity. This can be interpreted to mean that turbulent eddies having characteristic size η_K will be dissipated into heat in a single time scale τ_K. These *Kolmogorov scales* thus represent the smallest scales of turbulence; Scales smaller than these in a given flow will involve molecular motion.

The non-dimensional equation obtained for the spectrum function (Eq. 4.84) can be developed further, following Kolmogorov's second hypothesis. This states that the peaks of the energy spectrum and that of the dissipation spectrum are sufficiently apart. This is illustrated in Fig. 4.15. This implies that there is a dominating wave number range in the energy spectrum (small wave numbers i.e. large eddies) which contain most of the turbulent energy, while there is another dominating wave number range (large wave numbers i.e. small eddies) which are associated with the viscous dissipation. According to Kolmogorov's second hypothesis, these wave number ranges are sufficiently apart, meaning that there must be a wave number range in between these which does not contain significant energy, and where viscous dissipation is also insignificant.

First of all, this range should clearly be a part of the universal equilibrium range, because it does not house a significant portion of the turbulent energy (i.e. the bulk of the energy is assumed to be in the energy containing range). Second, in this range, the only significant activity is that the turbulent energy is transferred from small wave numbers to large wave numbers. This process is governed by the inertial (convective acceleration) terms in the N.-S. equation, as described in Example 4.1 above. Therefore, this particular range of wave numbers is called the *inertial subrange.* ("inertial" because of the previously mentioned inertia terms; and "subrange" because this range is a part of the larger universal equilibrium range, as pointed out in the preceding paragraphs).

Now, in the inertial subrange, the energy spectrum should be independent of the viscosity because no significant dissipation takes place in this range, as mentioned previously. This is an important statement: For the viscosity to drop out in Eq. 4.84, the function $\theta(\eta_K k)$ must take the following form

$$\theta(\eta_K k) = \text{Constant} \times (k\eta_K)^{-\frac{5}{3}} \tag{4.89}$$

Thus, inserting Eq. 4.89 in Eq. 4.84, the spectrum function is found to be

$$E(k) = \text{Constant} \;\times \varepsilon^{\frac{2}{3}}\; k^{-\frac{5}{3}} \tag{4.90}$$

This is known as Kolmogorov's $-5/3$ law (Kolmogorov, 1941). Experiments widely confirm the presence of this subrange where the spectrum is given by the $-5/3$ law (provided that the Reynolds number is sufficiently large

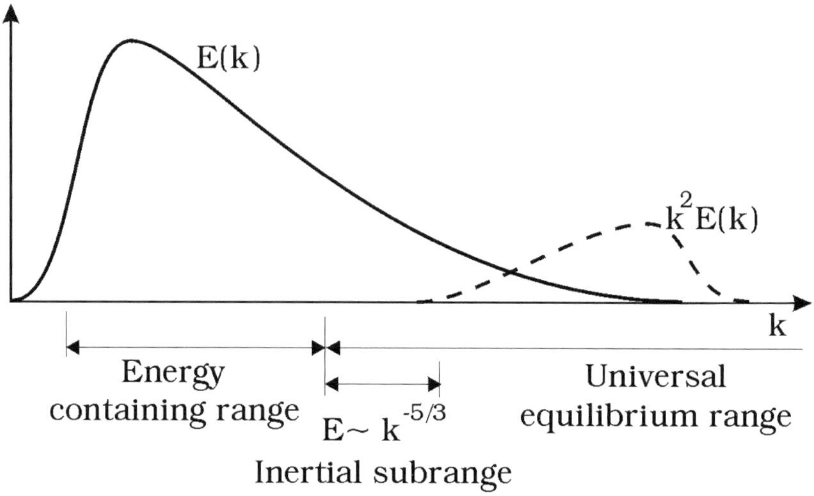

Fig. 4.17 General picture of energy spectrum.

such to yield the assumed wide range of scales). Finally, Fig. 4.17 gives a schematic picture of the spectral representation of the process.

Note finally that from Eq. 4.83 it is seen that the rate of dissipation is governed by the micro scales. This should not be taken, however, to mean that most energy is dissipated in eddies having this size. To see this, following similar arguments as e.g. in Tennekes and Lumley (1972) (their p. 21), let us consider an arbitrary (but large) length scale $\mathcal{L}$ defined on dimensional grounds as:

$$\mathcal{L} = \frac{\sigma_u^3}{\varepsilon}, \qquad Re_{\mathcal{L}} = \frac{\mathcal{L}\sigma_u}{\nu}, \qquad \sigma_u = \sqrt{\overline{u_1'^2}} \tag{4.91}$$

and with associated Reynolds number $Re_{\mathcal{L}}$ defined above. Invoking Eq. 4.86, following the steps below, leads to the ratio

$$\frac{\mathcal{L}}{\eta_K} = \frac{\mathcal{L}}{(\nu^3/\varepsilon)^{1/4}} = \mathcal{L}\left(\frac{\sigma_u^3}{\mathcal{L}\nu^3}\right)^{1/4} = \left(\frac{\sigma_u\mathcal{L}}{\nu}\right)^{3/4} = Re_{\mathcal{L}}^{3/4} \tag{4.92}$$

Similarly, invoking Eq. 4.83 yields the ratio

$$\frac{\mathcal{L}}{\lambda_f} = \frac{\mathcal{L}\varepsilon^{1/2}}{\sqrt{30}\nu^{1/2}\sigma_u} = \frac{\mathcal{L}\left(\sigma_u^3/\mathcal{L}\right)^{1/2}}{\sqrt{30}\nu^{1/2}\sigma_u} = \frac{1}{\sqrt{30}}\left(\frac{\sigma_u\mathcal{L}}{\nu}\right)^{1/2} = \frac{1}{\sqrt{30}}Re_{\mathcal{L}}^{1/2} \tag{4.93}$$

Hence, the ratio of the micro and Kolmogorov length scales is

$$\frac{\lambda_f}{\eta_K} = \frac{\lambda_f}{\mathcal{L}}\frac{\mathcal{L}}{\eta_K} = \frac{\sqrt{30}}{Re_{\mathcal{L}}^{1/2}}Re_{\mathcal{L}}^{3/4} = \sqrt{30}Re_{\mathcal{L}}^{1/4} \tag{4.94}$$

The factor $\sqrt{30}$ in the above is dependent on the definition of $Re_{\mathcal{L}}$ as well as which micro scale (λ_f or λ_g) is utilized. Regardless, from the above it is clear that for large Reynolds numbers typical of turbulent flows, the Taylor micro scale will be much larger than the Kolmogorov scale η_K, and that the large and small scales in a turbulent flow will become more widely separated as the Reynolds number is increased. Being intermediate in size (i.e. smaller than large energy containing eddies yet larger than Kolmogorov scales) this likewise explains why the micro scales λ_f and λ_g typically lie in the inertial subrange (see also Wilcox, 2006, p. 49), as was mentioned previously in Section 4.2.1. That the dissipation in Eq. 4.83 is governed by the micro scales is therefore likewise consistent with the notion of *equilibrium* within the universal equilibrium range i.e. that turbulent kinetic energy cascaded into the inertial subrange is equal to that dissipated at the much smaller Kolmogorov scales.

The preceding section has given a brief synopsis on some of the major contributions of Kolmogorov, which helped form the origin of modern turbulence theory. For a more detailed account of Kolmogorov's contributions to turbulence, see e.g. Jimenez (2004).

4.3.4 *Vortex stretching*

The physical mechanism resulting in the energy cascade is so-called *vortex stretching*. To explain this, let us consider a simple, but enlightening, flow situation where the velocity field is

$$u = sx, \qquad v = -sy, \qquad w = \omega_{01}y - \omega_{02}x \tag{4.95}$$

The corresponding initial vorticity vector, e.g. Eq. 2.66 or Eq. 6.89, is

$$\omega_i = (\omega_1, \omega_2, \omega_3) = (\omega_{01}, \omega_{02}, 0) \tag{4.96}$$

whereas the normal components of the strain rate tensor, Eq. 2.64, reduce to

$$s_{11} = \frac{\partial u}{\partial x} = s, \qquad s_{22} = \frac{\partial v}{\partial y} = -s, \qquad s_{33} = \frac{\partial w}{\partial z} = 0 \tag{4.97}$$

Hence, we have an idealized condition where the first vorticity component ω_{01} is in a direction having positive strain rate s, whereas the component ω_{02} is in the direction of negative strain rate $-s$.

Let us now assume that the scales being considered are within the inertial subrange, such that viscous effects can be neglected. The vorticity equation, Eq. 6.95 (derived in full in the Appendix to Chapter 6), here written in terms of instantaneous flow quantities, then reduces to

$$\frac{d\omega_i}{dt} = \omega_j \frac{\partial u_i}{\partial x_j} = \omega_1 \frac{\partial u_i}{\partial x_1} + \omega_2 \frac{\partial u_i}{\partial x_2} \tag{4.98}$$

The single tensor term above $\omega_j \partial u_i / \partial x_j$ is known as the *vortex stretching term* for reasons that will be made clear. For the conditions considered Eq. 4.98 reduces to the components

$$\frac{d\omega_1}{dt} = s\omega_1, \qquad \frac{d\omega_2}{dt} = -s\omega_2, \qquad \frac{d\omega_3}{dt} = \omega_1\omega_2 - \omega_2\omega_1 = 0 \tag{4.99}$$

This system yields the general solution:

$$\omega_1 = \omega_{01} e^{st}, \qquad \omega_2 = \omega_{02} e^{-st}, \qquad \omega_3 = 0 \tag{4.100}$$

where ω_{01} and ω_{02} are the initial vorticities in the i = 1,2 directions.

From Eq. 4.100 it is now clear that the vorticity component in the direction of negative strain rate will exponentially decay, whereas *the component in the direction of positive strain rate will exponentially increase.* Or in other words, any initial vorticity vector with component $\omega_1 > 0$ will evolve such that it eventually points in the direction of positive strain. As such, over time, this example demonstrates that the vorticity vector will tend to become aligned with the direction of positive strain rate, i.e. such that the vector is *stretched* rather than compressed.

Let us now simplify the situation yet further by considering a scenario in which the initial vorticity vector has its preferred alignment from the outset (e.g. as considered in Tennekes and Lumley, 1972, p. 256), which corresponds to setting $\omega_{02} = 0$ in the model above. Now the general solution from Eq. 4.100 simplifies to

$$\omega_1 = \omega_{01} e^{st}, \qquad \omega_2 = \omega_3 = 0 \tag{4.101}$$

i.e. the lone initial vorticity component growing exponentially in time. This is the simplest model that can be utilized to explain the phenomenon of vortex stretching. Note that even in this highly idealized situation the reduced initial velocity field from Eqs. 4.95:

$$u = sx, \qquad v = -sy, \qquad w = \omega_{01} y e^{st} \tag{4.102}$$

remains three dimensional.

Before proceeding further, we must first define a *vortex line* as one that is everywhere tangent to the the local vorticity vector (analogous to a *stream line* within a velocity field). Additionally, a *vortex tube* is defined as the surface formed by all vortex lines passing through a closed curve in

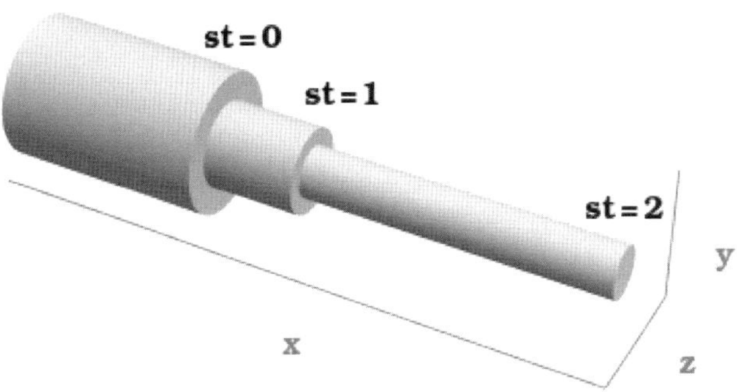

Fig. 4.18 Stretching of a vortex tube segment aligned with the direction of positive strain, at various non-dimensional times st.

the continuum (analogous to a *stream tube* in a velocity field). Let us now consider a small segment of a cylindrical vortex tube with radius R and length λ, with one end initially placed at the origin, such as depicted in Fig. 4.18 at time $st = 0$. Conservation of angular momentum requires that

$$\omega_1 A_v \sim \omega_1 R^2 = \text{Constant}, \qquad A_v = \pi R^2 \tag{4.103}$$

where A_v is the cross-sectional area of the vortex tube. Now, as a vortex tube occupies physical space, it is self evident that it is likewise a *material tube* i.e. it will move with the fluid. Hence, as it deforms, it must also conserve its mass (volume) i.e.

$$A_v \lambda \sim R^2 \lambda = \text{Constant} \tag{4.104}$$

Therefore,

$$\frac{\omega_1 R^2}{R^2 \lambda} = \frac{\omega_1}{\lambda} = \text{Constant} \tag{4.105}$$

and thus $\lambda \sim \omega_1 \sim e^{st}$, i.e. the length of the vortex tube segment (like the vorticity itself) will grow exponentially. Notice that, as we have conveniently placed one end at the origin (where $u = v = w = 0$), it will remain fixed in our example, though this would not generally be the case.

Now, as many a skeptical student has asked: "If the length is growing exponentially, and wave numbers are inversely proportional to length, how is this process contributing to the energy cascade to *higher* wave numbers?" To answer this question, we must first recognize from Eq. 4.104, that the exponentially growing length implies that the radius must instead be decaying as $R \sim e^{-st/2}$. The corresponding evolution of the vortex tube under consideration is depicted within Fig. 4.18 for various non-dimensional times st. While not sinusoidal, we can still view the segment of the vortex tube under consideration as a three-dimensional body, with characteristic wave numbers

$$k_x \sim \lambda^{-1} \sim e^{-st}, \qquad k_y = k_z = k_R \sim R^{-1} \sim e^{st/2} \tag{4.106}$$

Hence the modulus of the wave number vector (the x-axis on the energy spectrum $E(k)$) is

$$k = \sqrt{k_x^2 + k_y^2 + k_z^2} = \sqrt{k_x^2 + 2k_R^2} \sim \sqrt{e^{-2st} + 2e^{st}} \tag{4.107}$$

such that for large non-dimensional time st

$$k \sim \sqrt{2e^{st}} \tag{4.108}$$

From this, it is now clear that it is the decay in the radius of the vortex being stretched that causes the wave number modulus k to effectively increase via this process. This corresponds to the transfer of energy to higher wave numbers in the turbulent energy cascade.

The simple model utilized above neglects viscosity, and hence describes this process as continuing indefinitely; In reality, of course, the stretching process will only continue until a vortex tube becomes so thin (hence R so small, and k so large) that it is effectively "pinched" by viscosity once Kolmogorov wave number scales ($\sim \eta_K^{-1}$) are approached, representing the end of the inertial subrange. As established previously via Eq. 4.100, the preference for the vorticity vector (hence vortex tubes) to become aligned in the stretching direction can likewise explain why this *superharmonic* transfer of energy to higher wave numbers (corresponding to the exponential growth of ω_{01}) dominates the *subharmonic* transfer to lower wave numbers. The latter would correspond to exponential decay in vorticity, and hence *vortex compression*, which could similarly be demonstrated using the present general model, Eq. 4.100, if we had instead set $\omega_{01} = 0$ and retained ω_{02}.

An example demonstrating the energy cascade through the process of vortex stretching, in a more complicated situation can be seen in the series of plots depicted in Fig. 4.19. This figure depicts results from a direct numerical simulation (DNS) of a three-dimensional Taylor-Green vortex (see Taylor and Green, 1937), wherein the initial condition consists of series of triply-periodic large scale vortices, which then evolve in time based on a solution to Navier-Stokes equations. As can be seen, due to the process described above, the initially large scale vortices quickly become stretched and broken into smaller and smaller vortices, such that after sufficient time the initial large scale motion can hardly be recognized.

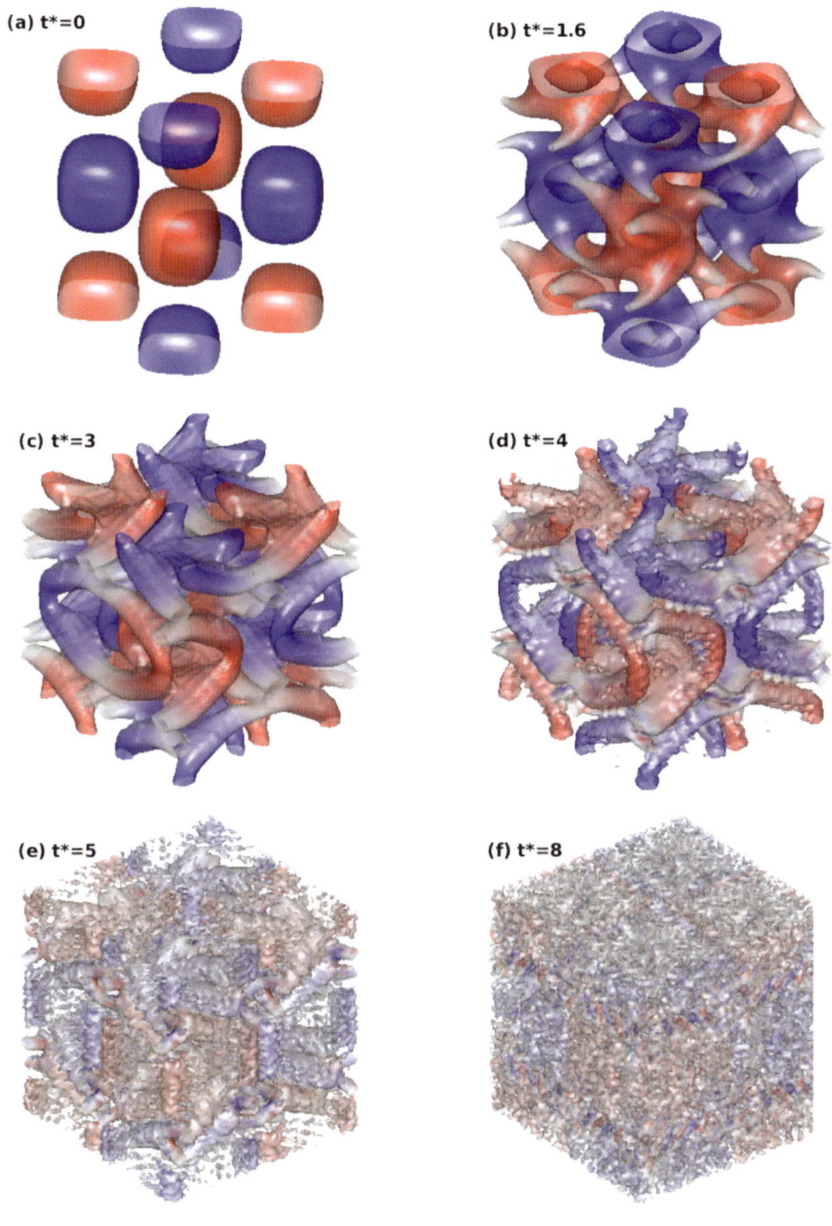

Fig. 4.19 Results from a direct numerical simulation (DNS) of a Taylor-Green vortex with $Re = 1600$. The plots depict isosurfaces of the so-called Q-criterion (Hunt et al., 1988), with red/blue colors indicating positive/negative vertical vorticity.

4.3.5 *One-dimensional spectra*

$E(k)$ is not an easily measurable quantity. To get around the problem, we define the following general one-dimensional spectrum for isotropic turbulence

$$\phi_{ij}(k) = \frac{1}{2\pi}\int_{-\infty}^{\infty} Q_{ij}(r)e^{-ikr}dr \tag{4.109}$$

where, due to symmetry about $r = 0$ this simplifies to

$$\phi_{ij}(k) = \frac{1}{\pi}\int_{0}^{\infty} Q_{ij}(r)\cos(kr)dr \tag{4.110}$$

In practice, the following special cases are of primary interest:

$$F(k) = \phi_{11}(k) = \frac{1}{\pi}\int_{0}^{\infty} Q_{11}(r)\cos(kr)dr \tag{4.111}$$

$$= \frac{\overline{u'^2}}{\pi}\int_{0}^{\infty} f(r)\ \cos(kr)\ dr \tag{4.112}$$

The inverse transform is

$$\overline{u'^2}f(r) = 2\int_{0}^{\infty} F(k)\ \cos(kr)\ dk \tag{4.113}$$

Note that setting $r = 0$ in the above immediately gives

$$\frac{1}{2}\overline{u'^2} = \int_{0}^{\infty} F(k)dk \tag{4.114}$$

In other words, analogous to Eq. 4.63 for the three-dimensional spectrum, the integral of the one-dimensional spectrum over all wave numbers is equal to half the variance of the respective fluctuating component. Analogous to the above, we similarly have:

$$G(k) = \phi_{22}(k) = \frac{\overline{v'^2}}{\pi}\int_{0}^{\infty} g(r)\ \cos(kr)\ dr \tag{4.115}$$

$$\overline{v'^2}g(r) = 2\int_{0}^{\infty} G(k)\ \cos(kr)\ dk \tag{4.116}$$

$$\frac{1}{2}\overline{v'^2} = \int_{0}^{\infty} G(k)dk \tag{4.117}$$

As proved in Appendix 4.E, for *isotropic turbulence*, the two one-dimensional spectra defined above, $F(k)$ and $G(k)$, are related by

$$G(k) = \frac{1}{2}\left[F(k) - kF'(k)\right] \tag{4.118}$$

Additionally, as proved in Appendix 4.F, the one-dimensional spectrum $F(k)$ is related to the three-dimensional energy spectrum function $E(k)$

$$E(k) = k^2 F''(k) - kF'(k) \tag{4.119}$$

Incidentally, inserting Eq. 4.90 (which is satisfied in the inertial subrange) into Eq. 4.119 gives

$$\frac{\varepsilon^{\frac{2}{3}}}{k^{\frac{5}{3}}} = k^2 F''(k) - kF'(k) \tag{4.120}$$

This has a general solution:

$$F(k) = \frac{9}{55}\varepsilon^{\frac{2}{3}} k^{-\frac{5}{3}} + \frac{k^2}{2} C_1 + C_2 \tag{4.121}$$

where C_1 and C_2 are constants. The second and third terms on the right-hand-side of the preceding equation should drop (hence $C_1 = C_2 = 0$) because $F(k)$ must go to zero for large values of k, and therefore

$$F(k) = \text{ Constant} \times \varepsilon^{\frac{2}{3}} k^{-\frac{5}{3}} \tag{4.122}$$

Invoking this result within Eq. 4.118 identically yields

$$G(k) = \text{ Constant} \times \varepsilon^{\frac{2}{3}} k^{-\frac{5}{3}}. \tag{4.123}$$

Hence, these results demonstrate that the local wave number region where the $-5/3$ law is evident in a three-dimensional spectrum $E(k)$ will similarly manifest in the one-dimensional spectra $F(k)$ and $G(k)$. While this result is formally demonstrated above for the special case of isotropic turbulence, the ability to detect the $-5/3$ law within experimentally obtained one-dimensional spectra is remarkably consistent.

Note that, similar to before, one-dimensional wave number spectra may be linked to analogous frequency-domain spectra by requiring equivalence in the variance. For example, for the u' fluctuating component we may link these according to

$$\overline{u'^2} = 2\int_0^\infty F(k)dk = \int_0^\infty S(f)df = \int_0^\infty S(\omega)d\omega \tag{4.124}$$

where it is seen that we have adopted the usual (though not universal) convention in which frequency-domain spectra are "one-sided," as it is redundant to allow both wave number components (k_1, k_2, k_3) and frequencies to be both positive and negative. After invoking the frozen turbulence approximation, Eq. 4.49, these ultimately lead to the relations

$$F(k) = \frac{\overline{u}}{4\pi} S(f) = \frac{\overline{u}}{2} S(\omega) \tag{4.125}$$

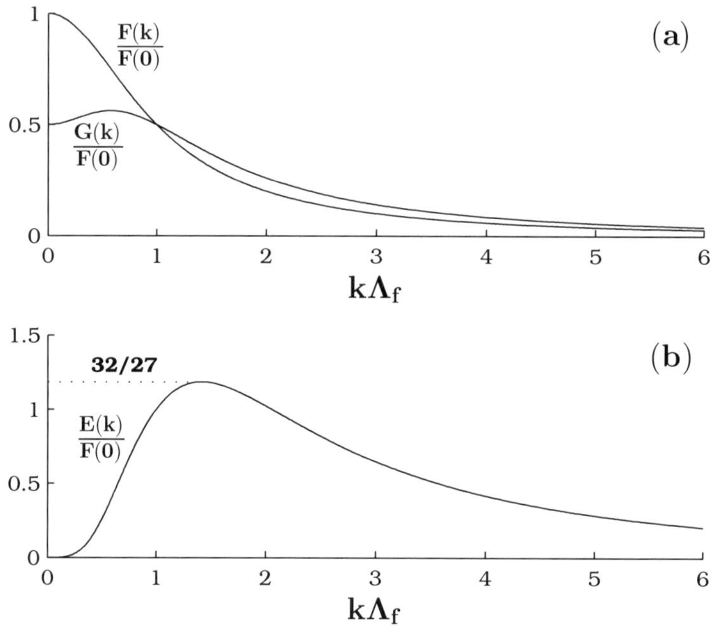

Fig. 4.20 One-dimensional and three-dimensional energy spectra for the numerical example.

where

$$S(\omega) = \frac{2\overline{u'^2}}{\pi} \int_0^\infty R_E(\tau)\ \cos(\omega t)\ dt. \tag{4.126}$$

defines the one-dimensional energy density spectrum in the frequency domain, and where the angular and ordinary frequencies are again related according to $\omega = 2\pi f$.

Example 4.2. Some simple model one-dimensional spectra.

To illustrate the development of an analytical model spectrum, as a first approximation, assume that the correlation function $f(r)$ may be approximated by

$$f(r) = \exp\left(-\frac{r}{\Lambda_f}\right) \tag{4.127}$$

From this, we may find the corresponding one-dimensional and three-dimensional spectrum functions. The one-dimensional spectrum function may be found by inserting Eq. 4.127 into Eq. 4.112, which leads to:

$$F(k) = \frac{\Lambda_f \; \overline{u'^2}}{\pi(1+(k\Lambda_f)^2)} = \frac{F(0)}{1+(k\Lambda_f)^2}, \qquad F(0) = \frac{\Lambda_f \; \overline{u'^2}}{\pi} \tag{4.128}$$

Assuming that the turbulence is isotropic, the corresponding $G(k)$ function can then be found from Eq. 4.118 to be

$$\frac{G(k)}{F(0)} = \frac{1+3(k\Lambda_f)^2}{2(1+(k\Lambda_f)^2)^2} \tag{4.129}$$

These resulting spectral functions are plotted in Fig. 4.20a. The corresponding three-dimensional energy spectrum function $E(k)$, can similarly be found using Eq. 4.119, yielding:

$$E(k) = 8F(0)\frac{(k\Lambda_f)^4}{(1+(k\Lambda_f)^2)^3} \tag{4.130}$$

This three-dimensional spectrum for $E(k)$ is plotted in Fig. 4.20b.

Eq. 4.128 above is known as the Dryden (1943) spectrum, and represents an early attempt at a simple analytical *model spectrum.* (This spectrum is again based on the assumed correlation function Eq. 4.127; As this does not satisfy $\partial f/\partial r = 0$ at $r = 0$, it is inaccurate for small r!) Note that for large k this spectrum yields $F(k) \sim k^{-2=-6/3}$, which is quite similar, though not equivalent to, Kolmogorov's inertial subrange where $F(k) \sim k^{-5/3}$.

Two additional simple one-dimensional model spectra having similar form to Eq. 4.128, but which do asymptotically satisfy Kolmogorov's $-5/3$ law for large k, are the von Karman (1948) spectrum

$$F(k) = \frac{\Lambda_f \overline{u'^2}}{\pi} \frac{1}{\left[1+70.78\left(\frac{k\Lambda_f}{2\pi}\right)^2\right]^{5/6}} \tag{4.131}$$

and the Kaimal (Kaimal et al., 1972) spectrum

$$F(k) = \frac{\Lambda_f \overline{u'^2}}{\pi} \frac{1}{\left[1+\frac{6}{2\pi}k\Lambda_f\right]^{5/3}} \tag{4.132}$$

These are both popular in wind engineering and atmospheric turbulence.

Obviously, none of the simple model spectra above attempt to describe the dissipation range, which will deviate from the $-5/3$ law once high Kolmogorov-scale wave numbers are reached. Accounting for the dissipation range, in fact, requires further modification to the the correlation function $f(r)$ for small r, as will be conceptually demonstrated in the next example.

Example 4.3. Extension of the Dryden one-dimensional model spectrum.

As a reasonably simple follow up example to the above, which can still be treated analytically, let us now generalize the correlation function from Eq. 4.127 to:

$$f(r) = (1-\alpha) + \alpha \exp\left(-\frac{\gamma r}{\Lambda_f}\right) \tag{4.133}$$

where

$$\alpha = 1 - \exp\left(-\frac{\gamma \beta^2 r}{\Lambda_f}\right) \tag{4.134}$$

and where β and γ are coefficients to be determined in what follows. Unlike Eq. 4.127, this correlation function now satisfies that $\partial f/\partial r = 0$ at $r = 0$, while also having a well-defined second derivative, which can in turn be easily related to the micro scale λ_f:

$$\left.\frac{\partial^2 f(r)}{\partial r^2}\right|_{r=0} = -\frac{2\beta^2\gamma^2}{\Lambda_f} \qquad \Leftrightarrow \qquad \frac{\Lambda_f}{\lambda_f} = \beta\gamma \tag{4.135}$$

Inserting $f(r)$ from the above into Eq. 4.112 leads to the one-dimensional spectrum

$$\frac{F(k)\pi}{\overline{u'^2}\Lambda_f} = \frac{\beta^2\left(\beta^2+1\right)\gamma^3\left(\left(\beta^4+\beta^2+1\right)\gamma^2+3k^2\Lambda_f^2\right)}{\left(\gamma^2+k^2\Lambda_f^2\right)\left(\beta^4\gamma^2+k^2\Lambda_f^2\right)\left(\left(\beta^2+1\right)^2\gamma^2+k^2\Lambda_f^2\right)} \tag{4.136}$$

The relationship between β and γ can finally be established by requiring that Eq. 4.34 be satisfied, leading to

$$\gamma = \frac{\beta^4+\beta^2+1}{\beta^2\left(\beta^2+1\right)} \tag{4.137}$$

As seen from the above, the resulting one-dimensional model spectrum $F(k)$ and correlation function $f(r)$ can be explicitly evaluated for a given macro scale Λ_f and β. Typically, γ from Eq. 4.137 is very near unity, such that $\beta \approx \Lambda_f/\lambda_f$, and hence the required input effectively (and conveniently) corresponds to the macro scale Λ_f and what amounts to a very close approximation to the micro scale λ_f. Similar to Example 4.2, it is relatively straight forward (at least if symbolic manipulator software is used!) to derive counterpart functions for $G(k)$ and $E(k)$ from $F(k)$ in Eq. 4.136 if isotropic turbulence is assumed. This is left as an additional exercise for the interested reader.

Comparison of the correlation functions $f(r)$ from Eqs. 4.127 and 4.133 (Fig. 4.21a) and the corresponding one-dimensional spectra $F(k)$ from

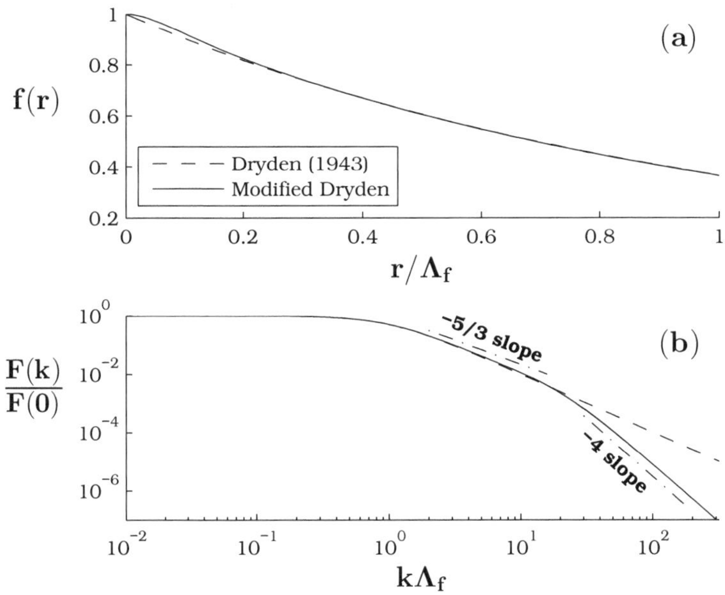

Fig. 4.21 Comparison of (a) the correlation function $f(r)$ and (b) the corresponding one-dimensional spectrum $F(k)$ for the Dryden and modified Dryden (with $\beta = 4 \approx \Lambda_f/\lambda_f$) models.

Eqs. 4.128 and 4.136 (Fig. 4.21b) are depicted in Fig. 4.21. A line having $-5/3$ slope is also added to Fig 4.21b as a reference. Comparison of the dashed (Dryden model) and full lines (modified Dryden model, described in the present example) in this figure graphically demonstrates how the increased detail in $f(r)$ near $r = 0$, as introduced in Eq. 4.133, manifests in the eventual break from the characteristic inertial subrange behavior in $F(k)$ (in this case reasonably close to the $-5/3$ law) once sufficiently large k is reached. The resulting tail having steeper slope at the high wave number end in Fig. 4.21b (full line) can be taken as conceptually representing the dissipation range of the spectrum.

The modified Dryden model spectrum above asymptotically leads to decay as $F(k) \sim k^{-4}$ in the high wave number dissipation range (an additional line having -4 slope is also added to Fig. 4.21b as a reference), whereas experiments suggest that this range would be better characterized by exponential decay in the spectrum once Kolmogorov scales are reached. A model spectrum describing the dissipation range in this fashion can be found e.g. in the book of Pope (2003). In contrast to the examples above, this model does not stem from a known correlation function, but is rather based on empirical fit to a wide range of experimental data.

Example 4.4. Comparison of one-dimensional spectra from a boundary layer in an open channel and from a jet.

Fig. 4.22a shows the one-dimensional spectrum function obtained for an open channel flow by Raichlen (1967). (The results are normalized as indicated in the figure. The Reynolds number is defined as $Re = 4r_h V/\nu$ in which r_h is the hydraulic radius). Fig. 4.22b, on the other hand, shows the one-dimensional spectrum measured in a jet, from Champagne (1978).

It may be noticed that, over a wave number interval, namely $1 < kh/(2\pi) < 6$, in Fig. 4.22a, $F(k)$ is apparently

$$F(k) \sim k^{-\frac{5}{3}}$$

revealing the previously mentioned Kolmogorov's $-5/3$ law. The same holds true for the second example (Fig. 4.22b) for the range $4\times10^{-2}\ \text{cm}^{-1} < k_1 < 10\ \text{cm}^{-1}$. Notice also that Fig. 4.22b actually contains both one-dimensional spectra $F(k)$ (top line) and $G(k)$ (bottom line). These coincide at more or less the same wave number where the $-5/3$ law begins. This can be taken as a demonstration of local isotropy once the inertial subrange is reached, in accordance with Kolmogorov's theory.

For larger wave numbers representing the dissipation range of the spectrum, there is likewise a clear deviation, resulting in a steeper negative slope when compared to the $-5/3$ law. Note the qualitative similarity with these measured one-dimensional spectra with the full line in Fig. 4.21b, based on the derived model spectrum. The example one-dimensional spectra presented in Fig. 4.22 are quite typical of those obtained in turbulent flows.

4.3.6 *Aliasing in one-dimensional spectra*

As seen above (Fig. 4.20), the one-dimensional spectral functions $F(k)$ and $G(k)$ are shaped rather differently than the three-dimensional spectrum $E(k)$. This is the result of *aliasing*. To demonstrate this formally, let us consider a hypothetical scenario where a three-dimensional isotropic turbulence field consists of a single wave number $k_0 > 0$. Mathematically this corresponds to the spectrum

$$E(k) = \frac{3\overline{u'^2}}{2}\delta(k - k_0) \tag{4.138}$$

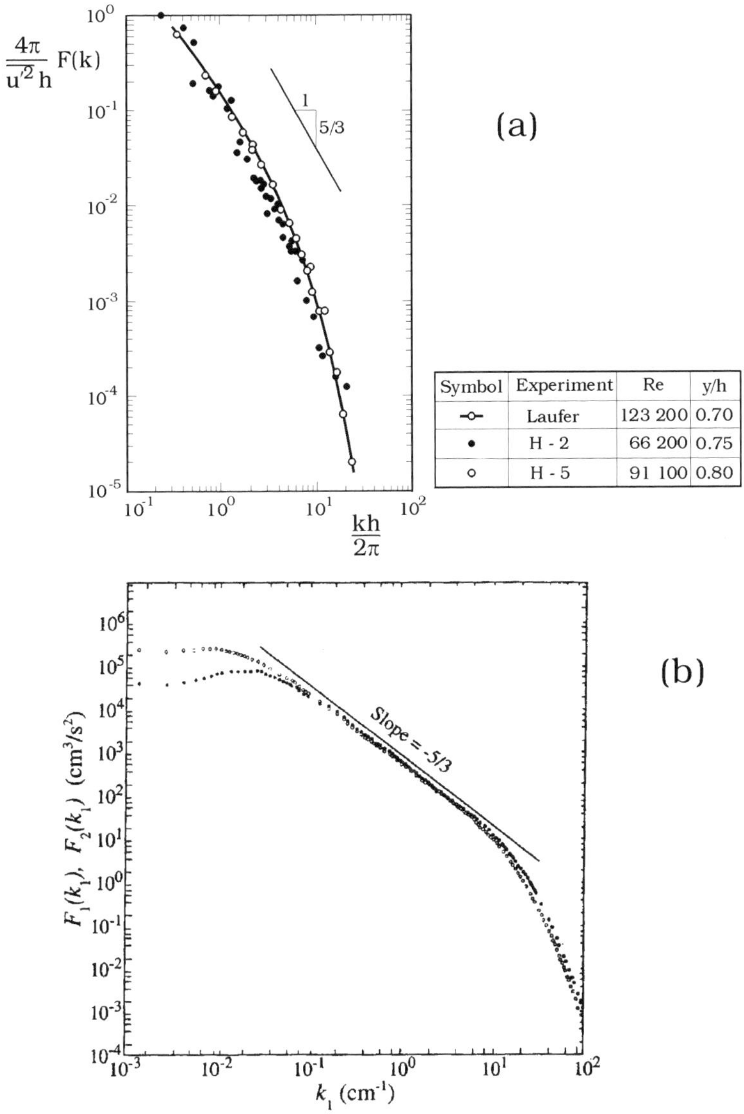

Fig. 4.22 One-dimensional energy spectra, measured in (a) an open channel (Raichlen, 1967), and (b) a jet (Champagne, 1978). Adapted from Raichlen (1967) and Champagne (1978).

where $\delta(k)$ is the Dirac delta function, consisting of a single pulse having unit area where the argument is zero i.e. at $k = k_0$. Notice that the lead constant has been chosen such that this spectrum satisfies

$$\int_0^\infty E(k)dk = \frac{3\overline{u'^2}}{2} \tag{4.139}$$

since $\delta(k)$ has unit area, by definition. This is plotted in Fig. 4.23a, where the arrow indicates the isolated pulse of energy at wave number $k = k_0$. Inserting this within Eq. 4.119 gives

$$\frac{3\overline{u'^2}}{2}\delta(k - k_0) = -kF'(k) + k^2F''(k) \tag{4.140}$$

with solution

$$\frac{k_0F(k)}{\overline{u'^2}} = \begin{cases} \dfrac{k_0}{\overline{u'^2}}\left[\dfrac{k^2}{2}C_1 + C_2\right], & k \leq k_0 \\ -\dfrac{3}{4} + \dfrac{3k^2}{4k_0^2} + \dfrac{k^2k_0}{2\overline{u'^2}}C_1 + \dfrac{k_0}{\overline{u'^2}}C_2, & k > k_0 \end{cases} \tag{4.141}$$

Let us first consider the solution in the region $k > k_0$. Grouping like terms, and requiring that $F(k)$ vanishes for large k requires that the integration constants are

$$C_1 = \frac{3\overline{u'^2}}{2k_0^3}, \qquad C_2 = \frac{3\overline{u'^2}}{4k_0} \tag{4.142}$$

Inserting these back into Eq. 4.141 yields the final solution

$$\frac{k_0F(k)}{\overline{u'^2}} = \begin{cases} \dfrac{3}{4}\left[1 - \left(\dfrac{k}{k_0}\right)^2\right], & k \leq k_0 \\ 0, & k > k_0 \end{cases} \tag{4.143}$$

Inserting this solution for $F(k)$ back into Eq. 4.118 likewise leads to

$$\frac{k_0G(k)}{\overline{u'^2}} = \begin{cases} \dfrac{3}{8}\left[1 + \left(\dfrac{k}{k_0}\right)^2\right], & k \leq k_0 \\ 0, & k > k_0 \end{cases} \tag{4.144}$$

These results for $E(k)$, $F(k)$, and $G(k)$ are plotted in Fig. 4.23. Inspection of this plot demonstrates how energy from a three-dimensional spectrum $E(k)$ consisting of a single pulse at $k = k_0$ is *aliased* to all wave numbers $k \leq k_0$ in the one-dimensional spectrum functions $F(k)$ and $G(k)$. This explains, in a mathematical sense, their qualitative difference in shape, as observed in Figs. 4.20, 4.21 and 4.22.

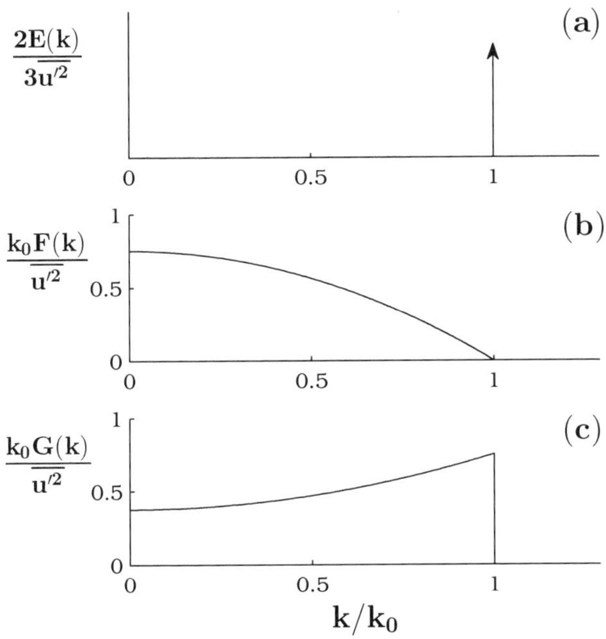

Fig. 4.23 Example showing the resulting one-dimensional spectra (b) $F(k)$ and (c) $G(k)$ resulting from a (a) three-dimensional spectrum $E(k)$ from Eq. 4.138.

Physically speaking, such aliasing occurs because in a one-dimensional measurement (e.g. along a line in the pure x-direction), there is simply no way to distinguish between a fluctuating component whose wave number vector is perfectly aligned with the measurement direction e.g. $\mathbf{k}_a = (k_x, 0, 0)$ having wave number modulus $k_a = k_x$, and one that is oblique but sharing the same component in the measurement direction e.g. $\mathbf{k}_b = (k_x, k_y, k_z)$ (such that modulus $k_b > k_a$). As a consequence, a one-dimensional spectrum obtained in a three-dimensional velocity field will, at a given wave number k, inevitably contain contributions from all wave number vectors having modulus $\geq k$.

4.3.7 *Matlab example: The discrete Fourier transform*

Let us finally close the present chapter with two Matlab examples, demonstrating use of the built-in discrete Fourier transform function `fft`. The time/frequency domain analogue to the continuous Fourier transform,

Eq. 4.109, applied here to an arbitrary time signal $u(t)$ is:

$$\hat{u}(\omega) = \frac{1}{2\pi}\int_{-\infty}^{\infty} u(t)e^{-i\omega t}dt \tag{4.145}$$

Note that from the Eulerian identity

$$e^{-i\omega t} = \cos(\omega t) - i\sin(\omega t), \qquad \omega = 2\pi f \tag{4.146}$$

Note also that if the time signal takes the simple harmonic form

$$u(t) = a_i\cos(\omega_i t) - b_i\sin(\omega_i t) \tag{4.147}$$

the resulting Fourier transform is

$$\hat{u}(\omega) = \left(\frac{a_i}{2} + i\frac{b_i}{2}\right)\delta(\omega_i - \omega) + \left(\frac{a_i}{2} - i\frac{b_i}{2}\right)\delta(\omega_i + \omega) \tag{4.148}$$

from which it is explicitly clear that the Fourier transform of a periodic function with angular frequency ω_i yields complex conjugate pulses (with half the input amplitudes a_i and b_i on the real and imaginary parts, respectively) at frequencies $\omega = \pm\omega_i$. This, in the simplest of terms, demonstrates how the Fourier transform converts from the time to the frequency domain!

In practice, rather than dealing with continuous turbulent time signals, we are instead typically working with discrete series with sampling interval Δt, as considered in the previous examples. Hence, to illustrate use of the FFT in practical Fourier decomposition, Program 4: `ExampleFFT.m` demonstrates a simple example using Matlab's discrete fast Fourier transform (FFT) function `fft` on such a discrete time series. For the sake of simplicity, this example considers an intentionally short time series of (even) length `N`$= 6$, created through simple summation of several harmonic components of the form from Eq. 4.147

$$u = \sum_{i=1}^{N} a_i\cos(2\pi f_i t) - b_i\sin(2\pi f_i t), \tag{4.149}$$

as seen in line 22. Here the harmonic amplitudes a_i and b_i are specified as input (line 2), corresponding to the discrete positive frequencies f_i given in line 6, ranging from $f = 0$ to the Nyquist frequency $f_N = 1/(2\Delta t)$ in discrete frequency increments $\Delta f = f_N/(N/2)$. The resulting signal is plotted in Fig. 4.24 (bottom), formed from the superposition of the cosine and sine harmonic components plotted in the two sub-plots immediately above. These are shown both continuously (dashed lines) and discretely (open circles) i.e. corresponding to the values at the discrete sample times t_i (stored in variable `t` in the script).

Once the signal (variable `u`) is constructed, this script uses Matlab's `fft` function to demonstrate how the input a_i and b_i coefficients can be detected from analysis of the input signal alone. The final two script lines result in the screen output below (some decimal points, zeros, and zero imaginary parts are omitted for clarity and conciseness):

```
out1 =

    0        10         20          30         -20          -10
    6 + 0i   0 + .5i    1 + 1.5i    4 + 0i     1 - 1.5i     0 - .5i

out2 =

    0        10         20          30
    6 + 0i   0 + 1i     2 + 3i      4 + 0i
```

In each of the output variables above (`out1` and `out2`) the top row corresponds to the frequencies (in Hz) and the bottom row to the corresponding complex amplitudes (in m/s i.e. the same units as the original signal). It is seen from the first output variable (`out1`) that direct use of the FFT (line 29) on the original signal, yields harmonic amplitudes corresponding to both positive and negative frequencies, in line with the previous analytic result from Eq. 4.148. These are stored in so-called reverse wrap-around (RWA) order, which in Matlab notation corresponds to: `[0:df:f_N df-f_N:df:-df]` where `df`=10 Hz and `f_N`=30 Hz in this example. While these may be re-ordered e.g. using the built-in `fftshift` command, we refrain from doing so here.

As mentioned previously, we have adopted the usual "one-sided" frequency domain convention such that frequencies are regarded as strictly positive, whereas Matlab's FFT function is generically "two-sided" (a requirement e.g. to properly handle complex signals). Hence, to account for this difference and detect the input a_i and b_i coefficients directly from the time series, we must multiply several of the resulting harmonic amplitudes (omitting the zero and Nyquist frequencies, which are only represented once) by a factor 2, effectively re-distributing contributions from the negative frequencies back to their positive counterparts. This procedure is performed in line 30, resulting in output variable `out2` above. From this result, we now recognize the detected a_i coefficients (input as `a = [6 0 2 4]`, see line 2) as the resulting real parts of the harmonic amplitudes, and several of the b_i coefficients (input as `b = [99 1 3 5]`, line 2) as the

Program 4: ExampleFFT.m

```
T = 0.1; N = 6; % Series duration (s) and length
a = [6 0 2 4]; b = [99 1 3 5]; % cos,sin (a,b) amps.

% Create discrete time & frequency vectors
dt = T/N; t = [0:dt:T-dt];
f_N = 1/(2*dt); df = f_N/(N/2); f = [0:df:f_N];

% Time series by superimposing cosines and sines
u = zeros(1,N); tres = [0:T/100:T]; % High res. t
for i = 1:length(f) % Loop & plot cos,sin harmonics
    subplot(3,1,1); hold on; % cos harmonics
    plot(t,a(i)*cos(2*pi*f(i)*t),'bo','MarkerSize',4);
    plot(tres,a(i).*cos(2.*pi.*f(i).*tres),'b--');
    ylim([-1 1].*max(a)); xlim([0 T]); box on;
    ylabel('a_icos(2\pif_it)')
    subplot(3,1,2); hold on; % sin harmonics
    plot(t,-b(i)*sin(2*pi*f(i)*t),'ro','MarkerSize',4);
    plot(tres,-b(i).*sin(2.*pi.*f(i).*tres),'r--');
    xlim([0 T]); box on; ylabel('-b_isin(2\pif_it)')

    % Add new harmonics to signal
    u = u+a(i)*cos(2*pi*f(i)*t)-b(i)*sin(2*pi*f(i)*t);
end
subplot(3,1,3); plot(t,u,'k-o','MarkerSize',4);
xlim([0 T]); xlabel('t (s)'); ylabel('u (m/s)');

% Use FFT to detect harmonic amplitudes, a & b
f_all = [0:df:f_N df-f_N:df:-df]; % RWA order
U = fft(u)/N; % Take Fourier transform
ab = [U(1) 2*U(2:N/2) U(N/2+1)]; % Harmonic amps.

% Output to screen
out1 = [f_all; U] % Direct from FFT, pos. & neg. f
out2 = [f; ab] % Gives (a + b*i) for pos. f
```

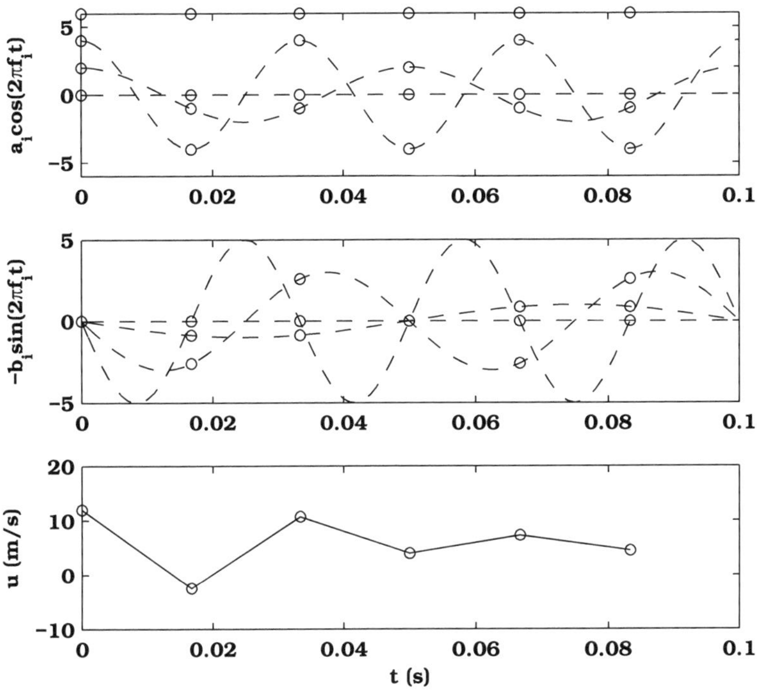

Fig. 4.24 Output from Matlab script `ExampleFFT.m`.

imaginary parts.

The lone exceptions are those components corresponding to the zero and Nyquist frequencies. The zero frequency component (originally input as $b_1 = 99$ m/s) is detected as zero, simply because $\sin(0) = 0$; this coefficient was input intentionally large to emphasize this point! The Nyquist component (input as $b_4 = 5$ m/s in the present example) is likewise detected as zero, not because $\sin(2\pi f_N t) = 0$ for all t, but because it happens to be zero at each of the discrete sample times t_i. This particular mode is seen as the dashed line having largest amplitude in Fig. 4.24 (middle), where each of the open circles (again corresponding to the discrete sample times) lands precisely at a zero crossing. Hence, while the input at the Nyquist frequency is properly detected for the cosine contribution, it cannot be detected for the sine contribution.

4.3.8 *Matlab example: Creating a one-dimensional energy density spectrum*

The preceding example demonstrates how the individual harmonic components comprising a given discrete signal can be detected directly using the FFT algorithm. However, when analyzing a turbulent velocity signal, we are more often interested in computing its one-dimensional energy spectrum function. This is illustrated in the final Matlab example script `ExampleSpectrum.m` given in Program 5.

In this script a synthetic turbulent velocity signal is first generated using `turbgen.m` (line 4), similar to several of the previous examples. The resulting discrete positive frequencies and corresponding harmonic amplitudes are then determined (lines 7–9) in the same manner as in Program 4, except that rather than determining the cosine a_i and sine amplitudes b_i separately (as real and imaginary parts), these are instead converted into the total harmonic amplitude using

$$A_i = \sqrt{a_i^2 + b_i^2} \tag{4.150}$$

which is accomplished using the `abs` function (line 9). Note that for each component $A_i^2/2$ represents the variance for the corresponding frequency f_i. The one-dimensional spectrum is then determined for each component through the operation (line 10):

$$S_i = \frac{1}{\Delta f}\frac{A_i^2}{2} \tag{4.151}$$

representing the energy density, i.e. kinetic energy (per unit mass) per discrete frequency, for each of the represented discrete frequencies (hence the division by Δf). On line 13 the sample variance of the original time series is calculated (as in Program 2), and compared with the integral of the resulting one-dimensional energy density spectrum, resulting in the screen output (both having units m^2/s^2):

```
sigma2 =

    0.0040

area =

    0.0040
```

This is an important (and always recommended) check: These must be equivalent, in accordance with the convention defined in Eq. 4.124. As such, while more commonly called the one-dimensional energy density spectrum, it may be equivalently thought of as the *variance density spectrum*, as it describes how energy at different frequencies contributes to the sample variance!

The graphical output from executing `ExampleSpectrum.m` is presented in the one-dimensional energy density spectrum depicted in Fig. 4.25 (top). The computed spectrum is compared with the von Karman spectrum, where the frequency domain analogue to Eq. 4.131 is:

$$S(f) = \frac{4\pi T_E}{\Lambda_f} F(k) = \frac{4T_E\overline{u'^2}}{(1 + 70.78(fT_E)^2)^{5/6}} \tag{4.152}$$

demonstrating an essentially perfect match. This is no surprise: The task performed in Program 5 (computing the spectrum from a time series) is in fact the inverse of what is done in the function `turbgen.m`, which creates a time series from a prescribed spectrum! (In the present example `spec=2` corresponds to utilizing the von Karman spectrum.)

Further inspection of Program 1: `turbgen.m` (lines 18 and 20) reveals that the two other simple model spectra discussed previously in Example 4.2 are also implemented as options. That is, if the input is set such that `spec=1` then the Dryden spectrum is utilized, where the frequency domain analogue to Eq. 4.128 is

$$S(f) = \frac{4\overline{u'^2}T_E}{1 + (2\pi f T_E)^2} \tag{4.153}$$

This was utilized previously e.g. for the signal analyzed in Program 3, as it is again conveniently based on a known correlation function. Alternatively, if input `spec=3` is utilized, then the resulting time series would be based on the Kaimal frequency spectrum

$$S(f) = \frac{4\overline{u'^2}T_E}{(1 + 6fT_E)^{5/3}} \tag{4.154}$$

which is the the frequency domain analogue to Eq. 4.132. Program 5 can easily be modified to utilize these other options, and other model spectra options may easily be added by the interested reader to `turbgen.m`, should this be of interest.

A corresponding one-dimensional wave number spectrum $F(k)$ can likewise easily be converted from the frequency spectrum $S(f)$ by invoking Eq. 4.125 (for the y-axis) and $k = 2\pi f/\overline{u}$ from Eq. 4.55 (for the x-axis), both of which stem from Taylor's frozen turbulence approximation.

Program 5: ExampleSpectrum.m

```
% Generate time series using turbgen.m
N = 1e6; T_E = 0.2; dt = 0.001; sigma2 = 0.004;
noise = 0; % Noise, try e.g. 0 (off) and 0.3
[u,t] = turbgen(N,dt,sigma2,T_E,2,noise,1842);

% Compute and plot the 1D spectrum
f_N = 1/(2*dt); df = f_N/(N/2); f = [0:df:f_N];
U = fft(u)/N; % Discrete Fourier transform (m/s)
A = abs([U(1) 2*U(2:N/2) U(N/2+1)]); % Harm amps (m/s)
S = 0.5*A.^2/df; % Energy density (m^2/s^2/Hz=m^2/s)
loglog(f,S,'.','MarkerEdgeColor',[1 1 1].*0.8);
xlabel('f (Hz)'), ylabel('S(f) (m^2/s)')
sigma2 = var(u), area = trapz(f,S) % These must match!

% Compare with the analytical spectrum
S_an = sigma2*4*T_E./(1 + 70.78*(f*T_E).^2).^(5/6);
    % Compare line above with turbgen.m, lines 19 & 23
hold on; plot(f,S_an,'r--'); xlim([f(2) f_N]);

% If noise is added, also use Matlab's PWELCH function
if noise, nw = 100; % No. of windows, modif. as needed
    lwin = round(length(u)./nw); % Window length
    [Sw,fw] = pwelch(u,lwin); % Normalized results
    fw = fw*f_N/pi; Sw = Sw*pi/f_N; % Un-normalized
    plot(fw,Sw,'k-');
end;
legend('Computed - FFT','von Karman spectrum', ...
      'Computed - PWELCH','Location','SouthWest', ...
      'AutoUpdate','off'); drawnow;

% Add reference -5/3 slope
plot(f(1e4:20e4),0.01.*f(1e4:20e4).^(-5/3),'k-.');
text(25,1.4e-4,'-5/3 slope','rot',-39)
ylim(10.^[-10 -1]); hold off;
```

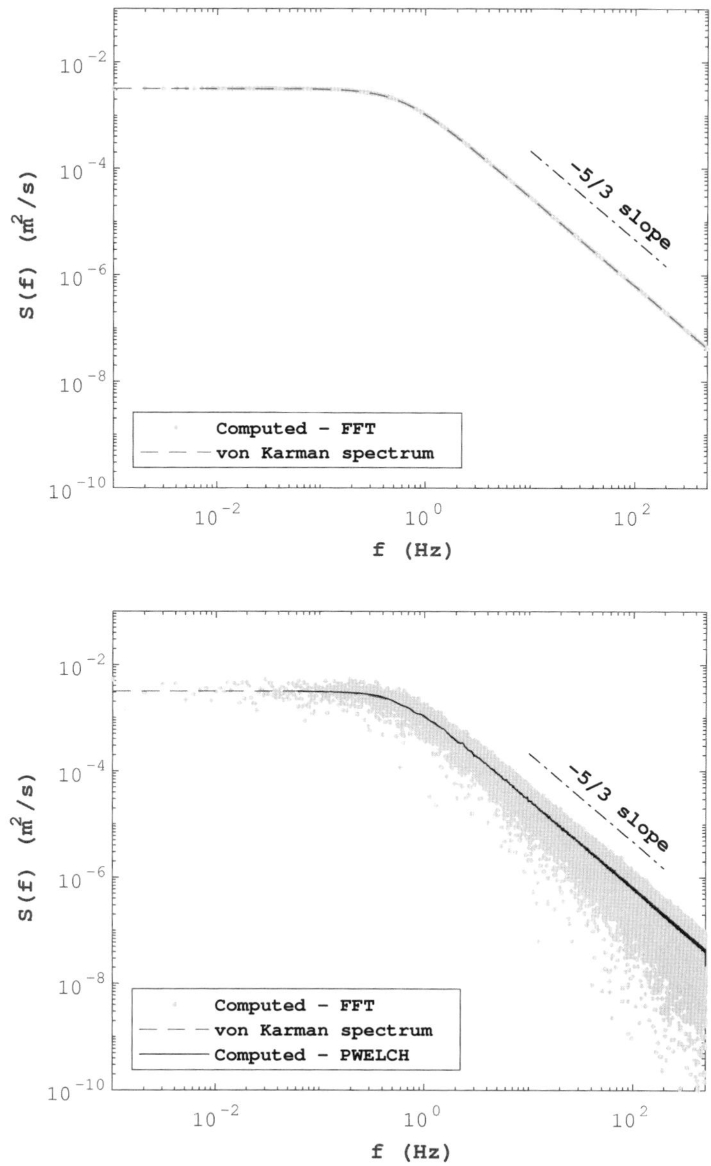

Fig. 4.25 Output from Matlab script `ExampleSpectrum.m` with (top) `noise=0` and (bottom) `noise=0.3`.

Fig. 4.25 (bottom) finally depicts the energy density spectrum that will be generated by executing Program 5 if artificial "noise" is added to the spectrum, accomplished by setting `noise=0.3` in line 3, rather than the default value `noise=0` (see Program 1, line 22 to see how noise is added to the analytical spectrum). The spectrum computed using the FFT algorithm (grey dots) is now seen to itself be rather noisy, fluctuating erratically about the smooth analytic shape. This is problematic, as it can e.g. be difficult to estimate the slope of the spectrum when it is so erratic, as would be required to identify e.g. a convincing match with the $-5/3$ law. This result resembles a spectrum that is typically achieved when FFT is applied to a real turbulent fluctuating signal. It is emphasized that this is not because real turbulent velocity signals are necessarily "noisy". Rather, this is a natural consequence of ongoing nonlinear interactions between the various modes, and is typical of non-periodic signals. (Similar spectra can even be generated from results based on solutions to nonlinear ordinary differential equations, see e.g. Fig. 3.2 of Abarbanel (1996), which stems from the well known Lorenz system.)

To combat this, it is often useful in practice to smooth out the spectrum, which can be done e.g. using Matlab's `pwelch` function, which effectively sub-divides the signal into several windows and takes a representative average. While we do not go into full detail about this algorithm (please read the Matlab help), an example correctly demonstrating its use is provided in Program 5, lines 21–25. This results in the smooth black line in Fig. 4.25 (bottom), which now again closely matches the analytic curve. Using this result, the inertial subrange could again be identified, similar to Fig. 4.25 (top). In practice, computed spectra are usually presented as smoothed (see e.g. the examples presented previously in Fig. 4.22), commonly accomplished using this or similar windowing techniques.

The savvy reader will have noticed that in the example just considered the spectrum was created by taking the Fourier transform of the velocity signal itself, whereas in the theory covered, e.g. Eq. 4.109, this was based on the Fourier transform of the correlation function. To explain why these are equivalent, let us define the general time correlation tensor

$$R_{ij} = \overline{u_i'(\tau)u_j'(\tau + t)} \tag{4.155}$$

such that

$$R_{11} = \overline{u'(\tau)u'(\tau + t)} \tag{4.156}$$

Now consider a single harmonic component, corresponding to the i-th Fourier component

$$u_i = A_i \sin(2\pi f_i t + \phi_i) \tag{4.157}$$

having amplitude A_i, frequency f_i, and arbitrary phase shift ϕ_i. The component above, again, has variance

$$\overline{u_i'^2} = \frac{A_i^2}{2} \tag{4.158}$$

Invoking Eq. 4.157 within Eq. 4.156 gives

$$R_{11} = \frac{A_i^2}{2}\cos(2\pi f_i \tau) = \overline{u_i'^2}\cos(2\pi f_i \tau) \tag{4.159}$$

From the above, it is now evident that the two-point time correlation of a harmonic signal (regardless of the phase shift ϕ_i) takes a cosine variation (remember, it must start at unity!), with amplitude equal to the signal variance. Hence, considering the spectrum resulting from the Fourier transform of the correlation function is equivalent to that based on the signal itself after squaring the harmonic amplitudes and dividing by two (i.e. $A_i^2/2$), which is precisely the operation seen in the script `ProgramSpectrum.m`, line 10.

As is now familiar, the whole basis of Fourier analysis is simply decomposing a signal into many such components, hence having explained this equivalence for a single harmonic, it likewise holds in general. While spectra are traditionally presented in terms of the correlation function in theoretical analysis, in practice when computing a spectrum from a turbulent velocity signal, it is usual to simply compute the one-dimensional spectrum based on the discrete Fourier transform of of the signal itself, in the manner demonstrated in `ProgramSpectrum.m`.

This completes the Matlab examples presented in this chapter. By design, these have demonstrated applications on synthetic turbulent time series, enabling results to be easily validated. Thus, the methods demonstrated may now be readily and confidently adapted to similarly analyze real life turbulent signals in practice. Such an opportunity is provided directly in hands-on Exercise 3 in Chapter 10, which involves the statistical, correlation, and spectral analysis of velocities measured in real turbulent air jets.

4.4 References

(1) Abarbanel, H.D.I. (1996): Analysis of Observed Chaotic Data. Springer.

(2) Buchhave, P. and Velte, C.M. (2017): Measurement of turbulent spatial structure and kinetic energy spectrum by exact temporal-to-spatial mapping, Phys. Fluids, vol. 29, article no. 085109.

(3) Champagne, F.H. (1978): The fine-scale structure of the turbulent velocity field. J. Fluid Mech., vol. 86, 67–108.
(4) Dryden, H.R. (1943): A review of the statistical theory of turbulence. Q. Appl. Mech., vol. 1, 7–42.
(5) Eckelmann, H. (1974): The structure of the viscous sublayer and the adjacent wall region in a turbulent channel flow. J. Fluid Mech., vol. 65, 439–459.
(6) Grass, A.J. (1971): Structural features of turbulent flow over smooth and rough boundaries, J. Fluid Mech., vol. 50, 233–255.
(7) Hinze, J.O. (1959): Turbulence: An Introduction to its Mechanism and Theory. McGraw-Hill.
(8) Hunt, J.C.R, Wray, A.A. and Moin, P. (1988): Eddies, stream, and convergence zones in turbulent flows. Center for Turbulence Research Report CTR-S88, pp. 193–208
(9) Jimenez, J. (2004): The contributions of A.N. Kolmogorov to the theory of turbulence. Arbor, vol. CLXXVIII, Issue 704, 589–606.
(10) Kaimal, J.C., Izumi, Y., Wyngaard, J.C. and Cote, R. (1972): Spectral characteristics of surface-layer turbulence. Quart. J. R. Met. Soc., vol. 98, 563–589.
(11) Kolmogorov, A.N. (1941): The local structure of turbulence in incompressible viscous fluids at very large Reynolds numbers (in Russian). Dokl. Adak. Nauk. SSSR, vol. 30, 299–303. English translation reprinted in Proc. Roy. Soc. Lond. A, vol. 434, 9–13 (1991).
(12) Kreplin, H.P. (1973): Mpi fur stromungsforschung. Tech. Rep. Rep. no. 2/1973, Gottingen.
(13) Pope, S.B. (2003): Turbulent Flows. Cambridge University Press.
(14) Raichlen, F. (1967): Some turbulence measurements in water. J. Eng. Mech. Div., vol. 93, 73–98.
(15) Raudkivi, A.J. and Callander, R.A. (1975): Advanced Fluid Mechanics. Edward Arnold.
(16) Richardson, L.F. (1926): Atmospheric diffusion shown in a distance-neighbour graph. Proc. Roy. Soc. Lond. A, vol. 110, 709–737.
(17) Sumer, B.M. and Deigaard, R. (1981): Particle motions near the bottom in turbulent flow in an open channel - Part 2. J. Fluid Mech., vol. 109, 311–337.
(18) Sumer, B.M. and Oguz, B. (1978): Particle motions near the bottom in turbulent flow in an open channel. J. Fluid Mech., vol. 86, 109–127.
(19) Taylor, G.I. (1935): Statistical theory of turbulence. Proc. R. Soc. Lond. A, vol. 151, 421–444.

(20) Taylor, G.I. (1938): The spectrum of turbulence. Proc. R. Soc. Lond. A, vol. 132, 476–490.
(21) Taylor, G.I. and Green, A.E. (1937): Mechanism of the production of small eddies from large ones. Proc. Roy. Soc. Lond. A, vol. 158, 499–521.
(22) Tennekes, H. and Lumley, J.L. (1972): A First Course in Turbulence. MIT Press.
(23) von Karman, T. (1948): Progress in the statistical theory of turbulence. Proc. Nat. Acad. Sci., vol. 34, 530–539.
(24) von Karman, T. and Howarth, L. (1938): On the statistical theory of isotropic turbulence. Proc. Roy. Soc. Lond. A, vol. 164, 192–215.
(25) Westfall, P.H. (2014): Kurtosis as peakedness, 1904–2014, R.I.P. The American Statistician, vol. 68, 191–195.
(26) Wilcox, D.C. (2006): Turbulence Modeling for CFD, 3rd edn. DCW Industries.

Appendix to Chapter 4: Proofs of some common isotropic turbulence relationships

This appendix serves as a companion to Chapter 4. Within, proofs are offered for any of the previously presented (but not yet formally established) relationships valid for isotropic turbulence, such that the entirety of this chapter is self contained.

4.A Proof of the isotropic relationship between the correlation functions $f(r)$ and $g(r)$, Eq. 4.24

Consider an isotropic, homogeneous turbulence in an incompressible fluid. From local continuity we have

$$\frac{\partial u_i'}{\partial x_i} = \frac{\partial u_i'}{\partial r_i} = 0 \tag{4.160}$$

where the vector $\mathbf{r}$ defines the position of a point relative to another having position $\mathbf{x}$. The two-point velocity correlation tensor may be written as

$$Q_{ij}(\mathbf{r}) = \overline{u_i'(\mathbf{x})u_j'(\mathbf{x}+\mathbf{r})} \tag{4.161}$$

with derivative

$$\begin{aligned}\frac{\partial Q_{ij}(\mathbf{r})}{\partial r_j} &= \overline{u_i'(\mathbf{x})\frac{\partial u_j'(\mathbf{x}+\mathbf{r})}{\partial r_j}} + \overline{u_j'(\mathbf{x}+\mathbf{r})\frac{\partial u_i'(\mathbf{x})}{\partial r_j}} \\ &= \overline{u_i'(\mathbf{x})\frac{\partial u_j'(\mathbf{x}+\mathbf{r})}{\partial r_j}} = 0\end{aligned} \tag{4.162}$$

since $u_i'(\mathbf{x})$ is not a function of $\mathbf{r}$, and and where local continuity has been invoked. Now let $Q_{ij}(\mathbf{r})$ be a general isotropic tensor of second order, which is a function of $\mathbf{r}$ alone (i.e. involving velocities at two points only) (Batchelor, 1960, p. 42)

$$Q_{ij}(\mathbf{r}) = Ar_ir_j + B\delta_{ij} \tag{4.163}$$

where the scalar coefficients A and B are even functions of r. This gives, for example

$$Q_{11}(r) = \overline{u_1'(x)u_1'(x+r)} = Ar_1r_1 + B = Ar^2 + B \tag{4.164}$$

$$Q_{22}(r) = \overline{u_2'(x)u_2'(x+r)} = Ar_2r_2 + B = B \tag{4.165}$$

where the coordinate system has been rotated such that $r_1 = r$ and $r_2 = 0$ i.e. the vector r_i is aligned with the x_1 axis. This freedom is allowed due to the rotational invariance inherent within isotropic turbulence, whereby any convenient orientation may be assumed. Similarly, differentiating Eq. 4.163 leads to

$$\begin{aligned}\frac{\partial Q_{ij}(\mathbf{r})}{\partial r_j} &= \frac{\partial}{\partial r_j}\left(Ar_ir_j + B\delta_{ij}\right)\\ &= A\left(r_i\frac{\partial r_j}{\partial r_j} + r_j\frac{\partial r_i}{\partial r_j}\right) + r_ir_j\frac{\partial A}{\partial r_j} + \delta_{ij}\frac{\partial B}{\partial r_j} = 0\end{aligned} \tag{4.166}$$

Note that

$$\frac{\partial r_j}{\partial r_j} = \frac{\partial r_1}{\partial r_1} + \frac{\partial r_2}{\partial r_2} + \frac{\partial r_3}{\partial r_3} = 3, \qquad \frac{\partial r_i}{\partial r_j} = \delta_{ij} \tag{4.167}$$

Inserting these, and letting $i = j = 1$ then gives

$$A(3r + r) + r^2\frac{\partial A}{\partial r} + \frac{\partial B}{\partial r} = 0 \tag{4.168}$$

which after division by r reads

$$4A + r\frac{\partial A}{\partial r} + \frac{1}{r}\frac{\partial B}{\partial r} = 0 \tag{4.169}$$

Consider now the correlation functions

$$f(r) = \frac{\overline{u_1'(x)u_1'(x+r)}}{\overline{u'^2}}, \qquad g(r) = \frac{\overline{u_2'(x)u_2'(x+r)}}{\overline{u'^2}} \tag{4.170}$$

Relations Eq. 4.164 and Eq. 4.165 yield the following equalities

$$Ar^2 + B = \overline{u'^2}f(r) \tag{4.171}$$

$$B = \overline{u'^2}g(r) \tag{4.172}$$

Eliminating B and solving for A gives

$$A = \frac{\overline{u'^2}}{r^2}\left(f(r) - g(r)\right) \tag{4.173}$$

From Eq. 4.172 and Eq. 4.173 we have

$$\frac{\partial B}{\partial r} = \overline{u'^2}g'(r) \tag{4.174}$$

$$\frac{\partial A}{\partial r} = \frac{\overline{u'^2}}{r^2}\left(f'(r) - g'(r)\right) - \frac{\overline{2u'^2}}{r^3}\left(f(r) - g(r)\right) \tag{4.175}$$

Inserting these and Eq. 4.173 into Eq. 4.169 gives

$$\frac{\overline{4u'^2}}{r^2}\left(f(r) - g(r)\right) - \frac{\overline{2u'^2}}{r^2}\left(f(r) - g(r)\right) + \frac{\overline{u'^2}}{r}\left(f'(r) - g'(r)\right) + \frac{\overline{u'^2}}{r}\left(g'(r)\right) = 0 \tag{4.176}$$

Simplifying further leads to

$$\frac{2}{r}\left(f(r) - g(r)\right) + f'(r) = \frac{2}{r}f(r) - \frac{2}{r}g(r) + f'(r) = 0 \tag{4.177}$$

Finally, solving for $g(r)$ gives

$$g(r) = f(r) + \frac{r}{2}f'(r) \tag{4.178}$$

This matches the isotropic relation Eq. 4.24, which was used extensively in Chapter 4. This relationship was originally proved by von Karman and Howarth (1938).

4.B Proof of Eqs. 4.30 and 4.31, relating λ_f and λ_g to the variance of fluctuating velocity gradients

From Eq. 4.26, the micro length scale λ_f is defined via

$$\frac{1}{\lambda_f^2} = -\frac{1}{2}\left[\frac{\partial^2 f}{\partial r^2}\right]_{r=0} = -\frac{1}{\overline{2u'^2}}\left[\frac{\partial^2}{\partial r^2}\overline{u'(x)u'(x+r)}\right]_{r=0} \tag{4.179}$$

Since $u'(x)$ is independent of r, we may further reduce this to

$$\frac{1}{\lambda_f^2} = -\frac{1}{\overline{2u'^2}}\left[\overline{u'(x)\frac{\partial^2 u'(x+r)}{\partial r^2}}\right]_{r=0} = -\frac{1}{\overline{2u'^2}}\overline{u'\frac{\partial^2 u'}{\partial r^2}} \tag{4.180}$$

Assuming homogeneous turbulence

$$\frac{\partial}{\partial r}\left(\overline{u'\frac{\partial u'}{\partial r}}\right) = \overline{\left(\frac{\partial u'}{\partial r}\right)^2} + \overline{u'\frac{\partial^2 u'}{\partial r^2}} = 0 \tag{4.181}$$

since there is no change of mean properties with distance. Whence

$$\overline{u'\frac{\partial^2 u'}{\partial r^2}} = -\overline{\left(\frac{\partial u'}{\partial r}\right)^2} \tag{4.182}$$

and we may write Eq. 4.180 as

$$\frac{1}{\lambda_f^2} = \frac{1}{\overline{2u'^2}}\overline{\left(\frac{\partial u'}{\partial r}\right)^2} \tag{4.183}$$

This matches Eq. 4.30. This result was originally shown by Taylor (1935).

Through analogous manipulations of Eq. 4.29, one may likewise derive Eq. 4.31. Note that the proof above has assumed homogeneous, but not necessarily isotropic, turbulence. The proof above is similar in nature to that worked out by Raudkivi and Callander (1975), p. 167.

4.C Proof of isotropic relation Eq. 4.66, the governing equation for the correlation function $Q_{ij}(\mathbf{r})$

Consider a flow field of an incompressible fluid, involving homogeneous and isotropic turbulence. The Navier-Stokes equation written at a point A is

$$\left[\frac{\partial u_i'}{\partial t} = -\left(\bar{u}_k + u_k'\right)\frac{\partial u_i'}{\partial x_k} - \frac{1}{\rho}\frac{\partial p}{\partial x_i} + \nu\frac{\partial^2 u_i'}{\partial x_l \partial x_l}\right]_A \tag{4.184}$$

After multiplication by $(u_j')_B$ this becomes

$$\begin{aligned}(u_j')_B\frac{\partial (u_i')_A}{\partial t} &= -\left(\bar{u}_k + (u_k')_A\right)\left(\frac{\partial}{\partial x_k}\right)_A (u_i')_A(u_j')_B \\ &\quad -\frac{1}{\rho}\left(\frac{\partial}{\partial x_i}\right)_A p_A(u_j')_B + \nu\left(\frac{\partial^2}{\partial x_l\partial x_l}\right)_A (u_i')_A(u_j')_B\end{aligned} \tag{4.185}$$

Analogously, multiplying the Navier-Stokes equation at a point B by $(u_i')_A$ leads to

$$\begin{aligned}(u_i')_A\frac{\partial (u_j')_B}{\partial t} &= -\left(\bar{u}_k + (u_k')_B\right)\left(\frac{\partial}{\partial x_k}\right)_B (u_j')_B(u_i')_A \\ &\quad -\frac{1}{\rho}\left(\frac{\partial}{\partial x_j}\right)_B p_B(u_i')_A + \nu\left(\frac{\partial^2}{\partial x_l\partial x_l}\right)_B (u_j')_B(u_i')_A\end{aligned} \tag{4.186}$$

Due to local continuity

$$(u_j')_B\left(u_i'\frac{\partial u_k'}{\partial x_k}\right)_A = (u_i')_A\left(u_j'\frac{\partial u_k'}{\partial x_k}\right)_B = 0 \tag{4.187}$$

Therefore

$$\begin{aligned}(u_k')_A\left(\frac{\partial}{\partial x_k}\right)_A (u_i')_A(u_j')_B =& (u_k')_A\left(\frac{\partial}{\partial x_k}\right)_A (u_i')_A(u_j')_B \\ &+ (u_j')_B\left(u_i'\frac{\partial u_k'}{\partial x_k}\right)_A \\ =& (u_j')_B\left[u_k'\frac{\partial u_i'}{\partial x_k} + u_i'\frac{\partial u_k'}{\partial x_k}\right]_A \\ =& \left(\frac{\partial}{\partial x_k}\right)_A (u_i')_A(u_k')_A(u_j')_B\end{aligned} \tag{4.188}$$

Analogously

$$(u_k')_B\left(\frac{\partial}{\partial x_k}\right)_B (u_j')_B(u_i')_A = \left(\frac{\partial}{\partial x_k}\right)_B (u_i')_A(u_k')_B(u_j')_B \tag{4.189}$$

Now note that

$$\frac{\partial}{\partial t}(u_i')_A(u_j')_B = (u_i')_A\frac{\partial}{\partial t}(u_j')_B + (u_j')_B\frac{\partial}{\partial t}(u_i')_A \tag{4.190}$$

Inserting Eq. 4.185 and Eq. 4.186 into the above then yields

$$\begin{aligned}\frac{\partial}{\partial t}(u_i')_A(u_j')_B = &-\bar{u}_k\left[\left(\frac{\partial}{\partial x_k}\right)_A(u_i')_A(u_j')_B + \left(\frac{\partial}{\partial x_k}\right)_B(u_j')_B(u_i')_A\right]\\ &-\left[\left(\frac{\partial}{\partial x_k}\right)_A(u_i')_A(u_k')_A(u_j')_B + \left(\frac{\partial}{\partial x_k}\right)_B(u_i')_A(u_k')_B(u_j')_B\right]\\ &-\frac{1}{\rho}\left[\left(\frac{\partial}{\partial x_i}\right)_A p_A(u_j')_B + \left(\frac{\partial}{\partial x_j}\right)_B p_B(u_i')_A\right]\\ &+\nu\left[\left(\frac{\partial^2}{\partial x_k\partial x_k}\right)_A + \left(\frac{\partial^2}{\partial x_k\partial x_k}\right)_B\right](u_i')_A(u_j')_B\end{aligned} \tag{4.191}$$

Time-averaging leaves

$$\begin{aligned}\frac{\partial}{\partial t}\overline{(u_i')_A(u_j')_B} = &-\left[\left(\frac{\partial}{\partial x_k}\right)_A\overline{(u_i')_A(u_k')_A(u_j')_B}\right.\\ &\left.+\left(\frac{\partial}{\partial x_k}\right)_B\overline{(u_i')_A(u_k')_B(u_j')_B}\right]\\ &-\frac{1}{\rho}\left[\left(\frac{\partial}{\partial x_i}\right)_A\overline{p_A(u_j')_B} + \left(\frac{\partial}{\partial x_j}\right)_B\overline{p_B(u_i')_A}\right]\\ &+\nu\left[\left(\frac{\partial^2}{\partial x_k\partial x_k}\right)_A + \left(\frac{\partial^2}{\partial x_k\partial x_k}\right)_B\right]\overline{(u_i')_A(u_j')_B}\end{aligned} \tag{4.192}$$

Note that, following Hinze (1959), p. 32, the terms involving $\bar{u}_k$ have now vanished. The resulting expression Eq. 4.192 therefore describes the relations between the turbulence motions as seen from a coordinate system moving with mean velocity $\bar{u}_k$. Now let

$$r_k = (x_k)_B - (x_k)_A \tag{4.193}$$

hence

$$\left(\frac{\partial}{\partial x_k}\right)_A = -\frac{\partial}{\partial r_k}, \qquad \left(\frac{\partial}{\partial x_k}\right)_B = \frac{\partial}{\partial r_k} \tag{4.194}$$

$$\left(\frac{\partial^2}{\partial x_k\partial x_k}\right)_A = \left(\frac{\partial^2}{\partial x_k\partial x_k}\right)_B = \frac{\partial^2}{\partial r_k\partial r_k} \tag{4.195}$$

We then obtain

$$\frac{\partial}{\partial t}\overline{(u_i')_A\,(u_j')_B} = \left[\frac{\partial}{\partial r_k}\overline{(u_i')_A(u_k')_A(u_j')_B} - \frac{\partial}{\partial r_k}\overline{(u_i')_A(u_k')_B(u_j')_B}\right]$$
$$+\frac{1}{\rho}\left[\frac{\partial}{\partial r_i}\overline{p_A(u_j')_B} - \frac{\partial}{\partial r_j}\overline{p_B(u_i')_A}\right]$$
$$+ 2\nu\frac{\partial^2}{\partial r_k \partial r_k}\overline{(u_i')_A(u_j')_B} \qquad (4.196)$$

We will now focus specifically on the terms like $\overline{p_A(u_j')_B}$ and $\overline{p_B(u_i')_A}$. As an example, the first of these is comprised of the following three components

$$\overline{p_A(u_1')_B}, \qquad \overline{p_A(u_2')_B}, \qquad \overline{p_A(u_3')_B} \qquad (4.197)$$

Let us now consider a situation where r_i is aligned with the x_1 axis, such that $r_1 = r$, $r_2 = r_3 = 0$. Under this orientation, a rotation of 180° about the x_1 axis would imply that

$$\overline{p_A(u_2')_B} = \overline{-p_A(u_2')_B}, \qquad \overline{p_A(u_3')_B} = \overline{-p_A(u_3')_B} \qquad (4.198)$$

which can only hold if

$$\overline{p_A(u_2')_B} = \overline{p_A(u_3')_B} = 0 \qquad (4.199)$$

Hence, the last two components of Eq. 4.197 can be neglected due to the assumed rotational invariance inherent within an isotropic turbulence field. The first component of Eq. 4.197, however, can apparently not be eliminated due to isotropy alone. To investigate this component further, we will consider $\overline{p_A(u_j')_B}$ as a general first-order isotropic tensor that is a function of r_i alone (Batchelor, 1960, p. 42, 51)

$$Q_j = \overline{p_A(u_j')_B} = Ar_j \qquad (4.200)$$

where A is an even function of r. Invoking the product rule

$$\left(\frac{\partial Q_j}{\partial r_j}\right)_B = \left(\frac{\partial}{\partial r_j}\right)_B \overline{p_A(u_j')_B} = \overline{p_A\left(\frac{\partial u_j'}{\partial r_j}\right)_B} + \overline{(u_j')_B\left(\frac{\partial p_A}{\partial r_j}\right)_B} = 0 \qquad (4.201)$$

It is emphasized that the first of the expanded terms in the above expression is zero only due to assumed incompressibility of the fluid. Equivalently, using the product rule on the right hand side of Eq. 4.200, we have

$$\frac{\partial Q_j}{\partial r_j} = \frac{\partial (Ar_j)}{\partial r_j} = A\frac{\partial r_j}{\partial r_j} + r_j\frac{\partial A}{\partial r_j} = 3A + r\frac{\partial A}{\partial r} = 0 \qquad (4.202)$$

where the above-mentioned orientation has again been invoked. The right-most differential equation has the general solution $A = C/r^3$, where C is a constant, implying that

$$\overline{p_A(u_1')_B} = Ar = \frac{C}{r^2} \tag{4.203}$$

Because $\overline{p_A(u_1')_B}$ cannot become infinite when $r = 0$, the constant C must therefore be zero. Hence

$$\overline{p_A(u_1')_B} = 0 \tag{4.204}$$

Combining this result with Eq. 4.199, along with analogous reasoning for the other like terms, we can now generally state that, under the assumed conditions

$$\overline{p_A(u_j')_B} = \overline{p_B(u_i')_A} = 0 \tag{4.205}$$

Terms involving these quantities can therefore be eliminated from Eq. 4.196, which then reduces to

$$\frac{\partial(Q_{ij})_{AB}}{\partial t} = \frac{\partial(S_{ik,j})_{AB}}{\partial r_k} - \frac{\partial(S_{i,kj})_{AB}}{\partial r_k} + 2\nu\frac{\partial^2}{\partial r_k \partial r_k}(Q_{ij})_{AB} \tag{4.206}$$

where

$$(Q_{ij})_{AB} = \overline{(u_i')_A(u_j')_B} \tag{4.207}$$

$$(S_{ik,j})_{AB} = \overline{(u_i')_A(u_k')_A(u_j')_B} \tag{4.208}$$

$$(S_{i,kj})_{AB} = \overline{(u_i')_A(u_k')_B(u_j')_B} \tag{4.209}$$

Furthermore, due to the assumed invariance under reflection with respect to point A (Hinze, 1959, p. 158)

$$\overline{(u_i')_A(u_k')_B(u_j')_B} = -\overline{(u_k')_A(u_j')_A(u_i')_B} \tag{4.210}$$

or equivalently

$$(S_{i,kj})_{AB} = -(S_{kj,i})_{AB} \tag{4.211}$$

Hence, we may write

$$\frac{\partial(Q_{ij})_{AB}}{\partial t} = \frac{\partial(S_{ik,j})_{AB}}{\partial r_k} + \frac{\partial(S_{kj,i})_{AB}}{\partial r_k} + 2\nu\frac{\partial^2}{\partial r_k \partial r_k}(Q_{ij})_{AB} \tag{4.212}$$

Defining

$$(S_{i,j})_{AB} = \frac{\partial(S_{ik,j})_{AB}}{\partial r_k} + \frac{\partial(S_{kj,i})_{AB}}{\partial r_k} \tag{4.213}$$

we finally obtain

$$\frac{\partial(Q_{ij})_{AB}}{\partial t} = (S_{i,j})_{AB} + 2\nu\frac{\partial^2}{\partial r_k \partial r_k}(Q_{ij})_{AB} \tag{4.214}$$

This matches the isotropic relation Eq. 4.66!

4.D Proof that the integral of $T(k)$ over all wave numbers is zero, Eq. 4.77

We will here prove Eq. 4.77, which states:

$$\int_0^\infty T(k)dk = 0, \qquad T(k) = 2\pi k^2 \Omega_{ii}(k) \tag{4.215}$$

where the definition Eq. 4.75 has been invoked. Setting $i = j$ in Eq. 4.72, and accounting for spherical symmetry in the wave number domain, leads to the single integral

$$S_{ii}(r) = \int_{-\infty}^\infty 4\pi k^2 \Omega_{ii}(k) e^{-ikr} dk \tag{4.216}$$

From the above, it is now clear that

$$\int_0^\infty T(k)dk = \int_0^\infty 2\pi k^2 \Omega_{ii}(k)dk = \frac{1}{2} S_{ii}(0) = \text{Constant} \tag{4.217}$$

Momentarily neglecting viscous effects, we may write Eq. 4.76 as:

$$\frac{\partial}{\partial t}\frac{K_t}{\rho} = \text{Constant} \tag{4.218}$$

from which is now clear that the constant must, in fact, be zero, as any other value would violate energy conservation. Hence, this proves Eq. 4.77 $\equiv$ Eq. 4.215!

It can be noted that Hinze (1959), p. 178, has also proved Eq. 4.215, albeit in a slightly more formal way.

4.E Proof of the isotropic relationship between the one-dimensional spectra $F(k)$ and $G(k)$, Eq. 4.118

In isotropic turbulence, from Eqs. 4.113 and 4.116 we have

$$\overline{u_1'^2} f(r) = 2\int_0^\infty F(k)\cos(kr)dk \tag{4.219}$$

$$\overline{u_1'^2} g(r) = 2\int_0^\infty G(k)\cos(kr)dk \tag{4.220}$$

Inspired by the last term in Eq. 4.178, let us first focus on the term

$$\overline{u_1'^2}\frac{r}{2} f'(r) = \int_0^\infty kF(k)\cdot(-r\sin(kr))dk \tag{4.221}$$

Applying definite integration by parts

$$\int_a^b f(k)g'(k)dk = [f(k)g(k)]_a^b - \int_a^b g(k)f'(k)dk \qquad (4.222)$$

with local functions (for clarity, it is emphasized that $f(k)$ and $g(k)$ above have nothing to do with the correlation functions $f(r)$ and $g(r)$!),

$$f(k) = kF(k), \qquad g'(k) = -r\sin(kr) \qquad (4.223)$$

$$f'(k) = F(k) + kF'(k), \qquad g(k) = \cos(kr) \qquad (4.224)$$

we obtain

$$\begin{aligned}\overline{u_1'^2}\frac{r}{2}f'(r) &= [kF(k)\cos(kr)]_0^\infty - \int_0^\infty \cos(kr)\left[F(k) + kF'(k)\right]dk \\ &= -\int_0^\infty \cos(kr)\left[F(k) + kF'(k)\right]dk \qquad (4.225)\end{aligned}$$

where it has been utilized that

$$[kF(k)\cos(kr)]_0^\infty = kF(k)\cos(kr)|_{k=\infty} = 0 \qquad (4.226)$$

It is now seen from Eqs. 4.219, 4.220, and 4.225 that all three terms within Eq. 4.178 have identical form, whence

$$2G(k) = 2F(k) - F(k) - kF'(k) \qquad (4.227)$$

which readily simplifies to

$$G(k) = \frac{1}{2}\left[F(k) - kF'(k)\right] \qquad (4.228)$$

This proves the relationship between $G(k)$ and $F(k)$ in isotropic turbulence, previously given as Eq. 4.118.

4.F Proof of the isotropic relationship between $E(k)$ and $F(k)$, Eq. 4.119

We shall now prove the isotropic relationship between the one-dimensional spectrum $F(k)$ and the three-dimensional energy spectrum function $E(k)$, previously given as Eq. 4.119, which reads:

$$E(k) = k^2F''(k) - kF'(k) \qquad (4.229)$$

Integrating both sides over all wave numbers, we obtain

$$\frac{1}{2}\overline{u_i'u_i'} = \int_0^\infty E(k)dk = \int_0^\infty \left(k^2F''(k) - kF'(k)\right)dk \qquad (4.230)$$

Making use of definite integration by parts, Eq. 4.222, on both terms on the right hand side, this becomes

$$\frac{1}{2}\overline{u_i'u_i'} = \left[k^2F'(k)\right]_0^\infty - \int_0^\infty 2kF'(k)dk - [kF(k)]_0^\infty + \int_0^\infty F(k)dk \quad (4.231)$$

Employing definite integration by parts once again, the second term on the right hand side is equivalently

$$-\int_0^\infty 2kF'(k)dk = -\left[2kF(k)\right]_0^\infty + \int_0^\infty 2F(k)dk \quad (4.232)$$

Inserting this and simplifying, we then have

$$\frac{1}{2}\overline{u_i'u_i'} = \left[k^2F'(k)\right]_0^\infty - 3\left[kF(k)\right]_0^\infty + 3\int_0^\infty F(k)dk \quad (4.233)$$

Since $F(\infty) = F'(\infty) = 0$, this reduces to

$$\frac{1}{2}\overline{u_i'u_i'} = 3\int_0^\infty F(k)dk \quad (4.234)$$

Now, as demonstrated in Eq. 4.114:

$$\int_0^\infty F(k)dk = \frac{1}{2}\overline{u_1'^2} \quad (4.235)$$

Thus, we have simply

$$\frac{1}{2}\overline{u_i'u_i'} = \frac{1}{2}\left(\overline{u_1'^2} + \overline{u_2'^2} + \overline{u_3'^2}\right) = \frac{3}{2}\overline{u_1'^2} \quad (4.236)$$

which obviously holds for isotropic turbulence, where all fluctuating components have identical variance. Hence, we have proved the isotropic relation Eq. 4.119 $\equiv$ Eq. 4.229!

Note that an alternative proof can be found e.g. in Raudkivi and Callander (1975), p. 174.

4.G Proof of the isotropic dissipation rate ε, Eq. 4.83

Consider again an isotropic, homogeneous turbulence field in an incompressible fluid. We begin from the general expression for the dissipation rate (per unit mass), as given previously in Eq. 4.82, which reads:

$$\varepsilon = \nu\overline{\left(\frac{\partial u_i'}{\partial x_j} + \frac{\partial u_j'}{\partial x_i}\right)\frac{\partial u_i'}{\partial x_j}} = \nu\left[\overline{\left(\frac{\partial u_i'}{\partial x_j}\right)^2} + \overline{\frac{\partial u_j'}{\partial x_i}\frac{\partial u_i'}{\partial x_j}}\right] \quad (4.237)$$

Consider now the derivatives in the last term; Freely adding an additional term, equal to zero due to local continuity, and simplifying leads to:

$$\overline{\frac{\partial u_j'}{\partial x_i}\frac{\partial u_i'}{\partial x_j}} = \overline{\frac{\partial u_j'}{\partial x_i}\frac{\partial u_i'}{\partial x_j}} + \overline{u_j'\frac{\partial^2 u_i'}{\partial x_i \partial x_j}} = \overline{\frac{\partial}{\partial x_i}\left(u_j'\frac{\partial u_i'}{\partial x_j}\right)} \tag{4.238}$$

We may similarly add an additional zero quantity within the final term above, and find

$$\overline{\frac{\partial}{\partial x_i}\left(u_j'\frac{\partial u_i'}{\partial x_j}\right)} = \frac{\partial}{\partial x_i}\left(\overline{u_j'\frac{\partial u_i'}{\partial x_j}} + \overline{u_i'\frac{\partial u_j'}{\partial x_j}}\right) = \frac{\partial}{\partial x_i}\frac{\partial \overline{\left(u_i'u_j'\right)}}{\partial x_j} \tag{4.239}$$

Due to assumed homogeneity, this final term is obviously zero, as the turbulence quantities cannot vary in space. Hence, we have now proved that

$$\overline{\frac{\partial u_j'}{\partial x_i}\frac{\partial u_i'}{\partial x_j}} = \frac{\partial}{\partial x_i}\frac{\partial \overline{\left(u_i'u_j'\right)}}{\partial x_j} = 0 \tag{4.240}$$

whereby Eq. 4.237 reduces to

$$\varepsilon = \nu\overline{\left(\frac{\partial u_i'}{\partial x_j}\right)^2} = \nu\overline{\frac{\partial u_i'}{\partial x_j}\frac{\partial u_i'}{\partial x_j}} \tag{4.241}$$

Expanding this yields a summation of nine quantities:

$$\begin{aligned}\varepsilon = \nu\Bigg[&\overline{\left(\frac{\partial u_1'}{\partial x_1}\right)^2} + \overline{\left(\frac{\partial u_2'}{\partial x_1}\right)^2} + \overline{\left(\frac{\partial u_3'}{\partial x_1}\right)^2}\\ &+\overline{\left(\frac{\partial u_1'}{\partial x_2}\right)^2} + \overline{\left(\frac{\partial u_2'}{\partial x_2}\right)^2} + \overline{\left(\frac{\partial u_3'}{\partial x_2}\right)^2}\\ &+\overline{\left(\frac{\partial u_1'}{\partial x_3}\right)^2} + \overline{\left(\frac{\partial u_2'}{\partial x_3}\right)^2} + \overline{\left(\frac{\partial u_3'}{\partial x_3}\right)^2}\Bigg]\end{aligned} \tag{4.242}$$

For isotropic turbulence, it follows immediately that

$$\overline{\left(\frac{\partial u_1'}{\partial x_1}\right)^2} = \overline{\left(\frac{\partial u_2'}{\partial x_2}\right)^2} = \overline{\left(\frac{\partial u_3'}{\partial x_3}\right)^2} \tag{4.243}$$

Now consider the isotropic relations for the micro scales, corresponding to Eqs. 4.30 and 4.31. Taking $r = x_1$, and utilizing that $\overline{u_1'^2} = \overline{u_2'^2}$, these may be written as:

$$\frac{1}{\lambda_f^2} = \frac{1}{2\overline{u_1'^2}}\overline{\left(\frac{\partial u_1'}{\partial x_1}\right)^2} \tag{4.244}$$

$$\frac{1}{\lambda_g^2} = \frac{1}{2\overline{u_1'^2}}\overline{\left(\frac{\partial u_2'}{\partial x_1}\right)^2} \tag{4.245}$$

As already established, the two micro scales are related; invoking Eq. 4.33, or equivalently

$$\lambda_f^2 = 2\lambda_g^2 \tag{4.246}$$

within Eq. 4.244 and Eq. 4.245 leads, after simplification, to the equality

$$\overline{\left(\frac{\partial u_1'}{\partial x_1}\right)^2} = \frac{1}{2}\overline{\left(\frac{\partial u_2'}{\partial x_1}\right)^2} \tag{4.247}$$

It then follows, due to isotropy, that similarly

$$\begin{aligned}\overline{\left(\frac{\partial u_1'}{\partial x_1}\right)^2} &= \frac{1}{2}\overline{\left(\frac{\partial u_1'}{\partial x_2}\right)^2} = \frac{1}{2}\overline{\left(\frac{\partial u_1'}{\partial x_3}\right)^2} = \frac{1}{2}\overline{\left(\frac{\partial u_2'}{\partial x_1}\right)^2}\\ &= \frac{1}{2}\overline{\left(\frac{\partial u_2'}{\partial x_3}\right)^2} = \frac{1}{2}\overline{\left(\frac{\partial u_3'}{\partial x_1}\right)^2} = \frac{1}{2}\overline{\left(\frac{\partial u_3'}{\partial x_2}\right)^2}\end{aligned} \tag{4.248}$$

After invoking the relations Eq. 4.243 and Eq. 4.248, Eq. 4.242 may be simplified to a single term, involving only the derivative of velocity u_1' with respect to x_1:

$$\varepsilon = 15\nu\overline{\left(\frac{\partial u_1'}{\partial x_1}\right)^2} \tag{4.249}$$

Utilizing Eq. 4.244, or Eq. 4.245 combined with Eq. 4.247, the dissipation rate (per unit mass) ε may be equivalently expressed in terms of either Taylor micro scale

$$\varepsilon = 30\nu\frac{\overline{u_1'^2}}{\lambda_f^2} = 15\nu\frac{\overline{u_1'^2}}{\lambda_g^2} \tag{4.250}$$

These match the previously-provided expressions found in Eq. 4.83! This was originally proved by Taylor (1935).

Likewise, in terms of the turbulent kinetic energy $K_t = 3/2\rho\overline{u_1'^2}$ (or alternatively, the turbulent kinetic energy per unit mass $k = K_t/\rho$, as used in several forthcoming chapters), these are equivalently

$$\varepsilon = \frac{20\nu}{\lambda_f^2}\frac{K_t}{\rho} = \frac{10\nu}{\lambda_g^2}\frac{K_t}{\rho}\left(= \frac{20\nu}{\lambda_f^2}k = \frac{10\nu}{\lambda_g^2}k\right) \tag{4.251}$$

This completes the proof of the dissipation rate (per unit mass) for homogeneous, isotropic turbulence in an incompressible fluid!

It can again be noted that these isotropic relations were originally proved by Taylor (1935). Note that the proof above follows more closely that provided by Raudkivi and Callander (1975), p. 185–186.

4.H References

(1) Batchelor, G.K. (1960): The Theory of Homogeneous Turbulence. Cambridge University Press.
(2) Hinze, J.O. (1959): Turbulence: An Introduction to its Mechanism and Theory. McGraw-Hill.
(3) Raudkivi, A.J. and Callander, R.A. (1975): Advanced Fluid Mechanics. Edward Arnold.
(4) Taylor, G.I. (1935): Statistical theory of turbulence. Proc. R. Soc. Lond. A, vol. 151, 421–444.
(5) von Karman, T. and Howarth, L. (1938). On the statistical theory of isotropic turbulence. Proc. Roy. Soc. Lond. A, vol. 164, 192–215.

Chapter 5

Wave boundary layers

In shallow waters, the orbital motion of water particles under a progressive wave degenerates into a bottom-parallel, straight line, oscillatory motion at the seabed (expressed as $U_0 = U_{0m}\sin(\omega t)$ in Fig. 5.1 for a small amplitude linear (sinusoidal) wave, where ω is the angular frequency, U_0 the streamwise velocity, and t the time). For each half cycle of this oscillatory motion, a "new" time-dependent boundary layer develops over the seabed. This cyclic boundary layer is called the *wave boundary layer*.

Wave boundary layers are of great importance for hydrodynamic processes associated with the sea bottom, such as turbulent mixing of mass and momentum; frictional dissipation and flow resistance; sediment transport processes; exchange of chemicals and organisms between the seabed and the main body of the water; dispersion of dredged and subsequently stockpiled material; etc. Therefore, understanding of the wave boundary layer is essential to develop predictive tools and capabilities related to these processes.

To gain a sense of where real-life wave boundary layers will be present, consider the following figures. First of all, to get an idea of the water depths where a wave boundary layer will exist in practice, we can work out from small amplitude linear wave theory (see the Appendix to the present chapter), that the flow beneath a wind wave with a period of $T = 12$ s will penetrate down to the bed in about 110 m of water or shallower (taking this wave period and depth $h = 110$ m and solving the linear dispersion relation, Eq. 5.182, yields $kh \approx 3$ i.e. near the practical deep water limit, where k is the wave number). This figure is about 30 m (or shallower) for a wave with a period of 6 s. Hence, wave boundary layers can be expected to develop out in the ocean for water depths smaller than say $O(100)$ m. The orbital velocity at the bed can be from nil to $O(2–3)$ m/s, see e.g. Appendix

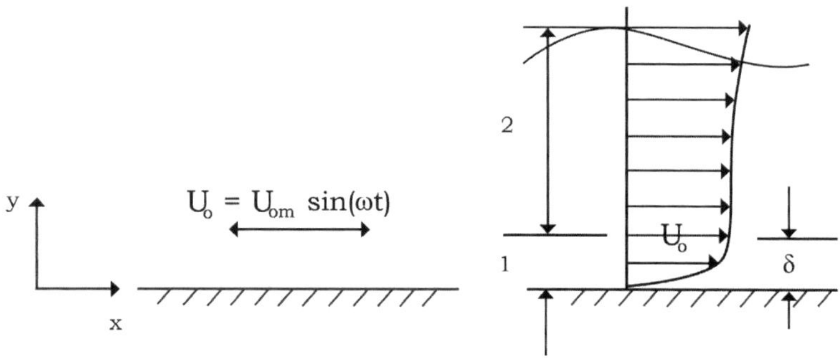

Fig. 5.1 Definition sketch. 1: Wave boundary layer, and 2: Free-stream region.

5.J. The stroke of particle motion at the bed (i.e., two times the amplitude of the particle motion) can be from nil to $O(10)$ m. As will be shown later in the chapter, the thickness of wave boundary layers can be from nil to $O(20-30)$ cm. These figures are for wind-generated-wave boundary layers, and do not cover those related to other types of wave boundary layers (such as tides or tsunami boundary layers). Tsunami wave boundary layers will be discussed separately in Section 5.11.

The wave boundary layers may occur over smooth beds (e.g., fine-sediment beds, clay beds, and freshly-deposited muddy beds), or over rough beds (e.g., sand-covered plane beds, ripple-covered beds, or beds covered with stones/armour blocks for protection work).

The quantities describing the wave boundary layer over a smooth bed depend on the wave-boundary-layer Reynolds number

$$Re = \frac{aU_{0m}}{\nu} \tag{5.1}$$

in which a is the amplitude of the orbital motion of water particles just outside the boundary layer (the so-called free stream region), U_{0m} is the amplitude of the orbital velocity associated with a (Fig. 5.1), and ν is the kinematic viscosity. For a small amplitude progressive linear (sinusoidal) wave, a and U_{0m} are related with $a = U_{0m}/\omega$.

The wave boundary layer undergoes tremendous changes as the Reynolds number is increased from zero. For small Reynolds numbers, the boundary-layer flow will be in the laminar regime. For higher Reynolds numbers, the boundary layer will turn into the so-called transitional boundary layer where signs of turbulence begin to emerge. This occurs at the critical value of the Reynolds number, $Re_{cr} = 1.5 \times 10^5$. For further increase of the Reynolds number, the flow in the boundary layer will be more and more turbulent. Unique to the wave boundary layers, even for very high Reynolds numbers, e.g., such as $O(10^6)$, the flow remains laminar for a small, initial portion of the half cycle until the transition to turbulence sets in, as the flow half cycle progresses, as will be discussed in greater detail later in the chapter.

In the case of wave boundary layers over rough beds, there will be an additional parameter, namely the roughness parameter

$$\frac{a}{k_s} \tag{5.2}$$

in which k_s is Nikuradse's equivalent sand roughness of the bed.

Waves in the field may co-exist with currents, and therefore the wave boundary layers may be influenced by the presence of currents. As will be seen, the wave boundary layer experiences considerable changes in the presence of current, or vice versa. For instance, turbulence in a current may diminish, partially or completely, by the superposition of waves in the combined flow situation, called the re-laminarization, which, along with other changes, will be discussed extensively in the present chapter.

The final section of this chapter is dedicated to the practical turbulence modelling of wave boundary layer flows. In this section the general description of turbulence modelling based on Reynolds-averaged Navier-Stokes (RANS) equations coupled with two-equation turbulence closures presented in Chapter 9, Section 9.4, is simplified to a so-called one-dimensional-vertical (1DV) framework, which is an appropriate description of wave boundary layer flows in many circumstances.

Some familiarity with basic linear potential flow theory is assumed in much of what follows. However, for the un-initiated, some essentials of linear potential flow wave theory are outlined in the present chapter's Appendix.

5.1 Laminar wave boundary layers

5.1.1 *Velocity distribution across the boundary layer depth*

We will in this section find the theoretical laminar solution for the boundary layer that will develop beneath an oscillatory (sinusoidally varying) free stream flow. The streamwise component of the N.-S. equations (Eq. 2.91) in the boundary layer (Fig. 5.1) reads:

$$\frac{\partial u}{\partial t} + u\frac{\partial u}{\partial x} + v\frac{\partial u}{\partial y} = -\frac{1}{\rho}\frac{\partial p}{\partial x} + \frac{1}{\rho}\frac{\partial \tau}{\partial y} \tag{5.3}$$

in which x is the coordinate axis in the horizontal direction, and y the coordinate axis in the vertical direction, u and v are the horizontal and vertical components of the velocity, respectively, p the pressure, τ the shear stress (Eq. 2.83), and ρ the fluid density. See the definition sketch in Fig. 5.1 in which the horizontal component of the orbital velocity just outside the boundary layer is also indicated for a small amplitude progressive linear wave, namely

$$U_0 = U_{0m}\sin(\omega t) \tag{5.4}$$

in which $\omega = 2\pi/T$ is the angular frequency of the wave and T the wave period.

First of all, the vertical component of the velocity v, in the boundary layer is practically zero, and therefore, the third term on the left hand side of Eq. 5.3 (as a first approximation) drops out. Secondly, the x distances can be assumed to be large compared to the y distances in the boundary layer, Fig. 5.2 – the wave length in the laboratory is like $L = O(2\text{–}3\text{ m})$, while the boundary layer thickness is like $\delta = O(1\text{ mm–}1\text{ cm})$; and, likewise, $L = O(100\text{ m})$ in the sea while $\delta = O(10\text{ cm})$. Therefore, the x-dependence in Eq. 5.3 can likewise be neglected as a first approximation. (In fact, one of these results implies the other, simply from inspection of the local continuity equation $\partial u/\partial x + \partial v/\partial y = 0$ i.e. if u varies weakly in x then the vertical velocity v will be small.) Hence, the second term on the left hand side of Eq. 5.3, $u\partial u/\partial x$, is small and can be neglected, again as a first approximation. Note, however, that the pressure gradient term on the right hand side of Eq. 5.3, $-(1/\rho)\partial p/\partial x$, is maintained, as this term actually drives the boundary layer flow. (We will return to this shortly.) Therefore, under the assumptions above, Eq. 5.3 reduces to:

$$\frac{\partial u}{\partial t} = -\frac{1}{\rho}\frac{\partial p}{\partial x} + \frac{1}{\rho}\frac{\partial \tau}{\partial y} \tag{5.5}$$

Now, the boundary layer is driven by the pressure gradient $\partial p/\partial x$, as noted above. This quantity can be calculated from the equation of motion outside the boundary layer, i.e., in the free-stream region, Fig. 5.1 (for a more formal justification of this see Section 5.12.1). The equation of motion in the free-stream region, by setting $\tau = 0$ in Eq. 5.5 reads:

$$\frac{\partial U_0}{\partial t} = -\frac{1}{\rho}\frac{\partial p}{\partial x} \tag{5.6}$$

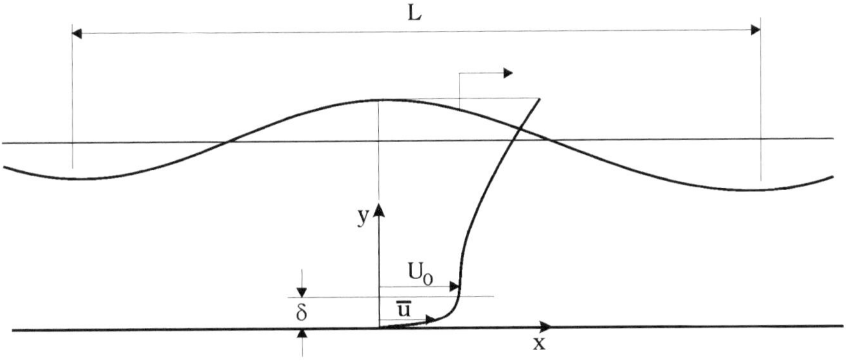

Fig. 5.2 The boundary layer thickness, δ, is small compared with the wave length, L.

in which U_0 for a small amplitude progressive linear wave is given by Eq. 5.4.

From Eqs. 5.5 and 5.6, one gets

$$\frac{\partial u}{\partial t} = \frac{\partial U_0}{\partial t} + \frac{1}{\rho}\frac{\partial \tau}{\partial y} \tag{5.7}$$

in which τ, the shear stress, is (Eq. 2.83)

$$\tau = \mu \frac{\partial u}{\partial y} \tag{5.8}$$

Therefore Eq. 5.7 becomes

$$\frac{\partial u}{\partial t} = \frac{\partial}{\partial t}(U_{0m}\sin(\omega t)) + \nu \frac{\partial^2 u}{\partial y^2} \tag{5.9}$$

As boundary conditions we require that the flow will asymptotically approach the specified free stream velocity:

$$y \to \infty: \qquad u \to U_{0m}\sin(\omega t) \tag{5.10}$$

(in this context $y \to \infty$ represents the location outside the boundary layer). In addition, we require that there will be no slip at the bed:

$$y = 0: \qquad u = 0 \tag{5.11}$$

The solution to Eq. 5.9 with the boundary conditions Eqs. 5.10 and 5.11 is (Batchelor, 1967, p. 354):

$$u(y,t) = U_{0m}\sin(\omega t) - U_{0m}\exp\left(-\frac{y}{\delta_1}\right)\sin\left(\omega t - \frac{y}{\delta_1}\right) \tag{5.12}$$

in which

$$\delta_1 = \left(\frac{2\nu}{\omega}\right)^{\frac{1}{2}} \tag{5.13}$$

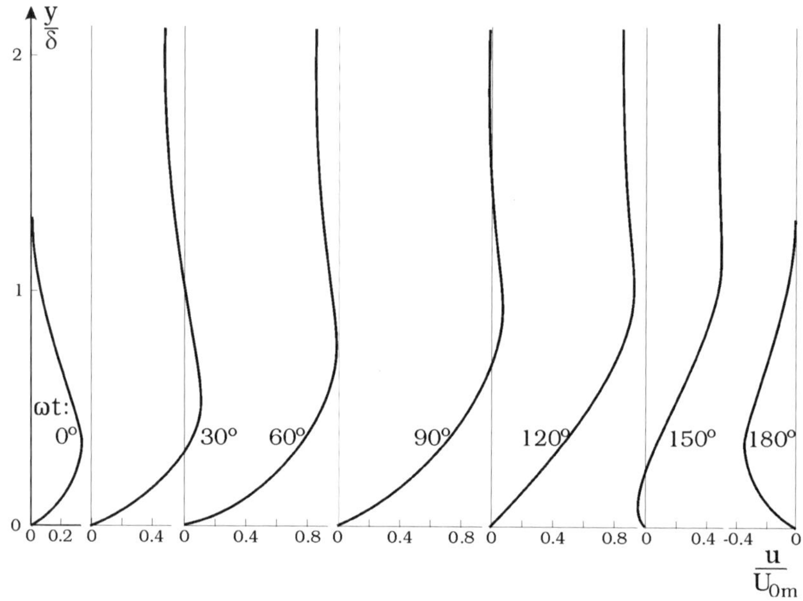

Fig. 5.3 Time development of velocity profiles in laminar wave boundary layer.

is the so-called Stokes length, which can be taken as a scale for the thickness of the boundary layer.

The time development of the velocity distribution is plotted from Eq. 5.12 in Fig. 5.3. In the figure, δ is the boundary layer thickness, defined slightly different from that in Eq. 5.13; namely, δ is the vertical distance from the bed to where the velocity is maximum at the phase value $\omega t = 90°$ (Fig. 5.4); this distance is found from the above solution as

$$\delta = \frac{3\pi}{4}\left(\frac{2\nu}{\omega}\right)^{\frac{1}{2}} = \frac{3\pi}{4}\delta_1 \tag{5.14}$$

or, in terms of the Reynolds number defined by Eq. 5.1,

$$\frac{\delta}{a} = \frac{3\pi}{4}\left(\frac{2}{Re}\right)^{\frac{1}{2}} \tag{5.15}$$

in which a is the amplitude of the orbital motion of water particles just outside of the boundary layer. Taking the position of such a particle as defined by: $x = -a\cos(\omega t)$ for a sinusoidal wave, then the velocity would be: $dx/dt = a\omega \sin(\omega t) = U_{0m}\sin(\omega t)$. It thus follows that the amplitude

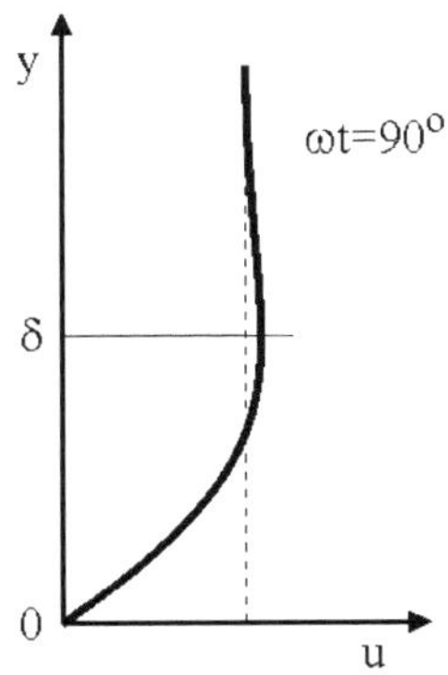

Fig. 5.4 Definition of the boundary layer thickness, δ, the distance from the bed where the velocity becomes maximum at the phase value $\omega t = 90°$.

a corresponds to

$$a = \frac{U_{0m}}{\omega} \tag{5.16}$$

We note that the Reynolds number for oscillatory boundary layers may also be defined in terms of the Stokes length, δ_1:

$$Re_\delta = \frac{\delta_1 U_{0m}}{\nu} \tag{5.17}$$

This Reynolds number is also widely used. The relationship between the two definitions of the Reynolds numbers in Eqs. 5.1 and 5.17 is

$$Re_\delta = \sqrt{2Re} \tag{5.18}$$

Returning to Fig. 5.3, as seen from the figure, the flow near the bed leads over the flow in the free stream region, meaning that the near-bed flow reverses earlier than the free stream flow.

This is related to the driving pressure gradient. From Eqs. 5.6 and 5.4, it can readily be seen that $\partial p/\partial x > 0$ after $\omega t = 90°$ in the $0° < \omega t < 180°$ half period of the motion. This means that the boundary layer will experience an *adverse pressure gradient* (i.e. increasing pressure in the direction of flow) during $90° < \omega t < 180°$ (Fig. 5.5), corresponding the the deceleration stage of the outer flow. This pressure gradient dictates the boundary-layer flow to reverse. However, for the flow to reverse, it will first have to "dissipate" its momentum. Clearly, the near-bed flow dissipates its momentum earlier than the outer flow, meaning that the near-bed flow will respond to this pressure gradient earlier, and therefore will reverse earlier than the outer flow, as revealed by the velocity profiles in Fig. 5.3.

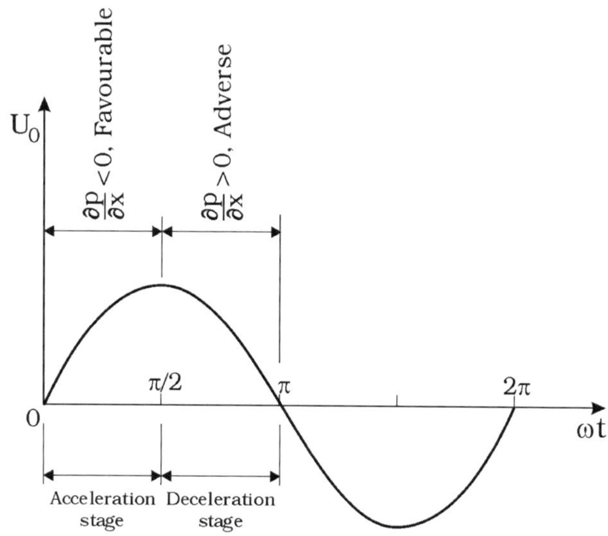

Fig. 5.5 Two different pressure gradients over one half cycle.

It may be noted that the preceding description may also explain the overshooting exhibited by the plotted velocity profiles in Fig. 5.3.

Analytical solutions have been developed for various types of wave boundary layers, such as cnoidal-wave boundary layers (Tanaka et al., 1998), and solitary-wave boundary layers (Tanaka et al., 1998, and Liu et al., 2007). We will return to the latter work under solitary-wave boundary layers in Section 5.10.

5.1.2 *Bed shear stress*

The bed shear stress from the solution, Eq. 5.12, corresponds to:

$$\tau_0 = \mu \left(\frac{\partial u}{\partial y}\right)_{y=0} = \frac{\mu U_{0m}}{\delta_1}[\sin(\omega t) + \cos(\omega t)] \tag{5.19}$$

or

$$\tau_0 = \frac{\sqrt{2}\mu U_{0m}}{\delta_1}\left[\sin(\omega t + \frac{\pi}{4})\right] \tag{5.20}$$

This equation shows that: (1) the bed shear stress leads over the free-stream flow, revealing the previous description of the near-bed boundary-layer process, and (2) the phase lead appears to be equal to $\pi/4$ radians, or

$$\phi = 45^\circ \tag{5.21}$$

meaning that the flow at the bed reverses $\phi = 45^\circ$ earlier than the free stream. We will return to this issue in Sections 5.2.4 and 5.2.6.

5.2 Laminar-to-turbulent transition

5.2.1 *General description: Turbulent spots*

Observations show that an oscillatory (or wave) boundary layer, beneath a free stream velocity given by Eq. 5.4, remains laminar for Reynolds numbers smaller than $Re_{cr} = 1.5 \times 10^5$. Fig. 5.6 gives a definition sketch for such a flow, with the coordinate axes as previously described. Fig. 5.7 shows a time sequence of flow pictures near the bed in a horizontal (x, z) plane at a distance from the bed of $y/\delta_1 = 0.24$ for the Reynolds number $Re = 4.5 \times 10^5$, the flow being visualized by plotting the wall-normal pressure gradient, a good indicator of the presence of turbulence. The wall-normal pressure gradient was obtained from Direct Numerical Simulation (DNS), Jensen et al. (2019). The sequence illustrates how the transition to turbulence develops at various stages of the flow half cycle, namely at $\omega t = 30^\circ$, 45° and 60°.

When inspected closely, it is seen that turbulence first emerges in the form of tiny turbulent patches, the tiny yellow areas in Fig. 5.7, at the phase $\omega t = 30^\circ$ (Fig. 5.7a). It may be noted that Jensen et al.'s (2019) DNS flow visualization at the phase value $\omega t = 15^\circ$ prior to the phase $\omega t = 30^\circ$ shows no sign of turbulence at the phase $\omega t = 15^\circ$ (this is not plotted here for the sake of brevity).

Fig. 5.7 indicates that the aforementioned turbulent patches grow in size, and increase in population, as the flow develops (Fig. 5.7b). They then merge (Fig. 5.7c), eventually forming a fully developed turbulent boundary layer for further phases during the half cycle of the oscillatory flow.

Fig. 5.8 gives a close-up picture (with an accompanying sketch, the bottom panel) of a single turbulent patch in a horizontal $(x,\ z)$ plane near the bed. This picture was obtained in a laboratory experiment by Carstensen et al. (2010). The experiments were carried out in an oscillating water tunnel (see Section 5.3 for a detailed description of this experimental facility). In the experiments, a thin-layered milk (skim milk diluted with water, with ordinary food coloring added, laid on the bed of the tunnel) was used to

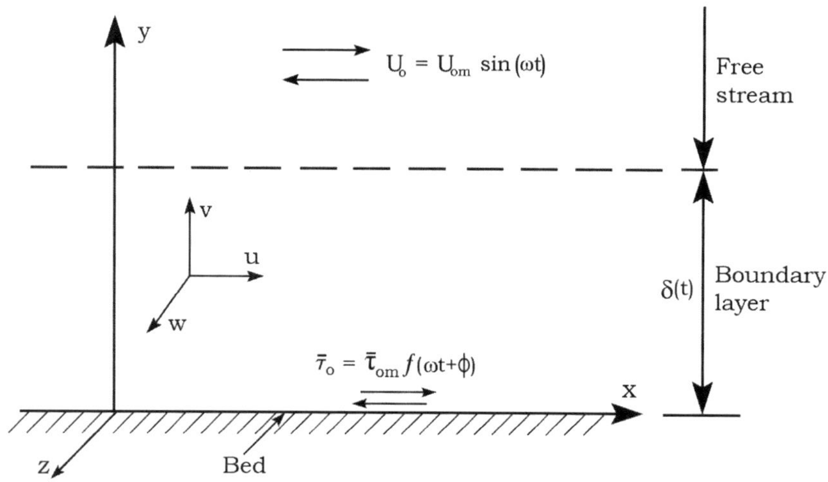

Fig. 5.6 Definition sketch.

make the turbulent areas visible. The Reynolds number corresponding to the flow visualization in Fig. 5.8 was $Re = 3 \times 10^5$, and the video frame in the figure corresponds to the phase value $\omega t = 143°$.

Fig. 5.8 shows that a turbulent patch is actually an isolated area of turbulence close to the bed in an otherwise laminar boundary layer flow. This arrowhead shaped flow feature is called the turbulent spot, first observed by Emmons, 1951, for steady boundary layers; by Carstensen et al., 2010, for oscillatory boundary layers; and by Sumer et al., 2010, for solitary wave boundary layers. The flow in this isolated area "bursts" with violent oscillations. Fig. 5.9 shows a snapshot illustrating multiple turbulent spots, obtained by DNS simulations (Mazzuoli, 2013, p. 51), with $Re = 4.5 \times 10^5$ and $\omega t \cong 140°$, the flow being visualized at $y/\delta_1 = 0.24$.

Fig. 5.10 shows a sequence of video frames (with the accompanying sketches), illustrating the time development of an individual turbulent spot from Carstensen et al.'s (2010) study. (The reader is also referred to the video accompanying Carstensen et al.'s paper, 2010, with the title "supplementary movie 2".) The Reynolds number is $Re = 4.5 \times 10^5$. The flow is from right to left. The top sketch in the figure (Fig. 5.10a) describes various probes appearing in the video frames. The middle, circular-shaped mark represents the brass housing of a hot-film probe, the size 0.2 mm ×

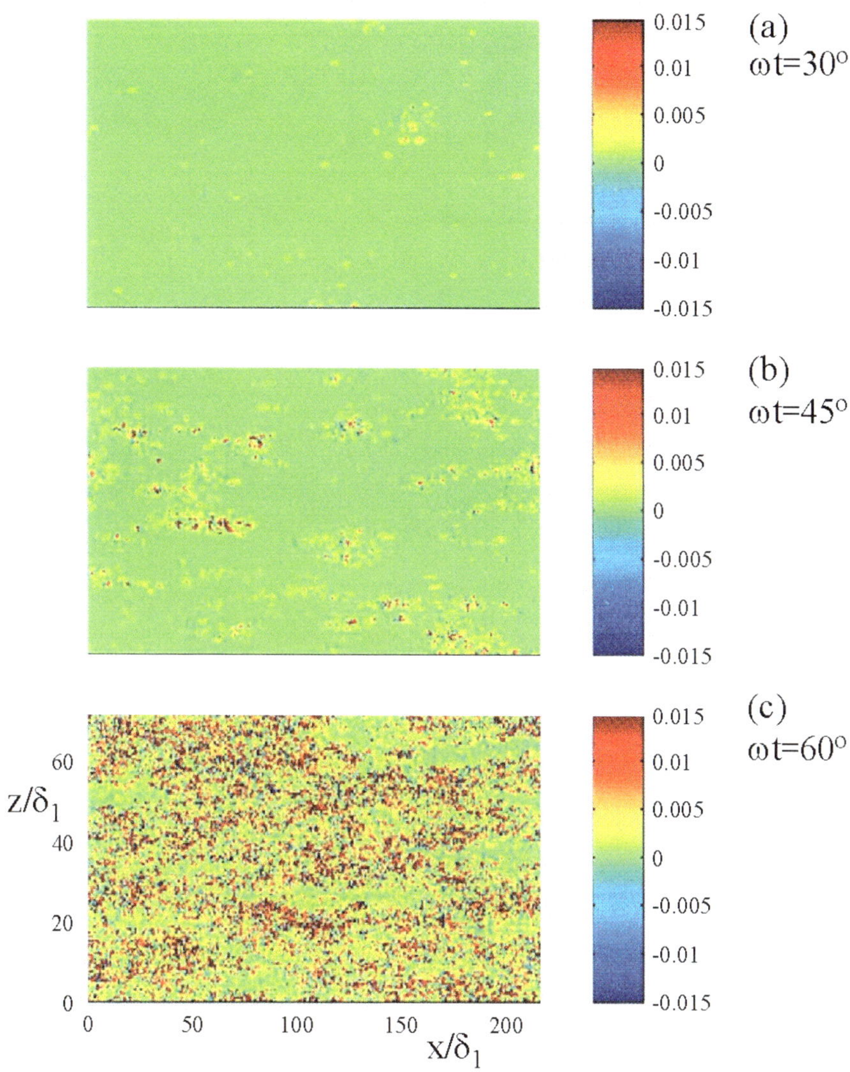

Fig. 5.7 Sequence of flow at horizontal (x, z) plane at a distance from the bed of $y/\delta_1 = 0.24$ for $Re = 4.5 \times 10^5$. Flow is from left to right. The flow is visualized by plotting the wall-normal pressure gradient obtained by DNS simulations. Jensen et al. (2019).

0.75 mm, mounted flush with the bed to measure the bed shear stress, the thick dot at the center of the circular mark. The rest of the circular marks in the definition sketch (Fig. 5.10a), the black circular areas in the video

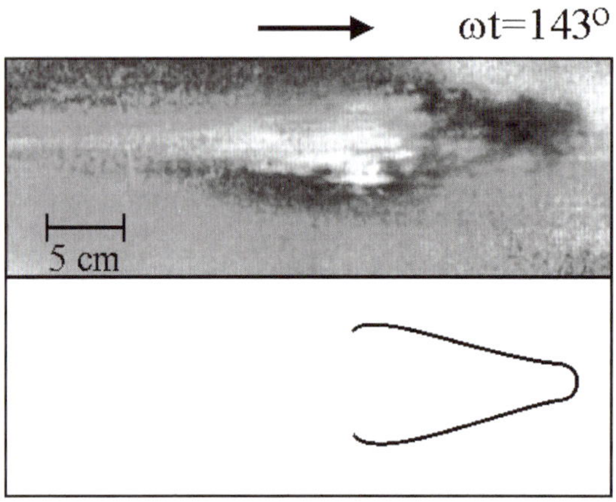

Fig. 5.8 A close-up picture (top panel) of a single turbulent patch with an accompanying sketch (bottom panel) in a horizontal (x, z) plane near the bed, obtained in a laboratory experiment (with $Re = 3 \times 10^5$). Carstensen et al. (2010).

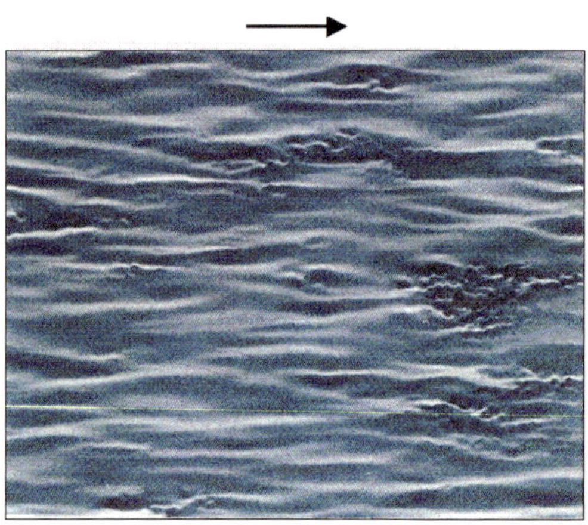

Fig. 5.9 A snapshot illustrating multiple turbulent spots, obtained by DNS simulations. Mazzuoli (2013).

frames, are dummy probes. They are mounted flush with the bottom with zero protrusion. Therefore, the reader should not take any account of these features. We will return to the bed shear stress measurements with the hot-film probe later.

Now, at the phase value $\omega t = 125°$ (Fig. 5.10b), the spot, which first

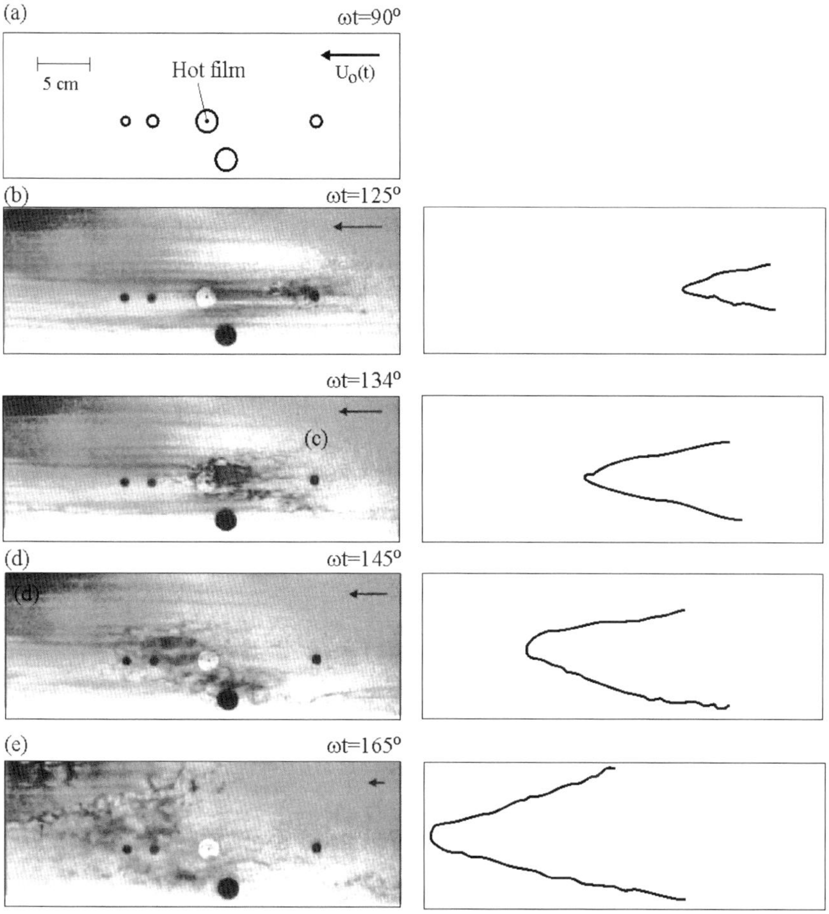

Fig. 5.10 Sequence of video frames illustrating the time development of a turbulent spot. $Re = 4.5 \times 10^5$. Carstensen et al. (2010).

emerged at an earlier phase, is seen to the right of the video frame. It travels to the left as it grows in size during the course of the half cycle (Fig. 5.10b-e). Its life time comes to an end by the time the flow reverses inside the spot. With this, the spot is "destructed" whereby it turns into an incoherent field of "background" turbulence as the boundary layer progresses into the next half-period of the motion. Note that the flow outside the spot reverses not at the same time but somewhat earlier. The delay in the flow reversal inside the turbulent spot is due to turbulence; the near-bed flow is supplied with the momentum-rich fluid continuously from the outer flow due to turbulence, and this delays the flow reversal inside the turbulent spot.

The way in which the spots come into existence can be described with reference to Fig. 5.11. The spot first emerges in the form of two finite-length streaks (the streaky white areas in Fig. 5.11a). Subsequently, the streaks develop twisting and turning motions (Fig. 5.11b and c), which are followed by the appearance of additional longitudinal streaks. The new streaks themselves develop turning and twisting motions and break into smaller structures, presumably resulting in the arrowhead-shaped turbulent spot. This observation is analogous to that made for steady boundary layers, e.g., Gad-el-Hak et al. (1981), Perry et al. (1981), Chambers and Thomas (1983), Browand and Plocher (1985), Seifert and Wygnanski (1995), and Matsubara and Alfredsson (2001).

As already cited in conjunction with Fig. 5.9, recently Mazzuoli et al. (2011) (see also Mazzuoli, 2013) made DNS simulations of the oscillatory boundary layer flow. The results of this latter study confirmed the experimental findings of Carstensen et al. (2010). In the numerical simulations, the turbulent spots appeared at about the end of accelerating phases at the Reynolds numbers investigated in the study, namely $Re = 3 \times 10^5$ and $Re = 4.5 \times 10^5$, in precisely the same manner as in Carstensen et al.'s (2010) observations. Carstensen et al.'s (2010) observations were further confirmed by the study of Scandura et al. (2016) where the latter authors numerically integrated the N.-S. equations for acceleration-skewed oscillatory flows.

Fig. 5.12 illustrates the side view of a turbulent spot at various phase values. The side-view of flow visualization was achieved using fluorescence dye and a vertical sheet of laser. The fluorescence dye was cooled, injected and allowed time to gain the same temperature as the surrounding water in the same way as the dye for the plan-view flow visualizations.

It may be noted that the vertical extent of turbulent spots observed in Carstensen et al.'s (2010) tests was $O(10 \text{ mm})$. Supplementary movie 3 in Carstensen et al. (2010) illustrates the time development of a turbulent spot in side view.

5.2.1.1 *Vortex tubes*

Finally, we note the following in conjunction with Carstensen et al.'s (2010) study. Carstensen et al. (2010) identified a second type of organized structure, which they called vortex tubes. These flow structures are essentially an array of cross-flow, two-dimensional vortices close to the bed extending

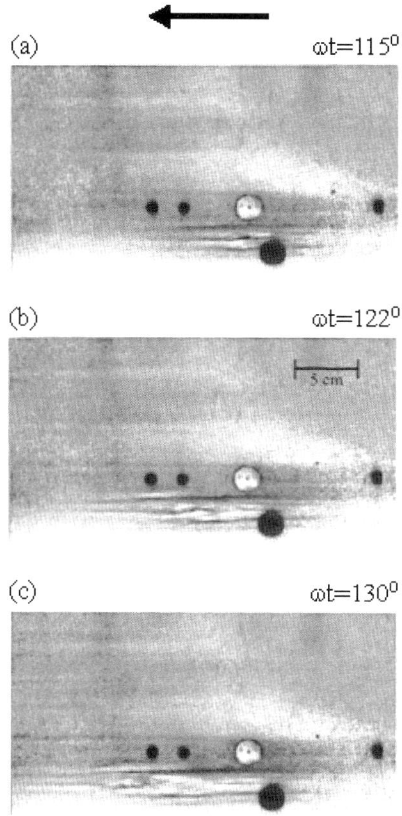

Fig. 5.11 The way in which a turbulent spot comes into existence, first emerging in the form of two finite-length streaks (the streaky white areas in a), then developing twisting and turning motions (b and c). Carstensen et al. (2010).

across the width of the boundary-layer flow (see the definition sketch in the inset in Fig. 5.13), caused by an inflectional-point shear-layer instability.

Fig. 5.13 illustrates the time evolution of the vortex tubes (in plan view) from Carstensen et al.'s (2010) study. These vortex tubes first emerge when $Re = 7 \times 10^4$, and no vortex tubes are observed for Re larger than 3×10^5. Fig. 5.14 illustrates the phase values at which these vortex tubes first emerge in the phase space (Fig. 5.14a) along with those at which they disappear (Fig. 5.14b).

The imprints of these vortices in the bed shear stress is a series of small, insignificant kinks and dips. These high-frequency oscillations in the bed

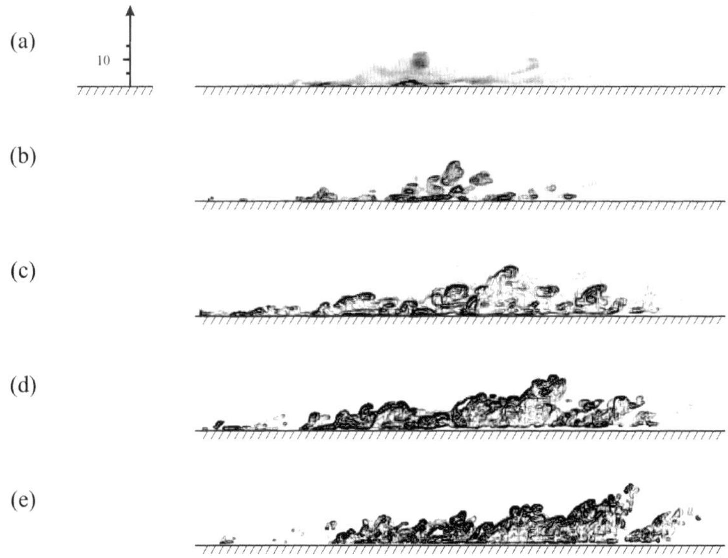

Fig. 5.12 Turbulent spot (side view). Flow direction from left to right: (a) $\omega t = 151°$, (b) $\omega t = 156°$, (c) $\omega t = 160°$, (d) $\omega t = 164°$ and (e) $\omega t = 173°$. Video frames from test 10, half-cycle 7, where $Re_w = 3.3 \times 10^5$. Vertical unit in millimeters. Carstensen (2006).

shear stress may be viewed as "turbulence". However, this turbulence is limited to the life time of the vortex tubes, and its magnitude is significantly (an order of magnitude) smaller than that introduced by turbulent spots and eventually by the boundary-layer process. See Carstensen et al. (2010) for further discussion.

The aforementioned flow features (vortex tubes) were later obtained by DNS simulations in the studies of Scandura (2013), Ozdemir et al. (2014) and Mazzuoli and Vittori (2019). Note that the wall was rough in the study of Mazzuoli and Vittori.

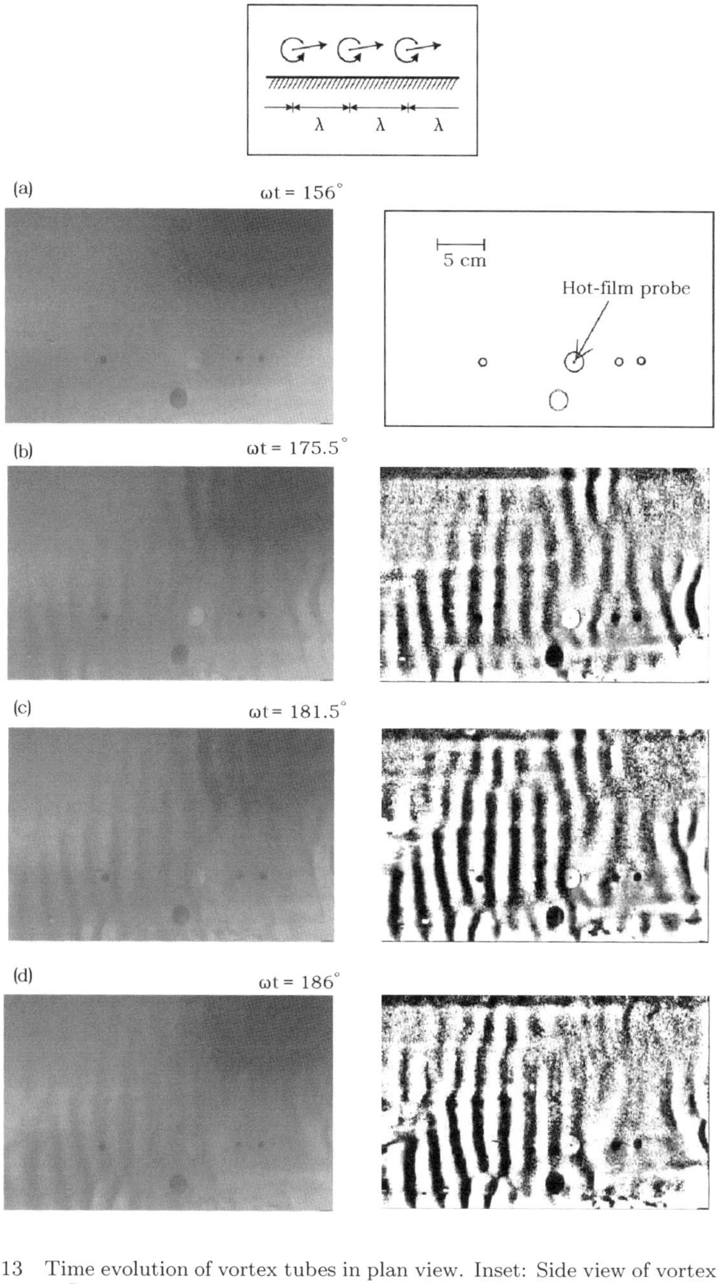

Fig. 5.13 Time evolution of vortex tubes in plan view. Inset: Side view of vortex tubes. $Re = 2.9 \times 10^5$. Flow where the vortex tubes are located is from left to right. Carstensen et al. (2010).

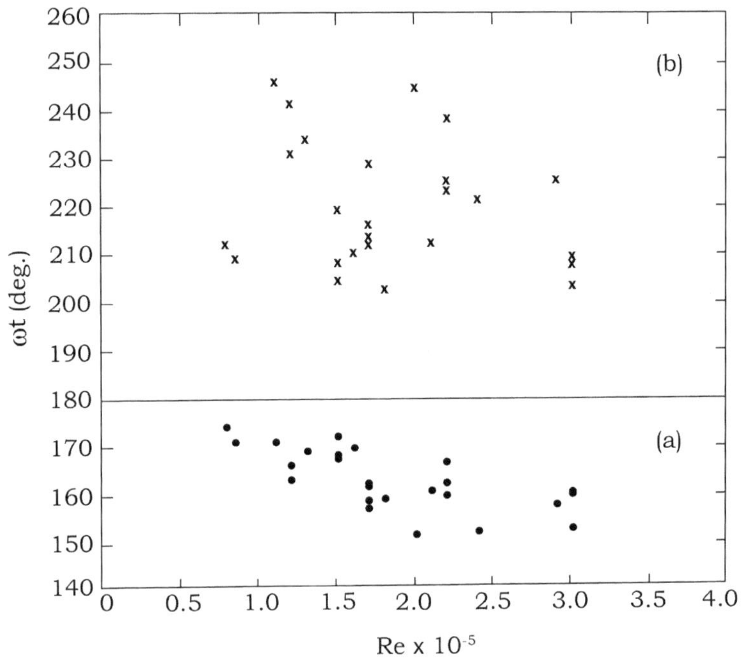

Fig. 5.14 Phase values of inception (a) and disappearance (b) of vortex tubes.

5.2.2 *Critical Reynolds number for transition*

In Carstensen et al.'s (2010) experiments, no turbulent spots were observed for Reynolds numbers smaller than 1.5×10^5, marking the critical value of the Reynolds number for transition to turbulence. Carstensen et al.'s (2010) observations have revealed for the first time that the inception of turbulence is associated with the generation of turbulent spots, an important feature not disclosed for oscillatory boundary-layer flows by the previous experimental and theoretical investigations.

The previously-mentioned critical value of the Reynolds number, $Re_{cr} = 1.5 \times 10^5$, associated with the emergence of turbulent spots appears to be in good agreement with the experimental data for laminar-to-turbulent transition reported in the literature (e.g., Sleath, 1984; Jensen et al., 1989; Lodahl et al., 1998), and also in very good agreement with the value $Re_{cr} = 1.58 \times 10^5$, obtained by a stability analysis (see Sleath, 1984) as well as the asymptotic result $Re_{cr} = 1.5 \times 10^5$ of Tromans (1978) as $R/\delta_1 \rightarrow$

∞ obtained from a stability analysis carried out for oscillatory flows in a circular pipe, R being the radius of the pipe. The results also agree well with the DNS simulations of laminar-to-turbulent transition of Akhavan et al. (1991), Vittori and Verzicco (1998) and Costamagna et al. (2003).

We note that, historically, some researchers (e.g. Foster et al. 1994; Foster 1997) have suggested that turbulence is caused by an inflection-point shear-layer instability, as the phase resolved velocity profiles immediately after the near-wall flow reversal develop inflectional-point instabilities, as seen in the preceding paragraphs. The inflection-point instability and the controversy it caused was brought into focus already in 1991 by Akhavan et al. (1991); These authors pointed out that a quasi-steady instability analysis for an oscillatory Stokes flow predicts the most unstable profiles to occur near the start of the acceleration stage, and the most unstable eigenmodes have their peak energies around inflectional points. They emphasized that these predictions do not agree with experimental observations which show that the flow becomes explosively unstable during the deceleration stage, and also that the predicted critical Reynolds number for the former type of the instability, $Re_{cr} \sim 3600$, is not in accord with $Re_{cr} = 1.5 \times 10^5$, the critical Reynolds number associated with the transition to turbulence (the explosive instability during deceleration) observed in the experiments.

5.2.3 *Imprint of turbulent spot in bed shear stress signal*

Fig. 5.15 displays two time series: the free-stream velocity (the bottom time series in the figure), and bed shear stress (the top time series), from Carstensen et al.'s (2010) oscillating-tunnel study. The bed shear stress was measured by a hot-film probe, as mentioned previously (Fig. 5.10). The Reynolds number of this test was $Re = 4.5 \times 10^5$. Carstensen et al. (2010) videotaped the flow, synchronized with the recording of the bed shear stress and the free stream velocity, displayed in Fig. 5.15. The video sequences are not shown here for reasons of space, but can be found in Carstensen et al. (2010). This synchronized recording of the flow along with the bed shear stress showed that the spikes in the bed shear stress time series in Fig. 5.15 (A and B) are generated by two separate turbulent spots passing over the hot-film probe, one after the other. Similar spikes in near-bed velocity and wall shear stress signals have been observed in previous research in conjunction with laminar-to-turbulent transition in oscillatory boundary layers (Hino et al., 1976, Jensen et al., 1989 and Lodahl et al., 1998). However, Carstensen et al.'s (2010) work related the previously

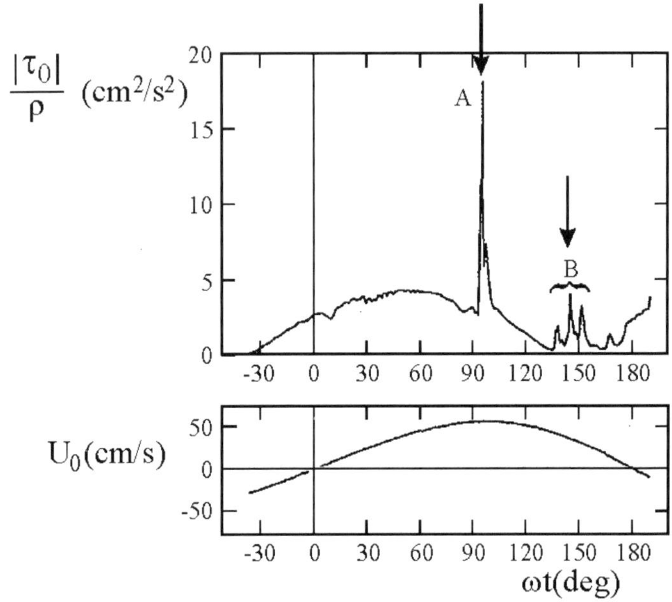

Fig. 5.15 Time series of bed shear stress and free-stream velocity. $Re = 4.5 \times 10^5$. The spikes in the bed shear stress time series (A and B) are generated by two separate turbulent spots passing over the hot-film probe, one after the other. Carstensen et al. (2010).

observed spikes in the wall shear stress signal (and in the near-bed velocity signal) to turbulent spots.

The turbulence associated with the turbulent spot brings momentum-rich fluid from the outer flow to the near-bed area. Therefore, the higher the momentum of the outer flow, the larger the magnitude of the spike in the bed shear stress. The momentum of the outer flow is highest at $\omega t = 90°$, and therefore the spikes should be largest around this phase value, as revealed in Fig. 5.15.

As mentioned earlier, Mazzuoli et al. (2011) made DNS simulations of the oscillatory boundary layer flow. The results of Mazzuoli et al.'s study confirmed that the turbulent spots do cause spikes in the wall shear stress and the velocity fields.

Fig. 5.16 displays bed shear stress time series reproduced from Jensen et al.'s (1989) oscillating-tunnel experiments for three different values of

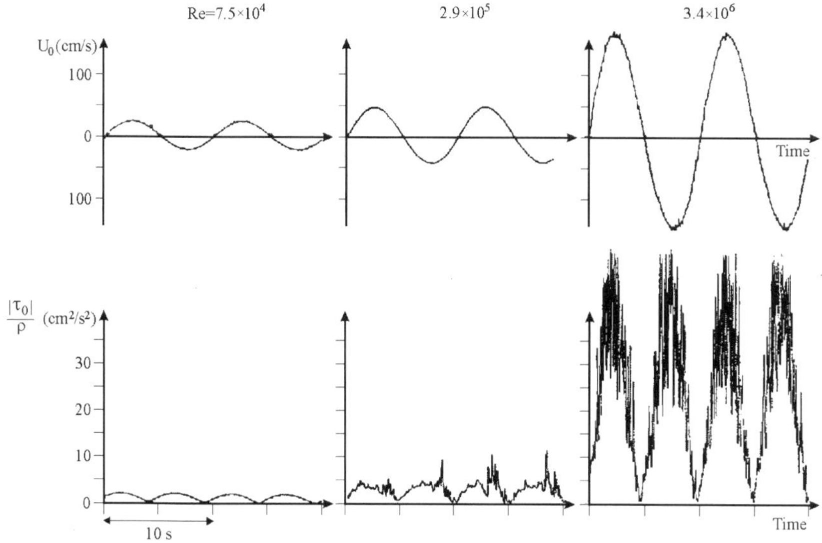

Fig. 5.16 Sample records of free-stream velocity and bed shear stress. Smooth bed. Jensen et al. (1989).

the Reynolds number, namely $Re = 7.5 \times 10^4$, 2.9×10^5 and 3.4×10^6, characterizing the three different flow regimes: the laminar regime, the transitional regime, and the fully-developed turbulent regime, respectively. The spikes characterizing turbulent spots, and described in the preceding paragraphs, can easily be recognized for the transitional Reynolds number $Re = 2.9 \times 10^5$.

5.2.4 *Laminar-to-turbulent transition in terms of friction coefficient*

One way of illustrating the laminar-to-turbulent transition is to plot the friction coefficient as a function the Reynolds number and the phase.

To this end, we define the friction coefficient by

$$f_w^\star = \frac{2\tau_0/\rho}{U_{0m}^2 \sin(\omega t + \frac{\pi}{4})} \tag{5.22}$$

This definition of the friction coefficient is slightly different from the traditional definition of the friction coefficient f_w

$$f_w = \frac{2\tau_{0m}/\rho}{U_{0m}^2} \tag{5.23}$$

in which the subscript m indicates the maximum value over a period. We will come back to Eq. 5.23 later.

As is seen, the bed shear stress in Eq. 5.22 is not the maximum value of the bed shear stress, but rather the phase-resolved bed shear stress, $\tau_0 = f(\omega t)$. The advantage of this representation is that, when divided by $\sin(\omega t + \frac{\pi}{4})$, the bed shear stress in the case of laminar flow,

$$\tau_0 = \frac{\rho U_{0m}^2}{Re^{\frac{1}{2}}} \sin\left(\omega t + \frac{\pi}{4}\right) \tag{5.24}$$

will be independent of the phase ωt and will depend only on *Re:*

$$f_w^\star = \frac{2}{Re^{\frac{1}{2}}} \tag{5.25}$$

The latter expression represents a straight line when plotted on a log-log graph.

This kind of representation enables us to identify the flow regimes in an experimental determination of the bed shear stress for various Reynolds numbers. The procedure may be summarized as follows: (1) Measure the bed shear stress, $\tau_0 = f(\omega t)$; (2) Calculate the friction coefficient $f_w^\star$ from Eq. 5.22; (3) Plot the latter quantity versus Re on a log-log graph, with ωt as a parameter; (4) If $f_w^\star$ versus Re on this plot collapses onto the straight line represented by Eq. 5.25, then the flow will be in the laminar regime; otherwise it will be either in the transitional regime, or in the turbulent regime.

Jensen et al. (1989) did this exercise. Their results are depicted in Fig. 5.17. This figure is an important figure. It implies that, even for very large *Re,* the flow regime may remain laminar for quite some time during the boundary layer development in the half cycle. For example, for $Re = 5 \times 10^5$, the flow will remain laminar until $\omega t \simeq 40°$, while for $Re = 10^6$, it will remain laminar until $\omega t \simeq 20°$–$25°$.

Fig. 5.17 further shows that the pattern of the variation of $f_w^\star$ with Re is not the same outside the laminar line; evidently there are two distinct regions, implying a change in the flow regime. These two regions are interpreted by Jensen et al. as the transitional and fully developed turbulent regimes. From the figure, it can be seen that, even for Re as high as 10^6, the flow becomes a fully developed turbulent flow only after the phase value $\omega t \simeq 60°$.

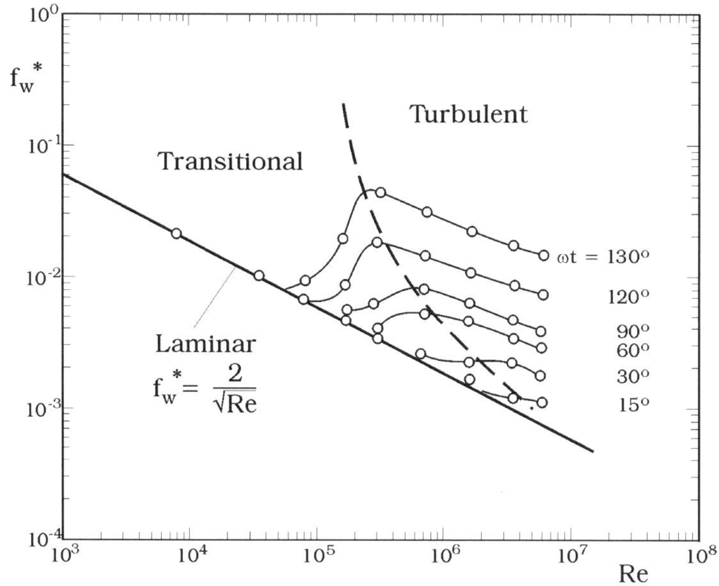

Fig. 5.17 Normalized friction coefficient at different phase values. Jensen et al. (1989).

It is seen that turbulence first occurs just prior to the bed shear stress reversal because the adverse pressure gradient becomes relatively large and the velocity of near-bed fluid particles becomes relatively small at this phase value, so that a very favourable environment is established for the initiation of turbulence, an environment opposite to that associated with the re-laminarization of flow. With increasing Re, transition to turbulence spreads towards smaller and smaller values of phase.

Carstensen et al. (2010) demonstrated that all the turbulent spots they observed in their tests fall in the "Transitional" region in Fig. 5.17 (see Fig. 10 in Carstensen et al., 2010), further validating Jensen et al.'s (1989) conjecture about the "Transitional" and "Turbulent" flow regimes based on their bed shear stress measurements.

5.2.5 *Turbulent spots over a rough bed*

Carstensen et al. (2012) continued the investigation of turbulent spots, extending their study to a rough wall. In the experiments, a fixed sand-grain bed was achieved by gluing sand of fairly uniform size one-layer deep on the bottom wall of the oscillating tunnel. The roughness height was $k = 0.25$

cm. This resulted in Nikuradse's equivalent sand roughness of $k_s = 0.5$ cm (see Carstensen, 2012, for details). The grain Reynolds number in the tests, $k_s^+ = k_s U_{fm}/\nu$, has been determined, and found to be $k_s^+ > O(60)$ when turbulent spots first emerge, implying that the boundary in Carstensen et al.'s (2012) study behaved as a nearly fully rough boundary.

Fig. 5.18 displays video sequences of the flow over the rough bed in a test with $Re = 7.8 \times 10^4$. The flow is from right to left. The figure clearly shows that turbulent spots form over the sand-grain bed and develop in the same way as in the smooth-bed oscillatory boundary layers. The reader can consult the video accompanying the paper of Carstensen et al. (2012) at the URL address: `http://dx.doi.org/10.1063/1.4767536`, to get a more complete view of the turbulent spots over a rough boundary.

It is interesting to note that the critical Reynolds number at which the turbulent spots first emerge is reduced considerably in the case of the rough bed: In Fig. 5.19 two sets of data are plotted for the onset of the turbulent spots in both the smooth-wall experiments (the empty circles, Carstensen et al., 2010), and the rough-bed experiments (the filled circles, Carstensen et al., 2012), plotted as function of the phase and the Reynolds number. As seen, the critical Reynolds number in the case of the rough wall is much smaller, namely $Re_{cr} \approx 0.5 \times 10^5$, than that of the smooth-wall case, $Re_{cr} = 1.5 \times 10^5$. This is consistent with the steady-boundary layer knowledge (e.g., Schlichting, 1979, p. 536). Also, the trend that the phase of the onset of turbulent spots decreases with Re appears to be much the same as in the smooth-bed case.

In a recent DNS study, Mazzuoli and Vittori (2019) have beautifully demonstrated the formation of turbulent spots over rough boundaries, thus confirming Carstensen et al.'s (2012) observations. In their DNS experiment, Mazzuoli and Vittori considered a rough bed made of spheres arranged in a honeycomb pattern. The sphere diameter was chosen to be 5 mm, "mimicking" Carstensen et al.'s (2012) 5 mm Nikuradse's equivalent sand roughness. The Reynolds number of the experiment (that showed the formation of turbulent spots) was selected to be the same as that of the experiment shown in Fig. 4 of Carstensen et al. (2012), reproduced here in Fig. 5.18. The numerical results of Mazzuoli and Vittori (2019) provide further information on transition to turbulence, including the speed of the extreme points of the spots, in general agreement with Carstensen et al.'s (2012) experiments.

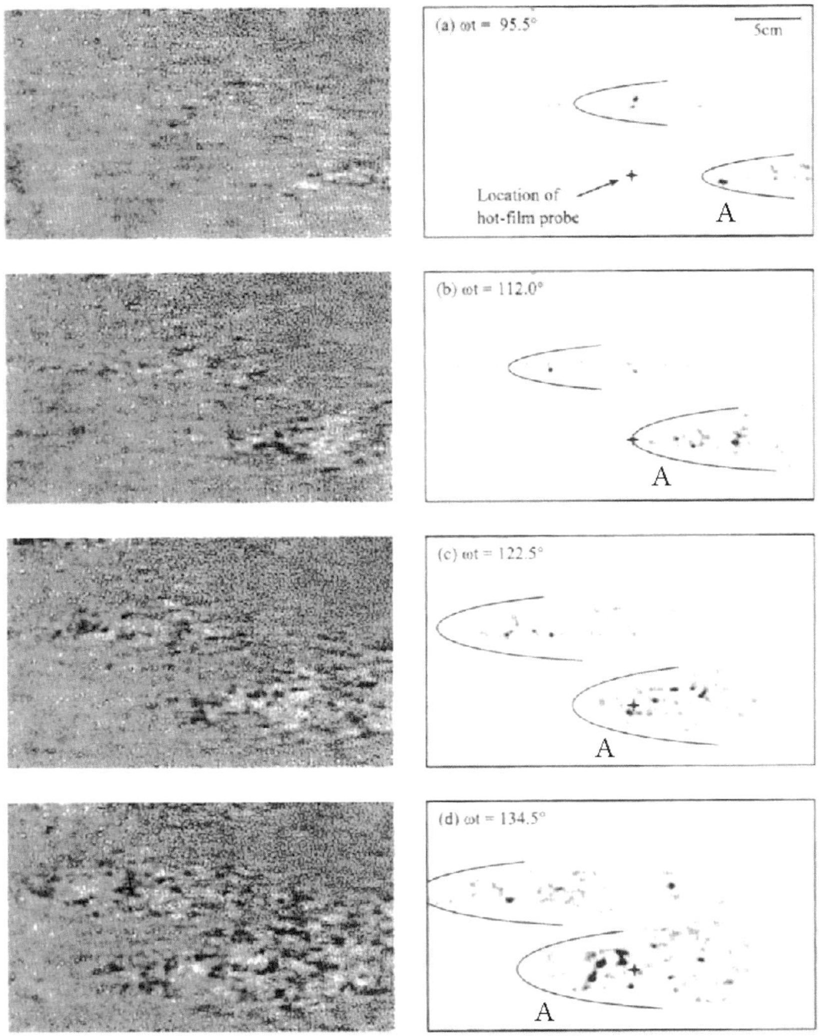

Fig. 5.18 Turbulent spots over the sand-grain bed and accompanying sketches. $Re = 7.8 \times 10^4$. Plus sign marks the location of the hot-film probe. Carstensen et al. (2012).

5.2.6 *Transition in terms of flow resistance and phase*

Fig. 5.20 presents the flow resistance and phase data for the smooth-bed tests where the data of Jensen et al. (1989) and Carstensen et al. (2010)

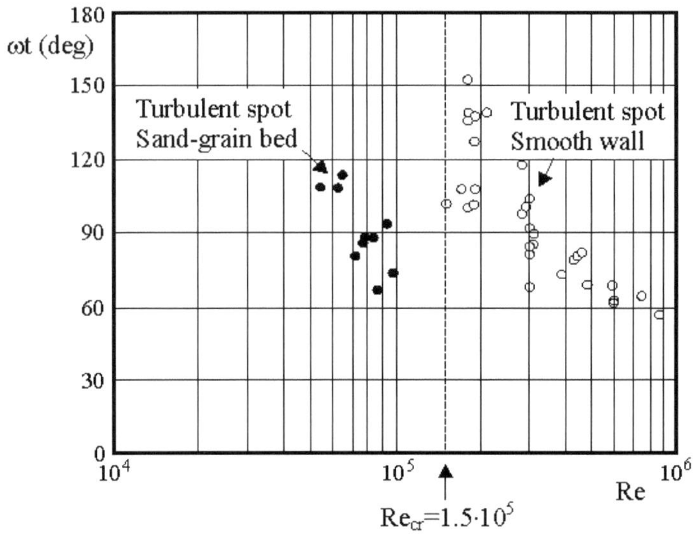

Fig. 5.19 Birth of turbulent spots. Filled circles: Carstensen et al. (2012), sand-grain bed. Empty circles: Carstensen et al. (2010), smooth bed. Bed roughness reduces the critical Reynolds number for the formation of turbulent spots, from 1.5×10^5 to approximately 0.5×10^5. Carstensen et al. (2012).

are plotted. No other data are plotted to keep the figure relatively simple. Incidentally, Jensen et al. (1989, Fig. 11) (and later Sarpkaya, 1993, Figs. 4–6, including his own experimental data) made an extensive comparison of their flow resistance and phase data with the existing experimental and theoretical work including Spalart and Baldwin's (1987) DNS simulation, Fredsøe's (1984) theoretical solution, Kamphuis' (1976), Hino et al.'s (1983) and Sleath's (1987) experiments, and demonstrated that all the data are consistent with the exception that Spalart and Baldwin's numerical data and Kamphuis' experimental data deviate somewhat from the other data in the transitional Reynolds-number range (Jensen et al., 1989, Fig. 11). We will return to this issue with the flow resistance and phase data reproduced from Jensen et al. (1989) in Section 5.5.

In Fig. 5.20, f_w is the wave friction coefficient defined in Eq. 5.23

$$f_w = \frac{2\tau_{0m}/\rho}{U_{0m}^2} \tag{5.26}$$

and ϕ is the phase lead of the bed shear stress over the free-stream velocity defined by

$$\overline{\tau}_0 = \tau_{0m}\, f(\omega t + \phi) \tag{5.27}$$

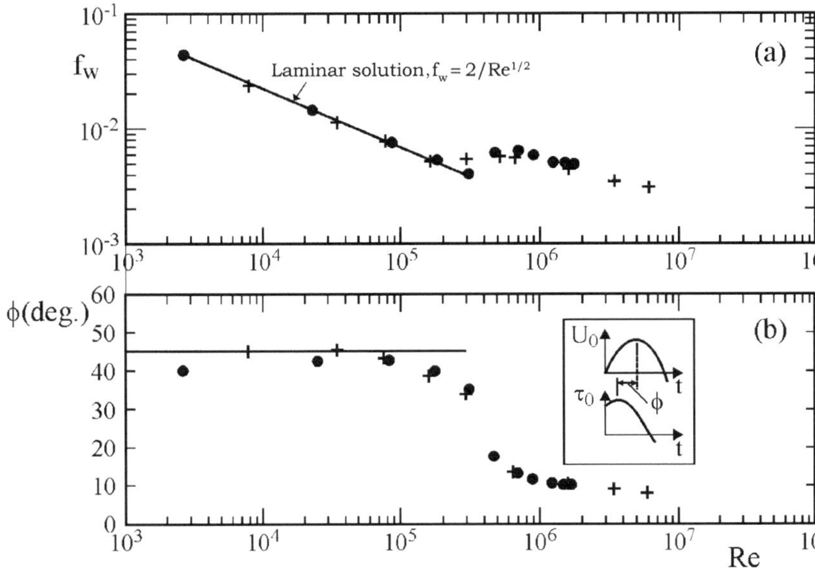

Fig. 5.20 Wave friction coefficient (a), and phase lead (b). Crosses: Jensen et al. (1989). Circles: Carstensen et al. (2010). Solid lines: Laminar solution. Carstensen et al. (2010).

Eq. 5.26 will, in terms of the friction velocity, read:

$$U_{fm} = \sqrt{\frac{f_w}{2}}U_{0m} \qquad (5.28)$$

in which U_{fm} is the maximum value of the friction velocity defined by

$$U_{fm} = \sqrt{\frac{\tau_{0m}}{\rho}} \qquad (5.29)$$

and τ_{0m} is the maximum value of the bed shear stress.

For the laminar case, from Eqs. 5.20 and 5.26, f_w

$$f_w = \frac{2}{Re^{\frac{1}{2}}} \qquad (5.30)$$

and the phase lead, from Eq. 5.21 is:

$$\phi = 45^\circ \qquad (5.31)$$

As seen from Fig. 5.20, the transition to turbulence occurs at $Re \simeq 1.5 \times 10^5$, consistent with the analysis of the preceding paragraphs.

It is also seen (Fig. 5.20) that the phase lead ϕ decreases with the transition. In the transitional/turbulent flows, the momentum exchange between the layers of fluid will be enhanced. Therefore, the near-bed flow will be supplied with the new, momentum-rich fluid from the upper layers of the boundary layer, and as a result, it will dissipate its momentum not as fast as in the case of the laminar flow, as mentioned previously in conjunction with turbulent spots, Section 5.2.1. Hence, its response to the adverse pressure gradient will be delayed, leading to the decrease in the phase lead, as indicated in Fig. 5.20b. In the fully-developed turbulent case (very large Re), the phase lead is only about 5°.

5.3 Turbulent wave boundary layers: Smooth bed

Turbulent wave boundary-layer flows have been the subject of a great many investigations for many years. With the pioneering work of Lundgren and Jonsson (1961), much research has been done in the past three decades or so, e.g., Hino et al. (1976 and 1983), Kemp and Simons (1982 and 1983), Sleath (1987), Jensen et al. (1989), Lodahl et al. (1998), Dixen et al. (2008) on the experimental side, and Grant and Madsen (1979), Fredsøe (1984), Justesen (1988a), Vittori and Verzicco (1998), Costamagna et al. (2003), Lohmann et al. (2006), Salon et al. (2007), Fornarelli and Vittori (2009) on the theoretical side, to name but a few. Major reviews on the subject have been presented by, among others, Soulsby et al. (1993), Blondeaux and Vittori (1999) and Sumer (2003).

The wave boundary layers may be simulated in the laboratory, using various techniques such as wave flumes, wave basins, oscillating plates or trays, and oscillating tunnels. Unless it is a large-scale facility, turbulent wave boundary-layer flows over smooth beds cannot be achieved in wave flumes simply because of the small scales involved. To circumvent this problem, oscillating water tunnels have been introduced in the research of turbulent wave boundary layers, simulating the full-scale wave boundary layer by the oscillatory boundary-layer flow that develops over the bottom of the oscillating tunnel.

The Danish oscillating tunnel (Lundgren and Sørensen, 1958), located at the Technical University of Denmark, was the first to enable experimental simulations of turbulent full-scale wave boundary layers. Jensen et al. (1989) used this facility to conduct an extensive study of turbulent wave boundary layers over a smooth bottom with high Reynolds numbers, complementary to two previous major studies with relatively smaller Reynolds numbers, namely Hino et al. (1983) and Sleath (1987). The studies of

Jensen et al. (1989), Sleath (1987) and Hino et al. (1983) formed a complementary set of information on turbulent wave boundary layers, which shed considerable light on the understanding of turbulent wave boundary-layer flows.

The present section (turbulent wave boundary layers over smooth beds) as well as the following section studying turbulent wave boundary layers over rough beds are mainly based on the work of Jensen et al. (1989).

5.3.1 *Oscillating tunnel in Jensen et al. (1989) study*

Fig. 5.21 displays a schematic description of the oscillating tunnel used in Jensen et al.'s (1989) experiments. This is a U-shaped oscillatory-flow water tunnel. It is essentially the same tunnel as described by Lundgren and Sørensen (1958) and later by Jonsson (1963). The working section was 10 m long, and 0.39 m wide. Its height was 0.29 m in most of the Jensen et al. tests, and 0.30 m in their sand-grain-bed experiments. (We note that the paper by Jensen et al., 1989, inadvertently reported that the height of the tunnel was 0.28 m.) The top and side walls of the working section were made of smooth, transparent Perspex plates. For the smooth-bed experiments, PVC plates were fixed rigidly to the bottom of the tunnel.

The oscillatory flow in the tunnel was driven by an electronically controlled pneumatic system. In the tests, the flow was maintained near its resonant period, 9.72 s. The spectra of the velocity at the centreline of the tunnel indicated that the contributions from the second and third harmonics to the motion is negligible for all practical purposes.

The tunnel was designed such that the full-scale wave boundary-layer flow under storm conditions could be replicated, with velocity amplitudes like $O(2\text{–}3)$ m/s, orbital-motion amplitudes like $O(5)$ m, and periods like $O(10)$ s. This enabled Jensen et al. (1989) to attain Reynolds numbers as high as $Re = 6 \times 10^6$.

In Jensen et al.'s (1989) experiments, the velocities were measured by laser-Doppler anemometer (LDA). In addition to the velocity measurements, the bed shear stress was also measured with a hot-film probe. Jensen et al. (1989) report that the dimensions of the measuring volume ($d_x \times d_y \times d_z$) are 0.3 mm $\times$ 0.3 mm $\times$ 3.2 mm for One-Component LDA, and 0.15 mm $\times$ 0.15 mm $\times$ 2.5 mm for Two-Component LDA for the smooth-bed experiments (Jensen et al. used both One- and Two-component LDAs; see Table 1 in Jensen et al., 1989), and ($d_x \times d_z$) are 0.2 mm $\times$ 0.75 mm for the hot-film probe, z being the spanwise direction. In one test, the

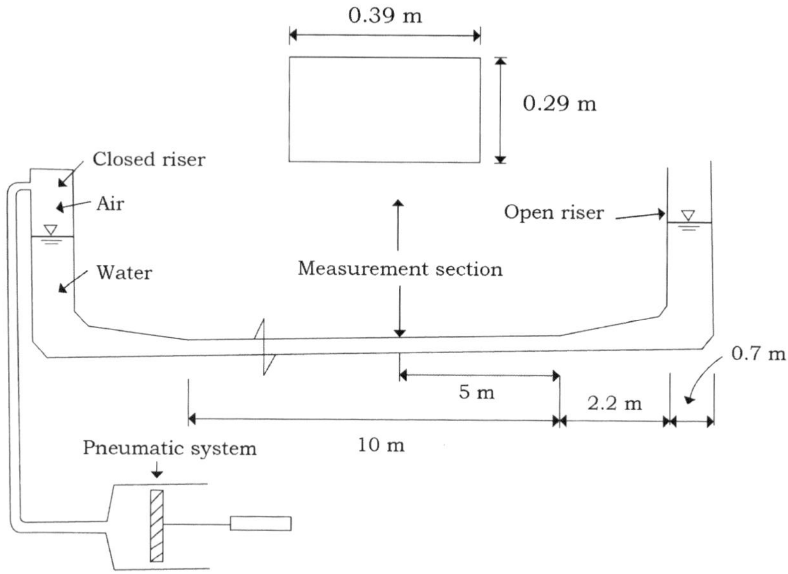

Fig. 5.21 Schematic description of the oscillating tunnel of Jensen et al.'s (1989) study.

spanwise component of the velocity was also measured. This was achieved by measuring the spanwise velocity in the boundary layer developing over the sidewall of the tunnel, with ($d_x \times d_y \times d_z$) are 0.15 mm × 2.5 mm × 0.15 mm.

The bed friction velocity was determined in two ways in Jensen et al.'s (1989) smooth-bed experiments: (1) from direct measurements of the bed shear stress by hot film; and (2) by fitting straight lines to the logarithmic-layer portion of the mean velocity distribution (plotted on a semi-logarithmic graph) for those tests where the velocity distribution was measured.

Using a wave gauge, the water level in the open riser of the U-tube (Fig. 5.21) was recorded simultaneously with the velocity measurements. This served as a reference signal in data processing.

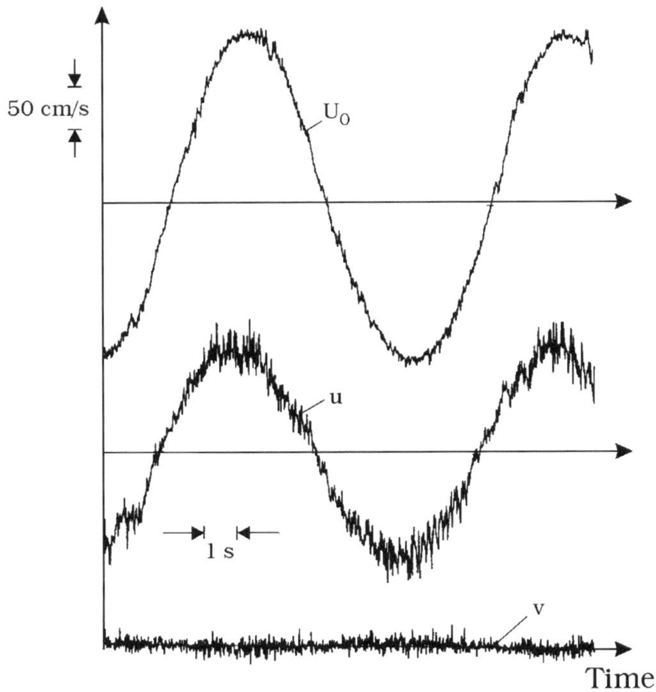

Fig. 5.22 Sample velocity records. $Re = 6 \times 10^6$. U_0 was measured at $y = 140$ mm, and u and v at $y = 1$ mm. Jensen et al. (1989).

5.3.2 *Ensemble averaging*

The velocity measurements in Jensen et al.'s (1989) smooth-bed experiments were made at sampling intervals of 48 ms, while the bed shear stress measurements were made at sampling intervals of 24 ms. The sampling interval of 48 ms corresponds to approximately 200 samples per flow cycle, large enough to be able to trace the time development of the measured statistical quantities over the flow cycle. Fig. 5.22 displays sample time series of the horizontal and vertical velocity components of the velocity in the boundary layer (the bottom two time series). The top time series is the free-stream velocity. This is from an experiment with $Re = 6 \times 10^6$.

In turbulent wave (oscillatory) boundary layers, mean values of the quantities are calculated through ensemble averaging. Jensen et al. (1989) also used the ensemble averaging to calculate the mean and the root-mean-square (r.m.s.) values of the hydrodynamic quantities. For example, the

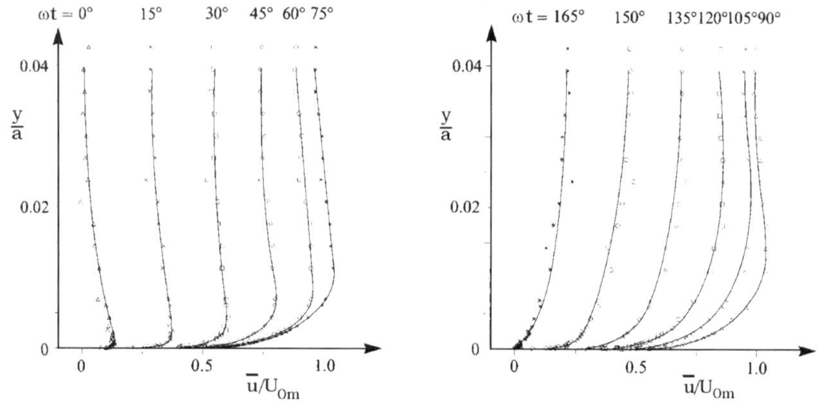

Fig. 5.23 Mean velocity distributions. Smooth bed. $Re = 6 \times 10^6$. Jensen et al. (1989).

mean value of the velocity component u is calculated by

$$\overline{u}(y, \omega t) = \frac{1}{N} \sum_{j=1}^{N} u[y,\ \omega(t + (j-1)T)] \tag{5.32}$$

Likewise, the r.m.s. value of the fluctuating velocity $u' = u - \overline{u}$,

$$\sqrt{\overline{u'^2}}(y, \omega t) = \left\{ \frac{1}{N-1} \sum_{j=1}^{N} [u[y,\ \omega(t + (j-1)T)] - \overline{u}(y, \omega t)]^2 \right\}^{\frac{1}{2}} \tag{5.33}$$

in which N is the total number of cycles sampled. N was 50 for the bed-shear-stress tests, and 80 for the other tests for the smooth bed experiments in Jensen et al. (1989). Sleath (1987) reported that, for record lengths larger than 50, no significant improvement was obtained with increase in the number of cycles sampled. Jensen et al.'s (1989) results confirmed this finding.

5.3.3 *General description of turbulent wave boundary layer*

Fig. 5.23 shows the time development of the velocity profiles for $Re = 6 \times 10^6$. The data is presented in a normalized form, using the outer-flow parameters, U_{0m} and a. In this outer-flow parameter normalization, the boundary-layer thickness is abandoned in favour of the amplitude of free stream orbital motion, a, to avoid the uncertainty involved in the experimental determination of the boundary-layer thickness.

The way in which the velocity profiles evolve with time is similar to that in the case of the laminar wave boundary layer (Fig. 5.3). Features such as the early reversal of the flow near the bed, and the overshooting in the velocity profiles are observed here, too.

Fig. 5.24 illustrates how the turbulence evolves as the flow progresses in the phase space: (1) the turbulence builds up near the bed and constantly diffuses away from the bed across the boundary layer, as the boundary layer develops in time; (2) by the time of approximately $\omega t = 140°$, the near-bed buildup of turbulence almost comes to an end; and (3) the turbulence becomes almost uniformly distributed across the depth, as the time of free-stream flow reversal is reached.

5.3.4 *Mean flow velocity*

The velocity profiles of Fig. 5.23 are plotted in semi-logarithmic plots in Fig. 5.25 in terms of the inner-flow parameters,

$$U_f = \sqrt{\frac{\overline{\tau}_0}{\rho}} \quad \text{and} \quad \nu \tag{5.34}$$

in which $\overline{\tau}_0$ is the temporal (or the phase-resolved) value of the mean bed shear stress, defined by $\overline{\tau}_0 = \tau_{0m} f(\omega t + \phi)$ with τ_{0m} being the maximum value of the mean bed shear stress, ϕ the phase lead, and ν the kinematic viscosity. The normalized distance from the bed y^+ in the figure is

$$y^+ = \frac{yU_f}{\nu} \tag{5.35}$$

There are two advantages of the inner-flow representation: (1) it gives a better picture of the boundary-layer behaviour close to the bed; and (2) the fact how the oscillatory boundary layer relates to the familiar steady boundary layer could possibly be best explained by reference to the inner-flow representation.

Fig. 5.25 also includes the van Driest profile as a reference line:

$$\frac{\overline{u}}{U_f} = 2\int_0^{y^+} \frac{dy^+}{1 + \left\{1 + 4\kappa^2 y^{+2}\left[1 - \exp(-y^+/A)\right]^2\right\}^{1/2}} \tag{5.36}$$

where κ is the von Karman constant (=0.4), and A the van Driest damping factor (=25) (van Driest, 1956). Recall the van Driest velocity profile, Eq. 3.108, obtained for steady boundary layers (Chapter 3), which is valid close to the wall. As mentioned previously in conjunction with Eq. 3.108 in Chapter 3, the van Driest profile tends to the logarithmic distribution

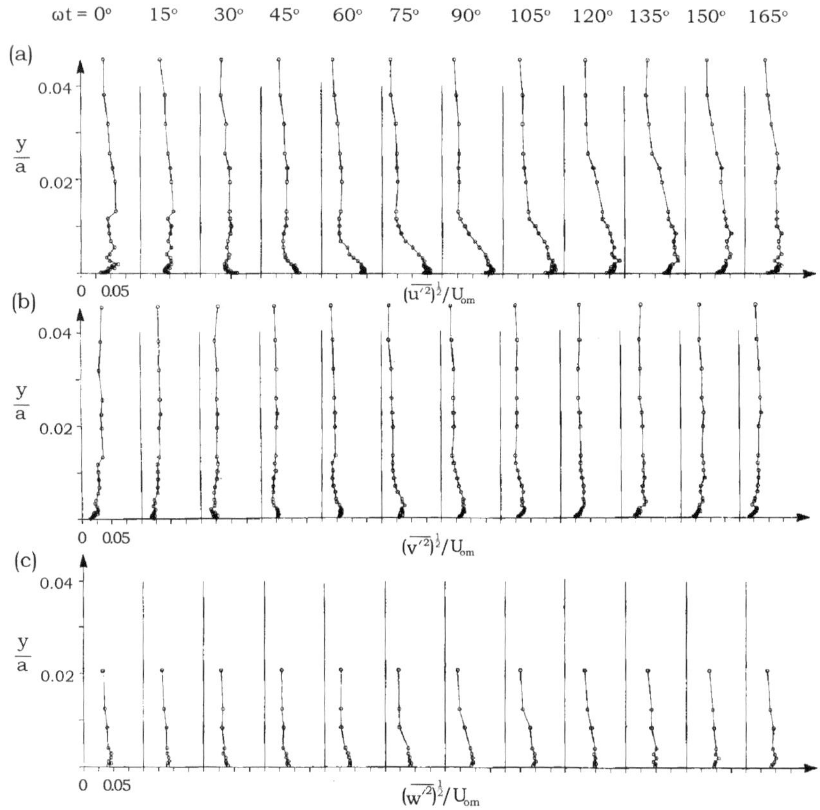

Fig. 5.24 Distribution of r.m.s. values of velocity fluctuations. $Re = 6 \times 10^6$. Smooth bed. Jensen et al. (1989).

for large values of y^+ (Eqs. 3.108 and 3.110), as seen from the figure (the dashed lines):

$$\frac{\overline{u}}{U_f} = \frac{1}{\kappa} \ln y^+ + 5 \tag{5.37}$$

Fig. 5.25 shows the following:

1. The oscillatory-flow velocity distributions agree remarkably well with the steady-flow ones in the range of $\omega t = 70°$–$110°$ where the flow experiences almost zero pressure gradient.

2. The logarithmic layer exists already at $\omega t = 15°$ and is maintained

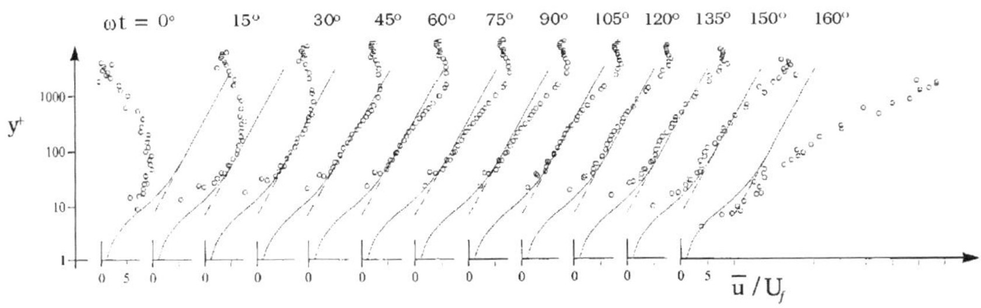

Fig. 5.25 Mean velocity distribution in semi-log plot. Smooth bed. $Re = 6 \times 10^6$. Dashed line: Log law. Solid curve: van Driest profile. Jensen et al. (1989).

throughout the ωt range from $\omega t \simeq 15°$ up to almost $160°$ where the near-bed flow reversal takes place.

3. At earlier phases, the logarithmic layer does not exist, however. This is simply because the boundary layer at these stages is not thick enough to house the logarithmic layer.

4. It may also be noted that the velocity profiles in the early phases of the acceleration stage (where the pressure gradient is favourable) resemble quite closely those obtained in favourable pressure-gradient steady boundary layers (Coles, 1956). Likewise, the velocity profiles in the later phases of the deceleration stage (where the pressure gradient is adverse) resemble, again, quite closely the velocity distributions obtained in adverse pressure-gradient steady boundary layers (Coles, 1956).

Finally, we note the following. Fig. 5.26 shows the semi-log plots of the velocity distributions obtained in another study (Sumer et al., 1987), with the corresponding shear stress distributions, Fig. 5.27. As seen from Fig. 5.27, the constant stress layer is displaced to higher elevations in the later phases (starting from $\omega t = \frac{9\pi}{12}$) of the deceleration stage. This means that the logarithmic layer should also be displaced to these higher elevations. The velocity distributions may still be represented by the logarithmic law. However, U_f in such a representation cannot be the bed shear velocity, but rather a velocity defined by $(\frac{\overline{\tau}}{\rho})^{\frac{1}{2}}$ where $\overline{\tau}$ is the shear stress occurring at these higher elevations rather than at the bed.

The above point is important because researchers have, in the past, mistakenly fitted straight lines to experimental data at such higher elevations, and treated the shear velocity they have obtained from this exercise as the bed shear velocity (i.e. the friction velocity), which is clearly not correct. To avoid this problem, the following criteria should be strictly observed

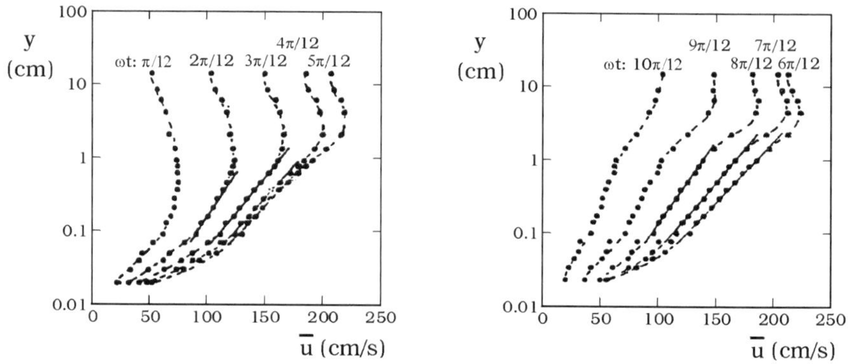

Fig. 5.26 Mean velocity distributions. Sumer et al. (1987).

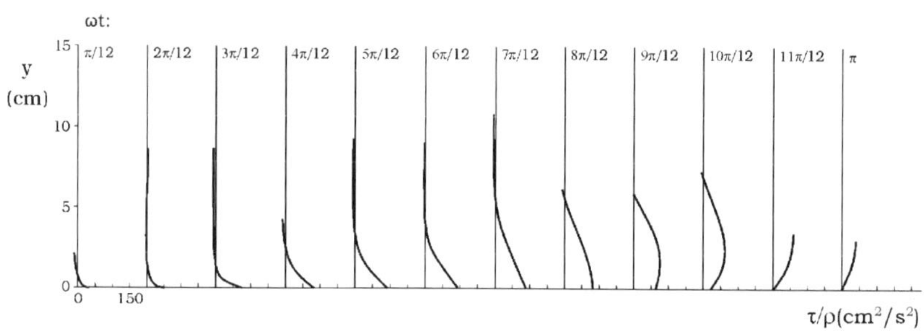

Fig. 5.27 Shear stress distributions corresponding to the velocity distributions in the previous figure. Sumer et al. (1987).

when fitting the logarithmic law to a measured set of velocity data:

1. The distance from the bed y should be in the "real" constant stress layer near the bed, defined by

$$y \ll \delta(t) \tag{5.38}$$

in which $\delta(t)$ is the phase-resolved boundary-layer thickness at the phase in consideration. $\delta(t)$ should not be confused in this context with the boundary layer thickness, δ, defined in conjunction with Fig. 5.4.

As discussed in the case of the steady boundary layer flows in Chapter 3, the condition in Eq. 5.38 can be relaxed to $y < (0.2\text{–}0.3)\delta(t)$.

2. At the same time, as discussed in Chapter 3, the distance y should be sufficiently large, so large that the variation of the velocity, $\frac{d\overline{u}}{dy}$, is inde-

pendent of the viscosity. For this, y^+ should be

$$y^+ > 30 - 70 \tag{5.39}$$

in which $y^+ = \frac{yU_f}{\nu}$ with U_f being the phase-resolved friction velocity at the phase in consideration.

The issue of fitting the logarithmic law to a measured velocity distribution has been discussed in greater detail with regard to applications in steady boundary layers (Chapter 3) in conjunction with Fig. 3.5 and Eqs. 3.45 and 3.46, and the reader is referred to that discussion for further details.

5.3.5 *Turbulence quantities*

Fig. 5.28 and 5.29 presents the turbulence data in normalized forms:

$$u'^+ = \frac{\sqrt{\overline{u'^2}}}{U_f}, \quad v'^+ = \frac{\sqrt{\overline{v'^2}}}{U_f}, \quad w'^+ = \frac{\sqrt{\overline{w'^2}}}{U_f}, \quad \overline{u'v'}^+ = \frac{-\overline{u'v'}}{U_f^2} \tag{5.40}$$

versus y^+ in terms of the inner-flow parameters. Also plotted in the figures are the known steady boundary-layer distributions (the solid lines) (see, for example, Fig. 3.7 for comparison). The turbulence quantities $\sqrt{\overline{u'^2}}$, $\sqrt{\overline{v'^2}}$, etc. are defined by Eq. 5.33. The quantity U_f on the other hand is the phase-resolved friction velocity, defined in Eq. 5.35.

Fig. 5.28a gives the $u'^+ (= \sqrt{\overline{u'^2}}/U_f)$ profiles. The dashed line in the figure represents the line tangent to the curve $u'^+(y^+)$ at $y = 0$. It was obtained from direct measurements of the bed shear stress through the equation

$$\sqrt{\overline{\tau_0'^2}} = \mu \frac{\partial \sqrt{\overline{u'^2}}}{\partial y} \tag{5.41}$$

in which τ_0' is the fluctuating component of the bed shear stress, $\tau_0' = \tau_0 - \overline{\tau_0}$. We note that there was a typo in the original publication in Jensen et al. (1989), and that the above expression is the correct version. Since the velocities could not be measured very close to the bed (for y values of $O(0.3$ mm$)$) owing to experimental constraints, plotting of the previously mentioned tangent lines is quite advantageous, as they will illustrate how the r.m.s. values of u' are distributed very close to the bed.

Except at very early phases of the acceleration stage, and also late phases of the deceleration stage, the u'^+ profiles appear to be in reasonable accord with the steady boundary-layer distributions throughout the phase

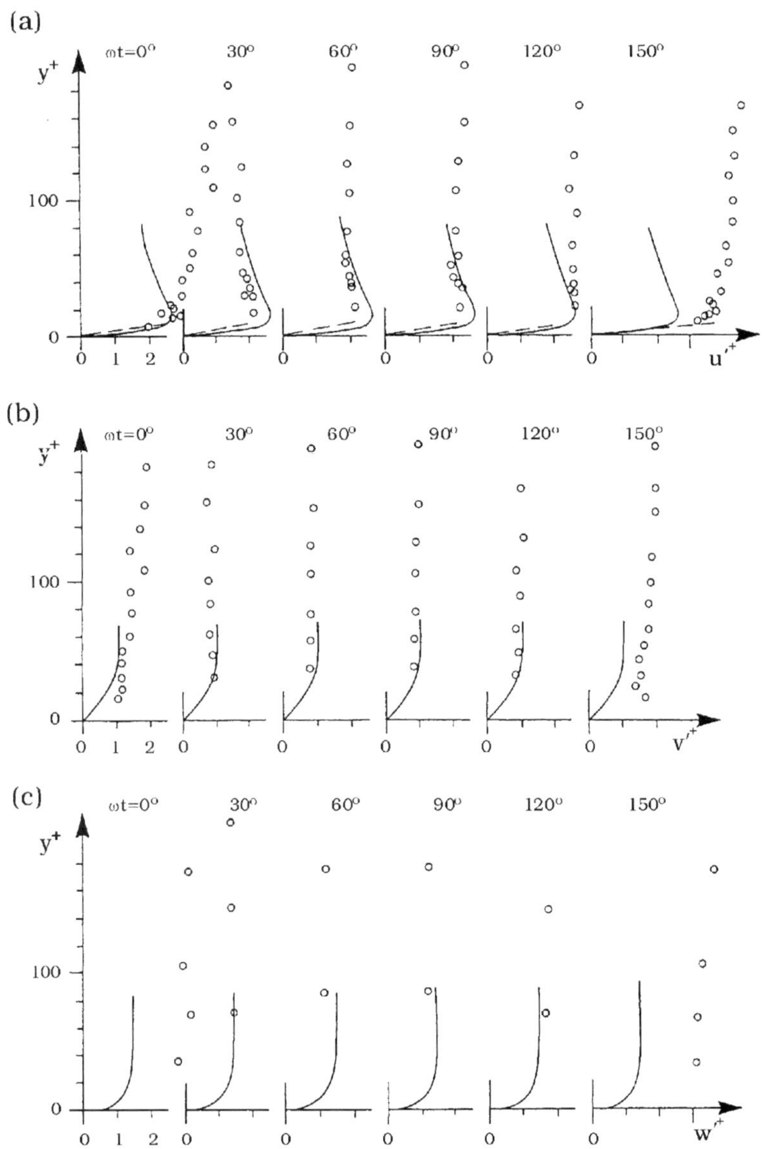

Fig. 5.28 R.m.s. values of u', v' and w'. Smooth bed. $Re = 6 \times 10^6$. Jensen et al. (1989).

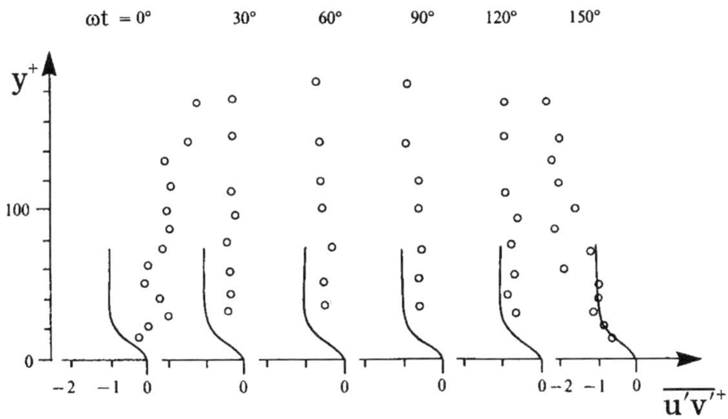

Fig. 5.29 Reynolds stresses. Smooth bed. $Re = 6 \times 10^6$. Jensen et al. (1989).

space. Jensen et al. (1989) argue that the slightly under-predicted values of u'^+ in the acceleration stage and slightly over-predicted values of u'^+ in the deceleration stage may be attributed to the "history effect" where the response of $\overline{u'^2}$ to the wall quantity U_f is not instantaneous.

Fig. 5.28b and Fig. 5.28c give the $v'^+(= \sqrt{\overline{v'^2}}/U_f)$ and $w'^+(= \sqrt{\overline{w'^2}}/U_f)$ profiles. The same arguments given above for the u'^+ component apply also for v'^+ and w'^+.

Fig. 5.29 gives the Reynolds-stress profiles normalized with the inner-flow parameters. One would expect that the Reynolds-stress profiles should approach to that for steady boundary-layer flows. Fig. 5.29 shows, however, that the Reynolds stresses in the oscillatory-flow boundary layer were apparently under-predicted. Jensen et al. (1989) attributed this to the high drop-out rate in the laser measurements in the region very close to the bottom. They further pointed out that also Sleath (1988) reported unexpectedly low values of laser-measured Reynolds stresses near the wall.

5.3.6 *Effect of Reynolds number*

Fig. 5.30 gives the mean velocity profiles for four different Reynolds numbers in terms of the inner-flow parameters. The characteristic features of the time development of the velocity profiles shown in Fig. 5.30 are as follows:

1. The velocity profile eventually reaches a state where the logarithmic layer comes into existence;

2. This logarithmic layer grows in thickness, as one proceeds further in the phase space ωt; and

3. This continues to be the case until practically the point of reversal near the bed is reached.

From the figure, it is seen that the higher the Re number, the earlier the logarithmic layer comes into existence. This can be explained by reference to Fig. 5.17; from the latter figure, it is seen that, for the establishment of the logarithmic layer, the flow has to reach the fully developed turbulent state. Jensen et al. (1989) also give the effect of Re on the turbulence quantities.

5.3.7 *Wall-normal pressure gradient*

In Chapter 3, we have seen that the lift force on a particle at the time of suspension at/near the bottom of an open channel essentially comes from the upward-directed pressure gradient. In Section 3.6.2, it was argued that other contributions such as the drag force and the apparent-mass force are not very significant. These results were based on Sumer's (1984) single particle experiments. It was further noted in Section 3.6.2 that the agreement between Sumer's (1984) lift force measurements and Jensen and Sumer's (2019) measurements of the vertical pressure gradient (obtained in Karsten Jensen's PhD study) supports the argument that the major contribution to the measured lift force in Sumer (1984) is from the upward-directed pressure gradient.

The latter quantity, namely the wall-normal pressure gradient, has been used earlier in conjunction with the laminar-to-turbulent transition (Fig. 5.7) to visualize turbulent spots in the laminar-to-turbulent transition in oscillatory boundary layers.

Although the wall-normal pressure gradient emerges to be an important quantity in the study of turbulence, not much attention has been given to this quantity so far. At this juncture, we emphasize that the wall-normal pressure gradient, $-\partial p/\partial y$, represents a so-called pressure-gradient force (in the y-direction) per unit volume of fluid in which p is the pressure, and y is the distance normal to the wall or the bed; see Fig. 5.6 for the definition sketch for the coordinate axes. The negative sign is introduced in front of $\partial p/\partial y$ to take care of the direction of the force i.e. such that positive values represent an upward force, and vice versa. In laminar-regime flows, no such force exists, and therefore $-\partial p/\partial y = 0$. This can readily be seen from the

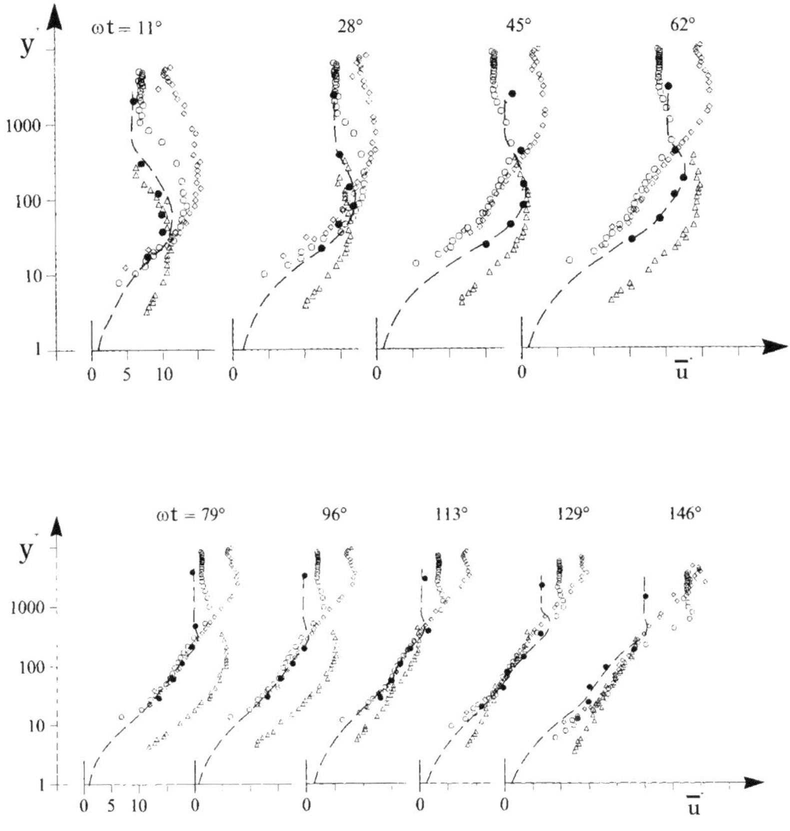

Fig. 5.30 Mean velocity distributions. Smooth bed. Diamonds: $Re = 6 \times 10^6$; Empty circles: $Re = 1.6 \times 10^6$; Filled circles: $Re = 5 \times 10^5$, all from Jensen et al. (1989). Dashed lines: $Re = 5 \times 10^5$, Spalart and Baldwin's (1987) direct N.-S. simulation. Triangles: $Re = 2.8 \times 10^5$, Hino et al. (1983).

wall-normal component of the Navier-Stokes equations.

Jensen et al. (2014a) were the first to study the wall-normal pressure gradient in a turbulent oscillatory boundary layer. In this section, we will summarize the results of Jensen et al.'s study (2014a).

Jensen et al. (2014a) treated the wall-normal pressure gradient as any other turbulence quantity. They calculated the wall-normal pressure gradient, using direct numerical simulation (DNS), and obtained the statistical properties associated with this quantity from the DNS simulations.

We note that, although the results were not included in their paper for reasons of space, Jensen et al. (2014a) compared their results for classic turbulence quantities such as $\overline{u'^2}$, $\overline{v'^2}$, $\overline{w'^2}$ and $\overline{u'v'}$ with Spalart and Baldwin (1987) and Jensen et al. (1989), and found that their results agree well with those of Spalart and Baldwin (1987) and Jensen et al. (1989). This gave confidence for the use of their DNS data in the calculations of the wall-normal pressure gradient.

Fig. 5.31 presents a sequence of the mean wall-normal pressure gradient versus the distance from the wall from Jensen et al.'s (2014a) study at four successive phase values, $\omega t = 0°$, $45°$, $90°$ and $135°$. The Reynolds number is $Re = 4.5 \times 10^5$, or the Reynolds number based on the Stokes length (Eq. 5.17) is $Re_\delta = 948$. The average is taken over the entire horizontal plane at a given distance from the wall. (Note that the authors calculated the previously mentioned turbulence quantities in the same fashion.) In Fig. 5.31, the first set of data (solid lines in the figure) represents the conditional averaged wall-normal pressure gradient corresponding to $-\partial p/\partial x_2 > 0$ (the upward-directed pressure gradient). The second set of data in Fig. 5.31 (the dashed lines in the figure) represents the conditional averaged wall-normal pressure gradient corresponding to $-\partial p/\partial x_2 < 0$ (the downward-directed pressure gradient). x_2 is the non-dimensional distance from the wall, and p the non-dimensional pressure. These non-dimensional quantities are defined in Eqs. 5.42–5.44.

The conditions described above imply that, for upward-directed pressure gradients, the fluid particles with zero vertical momentum will move upward, and for downward-directed pressure gradients, the fluid particles with zero vertical momentum will move downward. However, for fluid particles with finite vertical momentum in the opposite direction, first the aforementioned vertical pressure gradient is used to dissipate the momentum of the fluid particles, before the particles begin to move in the direction dictated by the vertical pressure gradient, upward or downward.

The non-dimensional quantities in Fig. 5.31 are defined as follows. The non-dimensional distance from the wall x_2

$$x_2 = \frac{y}{\delta_1} \tag{5.42}$$

in which δ_1 is the Stokes length, Eq. 5.13,

$$\delta_1 = \frac{\nu}{U_{0m}} Re_\delta \tag{5.43}$$

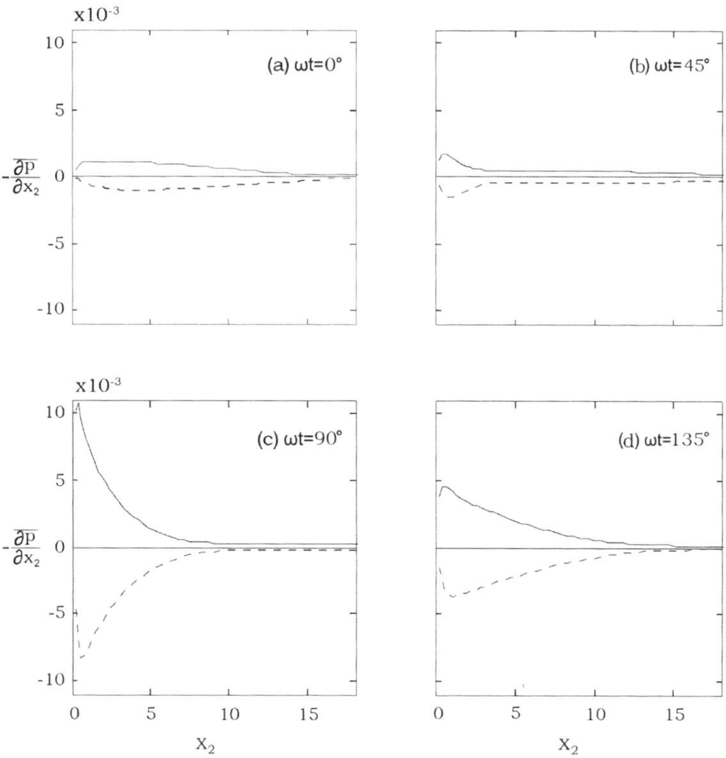

Fig. 5.31 Conditional averaged value of wall-normal pressure gradient. Solid line: $-\partial p/\partial x_2 > 0$. Dashed line: $-\partial p/\partial x_2 < 0$. Jensen et al. (2014a).

in which Re_δ is the Reynolds number based on the Stokes length, Eq. 5.17. The quantity p, the non-dimensional pressure, in Fig. 5.31 on the other hand is defined as

$$\frac{p}{\rho U_{0m}^2} \tag{5.44}$$

in which p is the actual pressure.

Returning to Fig. 5.31, first of all, the figure shows that, regardless of the phase value, the vertical pressure gradient appears to be largest very near the wall, in the region associated with the generation of turbulence, $x_2 = O(0.5)$, or $y^+ = O(20)$ from $y^+ = (Re_\delta/\sqrt{\frac{2}{f_w}})x_2$, taking $Re_\delta = 948$, and $f_w = 0.0040$, corresponding to the non-dimensional bed shear stress $\overline{\tau_0} = 1.9$ reported in Jensen et al. (2014a, p. 5).

Secondly, Fig. 5.31 shows that the pressure gradient increases tremendously with increasing phase, attaining very large values at $\omega t = 90°$ before it begins to fall off.

A third observation from the figure is that the upward-directed pressure gradient appears to be larger than the downward-directed pressure gradient, and this appears to be the case throughout the phase space.

Fig. 5.32 shows the standard deviation of the conditionally-averaged pressure gradients versus the distance from the wall

$$\sigma_{(-\frac{\partial p}{\partial x_2})} = \left\{ \overline{\left[(-\frac{\partial p}{\partial x_2}) - \overline{(-\frac{\partial p}{\partial x_2})} \right]^2} \right\}^{1/2} \tag{5.45}$$

Averaging (overbar) is, again, over the entire horizontal plane. The solid lines in the figure represent the condition $-\partial p/\partial x_2 > 0$ (the upward-directed pressure gradient), and the dashed lines $-\partial p/\partial x_2 < 0$ (the downward-directed pressure gradient). Two observations can be made. First, the magnitude (i.e., the standard deviation) for the upward-directed pressure gradient is larger than that of the downward-directed pressure gradient. Second, the time development observed for the standard deviation in Fig. 5.32 is similar to the conditionally-averaged gradients in Fig. 5.31. Furthermore, it may also be noted that the standard deviation values are as large as the mean values.

We note that the wall-normal pressure gradient data from the DNS simulations in Fig. 5.31 at $\omega t = 90°$ (corresponding to the condition $-\partial p/\partial x_2 > 0$, the upward-directed pressure gradient) agrees quite well with the upward-directed pressure gradient obtained from Sumer's (1984) single particle experiments. The latter has been discussed in greater detail in Section 3.6.2.

In the paper of Jensen et al. (2014a), they discussed the implications of wall-normal pressure gradient for sediment transport, and concluded that the large fluctuations of the wall-normal pressure gradient have significant implications for the latter.

Finally, it is believed that the wall-normal pressure gradient may have a significant role in the onset of turbulence, already recognized by the flow visualization of turbulent spots illustrated in Fig. 5.7. When the Reynolds number is larger than the critical value, and when the right value of the phase is reached, local generation of upward-directed pressure gradient momentarily near the bed will cause an impulsive movement of the near-bed fluid upward, in the way a turbulent spot is produced, as described in Section 5.2.1. Similar to the onset of turbulence, it is also believed that the

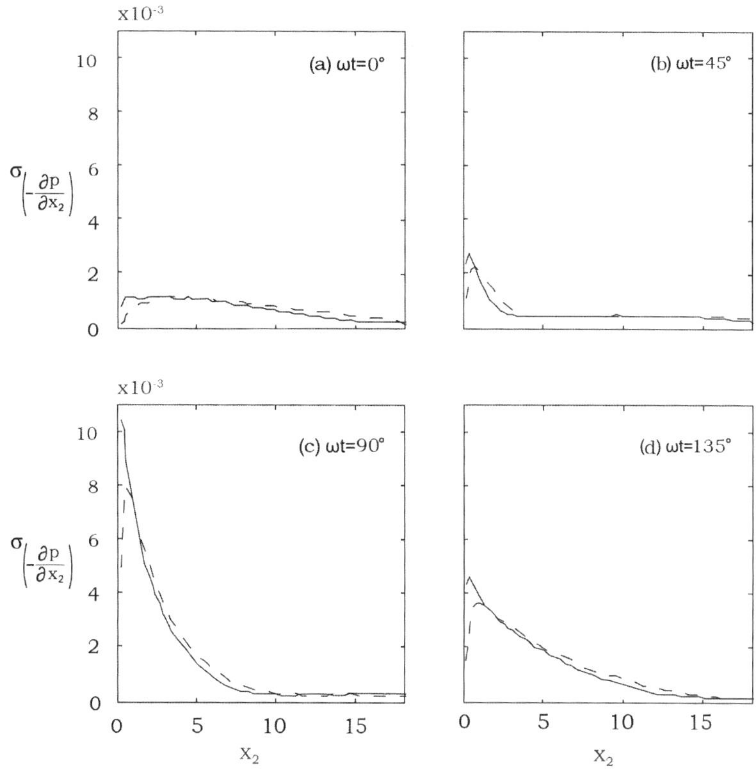

Fig. 5.32 Standard deviation of the conditional-averaged wall-normal pressure gradient. Solid line: $-\partial p/\partial x_2 > 0$. Dashed line: $-\partial p/\partial x_2 < 0$. Jensen et al. (2014a).

wall-normal pressure gradient may also have a significant role in the bursting process in a fully-developed turbulent boundary layer with regard to the ejection event in which a low-speed wall streak is subject to an impulsive "suction", resulting in the rigorous entrainment of low-speed fluid into the main body of the flow, presumably caused by locally- and momentarily-generated upward-directed pressure gradient. See Section 3.5 for a detailed description of the bursting process.

5.3.8 *Boundary-layer thickness*

Fig. 5.33 presents the normalized boundary-layer thickness $\frac{\delta}{a}$ plotted against Re. Here δ is the distance from the bed where the velocity be-

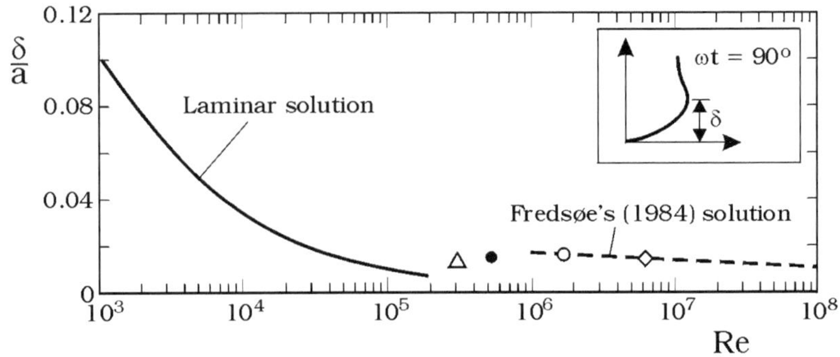

Fig. 5.33 Boundary-layer thickness. Smooth bed. Jensen et al. (1989).

comes maximum at the phase value $\omega t = 90°$ (Fig. 5.4). The laminar-flow solution in the figure is (from Eq. 5.14)

$$\frac{\delta}{a} = \frac{3\pi}{4}\left(\frac{2}{Re}\right)^{\frac{1}{2}} \tag{5.46}$$

Fredsøe's (1984) theoretical solution in the figure, obtained by integration of the momentum equation (dashed line in Fig. 5.33), can be approximated by:

$$\frac{\delta}{a} = 0.086 Re^{-0.11} \tag{5.47}$$

(see also Fredsøe and Deigaard, 1992, p. 29).

From the figure, the boundary-layer thickness for storm conditions (i.e., for very large Re)

$$\frac{\delta}{a} = O(0.01) \tag{5.48}$$

This corresponds to a boundary-layer thickness of $\delta = O(10 \text{ cm})$.

5.4 Turbulent wave boundary layers: Rough bed

Similar to steady boundary layers, there are three kinds of bed roughness in wave boundary layers: Hydraulically smooth bed when $k_s^+ < 5$; Completely rough bed (i.e. hydraulically rough) when $k_s^+ > 70$; and Transitional bed when $5 < k_s^+ < 70$ in which k_s^+:

$$k_s^+ = \frac{k_s U_{fm}}{\nu} \tag{5.49}$$

with k_s being Nikuradse's equivalent sand roughness of the bed, and U_{fm} the maximum value of the friction velocity defined in Eqs. 5.27–5.29.

The focus in this section will be turbulent wave boundary layers over completely rough beds, satisfying $k_s^+ > 70$.

On dimensional grounds, the quantities describing the wave boundary layer over a completely rough bed depend on the roughness parameter (Eq. 5.2):

$$\frac{a}{k_s} \tag{5.50}$$

The description in this section is essentially based on Jensen et al.'s (1989) study, already introduced in the previous section in conjunction with turbulent wave boundary layers over smooth beds.

We will first describe Jensen et al.'s (1989) setup for the rough bed experiments. Subsequently, we will compare the results of Jensen et al.'s rough bed experiments with their counterpart in their smooth-bed experiments. The discussion will end with the boundary layer thickness in the case of the rough-bed wave boundary layers.

The flow resistance in connection with the rough bed case will be discussed in Section 5.5, along with the smooth bed and transitional bed categories.

5.4.1 *Experimental setup in Jensen et al. (1989) rough-bed experiments*

As mentioned previously, Jensen et al. (1989) used the same oscillating tunnel as that described in the previous section. They experimented with two kinds of bed roughness. In one, they achieved the rough bed by gluing a sheet of sand paper on the bed. In the second, they glued sand of fairly uniform size one layer deep to the actual bottom of the tunnel. Nikuradse's equivalent sand roughness for these two rough beds was determined to be $k_s = 0.84$ mm for the former, and $k_s = 2.6$ mm and 3.7 mm for the latter. The roughness Reynolds number corresponding to the aforementioned rough beds was $k_s^+ = 44$ and 84 for the sand-paper experiments while $k_s^+ = 142$ and $k_s^+ = 460$ for the experiments where sand was glued to the bottom. As seen, except the case where $k_s^+ = 44$, the bed in Jensen et al.'s (1989) experiments behaved as a completely rough boundary.

In the case of rough beds, the measurement of the bed shear stress (and therefore the friction velocity) cannot be achieved by the hot-film technique as the latter requires the presence of the viscous sublayer (Sumer

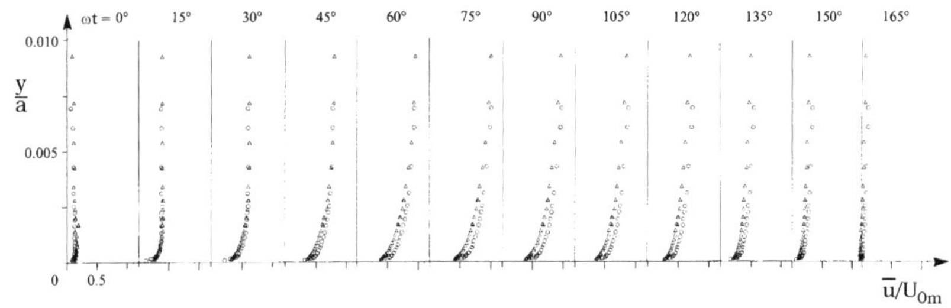

Fig. 5.34 Mean velocity $\overline{u}$. Circles: Smooth bed. Triangles: Rough bed. $Re = 6 \times 10^6$. Jensen et al. (1989).

et al., 1993a). Therefore the friction velocity U_f in the case of the rough bed in Jensen et al.'s (1989) experiments was determined by fitting straight lines to the logarithmic-layer portion of the phase-resolved mean velocity distributions plotted on a semi-logarithmic plane. This method gave also Nikuradse's equivalent sand roughness. The values given in the preceding paragraph are those obtained this way.

The LDA velocity measurements in Jensen et al.'s (1989) rough-bed experiments were made at sampling intervals of 48 ms for the sand-paper-roughness experiments, and 14 ms for the sand-grain-bed-roughness experiments. The sampling interval of 48 ms corresponds to approximately 200 samples per flow cycle, and 14 ms corresponds to about 690 samples per flow cycle; both are large enough to be able to trace the time development of the measured statistical quantities over the flow cycle. The total number of cycles sampled N was 50 for the sand-grain-roughness experiments, and 80 for the sand-paper-roughness experiments.

5.4.2 *Comparison of rough-bed and smooth-bed results*

Figs. 5.34–5.37 compare the rough-bed results with the corresponding smooth-bed ones in the tests of Jensen et al. (1989). Note that the free-stream conditions were maintained exactly the same in these two tests. Therefore comparison can be made on the same basis. The Reynolds number of the experiments was $Re = 6 \times 10^6$, and the roughness Reynolds number for the rough-bed experiment was $k_s^+ = 84$, and the roughness parameter $\frac{a}{k_s}$ was $\frac{a}{k_s} = 3700$.

From these figures, the following points appear to be noteworthy:

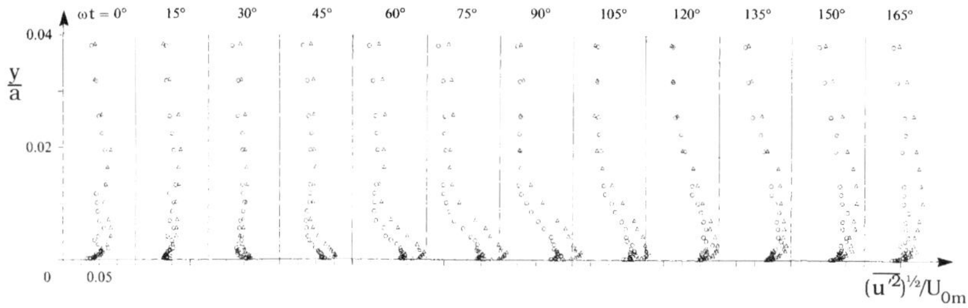

Fig. 5.35 R.m.s. value of u'. Circles: Smooth bed. Triangles: Rough bed. $Re = 6 \times 10^6$. Jensen et al. (1989).

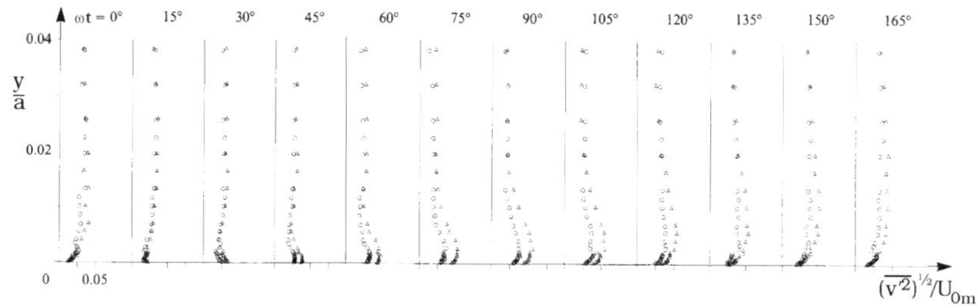

Fig. 5.36 R.m.s. value of v'. Circles: Smooth bed. Triangles: Rough bed. $Re = 6 \times 10^6$. Jensen et al. (1989).

1. The description given in the previous section of how the flow and turbulence evolve during one half-cycle is qualitatively the same for the rough-bed flows, too.

2. The introduction of the roughness, however, results in a decrease in the mean flow velocity and an increase in the turbulence quantities.

3. However, this effect tends to disappear with the distance from the bed; the mean flow velocity is practically the same for the two cases for distances $\frac{y}{a} \gtrsim 0.01$. The same is also true for the turbulence quantities, although this time the effect of roughness disappears at somewhat larger distances from the bed, namely for $\frac{y}{a} \gtrsim 0.03$. The decrease in the mean velocity is due to the retarding effect of the rough boundary, while the increase in the turbulence quantities is due to the enhancement of the momentum transfer by the presence of the roughness elements.

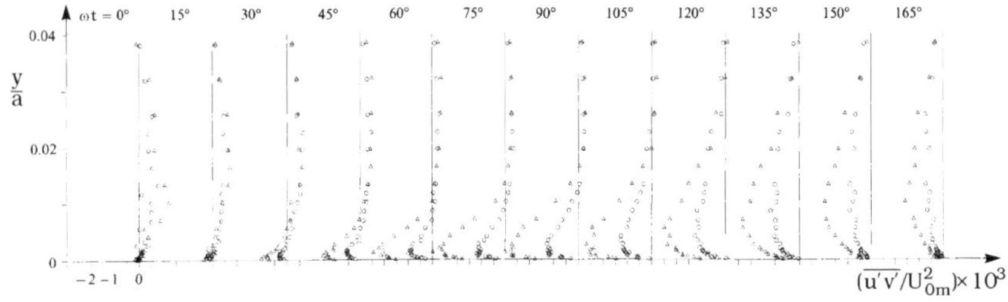

Fig. 5.37 Reynolds stress $\overline{u'v'}$. Circles: Smooth bed. Triangles: Rough bed. $Re = 6 \times 10^6$. Jensen et al. (1989).

4. The flow begins to feel the effect of the roughness only after ωt reaches the value of about 15°. This is because the temporal value of the roughness Reynolds number ($k_s^+ = k_s U_f/\nu$) is still far too small for these phases for the bed to act as a completely rough boundary and, as a result, the flow quantities for this short interval of time remain practically unchanged, regardless of the boundary category.

As mentioned previously, the rough turbulent boundary layers are governed solely by the parameter $\frac{a}{k_s}$. This parameter plays a role similar to that of Re in the case of the smooth turbulent wave boundary layers. Jensen et al. (1989) discussed the effect of this parameter on the mean and turbulence flow quantities in greater detail, and the reader is referred to this publication for further discussion.

It may also finally be noted that the flow structure within oscillatory wave boundary layers can alternatively, and rather generally (i.e. on smooth or rough beds), be viewed in terms of:

$$\overline{u}(y,t) = U_{0m}(1 - D_u(y))e^{i\omega t}, \qquad e^{i\omega t} = \cos(\omega t) + i\sin(\omega t) \tag{5.51}$$

where the physical value for $\overline{u}$ corresponds to the real part of the above. Here $D_u(y)$ is the so-called velocity defect function, see e.g. Sleath (1970, 1982) and Nielsen (1992, his Section 1.3.3). Based on comparison with experimental measurements, Nielsen (1985) proposed a velocity defect function of the form

$$D_u(y) = \exp\left[-(1+i)\left(\frac{y}{\delta_N}\right)^{p_N}\right] \tag{5.52}$$

Working within this same framework, Nielsen (2016) likewise proposed the following universal length scale:

$$\delta_N = \sqrt{0.0081ak_s + 2\frac{\nu}{\omega}} \tag{5.53}$$

which tends naturally to $\delta_N = \delta_1 = \sqrt{2\nu/\omega}$ on smooth beds (taking $k_s = 0$, i.e. the length scale becomes equivalent to the Stokes length) and to $\delta_N = 0.09\sqrt{ak_s}$ for large roughness. The parameter p_N above serves to effectively stretch the boundary layer structure vertically, relative to the laminar solution where $p_N = 1$. Taking this value for p_N, combined with $k_s = 0$, indeed leads to the laminar solution, analogous to Eq. 5.12 (but for a free stream velocity now assumed to vary as $U_0 = U_{0m}\cos(\omega t)$). For describing oscillatory boundary layers on fairly rough beds ($k_s/a > 0.06$ i.e. $a/k_s < 16.7$) data suggests that the flow structure is analogous to the laminar case, such that $p_N = 1$ may be retained, in conjunction with the physical value for k_s. For vanishing roughness (i.e. smooth beds with $k_s = 0$) and high Reynolds numbers the asymptotic value corresponds to $p_N = 1/3$. As pointed out by Nielsen (1985), this value corresponds to maximum energy dissipation for a given vertical scale. For further details and additional discussion of this approach, see Nielsen (2016) and Nielsen (1992).

5.4.3 *Boundary-layer thickness*

Fig. 5.38 depicts the boundary-layer thickness in the case of the rough bed. In the figure, Diamond, Circle and Square symbols are from Jensen et al. (1989); Triangles (tip pointing upward) from Jonsson and Carlsen (1967); Triangles (tip pointing downward) from Sleath (1987); and Solid lines: Fredsøe's (1984) theoretical solution obtained by integration of the momentum equation, approximated by (see also Fredsøe and Deigaard, 1992, p. 25)

$$\frac{\delta}{k_s} = 0.09\left(\frac{a}{k_s}\right)^{0.82} \tag{5.54}$$

The figure shows that, the boundary-layer thickness for storm conditions (very large $\frac{a}{k_s}$) is

$$\frac{\delta}{a} = O(0.02) \tag{5.55}$$

which corresponds to a boundary-layer thickness of $\delta = O(20\text{ cm})$.

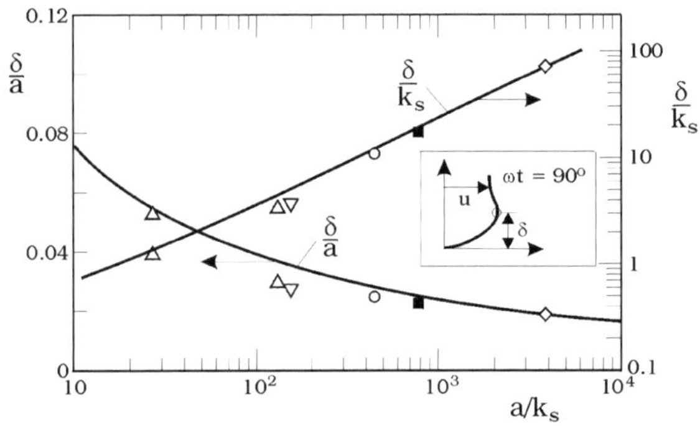

Fig. 5.38 Boundary-layer thickness. Rough bed. Jensen et al. (1989).

5.5 Flow resistance in wave boundary layers

Flow resistance is traditionally described in terms of the so-called wave friction coefficient, f_w:

$$f_w = \frac{2\tau_{0m}/\rho}{U_{0m}^2} \tag{5.56}$$

or

$$U_{fm} = \sqrt{\frac{f_w}{2}} U_{0m} \tag{5.57}$$

in which U_{fm} is the maximum value of the bed friction velocity defined by

$$U_{fm} = \sqrt{\frac{\tau_{0m}}{\rho}} \tag{5.58}$$

with τ_{0m} being the maximum value of the bed shear stress; see Eqs. 5.27–5.29.

On dimensional grounds, the friction coefficient is a function of the Reynolds number, $f_w = f(Re)$, in the case of smooth bed; it is a function of the roughness parameter, $f_w = f(\frac{a}{k_s})$, in the case of rough bed; and it is a function of both parameters, $f_w = f(Re, \frac{a}{k_s})$ in the case of transitional bed.

We note that the wave friction coefficient is traditionally accompanied by the phase information because the wave friction coefficient is defined on the basis of the maximum bed shear stress, and clearly it is important to determine where in the phase space this maximum value occurs; see the inset in Fig. 5.20, bottom panel. Similar to the wave friction coefficient, the phase difference ϕ is $\phi = f(Re)$ in the case of smooth bed; $\phi = f(\frac{a}{k_s})$ in the case of rough bed, and $\phi = f(Re, \frac{a}{k_s})$ in the case of transitional bed.

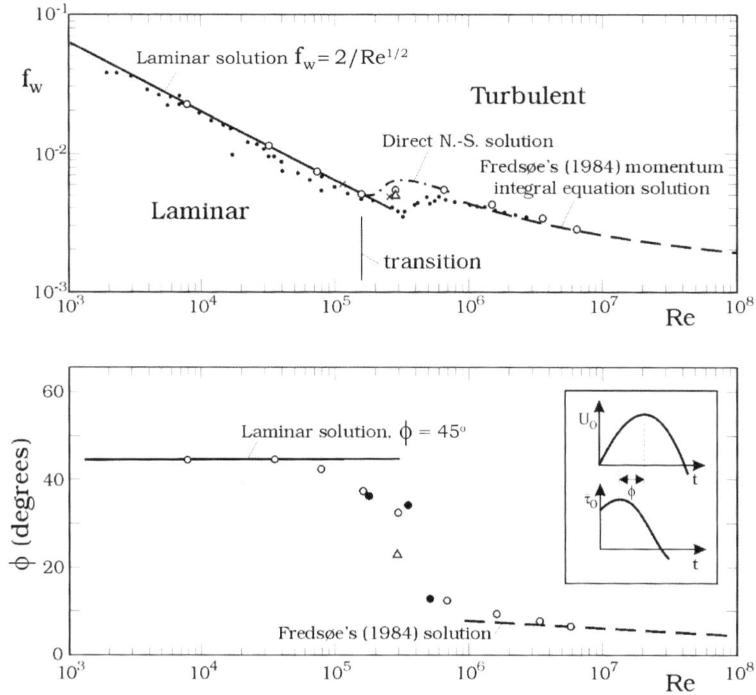

Fig. 5.39 (a) Wave friction coefficient and (b): Phase difference. See the text for the details of the data presented in this figure. Smooth bed. Jensen et al. (1989).

5.5.1 *Wave friction coefficient for smooth-bed wave boundary layer*

The flow resistance and phase data have been discussed previously in conjunction with the laminar-to-turbulent transition in Section 5.2.6 with Fig. 5.20. In that context, only the data from Carstensen et al. (2010) and Jensen et al. (1989) have been plotted to keep the figure relatively simple. We now reproduce in Fig. 5.39 the entire set of flow-resistance data published in Jensen et al. (1989).

In Fig. 5.39, top panel, (i): the laminar solution (solid line), $f_w = 2/\sqrt{Re}$; (ii): Spalart and Baldwin's (1987) DNS simulation; (iii): Fredsøe's (1984) theoretical solution obtained by integration of the momentum equation (dashed line); Empty circles: Jensen et al.'s (1989) experiments; Filled circles: Kamphuis' (1975) experiments; Triangles: Hino et al.'s (1983) experiments; Crosses: Sleath's (1987) 0.2 mm sand-bed experiments where the roughness Reynolds number corresponded to $k_s^+ = k_s U_f/\nu \approx 10$.

In Fig. 5.39, bottom panel, (i): Laminar solution (solid line), namely 45°; (ii): Fredsøe's (1984) theoretical solution obtained by integration of the momentum equation (dashed line); Filled circles: Spalart and Baldwin's (1987) DNS simulation; Empty circles: Jensen et al.'s (1989) experiments; Triangles: Hino et al.'s (1983) experiment.

From Fig. 5.39, the following expressions are given for the wave friction coefficient:

1. For laminar wave boundary layers ($Re \lesssim 1.5 \times 10^5$), from the laminar solution:

$$f_w = \frac{2}{Re^{\frac{1}{2}}} \tag{5.59}$$

2. For turbulent wave boundary layers ($Re \gtrsim 5 \times 10^5$), from Fredsøe's (1984) theoretical solution, which can be approximated to (see also Fredsøe and Deigaard, 1992, p. 29):

$$f_w = \frac{0.035}{Re^{0.16}} \tag{5.60}$$

3. For transitional wave boundary layers ($1.5 \times 10^5 \lesssim Re \lesssim 5 \times 10^5$), from Fig. 5.39, to a first approximation:

$$f_w = 0.004 - 0.005 \tag{5.61}$$

Regarding the phase difference in Fig. 5.39, bottom panel, the way in which ϕ varies with the Reynolds number has already been discussed in detail in conjunction with the laminar-to-turbulent transition in Section 5.2.6, and the reader is therefore referred there.

5.5.2 *Wave friction coefficient for rough-bed wave boundary layer*

Dixen et al. (2008) presented the flow resistance data where the authors compiled the existing data (from Bagnold, 1946, Kamphuis, 1975, Jonsson and Carlsen, 1976, Sleath, 1987, Sumer et al., 1987, Jensen et al., 1989, Simons et al., 2000), including theirs for small values of the roughness parameter a/k_s, covering a very broad range of a/k_s, namely $0.2 < a/k_s < 4 \times 10^3$. Figs. 5.40 and 5.41 are reproduced from Dixen et al. (2008). Fig. 5.40b gives a close-up picture of the wave friction coefficient versus a/k_s for very small values of the roughness parameter (very large roughnesses), namely for $0.2 < a/k_s < 4$.

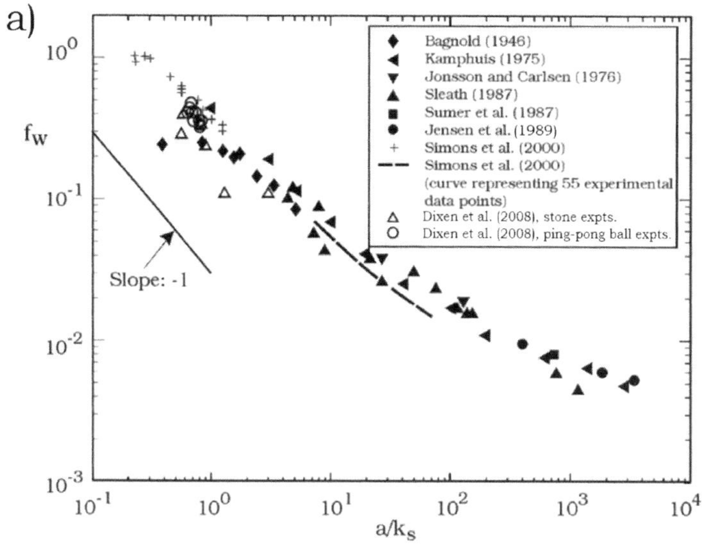

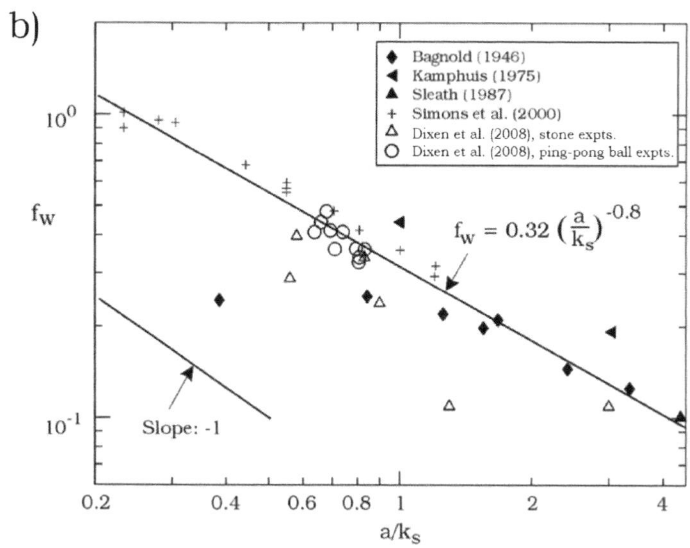

Fig. 5.40 (a) Wave friction coefficient. (b) Wave friction coefficient, close-up for small values of a/k_s. Rough bed. Dixen et al. (2008).

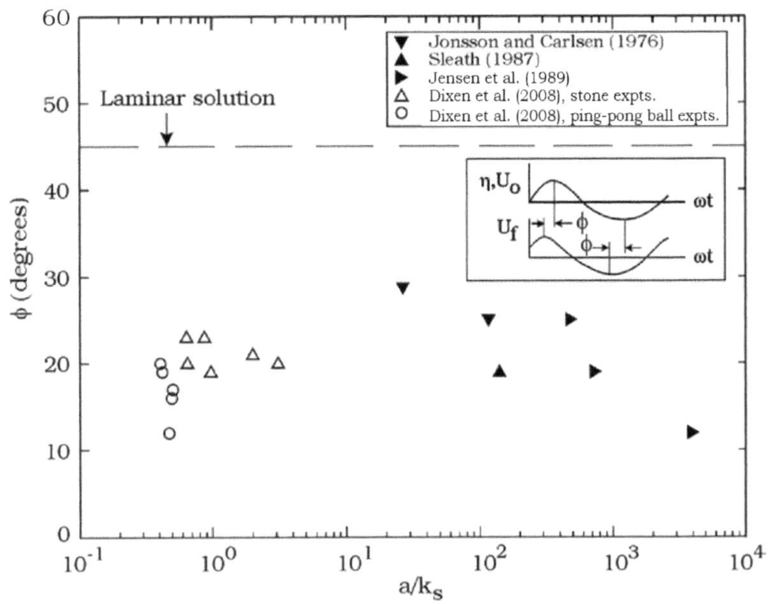

Fig. 5.41 Phase lead of the friction coefficient over the free-surface elevation. Rough bed. Dixen et al. (2008).

It would be interesting to discuss at this point the asymptotic behaviour of the variation of the wave friction coefficient for small values of $a/k_s \to 0$. Earlier, Kajiura (1968) and Jonsson (1975, 1980) suggested that f_w should go to a constant value as $a/k_s \to 0$, which was based on Bagnold (1946) data in Fig. 5.40. However, the data of Simons et al. (2000) and Dixen et al. (2008) in Fig. 5.40 do not seem to tend to a constant value as $a/k_s \to 0$.

Dixen et al. (2008) discussed this issue at some length. They showed that the bed shear stress can be written as

$$\tau_0 = \frac{1-n}{A_{xz}}\rho C_M V \frac{d}{dt}\langle U_0 \rangle \tag{5.62}$$

as $a/k_s \to 0$. Here n is the porosity of the rough bed, C_M the inertia coefficient $(= 1 + C_m)$ (see Sumer and Fredsøe, 2006), V the volume of the individual roughness element below the theoretical bed, and A_{xz} the area of the same volume projected on the plane perpendicular to the flow direction. Taking $< U_0 >= U_{0m} \sin(\omega t)$ and $V/A_{xz} = (\pi d^3/6)/(\pi d^2/4) = 2d/3$ (to a first approximation, where d is the particle diameter), the maximum value of the bed shear stress is found to be

$$\tau_{0m} = \frac{2}{3}(1-n)\rho C_M U_{0m}\omega d \tag{5.63}$$

Dixen et al. (2008) remark that although the entire concept of wave boundary layer breaks down for such small values of a/k_s, they, to a first approximation, define a wave-friction coefficient, to observe the way in which the friction coefficient varies asymptotically as $a/k_s \to 0$. Taking $d = k_s/2.5$, and recalling that $a = U_{0m}/\omega$, they obtain the wave friction coefficient

$$f_w = \frac{2\tau_{0m}/\rho}{U_{0m}^2} = \frac{8}{15}(1-n)C_M\left(\frac{a}{k_s}\right)^{-1} \tag{5.64}$$

For small values of a/k_s (i.e., for small values of the so-called Keulegan-Carpenter number), the inertia coefficient, C_M, should go to a constant value (see, e.g., Sumer and Fredsøe, 2006, p. 142). Therefore, for small values of a/k_s, the wave friction coefficient should be

$$f_w \propto \left(\frac{a}{k_s}\right)^{-1} \quad \text{as} \quad \frac{a}{k_s} \to 0 \tag{5.65}$$

Sleath (1984, p. 200) reached the same conclusion from a force balance analysis considering a small segment of the bed.

The line with the slope equal to -1 in Fig. 5.40 represents the variation in Eq. 5.65. The data in Fig. 5.40 do seem to reveal that f_w tends to go asymptotically to the $(\frac{a}{k_s})^{-1}$ variation. Dixen et al. (2008) furthermore explain why the earlier data was perceived as $f_w \to$ constant as $a/k_s \to 0$.

The following expressions are given for the wave friction coefficient displayed in Fig. 5.40:

1. For small values of the roughness parameter (large roughnesses), $0.2 < a/k_s < 10$, Dixen et al. (2008) give the following expression obtained from a curve fit exercise to the data in Fig. 5.40:

$$f_w = 0.32\left(\frac{a}{k_s}\right)^{-0.8} \tag{5.66}$$

This expression is slightly different from $f_w = 0.33(\frac{a}{k_s})^{-0.84}$, given by Simons et al. (2000).

2. For moderate values of the roughness parameter, $10 < a/k_s < 50$, the following flow-resistance equation (Kamphuis, 1975, which was based on his own data) can be used:

$$f_w = 0.4\left(\frac{a}{k_s}\right)^{-0.75} \tag{5.67}$$

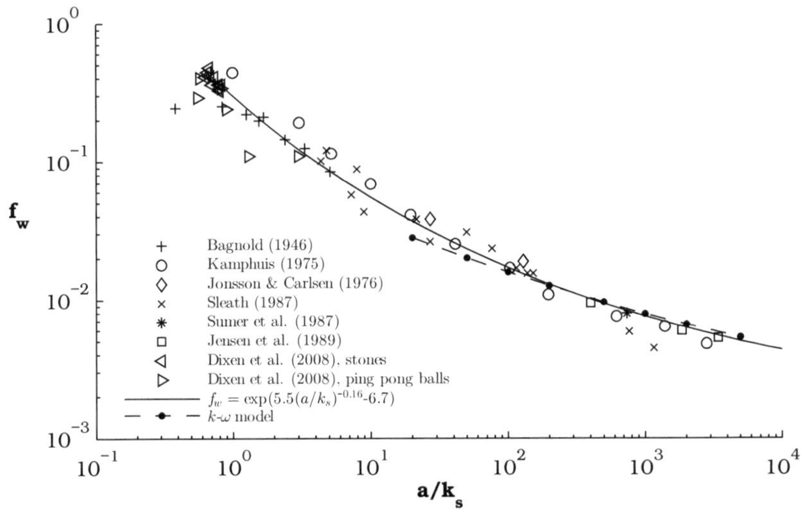

Fig. 5.42 Rough turbulent wave friction factors. Adapted from Fuhrman et al. (2013).

3. For large values of the roughness parameter, $a/k_s > 50$, Fredsøe and Deigaard (1992, p. 25) suggest the following equation which was based on Fredsøe's (1984) theoretical solution obtained by integration of the momentum equation, and shown to be in good agreement with the experimental data of Jensen et al. (1989), reproduced in Fig. 5.40:

$$f_w = 0.04 \left(\frac{a}{k_s} \right)^{-0.25} \tag{5.68}$$

Fuhrman et al. (2013) considered several of the data sets mentioned previously, spanning a wide range of a/k_s. They proposed the single formula, similar in form to those of Swart (1974) and Nielsen (1992), for estimating the wave friction coefficient f_w for rough beds:

$$f_w = \exp \left[5.5 \left(\frac{a}{k_s} \right)^{-0.16} - 6.7 \right] \tag{5.69}$$

The data sets and this curve (full line) are plotted in Fig. 5.42, where it is seen that this expression maintains reasonable accuracy for the entire range of a/k_s considered, say $1 \leq a/k_s \leq 10^4$. (Note that further extension of the above to predict the time variation of the bed shear stress $\tau_0(t)$ based on a free stream velocity signal $U_0(t)$ is presented in Section 5.11.3.)

Regarding the phase difference, Dixen et al. (2008) compiled all the existing data including their own and plotted ϕ versus the roughness parameter a/k_s. This data is reproduced from Dixen et al. (2008) in Fig. 5.41. The figure also includes the laminar solution, $\phi = 45°$, as a reference line.

Although there is a considerable amount of scatter, it seems that the phase difference does not change radically with the parameter a/k_s. This may be due to the fact that the mechanism responsible for the observed phase lead of friction velocity remains the same, irrespective of the value of a/k_s. This mechanism is closely related to the process where the lee-wake water is washed over the roughness elements prior to the flow reversal in the free-stream region. As this mechanism does not change with changing a/k_s, the phase lead ϕ will therefore remain practically unchanged. Dixen et al. (2008) further discuss the behaviour as $a/k_s \to 0$ in their publication, and the reader is referred to the original publication.

5.5.3 *Wave friction coefficient for transitional-bed wave boundary layer*

Fig. 5.43 displays the wave friction data plotted as function of the Reynolds number and the roughness parameter, reproduced from Justesen (1988a). The majority of the data in the figure for the transitional bed come from Kamphuis' (1975) experiments. Other experimental data in the figure are from Hino et al. (1983) (smooth bed), Kamphuis (1975) (also smooth bed), Jonsson (1963) (rough bed) and Sumer et al. (1987) (transitional bed). Fig. 5.43 also presents the friction-coefficient results of Justesen's (1988a) one-equation turbulence-modelling calculations (solid lines).

Based on the experimental data in Fig. 5.43, the following empirical (implicit) expression can be given for the wave friction coefficient, f_w:

$$f_w = f_{w,s} + (f_{w,r} - f_{w,s})\left(1 - \exp\left(-\frac{\sqrt{f_w/2}}{10}\frac{Re}{a/k_s}\right)\right)^3 \tag{5.70}$$

in which $f_{w,s}$ is the smooth-bed wave friction coefficient given in Eq. 5.60, and $f_{w,r}$ is the rough-bed friction coefficient given in Eq. 5.69. In the above curve-fitting exercise, the preceding expression is constructed such that $f_w \to f_{w,s}$ when k_s is small (smooth bed); and $f_w \to f_{w,r}$ when it is large (rough bed). This empirical expression, when plotted in Fig. 5.43, appears to agree fairly well with Justesen's (1988a) one-equation turbulence-modelling results (solid lines in the figure). It also appears to be in accord with Myrhaug's (1989) semi-empirical formula for the wave friction in transitional beds. The authors acknowledge help from Dr. Ozgur Kirca in the development of the above expression.

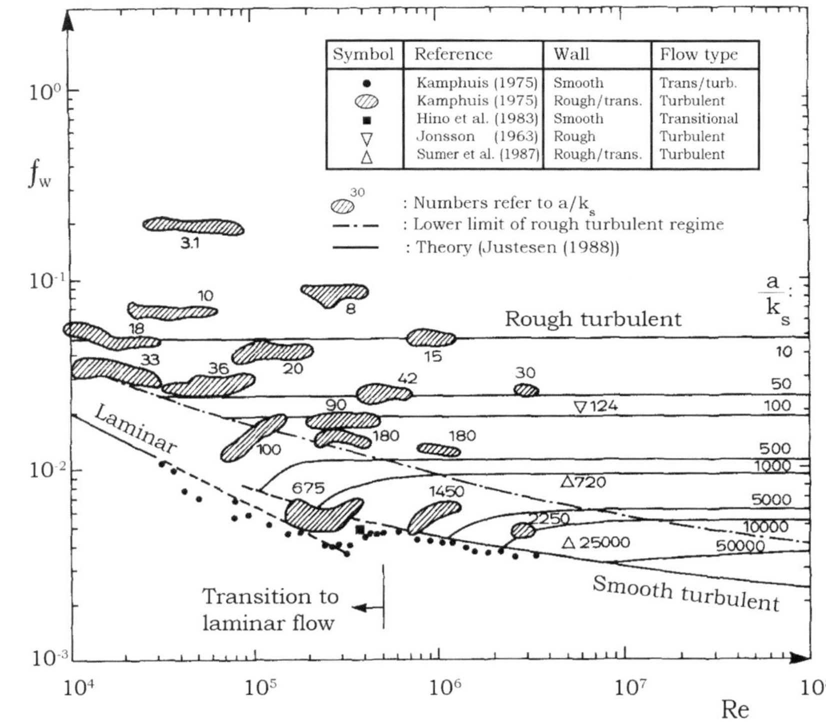

Fig. 5.43 Wave friction coefficient. Justesen (1988a).

5.6 Combined wave and current boundary layers

Waves in a coastal environment usually co-exist with currents. The waves and currents interact through a number of mechanisms such as modification of the wave kinematics by the current, modification of the current (over the entire flow depth) by the presence of the waves, and the interaction of the wave and current boundary layers. This section concerns the wave-current interaction within the bed boundary layer.

The wave-current boundary layer interaction has been the subject of a great many investigations in the last two decades or so; Bakker and van Doorn (1978), Kemp and Simons (1982, 1983), Asano et al. (1986), Myrhaug et al. (1987), Simons et al. (1988), Sleath (1990), Simons et al. (1992), Arnskov et al. (1993), and Simons et al. (1994), among others, on the experimental side; Bijker (1967), Grant and Madsen (1979),

Fredsøe (1984), van Kesteren and Bakker (1984), Christoffersen and Jonsson (1985), Justesen (1988a), Davies et al. (1988), Myrhaug and Slaattelid (1990), Huynh-Thanh and Temperville (1991) and Sleath (1991), among others, on the theoretical side. Field work has also been carried out, notably by Grant et al. (1983), Huntley and Hazen (1988), Myrhaug and Slaattelid (1989), and Soulsby and Humphery (1990), among others. Major reviews on the subject also appeared, e.g., Soulsby et al. (1993) and Lodahl et al. (1998).

The previously mentioned research has shown that a wave field superimposed on a current may change the mean and turbulence properties of the boundary layer significantly. The aforementioned changes in the flow may include the following processes: (1) The mean bed shear stress may increase, or remain constant, or even decrease with respect to the steady-current value. (2) Likewise, the maximum bed shear stress may increase, or it may retain its steady-current value when the waves are introduced. (3) The turbulence in the bed shear stress and in the near-bed flow may be suppressed, or it may even completely disappear when waves are superimposed on a current.

Although the previously mentioned research had thrown considerable light on the wave-current boundary layer interaction, comparatively little was known about the mechanics of the complex flow processes in this interaction mechanism, and a clear, comprehensive understanding of these processes was not available in early and mid 1990s.

This was the motivation of an extensive experimental investigation carried out by Lodahl et al. (1998). The purpose of the latter study was to investigate the flow processes in a "clean" wave-current environment. In Lodahl et al.'s (1998) study, the oscillating component of the combined flow (induced by real waves) was simulated by an oscillatory flow in an oscillating water tunnel, thereby eliminating the additional effects such as wave nonlinearity and wave asymmetry, plus more importantly enabling the "full scale" wave-induced velocities (hence, the fully developed turbulent wave boundary-layer conditions) to be produced under laboratory conditions, a condition which could not be fully achieved in the previous research referred to in the preceding paragraphs. Furthermore, Lodahl et al. (1998) adopted a circular pipe in favour of a tunnel with a rectangular cross section, to experiment with a truly two-dimensional wave-current flow, without any kind of secondary currents which are unavoidable in a rectangular cross-section flow environment.

There exist quite a substantial literature on what is called pulsating flows (the term used in the literature) in a circular pipe, the flow environment in the Lodahl et al. (1998) work: Gilbrech and Combs (1963), Sarpkaya (1966), Tromans (1978), Ramaprian and Tu (1980, 1983), Tu and Ramaprian (1983), Shemer et al. (1985), Mao and Hanratty (1986 and 1994), Stettler and Hussain (1986), and Finnicum and Hanratty (1988), among others. The researchers Tardu et al. (1994) studied pulsating flows in a channel.

The common feature of these earlier studies was that the combined flow was always in the current-dominated regime, and the oscillatory component of the combined flow was in the laminar- or, at most, transitional-flow regime, and therefore various characteristics of the combined flow associated with the wave-dominated fully-developed turbulent wave boundary-layer flows could not be captured. Lodahl et al.'s (1998) study covered practically the entire range of wave-current flows, namely from the current-dominated flow regime to the wave-dominated flow regime, and from the laminar to the fully-developed-turbulent wave boundary-layer regime.

This section is essentially based on Lodahl et al's (1998) study.

5.6.1 *Experimental setup in Lodahl et al. (1998) study*

Lodahl et al. (1998) conducted their experiments in the same oscillating tunnel as described in Section 5.3.1 (Fig. 5.21). The flow environment in Lodahl et al.'s study was a circular pipe, as mentioned previously. The pipe was placed in the tunnel, and rigidly fixed to the tunnel (Fig. 5.44). Three kinds of pipes were used, with diameters $D = 19$ cm, 9 cm and 1.1 cm. This enabled the researchers to experiment with a broad range of the parameter R/δ_1, one of the parameters of the study. Here R is the pipe radius and δ_1 the Stokes length (or the Stokes layer thickness) (Eq. 5.13). These transparent pipes had trumpet-mouth inlets at both ends. The pipes' surface was smooth. No water flow was allowed in the space between the pipe and the tunnel. The length of the 19 cm pipe was 10 m, the same as that of the working section of the tunnel (Fig. 5.21). This figure was 7.5 m for the 9 cm pipe and 5 m for the 1.1 cm pipe.

The current in the tunnel was provided by a constant head tank, and the desired flow rate was obtained by simultaneous adjustments of the inlet and outlet valves of the tunnel. The oscillatory flow in the tunnel is driven by an electronically controlled pneumatic system in exactly the same fashion as in Jensen et al. (1989), as described in Section 5.3.1.

Two kinds of measurements were made in Lodahl et al.'s (1998) study:

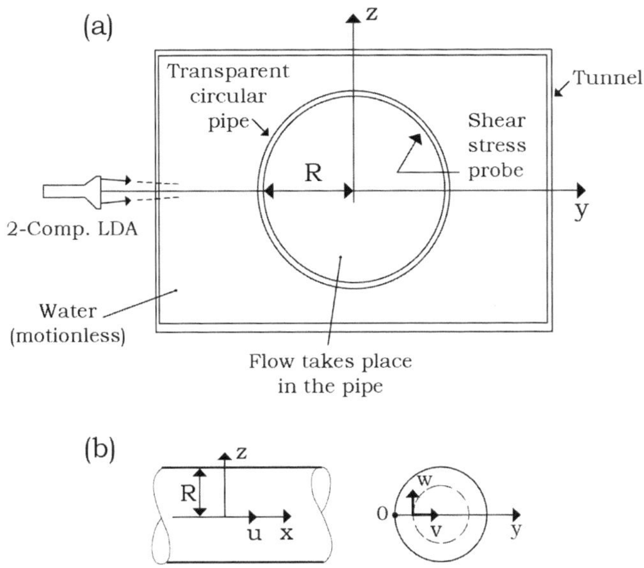

Fig. 5.44 Experimental setup in Lodahl et al.'s (1998) study. (a) Cross-section of the oscillating tunnel. (b) Definition sketch.

wall shear stress measurements, and velocity profile measurements.

The wall shear stress was measured with a hot film, described previously in conjunction with turbulent smooth-bed wave boundary layers in Section 5.2. The wall shear stress was not measured for the smallest 1.1 cm diameter pipe, because the housing of the hot-film probe was too big for this pipe.

The velocity profile measurements were made in the 19 cm diameter pipe, using a two-component LDA system. The measured components are x- (the streamwise) and z- (the tangential) velocity components (see Fig. 5.44). The LDA probe was traversed along the y-axis over the entire radius of the pipe (Fig. 5.44) for these measurements. The dimensions of the measuring volume were $(d_x \times d_y \times d_z) = 0.15$ mm $\times$ 2.5 mm $\times$ 0.15 mm for the two-component LDA measurements, and $(d_x \times d_z)$ are 0.2 mm $\times$ 0.75 mm for the hot-film probe.

The measurements were made with a sampling interval of 24 ms, corresponding to 416 samples per cycle. In the turbulent regime, at least 50 cycles were sampled in each test. In the laminar oscillatory flow case, typically 5 cycles were sampled. For the case of current alone, mean values were obtained through time averaging for which the sampling time was about 5 minutes, and, in the case of oscillatory flow and combined oscillatory flow and current, mean values were calculated through ensemble averaging (similar to Eqs. 5.32 and 5.33).

The experimental procedure adopted in a typical experiment to measure the wall shear stress in combined flows was as follows: (1) Calibrate the hot-film probe in situ (see the original paper for details). (2) Establish the desired current in the pipe, and make the "current alone" measurement (the wall shear stress and the centre line velocity). (3) Superimpose the first oscillatory flow on the current, and make the the wall shear stress measurement. (Here, the desired oscillatory flow was achieved in the combined flow by driving the piston of the pneumatic system at the desired amplitude. Also, the current flow rate was monitored continuously to check if it was maintained constant.) (4) Repeat the previous step for further increases in the velocity of the oscillatory flow. The same procedure was adopted also for the velocity profile measurements. The reader is referred to the original publication for details.

5.6.2 *Laminar-to-turbulent transition in combined oscillatory flow and current*

5.6.2.1 *Oscillatory flow alone in a pipe*

As seen in Section 5.2, laminar-to-turbulent transition in the case of oscillatory flow in wave boundary layers over a smooth bed is governed by the Reynolds number, Re_w. (Note that, in this section, we will use the symbol Re_w for the Reynolds number associated with the oscillatory component of the combined flow, and Re_c for the current component of the combined flow.) In the case of an oscillatory flow which takes place in a circular pipe, there will be an additional parameter, namely R/δ_1 in which R is the pipe radius, and δ_1 the Stokes length (or the Stokes layer thickness) (Eq. 5.13). Therefore, the parameters governing the laminar-to-turbulent transition in the case of pure oscillatory flow in a circular pipe are

$$Re_w \text{ and } \frac{R}{\delta_1} \tag{5.71}$$

in which

$$Re_w = \frac{aU_{0m}}{\nu} \tag{5.72}$$

with U_{0m}, the maximum value of the sinusoidal oscillatory velocity at the centre line of the pipe

$$U_0 = U_{0m} \sin(\omega t) \tag{5.73}$$

ω the angular frequency of the oscillatory flow ($\omega = 2\pi/T$), T the period of the oscillatory flow, a the amplitude of the oscillatory flow at the same location (i.e., at the centre line of the pipe), calculated as $a = U_{0m}T/(2\pi)$, and ν the kinematic viscosity. The quantity δ_1 in the parameter R/δ_1 in Eq. 5.71 is the Stokes length, again defined by (Eq. 5.13)

$$\delta_1 = \left(\frac{2\nu}{\omega}\right)^{\frac{1}{2}} \tag{5.74}$$

Fig. 5.45 displays the transition-to-turbulence data in the (R/δ_1, Re_w) plane along with Tromans' (1978) result obtained from the linear stability analysis. Transition in Lodahl et al.'s (1998) study was judged to occur when there was any sign of imperfection in the wall shear stress signal, which emerged in the form of spikes, as described in Section 5.2.3 (e.g., Fig. 5.15). Note the good agreement between Lodahl et al.'s data and the Tromans' stability curve.

Fig. 5.45 shows that the critical value of the Reynolds number Re_w approaches the familiar value 1.5×10^5, as $R/\delta_1 \to \infty$, the plane-wall oscillatory flow case (see Section 5.2). Details regarding various discrepancies and also the interesting asymptotic case for small values of R/δ_1 are discussed in detail in Lodahl et al. (1998).

Next, we will turn our attention to the combined oscillatory flow and current case.

5.6.2.2 *Combined oscillatory flow and current in a pipe*

In the case of combined flows, in addition to the parameters in Eq. 5.71, the current Reynolds number Re_c should also be involved. Hence, the transition to turbulence is governed by the following three parameters:

$$Re_w, \quad \frac{R}{\delta_1} \quad \text{and } Re_c \tag{5.75}$$

in which Re_c is defined by

$$Re_c = \frac{DV}{\nu} \tag{5.76}$$

with $D = 2R$, the pipe diameter, and V the cross-sectional mean velocity, obtained in Lodahl et al.'s (1998) study from the measured velocity profiles.

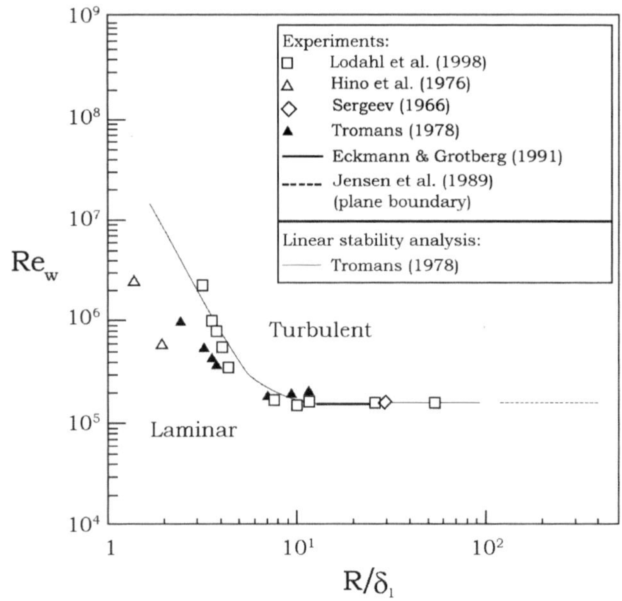

Fig. 5.45 Transition to turbulence in oscillatory flow alone in a circular pipe. Lodahl et al. (1998).

Fig. 5.46 presents the transition data of Lodahl et al. (1998) in the (Re_w, Re_c) plane, obtained for the 19 cm diameter pipe (corresponding to $R/\delta_1 = 53$). Similar to the oscillatory-flow-alone case, transition was judged to occur when there was any sign of imperfection in the wall shear stress signal. Fig. 5.46 also contains the contour plot of V/U_{0m}. We shall return to the latter shortly.

Let us first discuss Fig. 5.46:

1. The data indicate that (1) the critical value of Re_c approaches $(Re_c)_{cr} \approx 2 \times 10^3$, as $Re_w \to 0$, the value known from the steady pipe flow research (Schlichting, 1979, p. 451); and (2) the critical value of Re_w approaches $(Re_w)_{cr} \approx 1.5 \times 10^5$, as $Re_c \to 0$, the value obtained from the oscillatory-flow-alone tests and from the linear stability analysis (Fig. 5.45).

2. The transition to turbulence in current flow is delayed by superimposing an oscillatory flow on the current. While transition occurs at $(Re_c)_{cr} \approx 2 \times 10^3$ in the case of the pure current, it occurs only after Re_c reaches a value of about 7.5×10^3 in the case of the combined flow with

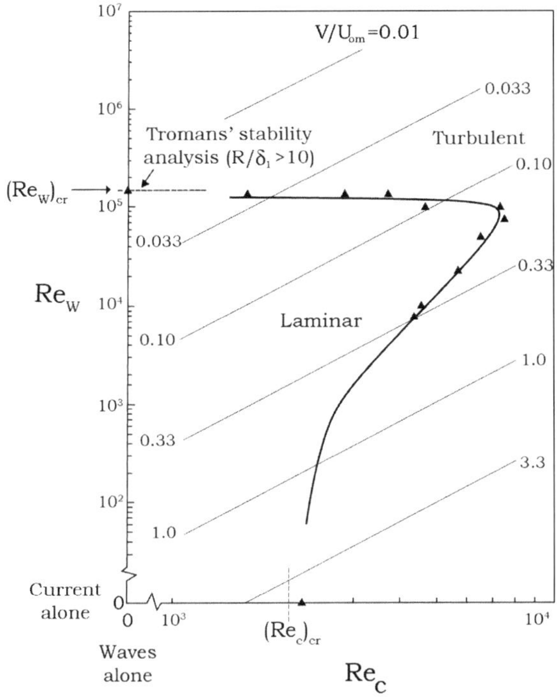

Fig. 5.46 Transition to turbulence in a circular pipe in combined flow $R/\delta_1 = 53$. Lodahl et al. (1998).

$Re_w = 8 \times 10^4$, a factor of 3.5 increase in the critical value of Re_c.

Lodahl et al. (1998, Figs. 6 and 7) present similar transition-to-turbulence diagrams for the other cases with respect to the parameter R/δ_1 tested in the experiments. They also compare their experimental results with the results of Tromans' (1978) stability analysis as well as the experimental data from Gilbrech and Combs (1963) and Sarpkaya (1966) studies.

5.6.2.3 *Re-laminarization*

A turbulent current flow can be laminarized by introducing an oscillatory flow, as revealed by Fig. 5.46. For example, the current with a Reynolds number of $Re_c = 3 \times 10^3$ in Fig. 5.46 is laminarized with the introduction of an oscillatory flow with a Reynolds number Re_w in the range of $Re_w = 10^3$–1.5×10^5. This process, termed the re-laminarization, has been recognized

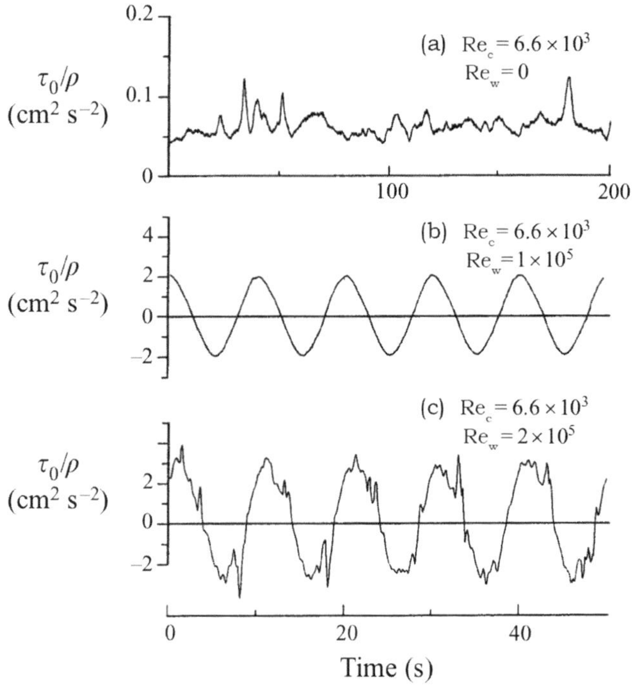

Fig. 5.47 Time series of the wall shear stress. $R/\delta_1 = 53$. Lodahl et al. (1998).

by earlier researchers such as Gilbrech and Combs (1963), Sarpkaya (1966), and Ramaprian and Tu (1980). Fig. 5.46 shows that the range of Re_w over which re-laminarization occurs varies with the current Reynolds number, Re_c.

Fig. 5.47 presents the time series of the wall shear stress from Lodahl et al. (1998). Fig. 5.47a depicts the pure current case with $Re_c = 6.6 \times 10^3$; from the time series, it is clear that the flow is in the turbulent regime. In Fig. 5.47b, an oscillatory flow with a Reynolds number of $Re_w = 10^5$ is introduced. As seen from Fig. 5.47b, there is no sign of turbulence in the wall shear stress time series; the flow is laminarized by the introduction of this oscillatory flow. In Fig. 5.47c, the oscillatory flow velocity is increased by a factor of 2 (giving a Reynolds number of $Re_w = 2 \times 10^5$). As seen from the wall shear stress time series in Fig. 5.47c, turbulence "re-emerges", in the form of small spikes just prior to the reversal of the wall shear stress.

As mentioned previously, Fig. 5.46 depicts the V/U_{0m} = constant lines.

These lines are obtained from

$$\frac{V}{U_{0m}} = \frac{1}{2\sqrt{2}} \frac{Re_c}{(\frac{R}{\delta_1})(Re_w)^{1/2}} \tag{5.77}$$

Small values of V/U_{0m} such as $V/U_{0m} = O(0.1)$ mean that the combined flow will be dominated by its oscillatory-flow component, hereafter called "wave-dominated flow".

A close inspection of Fig. 5.46 together with Fig. 5.45 indicates that, in the wave-dominated case, the flow regime is actually determined by the regime of the oscillatory-flow component itself: if the oscillatory component is in the laminar regime, then the combined flow will also be in the laminar regime, and vice versa, regardless of the regime of the current component of the combined flow. So, for the re-laminarization to occur, first and foremost, the velocity ratio V/U_{0m} should be sufficiently small so that the flow is wave dominated, and secondly the Reynolds number of the oscillatory component of the flow, Re_w, should be sufficiently small so that the oscillatory component of the flow is in the laminar regime.

We further note that Lodahl et al. (1998) also report that other tests with different values of the parameter R/δ_1 all indicated the same behaviour. The reader is referred to Lodahl et al. (1998) for further discussions.

From the preceding paragraphs, it is evident that current turbulence can completely disappear if the superimposed waves (oscillatory flow) are large, so large that the combined flow is wave dominated, and, obviously, when the wave component of the combined flow is in the laminar regime.

Now, Fig. 5.48 gives the definition sketch. One relevant question here is: What if the waves are not tremendously large? In this case, the current turbulence will be suppressed, but not completely disappear. Now, Fig. 5.49 displays the counter plot of turbulence in terms of the wall shear stress

$$\frac{(\sqrt{\overline{\tau_0'^{\,2}}})_{\max}}{(\overline{\tau_0})_{\max}} \tag{5.78}$$

as a function of Re_c and Re_w with the parameter V/U_{0m}. Here, τ_0' is the fluctuating component of the wall shear stress, $\tau_0' = \tau_0 - \overline{\tau_0}$, where the overbar signifies ensemble averaging, and the subindex max indicates the maximum value of the quantity in question over one cycle (Fig. 5.48). The case that the preceding question refers to corresponds to the area where the wave component is not tremendously large, corresponding to the values of the parameter like $V/U_{0m} > O(0.3)$ in Fig. 5.49. As seen, the current

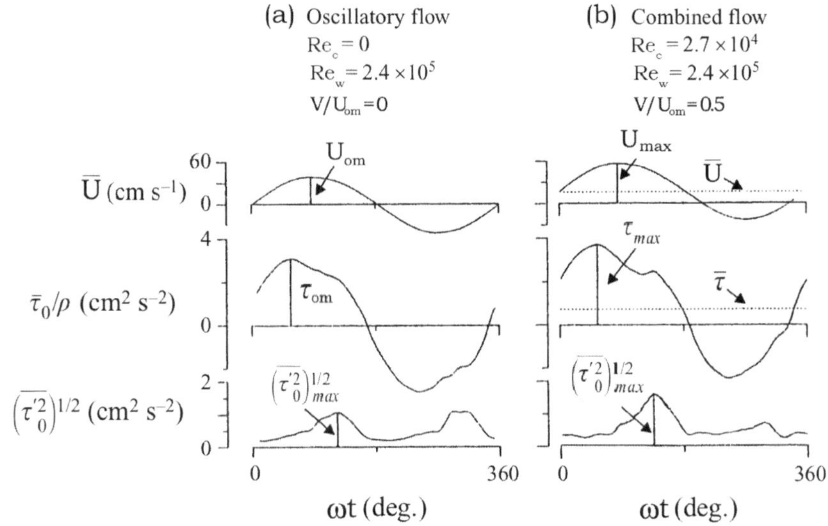

Fig. 5.48 Definition sketch.

turbulence in this area is suppressed, with a reduction in the values of $(\sqrt{\overline{\tau_0'^{\,2}}})_{\max}/(\overline{\tau_0})_{\max}$ from 0.25–0.35 to 0.1-0.25 and eventually to 0–0.1 in a band adjacent to the neutral curve separating the laminar and turbulent regions.

Fig. 5.50 presents a similar diagram obtained in the case of a combined-waves-and-current flow environment with the waves propagating perpendicular to the current (Arnskov et al., 1993). The latter authors conducted their experiments in a wave basin with a current facility where the waves propagate perpendicular, or at an oblique angle to the current. In Fig. 5.50, on the vertical axis, the same quantity as in Eq. 5.78 is plotted versus

$$X = \frac{\tau_c}{\tau_c + \tau_{0m}} \tag{5.79}$$

in which τ_c is the mean bed shear stress corresponding to the current component of the combined flow, and τ_{0m} the maximum value (the amplitude) of the bed shear stress corresponding to the wave component of the combined flow. $X = 0$ represents the waves-alone case, and $X = 1$ the current alone case. The figure clearly indicates that the current turbulence is suppressed gradually with decreasing values of the parameter X (i.e., with larger and larger waves). An important point which Arnskov et al. (1993)

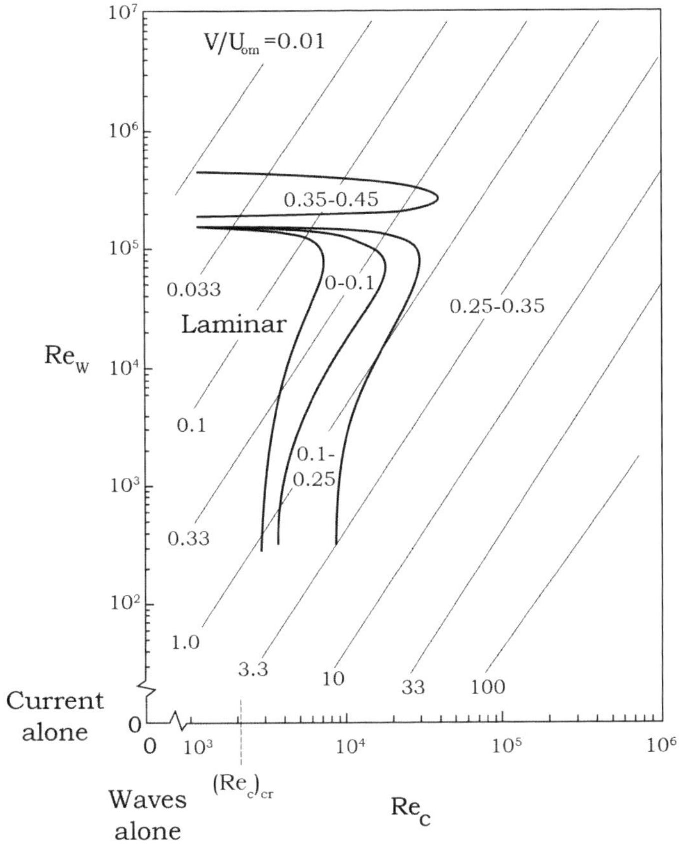

Fig. 5.49 Contour plot of $(\overline{\tau_0'^2})^{1/2}_{\max}/\tau_{\max}$. $R/\delta_1 = 53$. Lodahl et al. (1998).

note is that this finding has large implications for physical model tests with sediment transport, as it suggests that turbulence will be suppressed in a small scale laboratory facility.

5.6.3 *Mean wall shear stress*

The time-averaged mean wall shear stress, $\overline{\tau}$, is defined as the period-averaged value of the ensemble-averaged wall shear stress

$$\overline{\tau} = \frac{1}{T}\int_0^T \overline{\tau_0}(\omega t)dt \tag{5.80}$$

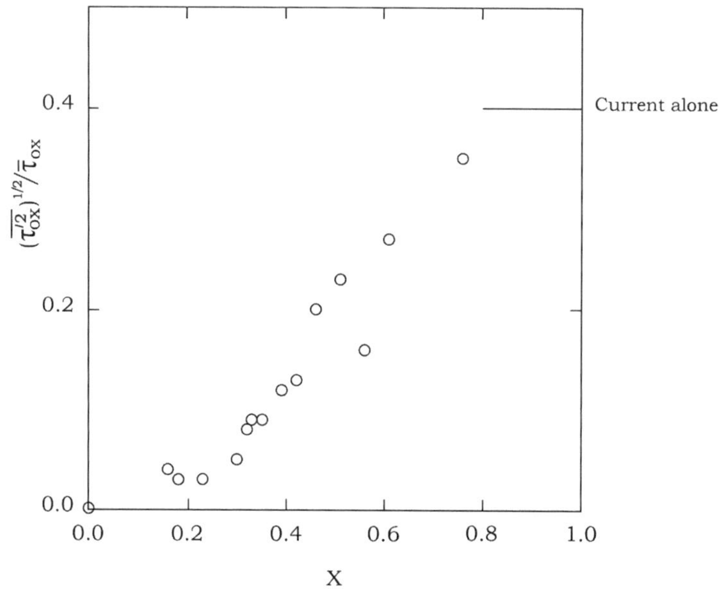

Fig. 5.50 Suppression of current turbulence by superimposed waves. Arnskov et al. (1993).

in which $\overline{\tau_0}(\omega t)$ is the ensemble-mean wall shear stress. $\overline{\tau}$ represents an overall mean (Fig. 5.48).

Fig. 5.51 presents the data on this quantity for $R/\delta_1 = 53$, normalized by τ_c, the mean wall shear stress corresponding to the current component of the combined flow. The vertical arrows in the plot indicate that V/U_{0m} is unity at these points, $V/U_{0m} = 1$. The regions to the left of these points correspond to the current-dominated regime, and the regions to the right of these points correspond to the wave-dominated flow regime. (Obviously, there is a range of V/U_{0m} around $V/U_{0m} = 1$ where both components of the flow are equally significant.)

Clearly, the ratio $\overline{\tau}/\tau_c$ will have to be unity if a linear interaction takes place between the current component and the oscillatory-flow component of the combined flow.

Fig. 5.51 shows, however, the following:

1. $\overline{\tau}/\tau_c$ can be unity, meaning that there is no change in the current mean wall shear stress when an oscillatory flow is superimposed.

2. $\overline{\tau}/\tau_c$ can be smaller than unity, meaning that the mean wall shear

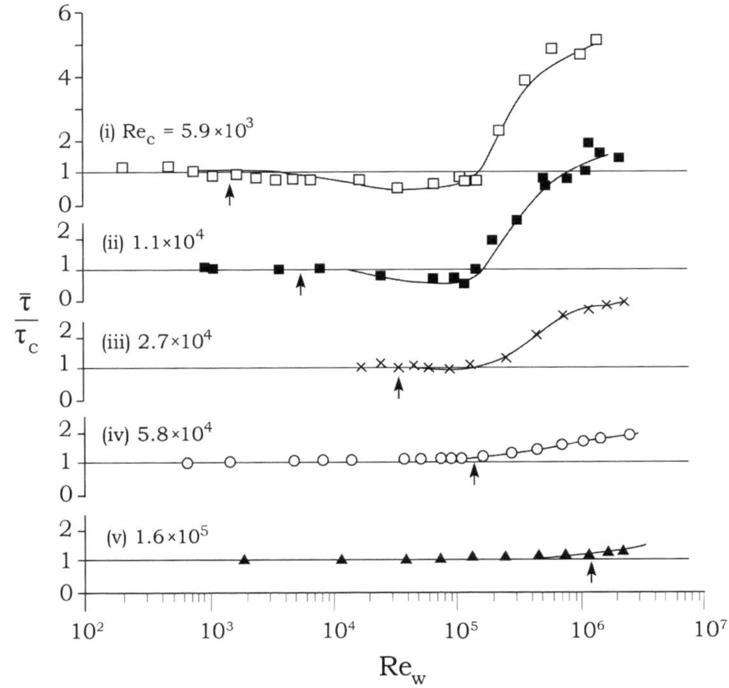

Fig. 5.51 Mean wall shear stress normalized by the current wall shear stress. Vertical arrows indicate that V/U_{0m} is unity at these points. Lodahl et al. (1998).

stress decreases with respect to τ_c.

3. $\overline{\tau}/\tau_c$ can be larger than unity, meaning that the mean wall shear stress increases with respect to τ_c.

The way in which $\overline{\tau}/\tau_c$ varies with Re_w in Fig. 5.51 is shown schematically in Fig. 5.52.

There are two distinct patterns:

1. Fig. 5.52a: $\overline{\tau}/\tau_c$ first decreases from the value $\overline{\tau}/\tau_c = 1$, goes through a minimum, and then increases with increasing Re_w. See Fig. 5.51, curves (i) and (ii), and to a smaller degree curve (iii); and

2. Fig. 5.52b: $\overline{\tau}/\tau_c$ increases monotonically with increasing Re_w. See Fig. 5.51, curves (iv) and (v).

These two patterns are explained as follows.

First, Fig. 5.52a will be discussed. The decrease in the mean wall shear stress occurs because of the combined action of the following two effects: (1)

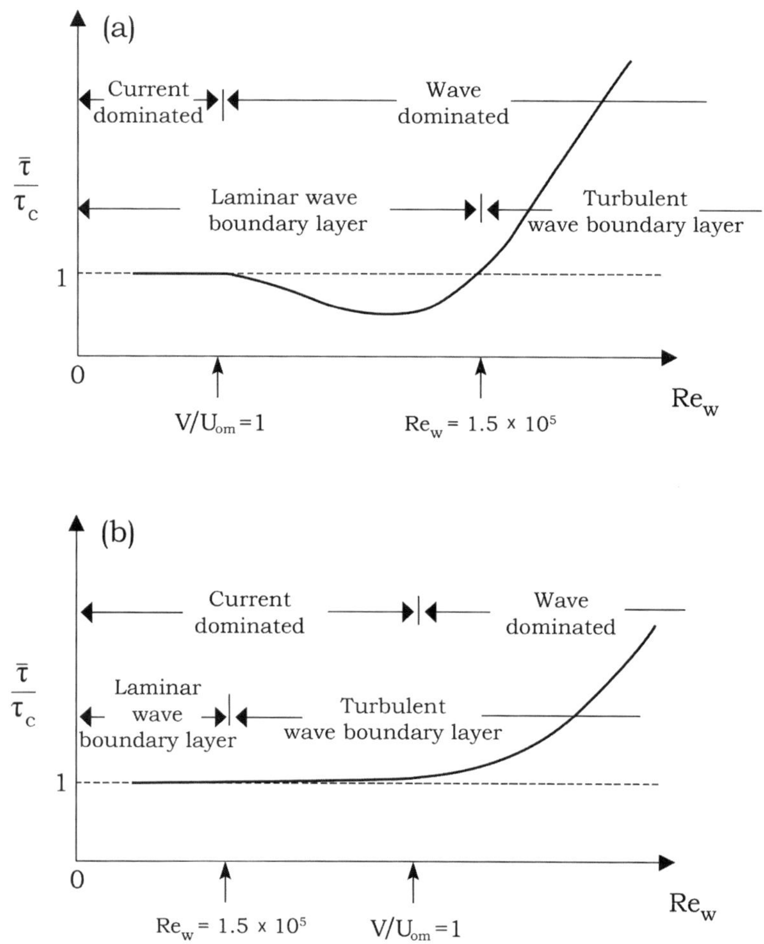

Fig. 5.52 Schematic description of the variation of the normalized mean wall shear stress with wave Reynolds number for a given current Reynolds number. Two distinct patterns appear: (a) and (b). Lodahl et al. (1998).

the flow becomes wave-dominated; and (2) Re_w is such that the oscillatory-boundary-layer component of the combined flow is in the laminar regime ($Re_w < 1.5 \times 10^5$). This means that the momentum-rich, high speed flow from the upper portion of the boundary layer is no longer brought into the neighbourhood of the wall by the flow turbulence because the turbulence is suppressed by the re-laminarization process; the end result is therefore a decrease in the mean wall shear stress. Fig. 5.51 shows that the mean wall shear stress in these regions can easily take values as small as $\overline{\tau}/\tau_c = O(0.5)$, a factor of 2 decrease in $\overline{\tau}$.

Next, Fig. 5.52b will be discussed. In contrast to Fig. 5.52a, the mean wall shear stress increases when the flow becomes wave-dominated, but the oscillatory boundary layer of the flow is in the turbulent regime ($Re_w > 1.5 \times 10^5$). This is evidently due to the turbulence generated by the oscillatory boundary layer of the flow. The momentum-rich, high-speed flow (brought into the neighbourhood of the wall by this turbulence) in this case penetrates much closer to the wall owing to the much smaller wave boundary-layer thickness. Hence, the end result will be an increase in the mean wall shear stress. Incidentally, Lundgren (1972), Grant and Madsen (1979) and Fredsøe (1984) have introduced the concept of "wave roughness" felt by the current to include this increase in the bed shear stress in modelling of the combined flow. The concept of wave roughness will be discussed in detail in Section 5.6.5.

It is important to note that even the fact that the oscillatory boundary layer becomes turbulent is not enough for the mean wall shear stress to be affected, unless the flow becomes wave-dominated. As long as the flow remains current-dominated, no change will occur in the mean wall shear stress.

Fig. 5.53 presents the entire data on the (Re_w, Re_c) plane in the form of a contour plot for $R/\delta_1 = 53$. Notice that the increase in the wall shear stress with respect to its steady-current value can be as much as by a factor of 4. Lodahl et al. (1998) present a similar contour plot for $R/\delta_1 = 25$, which shows similar behaviour.

Lodahl et al. (1998) note that their findings explain qualitatively the results obtained by the previous researchers experimenting with actual wave boundary layers in wave flumes. The then existing data have been plotted by R.R. Simons in the paper by Soulsby et al. (1993): Kemp and Simons' (1982, 1983) data indicate that, while the mean bed shear stress increases quite substantially in the case of a rough bed (the transitional/turbulent regime wave boundary layer, it is not radically different from the current-alone value in the case of a smooth bed (the laminar-regime wave boundary layer, $Re_w = O(10^3)$). Likewise, Simons et al.'s (1988) data indicate a substantial increase in the bed shear stress; this is because the bed of the wave flume was covered by gravel in their experiments, and the wave boundary layer was in the transitional/turbulent regime.

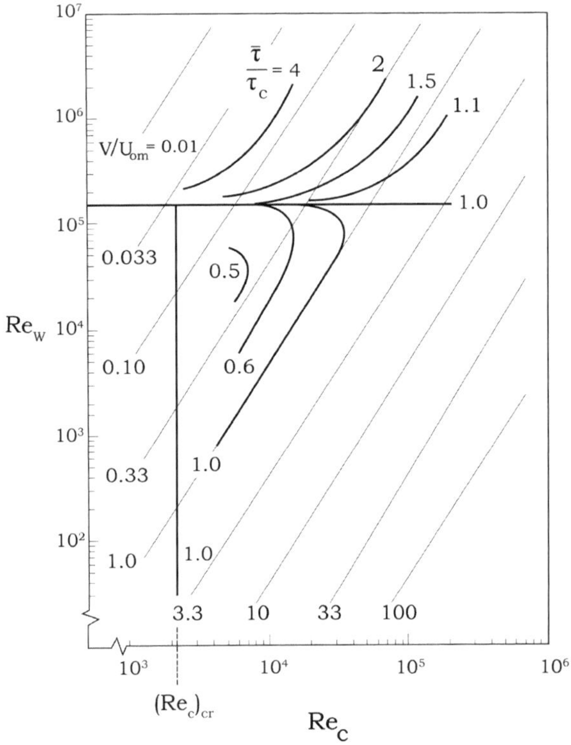

Fig. 5.53 Increase in the mean wall shear stress with respect to its steady-current value, based on the same experiment as shown in Fig. 5.48. Lodahl et al. (1998).

5.6.4 *Friction coefficient for combined flow*

The friction coefficient for the combined flow, corresponding to the mean wall shear stress can be defined by

$$\overline{f} = 2\frac{\overline{\tau}}{\rho V^2} \tag{5.81}$$

Fig. 5.54 displays the friction-coefficient data plotted against the velocity ratio U_{0m}/U_{fc} for $R/\delta_1 = 53$, adopting the representation used by Fredsøe (1984). Here U_{fc} is the friction velocity of the current component of the combined flow:

$$U_{fc} = \sqrt{\frac{\tau_c}{\rho}} \tag{5.82}$$

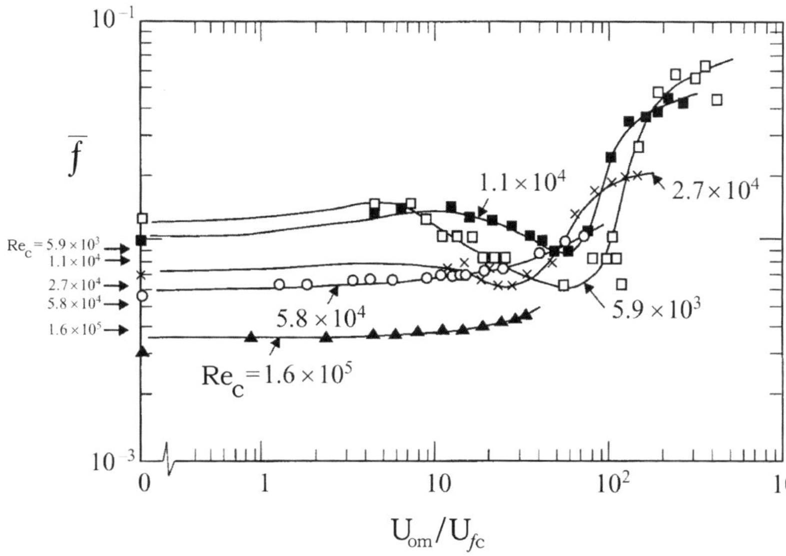

Fig. 5.54 Friction coefficient corresponding to the mean wall shear stress. The arrows on vertical axis: Friction coefficient found from Nikuradse diagram. $R/\delta_1 = 53$. Lodahl et al. (1998).

Fig. 5.54 also includes the values of the friction coefficient found from the Nikuradse diagram (the arrows on the vertical axis) (Schlichting, 1979, p. 598) when $U_{0m}/U_{fc} = 0$, the current case.

Fig. 5.54 shows that the friction coefficient varies with U_{0m}/U_{fc} in much the same way as in Fig. 5.51. The dips experienced by the friction coefficient correspond to the fall in $\overline{\tau}/\tau_c$ in Fig. 5.51 where $\overline{\tau}/\tau_c < 1$. Lodahl et al. (1998) also present a plot for $R/\delta_1 = 25$ as well, which shows a similar behaviour.

It may be noted that the way in which the friction coefficient changes with U_{0m}/U_{fc} is similar to that obtained theoretically by Fredsøe (1984), using the depth-integrated momentum equation with a logarithmic velocity distribution both inside and outside the wave boundary layer. Lodahl et al. (1998) remark, however, that neither in Fredsøe's (1984) work nor in more refined flow modelling studies (such as Grant and Madsen's 1979 eddy viscosity model, or Davies et al.'s (1988) turbulent-energy closure model) could the previously mentioned dips be captured, because these are related to the wave boundary layer processes where the turbulence is suppressed, as described in the preceding paragraphs.

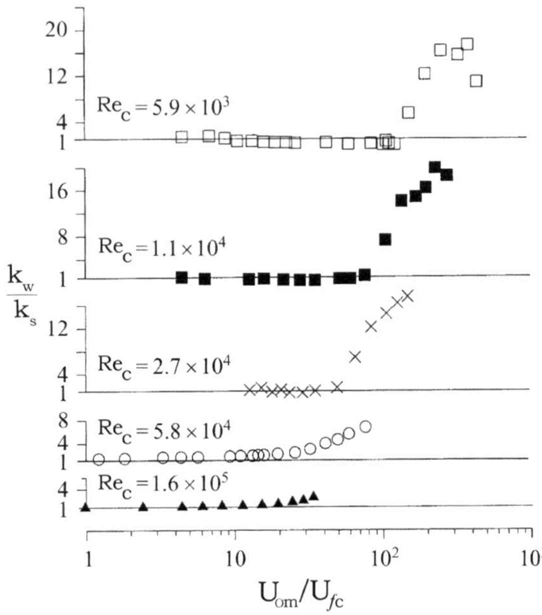

Fig. 5.55 Apparent roughness (caused by the oscillatory component of the flow) normalized by the Nikuradse equivalent roughness experienced in the case of the steady current. $R/\delta_1 = 53$. Lodahl et al. (1998).

5.6.5 *Wave-induced apparent roughness*

Fig. 5.55 presents the flow-resistance data associated with the mean wall shear stress (for $R/\delta_1 = 53$) in terms of the apparent roughness, k_w, which is defined by the familiar resistance formula

$$\frac{V}{\overline{U_f}} = 2.46 \ln\left(\frac{7.4R}{k_w}\right) \tag{5.83}$$

in which k_w is Nikuradse's equivalent sand roughness (Schlichting, 1979, p. 621), and represents the roughness caused by the wave boundary layer, and $\overline{U_f}$ is the friction velocity corresponding to the mean wall shear stress:

$$\overline{U_f} = \sqrt{\frac{\overline{\tau}}{\rho}} \tag{5.84}$$

Eq. 5.83 is precisely the expression used by Lodahl et al. (1998), and it actually corresponds to the flow resistance equation presented in Chapter 3 in conjunction with steady boundary layers (Eq. 3.134). Lodahl et al. (1998) obtained k_w via Eq. 5.83, using the data on V and $\overline{U_f}$.

In the plot in Fig. 5.55, for convenience, k_w is normalized by k_s, the roughness obtained from Eq. 5.83 in the case of the steady current (where k_w is replaced by k_s, and $\overline{U_f}$ by U_{fc}). Lodahl et al. (1998) point out that the actual values of k_s were found in the range $2 \lesssim k_s U_{fc}/\nu \lesssim 8$, which is not radically different from $k_s U_{fc}/\nu \approx 4$, the value corresponding to a smooth wall, the wall in Lodahl et al.'s experiments.

As seen from Fig. 5.55, the wave-induced apparent roughness experiences a variation similar to the friction coefficient (cf. Fig. 5.54). The figure further shows that the apparent roughness increases tremendously with increasing U_{0m}/U_{fc} in the wave-dominated turbulent regime.

5.6.6 *Oscillating component of the wall shear stress in combined flow*

One of the quantities related to the oscillating component of the wall shear stress is $\tau_{\max}$, the maximum value (Fig. 5.48 for the definition). Fig. 5.56 displays the variation of this quantity normalized by $(\tau_c + \tau_{0m})$ as a function of the wave Reynolds number and the current Reynolds number for the value of 53 of the parameter R/δ_1. Recall that τ_c is the mean bed shear stress corresponding to the current component of the combined flow, and τ_{0m} the maximum value (the amplitude) of the bed shear stress corresponding to the wave component of the combined flow. In the case of linear interaction, the quantity $\tau_{\max}/(\tau_c + \tau_{0m})$ will be equal to unity. Fig. 5.56 indicates that, for the majority of cases, this interaction is nonlinear.

Consider the wall shear stress difference between the maximum wall shear stress and the mean wall shear stress, $\tau_{\max} - \overline{\tau}$ (Fig. 5.48). This is actually the "amplitude" of the oscillating wall shear stress in the combined flow. One can define a friction coefficient associated with this quantity:

$$f_w = 2\frac{\tau_{\max} - \overline{\tau}}{\rho U_{0m}^2} \tag{5.85}$$

This friction coefficient reduces to the familiar wave friction coefficient when the current component of the combined flow becomes zero ($\overline{\tau} = 0$), Eq. 5.56. The data regarding the friction coefficient defined in the above equation are given in Lodahl et al. (1998), first for the oscillatory-flow-alone case and subsequently for two combined flow cases ($R/\delta_1 = 53$ and 25) studied in Lodahl et al. (1998); see Figs 16 and 17 in the latter publication.

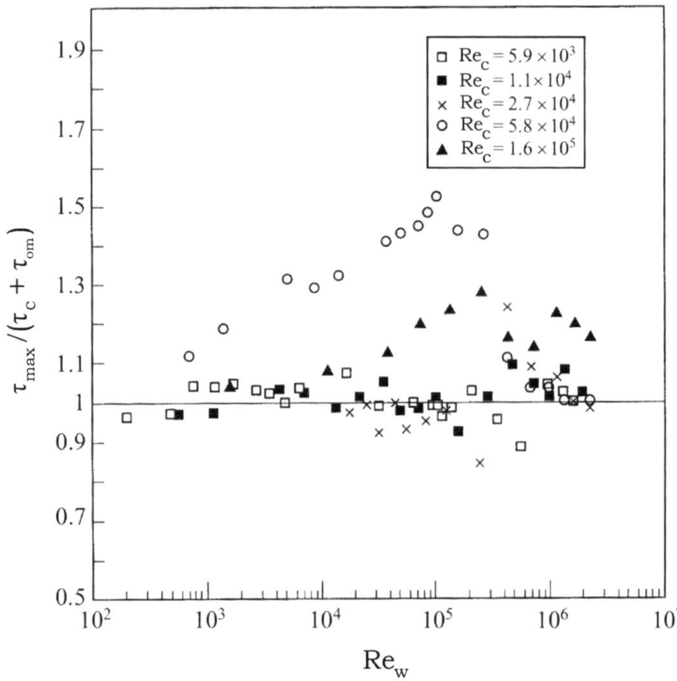

Fig. 5.56 Maximum wall shear stress. $R/\delta_1 = 53$. Lodahl et al. (1998).

5.6.7 *Flow resistance in combined wave and current boundary layers*

In this section, parallel to Section 5.5, we will summarize the flow resistance relations in the case of combined wave and current boundary layers, based on Soulsby et al. (1993) and Soulsby (1997).

Fig. 5.57 gives the definition sketch, adapted from Soulsby et al. (1993). The quantity τ_c is the bed shear stress in the case of current alone (Fig. 5.57a), and τ_w the maximum value of the bed shear stress in the case of wave alone, which is at an angle ϕ to the current (Fig. 5.57b). (We note that, throughout this section, we use the same symbols as in Soulsby et al. for convenience; variable ϕ should therefore not be confused with the phase lead of the bed shear stress discussed previously.)

The question is: What is the mean (period averaged) bed shear stress, and, likewise, what is the maximum value of the bed shear stress when the

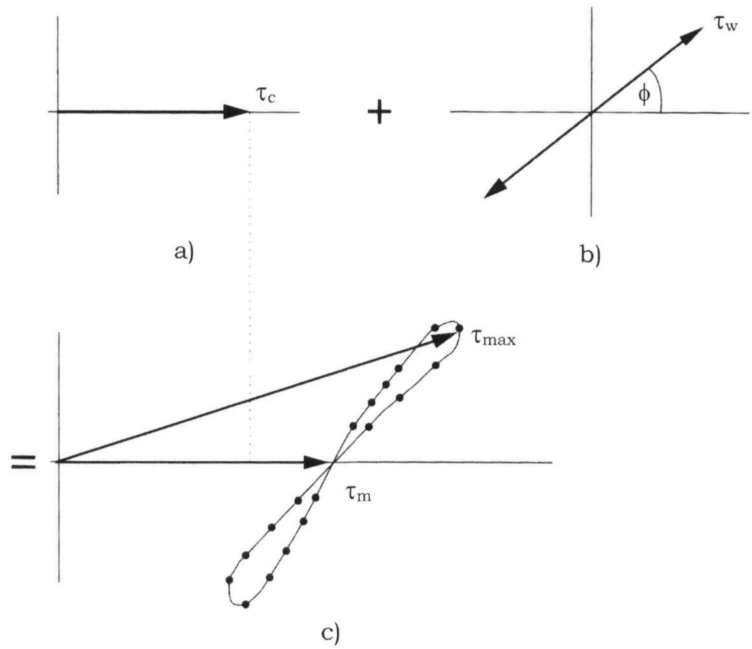

Fig. 5.57 Schematic diagram of interaction of wave and current bed shear stresses. Plan view. Adapted from Soulsby et al. (1993).

aforementioned two flows are superimposed?

First of all, we know from our previous analysis that the mean bed shear stress, τ_m, will be larger than the bed shear stress in the current alone case, τ_c, if the wave component of the combined flow is in the turbulent regime (Fig. 5.52a,b), a situation that normally exists in the field. This is reflected in Fig. 5.57c where $\tau_m > \tau_c$. Recall, however, that the opposite is true when (1) the wave component of the combined flow is in the laminar regime, and (2) the combined flow is in the wave-dominated regime (Fig. 5.52a). We will in the following consider that $\tau_m > \tau_c$, as we are, in this treatment, interested in field situations.

Now, with the "introduction" of the wave component, the tip of the bed shear stress vector will move in the way as illustrated in Fig. 5.57c. As seen in this schematic representation, the trajectories of the tip of the bed shear stress vector do not collapse during one half cycle of the motion, and this is due to the assumed asymmetry in the variation of the bed shear stress during the half cycle.

Clearly, the bed shear stress vector will, during the wave cycle, experience a maximum value. This maximum value is designated τ_{max} (Fig. 5.57c).

Soulsby (1997) notes that more than 20 different theories and models have been proposed to describe this process. A comparison between predictions of the mean and maximum bed shear-stress during a wave cycle by eight of these models is shown in Soulsby et al. (1993), also reproduced in Soulsby (1997, Fig. 17). Soulsby notes that differences between the models of 30%–40% are commonly found, and differences of up to a factor of 3 occur for strongly wave-dominated conditions.

An algebraic approximation (accurate to ±5% in most cases) to the models was also derived by Soulsby et al. (1993). A test of the parameterized models against a large data set of 61 laboratory values and 70 field values of τ_m was performed by Soulsby (1995). Various criteria for goodness of fit were used. Soulsby (1997) notes that no one model performed best according to all the criteria. He adds, however, that four models which gave a good all-round performance and/or are widely used are the analytical models of Grant and Madsen (1979) and Fredsøe (1984), and the numerical models of Huynh-Thanh and Temperville (1991) and Davies et al. (1988). Parametric fitting coefficients (thirteen of them) for these models are given in Soulsby (1997, Table 9).

A data-based method was also derived by Soulsby (1995) by optimizing the aforementioned thirteen coefficients in the parameterized expression for τ_m which had been used for fitting the theoretical models cited in the preceding paragraph. The same set of 131 data points was used. This approach gave a significantly better fit to the data than any of the models (although, Soulsby states, it is a mathematical necessity that it cannot be worse). The coefficients for this method are given also in Soulsby (1997, Table 9). More importantly, it was also found that a two-coefficient optimization gave almost as good a fit to the data as the best theoretical models. This method reduces to a simple equation, which is reproduced here:

$$\tau_m = \tau_c \left[1 + 1.2 \left(\frac{\tau_w}{\tau_c + \tau_w} \right)^{3.2} \right] \tag{5.86}$$

with τ_c and τ_w being the bed shear stresses which would occur due to the current alone and to the wave alone, respectively, and τ_m the mean bed shear stress (Fig. 5.57c).

The corresponding equation for the maximum bed shear stress, τ_{max}, (Fig. 5.57c) is given by a vector addition of τ_m from the preceding equation, and τ_w obtained through $f_w = 0.237(a/k_s)^{-0.52}$ (similar to Eqs. 5.66–5.68 or Eq. 5.69) and Eq. 5.56:

$$\tau_{max} = \left[(\tau_m + \tau_w \cos\phi)^2 + (\tau_w \sin\phi)^2\right]^{1/2} \qquad (5.87)$$

In the case of a co-directional, combined wave and current, i.e. $\phi = 0$, the above equation becomes

$$\tau_{max} = \tau_m + \tau_w \qquad (5.88)$$

An example demonstrating application of the above formulation can be found in Section 5.11.4, in the context of tsunami-wind wave boundary layer interaction, where the long tsunami is demonstrated to effectively play the role of the current.

5.7 Boundary layers in wave flumes

As discussed earlier, turbulent wave boundary layers over smooth beds cannot be achieved in wave flumes because of the small scales involved (unless large-scale wave flumes are used). To circumvent this problem, oscillating water tunnels have been used in the research of wave boundary layers. We have discussed the wave boundary layers developing in oscillating water tunnels extensively in the previous sections.

Although turbulent wave boundary layers are hardly achieved in the laboratory wave flumes, wave boundary layers are also studied, using wave flumes for academic purposes or otherwise. Real-life turbulent wave boundary layers over smooth beds can be achieved in large-scale wave flume facilities without any problem. Real-life turbulent wave boundary layers over rough beds, on the other hand, can be achieved even using small, or medium-scale wave flumes without any major problem. However, wave boundary layers developing in wave flumes differ in several aspects from those simulated in oscillating tunnels. For example, the steady streaming in the wave boundary layer that comes into existence as a result of the presence of the free surface (hence active convective terms, this will be discussed in detail in Chapter 6) does not exist in the wave boundary layer developing in an oscillating tunnel. Another example is that there exists a return flow (see Appendix 5.D) that affects the wave boundary layer in the wave flume, whereas this effect is absent in the case of the oscillating tunnel. A third example is that, in the case of combined waves and current, the flow takes place in an extended depth (the water depth plus the surface-elevation amplitude) in wave flumes whereas the latter is nonexistent in wave boundary

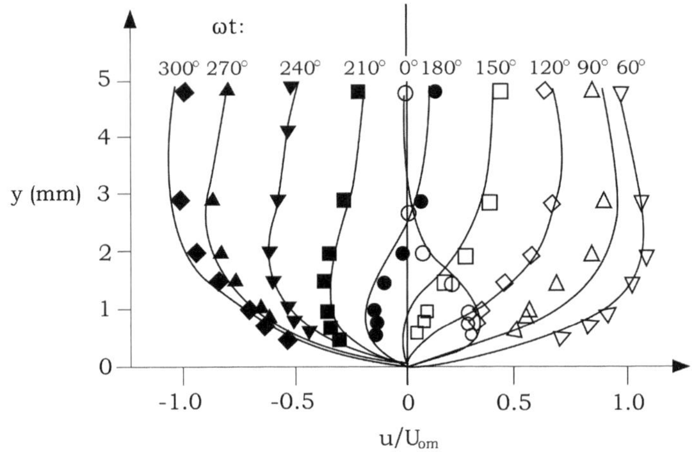

Fig. 5.58 Velocity profiles at the measurement section. Waves alone. Sumer et al. (1997).

layers developing in an oscillating tunnel. Therefore, it is imperative also to consider wave boundary layers that develop in wave flumes in order to get a more complete view of wave boundary-layer processes.

5.7.1 *Waves alone*

Fig. 5.58 illustrates the time development of the velocity profiles in real waves in a wave flume close to the bed, reproduced from Sumer et al. (1997). This is Test 14 in Sumer et al.'s (1997) experiments. The boundary layer is in the laminar regime with a Reynolds number of $Re_w = 2.3 \times 10^4$. The solid lines in the figure are the laminar solution represented by

$$\frac{u}{U_{0m}} = a\sin(\omega t) + b\sin(\omega t) \tag{5.89}$$

in which a and b are coefficients determined by the least square method. The boundary layer thickness is $\delta = 2.8$ mm.

Although there is an apparent asymmetry in the profiles with respect to the phase due to the asymmetry in the waves, the way in which the velocity profiles evolve resembles that obtained in the oscillating tunnel experiment of Jensen et al. (1989), Fig. 5.23.

In wave boundary layers in real waves, period-averaged velocity profiles are of large interest. Fig. 5.59 presents two such period-averaged velocity-profile data obtained in a wave-flume with a ripple-covered bed, Fredsøe et

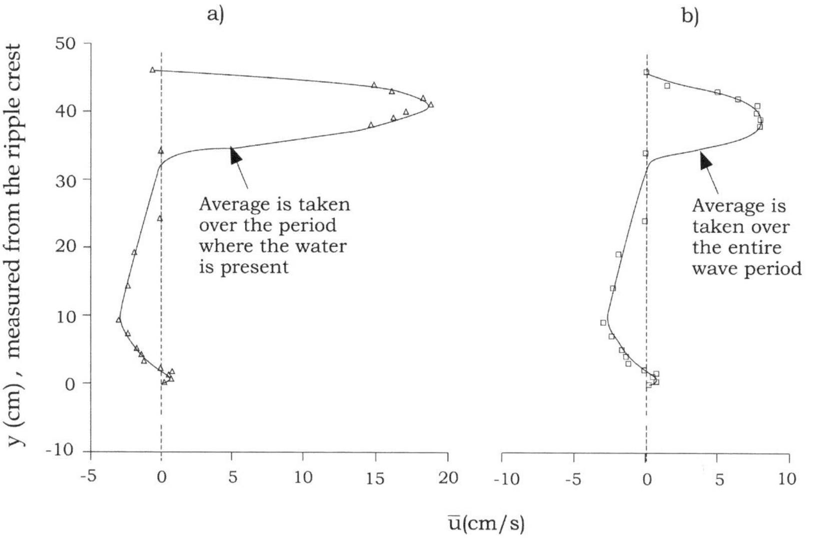

Fig. 5.59 Period-averaged velocity distribution at the ripple crest. Waves alone. Fredsøe et al. (1999).

al. (1999). The ripples (22 cm in length and 3.5 cm in height), cast from concrete, had a hydraulically smooth surface. The water depth was 42 cm, the wave height 13 cm, the wave period 2.5 s, and the roughness parameter for the test exhibited in the figure was $a/k_s = 1.2$. The data presented in the figure is the period-averaged velocity profile in Test W1, measured at the ripple crest section, in Fredsøe et al.'s (1999) experiments.

Two profiles are displayed in Fig. 5.59. In one, the period-averaging is taken over the time where the water is present (Fig. 5.59a), and in the other, the period-averaging is taken over the entire wave period (Fig. 5.59b).

The top portion of the velocity profile in Fig. 5.59b represents the net discharge of water (in the direction of wave propagation) between the wave crest and the wave trough, called the wave drift. From Fig. 5.59b, this is found to be $q_{drift} = 61\ \mathrm{cm}^2/\mathrm{s}$. Fredsøe et al. (1999) report that this value is 44% smaller than the theoretical value obtained from the sinusoidal wave theory (Fredsøe and Deigaard, 1992, p. 8; see also Eq. 5.188 in Appendix 5.D),

$$q_{drift} = \frac{H^2\pi}{4T}\frac{1}{\tanh(kh)} \tag{5.90}$$

in which H is the wave height, T the wave period, h the water depth and k

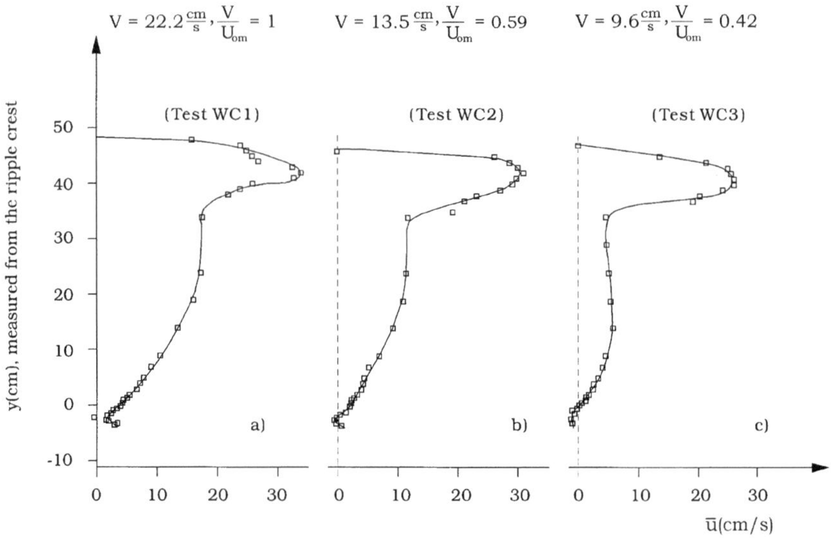

Fig. 5.60 Period-and-space-averaged velocity profiles. Combined waves and current. Fredsøe et al. (1999).

the wave number. Fredsøe et al. (1999) attributed this discrepancy to the asymmetry in the waves.

Fig. 5.59b further indicates that the water discharge caused by the wave drift is balanced by a return flow in the opposite direction in the lower depths, except the small kink near the bed. Fredsøe et al. (1999) relates this latter small kink to the mass transport near the bed in the direction of wave propagation (i.e. conventional wave boundary layer streaming, see Section 6.1).

5.7.2 *Combined waves and current*

Fig. 5.60 presents three period-averaged velocity profiles from Fredsøe et al.'s (1999) experiments for the case of combined waves and current. Wave drift is quite clear even in the case of $V/U_{0m} = 1$, with the top portion of the velocity profiles.

Fig. 5.61 compares two velocity profiles, one the velocity profile in the current case (dashed line), and the other that when this particular current is superimposed with a wave, representing the period-averaged velocity profile for the combined flow (solid line). These profiles are from the current test C1 (the current-alone case), and the combined waves and current test WC1 from the Fredsøe et al. (1999) experiments. The superimposed wave (designated as W1 in Fredsøe et al., 1999) has a wave height of 13 cm and wave period of 2.5 s, described in the previous section in conjunction with the period-averaged velocity profiles in real waves (Fig. 5.59). The current-velocity-wave-velocity parameter is $V/U_{0m} = 1$. The current-alone profile in the figure is reproduced from Fig. 2 of Fredsøe et al. (1999), and Waves+Current profile from Fig. 8a of Fredsøe et al. (1999) (also reproduced in Fig. 5.60a here). We note that Fredsøe et al. (1999) point out that the water discharge over the depth in the case of the combined flow, i.e., $q = \int_0^{wave\ crest} u\, dy$, and that in the case of the current alone, $q = \int_0^{depth} u\, dy$, are not entirely equal to each other, but close, within 10%.

Now, Fig. 5.61 shows that the velocity profile in the case of the combined flow is "elevated" quite substantially with respect to that in the case of current alone. This is because, in the W+C case, (1) the flow in the crest half period takes place over a depth substantially larger than that of the current-alone case; and (2) the flow in the trough half period is rather small or negative (see Figs. 7i-7r in Fredsøe et al., 1999).

When plotted on a semi-logarithmic plane (Fig. 5.62), the two velocity profiles in Fig. 5.61 exhibit some important features. First of all, the displacement of the velocity profile to upper elevations is even more clear in this representation. Secondly, Nikuradse's equivalent sand roughnesses found from the log-fit exercise for the two cases (namely, the current alone case, and the combined waves and current case, k_s and k_w, respectively) are an order of magnitude different, k_s being an order of magnitude smaller than k_w. Clearly, Nikuradse's equivalent sand roughness for the combined flow, k_w, corresponds to the wave-induced roughness, discussed earlier, while that in the case of the current alone, k_s, corresponds to the bed roughness, the ripple roughness in the present case.

On top of the previously mentioned mechanism, there is an additional mechanism to increase the period-averaged bed shear stress due to Lodahl et al. (1998) in the case of wave-dominated combined W+C with the wave component being in the turbulent regime (see the discussion in Section 5.6.3).

(Although there are no significant implications on the end results, we note that there are two combined flow profiles shown in Fig. 5.62; one without the wave drift modification (the filled circles), and the other with

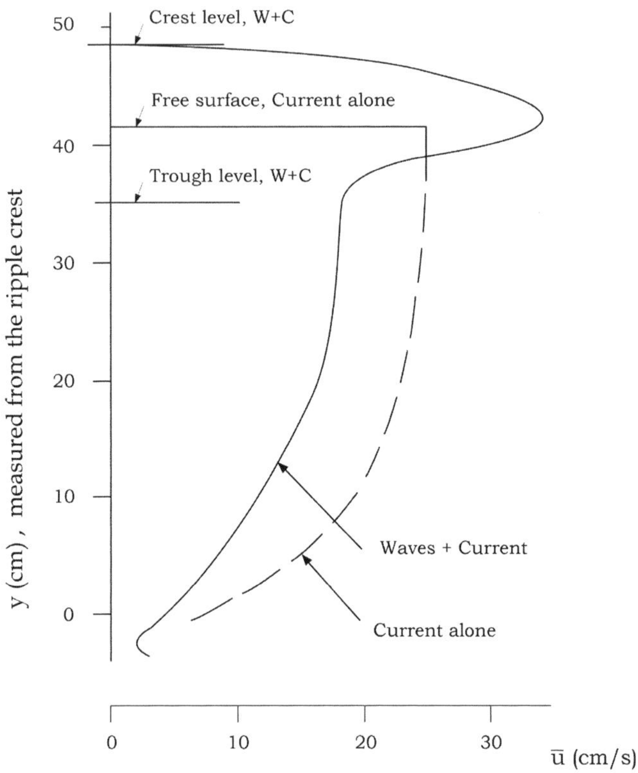

Fig. 5.61 Comparison of velocity profiles. Combined waves and current, and current alone. Adapted from Fredsøe et al. (1999).

the wave drift modification (the open circles). A detailed discussion of the wave drift modification is given in Fredsøe et al. (1999, pp. 195–196).)

Fredsøe et al. (1999, Fig. 13) presented the data on wave-induced roughness k_w/k_s together with an additional set of data they compiled from the literature as a function of U_{0m}/U_f. They found that k_w/k_s increases with increasing U_{0m}/U_f, in agreement with the experimental and theoretical results reported in the literature, notably Fredsøe (1984), Nielsen (1992) and Fredsøe et al. (1999).

Finally, it may be noted in conjunction with Fig. 5.62 that the velocity profile in the combined flow case has the same shape as that observed in the study of Klopman (1994, Fig. 4.10) for the case where the waves follow

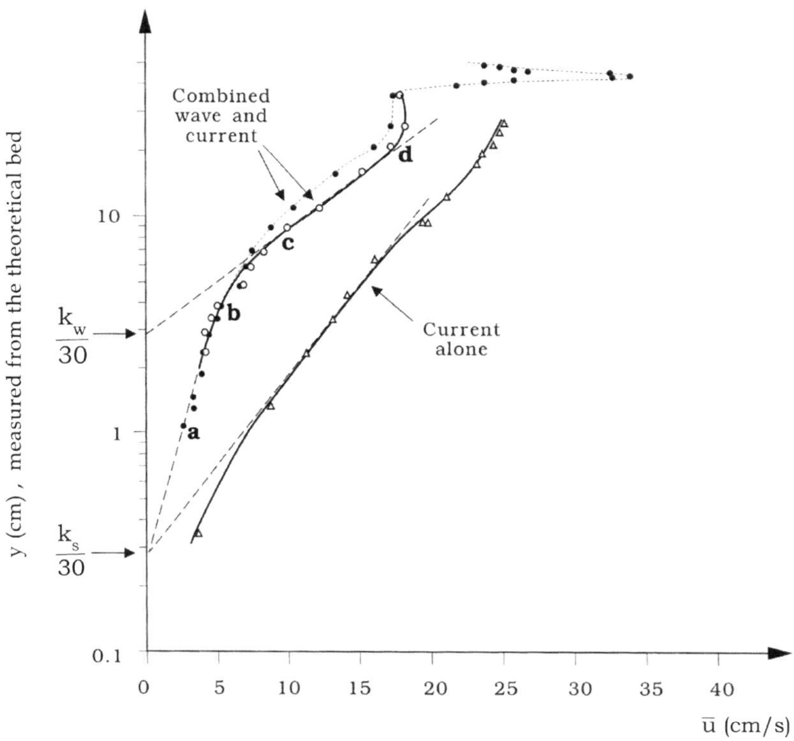

Fig. 5.62 Comparison of velocity profiles. Combined wave and current, and current alone. Fredsøe et al. (1999). Empty circles: wave-drift and its associated return flow are subtracted. Filled circles: not subtracted. Fredsøe et al. (1999).

the current. Klopman's study showed that the velocity profile changes substantially when the waves oppose the current. The latter study also showed that these changes occur over the entire depth of the flow; the velocity profile for the case of opposing current is displaced to even higher elevations.

5.8 Bursting process in wave boundary layers

Fig. 5.17 summarizes the flow regimes in wave boundary layers, namely, Laminar Regime, represented by the straight line $f_w^\star = 2/\sqrt{Re}$; Transitional Regime, the area lying between the line $f_w^\star = 2/\sqrt{Re}$ and the dashed line in

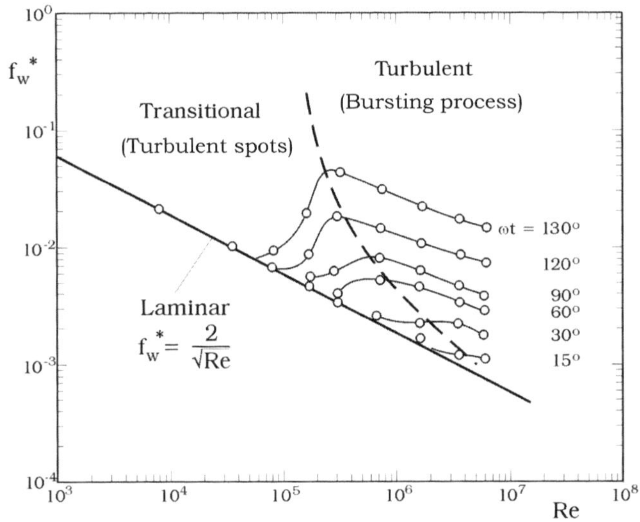

Fig. 5.63 Turbulent spots and bursting process areas.

the figure; and Turbulent Regime, the area lying to the right of the dashed line. It is expected that the bursting process, the quasi-cyclic process that takes place near the wall in turbulent boundary layers (see Section 3.5), occurs in the Turbulent-Regime area (Fig. 5.63, reproduced from Fig. 5.17). As implied by Fig. 5.63, the presence of the bursting process is not only a function of the Reynolds number but also a function of the phase; the bursting process cannot be observed for Reynolds numbers as high as 10^6, or even for 10^7, for small values of the phase ωt during the course of the wave cycle.

The bursting process in wave boundary layers has not been studied extensively, and the literature on the subject is sparse. To the authors' knowledge, Hino et al. (1983) were the first to cover the bursting process in their experimental study on turbulent wave boundary layers. They did the experiments in an oscillating wind tunnel of rectangular cross-section. They measured the velocity, using LDA and also a hot-wire anemometer. To aid the latter measurements, Hino et al. (1983) also made flow visualization experiments, using the smoke-wire technique. The authors produced diagrams illustrating the time evolution of the conditionally-averaged velocity profiles, satisfying four different conditions (1) $u' < 0$ and $v' > 0$; (2) $u' > 0$ and $v' < 0$; (3) $u' < 0$ and $v' < 0$; and (4) $u' < 0$ and $v' > 0$, similar to those depicted in Table 3.3, representing the ejection, sweep, wallward interaction, and outward interaction types of motion, respectively. Likewise, they also produced similar time variations for the Reynolds stress. These measurements revealed the presence of the bursting process in quite the same way as in steady boundary layers.

Hino et al. (1983, flow-visualization photographs in their Fig. 32) further were able to visualize the near-wall streaky pattern of the boundary layer. They reported that the spacing between the streaks was

$$\lambda^+ \approx 150 \tag{5.91}$$

in accordance with the corresponding data in steady boundary layers (see Section 3.5).

At this juncture, we recall the sequence of transition to turbulence illustrated in Fig. 5.7, obtained by DNS simulations. The Reynolds number in this numerical test is $Re = 4.5 \times 10^5$. For this Reynolds number, the bursting process comes into existence only after $\omega t \simeq 120°$, according to Fig. 5.63. Nevertheless, the streaky pattern, the signature of the bursting process, begins to appear already at $\omega t = 60°$ (Fig. 5.7c). When closely inspected, the spacing between the low-speed fluid zones (the elongated green areas) in Fig. 5.7c is worked out to be like $\lambda^+ = O(500)$. The boundary layer evolves into the fully developed stage for further phases (larger than $\omega t \simeq 120°$) with the fully-developed-turbulent-boundary-layer bursting process occurring with the spacing between the low speed streaks being like $\lambda^+ = O(100)$.

More recently, Zhang and Simons (2019) studied experimentally the turbulence structures in the bottom wave-current boundary layers, using Particle Image Velocimetry (PIV). They did their experiments in two flumes, a low turbulence flume (6 m in length, 49.3 cm in width and 27 cm in depth), and a medium scale flume (16 m in length, 45 cm in width and 60 cm in depth). The authors did additional hydrogen bubble experiments in the low turbulence flume. They did two kinds of experiments: Current-alone experiments; and Combined-waves-and-current experiments. Their main findings can be summarized as follows:

1. For the current-alone experiments, the mean spacing of the low-speed streaks was obtained to be $\lambda^+ = 118$ and 111 from the PIV and the hydrogen-bubble experiments, respectively, for the low-turbulence flume tests; and $\lambda^+ = 125$ from the PIV experiments for the 16 m flume tests, which is consistent with the steady-boundary-layer data (Section 3.5).

2. For the combined-waves-and-current experiments, the corresponding values were found to be $\lambda^+ = 141$–160 for the low-turbulence flume tests;

and $\lambda^+ = 124$–137 for the 16 m flume tests. As seen, the combined waves-and-current values are consistently larger than the current-alone values.

3. It is to be noted that the combined waves-and-current values above are consistent with Hino et al.'s (1983) oscillatory-boundary-layer value (Eq. 5.91), $\lambda^+ \approx 150$.

4. Zhang and Simons found that the spacing of the low-speed streaks varied periodically within a wave cycle.

5. Zhang and Simons further found that more streaks emerged during the deceleration stage (therefore the spacing decreased), and the opposite during the acceleration stage of the wave cycle.

6. They also observed "destruction" of the low speed streaks with the flow reversal of the wave component of the combined flow, depending on the strength of the flow reversal and the strength of the low speed streaks.

7. Finally, it was found that the shear stress induced by ejections is higher with waves of increasing amplitudes than in the current alone case. It was also found that the same trend existed for sweep events, but the increase was not as much.

5.9 Miscellaneous wave boundary layer examples

Example 5.1. Non-uniform wave boundary layers.

Fredsøe et al. (1993) have studied non-uniform wave boundary layers, where the non-uniformity is due to a sudden change in the bed roughness, while Sumer et al. (1993b) have studied non-uniform wave boundary layers, where the non-uniformity is due to the convergent/divergent geometry of the flow environment.

In the former study (Fredsøe et al., 1993), experiments have been conducted in the same oscillating tunnel as that used in Jensen et al. (1989). Sudden change in the roughness over the bed was introduced, with two kinds of change in the roughness: in one, the roughness changed from a smooth-wall roughness to a roughness equal to 4.8 mm, and in the other, it changed from a roughness equal to 0.35 mm to the same roughness as in the previous experiment (4.8 mm). The Reynolds number was 6×10^6 for the major part of the experiments. The response of the boundary layer to the sudden change in the roughness was found to occur over a transitional length of the flow. The bed shear stress over this transitional length attains a peak value over the bed section with the larger roughness. The amplification in the bed shear stress due to this peak could be up to 2.5 times its asymptotic value. Also, the turbulence is quantitatively different in the two half periods; a much stronger turbulence is experienced in the half period where the flow is toward the less-rough section. Of particular interest is that a constant steady streaming occurs near the bed in the neighborhood of the junction between the two bed sections. This streaming is directed toward the section with the larger roughness.

The experimental study of Fredsøe et al. (1993) involving sudden changes in bed roughness are considered in more detail in the next chapter, see Section 6.3.

In the study of Sumer et al. (1993b), similar experiments have been conducted in the same oscillating tunnel where a convergent-divergent section has been achieved with a convergent-divergent "ceiling" in the tunnel. Most of the measurements were conducted for three different values of the half angle of divergence, namely $0°$, $0.5°$ and $1°$. The Reynolds number was 6×10^6. The results indicated that there exists a constant steady streaming near the wall in the direction toward the convergent end of the tunnel. The magnitude of this streaming increases with the angle of divergence. Turbulence is largely suppressed in the convergent half period; this is felt even in the early stage of the following divergent half period. It recovers, however, shortly after, and an immense growth of turbulence is observed for the rest of the divergent half period of the flow.

The experiments of Sumer et al. (1993b) are considered in greater detail in the following chapter, see Section 6.2.2.

Example 5.2. Effect of externally generated turbulence on wave boundary layers.

The effect of turbulence generated externally on wave boundary layers has been studied by Fredsøe et al. (2003). The experiments have been conducted in the same oscillating tunnel as that used in the previous examples. The study deals with the effect of externally generated turbulence on the oscillatory boundary layer to simulate the turbulence in the wave boundary layer under broken waves in the swash zone. Turbulence was generated "externally" as the flow in the oscillator was passed through a series of grids that extended from the cover of the water tunnel to about mid-depth. Two different types of grid porosities were used. Direct measurements of the bed shear stress and velocity measurements were carried out. For the velocity measurements, mean and turbulence properties were measured in both the streamwise direction and in the direction perpendicular to the bed. A supplementary measurement for the undisturbed (without grids) case was also carried out, for comparison with the grid results. The mean and turbulence quantities in the outer flow region are increased substantially with the introduction of the grids. It is shown that the externally generated turbulence is able to penetrate the bed boundary layer, resulting in an increase in the bed shear stress, and therefore the friction coefficient. Other features related to the bed shear stress, such as transition, the friction factor and phase lead, are discussed. Of particular interest, the critical Reynolds number for transition to turbulence is reduced, as the level of externally generated turbulence is increased. The range of the Reynolds number investigated in the study was 1×10^4–6×10^6.

Example 5.3. Particle motions in oscillatory flow over a smooth bed.

Jensen et al. (2014b) studied this problem experimentally in the same oscillating tunnel as that described in the previous examples. As detailed in Chapter 3, the motion of sediment particles have been studied fairly extensively over the years, Grass (1974), Sumer and Oguz (1978), Sumer and Deigaard (1981), Nino and Garcia (1996). These investigations were undertaken in steady flows. Jensen et al.'s work (2014b) extended the previous work to oscillatory boundary layer flows. In the experiments, trajectories of single heavy particles were recorded in two dimensions (horizontal and vertical) and in time. The quantities related to particle motion such as the probability distribution of particle position in the vertical (equivalent to the sediment concentrations in the Eulerian approach), the horizontal and vertical particle velocities, and several other related statistics, now a function of phase, are determined from the particle trajectories data. The Reynolds number was 5.2×10^5. Comparison with steady-flow data was made. The vertical velocities normalized by the inner flow parameters around phase values $\omega t = 90°$ and $270°$ compare well with the results from steady flow research (referred to above, and discussed in greater detail in Section 3.6.1). It was found that, during the flow reversal, the particles stay near the bed. Otherwise, the particles can reach heights as large as $2.5d$–$3d$ from the bed (d being the particle size).

5.10 Solitary wave boundary layers

5.10.1 *Laminar solitary wave boundary layers*

Although much research has been carried out on boundary layers under harmonic progressive waves, reviewed in the previous sections, to the authors knowledge, Liu et al. (2007) were the first to study laminar boundary-layer flows under solitary waves (Fig. 5.64) in greater detail. They derived analytical solutions, based on (1) those of Liu and Orfila (2004) for viscous boundary-layer flows under transient long waves, and (2) the solutions for the nonlinear boundary-layer equations. They also carried out laboratory measurements that include the free surface displacement, particle image velocimetry (PIV)-resolved velocity fields of the boundary layer and the bottom shear stress. Liu et al. (2007) gave a detailed account of the existing work on the subject prior to their study, notably Keulegan (1948) and Mei (1983). We note that Park et al. (2014) later published a paper in which the authors addressed two shortcomings in Liu et al. (2007), namely (1) the authors clarified the notation of the outer velocity potential used in the formulae and derivations, which, the authors pointed out, was not consistently defined in Liu et al. (2007); and (2) some terms were missing in the nonlinear boundary layer equation, which was solved numerically. The authors presented the corrected nonlinear boundary layer equations and compared their numerical solutions with experiments.

Keulegan (1948) developed an analytical solution for the spatial variations of the velocity distribution and the bed shear stress under solitary waves. Tanaka et al. (1998), in their study of boundary layers under cnoidal waves, derived equations for the time variations of these quantities for solitary waves (which they viewed as the asymptotic case of cnoidal waves as the Ursell parameter goes to infinity), using Keulegan's (1948) analytical solution.

We will adopt the expressions given by Tanaka et al. (1998) for the aforementioned quantities to describe the laminar solitary wave boundary layers. The coordinate system is shown in the definition sketch in Fig. 5.64. The expression for the temporal variation of the velocity distribution in the boundary layer given by Tanaka et al. (1998) is:

$$\frac{u}{U_c} = \operatorname{sech}^2(-\xi) - \frac{2}{\sqrt{\pi}}\int_0^\infty \operatorname{sech}^2\left\{-\xi + (\frac{\beta' y}{\alpha})^2\right\} e^{-\alpha^2} d\alpha \tag{5.92}$$

and that for the temporal variation of the bed shear stress:

$$\frac{\tau_0}{\rho}\frac{\sqrt{\pi}}{4U_c\sqrt{ac\nu}} = \int_0^\infty \operatorname{sech}^2(-\xi + \alpha^2)\tanh(-\xi + \alpha^2)\, d\alpha \tag{5.93}$$

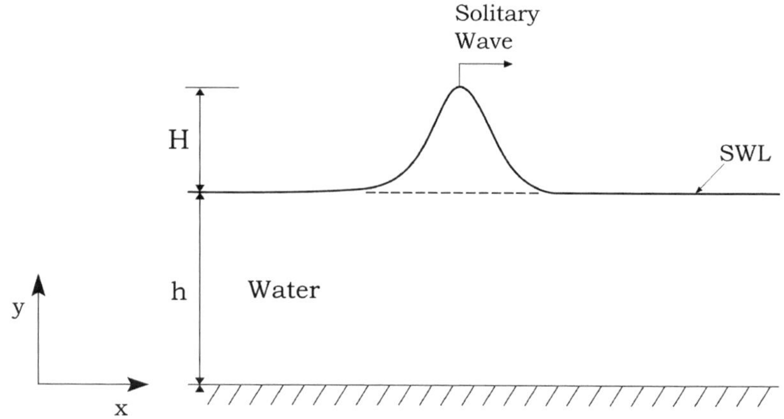

Fig. 5.64 Solitary wave. Wave height is grossly exaggerated for illustration purposes.

in which

$$a = \frac{1}{h}\sqrt{\frac{3H}{4h}} \tag{5.94}$$

$$\xi = act \tag{5.95}$$

$$c = \sqrt{g(h+H)} \tag{5.96}$$

$$U_c = c\frac{H}{h} \tag{5.97}$$

$$\beta' = \frac{1}{2}\sqrt{\frac{ac}{\nu}} \tag{5.98}$$

where h is the water depth, H the wave height, y the vertical distance from the bed (Fig. 5.64), t time, ν the kinematic viscosity, α a dummy variable, and the other quantities are as defined in the preceding equations. We note that the symbols used in Tanaka et al.'s (1998) paper have been adopted unchanged for convenience. (This is with the exception of the symbol used for the vertical distance, y, to be consistent with the rest of the chapter.) We also note that the quantity U_c should not be confused with the current velocity used in conjunction with combined waves and current boundary layers. One final note is that the independent variable ξ in the original formulation in Tanaka et al. (1998) has been replaced by $-\xi$, in order to recover Fig. 5.65, reproduced from Fig. 7 of Tanaka et al. (1998).

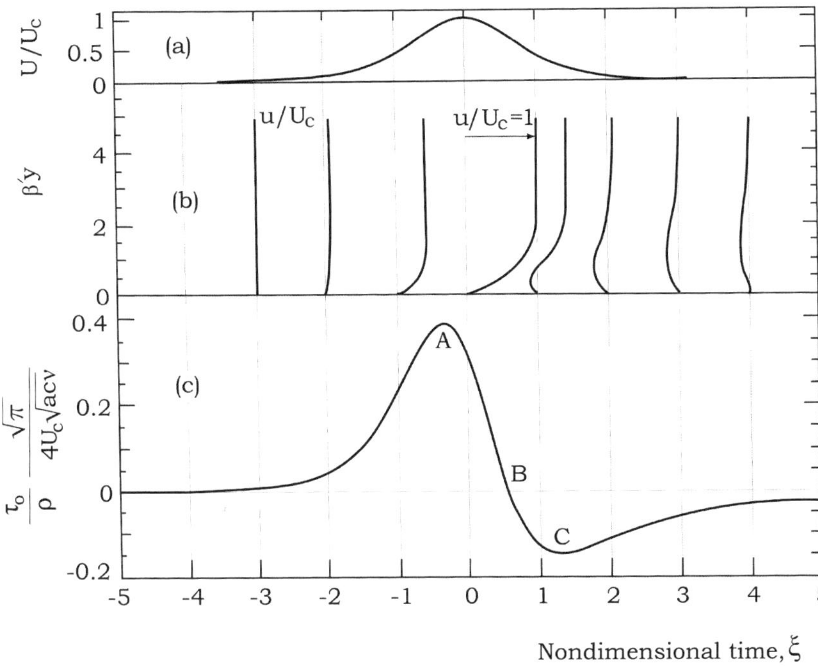

Fig. 5.65 Variations of the velocity distribution and the bed shear stress under solitary waves. Laminar solution. Tanaka et al. (1998).

The time variation of the velocity distribution across the depth is plotted in normalized form in Fig. 5.65b while that of the bed shear stress, again in normalized form, in Fig. 5.65c. Fig. 5.65a, on the other hand, illustrates the time variation of the velocity in the free-stream region, obtained from Eq. 5.92 for large values of $\beta' y$ like $\beta' y > O(4)$, or formally, as $\beta' y \to \infty$:

$$\frac{u}{U_c} = \operatorname{sech}^2(-\xi) \tag{5.99}$$

Firstly, the way in which the velocity distribution evolves in time (Fig. 5.65b) is similar to that in the wave boundary layer in oscillatory motion described in Fig. 5.3. As in the case of the oscillatory boundary layer, with the deceleration stage set in ($t > 0$), an adverse pressure gradient is generated. As a result, the flow in the boundary layer reverses, first at the bed, and gradually at larger and larger depths from the bed, as the flow progresses, as illustrated in Fig. 5.65b.

Secondly, it is seen from Fig. 5.65c, that the bed shear stress leads over the free-stream velocity by $\xi \simeq 0.35$ at time $\xi = 0$, Point A in Fig. 5.65c. This issue will be discussed in details later in the section.

Thirdly, the bed shear stress reverses at time $\xi \simeq 0.53$ (Point B), and the flow at the bed gradually increases, reaches a peak (Point C), and subsequently begins to decrease, as the flow continues.

As will be seen in the next section, there are two laminar-flow regimes in solitary wave boundary layers. The laminar solitary wave boundary-layer flow, studied in the preceding paragraphs, turns into a second kind of laminar solitary wave boundary-layer flow with an increase in the Reynolds number. This and further transitional regimes of solitary wave boundary layer flows will be studied in the next section.

5.10.2 *Transitional and turbulent solitary wave boundary layers*

Transitional and turbulent solitary wave boundary layers have only recently attracted some attention. Liu (2006) developed analytical solutions to the wave damping due to a turbulent boundary layer. In a relatively recent study, Vittori and Blondeaux (2008a and 2008b) presented the results of an investigation in which turbulent boundary layer under a solitary wave was studied by direct numerical simulation (DNS). The latter study was followed by further studies of solitary wave boundary layers by the Genoa group, Vittori and Blondeaux (2011), and Blondeaux et al. (2012). An extensive review of the subject, including the implications for sediment transport, has been published recently by Vittori and Blondeaux (2014).

In a PhD study at the Technical University of Denmark, wave boundary layer over a smooth bed was studied under oscillatory motion, described in the preceding sections, notably in Section 5.2, Carstensen et al. (2010). The focus was on the transitional regimes of the oscillatory boundary-layer flows. The work of Carstensen et al. (2010) was later extended to solitary wave boundary layers, Sumer et al. (2010). To this end, the solitary wave boundary-layer flow is simulated in the laboratory in an oscillating water tunnel. With this, the wave-induced velocities can be increased quite substantially, up to 1.2 m/s in the oscillating water tunnel facility, resulting in Reynolds numbers up to 2×10^6.

This section is mainly based on the work of Sumer et al. (2010).

5.10.3 *Experimental facility in Sumer et al. (2010) study*

The experiments were carried out in the same U-shaped oscillating water tunnel as in Carstensen et al. (2010). See Section 5.3.1 for a detailed description of the oscillating water tunnel. See also Fig. 5.21 for a schematic description of the tunnel including the pneumatic system. The solitary motion in the tunnel was driven by this electronically controlled pneumatic system. The piston of the pneumatic system was driven such that the free-stream velocity in the tunnel satisfied

$$U_0 = U_{0m}\text{sech}^2(\omega t) \tag{5.100}$$

the variation of the particle velocity at the bed at a given location under a small-amplitude solitary wave, Eq. 5.99 where the notation for the maximum velocity has changed to U_{0m}, and the argument of sech^2 has changed to ωt, adopting the same notation as in Sumer et al. (2010) for convenience. Here ω is the angular frequency

$$\omega = \frac{2\pi}{T} \tag{5.101}$$

with T, which can be interpreted as the effective period of the motion, a time scale that characterizes the width of the velocity time series, Fig 5.66a. We note that the effective period T is formally defined by Fig. 5.66a, namely, T is the length of the time window which covers $-180° \le \omega t \le 180°$.

Two kinds of measurements were performed in the experiments: bed shear stress measurements and velocity measurements. The bed shear stress was measured, using the same hot-film probe as in Carstensen et al. (2010). The streamwise velocity was measured with a Laser Doppler Anemometer (LDA). The details of these measurements are given in Carstensen et al. (2010) and Sumer et al. (2010).

Mean values of the quantities (the bed shear stress, or the velocity) are calculated through ensemble averaging according to

$$\overline{\phi}(\omega t) = \frac{1}{N}\sum_{j=1}^{N}[\phi(\omega t)]_j \tag{5.102}$$

in which $[\phi(\omega t)]_j$ is the time series of ϕ obtained in run j, and N is the total number of runs (i.e., the sample size). The root-mean-square (r.m.s.) value of the quantity in question, $\phi' = \phi - \overline{\phi}$, is calculated by

$$(\overline{\phi'^2})^{1/2}(\omega t) = \left\{\frac{1}{N-1}\sum_{j=1}^{N}\left\{[\phi(\omega t)]_j - \overline{\phi}(\omega t)\right\}^2\right\}^{1/2} \tag{5.103}$$

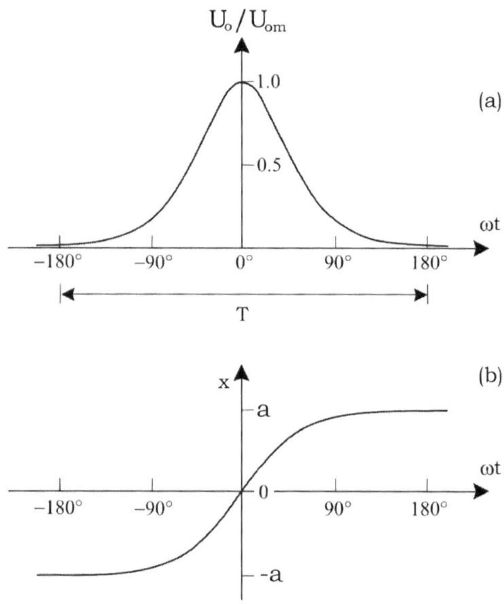

Fig. 5.66 (a) Free-stream velocity. (b) Water particle displacement in the free stream region in a solitary motion.

Although the main focus in Sumer et al. (2010) was the statistical properties of the hydrodynamic quantities such as the bed shear stress and the streamwise velocity, some limited number of flow visualization experiments were made to complement the bed shear stress and velocity measurements. See Section 5.2.1 and the original references (Carstensen et al., 2010, and Sumer et al., 2010) for the details of these flow visualization experiments.

5.10.4 *Flow regimes in solitary wave boundary layers*

The boundary layer developing over a smooth bed of a solitary motion in the half space $y > 0$ is described by a single parameter, namely the Reynolds number, similar to oscillatory boundary layers. The Reynolds number in solitary wave boundary layers is defined as

$$Re = \frac{aU_{0m}}{\nu} \tag{5.104}$$

in which U_{0m} is the maximum value of the free-stream velocity, defined by Eq. 5.100, and a is the half of the stroke of the water particle displacement in the free-stream region, defined by

$$x = a\tanh(\omega t) \tag{5.105}$$

and found from

$$a = \frac{U_{0m}}{\omega} \tag{5.106}$$

(The time derivative of both sides of Eq. 5.105 will give the expression for the free-stream velocity in Eq. 5.100 with the half of the stroke a given by Eq. 5.106.)

The boundary-layer flow undergoes tremendous changes as the Reynolds number is increased. The flow regimes experienced with increasing Reynolds number are summarized in Fig. 5.67 with reference to sampled bed shear stress traces. In the figures, τ_0 is the bed shear stress and ρ the density of water.

(We note that, in the case of a solitary-motion boundary layer developing in a finite-depth flow environment rather than the half space $y > 0$, the boundary layer is described by two parameters, the Reynolds number and the boundary-layer-thickness-to-depth ratio. Vittori and Blondeaux (2008b) chose the following set of parameters for real solitary wave boundary layers, namely H/h and δ_{VB}/h in which δ_{VB} is the boundary-layer thickness defined by

$$\delta_{VB} = \left(\frac{2\nu h}{\sqrt{gh}}\right)^{1/2} \tag{5.107}$$

Vittori and Blondeaux's set of parameters can be converted to the previously mentioned set of parameters, as will be done later in the chapter.)

For Re numbers up to 2×10^5, the flow is laminar (Fig. 5.67a); no sign of turbulence is observed in the bed shear stress signal. When Re is increased further, typically a strong decrease in magnitude (A, Fig. 5.67b), followed by a strong increase (B), emerges in the bed shear stress signal; this flow regime falls into the Reynolds number range $2 \times 10^5 \leq Re < 5 \times 10^5$. As will be seen below, this behaviour is related to the emergence of quasi-two-dimensional vortex tubes, the coherent flow features studied in Carstensen et al. (2010), see also the discussion in Section 5.2.1.1.

With a further increase in the Reynolds number, for Reynolds numbers larger than 5×10^5 up to 2×10^6 (the highest Reynolds number achieved in Sumer et al., 2010, experiments), the bed shear stress experiences single or multiple, or sometimes quite dense spikes, produced by the passage of turbulent spots over the measuring point (Fig. 5.67c), first emerging toward

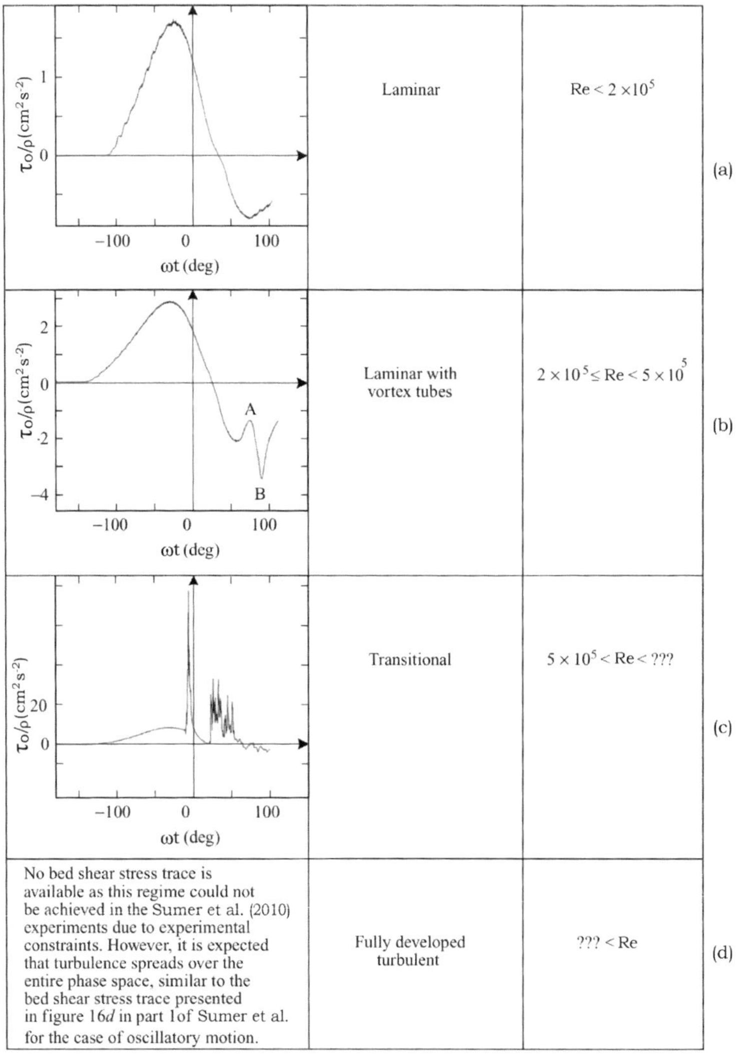

Fig. 5.67 Flow regimes in solitary wave boundary layer. Part 1 of Sumer et al. referred to in the bottom left box is the reference Carstensen et al. (2010). Sumer et al. (2010).

the end of the flow period, and gradually spreading to earlier phase of the motion with increasing Re, in the same way as in the oscillatory motion (see Section 5.2.4) This regime corresponds to the transitional flow regime.

For even further increase in the Reynolds number ($Re > 2 \times 10^6$), the previously described spikes are expected to spread over practically the entire phase space, and with this, the flow is expected to become fully developed turbulent, similar (qualitatively) to the case of oscillatory boundary layers (Jensen et al., 1989, Lodahl et al., 1998, and Carstensen et al., 2010). Further discussion will be given later in Section 5.10.4.3.

Each regime will now be considered individually.

5.10.4.1 *Laminar regime ($Re < 2 \times 10^5$)*

As pointed out earlier in conjunction with Fig. 5.65, the bed shear stress leads over the free-stream velocity (Fig. 5.67a) (in agreement with the oscillatory boundary-layer flows, Fig. 5.16), and eventually reverses at the phase $\omega t = 33°$, which agrees quite well with Liu et al.'s (2007) analytical solution for laminar boundary layers, as reported in Sumer et al. (2010, Fig. 6). We also note that Liu et al.'s (2007) laminar solution and Tanaka et al.'s laminar solution (Eq. 5.92) agree very well. We will, however, refer to Liu et al.'s (2007) laminar solution throughout the rest of the section, to be consistent with the paper by Sumer et al. (2010).

The flow reversal is related to the driving pressure gradient $\partial p/\partial x$, as mentioned previously; this pressure gradient becomes positive (i.e., the flow experiences an adverse pressure gradient) in the deceleration period ($\omega t > 0°$). However, it takes $\omega t = 33°$ in phase for the flow adjacent to the bed to "dissipate" its momentum and also to overcome the viscous resistance, to respond to the adverse pressure gradient. At this juncture, we further note that Liu et al.'s (2007) results show a very good agreement between their experimental data and their analytical prediction.

5.10.4.2 *Laminar regime with vortex tubes ($2 \times 10^5 \leq Re < 5 \times 10^5$)*

In the Reynolds number interval $2 \times 10^5 \leq Re < 5 \times 10^5$, typically a strong decrease (A, Fig. 5.67b), followed by a strong increase (B), emerges in the bed shear stress signal, as discussed above. This behaviour, which is clearly different from the pure laminar-regime flow, is related to the emergence of vortex tubes, the coherent structures generated as a result of the inflectional-point shear-layer instability. Fig. 5.68 shows a video frame illustrating the vortex tubes in plan view ($Re = 3.8 \times 10^5$). Fig. 5.69, on the other hand, shows another video frame for a side view of the vortex tubes obtained in another test ($Re = 2.8 \times 10^5$). See supplementary movies

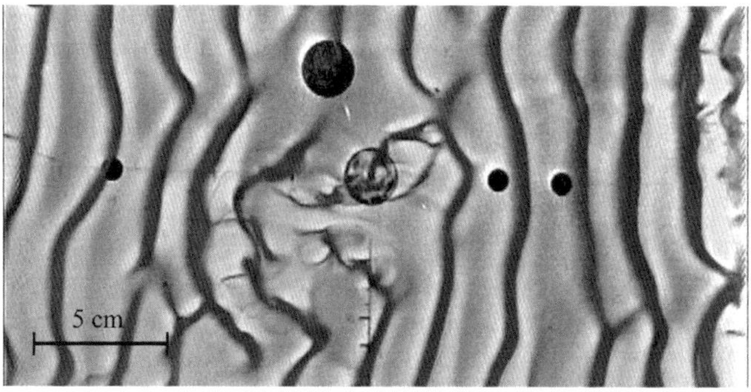

Fig. 5.68 Vortex tubes in plan view in solitary wave boundary layer. Sumer et al. (2010).

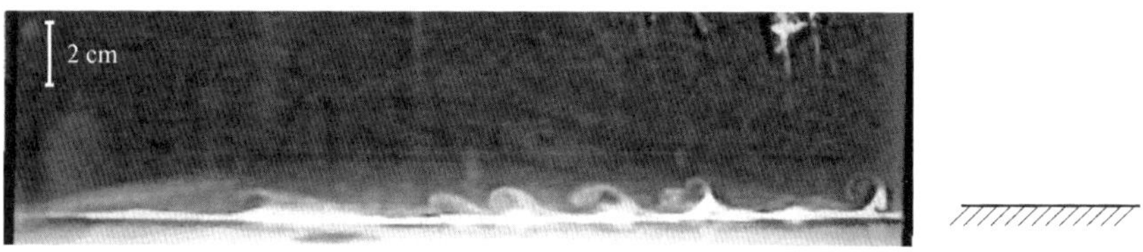

Fig. 5.69 Vortex tubes in solitary wave boundary layer, in side view. Sumer et al. (2010).

1 and 2 at `journals.cambridge.org/flm`.

The decrease (A) in the bed shear stress signal (Fig. 5.67b) corresponds to the situation where the bed shear stress probe experiences the flow between two successive vortices (marked A in Fig. 5.70), while the increase (B) in Fig. 5.67b corresponds to the situation where the wall shear-stress probe experiences the next vortex (marked B in Fig. 5.70) as the system of the vortex tubes is convected in the direction of the near-bed flow. As mentioned in Carstensen et al. (2010), these vortices are caused by the inflection-point instability.

The vortex tubes have been obtained by numerical simulations of Vittori and Blondeaux (2011), Blondeaux et al. (2012), Scandura (2013), and Ozdemir et al. (2013), thereby confirming the existence of these flow features, first discovered by Sumer et al. (2010) for boundary layers under solitary motion. Scandura (2013) report the emergence of an array of sec-

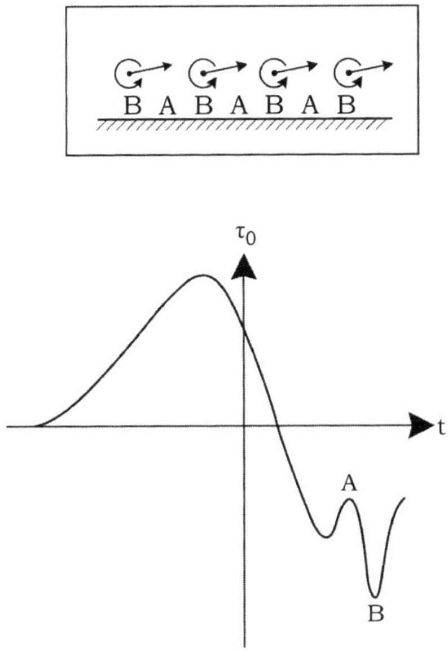

Fig. 5.70 Time history of the bed shear stress in the laminar regime with vortex tubes. Sumer et al. (2010).

ondary vortices (Vortices b in Fig. 5.71) associated with the vortices caused by the inflection-point instability (Vortices B in Fig. 5.71). These latter vortices come into existence relatively later in the process, also observed in the videos accompanying the paper by Sumer et al. (2010), the time t_2 in Fig. 5.71 corresponding to the late stages in the phase. These secondary vortices (Vortices b) are driven by the primary vortices (Vortices B). With the generation of these secondary vortices, Vortex B and Vortex b form a so-called vortex pair, which is subject to a self-induced motion in the direction normal to the line connecting the centers of the Vortices B and b. The generation of these secondary vortices, the formation of the vortex pairs, and the motion of the system of the vortex pairs are clearly revealed by the flow visualization results of Scandura's (2013, Figs. 11 and 12) numerical simulations. We also note that the generation of vortex pairs are also illustrated in the flow visualization results of Vittori and Blondeaux (2011) and Ozdemir et al.'s (2013) direct numerical simulations.

Four tests of Sumer et al. (2010), Tests 5–8, fall into the range $2 \times 10^5 \leq Re < 5 \times 10^5$ where the flow regime was laminar with vortex tubes. Of

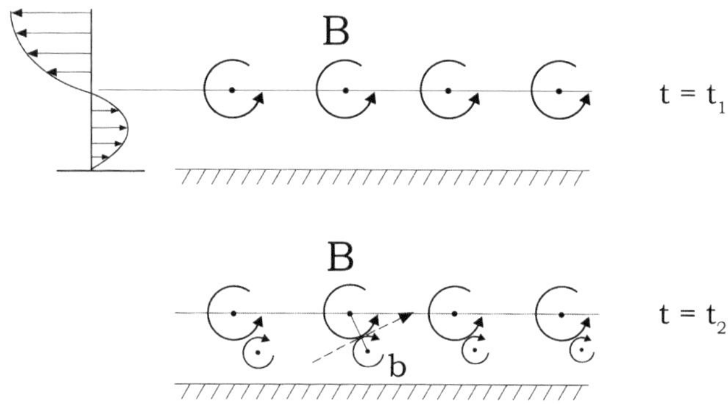

Fig. 5.71 Emergence of an array of secondary vortices, Vortices b.

these, Test 8 was presumably too close to the demarcation line between this regime and the laminar-to-turbulent transition regime, and therefore the bed shear stress did not behaved the same way in all runs as described in the preceding paragraphs, and there was sign of turbulence in the time histories of the bed shear stress in several cases. In the case of Tests 5–7 (with $Re = 2 \times 10^5$, 2.7×10^5 and 3.1×10^5), however, all the time histories showed the same behaviour as described above. It is also important to note the following observation: When plotted on top of each other, the bed shear stress traces collapsed practically on a single curve, meaning that, for a given Reynolds number, (1) the vortex tubes come into existence at exactly the same phase value; (2) they travel with exactly the same speed; and (3) they have exactly the same lifespan regardless of the test. The data regarding the latter have been given in Sumer et al. (2010). However, we note that A in Fig. 5.67b occurs at $\omega t = 74°–77°$, and B at $\omega t = 90.5°–100°$.

5.10.4.3 *Transitional regime* ($5 \times 10^5 < Re <?$)

With a further increase in the Reynolds number, for *Re* larger than 5×10^5 up to 2×10^6 (the highest Reynolds number achieved in Sumer et al. (2010) experiments), the bed shear stress experiences single, or multiple or, sometimes quite dense spikes, produced by the passage of turbulent spots over the measuring point, first emerging toward the end of the flow period, and gradually spreading to earlier phases of the motion with increasing *Re*, in the same way as in oscillatory boundary-layer flows, as mentioned earlier.

Fig. 5.72 displays four sample traces for the measured bed shear stress,

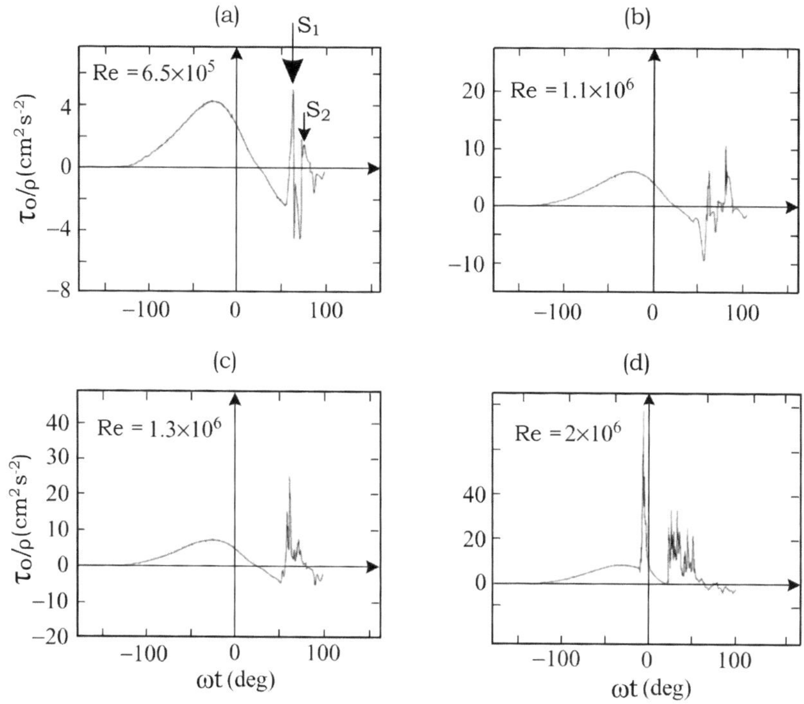

Fig. 5.72 Four sample time histories of bed shear stress in the laminar-to-turbulent transition regime. Sumer et al. (2010).

each corresponding to a different Reynolds number. The figure shows that the bed shear stress experiences spikes, sometimes very dense, similar to the case of oscillatory-motion boundary-layer flow described in the previous sections, Sections 5.2.1–5.2.3. These spikes are caused by turbulent spots, as detailed in Section 5.2.1. For example, in Fig. 5.72a, the flow at the bed reverses at around $\omega t = 25°$. However, it then experiences a very strong spike (marked S_1) starting at $\omega t = 55°$, indicating that a turbulent spot (forming upstream of the wall shear-stress probe, and traveling in the positive direction) hits the wall shear-stress probe, presumably changing the flow direction again at the bed. This is followed by a second, smaller spike (marked S_2), indicating that a second turbulent spot is passing overhead. It may be noted that there is a striking resemblance between the previously described bed-shear-stress time history of Sumer et al. (2010) and that given in Fig. 2 of Vittori and Blondeaux (2008b) of the longitu-

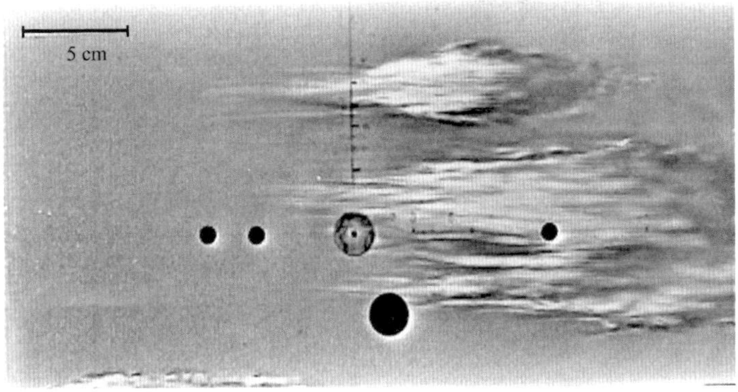

Fig. 5.73 Turbulent spots in solitary wave boundary layer. Sumer et al. (2010).

dinal velocity component near the bed in the turbulent regime, with the velocity experiencing positive and negative spikes in the same manner as in Fig. 5.72a. Fig. 5.73 shows a video frame illustrating turbulent spots in plan view. See also supplementary movie 3 accompanying the paper by Sumer et al. (2010).

Fig. 5.72 shows that the flow reversal does not occur in the bed shear stress where there is a turbulent spot passing overhead. This is linked to the fact that the turbulence in the turbulent spot brings the momentum-rich fluid from the upper elevations to the near-bed region This obviously offsets the effect of adverse pressure gradient, and therefore no flow reversal will occur near the bed.

Similar to the oscillatory-motion boundary layers (Carstensen et al., 2010), Section 5.2.1, Sumer et al. (2010) report that their plan-view flow visualization showed that turbulent spots and vortex tubes may occur concurrently.

Sumer et al. (2010) present the phase variations of the mean bed shear stress and that of the r.m.s. value of the fluctuations in the bed shear stress, not shown here for reasons of space, for various Reynolds numbers. From the latter data, it is seen that the occurrence of turbulence in the bed shear stress spreads toward the lower values of the phase ωt, as Re is increased. This is in good agreement with the results of Vittori and Blondeaux's (2008b) numerical simulations, as discussed by Sumer et al. (2010).

Although not captured in Sumer et al.'s (2010) study due to experimental constraints, Sumer et al. (2010) state that turbulence is expected to spread toward lower and lower values of the phase ωt, as Re is increased further, eventually covering the entire phase range. Sumer et al. (2010) note that this higher Reynolds number regime could not be captured in their study due to the finite stroke length of the piston of the pneumatic system to test this argument.

In the context of the above discussion, Sumer et al. (2010) add the following: One may note that, in the case of the oscillatory boundary layers, the role of the turbulence left from the previous half cycle is such that it will promote transition (trigger turbulence) in the boundary layer, as demonstrated by Fredsøe et al. (2003) and by DNS simulations of Costamagna et al. (2003). Clearly this "background" turbulence does not exist in the case of the solitary motion, and therefore transition to turbulence in the acceleration stage will be triggered by the regular boundary-layer instability mechanism. Considering the very strong favourable pressure gradient environment in the acceleration stage, the inception of turbulence in solitary wave boundary layers for earlier phases may differ quantitatively from that in oscillatory boundary layers.

In Fig. 5.74 the maximum value of the r.m.s. value of the fluctuating bed shear stress is plotted as a function of Re. As seen, turbulence grows explosively after Re reaches $Re \simeq 5 \times 10^5$. The time histories of the bed shear stress in combination with the flow visualization test of Sumer et al. (2010) indicate that this transition is related to the inception of turbulent spots.

Vittori and Blondeaux (2008b, Fig. 6) plotted their DNS results regarding the transition to turbulence in terms of two non-dimensional parameters, namely H/h and δ_{VB}/h in which δ_{VB} is the boundary-layer thickness defined in Eq. 5.107. The data were recast (using the small-amplitude solitary-wave theory) by Sumer et al. (2010), and are plotted together with their own experimental data on the plane $(Re, h/\delta_1)$ in Sumer et al. (2010) which is reproduced in Fig. 5.75. In plotting their experimental data, Sumer et al. (2010) have taken the water depth h as the half height of the tunnel of their experiments. The quantity δ_1 in the figure is the Stokes length, Eq. 5.13. The experimental data of Sumer et al. (2010) and Vittori and Blondeaux's (2008b) numerical data seem to be in good agreement. Sumer et al. (2010) note that Re_{cr} is expected to increase with decreasing h/δ_1, similar to oscillatory boundary layers (e.g., Lodahl et al., 1998, Fig. 5.45 here) and Tuzi and Blondeaux (2008, their Fig. 24).

It appears that there is a substantial delay in the laminar-to-turbulent transition in terms of the Reynolds number in the solitary-wave boundary

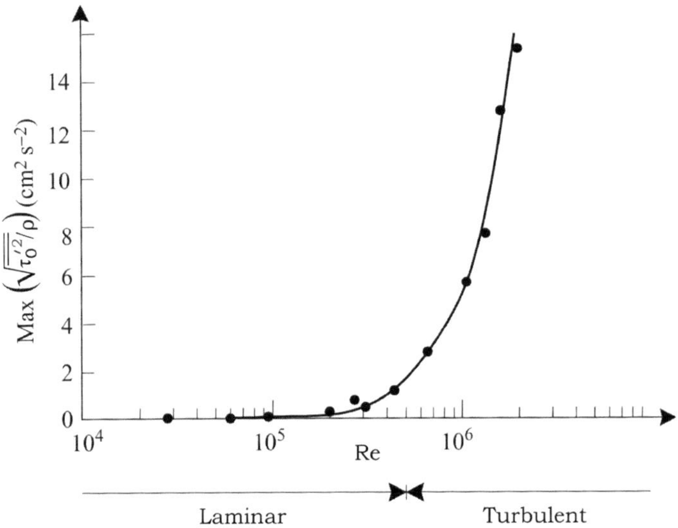

Fig. 5.74 Maximum value of the r.m.s. value of the fluctuating bed shear stress as function of the Reynolds number. Sumer et al. (2010).

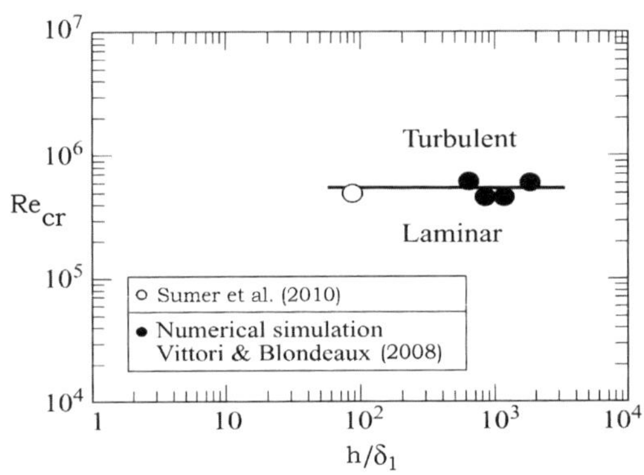

Fig. 5.75 Transition to turbulence in solitary wave boundary layer. Sumer et al. (2010).

layer (c.f., $Re_{cr} \simeq 5 \times 10^5$ of the solitary-wave boundary layer case, and $Re_{cr} \simeq 1.5 \times 10^5$ of the oscillatory boundary layer case, Section 5.2.2). This is mainly because the adverse pressure gradient toward the end of the motion, a key factor in the transition, is more gentle in the present case than in the case of the oscillatory motion.

5.10.5 *Flow resistance in solitary wave boundary layers*

The wave friction coefficient is a key quantity in the analysis of oscillatory wave boundary layers. Drawing an analogy to oscillatory wave boundary layers, two friction coefficients can be defined for solitary wave boundary layers: (1) a friction coefficient associated with the maximum bed shear stress in the acceleration stage where the near-bed flow is in the forward direction (Fig. 5.76a):

$$f_f = \frac{2\overline{\tau}_{0f}}{\rho U_{0m}^2} \tag{5.108}$$

and (2) that associated with the maximum bed shear stress in the deceleration stage where the near-bed flow is in the reverse direction (Fig. 5.76 b):

$$f_r = \frac{2\left|\overline{\tau}_{0r}\right|}{\rho U_{0m}^2} \tag{5.109}$$

The flow resistance data of Sumer et al. (2010) are plotted in Fig. 5.76 (circles) together with Liu et al.'s (2007) laminar solution (the solid lines). Jensen et al.'s (1989) oscillatory-flow data also are plotted in the figure as reference lines (the dashed lines). Sumer et al. (2010) data points 1–3 lie in the laminar-regime Re range; 4–6 in the laminar-regime Re range with vortex tubes being present; and 7–10 in the transitional-regime Re range.

The friction-coefficient diagrams are accompanied with the phase diagrams in Fig. 5.77 where Φ_f is the phase lead of the maximum bed shear stress in the acceleration stage over the maximum free-stream velocity, and Φ_r is the phase difference between the maximum free-stream velocity and the maximum bed shear stress associated with the reverse flow in the deceleration stage. Fig. 5.77 also includes Liu et al.'s (2007) laminar solutions for Φ_f and Φ_r.

From Fig. 5.76a, it is seen that the agreement between the experiments and Liu et al.'s (2007) laminar solution is very good, even for values of Re up to $Re = 1.3 \times 10^6$, a Reynolds number well beyond $Re_{cr} \simeq 5 \times 10^5$. This is because, even for such large Reynolds numbers, the flow remains in the laminar regime for the phase value where the shear stress attains its maximum value, and therefore the maximum bed shear stress (and therefore f_f) behaves like in the case of the laminar flow, revealing the observed agreement.

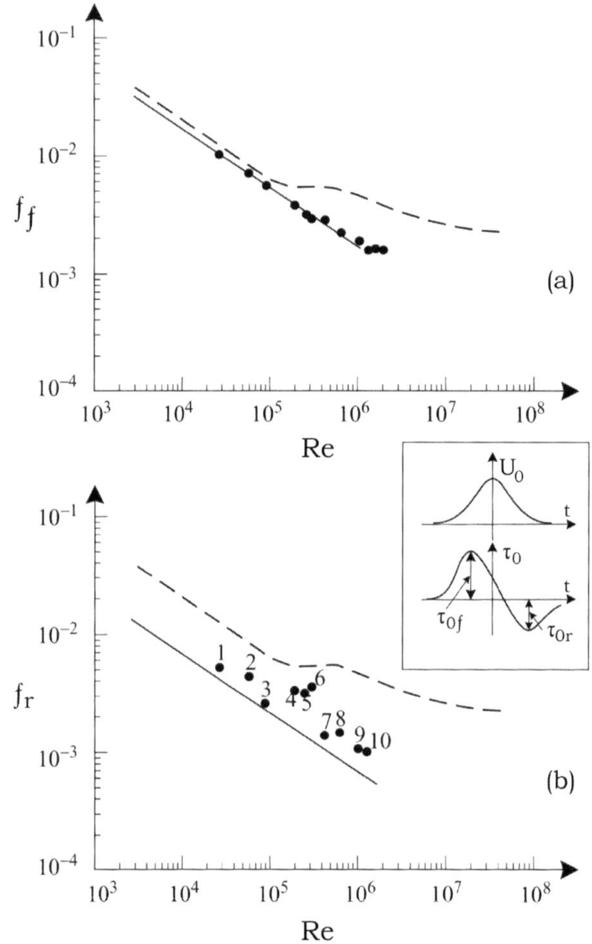

Fig. 5.76 Friction coefficient data in solitary wave boundary layer. Circles: Sumer et al.'s experiments (2010). Dashed lines: Jensen et al.'s (1989) oscillatory flow experiments plotted for reference. Solid lines: Liu et al.'s (2007) laminar solution. Sumer et al. (2010).

The friction coefficient data of Sumer et al. (2010) for solitary wave boundary layer in Fig. 5.76 are quite close to those of Jensen et al. (1989) oscillatory flow data, and this is for Reynolds numbers up to $Re = 1.5 \times 10^5$, the critical Reynolds number for the oscillatory-motion boundary layer. This is because the solitary-motion boundary layer flow and the oscillatory-motion boundary-layer flow are quite similar for ωt up to the point where the near-bed flow reverses, and therefore the corresponding friction coefficients appear to be quite close to each other until the point of turbulent transition for oscillatory-motion boundary layer is reached.

Regarding the friction coefficient for reversed flow (Fig. 5.76b), Sumer et al.'s (2010) data corresponding to the laminar regime (Points 1–3) agree fairly well with the laminar solution of Liu et al. (2007). For the three other data points in the laminar regime $Re < 5 \times 10^5$ (Points 4–6), the friction coefficient is predicted to be relatively larger. This is linked to the presence of the dips (B in Fig. 5.67b) in the bed shear stress traces, which are induced by vortex tubes as described in the preceding paragraphs. For higher Reynolds numbers ($Re > 5 \times 10^5$, the transitional regime), Points 7–10, the experimental data differ from the laminar solution, as expected. Sumer et al. (2010) note that the data for the two highest Re numbers ($Re = 1.6 \times 10^6$ and 2×10^6) of the Sumer et al. (2010) experiments are not included in the diagram on the grounds that the maximum value of $|\overline{\tau}_{0r}|$ could not be captured in the mean bed shear stress variations.

As for the phase diagrams (Fig. 5.77), Fig. 5.77a shows that Sumer et al.'s data for Φ_f in the laminar regime are in fairly good agreement with the laminar solution of Liu et al. (2007), although there is a tendency that the experimental values are, in general, slightly larger than predicted by the laminar theory.

Regarding the two data points corresponding to the highest Reynolds numbers, the phase becomes almost zero for $Re = 1.6 \times 10^6$, and even negative for $Re = 2 \times 10^6$. This is because of the turbulence (see Fig. 12f, left, in Sumer et al., 2010). Although not captured in Sumer et al.'s (2010) experiments due to experimental constraints, the latter authors note that Φ_f is expected to "recover" again with a further increase in Re as the turbulence spreads further toward the smaller values of ωt in the fully developed turbulent regime, and it is expected to settle at a positive value, albeit smaller than $\omega t = 20°$ (the laminar-regime phase lead), similar to the behaviour of the phase lead observed in the case of the oscillatory flow (e.g., Jensen et al., 1989, Fig. 11b).

As for Φ_r, the agreement between the experiment and Liu et al.'s (2007) laminar solution is very good in the laminar-regime Re number range for the first three data points. However, Φ_r data for the next three points in the laminar regime deviate from Liu et al.'s laminar solution because these data points correspond to the laminar regime with vortex tubes. For the transitional Reynolds number range ($Re > 5 \times 10^5$), Fig. 5.77b shows that

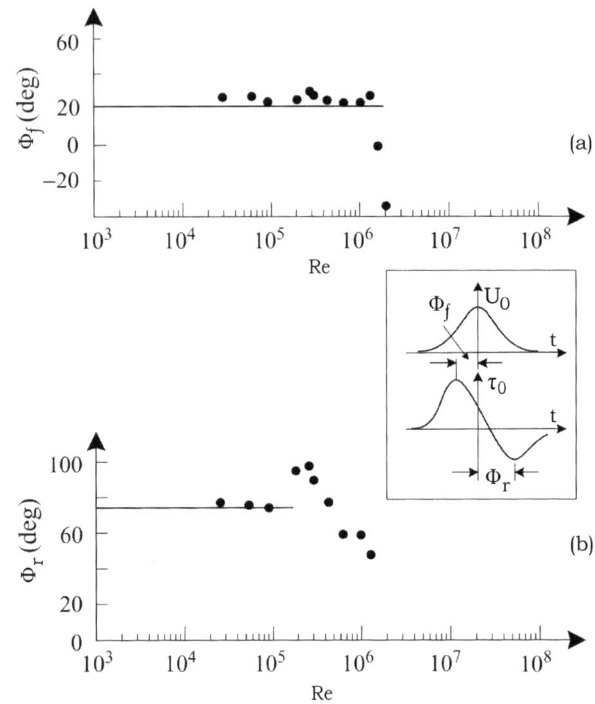

Fig. 5.77 Phase data for solitary wave boundary layer. Circles: Sumer et al.'s experiments (2010). Solid lines: Liu et al.'s (2007) laminar solution. Sumer et al. (2010).

Φ_r decreases with respect to the values experienced in the laminar regime. This is because of the introduction of turbulence via the appearance of turbulent spots at the phase values where the maximum bed shear stress in the deceleration stage is experienced. However, as *Re* is increased further, Φ_r is expected to pick up values larger than those experienced in the laminar and transitional regimes exhibited in Fig. 5.77b simply because the flow reversal in this case will be delayed due to turbulence, resulting in values of Φ_r larger than the laminar and transitional ones.

Fig. 5.78 displays the phase data regarding the phase at which the flow reversal takes place at the bed, Φ_c, from Sumer et al. (2010). The solid line is Liu et al.'s (2007) laminar solution. Again, the experimental data and the laminar solution are in very good agreement. The boundary layer reverses in the laminar regime until *Re* reaches the value $Re = 1.3 \times 10^6$ because only after this Reynolds number, the turbulence on average will

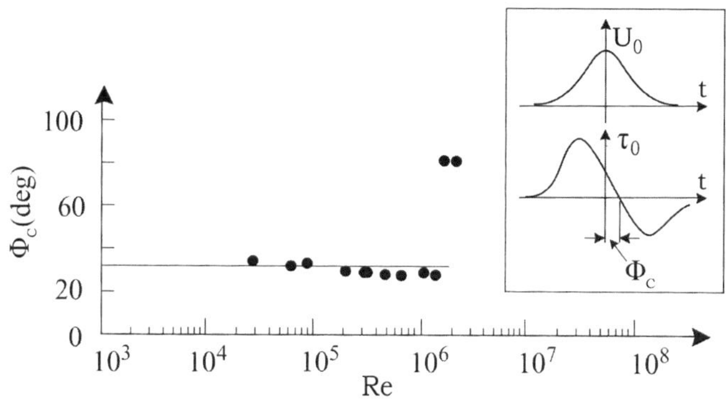

Fig. 5.78 Phase at which the flow reversal takes place. Circles: Sumer et al.'s experiments. Solid line: Liu et al.'s laminar solution. Sumer et al. (2010).

reach the phase where the flow reversal takes place, although transition to turbulence occurs much earlier, i.e. at a much lower Reynolds number $Re = 5 \times 10^5$. Once turbulence reaches that point, the reversal will occur in the turbulent regime, and this will delay the flow reversal, revealing the higher values of Φ_c around 80°, observed for the two highest Re numbers in Sumer et al's (2010) test. The flow reversal at the bed is delayed quite considerably in the case of the turbulent regime, similar to the oscillatory-flow boundary-layer flows (Section 5.2.6), Jensen et al. (1989), Carstensen et al. (2010); this is related to the continuous supply of momentum-rich fluid from the outer flow as discussed previously.

5.10.6 *Velocity profiles in solitary wave boundary layers*

Fig. 5.79 displays the phase evolution of the mean velocity profiles for three different Re numbers, $Re = 7.1 \times 10^4$, $Re = 4.8 \times 10^5$ and $Re = 1.8 \times 10^6$ from Sumer et al.'s (2010) study, the first being in the laminar regime, the second at the onset of turbulent transition, and the third in the transitional regime. The figure also includes Liu et al. 's (2007) laminar solution for the phase values $\omega t = -100°$, $-40°$, $100°$ and $40°$. The agreement between the experimental data and the laminar solution is fairly good.

From Fig. 5.79, it is seen that the flow reversal occurs at larger and larger ωt values with the distance from the bed, similar to oscillatory boundary layers (e.g. Fig. 5.3), see also Jensen et al. (1989). It is also seen that the flow reversal across the depth is, as expected, delayed quite substantially

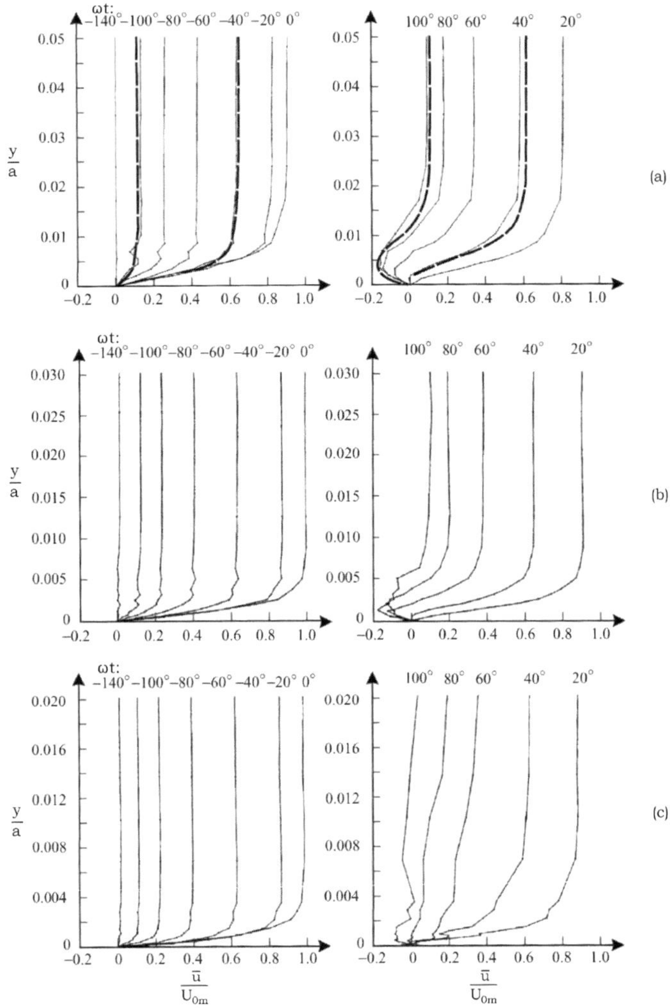

Fig. 5.79 Mean velocity profiles. Solitary wave boundary layer. (a) $Re = 7.1 \times 10^4$ (laminar). (b) $Re = 4.8 \times 10^5$ (onset of transition). (c) $Re = 7.8 \times 10^6$ (transition). Dashed lines: Liu et al.'s laminar solution. Sumer et al. (2010).

when Re is increased from $Re = 4.8 \times 10^5$ to 1.8×10^6, in agreement with the bed shear stress measurements.

Fig. 5.80 presents the phase evolution of the mean velocity profiles in

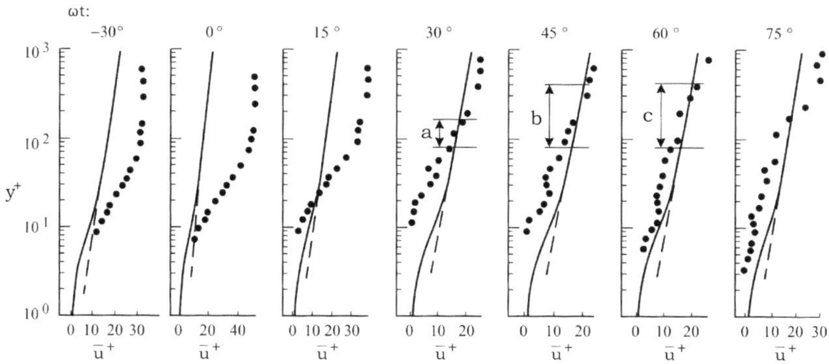

Fig. 5.80 Mean velocity profiles in semi-logarithmic plot. $Re = 1.8 \times 10^6$. Circles: Sumer et al.'s (2010) experiments. Solid lines: van Driest profile. Dashed lines: logarithmic velocity distribution. Sumer et al. (2010).

semi-logarithmic plots for $Re = 1.8 \times 10^6$, from Sumer et al. (2010). The velocity and the vertical distance from the bed are normalized in terms of the inner-flow parameters, $U_f(= \sqrt{\overline{\tau}_0/\rho})$, the phase-resolved mean friction velocity, and the kinematic viscosity ν:

$$\overline{u}^+(y, \omega t) = \frac{\overline{u}(y, \omega t)}{U_f(\omega t)} \quad \text{and} \quad y^+ = \frac{yU_f(\omega t)}{\nu} \tag{5.110}$$

In the figure is also plotted the van Driest velocity profile (van Driest, 1956), see e.g., Eq. 5.36 and the discussion related to the latter equation:

$$\frac{\overline{u}}{U_f} = 2\int_0^{y^+} \frac{dy^+}{1 + \left\{1 + 4\kappa^2 y^{+2}\left[1 - \exp(-y^+/A)\right]^2\right\}^{1/2}} \tag{5.111}$$

as a reference line (solid line). As is known from previous discussions (Section 5.3.4), the van Driest equation tends to the logarithmic distribution

$$\frac{\overline{u}}{U_f} = \frac{1}{\kappa}\ln y^+ + 5 \tag{5.112}$$

for large values of y^+ (the dashed lines in the figure).

The y^+ interval over which the measured velocity profile comes closest to the logarithmic distribution is marked a, b and c (over the phase interval $\omega t \simeq 30°$–$60°$) in the figure. Although the behaviour of the measured velocity distribution as a function of phase is qualitatively quite similar to that of oscillatory boundary-layer flows (Fig. 5.25), the agreement between the velocity data and the logarithmic distribution over the phase interval $\omega t \simeq 30°$–$60°$ is not very good. Sumer et al. (2010) attributes this to the fact that the boundary layer in the experiment is not in the fully developed turbulent regime.

Sumer et al. (2010) further present the time evolution of the turbulence in terms of the r.m.s. value of the fluctuating component of the velocity normalized by the inner-flow parameters.

Finally, we note that Vittori and Blondeaux (2011), via direct numerical simulations, obtained velocity profiles, bed shear stress variations including the flow resistance properties for solitary wave boundary layers, including transitional and turbulent regime flows. Their results are in accordance with Sumer et al.'s (2010) experimental findings.

5.10.7 *Remarks on practical applications*

For a small-amplitude solitary wave, the surface elevation and the horizontal (free stream) particle velocity are, respectively, given as

$$\eta = H \operatorname{sech}^2(\omega t) \tag{5.113}$$

$$U_0 = U_{0m} \operatorname{sech}^2(\omega t) \tag{5.114}$$

in which U_{0m} and ω are

$$U_{0m} = \sqrt{gh}\frac{H}{h} \tag{5.115}$$

$$\omega = \sqrt{\frac{3}{4}gH}\frac{1}{h} \tag{5.116}$$

in which H is the wave height, and h the water depth.

Similar to Eq. 5.101, an effective period can be defined by

$$T = \frac{2\pi}{\omega} = 2\pi\sqrt{\frac{4}{3gH}}h \tag{5.117}$$

where T can be interpreted as a time scale that characterizes the width of the free-stream velocity time series as in Fig. 5.66.

The boundary layer developing over a smooth bed under a solitary wave where the half-space approximation is valid ($y > 0$) is described by a single non-dimensional quantity, namely the Reynolds number, defined by Eq. 5.104 in which U_{0m} is given by Eq. 5.115, and a by Eqs. 5.106, 5.116 and 5.117.

In the case when T is large, so large that the boundary-layer thickness becomes comparable to the water depth, the parameter water-depth-to-boundary-layer-thickness ratio should also be involved, as pointed out earlier. This case normally involves turbulent-regime boundary-layer flows (not only in the field but also, to a large extent, in the laboratory). The boundary-layer thickness, δ, in this case may be estimated as follows.

δ corresponding to T may be written as

$$\delta = O\left(\sqrt{2\nu_t T}\right) \tag{5.118}$$

in which ν_t is the turbulent diffusion coefficient (eddy viscosity) and may, to a first approximation, be taken as

$$\nu_t = O\left(\frac{\kappa}{6}\delta U_{fm}\right) \tag{5.119}$$

in which κ is the von Karman constant ($\simeq 0.4$), and U_{fm} is the maximum value of the friction velocity.

(Eq. 5.119 is obtained by integrating the turbulent diffusion coefficient over the boundary-layer thickness, $\nu_t = \frac{1}{\delta}\int_0^\delta \epsilon(y)dy$, with $\epsilon(y)$ obtained from the Reynolds analogy (i.e., the transfer of momentum and mass is analogous) as $\epsilon(y) = \kappa U_{fm} y(1 - (y/\delta))$. See Section 8.4.1.

Now, solving for the boundary-layer thickness, δ, from Eqs. 5.118 and 5.119 yields

$$\delta = O(0.1 U_{fm} T) \tag{5.120}$$

In Sumer et al.'s (2010) experiments, the largest boundary-layer thickness was, from the preceding equation, $\delta = O(0.1U_{fm}T) = O(0.1 \times 2.8$ cm/s $\times$ 8.5 s$) = O(2.4$ cm$)$ in Test 15 corresponding to the highest Reynolds number, and hence $h/\delta = O(14.5$ cm / 2.4 cm$) = O(6)$ where h is taken as $h = 29/2 = 14.5$ cm, the half-height of the experimental tunnel. Sumer et al. (2010) note that the above estimate of the boundary-layer thickness agrees quite well with that found from the measured velocity profiles (Fig. 5.79).

So, the ratio h/δ in Test 15 of Sumer et al. (2010) was like $O(6)$, large enough for the parameter h/δ not to be very significant. (This parameter was even larger in the other tests in Sumer et al., 2010.) Therefore, the results of the Sumer et al. (2010) study are directly applicable to boundary layers developing under a solitary wave provided that the wave period is not very large, such that the water-depth-to-boundary-layer-thickness ratio remains sufficiently large. Otherwise, the boundary-layer properties should be dependent not only on Re but also on the water-depth-to-boundary-layer-thickness ratio, h/δ.

Sumer et al.'s (2010) tests cover the smooth-bed case. In the case of a rough bed, there will be a third parameter, a/k_s, in which k_s is Nikuradse's equivalent sand roughness. For the Sumer et al. (2010) results to be applicable for the rough-bed case, the bed should act as a hydraulically smooth boundary, i.e., $k_s U_{fm}/\nu$ should be sufficiently small, certainly smaller than $O(10)$. So far, no data are available for solitary motion/wave boundary layers over rough beds.

5.11 Tsunami wave boundary layers

Waves in coastal environments range from relatively short-scale wind waves and swell (periods typically less than 30 s) to much longer tidal waves (periods of approximately 12 h). Due to their extreme lengths such tidal flows can be reasonably treated as quasi-steady currents for most practical purposes. Alternatively, due to their relatively short periods, the bottom boundary layers induced by wind waves can be effectively reproduced experimentally at full scale within oscillatory tunnel environments (e.g. Jensen et al., 1989, see Fig. 5.21 herein), and have therefore been extensively studied in the literature, as indicated previously in the present chapter. Lying in between these extremes are infrequent, but potentially devastating, tsunami wave events, which are created via the displacement of large volumes of water e.g. due to earthquakes, landslides, volcanic eruptions, glacier calvings, meteorite impacts, and even underwater explosions.

Such tsunami waves are typically characterized by periods ranging from several minutes to hours and/or flow velocities that can be well in excess of typical wind-wave scales (obviously, depending on the generating conditions). Due to their long, yet unsteady and transient nature, tsunami scales are difficult (though not impossible, see e.g. the recent works of Goseberg et al., 2013; Schimmels et al., 2016) to achieve properly in controlled laboratory settings. As a result of this difficulty, the boundary layer dynamics beneath tsunami waves has received relatively little specific attention in the literature, though such knowledge is imperative for the understanding of induced sediment transport and morphological effects, including e.g. scour induced around coastal structures.

5.11.1 *Idealized tsunami signals*

To help fill this gap Williams and Fuhrman (2016) conducted a detailed numerical study of tsunami-scale wave boundary layers, utilizing a one-dimensional-vertical model based on the k-ω turbulence closure model of Wilcox (2006), as will be presented in the forthcoming Section 5.12 (see also Chapter 9). They considered three different idealized tsunami wave signals, where the tsunami free surface at some reference depth h_0 was envisioned to vary either (1) sinusoidally, (2) as a "single wave" (elongated soliton), or (3) as a so-called N-wave. For the sinusoidal description, they envisioned the time-varying flow beneath a free surface corresponding to

$$\eta_0(t) = A_0 \sin(\omega t), \qquad \omega = \frac{2\pi}{T} \tag{5.121}$$

where A_0 is the amplitude at some reference depth h_0. For the "single wave" description the tsunamis shape was envisioned as an elongated soliton (having independent wave amplitude and effective period)

$$\eta_0(t) = A_0 \text{sech}^2(\omega t), \qquad \omega = \frac{2\pi}{T} \tag{5.122}$$

Finally, the envisioned N-wave form corresponds to the superposition of two (one positive, one negative) single waves, as introduced by Madsen and Schäffer (2010):

$$\eta_0(t) = 1.165 A_0 \left[\text{sech}^2\left(\omega t - \frac{3\pi}{4}\right) - \text{sech}^2\left(\omega\left(t - \frac{\pi}{2\omega}\right) - \frac{3\pi}{4}\right)\right] \tag{5.123}$$

where the factor 1.165 was added to ensure that the maximum elevation corresponds to A_0 in the expression above. (We use this notation such that A_0 represents the maximum surface elevation for all three wave descriptions above.) These three idealized tsunami wave forms are depicted in Fig. 5.81. The sinusoidal and N-wave shapes are presented as leading elevation events, though they can readily be changed to leading depression events simply by using negative A_0. The single wave shape can likewise be converted from a purely elevation to a depression event in the same fashion.

To translate the envisioned surface elevation signals to water depths h other than the reference depth h_0, Williams and Fuhrman (2016) assumed that the surface would vary according to linear shallow water shoaling theory (Green's law, see Eq. 5.212):

$$\frac{\eta(t)}{\eta_0(t)} = \frac{A}{A_0} = \left(\frac{h_0}{h}\right)^{1/4} \tag{5.124}$$

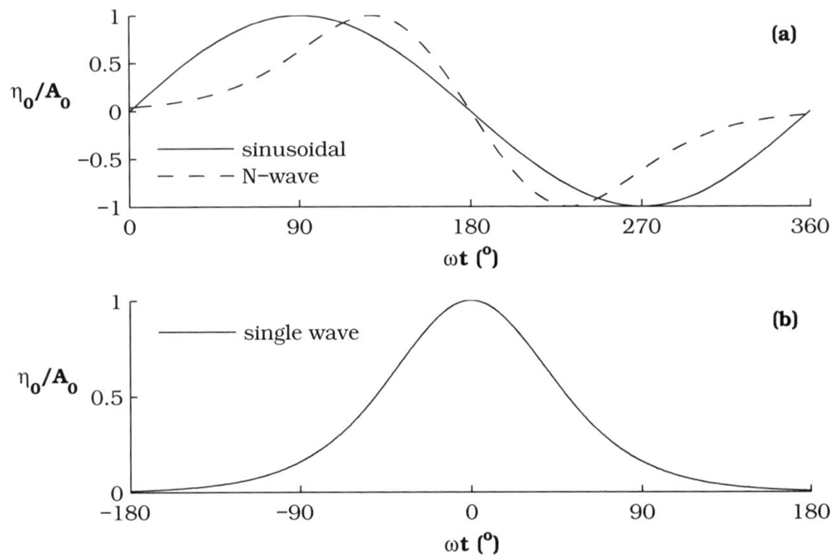

Fig. 5.81 Comparison of idealized tsunami wave shapes.

where η and A are, respectively, the free surface elevation and amplitude at depth h, and where t is treated independently at all depths. With emphasis on studying tsunami-induced boundary layers, the resulting surface elevations from Eq. 5.124 were converted to free stream velocities based on linear shallow water wave theory (see Eq. 5.200), according to:

$$U_0(t) = \sqrt{\frac{g}{h}}\eta(t) \tag{5.125}$$

These velocities (actually, their time derivatives) were then utilized to determine the vertically-constant pressure gradient (within the forthcoming Eq. 5.159) driving the tsunami-induced boundary layer flow, which in the present situation (neglecting steady as well as all convective acceleration effects) simplifies to

$$-\frac{1}{\rho}\frac{\partial p}{\partial x} = \frac{\partial U_0}{\partial t} \tag{5.126}$$

Note that the resulting free stream velocity magnitude from this procedure corresponds to

$$U_{0m} = \sqrt{\frac{gA^2}{h}} \tag{5.127}$$

which can be readily seen from Eq. 5.125, after substituting the surface elevation η and velocity signal U_0 with their respective amplitudes A and U_{0m}. This simple approach based on linear shallow water theory obviously neglects nonlinearity, any reflections, as well as three dimensional effects such as refraction and diffraction; It is still physically-based, however, and reasonable for systematic study, especially given the lack of other simple alternatives. Note that the adopted approach is generally consistent with the findings of Chan and Liu (2012), that the leading tsunami waves in both near- and far-field regions correspond to small amplitude long waves.

To achieve realistic scales, Williams and Fuhrman (2016) considered the induced idealized flow beneath tsunami waves having period $T = 15$ min, well-known to be within a range typical of tsunamis, and also reasonably resembling observations from the 2004 Indian Ocean tsunami (see e.g. Madsen and Fuhrman, 2008). Waves were considered to have a characteristic amplitude $A_0 = 0.5$ m (corresponding to wave height $H_0 = 2A_0 = 1$ m) at an offshore reference depth $h_0 = 4000$ m, approximately equal to the average depth of Earth's oceans. The resulting waves and corresponding velocities were then translated (shoaled) to other depths as shallow as $h = 10$ m, as described above. Tsunami-induced boundary layer flow (using all three idealized wave forms above) over smooth beds, as well as over three fixed grain sizes were systematically considered: $d = 0.3$ mm, 1.5 mm, and 3 mm, where Nikuradse's equivalent sand grain roughness was calculated according to $k_s = 2.5d$. The resulting simulations produced a wide range of scales, resulting in Reynolds numbers approaching $Re = aU_{0m}/\nu = 10^9$ and a/k_s values in excess of 40,000, well beyond those typical of wind waves. (Here $a = U_{0m}/\omega$ and $\omega = 2\pi/T$.)

The results on hydraulically smooth and rough beds were then collected, and Williams and Fuhrman (2016) used these to parameterize the tsunami wave boundary layer characteristics. They found that all three of the idealized wave forms resulted in similar results. In the hydraulically rough regime (likely of most practical interest), they proposed the following relationships for estimating the tsunami wave boundary layer thickness δ:

$$\frac{\delta}{a} = 0.05\left(\frac{a}{k_s}\right)^{-0.11}, \qquad a = \frac{U_{0m}}{\omega} \tag{5.128}$$

the maximum bed shear stress (parameterized as a wave friction factor):

$$f_w = \frac{2\tau_{0m}}{\rho U_{0m}^2} = \exp\left(5.5\left(\frac{a}{k_s}\right)^{-0.16} - 6.7\right) \tag{5.129}$$

and the maximum turbulent kinetic energy density:

$$\frac{k_m}{U_{0m}^2} = 0.068\left(\frac{a}{k_s}\right)^{-0.22} \tag{5.130}$$

Note that Eq. 5.129 matches the expression from Fuhrman et al. (2013), presented previously as Eq. 5.69, originally formulated for rough beds at wind wave scales. This expression was found to maintain accuracy even when extrapolated to tsunami-scale waves, suggesting that there is a continuum connecting these two scales. (Note that a generalization of the above, which can be utilized for predicting the time variation of the boundary layer thickness and bed shear stress is presented in Section 5.11.3.)

Notably, even for the cases having the smallest depth considered i.e. $h = 10$ m, the resulting tsunami wave boundary layer thickness was found to be at most $\delta \approx 5$ m. Hence, over the parametric range considered the tsunami-induced boundary layers did not span the entirety of the water depth, and consequently depth limitation effects were not significant. This is consistent with a notion that, for a wide range of geophysical water depths, the boundary layers beneath tsunami waves are fundamentally both current-like, due to their long durations, as well as wave-like, in the sense that the boundary layer is unsteady and may not extend over the entire water depth.

This finding is likewise consistent with later simulations of Larsen et al. (2017) as well as measurements made by Larsen et al. (2018), where properly scaled tsunami-scale wave boundary layers were approximately created utilizing long time-varying (pump-driven) currents of the form

$$U_0 = U_{0m}\mathrm{sech}^2(\omega t), \qquad \omega = \frac{2\pi}{T} \tag{5.131}$$

This signal was intended to approximate the expected shallow water flow beneath a long single wave having the form Eq. 5.122. While their study was primarily focused on tsunami-induced scour around a monopile, they conducted three tests measuring the induced pure boundary layer flow in the absence of any structure. The experimental conditions from these cases are presented in Table 5.1. As seen there, the maximum velocity is nearly held constant, with the effective period varied considerably. Note that the periods in each case are $T = O(1\text{ min})$, much longer than typical laboratory-scale flume studies involving wind scale waves, which have periods $T = O(1$ s).

The resulting experimentally measured velocity profiles beneath the peak flow (circles) for each of the three cases considered by Larsen et al.

Table 5.1 Experimental conditions involving tsunami-induced wave boundary layer flows from Larsen et al. (2018). All conditions utilize a depth $h = 0.27$ m and bottom sediment size $d = 0.18$ mm. The boundary layer thickness δ is estimated as the minimum of that calculated from Eq. 5.128 and the water depth h.

Case	T (s)	U_{0m} (m/s)	a (m)	δ (m)	$Re = aU_{0m}/\nu$	a/k_s	δ/h
1	80	0.49	6.2	0.11	3.1×10^6	14,000	0.40
2	123	0.51	10	0.17	5.1×10^6	22,000	0.61
3	497	0.49	39	0.27 (= h)	1.9×10^7	86,000	1.0

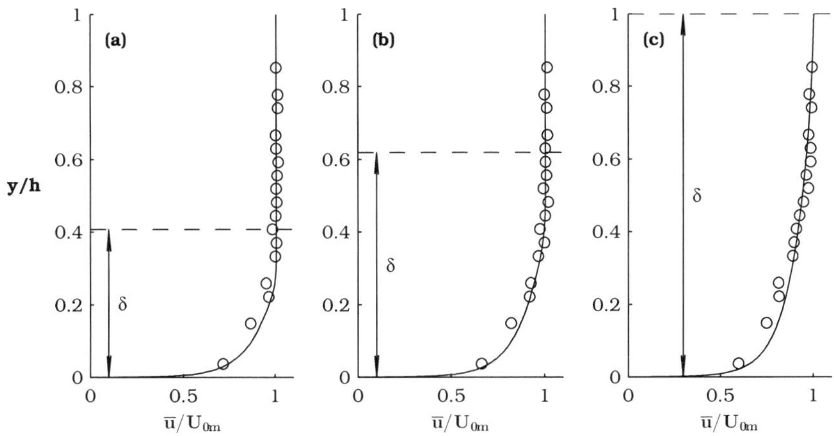

Fig. 5.82 Comparison of measured (circles) and simulated (full lines) velocity profiles at maximum phase for a laboratory-scale tsunami wave boundary layer, approximated as a time varying current. The results are based on a velocity varying approximately as Eq. 5.131 on a depth $h = 0.27$ m, with (a) $T = 80$ s, $U_{0m} = 0.49$ m/s; (b) $T = 123$ s, $U_{0m} = 0.51$ m/s; and (c) $T = 497$ s, $U_{0m} = 0.49$ m/s. The horizontal dashed lines represent the boundary layer thickness approximated via Eq. 5.128 in (a) and (b), and taken as $\delta = h$ in (c).

(2018) are compared with those computed using the k-ω boundary layer model of Fuhrman et al. (2013) (full lines, as described in Section 5.12) in Fig. 5.82. Also shown is the expected tsunami wave boundary layer thickness δ, here estimated as the the minimum of that calculated from Eq. 5.128 and the water depth h. In Cases 1 and 2 (Figs. 5.82a and 5.82b) it is clear that the boundary layer does not span the entirety of the water depth, a situation consistent with our previous discussions. This finding is adequately predicted by both the model results, as well as the empirical formula Eq. 5.128, giving further credence to their validity. In the third case (with very long period $T = 497$ s), the boundary layer thickness predicted by Eq. 5.128 corresponds to approximately $2.1h$, and hence is instead simply taken as $\delta = h$, as seen in Fig. 5.82c. The prediction that the boundary layer in this case would be depth-limited is clearly supported by both the experimental data and numerical results, which show a boundary layer spanning the entirety of the water depth. Hence, this situation can be taken to exemplify conditions further shoreward (or involving longer periods) than considered by Williams and Fuhrman (2016).

For further discussion of depth-limitation effects on wave boundary layers on rough beds the interested reader is referred to the experimental study and analysis of Tanaka et al. (1999).

5.11.2 *Signals based on tsunami observations*

In addition to the idealized tsunami wave signals presented in the previous sub-section, to gain further insight Williams and Fuhrman (2016) likewise simulated the boundary layers that could be expected from flow based directly on free surface observations during real geophysical tsunami events. Following their approach, two measured tsunami signals will be considered in what follows: one from the 2004 Indian Ocean tsunami (measured off the yacht Mercator at approximate depth $h_0 = 14$ m, see e.g. Rabinovich and Thomson, 2007; Madsen and Fuhrman, 2008; Schimmels et al., 2016), and one stemming from the 2011 Tohoku tsunami (measured by the GPS-based station Iwate South at a depth $h_0 = 204$ m, see e.g. Chan and Liu, 2012). The Indian Ocean tsunami occurred as a result of a magnitude 9.1–9.3 earthquake at December 26, 2004, with the epicenter off the west coast of Sumatra, Indonesia. The Tohoku tsunami was similarly the result of a magnitude 9.0–9.1 earthquake on March 11, 2011, with epicenter approximately 70 km off the Pacific coast of Tohoku, Japan.

Chan and Liu (2012) found that the leading wave of the measured free surface elevation $\eta(t)$ time series relating to the 2011 Tohoku tsunami could be well approximated by the superposition of three single waves of varying wave height and period, described by:

$$\eta_0(t) = \sum_{n=1}^{3} A_n \text{sech}^2\left(\omega_n(t - (t_0 + t_n))\right) \tag{5.132}$$

where t_0 is a reference time and t_n represents a time shift. Similarly, Schimmels et al. (2016) found that the leading wave surface elevation corresponding to the 2004 Indian Ocean tsunami could also be reasonably well described through this methodology. The corresponding input coefficients for their descriptions of these two signals based on Eq. 5.132 are given in

Table 5.2 Coefficients for the superposition of single waves used to approximate measured tsunami signals, and the corresponding water depth h_0 at which the signal was measured.

Tsunami	Indian Ocean (2004)	Tohoku (2011)
h_0 (m)	14	204
$\{A_1, A_2, A_3\}$ (m)	$\{-3.10, 3.80, -1.00\}$	$\{-0.80, 2.20, 5.85\}$
$\{\omega_1, \omega_2, \omega_3\}$ (1/s)	$\{0.00420, 0.00530, 0.00900\}$	$\{0.00298, 0.00330, 0.0109\}$
$\{t_1, t_2, t_3\}$ (s)	$\{600, 1000, 1399\}$	$\{580, 980, 1298\}$

Table 5.2, where $t_0 = 0$ in all cases. Also provided are the two reference observation depths h_0.

Comparisons between the measured (circles) and fitted (full lines) signals for these two events are presented in Fig. 5.83a and 5.84a, respectively, which illustrates the close match achieved for the description of the leading tsunami waves. From these, it can also be readily observed that the form of the two tsunami signals are rather different. In relation to the three idealized signals considered in Section 5.11.1, the Tohoku signal (Fig. 5.84a) bares most resemblance to a single elevation wave, while the Indian Ocean signal (Fig. 5.83a) is more reminiscent of a leading-depression N-wave, though it could likewise also be reasonably characterized as sinusoidal. These approximations (appropriately phase shifted) are added to these figures as either dashed or dashed-dotted lines to enable direct comparison.

Following the approach utilized by Williams and Fuhrman (2016), as detailed in Section 5.11.1, the measured surface elevation signals from Eq. 5.132 at depth h_0 can again be converted into expected free stream velocity signals at any other depth h utilizing Eq. 5.124 and Eq. 5.125. These can then be utilized to drive turbulent boundary layer models, such as that presented in Section 5.12. Based on this methodology, computed bed shear stress time series for the Indian Ocean (using $h = h_0 = 14$ m) and Tohoku (using $h = 20$ m) tsunami events are respectively presented in Figs. 5.83b and 5.84b for each of the relevant analytical free surface approximations. Note that all results presented here consider a bed assumed to be comprised of medium sand grains with diameter $d = 0.3$ mm, and with roughness calculated as $k_s = 2.5d = 7.5$ mm. Not surprisingly, the resulting bed shear stresses using the three approximations in these figures only show slight differences.

We will first discuss the results based on the boundary layer simulations of the 2004 Indian Ocean event, Fig. 5.83. It is seen that the bed shear stress magnitude in the trailing elevation wave from Fig. 5.83 is slightly larger

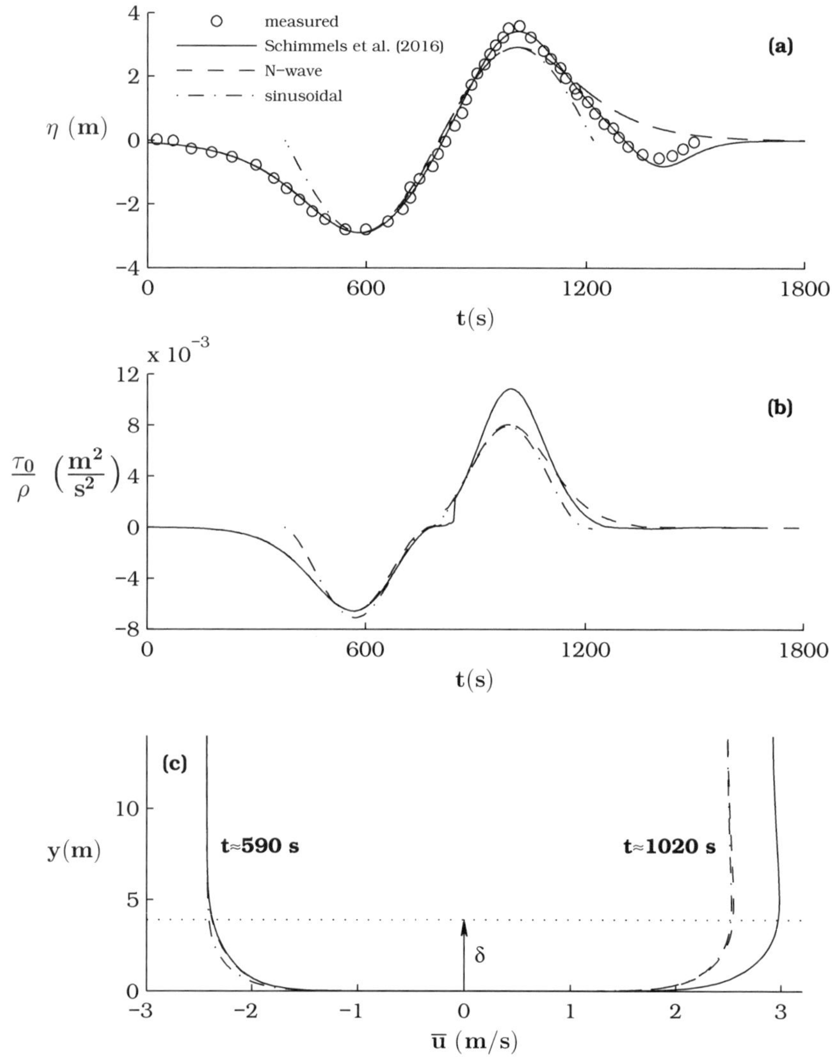

Fig. 5.83 (a) Measured (Mercator yacht signal at depth $h_0 = 14$ m, circles) and approximated (lines) free surface elevation time series from the 2004 Indian Ocean tsunami; The N-wave approximation utilizes $T = 25$ min, $A_0 = -2.905$ m, whereas the sinusoidal approximation utilizes $T = 14$ min, $A_0 = -2.905$ m. (b) Corresponding computed bed shear stress time series. (c) Computed velocity profiles when the flow is minimum and maximum; The depicted boundary layer thickness δ is calculated from Eq. 5.128 based on the sinusoidal period above with $U_{0m} = 2.43$ m/s calculated from Eq. 5.127.

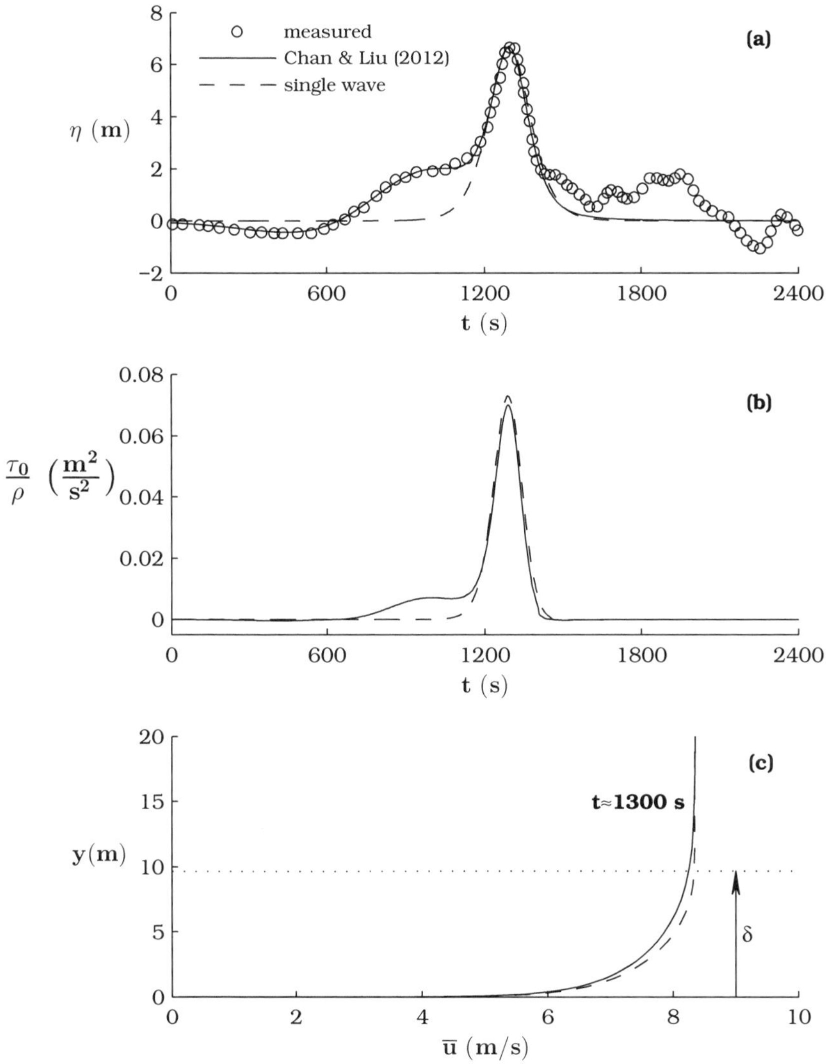

Fig. 5.84 (a) Measured (Iwate South signal at depth $h_0 = 204$ m, circles) and approximated (lines) free surface elevation time series from the 2011 Tohoku tsunami; The single wave approximation utilizes $T = 11.3$ min, $A_0 = 6.7$ m. (b) Corresponding computed bed shear stress time series at depth $h = 20$. (c) Computed velocity profiles when the flow is maximum; The depicted boundary layer thickness δ is calculated from Eq. 5.128 based on the period above with $U_{0m} = 8.35$ m/s calculated from Eq. 5.127.

when based on the approximation of Schimmels et al. (2016), compared to the other signals; This is simply due to the added flexibility inherent within Eq. 5.132, which can directly account for the asymmetric peak and trough amplitudes (hence velocity magnitudes) that occurred during this event. Such asymmetry in the velocity magnitudes is obviously not accounted for in the simpler N-wave and sinusoidal descriptions. Upon closer inspection, it can also be noticed that even under these simpler approximations, that the magnitude of the bed shear stress beneath the trailing elevation wave exceeds that predicted beneath the the leading depression wave. This is due to the transient nature of the tsunami boundary layer, which is undeveloped prior to the arrival of the leading depression wave.

Now turning attention to the simulations based on the 2011 Tohoku tsunami event, similar small deviations are seen when comparing the two computed bed shear stress time series in Fig. 5.84b. The maximum bed shear stress in Fig. 5.84b is slightly larger under the single wave approximation, owing to the somewhat more dramatic acceleration from zero velocity to the peak. Apart from the minor differences discussed, both of these examples based on real-life tsunami signals can be taken as demonstrations that the simple wave forms utilized in Section 5.11.1 are quite useful for idealizing actual geophysical tsunami events.

This conclusion is yet further demonstrated based on the comparisons of the computed instantaneous velocity profiles beneath the tsunami wave peaks and/or troughs, presented in Figs. 5.83c and 5.84c. Regardless of the approximation utilized, all results are quite close to one another, taking into account the previously discussed differences in the trailing elevation (hence free stream velocity) in the approximations of the 2004 Indian Ocean event. Concerning the leading wave in both events, it is seen that the sinusoidal (Fig. 5.83c) and single wave descriptions (Fig. 5.84c) result in slightly larger near-bed velocity gradients than do the other approximations, again the result of the more rapid acceleration from zero to the trough (Fig. 5.83c) and peak (Fig. 5.84c) flow. The boundary layer thickness, approximated from Eq. 5.128, is likewise depicted on both of these plots, and seemingly does a good job of predicting the extent of the boundary layer in both cases. Consistent with the findings of Williams and Fuhrman (2016), the model results suggest that the tsunami-induced boundary layers during these events at the depths considered herein would again only span a fraction of the total water depth. This is an important realization, as it implies that quasi-steady friction laws (which assume that the boundary layer spans the full depth) must be used with great care when describing tsunami-induced flows at depths $O(10 \text{ m})$ or larger.

We finally wish to emphasize that, even though they have commonly been utilized with the intention of studying tsunamis in theoretical and experimental studies dating back to the 1970s, that true solitary waves are a rather poor approximation for the wave forms that can be expected from geophysical tsunamis. As discussed by Madsen et al. (2008), use of the so-called solitary wave paradigm can in fact result in spatial and temporal scales that may be off by several orders of magnitude. Such large deviations are largely due to the fact that in a solitary wave the free surface elevation (adopting the leading-order theory) will be given by

$$\eta_0 = A_0 \text{sech}^2\left(\sqrt{\frac{3A_0}{4h_0^3}}(x - c(t - t_0))\right), \qquad c = \sqrt{g(h_0 + A_0)} \quad (5.133)$$

This represents a special situation in which dispersion and nonlinearity are in balance, resulting in a propagating wave having permanent form (according to potential flow theory). Taking $x = 0$, this expression is equivalent to our single wave signal Eq. 5.122, provided that

$$\omega = \frac{2\pi}{T} = \sqrt{\frac{3gA_0(A_0 + h_0)}{4h_0^3}} \quad (5.134)$$

From this it is clear that under a solitary wave approximation the effective period T is tied to the specific combination of the wave amplitude A_0 and water depth h_0, whereas in a real tsunami these would be independent parameters, largely determined by the generating source. As demonstration of the severe errors that can occur utilizing a solitary wave model to describe a tsunami, consider Fig. 5.85 which again depicts the previously considered Mercator 2004 Indian Ocean signal (from Fig. 5.83a), but now with a solitary wave added with amplitude $A = 3.42$ m, chosen to approximately match the maximum surface elevation. Clearly the duration of the resulting solitary wave ($T = 15.7$ s) is not close to that of the tsunami, and is in fact much closer to expected wind wave scales. This result is consistent with the concerns presented by Madsen et al. (2008), and demonstrates clearly the importance of maintaining independent A and T when utilizing the single wave approximation Eq. 5.122 to represent tsunamis.

5.11.3 *Time variation of the boundary layer thickness and bed shear stress*

Based on results from full-scale CFD simulations of the runup and inundation of tsunamis, Larsen and Fuhrman (2019a,b) extended the expressions

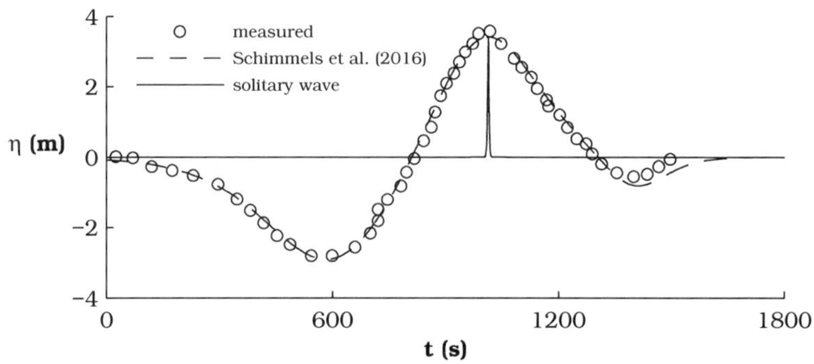

Fig. 5.85 Comparison of the Mercator 2004 Indian Ocean signal with a solitary wave having matching maximum surface elevation. A similar plot can be seen in Fig. 2b of Schimmels et al. (2016).

above for estimating the boundary layer thickness and bed shear stress near the time of maximum free stream flow (Williams and Fuhrman, 2016), Eqs. 5.128–5.129, to yield similar approximations for predicting their time variation.

Their method can quite generally be utilized to estimate the time variation of both the boundary layer thickness and bed shear stress for a given free stream velocity signal $U_0(t)$. In their method, the time variation of the friction velocity (hence, bed shear stress) is given by

$$U_f(t) = \sqrt{\frac{f(t)}{2}}U_0(t), \qquad \frac{\tau_0(t)}{\rho} = U_f(t)\,|U_f(t)| \tag{5.135}$$

where

$$f(t) = \max(f_w(t), f_c) \tag{5.136}$$

is a time-varying friction factor, taken as the maximum of a time-varying (wave-like) contribution, $f_w(t)$, and that from a steady current, f_c. The wave-like contribution is:

$$f_w(t) = \exp\left(5.5\left(\max\left(\frac{a(t)}{k_s}, 1\right)\right)^{-0.16} - 6.7\right) \tag{5.137}$$

which is an obvious extension of Eq. 5.129, with the time varying length scale given as:

$$a(t) = \int_{t_0}^{t} |U_0(t)|dt \tag{5.138}$$

where t_0 is the time corresponding to the most recent free stream flow reversal (i.e. the most recent zero crossing). This will coincide with the standard constant value $a = U_{0m}/\omega$ in the special case where the free stream velocity is sinusoidally varying at the phase $\omega t = 90°$. The steady current contribution is given by

$$f_c = \frac{2}{(C - 2.5\ln(k_s/h))^2} \tag{5.139}$$

where $C = 8.5$ and h is the water depth. (This is essentially the Colebrook-White formula for predicting the steady flow resistance above above a rough wall. The coefficient C is changed from the traditional $C = 6.5$ because the present formulation is based on the free stream velocity i.e. that at the top of the boundary layer, rather than a boundary-layer-averaged velocity.)

The time-varying boundary layer thickness was similarly estimated by

$$\delta(t) = \min\left(0.05a(t)^{0.89}k_s^{0.11}, h\right) \tag{5.140}$$

which is an obvious extension of Eq. 5.128. Similar to the friction factor above, the first argument represents the time-varying wave boundary layer thickness, which is not allowed to become larger than the water depth (and second argument) h.

The formulation above naturally defaults to utilize the wave-like expressions in situations where the boundary layer does not exceed the water depth, and similarly defaults to a steady current formulation at the limit where the boundary layer becomes depth-limited (both scenarios are realistic for tsunami-induced flows, depending on the water depth and duration of the tsunami).

While the above formulation was originally developed to match the computed variations of $\delta(t)$ and $\tau_0(t)$ from full-scale tsunami simulations, Larsen and Fuhrman (2019b) showed that it likewise works quite well e.g. in predicting the time variation of the bed shear stresses beneath short waves as well. See their appendix, where the simple formulation was successfully tested for predicting the time variation of the bed shear stress beneath free stream velocities for regular waves, which were both sinusoidal (Jensen et al., 1989) as well as acceleration-skewed (van der A et al., 2011). They likewise successfully tested the method for the time-varying bed shear stress induced by irregular wave free stream velocity signals, which were both velocity- and acceleration-skewed (Yuan and Dash, 2017). (For further discussion of the boundary layer beneath non-sinusoidal wave shapes, see Section 6.4.)

The formulation above is designed to be valid in coastal regions which are fully submerged. For a model describing the bed shear stress, surface shape and velocity field near the tip of a tsunami (or dam break, etc.) during runup, see the recent work of Nielsen (2018).

5.11.4 *Comparison with tsunami field observations in the presence of energetic wind waves*

As a final example, we will now compare and reconcile the predictions based on the numerical simulations of Williams and Fuhrman (2016), from Section 5.11.1, directly against field observations of Lacy et al. (2012), who described the currents and suspended sediment transport in Monterey Bay, California associated with the arrival of the 2010 tsunami originating due to a magnitude 8.8 earthquake with epicenter offshore of Maule, Chile. The earthquake occurred on the early morning of February 27, 2010, and the resulting tsunami arrived in Monterey Bay approximately 14 hours later. The data collected by Lacy et al. (2012) represents the first reported field observations of tsunami-generated currents that include the near-bed region, and are hence an excellent opportunity for comparison.

Lacy et al. (2012) report that the largest observed tsunami oscillation corresponded to a wave amplitude of approximately $A_0 = 0.55$ m and a period $T = 16$ min, on a depth of approximately $h_0 = 9$ m, resulting in a peak tsunami-induced velocity of $U_{0m} = 0.36$ m/s. From this event they estimated a maximum tsunami wave boundary layer thickness of $\delta = 0.79$ m, which is slightly less than 10% of the water depth, again representing conditions that are not depth limited, consistent with much of the preceding discussion. Additionally, based on the measured near-bed velocity profiles they estimated a peak tsunami-induced bed shear stress of $\tau_m = 0.4$ Pa. They also noted the presence of an energetic local wind wave climate at the time of tsunami arrival, resulting in an estimated maximum wind wave-induced shear stress $\tau_{0m} = 2.5$ Pa ($= \tau_w$ in the notation used in Section 5.6.7; This notation will be maintained in what follows). Based on these, Lacy et al. (2012) estimated a maximum total (wind wave plus tsunami) bed shear stress corresponding to $\tau_{max} = 2.9$ Pa.

To compare with their observations, we must first approximate the bed roughness. Lacy et al. (2012) report a sediment size $d = 0.315$ mm (medium sand) in the observation region. Although Lacy et al. (2012) discuss the presence of bed ripples in the vicinity of the observation point, for simplicity in the present example we will calculate the roughness as $k_s = 2.5d =$

0.788 mm. Utilizing their observed tsunami period $T = 2\pi/\omega$ and velocity magnitude U_{0m}, yields the length scale $a = U_{0m}/\omega = 55.0$ m (notice that this is much larger than conventional wind waves!). Invoking this directly within Eq. 5.128 then results in a predicted tsunami wave boundary layer thickness of $\delta = 0.81$ m, in excellent agreement with their estimate.

As a next step, let us estimate the bed shear stress that would be induced by the tsunami in the absence of any wind waves. For this purpose, direct application of Eq. 5.129 yields $f_w = 0.00310$. Utilizing the density $\rho = 1025$ kg/m^3, this corresponds to a peak bed shear stress of $\tau_c = f_w \rho U_{0m}^2/2 = 0.21$ Pa. This is only approximately half of that reported by Lacy et al. (2012) (called τ_m above, for reasons which will become clear).

It must be emphasized that the predicted peak bed shear stress from Eq. 5.129, resulting in τ_c above, strictly considers the idealized situation where the tsunami-induced flow is in isolation. However, as mentioned above, Lacy et al. (2012) likewise report that at the time of the tsunami arrival the surface wind waves were also quite energetic, with the maximum bed shear stress due to waves being approximately $\tau_w = 2.5$ Pa, nearly five times greater than that induced by the (in this case, locally relatively small) tsunami. Hence, in this context, we must in fact expect that the interaction of the tsunami flow with the short-duration wind waves will more closely resemble that within classical wave-current interaction (as treated throughout Section 5.6, and especially in Section 5.6.7), with the long-duration tsunami effectively playing the role of the current.

Maintaining this analogy, and treating the peak tsunami-induced flow in a quasi-steady fashion, we can therefore more properly re-estimate the tsunami-induced bed shear stress utilizing the empirical wave-plus-current decomposition of Soulsby (1995) (see Section 5.6.7)

$$\tau_m = \tau_c \left(1 + 1.2 \left(\frac{\tau_w}{\tau_c + \tau_w} \right)^{3.2} \right) \tag{5.141}$$

Here τ_m can be taken to represent the peak bed shear stress due to the tsunami-induced current in the presence of waves, whereas τ_c represents the peak bed shear stress expected due to the tsunami (current) in isolation, as calculated previously from Eq. 5.129, and τ_w is the maximum bed shear stress due to the waves (this notation is that same as that used in Section 5.6.7). Hence, taking $\tau_c = 0.21$ Pa from our original estimate above, and taking the reported value from the waves $\tau_w = 2.5$ Pa as given, yields a peak tsunami-induced bed shear stress of $\tau_m = 0.43$ Pa. This corrected result is now very close to the 0.4 Pa reported by Lacy et al. (2012)!

The significant increase in the bed shear stress predicted by Eq. 5.141, relative to that predicted naively by Eq. 5.129, can be interpreted as due to the additional apparent wave-induced roughness felt by the tsunami-induced flow. Based on the tsunami-induced bed shear stress $\tau_m = 0.43$ Pa calculated above, the corresponding wave friction factor would be $f_{w,m} = 2\tau_m/(\rho U_{0m}^2) = 0.0065$. Invoking this in the left-hand-side of Eq. 5.129 and solving for the roughness (replacing k_s with k_w) then yields the apparent wave-induced roughness $k_w = 0.032$ m, which is roughly 40 times that estimated based solely on the grain size, k_s. This also exceeds (by more than a factor 4) the value 0.0072 m estimated by Lacy et al. (2012) based solely on observed ripple dimensions, but not invoking a combined wave-current analogy.

The maximum bed shear stress in the combined tsunami current-plus-wind wave flow can finally generally be estimated through vector addition, according to

$$\tau_{max} = \sqrt{(\tau_m + \tau_w \cos\phi)^2 + (\tau_w \sin\phi)^2} \tag{5.142}$$

where ϕ defines the angle between the tsunami-current and wind wave propagation direction, see Eq. 5.87. Since, according to Lacy et al. (2012) these were effectively co-linear, we may simply take $\phi = 0°$, by which this reduces to

$$\tau_{max} = \tau_m + \tau_w \tag{5.143}$$

This yields $\tau_{max} = 2.9$ Pa, which again matches precisely the value reported by Lacy et al. (2012).

This example demonstrates how the tsunami-alone wave boundary layer predictions taken from the numerical study of Williams and Fuhrman (2016) can likewise be further utilized to predict the tsunami-induced shear stress in the presence of shorter wind waves, based on a wave-plus-current boundary layer flow analogy. In this example the bed shear stress from the waves (τ_w in the above) was again taken as given, though if this too was unknown it could likewise be calculated from Eq. 5.69 $\equiv$ Eq. 5.129, as this expression holds at both scales.

5.12 Mathematical modelling of turbulent wave boundary layers

5.12.1 *One-dimensional vertical (1DV) RANS equations*

While for more complicated problems involving complex domains require solution of full two- or three-dimensional RANS equations coupled with appropriate turbulence closure models, for a detailed account see Chapter 9. Many problems involving turbulent wave boundary layer flows can, however, be adequately treated utilizing so-called one-dimensional-vertical (1DV) models. Formally developing such an approach will be the focus of the present section.

Considering the horizontal component of the RANS equations, e.g. Eq. 9.2 while invoking the Boussinesq approximation Eq. 9.30, assuming uniformity in the transverse z-direction, and finally invoking local continuity e.g. Eq. 9.1, leads in full to:

$$\begin{aligned}\frac{\partial \overline{u}}{\partial t} + \overline{u}\frac{\partial \overline{u}}{\partial x} + \overline{v}\frac{\partial \overline{u}}{\partial y} = &-\frac{1}{\rho}\frac{\partial \overline{p}}{\partial x} + \nu\left(\frac{\partial^2 \overline{u}}{\partial x^2} + \frac{\partial^2 \overline{u}}{\partial y^2}\right) - \frac{2}{3}\frac{\partial k}{\partial x} \\ &+ 2\frac{\partial}{\partial x}\left(\nu_t \frac{\partial \overline{u}}{\partial x}\right) + \frac{\partial}{\partial y}\left(\nu_t\left(\frac{\partial \overline{u}}{\partial y} + \frac{\partial \overline{v}}{\partial x}\right)\right) \qquad (5.144)\end{aligned}$$

However, for typical problems involving wave boundary layers, many of the terms above will be rather insignificant. For a wave propagating in the horizontal x-direction, it is quite reasonable to assume that vertical y variations in the boundary layer are much stronger than those in the horizontal x direction, and hence that vertical velocities are likewise small relative to horizontal velocities. Moreover, it is likewise reasonable to assume that the turbulent kinetic energy density $k = 1/2\overline{u_i'u_i'}$ is typically small relative to $\overline{u}^2$. Formally, this can be achieved by invoking the following substitutions within Eq. 5.144:

$$\begin{aligned} x \to Lx^*, \quad y \to ay^*, \quad t \to \frac{a}{U_{0m}}t^*, \\ \overline{u} \to U_{0m}\overline{u}^*, \ \overline{v} \to \epsilon U_{0m}\overline{v}^*, \ \epsilon = a/L, \\ \overline{p} \to \frac{\rho U_{0m}^2}{\epsilon}\overline{p}^*, \ \nu_t \to U_{0m}a\nu_t^*, \ k \to f_w U_{0m}^2 k^* \qquad (5.145)\end{aligned}$$

which transforms the dimensional variables to properly scaled dimensionless counterparts having asterisk superscript. These substitutions, after some simplification, leads straightforwardly to:

$$\frac{\partial \overline{u}^*}{\partial t^*} = -\epsilon \overline{u}^* \frac{\partial \overline{u}^*}{\partial x^*} - \epsilon \overline{v}^* \frac{\partial \overline{u}^*}{\partial y^*} - \frac{\partial \overline{p}^*}{\partial x^*} + \frac{1}{Re}\left(\epsilon^2 \frac{\partial^2 \overline{u}^*}{\partial x^{*2}} + \frac{\partial^2 \overline{u}^*}{\partial y^{*2}}\right) - \epsilon f_w \frac{2}{3}\frac{\partial k^*}{\partial x^*}$$
$$+ \epsilon^2 2 \frac{\partial}{\partial x^*}\left(\nu_t^* \frac{\partial \overline{u}^*}{\partial x^*}\right) + \frac{\partial}{\partial y^*}\left(\nu_t^* \left(\frac{\partial \overline{u}^*}{\partial y^*} + \epsilon^2 \frac{\partial \overline{v}^*}{\partial x^*}\right)\right) \tag{5.146}$$

from which the relative importance of each of the terms is now readily apparent. In line with the previously stated assumptions, the parameter ϵ must be small, and we may also take $f_w = O(\epsilon)$ i.e. that the wave friction coefficient is also small, as is typically the case. Retaining only terms up to $O(\epsilon)$ in Eq. 5.146, and transforming back to the dimensional variables, leads to the simplified equation

$$\boxed{\frac{\partial \overline{u}}{\partial t} = -\frac{1}{\rho}\frac{\partial \overline{p}}{\partial x} + \nu \frac{\partial^2 \overline{u}}{\partial y^2} + \frac{\partial}{\partial y}\left(\nu_t \frac{\partial \overline{u}}{\partial y}\right) - \underline{\left(\overline{u}\frac{\partial \overline{u}}{\partial x} + \overline{v}\frac{\partial \overline{u}}{\partial y}\right)}} \tag{5.147}$$

This equation will be called the **1DV RANS equation**. [Note that the framing of the equation above with the "box" (in additional to several equations which follow), is used to highlight the resulting final governing equations for the 1DV model formulation; These framed governing equations are the basis e.g. for the turbulence modelling hands-on Exercise 4 in Chapter 10.]

Doing similarly in the vertical direction, the y-component of Eq. 9.2 may be written in full:

$$\frac{\partial \overline{v}}{\partial t} + \overline{u}\frac{\partial \overline{v}}{\partial x} + \overline{v}\frac{\partial \overline{v}}{\partial y} = -\frac{1}{\rho}\frac{\partial \overline{p}}{\partial y} + \nu\left(\frac{\partial^2 \overline{v}}{\partial x^2} + \frac{\partial^2 \overline{v}}{\partial y^2}\right) - \frac{2}{3}\frac{\partial k}{\partial y}$$
$$+ \frac{\partial}{\partial x}\left(\nu_t\left(\frac{\partial \overline{v}}{\partial x} + \frac{\partial \overline{u}}{\partial y}\right)\right) + 2\frac{\partial}{\partial y}\left(\nu_t \frac{\partial \overline{v}}{\partial y}\right) \tag{5.148}$$

Invoking the substitutions Eq. 5.145 into Eq. 5.148 leads, after tedious but straight-forward simplification, to

$$\epsilon^2 \frac{\partial \overline{v}^*}{\partial t^*} = -\epsilon^3 \overline{u}^* \frac{\partial \overline{v}^*}{\partial x^*} - \epsilon^3 \overline{v}^* \frac{\partial \overline{v}^*}{\partial y^*} - \frac{\partial \overline{p}^*}{\partial y^*} + \frac{1}{Re}\left(\epsilon^4 \frac{\partial^2 \overline{v}^*}{\partial x^{*2}} + \epsilon^2 \frac{\partial^2 \overline{v}^*}{\partial y^{*2}}\right) - \epsilon f_w \frac{2}{3}\frac{\partial k^*}{\partial y^*}$$
$$+ \frac{\partial}{\partial x^*}\left(\nu_t^*\left(\epsilon^4 \frac{\partial \overline{v}^*}{\partial x^*} + \epsilon^2 \frac{\partial \overline{u}^*}{\partial y^*}\right)\right) + \epsilon^2 2 \frac{\partial}{\partial y^*}\left(\nu_t^* \frac{\partial \overline{v}^*}{\partial y^*}\right) \tag{5.149}$$

Hence, to the same order of approximation as maintained previously, i.e. retaining terms up to $O(\epsilon)$, the vertical component of the RANS equations

Eq. 5.148 reduces dramatically to:

$$\boxed{\frac{1}{\rho}\frac{\partial \overline{p}}{\partial y} = 0} \tag{5.150}$$

That is, the pressure in the boundary layer will be vertically constant! As this results in a trivial governing equation, it does not require separate solution in its own right; This result, however, is very important as it implies that the pressure gradient term in Eq. 5.147 can be taken as vertically uniform i.e. the pressure in the boundary layer may be taken as equal to that just above in the free-stream region. This has been utilized previously throughout the present chapter.

Equation Eq. 5.147 incorporates, to leading-order, effects due to variations in the propagating x-direction (convective acceleration) and, as will be shown, adequately describes wave boundary layers for many situations of practical interest. Note that in Eq. 5.147 and Eq. 5.150 leading-order terms of $O(1)$ are unmarked, whereas higher-order terms of $O(\epsilon)$ are underlined. For conditions corresponding to oscillating tunnels, as commonly considered experimentally, the $O(\epsilon)$ terms naturally drop out due to uniformity in the x-direction. These higher-order terms are essential, however, for describing other effects e.g. the phenomenon of so-called boundary layer streaming, which is the focus of Chapter 6. Note that even though there are x-derivatives appearing in Eq. 5.147, these may be approximated (at least in the case of regular waves) in practice, and this equation can hence conveniently be solved numerically requiring resolution of only a single (vertical) direction, i.e. within a 1DV framework.

5.12.2 *Treatment of convective acceleration terms*

To solve Eq. 5.147 in a 1DV setting requires that the convective acceleration (single underlined) terms can be adequately approximated, when they are of interest. It can be noted that, while these convective terms are of no importance within (horizontally uniform) oscillatory tunnel flow environments, they will be non-zero whenever there is spatial non-uniformity in the x-direction. Such non-uniformity arises naturally e.g. due to: (1) spatial variations in the propagating wave itself (and therefore its boundary layer) as well as (2) due to sloping beds, which will create an oscillatory converging-diverging flow. Methods for approximating the convective acceleration terms due to both of these effects are described below.

For monochromatic (i.e. periodic with a single frequency) waves the convective terms can be evaluated by assuming constant form wave propagation. This allows x-derivatives to be linked to time derivatives via the relation

$$\frac{\partial}{\partial x} = -\frac{1}{c}\frac{\partial}{\partial t} \tag{5.151}$$

Here c is the wave celerity (i.e. the speed of wave propagation), which can typically be found by solution of the linear dispersion relation, e.g. Eq. 5.182. Similar treatment of the horizontal derivative has been utilized by many authors e.g. Trowbridge and Madsen (1984), Brøker (1985), Justesen (1988a), Deigaard et al. (1999), and Holmedal and Myrhaug (2009), among others.

Let us now also consider the effects of a sloping bed on an oscillatory wave boundary layer e.g. as would occur in an converging-diverging flow in an oscillating tunnel. Such a convergent tunnel (with local half-depth h at $x = 0$ and slope $S = \tan\beta = -dh/dx$ leads to the depth variation

$$h(x) = h - Sx \tag{5.152}$$

For a particular discharge (per unit width) q, the mean horizontal velocity (here taken as the free stream velocity U_0, assuming small boundary layer thickness) will then vary according to

$$U_0(x) = \frac{q}{h(x)} = \frac{q}{h - Sx} = U_0(x=0) + U_0\frac{Sx}{h} + O\left(\frac{x^2S^2}{h^2}\right). \tag{5.153}$$

Hence, at $x = 0$ (taken as our local point of interest) we have

$$\frac{\partial U_0}{\partial x} = \frac{SU_0}{h} \qquad \text{or} \qquad \frac{\partial}{\partial x} = \frac{S}{h}, \tag{5.154}$$

i.e. differentiating with respect to x can be locally approximated by multiplication with S/h.

Combining these two effects, we may then approximate x-derivatives appearing in Eq. 5.147 with

$$\boxed{\frac{\partial}{\partial x} = \underline{\underline{-\frac{1}{c}\frac{\partial}{\partial t}}} + \underline{\underline{\underline{\frac{S}{h}}}}.} \tag{5.155}$$

Here the double-underlined term stems from wave-propagation effects, whereas the triple-underlined term stems from sloping bed effects. This convention will be maintained in what follows.

Evaluation of the convective terms in Eq. 5.147 likewise requires determination of the induced vertical velocity $\overline{v}$. This may be approximated by solving the local continuity equation, while invoking Eq. 5.155, leading to

$$\frac{\partial \overline{u}}{\partial x} + \frac{\partial \overline{v}}{\partial y} \approx \underline{\underline{-\frac{1}{c}\frac{\partial \overline{u}^{(1)}}{\partial t}}} + \underline{\underline{\underline{\frac{S\overline{u}}{h}}}} + \frac{\partial \overline{v}}{\partial y} = 0 \tag{5.156}$$

Here $\partial \overline{u}^{(1)}/\partial t$ can conveniently be taken as the leading-order (i.e. non-underlined) part of $\partial \overline{u}/\partial t$ from Eq. 5.147. This is justifiable because all terms containing $\overline{v}$ in the governing equations are already of secondary importance, hence any errors made by this approximation will be at the same order of importance as the terms already neglected in Eq. 5.147. Based on Eq. 5.156, the vertical velocity field for all y can be determined through cumulative integration of

$$\boxed{\overline{v} = \int_0^y \left(\underline{\underline{\frac{1}{c}\frac{\partial \overline{u}^{(1)}}{\partial t}}} - \underline{\underline{\frac{S\overline{u}}{h}}} \right) dy} \tag{5.157}$$

From the approximations described above, given the instantaneous velocity field $\overline{u}$, convective terms appearing in Eq. 5.147 can be evaluated, while still only requiring discretization in the vertical y-direction, as required in a 1DV framework.

5.12.3 *Pressure gradient*

For simulation of wave boundary layers, the pressure gradient term in Eq. 5.147 can be utilized to drive a desired free stream velocity U_0. In an idealized free stream region taken as just outside the boundary layer shear stresses (both viscous and turbulent) can be neglected. Consequently, the velocity gradient $\partial U_0/\partial y$ may likewise be neglected, and in this region the governing 1DV RANS equation Eq. 5.147 reduces to

$$-\frac{1}{\rho}\frac{\partial \overline{p}}{\partial x} = \frac{\partial U_0}{\partial t} + \underline{U_0 \frac{\partial U_0}{\partial x}} \tag{5.158}$$

It is again emphasized that, due to Eq. 5.150, the driving pressure gradient is taken as vertically uniform i.e. its free stream value may be justifiably utilized throughout the boundary layer. After invoking Eq. 5.155, this becomes

$$\boxed{-\frac{1}{\rho}\frac{\partial \overline{p}}{\partial x} = \left(1 - \underline{\underline{\frac{U_0}{c}}}\right)\frac{\partial U_0}{\partial t} + \underline{\underline{\frac{SU_0^2}{h}}} - P_x} \tag{5.159}$$

where a constant pressure gradient P_x has also been added for generality, which can be utilized to drive additional steady currents, either in isolation, for combined wave-plus-current flows, or to counter induced period-averaged drifts that can arise within wave boundary layer simulations. Unless otherwise stated, $P_x = 0$ and $S = 0$ in what follows in the present chapter (these features will be utilized in Chapter 6).

From Eq. 5.159 it is seen that to drive a desired free stream flow U_0 likewise requires specification of the free stream acceleration $\partial U_0/\partial t$. For the simplest cases involving boundary layers beneath an oscillatory (sinusoidal) free stream signal, these would correspond simply to

$$U_0 = U_{0m}\sin(\omega t) \tag{5.160}$$

$$\frac{\partial U_0}{\partial t} = U_{0m}\omega\cos(\omega t) \tag{5.161}$$

where $\omega = 2\pi/T$ is the angular frequency of the wave. Other, more general, free stream flow and acceleration combinations useful for driving e.g. non-sinusoidal wave shapes are presented later e.g. in the pairs Eqs. 6.68, 6.69 and Eqs. 6.81, 6.82.

Note also that a steady current can be driven in the 1DV framework described above simply by setting $P_x = -U_f^2/h$, where U_f is the desired friction velocity that will be reached at equilibrium i.e. once the driving pressure gradient is balanced by shear stresses and there is no longer any acceleration.

5.12.4 *Two-equation k-ω turbulence closure*

To evaluate the right-hand-side of Eq. 5.147 one must resort to some form of turbulence model, in order to obtain the eddy viscosity ν_t. We will here focus on the (Wilcox, 2006) k-ω model, presented in full form in Section 9.4. Based on analogous scaling as described above and leading to Eq. 5.147, we may simplify Eq. 9.35 to the following 1DV transport equation for the turbulent kinetic energy (per unit mass), $k = (1/2)\overline{u_i'u_i'}$:

$$\boxed{\frac{\partial k}{\partial t} = \nu_t\left(\frac{\partial \overline{u}}{\partial y}\frac{\partial \overline{u}}{\partial y}\right) - \beta^* k\omega + \frac{\partial}{\partial y}\left[\left(\nu + \sigma^*\frac{k}{\omega}\right)\frac{\partial k}{\partial y}\right] - \underline{\left(\overline{u}\frac{\partial k}{\partial x} + \overline{v}\frac{\partial k}{\partial y}\right)}} \tag{5.162}$$

This form is appropriate for use in the 1DV modelling framework, with x-derivatives and vertical velocity again approximated as described above. Similarly, simplification of Eq. 9.36 yields the following transport equation for the specific rate of dissipation (also termed ω):

$$\boxed{\begin{aligned}\frac{\partial \omega}{\partial t} = \; & \alpha\frac{\omega}{k}\nu_t\left(\frac{\partial \overline{u}}{\partial y}\frac{\partial \overline{u}}{\partial y}\right) - \beta\omega^2 + \frac{\partial}{\partial y}\left[\left(\nu + \sigma\frac{k}{\omega}\right)\frac{\partial \omega}{\partial y}\right] \\ & + \frac{\sigma_d}{\omega}\frac{\partial k}{\partial y}\frac{\partial \omega}{\partial y} - \underline{\left(\overline{u}\frac{\partial \omega}{\partial x} + \overline{v}\frac{\partial \omega}{\partial y}\right)}\end{aligned}} \tag{5.163}$$

Simplification of Eq. 9.37 likewise yields the eddy viscosity ν_t:

$$\boxed{\nu_t = \alpha^* \frac{k}{\tilde{\omega}}, \qquad \tilde{\omega} = \max\left\{\omega, \lambda_1 \frac{|\partial \overline{u}/\partial y|}{\sqrt{\beta_0^*/\alpha^*}}\right\}} \tag{5.164}$$

The standard "high Reynolds number" model closure coefficients are as defined in Section 9.4.1, Eqs. 9.39–9.41, taking $f_\beta = 1$. Alternatively, a "low Reynolds number" model variation (capable of describing laminar to turbulence transition, obviously still working within the confines of a Reynolds-averaged flow description) can be obtained by simply re-defining selected closure coefficients to be flow dependent, as indicated in Eqs. 9.43–9.46. Results from both of these models in the present 1DV framework will be presented in what follows.

5.12.5 *Boundary conditions*

As boundary conditions, a frictionless rigid lid may be conceptually imposed at the top boundary, whereby vertical derivatives of $\overline{u}$, k, and ω are all set to zero. The top boundary must lie sufficiently far from the bottom to lie in the free stream region (typically placed several times the wave boundary layer thickness δ above the bed). The bottom boundary is taken as a friction wall, and thus a no-slip boundary condition is imposed, i.e. $\overline{u} = \overline{v} = 0$. Bottom boundary conditions for k and ω are applied directly on the wall surface, as outlined in Section 9.4.2. Results using the standard k-ω model utilize a $\partial k/\partial y = 0$ bottom boundary condition, whereas results utilizing the "low Reynolds number" model use $k = 0$. For further discussion of boundary conditions in such models, see Section 9.4.2.

5.12.6 *Numerical solution*

Notice that governing time-stepping equations Eq. 5.147, Eq. 5.162, and Eq. 5.163 have been presented with time derivatives on the left hand side, and all other terms grouped on the right hand side. The system of partial differential equations then lends itself readily to numerical solution via the so-called *method of lines*. Within this approach, spatial derivatives (all on the right hand sides) are discretized independently of the time derivatives (left hand sides). Vertical spatial derivatives are evaluated utilizing standard finite difference approximations. Starting from specified initial conditions, time marching of the PDEs can then be handled based on any suitable standard time integration method.

The present implementation has been developed within the Matlab environment, with time integration conveniently achieved using the built in `ode15s` function, designed for the solution of stiff equations, including built-in adaptive time stepping and error control. The resulting code is called the "MatRANS" model, originally described by Fuhrman et al. (2013), with subsequent developments and applications made by Williams and Fuhrman (2016), Kirca et al. (2016), and Caliskan and Fuhrman (2017). It is available at (filename: `MatRANS.zip`):

`https://doi.org/10.11583/DTU.c.4508648`

including several example cases already set up. This model is also intended to be utilized within (parts of) several of the hands on Exercises in Chapter 10, as described there.

When solving boundary layer problems, it is imperative that a computational grid be utilized that provides adequate resolution of the near-wall physics. Guidelines for ensuring proper resolution near walls are provided in Section 9.4.2, which are adhered to in what follows.

5.12.7 *Standard Wilcox (2006) k-ω closure model results*

To demonstrate applications based on the standard k-ω turbulence closure model (i.e. with closure coefficients taken as constants, as discussed in Section 9.4.1.1, Eq. 9.39–9.41), we will simulate conditions based on selected high Reynolds number experimental tests originally conducted by Jensen et al. (1989) in the oscillating tunnel facility at the Technical University of Denmark, Fig. 5.21. Specifically, two conditions will be considered, both having Reynolds number $Re = 6 \times 10^6$, with one on a smooth bed (corresponding to their Test 10) and one on a rough bed (sand paper, with $k_s = 0.18$ mm, corresponding to their Test 13). Dating back to Justesen (1988b) these cases have been widely utilized for testing the ability of various turbulence closure models to describe turbulent oscillatory boundary layer flows. Note that in these simulations convective acceleration (underlined) terms are switched off, due to the horizontal uniformity in oscillating tunnel facilities (they are included in several other applications of this model, described in Chapter 6).

Let us first compare the smooth bed results. Comparison is made for the resulting velocity (Fig. 5.86), turbulent kinetic energy density (Fig. 5.87),

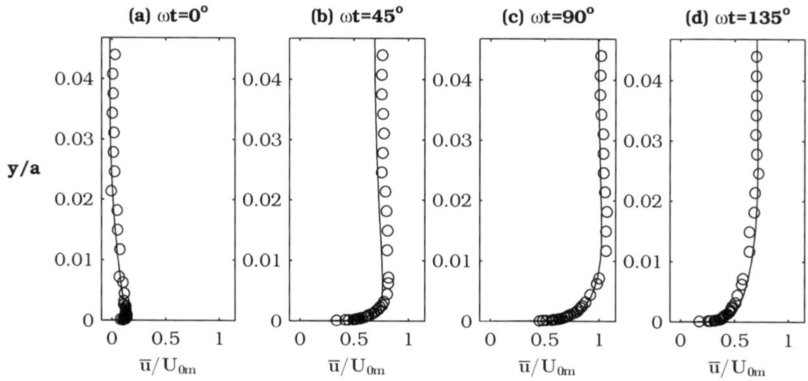

Fig. 5.86 Comparison of measured (circles, Test 10 of Jensen et al., 1989, with U_{0m} = 1.97 m/s, T = 9.72 s; $Re = 6 \times 10^6$, smooth wall) and computed (full lines) velocity profiles at selected phases within a turbulent oscillatory boundary layer.

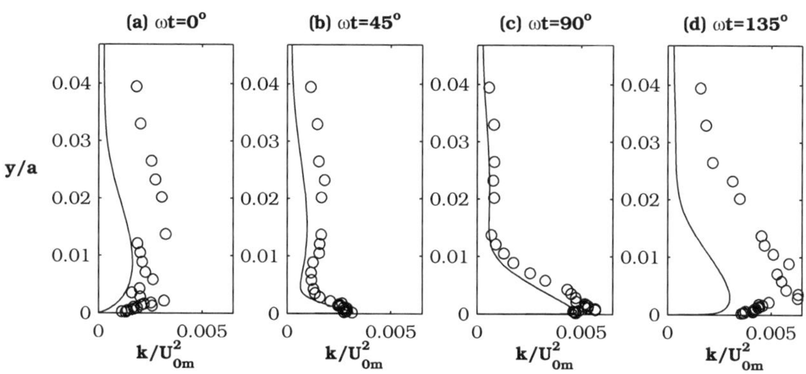

Fig. 5.87 Comparison of measured (circles, Test 10 of Jensen et al., 1989, smooth wall) and computed (full lines) turbulent kinetic energy density profiles at selected phases.

and Reynolds stress (Fig. 5.88) profiles. (These experiments have been discussed earlier in conjunction with Figs. 5.21–5.25). In each comparison, results are depicted just after near-bed flow reversal ($\omega t = 0°$), during the acceleration stage ($\omega t = 45°$), at peak free stream flow ($\omega t = 90°$), and during the deceleration stage ($\omega t = 135°$). It should be noted that the measurements of Jensen et al. (1989) consisted of two velocity components, hence for the sake of comparison these have been converted to turbulent

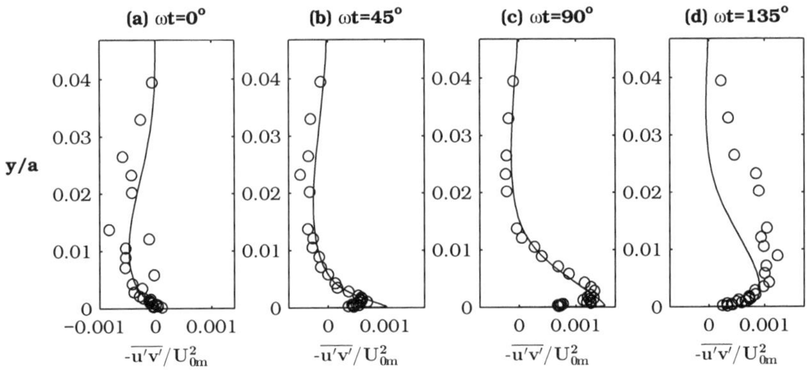

Fig. 5.88 Comparison of measured (circles, Test 10 of Jensen et al., 1989, smooth wall) and computed (full lines) Reynolds stress profiles at selected phases.

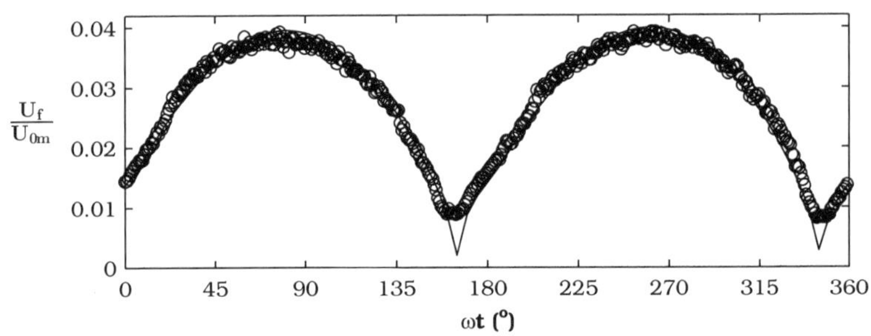

Fig. 5.89 Comparison of measured (circles, Test 10 of Jensen et al., 1989, smooth wall) and computed (full lines) friction velocity time series.

kinetic energy via the approximation

$$k = 0.65\left(\overline{u'^2} + \overline{v'^2}\right) \tag{5.165}$$

as suggested by Justesen (1991). From Fig. 5.86 it is seen that the model makes reasonable predictions of the velocities throughout all stages of the wave boundary layer. Furthermore, based on Figs. 5.87 and 5.88 it is seen that the turbulence properties are likewise quite well produced, with the exception of the deceleration stage ($\omega t = 135°$) where both the turbulent kinetic energy and Reynolds stresses are under-predicted. Importantly, even at this stage the model accurately predicts the Reynolds stresses very near the wall, as also illustrated by the excellent comparison between computed and directly measured friction velocities depicted in Fig. 5.89. (Direct measurement of friction velocities was achieved using the hot-film technique, see Section 5.3.1.)

Similar comparison will now be made for the rough bed case (Test 13, with $Re = 6 \times 10^6$, $a/k_s = 3700$). (These experiments have been discussed earlier in conjunction with Figs. 5.34–5.37.) As before, the measured and computed velocity (Fig. 5.90), turbulent kinetic energy density (Fig. 5.91), and Reynolds stress (Fig. 5.92) profiles are presented. Results for the rough wall case are of similar quality as in the smooth bed case, generally comparing quite well with measurements. During the deceleration stage it is again seen that the velocity is slightly over-predicted, whereas the turbulent kinetic energy k is under-predicted. This implies that during the adverse pressure gradient the model is not extracting enough energy from the mean flow, and is apparently hence indicative of poorer accuracy of the eddy viscosity at these times. It is again emphasized, that the Reynolds stress very near the wall remains quite accurate, which is again illustrated via the comparison with the measured friction velocities in Fig. 5.93. The friction velocity in the experiments (Jensen et al., 1989, their Fig. 7) was determined by fitting straight lines to the logarithmic-layer portion of the mean velocity distribution (see Section 5.4.1). Note that the experimental differences in the two half cycles seen there are apparently due to asymmetries that occurred in the experiments, and which are not present in the numerical simulations, which expectedly result in essentially identical results for both positive and negative half-cycles of the oscillatory flow. The close match in the friction velocity for the first half cycle (where the velocities are also in reasonable agreement) can hence be considered as acceptable validation.

Through comparison of the smooth and rough bed cases just presented, it is clear that the turbulence closure faithfully captures the effects associated with bottom roughness i.e. reduced Reynolds averaged velocities, increased near-wall turbulence, and increased bed shear stresses. See the discussion in Section 5.4.2.

5.12.8 *Low Reynolds number Wilcox (2006) k-ω closure model results*

To demonstrate the performance of the "low Reynolds number" k-ω turbulence model (i.e. now adopting selected flow-dependent closure coefficients as described in Eqs. 9.43–9.46) we will consider simulations of the boundary layer beneath transient solitary waves. Specifically, we will present results from Williams and Fuhrman (2016), who simulated numerically a series

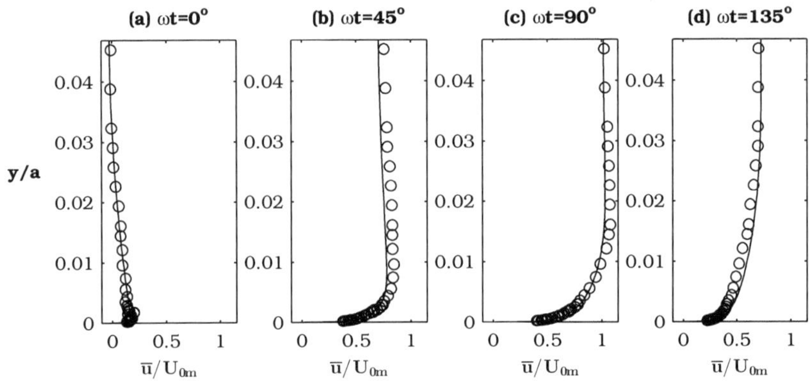

Fig. 5.90 Comparison of simulated (full lines, based on the standard k-ω turbulence closure) and measured (test 13 of Jensen et al., 1989, with $Re = 6 \times 10^6$, rough wall with $a/k_s = 3700$) ensemble averaged velocity profiles at selected phases.

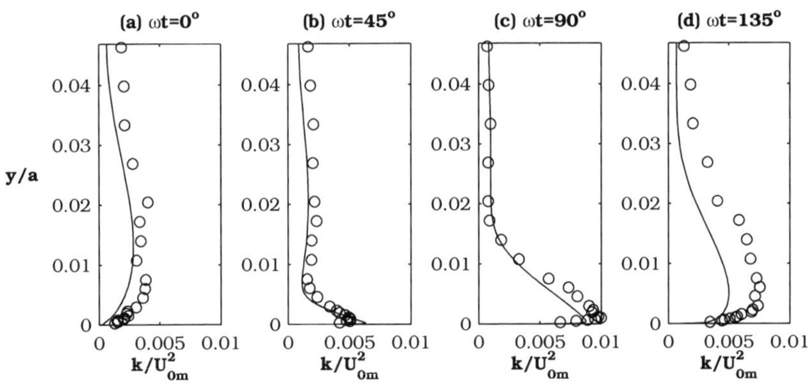

Fig. 5.91 Comparison of simulated (full lines, based on the standard k-ω turbulence closure) and measured (test 13 of Jensen et al., 1989, rough wall with $a/k_s = 3700$) turbulent kinetic energy density k profiles at selected phases.

of cases based on the smooth-wall oscillating tunnel experiments of Sumer et al. (2010), which are presented in detail in Section 5.10. [Note that the model version actually utilized by Williams and Fuhrman (2016) differred slightly from that presented in Section 5.12.4. Specifically, in the ω production term i.e. the first term in the right hand side of Eq. 5.163, the eddy viscosity was missing the lead α^* coefficient (see Eq. 5.164) and instead implemented as $\nu_t = k/\tilde{\omega}$. This lead to their use of the coefficient $R_k = 3$

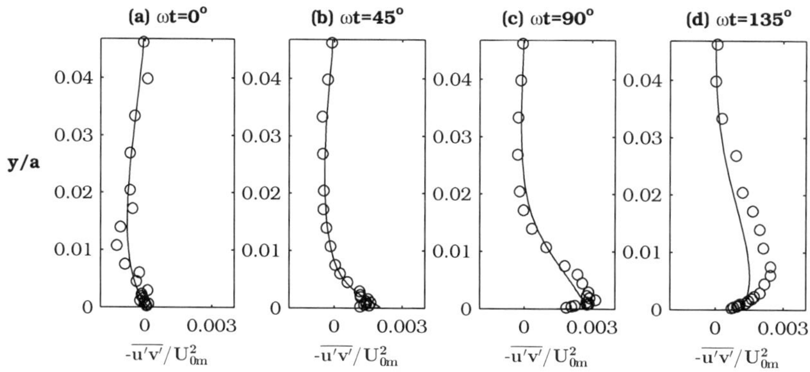

Fig. 5.92 Comparison of simulated (full lines, based on the standard Wilcox (2006) k-ω turbulence closure) and measured (test 13 of Jensen et al., 1989, rough wall with $a/k_s = 3700$) Reynolds stress $\overline{u'v'}$ profiles at selected phases.

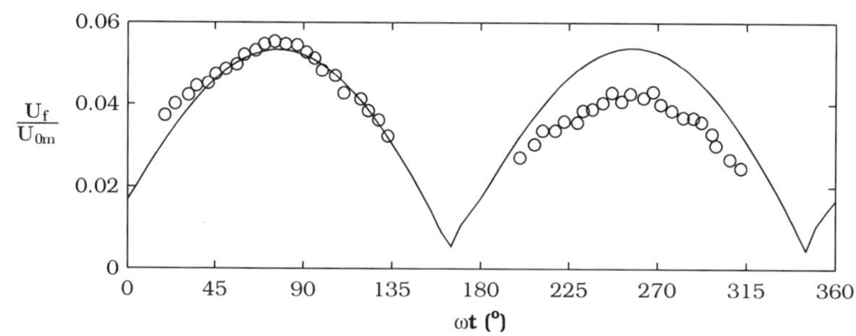

Fig. 5.93 Comparison of simulated (full line, based on the standard Wilcox (2006) k-ω turbulence closure) and measured (test 13 of Jensen et al., 1989, rough wall with $a/k_s = 3700$) time series of friction velocity.

rather than the standard value of 6, see Section. 9.4.1.2. Their resulting model was successfully validated, indicating that these two variations can give quite similar results.] To simulate the boundary layer beneath a free stream having the leading-order form of a solitary wave:

$$U_0 = U_{0m}\text{sech}^2(\omega t) \tag{5.166}$$

the driving pressure gradient can taken as equal to the corresponding time derivative

$$-\frac{1}{\rho}\frac{\partial p}{\partial x} = \frac{\partial U_0}{\partial t} = U_{0m}\omega\text{sech}^2(\omega t)\tanh(\omega t) \tag{5.167}$$

where $\omega = 2\pi/T$ is here the angular frequency.

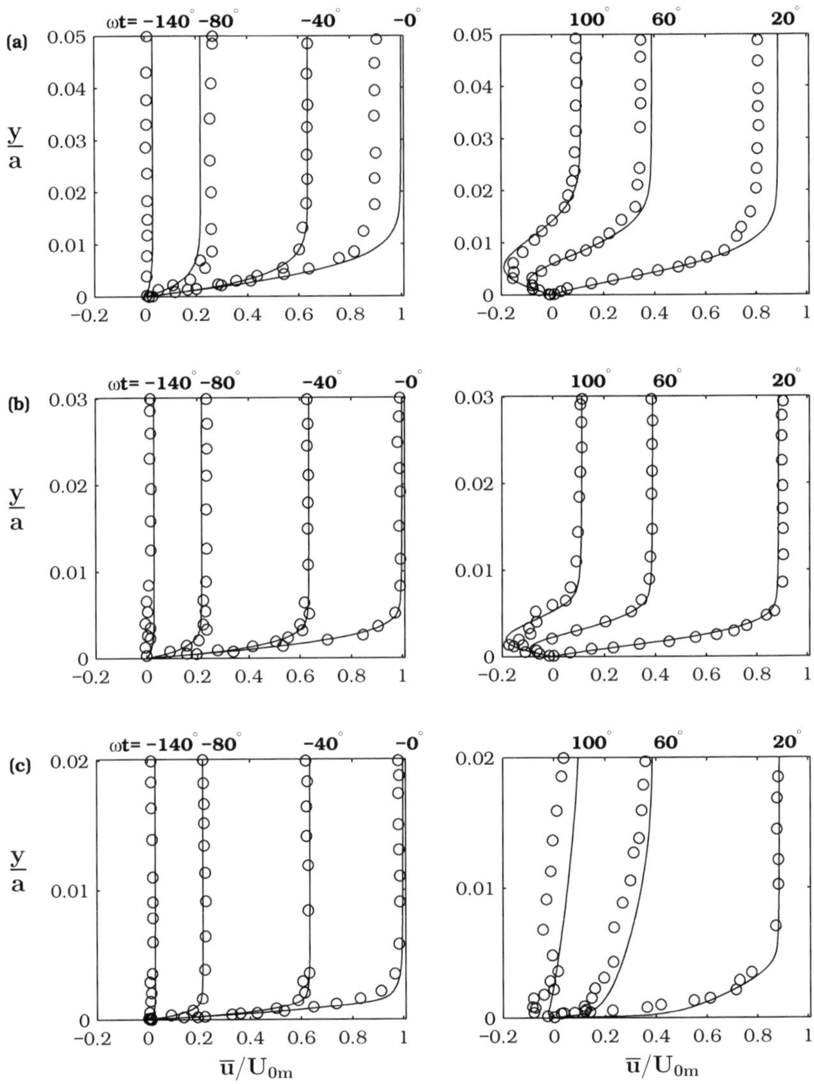

Fig. 5.94 Comparison of computed (lines, from Williams and Fuhrman, 2016, utilizing the low Reynolds number k-ω model) and measured (circles, from Sumer et al., 2010) velocity profiles for solitary wave boundary layers on smooth beds. The results correspond to (a) laminar conditions with $Re = 7.1 \times 10^4$, transitional conditions with $Re = 4.8 \times 10^5$, and (c) turbulent conditions with $Re = 1.8 \times 10^6$. The left- and right-hand sub-plots correspond, respectively, to times of accelerating and decelerating free stream flow.

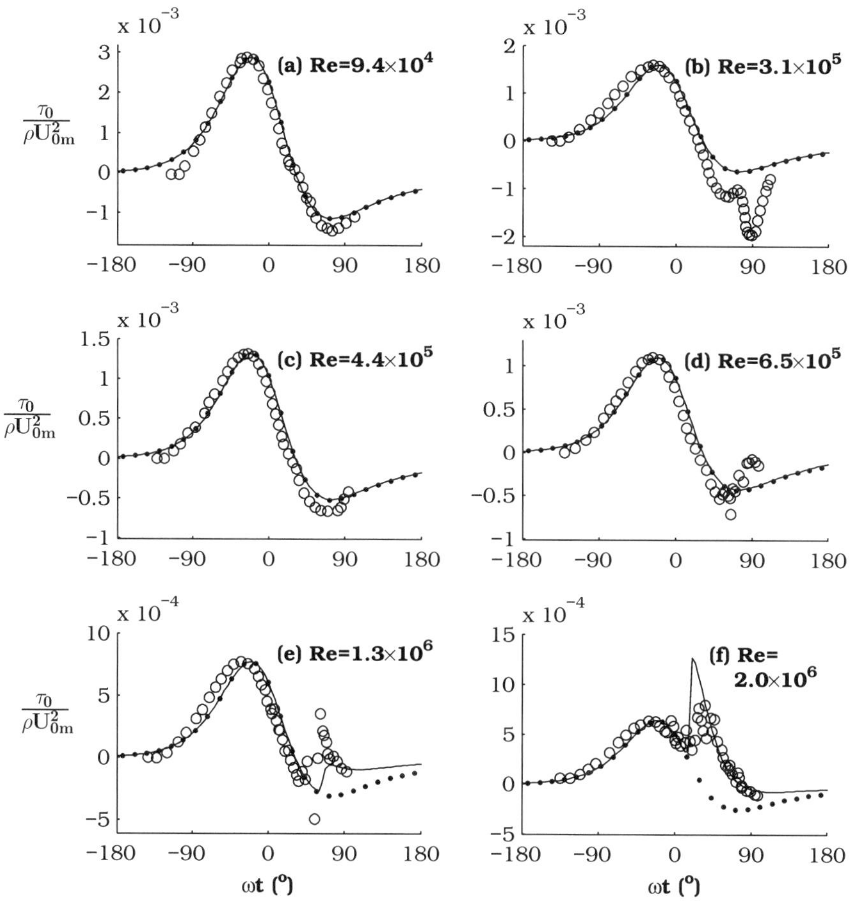

Fig. 5.95 Comparison of bed shear stress time series beneath a solitary wave free stream velocity. Open circles depict measurements made by Sumer et al. (2010), dots depict the laminar theory of Liu et al. (2007), and full lines depict numerical simulation results from Williams and Fuhrman (2016) made with a 1DV "low Reynolds number" k-ω turbulence model.

Comparison of the measured and simulated velocity profiles at several phases are presented in Fig. 5.94 for three cases having variable Reynolds numbers (a) $Re = aU_{0m}/\nu = 7.1 \times 10^4$, (b) 4.8×10^5, and (c) 1.8×10^6. The model is seen to capture well the differences between laminar and turbulent regimes, where the latter results in a significant reduction in phase differences throughout the boundary layer due to additional turbulent mixing of momentum.

The ability to simulate transition is perhaps more clearly evident in Fig. 5.95 which depicts comparison of computed (lines) and measured (circles) bed shear stresses from six different cases. For sufficiently low Reynolds number, the measured and model results essentially match the laminar theory of Liu et al. (2007), depicted as the dots. (It can be noted that e.g. for $Re = 3.1 \times 10^5$, Fig. 5.95b, the experimental measurements deviate from this theory at later phases due to the formation of 2D vortex tubes, though these can obviously not be simulated within a 1DV framework.) For the largest two Reynolds numbers considered, $Re = 1.3 \times 10^6$ (Fig. 5.95e) and $Re = 2.0 \times 10^6$ (Fig. 5.95f), the simulated transition to turbulence is very clear, as characterized by a sudden increase in the bed shear stress which clearly deviates from the laminar theory. The transition is likewise seen to occur at approximately the same phase as was observed. Hence, these cases in particular demonstrate the ability of the low Reynolds number k-ω turbulence model to reasonably simulate transition, while still utilizing a boundary layer model based on RANS equations. (Obviously, the transition in this context is only in terms of averaged flow properties, it does not directly model detailed aspects of this phenomenon, such as turbulent spots, etc., which would require Direct Numerical Simulation with much greater resolution).

It can finally be noted that the apparent over-prediction in the simulated turbulence-induced spike in the bed shear stress may, at least in part, be attributed to ensemble averaging of the experimental results. This ensemble averaging will tend to reduce the experimental spike, since the detected transition will not occur at precisely the same phase in a series of repeated experimental measurements. This is due to natural variations in the spatial development of turbulent spots, see e.g. the detailed discussion in the Appendix to Chapter 1.

5.13 References

(1) Akhavan, R., Kamm, R.D., and Shapiro, A.H. (1991): An investigation of transition to turbulence in bounded oscillatory Stokes flow. Part 2. Numerical simulations. J. Fluid Mech., vol. 225, 423–444.

(2) Arnskov, M.M., Fredsøe, J. and Sumer, B.M. (1993): Bed shear stress measurements over a smooth bed in three-dimensional wave-current motion. Coast. Eng., vol. 20, 277–316.

(3) Asano, T., Nakagawa, M. and Iwagaki, Y. (1986): Changes in current

properties due to wave superimposing. Proc. 20th Int. Conf. Coast. Eng. (Ed. B.L. Edge), vol. 1, ch. 70, 925–939, ASCE.
(4) Bagnold, R.A. (1946): Motion of waves in shallow water. Interaction between waves and sand bottom. Proc. Roy. Soc. Ser. A, vol. 187, 1–15.
(5) Bakker, W.T. and van Doorn, T. (1978): Near bottom velocities in waves and current. Proc. 16th Int. Conf. Coast. Eng., Hamburg (ed. B.L. Edge), vol 2, ch. 82, 1394–1413, ASCE.
(6) Batchelor, G.K. (1967): An Introduction to Fluid Dynamics. Cambridge University Press.
(7) Bijker, E.W. (1967): Some considerations about scales for coastal models with movable bed. Publ. 50, Delft Hydraulics Lab., Delft, 142 p.
(8) Blondeaux, P., Pralitis, J. and Vittori, G. (2012): Transition to turbulence at the bottom of a solitary wave. J. Fluid Mech., vol. 709, 396–407.
(9) Blondeaux, P. and Vittori, G. (1999): Boundary layer and sediment dynamics under sea waves. In: Adv. Coast. Ocean Eng. (ed. P.L.-F. Liu), vol. 4, pp. 133–190. World Scientific.
(10) Browand, F. and Plocher, D.A. (1985): Image processing for sediment transport. In 21st IAHR Congress, pp. 19–23, Melbourne, Australia.
(11) Brøker, I.H. (1985): Wave generated ripples and resulting sediment transport in waves. Series paper no. 36, ISVA, Technical University of Denmark.
(12) Caliskan, U. and Fuhrman, D.R. (2017): RANS-based simulation of wave-induced sheet-flow transport of graded sediments. Coast. Eng., vol. 121, 90–102.
(13) Carstensen, S. (2006): Experimental Investigation of Coherent Structures in Wave Boundary Layer. PhD Thesis. Principal Supervisor: B.M. Sumer. Technical University of Denmark, Coastal, Maritime and Structural Eng., January 31, 2006.
(14) Carstensen, S., Sumer, B.M. and Fredsøe, J. (2010): Coherent structures in wave boundary layers. Part 1. Oscillatory motion. J. Fluid Mech., vol. 646, 169–206.
(15) Carstensen S., Sumer, B.M. and Fredsøe, J. (2012): A note on turbulent spots over a rough bed in wave boundary layers. Phys. Fluids, vol. 24, article no. 115104.
(16) Chambers, F.W. and Thomas, A.S.W. (1983): Turbulent spots, wave packets and growth. Phys. Fluids, vol. 26, 1160–1162.
(17) Chan, I.-C. and Liu, P.L.-F. (2012): On the runup of long waves on a

plane beach, J. Geophys. Res., vol. 117, article no. C08006.
(18) Christoffersen, J.B. and Jonsson, I.G. (1985): Bed friction and dissipation in a combined current and wave motion. Ocean Eng., vol. 12, 387–423.
(19) Coles, D. (1956): The law of the wake in a turbulent boundary layer. J. Fluid Mech., vol. 1, 191–226.
(20) Costamagna, P., Vittori, G. and Blondeaux, P. (2003): Coherent structures in oscillatory boundary layers. J. Fluid Mech., vol. 474, 1–33.
(21) Davies, A.G., Soulsby, R.L. and King, H.L. (1988): A numerical model of the combined wave and current bottom boundary layer. J. Geophys. Res., vol. 93, 491–508.
(22) Deigaard, R., Jakobsen, J.B., and Fredsøe, J. (1999): Net sediment transport under wave groups and bound long waves. J. Geophys. Res., vol. 104(C6), 13559–13575.
(23) Dixen, M., Hatipoglu, F., Sumer, B.M. and Fredsøe, J. (2008): Wave boundary layer over a stone-covered bed. Coast. Eng., vol. 55, 1–20.
(24) Eckmann, D.M. and Grotberg, J.B. (1991): Experiments on transition to turbulence in oscillatory pipe flow. J. Fluid Mech., vol. 222, 329–350.
(25) Emmons, H.W. (1951): Laminar-turbulent transition in boundary layer. J. Aeronaut. Sci., vol. 18 (7), 490–498.
(26) Finnicum, D.S. and Hanratty, T.J. (1988): Influence of imposed flow oscillations on turbulence. Physochem. Hydrodyn., vol. 10, 585–598.
(27) Fornarelli, F. and Vittori, G. (2009): Oscillatory boundary layer close to a rough wall. Eur. J. Mech. B. Fluids, vol. 28, 283–295.
(28) Foster, D.L. (1997): Dynamics of the near shore wave bottom boundary layer. PhD Thesis, Oregon State University.
(29) Foster, D.L., Holman, R.A. and Beach, R.A. (1994): Sediment suspension events and shear instabilities in the bottom boundary layer. Coastal Dynamics '94. In: Proc. Int. Conf. on the Role of the Large Scale Experiments in Coastal Research, ASCE (Eds. A.S. Arcilla, M.F. Stive and N.C. Kraus), pp. 712–726. Universitat Politecnica de Catalunya, Barcelona, Spain.
(30) Fredsøe, J. (1984): Turbulent boundary layer in wave-current motion. J. Hydraul. Eng. ASCE., vol. 110, 1103–1120.
(31) Fredsøe, J., Andersen, K.H. and Sumer, B.M. (1999): Wave plus current over ripple-covered bed. Coast. Eng., vol. 38, 177–221.
(32) Fredsøe, J. and Deigaard, R. (1992): Mechanic of Coastal Sediment Transport. Advanced Series on Ocean Engineering. World Scientific.

(33) Fredsøe, J., Sumer, B.M., Kozakiewicz, A., Chua, L.H.C. and Deigaard, R. (2003): Effect of externally generated turbulence on wave boundary layer. Coast. Eng., vol. 49, 155–183.
(34) Fredsøe, J., Sumer, B.M., Laursen, T.S. and Pedersen, C. (1993): Experimental investigations of wave boundary layers with a sudden change in roughness. J. Fluid Mech., vol. 252, 117–145.
(35) Fuhrman, D.R., Schløer, S. and Sterner, J. (2013): RANS-based simulation of turbulent wave boundary layer and sheet-flow sediment transport processes. Coast. Eng., vol. 73, 151–166.
(36) Gad-el-Hak, M., Blackwelder, R.F. and Riley, J.J. (1981): On the growth of turbulent regions in laminar boundary layers. J. Fluid Mech., vol. 110, 73–95.
(37) Gilbrech, D.A. and Combs, G.D. (1963): Critical Reynolds numbers for incompressible pulsating flows in tubes. Developments in Theoretical and Applied Mechanics, vol. 1, 292–304. Plenum.
(38) Goseberg, N., Wurpts, A. and Schlurmann, T. (2013): Laboratory-scale generation of tsunami and long waves. Coast. Eng., vol. 79, pp. 57–74.
(39) Grant, W.D. and Madsen, O.S. (1979): Combined wave and current interaction with a rough bottom. J. Geophys. Res., vol. 84, 1797–1808.
(40) Grant, W.D., Williams, A.J., Glenn, S.M., Cacchione, D.A. and Drake, D.E. (1983): High frequency bottom stress variability and its prediction in the COSE region. WHOI Tech. Rep., 83–19, Woods Hole Oceanographic Institute.
(41) Grass, A.J. (1974): Transport of fine sand on a flat bed: turbulence and suspension mechanics. Euromech 48, Inst. Hydrodyn. and Hydraulic Engng., Tech. Univ. Denmark, p. 33.
(42) Hino, M., Kashiwayanagi, M., Nakayama, A. and Hara, T. (1983): Experiments on turbulence statistics and structure of reciprocating oscillatory flow. J. Fluid Mech., vol. 131, 363–399.
(43) Hino, M., Sawamoto, M. and Takasu, S. (1976): Experiments on transition to turbulence in an oscillatory pipe flow. J. Fluid Mech., vol. 75, 193–207.
(44) Holmedal, L.E. and Myrhaug, D. (2009): Wave-induced steady streaming, mass transport and net sediment transport in rough turbulent ocean bottom boundary layers. Cont. Shelf Res., vol. 29, 911–926.
(45) Huntley, D.A. and Hazen, D.G. (1988): Seabed stresses in combined wave and steady flow conditions on the Nova Scotia continental shelf - field measurements and predictions. J. Phys. Oceanogr., vol. 18,

347–362.
(46) Huynh-Thanh, S. and Temperville, A. (1991): A numerical model of the rough turbulent boundary layer in combined wave and current interaction. In Sand Transport in Rivers, Estuaries and the Sea (ed. R.L. Soulsby and R. Bettess), 93–100. Balkema.
(47) Jensen, B.L., Sumer, B.M. and Fredsøe, J. (1989): Turbulent oscillatory boundary layers at high Reynolds numbers. J. Fluid Mech., vol. 206, 265–297.
(48) Jensen, K.L. and Sumer, B.M. (2019): Experimental investigation of wall-normal pressure gradient in a turbulent boundary layer in an open channel. Paper in preparation. From the PhD study of Karsten L. Jensen supervised by B. M. Sumer at Technical University of Denmark, DTU Mechanical Engineering, "Vertical pressure gradient and particle motions in wave boundary layers", October 2014.
(49) Jensen, K.L., Sumer, B.M., Vittori, G. and Blondeaux, P. (2014a): Role of the vertical pressure gradient in wave boundary layers. Proc. 34th Int. Conf. Coast. Eng., Seoul, Korea (Ed. Patrick Lynett). pp. 1–11, https://doi.org/10.9753/icce.v34.waves.47
(50) Jensen, K.L., Sumer, B.M., Fredsøe, J. and Jensen, J.J. (2014b): Particle motions in oscillatory flow over a smooth bed. In Book of Proceedings, 2014, IAHR. Presented at 3rd IAHR Europe Congress, 2014, Porto, Portugal.
(51) Jensen, K.L., Sumer, B.M., Vittori, G., Blondeaux, P. (2019): Direct numerical simulation of turbulent wave boundary layer. Paper in preparation. From the PhD study of Karsten L. Jensen supervised by B. M. Sumer at Technical University of Denmark, DTU Mechanical Engineering, "Vertical pressure gradient and particle motions in wave boundary layers", October 2014.
(52) Jonsson, I.G. (1963): Measurements in the turbulent wave boundary layer. Proc. 10th Cong. IAHR, London, vol. 1, 85–92.
(53) Jonsson, I.G. (1975): The wave friction factor revisited. Technical University of Denmark, Institute of Hydrodynamics and Hydraulic Engineering, Lyngby, Denmark, Progress Report, vol. 37, 3–8.
(54) Jonsson, I.G. (1980): A new approach to oscillatory rough turbulent boundary layers. Ocean Eng., vol. 7, 109–152.
(55) Jonsson, I.G. and Carlsen, N.A. (1976): Experimental and theoretical investigations in an oscillatory turbulent boundary layer. J. Hydraul. Res., vol. 14, 45–60.
(56) Justesen, P. (1988a): Turbulent Wave Boundary Layers. Series Paper

43, Institute of Hydrodynamic and Hydraulic Engineering, Technical University of Denmark.
(57) Justesen, P. (1988b): Prediction of turbulent oscillatory flow over rough beds. Coast. Eng., vol. 12, 257–284.
(58) Justesen, P. (1991): A note on turbulence calculations in the wave boundary layer. J. Hydraul. Res., vol. 29, 699–711.
(59) Kajiura, K. (1968): A model of bottom boundary layer in water waves. Bull. Earthquake Res. Inst., vol. 46, 75–123.
(60) Kamphuis, J.W. (1975): Friction factor under oscillatory waves. J. Waterways, Port, Coastal and Ocean Eng. Div., ASCE, vol. 101 (WW2), 135–144.
(61) Kemp, P.H. and Simons, R.R. (1982): The interaction between waves and a turbulent current: waves propagating with the current. J. Fluid Mech., vol. 116, 227–250.
(62) Kemp, P.H. and Simons, R.R. (1983): The interaction between waves and a turbulent current: waves propagating against the current. J. Fluid Mech., vol. 130, 73–89.
(63) Keulegan, G.H. (1948): Gradual damping of solitary wave. J. Res. Natl. Bur. Stand., vol. 40, 607–614.
(64) Kirca, V.S.O., Sumer, B.M., Steffensen, M., Jensen, K.L. and Fuhrman, D.R. (2016): Longitudinal dispersion of heavy particles in an oscillating tunnel and application to wave boundary layers. J. Ocean Eng. Mar. Energy., vol. 2, 59–83.
(65) Klopman, G. (1994): Vertical structure of the flow due to waves and currents. Progress Report, H 840.30, Part II, Delft Hydraulics. Prepared for Rijkwaterstaat, Tidal Water Division and Commission of the European Communities, Directorate General XII.
(66) Lacy, J.R., Rubin, D.M. and Buscombe, D. (2012): Currents, drag, and sediment transport induced by a tsunami. J. Geophys. Res., vol. 117, article no. C09028.
(67) Larsen, B.E., Arbøll, L.K., Frigaard, S., Carstensen, S. and Fuhrman, D. R. (2018): Experimental study of tsunami-induced scour around a monopile foundation. Coast. Eng., vol. 138, 9–21.
(68) Larsen, B.E. and Fuhrman, D.R. (2019a): Full-scale CFD simulation of tsunamis. Part 1: Model validation and run-up. Coast. Eng., vol. 151, 22–41.
(69) Larsen, B.E. and Fuhrman, D.R. (2019b): Full-scale CFD simulation of tsunamis. Part 2: Boundary layers and bed shear stresses. Coast. Eng., vol. 151, 42–57.

(70) Larsen, B.E., Fuhrman, D.R., Baykal, C. and Sumer, B. M. (2017): Tsunami-induced scour around monopile foundations. Coast. Eng., vol. 129, 36–49.
(71) Liu, P.L.-F. (2006): Turbulent boundary-layer effects on transient wave propagation in shallow water. Proc. Roy. Soc. A, vol. 462, 3431–3491.
(72) Liu, P.L.-F. and Orfila, A. (2004): Viscous effects on transient long-wave propagation. J. Fluid Mech., vol. 520, 83–92.
(73) Liu, P.L.-F., Park, Y.S. and Cowen, E.A. (2007): Boundary layer flow and bed shear stress under a solitary wave. J. Fluid Mech., vol. 574, 449–463.
(74) Lodahl, C.R., Sumer, B.M. and Fredsøe, J. (1998): Turbulent combined oscillatory flow and current in a pipe. J. Fluid Mech., vol. 373, 313–348.
(75) Lohmann, I.P., Fredsøe, J., Sumer, B.M. and Christensen, E.D. (2006): Large eddy simulation of the ventilated wave boundary layer. J. Geophys. Res., vol. 111, article no. C06036.
(76) Lundgren, H. (1972): Turbulent currents in the presence of waves. Proc. 13th Coastal Engng. Conf. July 10–14, 1972, Vancouver, B.C., Canada, vol. 1, chap. 33, ASCE.
(77) Lundgren, H. and Jonsson, I.G. (1961): Bed shear stress induced by a wave motion. Coastal Engineering Laboratory, Technical University of Denmark. Basic Research Progress Report, vol. 1, 3–5.
(78) Lundgren, H. and Sørensen, T. (1958): A pulsating water tunnel. Proc. 6th Intl. Conf. on Coastal Engineering, Miami, Fl., pp. 356–358. Council on Wave Research, The Engng. Foundation, Engineering Field Station, University of California, Richmond, CA.
(79) Madsen, P.A. and Fuhrman, D.R. (2008): Run-up of tsunamis and long waves in terms of surf-similarity. Coast. Eng., vol. 55, 209–223.
(80) Madsen, P.A., Fuhrman, D.R. and Schäffer, H.A. (2008): On the solitary wave paradigm for tsunamis. J. Geophys. Res., vol. 113, article no. C12012.
(81) Madsen, P.A. and Schäffer, H.A. (2010): Analytical solutions for tsunami runup on a plane beach: single waves, N-waves and transient waves. J. Fluid Mech., vol. 645, 27–57.
(82) Mao, Z. and Hanratty, T.J. (1986): Studies of the wall shear stress in a turbulent pulsating pipe flow. J. Fluid Mech., vol. 170, 545–564.
(83) Mao, Z. and Hanratty, T.J. (1994): Influence of large-amplitude oscillations on turbulent drag. AIChE J., vol. 40, 1601–1610.
(84) Matsubara, M. and Alfredsson, P.H. (2001): Disturbance growth in boundary layers subjected to free-stream turbulence. J. Fluid Mech.,

vol. 430, 149–168.

(85) Mazzuoli, M. (2013): Transition to Turbulence in an Oscillatory Boundary Layer and its Effects on the Motion of a Rigid Particle. PhD Thesis. University of Genoa, Dept. of Civil, Chemical and Environmental Eng. Supervisors: G. Vittori, P. Blondeaux and G. Seminara, March 19th, 2013, 157 pages.

(86) Mazzuoli, M. and Vittori, G. (2019): Turbulent spots in an oscillatory flow over a rough wall. Eur. J. Mech. B Fluids, vol. 78, 161–168.

(87) Mazzuoli, M., Vittori, G. and Blondeaux, P. (2011): Turbulent spots in oscillatory boundary layers. J. Fluid Mech., vol. 685, 365–376.

(88) Mei, C.C. (1983): The Applied Dynamics of Ocean Surface Waves. John Wiley & Sons.

(89) Myrhaug, D., Reed, K. and Fyfe, A.J. (1987): Seabed boundary layer studies for pipelines: Large scale laboratory experiments. In Proc. 9th Intl. Symp. on Offshore Engineering, Rio de Janeiro (Ed. F.L.L. Carneiro et al.), 345–359, Wiley.

(90) Myrhaug, D. (1989): A rational approach to wave friction coefficients for rough, smooth and transitional turbulent flow. Coast. Eng., vol. 13, 11–21.

(91) Myrhaug, D. and Slaattelid, O.H. (1989): Combined wave and current boundary layer model for fixed rough beds. Ocean Eng., vol. 16, 119–142.

(92) Myrhaug, D. and Slaattelid, O.H. (1990): A rational approach to wave-current friction coefficients for rough, smooth and transitional turbulent flow. Coast. Eng., vol. 14, 265–293.

(93) Nielsen, P. (1985): On the structure of oscillatory boundary layers. Coast. Eng., vol. 9, 261–276.

(94) Nielsen, P. (1992): Coastal Bottom Boundary Layers and Sediment Transport. Advanced Series on Ocean Engineering. World Scientific.

(95) Nielsen, P. (2016): 1DV structure of turbulent wave boundary layers. Coast. Eng., vol. 112, 1–8.

(96) Nielsen, P. (2018): Bed shear stress, surface shape and velocity field near the tips of dam-breaks, tsunami and wave runup. Coast. Eng., vol. 138, 126–131.

(97) Nino, Y. and Garcia, M.H. (1996): Experiments on particle-turbulent interactions in the near wall region of an open channel flow: implications for sediment transport. J. Fluid Mech., vol. 326, 285–319.

(98) Ozdemir, C.E., Hsu, T.-J. and Balachandar, S. (2013): Direct numerical simulations of instability and boundary layer turbulence under a

solitary wave. J. Fluid Mech., vol. 731, 545–578.
(99) Ozdemir, C.E., Hsu, T.-J. and Balachandar, S. (2014): Direct numerical simulations of transition and turbulence in smooth walled Stokes boundary layer. Phys. Fluids, vol. 26, article no. 045108.
(100) Park, Y.S., Verschaeve, J. Pedersen, G.K. and Liu, P.L.F. (2014): Boundary-layer flow and bed shear stress under a solitary wave: revision. J. Fluid Mech., vol. 753, 554–559.
(101) Perry, A., Lim, T. and Teh, E. (1981): A visual study of turbulent spots. J. Fluid Mech., vol. 104, 387–405.
(102) Rabinovich, A.B. and Thomson, R.E. (2007): The 26 December 2004 Sumatra tsunami: Analysis of tide gauge data from the world ocean Part 1. Indian Ocean and South Africa. Pure Appl. Geophys., vol. 164, 261–308.
(103) Ramaprian, B.R. and Tu, S.W. (1980): An experimental study of oscillatory pipe flow at transitional Reynolds number. J. Fluid Mech., vol. 100, 513–544.
(104) Ramaprian, B.R. and Tu, S.W. (1983): Fully developed periodic turbulent pipe flow. Part 2. The detailed structure of the flow. J. Fluid Mech., vol. 137, 59–81.
(105) Salon, S., Armenio, V. and Crise, A. (2007): A numerical investigation of the Stokes boundary layer in the turbulent regime. J. Fluid Mech., vol. 570, 253–296.
(106) Sarpkaya, T. (1966): Experimental determination of the critical Reynolds number for pulsating Poiseuille flow. Trans. ASME: J. Basic Engng., Sept., 589–598.
(107) Sarpkaya, T. (1993): Coherent structures in oscillatory boundary layers. J. Fluid Mech., vol. 253, 105–140.
(108) Scandura, P. (2013): Two dimensional vortex structures in the bottom boundary layer of progressive and solitary waves. J. Fluid Mech., vol. 728, 340–361.
(109) Scandura, P., Faraci, C. and Foti, E. (2016): A numerical investigation of acceleration-skewed oscillatory flows. J. Fluid Mech., vol. 808, 576–613.
(110) Schimmels, S., Sriram, V. and Didenkulova, I. (2016): Tsunami generation in a large scale experimental facility. Coast. Eng., vol. 110, 32–41.
(111) Schlichting, H. (1979): Boundary Layer Theory. McGraw-Hill.
(112) Seifert, A. and Wygnanski, I.J. (1995): On turbulent spots in a laminar boundary layer subjected to a self-similar adverse pressure gradient. J.

Fluid Mech., vol. 296, 185–209.

(113) Sergeev, S.I. (1966): Fluid oscillations in pipes at moderate Reynolds numbers. Mekh. Zhid. i Gaza, vol. 1, no. 1, 21.

(114) Shemer, L., Wygnanski, I. and Kit, E. (1985): Pulsating flow in a pipe. J. Fluid Mech., vol. 153, 313–337.

(115) Simons, R.R., Grass, A.J. and Mansour-Tehrani, M. (1992): Bottom shear stresses in the boundary layer under waves and currents crossing at right angles. In Proc. 23rd Intl. Conf. on Coastal Engng., Venice, Italy (ed. B.L. Edge), vol. 1, ch. 45, 604–617. ASCE.

(116) Simons, R.R., Grass, A.J., Saleh, W.M. and Tehrani, M.M. (1994): Bottom shear stresses under waves with a current superimposed. Proc. 24th Intl. Conf. on Coastal Engng., Oct. 1994, Kobe, Japan (ed. B.L. Edge), vol. 1, Ch. 42, 565–578.

(117) Simons, R.R., Kyriacou, A., Soulsby, R.L. and Davies, A.G. (1988): Predicting the nearbed turbulent flow in waves and currents. Proc. I.A.H.R. Symposium on Mathematical Modelling of Sediment Transport in the Coastal Zone, Copenhagen, 33–47.

(118) Sleath, J.F.A. (1984): Sea Bed Mechanics. John Wiley & Sons.

(119) Sleath, J.F.A. (1987): Turbulent oscillatory flow over rough beds. J. Fluid Mech., vol. 182, 369–409.

(120) Sleath, J.F.A. (1970): Velocity measurements close to the bed in a wave tank. J. Flui Mech., vol. 42, 111–123.

(121) Sleath, J.F.A. (1982): The effect of jet formation on the velocity distribution in oscillatory flow over flat beds of sand or gravel. Coast. Eng., vol. 6, 151–177.

(122) Sleath, J.F.A. (1988): Transition in oscillatory flow over rough beds. J. Waterway, Port, Coastal, Ocean Engng. Div. ASCE, vol. 114, 18–33.

(123) Sleath, J.F.A. (1990): Velocities and bed friction in combined flow. In 22nd Intl. Conf. on Coastal Engineering, Delft, 1990, vol. 1, 450–463. ASCE.

(124) Sleath, J.F.A. (1991): Velocities and shear stresses in wave-current flows. J. Geophys. Res., vol. 96, 15237–15244.

(125) Soulsby, R. L. (1995): Bed shear-stresses due to combined waves and currents. In: M. J. F. Stive, H. J. de Vriend, J. Fredsøe, L. Hamm, R. L. Soulsby, C. Teisson and J. C. Winterwerp (Eds.), Advances in Coastal Morphodynamics (Delft Hydraulics, Netherlands), pp. 4-20–4-23.

(126) Soulsby, R. (1997): Dynamics of Marine Sands: A Manual for Practical Applications. Thomas Telford.

(127) Soulsby, R.L., Hamm, L., Klopman, G., Myrhaug, D., Simons R.R. and

Thomas, G.P. (1993): Wave-current interaction in the vertical. Coast. Eng., vol. 21, 41–69.
(128) Soulsby, R.L. and Humphery, J.D. (1990): Field observations of wave-current interaction at the seabed. In Water Wave Kinematics (ed. A. Tørum and O.T. Gudmester), 413–428, Kluwer.
(129) Spalart, P.R. and Baldwin, B.S. (1987): Direct simulation of a turbulent oscillating boundary layer. NASA Tech. Mem. 89460 Ames Research Center, Moffett Field, California.
(130) Stettler, J.C. and Hussain, A.K.M.F. (1986): On transition of the pulsating flow. J. Fluid Mech., vol. 170, 169–197.
(131) Sumer, B.M. (1984): Lift forces on moving particles near boundaries. J. Hydraul. Eng.-ASCE, vol. 110, 1272–1278.
(132) Sumer, B.M. (2003): Experimental investigation of wave boundary layer. EUROMECH Colloquium No. 451, 26–29. Taormina, Italy, General Lecture. Power Point presentation obtainable from the author (bms@bmsumer.com).
(133) Sumer, B.M., Arnskov, M.M., Christiansen, N. and Jørgensen, F.E. (1993a): Two-component hot-film probe for measurements of wall shear stress. Experiments in Fluids, vol. 15, 380–384.
(134) Sumer, B.M., Christiansen, N. and Fredsøe, J. (1997): Horseshoe vortex and vortex shedding around a vertical wall-mounted cylinder exposed to waves. J. Fluid Mech., vol. 332, 41–70.
(135) Sumer, B.M. and Deigaard, R. (1981): Particle motions near the bottom in turbulent flow in an open channel - Part 2. J. Fluid Mech., vol. 109, 311–337.
(136) Sumer, B.M., Jensen, B.L. and Fredsøe, J. (1987): Turbulence in oscillatory boundary layers. Advances in Turbulence. Proc. 1st European Conference, Lyon, France, published by Springer, pp. 556–567.
(137) Sumer, B.M., Jensen, P.M., Sørensen, L.B., Fredsøe, J., Liu, P.L.-F. and Carstensen, S. (2010): Coherent structures in wave boundary layers. Part 2. Solitary motion. J. Fluid Mech., vol. 646, 207–231.
(138) Sumer, B.M., Laursen, T.S. and Fredsøe, J. (1993b): Wave boundary layers in a convergent tunnel. Coast. Eng., vol. 20, 317–342.
(139) Sumer, B.M. and Fredsøe, J. (2006): Hydrodynamics Around Cylindrical Structures. Advanced Series on Ocean Engineering. World Scientific.
(140) Sumer, B.M. and Oguz, B. (1978): Particle motions near the bottom in turbulent flow in an open channel. J. Fluid Mech., vol. 86, 109–127.
(141) Swart, D.H. (1974) Offshore sediment transport and equilibrium beach

profiles. Delft Hydraulics Laboratory, Publication No. 131.
(142) Tanaka, H., Sana, A., Kawamura, I. and Yamaji, H. (1999): Depth-limited oscillatory boundary layers on a rough bottom. Coast. Eng. J., vol. 41, 85–105.
(143) Tanaka, H., Sumer, B.M. and Lodahl, C. (1998): Theoretical and experimental investigation on laminar boundary layers under cnoidal wave motion. Coast. Eng. J., vol. 40, 81–98.
(144) Tardu, S.F., Binder, G. and Blackwelder, R.F. (1994): Turbulent channel flow with large-amplitude velocity oscillations. J. Fluid Mech., vol. 267, 109–151.
(145) Tromans, P.S. (1978): Stability and Transition of Periodic Pipe Flows. PhD Thesis, Cambridge University.
(146) Trowbridge, J. and Madsen, O.S. (1984): Turbulent wave boundary layers 2. Second-order theory and mass transport. J. Geophys. Res., vol. 89, 7999–8007.
(147) Tu, S.W. and Ramaprian, B.R. (1983): Fully developed periodic turbulent pipe flow. Part 1. Main experimental results and comparison with predictions. J. Fluid Mech., vol. 137, 31–58.
(148) Tuzi, R. and Blondeaux, P. (2008): Intermittent turbulence in a pulsating pipe flow. J. Fluid Mech., vol. 599, 51–79.
(149) van der A, D.A., O'Donoghue, T., Davies, A.G. and Ribberink, J.S. (2011): Experimental study of the turbulent boundary layer in acceleration-skewed oscillatory flow. J. Fluid Mech., vol. 648, 251–283.
(150) van Driest, E.R. (1956): On turbulent flow near a wall. J. Aero. Sci., vol. 23, 1007–1011.
(151) van Kesteren, W.G.M. and Bakker, W.T. (1984): Near bottom velocities in waves with a current; analytical and numerical computations. In Proc. 19th Conf. Coastal Engng., Houston, TX, 1161–1177. ASCE.
(152) Vittori, G. and Blondeaux, P. (2008a): Boundary layer flow and bed shear stress under solitary wave. In Book of Abstract of the 31st Intl. Conf. on Coastal Engng., Hamburg, Germany, Abstract No. 062.
(153) Vittori, G. and Blondeaux (2008b): Turbulent boundary layer under a solitary wave. J. Fluid Mech., vol. 615, 433–443.
(154) Vittori, G. and Blondeaux (2011): Characteristics of the boundary layer at the bottom of a solitary wave. Coast. Eng., vol. 58, 206–213.
(155) Vittori, G. and Blondeaux (2014): The boundary layer at the bottom of a solitary wave and implications for sediment transport. Progress in Oceanography, vol. 120, 399–409.
(156) Vittori, G. and Verzicco, R. (1998): Direct simulation of transition in

an oscillatory boundary layer. J. Fluid Mech., vol. 371, 207–232.
(157) Wilcox, D.C. (2006): Turbulence Modeling for CFD, 3rd edn. DCW Industries.
(158) Williams, I.A. and Fuhrman, D.R. (2016): Numerical simulation of tsunami-scale wave boundary layers. Coast. Eng., vol. 110, 17–31.
(159) Yuan, J. and Dash, S.M. (2017): Experimental investigation of turbulent wave boundary layers under irregular coastal waves. Coast. Eng., vol. 128, 22–36.
(160) Zhang, X. and Simons, R. (2019): Experimental investigation on the structure of turbulence in the bottom wave-current boundary layers. Coast. Eng., vol. 152, article no. 103511.

Appendix to Chapter 5: Some essentials of linear potential flow wave theory

This appendix will first present some basic aspects of so-called linear (small amplitude) potential (i.e. irrotational) flow wave theory. While such theory inevitably breaks down in boundary layer regions near the sea bed, as well as in the turbulent surf zone where waves are breaking, it is still of great importance, as the bulk region beneath (non-breaking) surface waves is quite well described by potential flow theory. Hence, we will first review some basic and fundamental aspects, on which the preceding chapter has built.

While nonlinear wave theories will not be specifically covered in this book, it is still worth mentioning that the state of the art in nonlinear regular wave solutions corresponds to e.g. the stream function solution of Fenton (1988) as well as the recent fast method of Clamond and Dutykh (2018). These are capable of providing highly nonlinear (potential flow) regular wave solutions based on numerical solutions to the fully nonlinear potential flow wave equations (see the next section). For nonlinear irregular (potential flow) wave theories, see the second order solution of Sharma and Dean (1981), as well as the theory of Madsen and Fuhrman (2012) who extended this solution to third order. For a linear description of irregular waves simple superposition holds.

5.A Governing equations

Consider the irrotational flow with a free surface varying in the horizontal x direction, such that y is vertical, with the origin placed at the still water level for convenience. Uniformity is assumed in the transverse z direction, for simplicity. The flow is bounded by the bottom at $y = -h$ and the free surface at $y = \eta$, such that h is the still water depth. The governing equations and boundary conditions for the irrotational flow of an incompressible inviscid fluid with a free surface are well known and correspond to

$$\frac{\partial u}{\partial t}+g\frac{\partial \eta}{\partial x}+\frac{1}{2}\frac{\partial}{\partial x}\left(u^2+v^2\right)=0, \qquad y=\eta \tag{5.168}$$

$$\frac{\partial \eta}{\partial t}+u\frac{\partial \eta}{\partial x}-v=0, \qquad y=\eta \tag{5.169}$$

$$\frac{\partial u}{\partial x}+\frac{\partial v}{\partial y}=0, \qquad -h<y<\eta \tag{5.170}$$

$$v+u\frac{\partial h}{\partial x}=0, \qquad y=-h \tag{5.171}$$

(It can be noted that the above is presented in terms of velocity variables, rather than a so-called velocity potential, for the sake of consistency with the rest of this book.) In the above, Eq. 5.168 is the *dynamic free surface boundary condition*, which stems from the pressure at the free surface being constant. Eq. 5.169 is the *kinematic free surface boundary condition*, which states that a particle at the free surface will remain there. Eq. 5.170 is the familiar *local continuity equation*. Eq. 5.171 is the *kinematic bottom condition*, which states that the bed is impermeable.

Let us now assume that the sea bed is flat, such that $\partial h/\partial x=0$, and also that the variations of the free surface η are small (such that nonlinear terms at the surface can be neglected). Moving the origin from the still water level down to the sea bed (the usual convention adopted in the present book, such that $y=0$ is now the bed elevation, and the still water level is now at $y=h$), we may re-cast the governing equations above to the following linearized form:

$$\frac{\partial u}{\partial t}+g\frac{\partial \eta}{\partial x}=0, \qquad y=h \tag{5.172}$$

$$\frac{\partial \eta}{\partial t}-v=0, \qquad y=h \tag{5.173}$$

$$\frac{\partial u}{\partial x}+\frac{\partial v}{\partial y}=0, \qquad 0<y<h \tag{5.174}$$

$$v=0, \qquad y=0 \tag{5.175}$$

These provide a leading-order description for surface waves propagating on a flat bottom in the x-direction.

5.B Surface elevation and velocity kinematics beneath a progressive regular wave

Consider a sinusoidal progressive wave having free surface of the form

$$\eta = \frac{H}{2}\sin\theta, \qquad \theta = \omega t - kx \tag{5.176}$$

where H is the (crest-to-trough) wave height, and the angular frequency and wave number are respectively defined by

$$\omega = \frac{2\pi}{T}, \qquad k = \frac{2\pi}{L} \tag{5.177}$$

where T is the wave period and L is the wave length. According to linear wave theory, the velocity field beneath this wave will correspond to

$$u = \frac{H\omega}{2}\frac{\cosh(ky)}{\sinh(kh)}\sin\theta \tag{5.178}$$

$$v = \frac{H\omega}{2}\frac{\sinh(ky)}{\sinh(kh)}\cos\theta \tag{5.179}$$

where h is again the still water depth, and the sea bed is again assumed to be flat at $y = 0$. It is straight forward to show that the description above satisfies the linearized governing equations, Eqs. 5.172–5.175. Notice that the horizontal velocity u is in phase with the free surface η, whereas the vertical velocity v is in phase with $\partial\eta/\partial t$ i.e. v is positive when the free surface is rising and negative when it is falling. It is emphasized that as the flow description above stems from potential (irrotational) flow, it yields a slip condition at the sea bed $y = 0$ for the horizontal velocity u. Thus, in no way do these expressions attempt to describe the flow in the bottom boundary layer region.

When dealing with surface waves it is usual to separate the total pressure beneath the surface as

$$p = \tilde{p} + \rho g(h - y) \qquad \text{or} \qquad \tilde{p} = p + \rho g(y - h) \tag{5.180}$$

where from linear wave theory

$$\tilde{p} = \rho g\frac{H}{2}\frac{\cosh(ky)}{\cosh(kh)}\sin\theta \tag{5.181}$$

is the (wave-induced) dynamic pressure i.e. that in addition to the hydrostatic pressure from the mean water level (corresponding to the second term on the right hand side of 5.180, left). It is seen that $\tilde{p}$ is in phase with η such that the wave-induced pressure is largest (smallest) beneath the wave crest (trough).

5.C Linear dispersion relation

The angular frequency and wave number (hence the wave length and period) are not independent, but rather linked via the so-called *linear dispersion relation*:

$$\omega^2 = gk \tanh(kh) \tag{5.182}$$

This condition is necessary in order for the wave description above to satisfy the linearized dynamic free surface condition, Eq. 5.172. Invoking Eq. 5.182, the wave celerity c is given by

$$c = \frac{L}{T} = \frac{\omega}{k} = \sqrt{\frac{g}{k}\tanh(kh)} \tag{5.183}$$

Note for generality that in a combined (co-linear) wave-plus-current situation, Eq. 5.182 can be generalized to

$$(\omega - kV)^2 = gk \tanh(kh) \tag{5.184}$$

where V is the depth-averaged current velocity. In a non-dimensional form, this is equivalently

$$\left(\frac{c}{\sqrt{gh}} - Fr\right)^2 = \frac{\tanh(kh)}{kh} \tag{5.185}$$

where $Fr = V/\sqrt{gh}$ is the Froude number.

5.D Wave drift

The period-averaged wave drift (i.e. the discharge per unit width that occurs in the region between the wave trough and crest), here utilizing an Eulerian frame of reference, is defined by

$$q_{drift} = \frac{1}{T}\int_0^T \int_{h-H/2}^{h+\eta} u dy dt \tag{5.186}$$

As a first approximation, justifiable since the wave height is assumed small, it is sufficient to consider the horizontal velocity from Eq. 5.178 applied at the still water level $y = h$. This corresponds to

$$u(y = h) = \frac{H\omega}{2}\frac{\cosh(kh)}{\sinh(kh)}\sin\theta = \frac{H\omega}{2}\frac{1}{\tanh(kh)}\sin\theta \tag{5.187}$$

Invoking Eqs. 5.187 and 5.176 in Eq. 5.186, and evaluating the integral, leads to the wave drift

$$q_{drift} = \frac{H^2\pi}{4T}\frac{1}{\tanh(kh)} \tag{5.188}$$

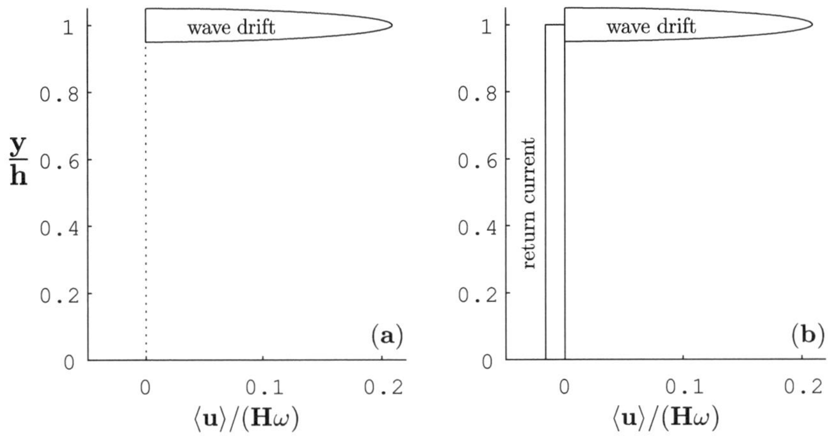

Fig. 5.96 Plot demonstrating the period-averaged velocity profile beneath a linear progressive wave (a) with wave drift only and (b) with a return current added such that there is zero net mass flux. The wave conditions considered correspond to $kh = 1$ and $H/h = 0.1$.

This matches Eq. 5.90. Notice that this is formally a second order $O(H^2)$ quantity forced from the first order (linear) $O(H)$ wave description.

If zero net mass flux is assumed over a given cross section then the above would imply the development of an additional uniform return current (again, assuming potential flow!) having strength

$$V = -\frac{q_{drift}}{h} = -\frac{H^2\pi}{4hT}\frac{1}{\tanh(kh)} \tag{5.189}$$

This would be the case in closed wave flumes. It is also commonly assumed to be valid in nearshore regions, taking the beach as impermeable (this obviously neglects any additional three-dimensional features such as rip currents, etc.). The resulting situation showing the period-averaged wave drift, both in the absence and presence of a return current, is depicted in Fig. 5.96.

It is also emphasized that as the return current V above is also an $O(H^2)$ quantity, it need not be incorporated into the generalized dispersion relation Eq. 5.184 if only a leading order description is desired i.e. the dispersion relation Eq. 5.182 is appropriate to leading order for pure wave problems i.e. in the absence of ambient currents.

5.E Energy flux

An exact expression for the instantaneous energy flux (per unit width) through a given vertical section beneath a surface wave is

$$E_f(t) = \int_0^{h+\eta} pudy + \int_0^{h+\eta} \left[\rho g(y-h) + \frac{\rho}{2}\left(u^2+v^2\right)\right] udy \qquad (5.190)$$

Here the first term accounts for the rate of work done by the pressure on this section, while the two quantities in the second integral respectively account for the flux of potential and kinetic energy through the section. From Eqs. 5.178 and 5.179 it is seen that both u and v are $O(H)$ quantities, meaning that the flux of kinetic energy will be $O(H^3)$, and can therefore be disregarded as small. Since the surface elevation η is also assumed small relative to h, we are left with the following leading-order approximation

$$E_f(t) = \int_0^h pudy + \int_0^h [\rho g(y-h)]\, udy = \int_0^h \tilde{p}udy \qquad (5.191)$$

where the definition of $\tilde{p}$ from Eq. 5.180 has been invoked. Inserting $\tilde{p}$ from Eq. 5.181 and u from Eq. 5.178 into the above and period averaging then leads to the period-averaged energy flux, which may be compactly written as

$$E_f = Ec_g, \qquad E = \frac{1}{8}\rho g H^2, \qquad c_g = \frac{c}{2}(1+G), \qquad G = \frac{2kh}{\sinh(2kh)} \qquad (5.192)$$

In the above E is the energy density and c_g is the group velocity; From the above (left-most equation) it is clear that c_g can be interpreted as the speed of wave energy propagation.

5.F Radiation stress

The radiation stress (Longuet-Higgins and Stewart, 1962, 1964) may be defined as the excess flow of momentum due to the presence of waves. [Note for historical reasons that the concept of a time-averaged momentum signature of short waves, termed the "wave thrust", was developed independently at the Technical University of Denmark by Prof. Helge Lundgren in the 1950s, eventually published as Lundgren (1963); See Svendsen (2006), p. 546.] The principal component in the direction of wave propagation corresponds to the total momentum flux due to the presence of waves, minus the mean flux in the absence of waves. This may be written as

$$S_{xx} = \left\langle \int_0^{h+\eta} (p+\rho u^2)dy \right\rangle - \int_0^h \rho g(h-y)dy \qquad (5.193)$$

where the hydrostatic pressure force from the mean water level (here assumed to be at $y = h$) corresponds to

$$\int_0^h \rho g(h-y)dy = \frac{1}{2}\rho g h^2 \tag{5.194}$$

To leading order, the momentum part can be found as

$$F_m = \left\langle \int_0^{h+\eta} (\rho u^2) dy \right\rangle \approx \left\langle \int_0^{h} (\rho u^2) dy \right\rangle = \frac{1}{2}E(1+G) \tag{5.195}$$

after invoking the velocity from Eq. 5.178. To leading order, it can be shown that the pressure part corresponds to

$$F_p = \left\langle \int_0^{h+\eta} p dy \right\rangle - \int_0^h \rho g(h-y)dy = \frac{1}{2}EG \tag{5.196}$$

It is noted that the formal derivation of this quantity requires going to second-order in the dynamic pressure $\tilde{p}$, see e.g. Svendsen (2006), p. 95–96 (an exception to this is in the shallow water regime, where the pressure is essentially hydrostatic, and this additional complication is not necessary). The total radiation stress is thus

$$S_{xx} = F_m + F_p = \frac{1}{2}E(1+2G) \tag{5.197}$$

where E and G are as defined in Eq. 5.192.

To see how the radiation stress S_{xx} arrises naturally within the period-averaged and depth-integrated momentum equation, see the Appendix to Chapter 7, Eq. 7.132.

5.G Shallow water approximations

It can be noticed that the dimensionless depth (i.e. wave number times depth) kh commonly appears in many of the expressions above. If kh is small, say $kh \leq \pi/10 \approx 0.3$, then the hyperbolic functions above are well approximated by their leading-order terms from their Taylor series:

$$\tanh(kh) = kh + O(kh)^3, \qquad \sinh(kh) = kh + O(kh)^3 \tag{5.198}$$

$$\cosh(kh) = 1 + O(kh)^2 \tag{5.199}$$

where kh can be replaced with any argument. Invoking these then leads to the following shallow water approximations:

$$u = \frac{H\omega}{2}\frac{1}{kh}\sin\theta = c\frac{\eta}{h} = \sqrt{\frac{g}{h}}\eta, \qquad v = \frac{H\omega}{2}\frac{y}{h}\cos\theta \tag{5.200}$$

$$\tilde{p} = \rho g \frac{H}{2} \sin\theta = \rho g \eta \tag{5.201}$$

$$\omega^2 = gk^2 h, \qquad G = 1, \qquad c_g = c = \sqrt{gh} \tag{5.202}$$

$$E_f = Ec, \qquad F_m = E, \qquad F_p = \frac{E}{2}, \qquad S_{xx} = \frac{3}{2}E \tag{5.203}$$

These are often reasonable in nearshore regions.

Notice that in shallow water the horizontal velocity u becomes uniform across the depth, whereas v grows linearly from the bottom. The dynamic pressure is equivalent to the hydrostatic pressure from the varying free surface.

5.H Deep water approximations

Conversely, if kh is large, say $kh \geq \pi$, then the following asymptotic approximations may be used:

$$\sinh(kh) = \frac{e^{kh} - e^{-kh}}{2} \to \frac{1}{2}e^{kh}, \qquad \cosh(kh) = \frac{e^{kh} + e^{-kh}}{2} \to \frac{1}{2}e^{kh} \tag{5.204}$$

such that:

$$\tanh(kh) = \frac{\sinh(kh)}{\cosh(kh)} \to 1, \qquad G \to 0 \tag{5.205}$$

These lead to the deep water approximations:

$$\omega^2 = gk_0, \qquad k_0 = \frac{\omega^2}{g}, \qquad L_0 = \frac{2\pi}{k_0} = \frac{gT^2}{2\pi}, \tag{5.206}$$

$$c_{g0} = \frac{c_0}{2}, \qquad c_0 = \sqrt{\frac{g}{k_0}} = \frac{g}{\omega} \tag{5.207}$$

$$E_{f0} = E_0 \frac{c_0}{2}, \qquad S_{xx0} = \frac{1}{2}E_0, \qquad E_0 = \frac{1}{8}\rho g H_0^2 \tag{5.208}$$

In the above a 0 subscript indicates a deep water quantity. An expression for the deep water wave height H_0 will be derived in the next sub-section, Eq. 5.211.

5.I Shoaling

Shoaling refers generally to changes in wave properties (primarily wave height H, wave length L and celerity c) as a wave propagates over varying (typically decreasing) depth, up to the point of incipient breaking. From offshore to the point just prior to incipient breaking, it is a good approximation to neglect energy losses that will occur in the boundary layer, and assume that the energy flux is conserved. Such an expression of conservation of energy flux from a hypothetical infinite depth (deep water, left hand side) to an arbitrary depth h (right hand side) may be written as

$$E_0 c_{g0} = E c_g \qquad \Rightarrow \qquad \frac{1}{8}\rho g H_0^2 \frac{c_0}{2} = \frac{1}{8}\rho g H^2 \frac{c}{2}(1+G) \tag{5.209}$$

where the 0 subscript again indicates a deep water quantity. Note that

$$\frac{c}{c_0} = \frac{\omega^2}{gk} = \tanh(kh) \tag{5.210}$$

where the dispersion relation, Eq. 5.182, has been utilized. Invoking this within Eq. 5.209 and simplifying leads to the following expression for the deep water wave height H_0

$$H_0^2 = H^2(1+G)\tanh(kh) \tag{5.211}$$

For given wave properties (e.g. H and T) at a particular depth h, the wave number can be readily found from a solution of the linear dispersion relation, Eq. 5.182. From this the deep water wave height H_0 can be computed from Eq. 5.211. Once this is known, the wave height H and wave number k at other depths h can be found from a solution of Eqs. 5.211 and 5.182. In this fashion linear shoaling problems (e.g. determining how H varies with h) may be solved.

It is very often useful to parameterize wave-induced processes in terms of deep water quantities H_0 and L_0 (e.g. the surf similarity parameter, Eq. 7.1), as these two parameters uniquely define the wave properties at any other depth h according to linear wave theory, as described above.

Note also that energy flux conservation between two locations (indicated by subscripts 1 and 2), both of which are in shallow water, becomes:

$$\frac{1}{8}\rho g H_1^2 \sqrt{gh_1} = \frac{1}{8}\rho g H_2^2 \sqrt{gh_2} \qquad \Rightarrow \qquad \frac{H_2}{H_1} = \left(\frac{h_1}{h_2}\right)^{1/4} \tag{5.212}$$

This is known as Green's law.

5.J Link to wave boundary layer quantities

From the above, wave boundary layer quantities utilized at length in Chapters 5 and 6 can now be linked to the basic surface wave parameters. For example, taking $y = 0$ in Eq. 5.178, the free-stream (near bed) horizontal velocity may be approximated from the linear potential flow theory applied at the bottom, leading to:

$$U_0 = u(y=0) = U_{0m}\sin\theta, \qquad U_{0m} = \frac{H\omega}{2\sinh(kh)} \tag{5.213}$$

since $\cosh(0) = 1$. Note that since potential flow admits a slip condition at the sea bed, there is little reason to account for the (typically small) wave boundary layer thickness δ in the approximation just above. Similarly, invoking the above, the amplitude of the free-stream orbital motion may be approximated as

$$a = \frac{U_{0m}}{\omega} = \frac{H}{2}\frac{1}{\sinh(kh)} \tag{5.214}$$

In terms of the basic wave parameters, the corresponding wave boundary layer Reynolds number is then

$$Re = \frac{aU_{0m}}{\nu} = \frac{H^2\omega}{4\sinh^2(kh)\nu} \tag{5.215}$$

where ν is again the kinematic fluid viscosity.

Hence, this appendix illustrates how wave boundary layer quantities (e.g. U_{0m} and a) may be easily estimated based on knowledge of local surface wave quantities (e.g. H and T), combined with the water depth h.

Numerous more complicated potential flow wave theories exist, which account for additional nonlinear effects. The essentials described above, however, are sufficient for a leading-order description of the flow and general behavior of surface waves and for understanding the contents of this book.

5.K References

(1) Clamond, D. and Dutykh, D. (2018): Accurate fast computation of steady two-dimensional surface gravity waves in arbitrary depth. J. Fluid Mech., vol. 844, 491–518.

(2) Fenton, J.D. (1988): The numerical solution of steady water wave problems. Comput. Geosci., vol. 14, 357–368.

(3) Longuet-Higgins, M.S. and Stewart, R.W. (1962): Radiation stress and mass transport in gravity waves with application to 'surf-beats'. J. Fluid Mech., vol. 8, 565–583.
(4) Longuet-Higgins, M.S. and Stewart, R.W. (1964): Radiation stress in water waves, a physical discussion with applications. Deep Sea Research, vol. 11, 529–563.
(5) Lundgren, H. (1963): Wave thrust and wave energy level. IAHR Proc. 10th Int. Congr. Assoc. Hydr. Res. London. vol. 1, Paper 1.20, 147–151.
(6) Madsen, P.A. and Fuhrman, D.R. (2012): Third-order theory for multi-directional irregular waves. J. Fluid Mech., vol. 698, 304-334.
(7) Sharma, J.N. and Dean, R.G. (1981): Second-order directional seas and associated wave forces. J. Soc. Petroleum Eng., vol. 21, 129-140.
(8) Svendsen, I.A. (2006): Introduction to Nearshore Hydrodynamics. World Scientific.

Chapter 6

Streaming in wave boundary layers

The preceding chapter has focused primarily on wave boundary layers which develop beneath spatially homogeneous oscillatory (typically, sinusoidal) free-stream flows. This corresponds precisely to the situation in conventional oscillating tunnel facilities, e.g. Fig. 5.21, and is a sufficiently accurate description of real wave boundary layers for many purposes. However, in reality, the flow beneath water waves will generally not be spatially homogeneous. This is due either to their progressive nature alone (i.e. variations over their wave length), or for other more deliberate reasons e.g. due to sloping beds or changing bed roughness. Moreover, while waves may, to leading-order, be described as sinusoidally shaped (see the Appendix to Chapter 5), in reality their shape will often deviate significantly from this idealized form due to nonlinear (finite amplitude) effects. This will also affect their corresponding near-bed free stream flows, and hence the boundary layers that develop beneath.

The present chapter extends the investigations made in the previous chapter to consider additional effects that arise either due to (1) spatial non-uniformity or (2) deviations from the pure sinusoidal description. As will be shown, whereas spatially-uniform sinusoidal wave boundary layers inevitably lead to zero net (period-averaged) boundary layer flux (obviously, as the forward and backward half cycles are simply mirrored), incorporation of these various additional effects will invariably lead to net period-averaged flows within the boundary layer, even in the absence of externally generated currents. Such phenomenon can be collectively described as *steady streaming* in wave boundary layers. The most common of these effects will be considered individually in this chapter, free of other effects for simplicity. Specifically, we will consider the streaming beneath progressive waves in Section 6.1, due to sloping beds (converging-diverging oscillatory flows) in Section 6.2, due to changes in bottom roughness in Section 6.3, and due to non-sinusoidal wave shapes in Section 6.4. Some examples of streaming measured beneath real (laboratory or field) waves will be presented in Section 6.5, and the importance of streaming and wave shape in coastal sediment transport is discussed in Section 6.6.

The chapter Appendix additionally derives the vector vorticity equation, and explains a vorticity source term important in promoting streaming in the form of secondary flows when there is a change in bed roughness, as explained in Section 6.3.

6.1 Streaming beneath sinusoidal progressive waves

Let us begin our discussion by considering the effects within a wave boundary layer associated with spatial non-uniformity due to the progressive nature of a sinusoidal wave train (see e.g. Fig. 5.2) i.e. now accounting for x-variations along the wave length, which were previously neglected as "small". We will first consider the laminar case analytically, which is itself an enlightening introduction to the phenomenon of streaming in wave boundary layers. This will in turn be followed by analogous presentation of boundary layer streaming beneath progressive waves in the turbulent regime (both in hydraulically smooth and rough conditions).

6.1.1 *Laminar theory*

6.1.1.1 *Eulerian analysis*

Recall from Section 5.1, Eq. 5.12, that the theoretical laminar solution for the boundary layer beneath a spatially-uniform sinusoidal free stream motion

$$U_0 = U_{0m} \sin(\theta) \tag{6.1}$$

where $\theta = \omega t$ and $\omega = 2\pi/T$ (T being the wave period), corresponds to

$$u = U_{0m} \sin\theta - U_{0m} e^{-\xi} \sin(\theta - \xi) \tag{6.2}$$

where here $\xi = y/\delta_1$, and $\delta_1 = \sqrt{2\nu/\omega}$ is again the Stokes length. Notice that upon inspection of the local continuity equation

$$\frac{\partial u}{\partial x} + \frac{\partial v}{\partial y} = 0 \tag{6.3}$$

that a lack of x-variation in u (coupled with a no-slip condition $u = v = 0$ at the bottom $y = 0$), implies that there would be no vertical velocity i.e.

$v = 0$. Hence, in the spatially uniform condition, the complete neglect of the convective acceleration terms in the Navier-Stokes equation is justified.

While the spatially-uniform situation is ideally satisfied within e.g. experimental oscillating tunnel facilities (e.g. Fig. 5.21), real life progressive waves are, of course, not spatially uniform. Rather, they vary periodically in the progressive x-direction, just as they do in time. To consider the consequences of these spatial variations, we may simply consider the solution above as the leading-order result, but with the phase function now replaced by

$$\theta = \omega t - kx, \tag{6.4}$$

where $k = 2\pi/L$ is the wave number, and L the wave length. Invoking Eq. 6.4, and then inserting Eq. 6.2 into Eq. 6.3 then leads, after integration, to the corresponding solution for the vertical velocity

$$v = ykU_{0m}\cos\theta - \frac{1}{2}U_{0m}k\delta_1(\cos\theta + \sin\theta) + \frac{1}{2}U_{0m}k\delta_1 e^{-\xi}\left[\cos(\theta - \xi) + \sin(\theta + \xi)\right] \tag{6.5}$$

Note that for large ξ, i.e. far outside the boundary layer, this reduces to

$$\frac{v_\infty}{kU_{0m}\delta_1} = \frac{1}{2}\left[(2\xi - 1)\cos\theta - \sin\theta\right] \tag{6.6}$$

Notice that the present solution, which incorporates boundary layer effects, yields both $\cos\theta$ and $\sin\theta$ contributions. This is in stark contrast to expectations based on a purely potential flow wave solution, which would yield vertical velocities varying as

$$\frac{v_\infty}{kU_{0m}\delta_1} = \xi\cos\theta \tag{6.7}$$

Now consider the individual terms in the x-component of the Navier-Stokes equation, which may be written as

$$\frac{\partial u}{\partial t} + \frac{\partial(uu)}{\partial x} + \frac{\partial(uv)}{\partial y} = -\frac{1}{\rho}\frac{\partial p}{\partial x} + \frac{1}{\rho}\frac{\partial \tau}{\partial y} \tag{6.8}$$

Notice that convective-terms have now been retained, as they will be active due to the spatially varying velocities. Upon period-averaging (denoted by $\langle\cdot\rangle$) each of these terms, after invoking the previous solutions for u and v, it is found that

$$\left\langle\frac{\partial u}{\partial t}\right\rangle = \left\langle\frac{\partial(uu)}{\partial x}\right\rangle = \left\langle\frac{1}{\rho}\frac{\partial p}{\partial x}\right\rangle = 0 \tag{6.9}$$

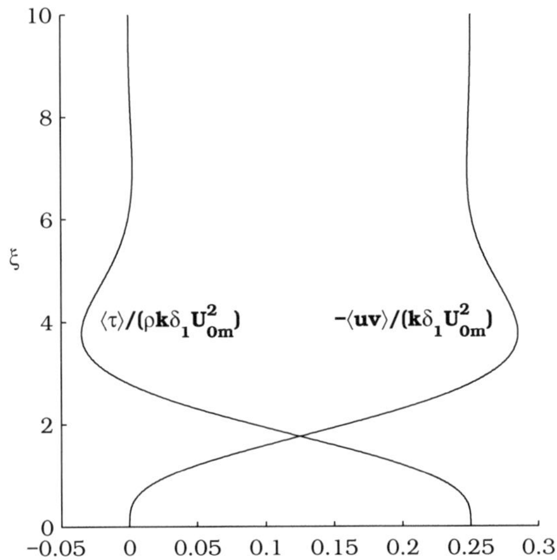

Fig. 6.1 Laminar solution for $-\langle uv\rangle$.

Hence, the period-averaged Navier-Stokes equation reduces to

$$\frac{\partial\langle uv\rangle}{\partial y}=\frac{\partial\langle\tau\rangle}{\partial y} \tag{6.10}$$

Considering the first term, after invoking Eq. 6.2 and Eq. 6.5, we find that

$$\frac{\langle uv\rangle}{k\delta_1 U_{0m}^2}=\frac{1}{2}e^{-\xi}[\cos\xi-\cosh\xi+\xi\sin\xi] \tag{6.11}$$

Integrating Eq. 6.10 leads immediately to

$$\langle\tau\rangle=\langle uv\rangle+c_1 \tag{6.12}$$

Invoking Eq. 6.11, and requiring that $\langle\tau\rangle$ goes to zero for large ξ (i.e. no shear stress in the free stream region outside the boundary layer) leads to the integration constant

$$c_1=\frac{1}{4}kU_{0m}^2\delta_1 \tag{6.13}$$

Hence, the period-averaged shear stress distribution may be written

$$\frac{\langle\tau\rangle}{\rho kU_{0m}^2\delta_1}=-\frac{1}{4}e^{-2\xi}+\frac{1}{2}e^{-\xi}(\cos\xi+\xi\sin\xi) \tag{6.14}$$

The expressions Eq. 6.11 and Eq. 6.14 are plotted in Fig. 6.1. From inspection of this figure, or for that matter Eq. 6.14, it is seen that the inclusion of the convective terms forces a period-averaged bed shear stress (taking $\xi = 0$) in the direction of the wave propagation corresponding to

$$\frac{\langle \tau_0 \rangle}{\rho k U_{0m}^2 \delta_1} = \frac{1}{4} \tag{6.15}$$

Moreover, it is clear that generally

$$\frac{\langle \tau_0 \rangle}{\rho} = \frac{\langle \tau \rangle}{\rho} - \langle uv \rangle \tag{6.16}$$

and therefore specifically

$$\frac{\langle \tau_0 \rangle}{\rho} = -\langle U_0 v_\infty \rangle \qquad \text{or} \qquad \langle \tau_0 \rangle = -\rho \langle U_0 v_\infty \rangle \tag{6.17}$$

since, again, it is assumed that $\langle \tau \rangle$ goes to zero for large ξ. The quantity $-\rho\langle U_0 v_\infty \rangle$ represents the period-averaged vertical flux of horizontal momentum through the top of a control volume, which is balanced by the period-averaged bed shear stress $\langle \tau_0 \rangle$ at the bottom of such a control volume. That $\langle U_0 v_\infty \rangle \neq 0$ can be surmised from Eqs. 6.1 and 6.6, both of which have $\sin\theta$ contributions. (This is again in contrast to the resulting v_∞ from potential flow, Eq. 6.7!)

Let us now solve for the corresponding period-averaged flow within the boundary layer. Noting that

$$\frac{\langle \tau \rangle}{\rho} = \nu \frac{\partial \langle u \rangle}{\partial y} \tag{6.18}$$

we may similarly integrate Eq. 6.14 to arrive at the following expression for the so-called steady streaming (period-averaged Eulerian velocity) profile within the boundary layer

$$\frac{\langle u \rangle c}{U_{0m}^2} = \frac{3}{4} + \frac{1}{4} e^{-2\xi} + e^{-\xi} \left[\frac{\sin\xi}{2} - \cos\xi - \frac{\xi}{2}(\cos\xi + \sin\xi) \right] \tag{6.19}$$

where $c = \omega/k$ is the wave celerity, and where the integration constant has been taken as zero in order to satisfy the no-slip condition at the sea bed. This solution is plotted in Fig. 6.2. It corresponds to the period-averaged velocity profile that would be measured from fixed-point measurements i.e. utilizing an Eulerian frame of reference. For large ξ, corresponding to the top of the boundary layer, only the first term remains and the solution tends to the so-called *streaming velocity*

$$\frac{u_\infty c}{U_{0m}^2} = \frac{3}{4} \tag{6.20}$$

These results were first found by Longuet-Higgins (1958). Hence, the existence of period-averaged flow in the boundary layer beneath progressive

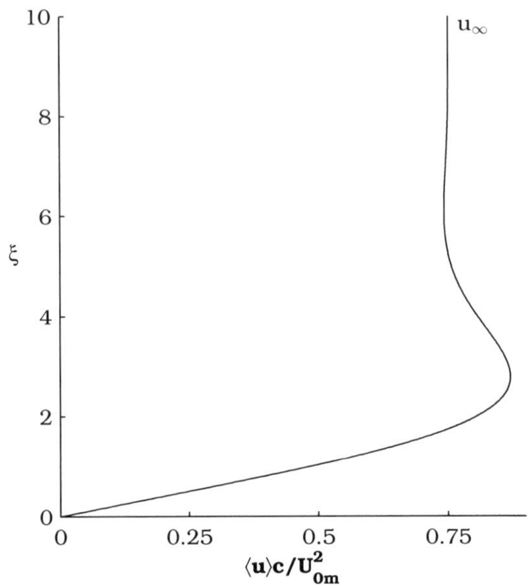

Fig. 6.2 Laminar solution for the period-averaged velocity profile in the wave boundary layer.

waves, in the manner described above, is commonly called Longuet-Higgins type streaming. Note that the streaming solution is a second-order phenomenon, with strength U_{0m}/c relative to the leading-order purely oscillatory flow. Note also that in a progressive sinusoidal wave, since $c = \omega/k$ and $a = U_{0m}/\omega$, this factor is equivalent to

$$\frac{U_{0m}}{c} = \frac{U_{0m}k}{\omega} = ak \tag{6.21}$$

Since $\delta \sim a$ and $k \sim L^{-1}$ it is seen that $ak \sim \delta/L$, hence treating the streaming phenomenon as a second-order phenomenon is in line with the boundary layer assumption that vertical flow variations are much more rapid than horizontal variations.

To conclude: taking into account the spatial variation inherent within a sinusoidal progressing wave results in (1) a period-averaged boundary layer drift known as *steady streaming*, as well as (2) a period-averaged bed shear stress in the direction of wave propagation. Neither of these phenomena would be predicted from spatially-uniform considerations; they are promoted as a second-order response (in terms of $U_{0m}/c = ak$, see Eq. 6.19) to the convective acceleration terms, which are in reality not completely negligible.

6.1.1.2 *Lagrangian analysis*

While the present chapter will primarily focus on analysis from an Eulerian perspective, as above, we may also note that the Lagrangian (particle following) velocity vector $\mathbf{u}_L = (u_L, v_L, w_L)$ is related to the Eulerian velocity vector $\mathbf{u} = (u, v, w)$ according to

$$\mathbf{u}_L = \mathbf{u}\left(\mathbf{x}_0 + \int_0^t \mathbf{u}_L dt, t\right) \tag{6.22}$$

where $\mathbf{x}_0 = (x_0, y_0, z_0)$ is the particle position at $t = 0$. Expressed as a Taylor series (keeping the first two terms), this may be written as

$$\mathbf{u}_L = \mathbf{u}\,(\mathbf{x}_0, t) + \int_0^t \mathbf{u}_L dt \cdot \nabla \mathbf{u}(\mathbf{x}_0, t) + \ldots \tag{6.23}$$

where $\nabla = (\partial/\partial x, \partial/\partial y, \partial/\partial z)$ is the gradient operator. Consider now a progressive wave propagating in the pure x direction, as before. The x component of the above is

$$u_L = u + \left(\int_0^t u_L dt\right)\frac{\partial u}{\partial x} + \left(\int_0^t v_L dt\right)\frac{\partial u}{\partial y} \tag{6.24}$$

Since the Lagrangian velocities u_L and v_L appear exclusively in the second-order terms in Eqs. 6.23 and 6.24 above, it is consistent to the order of accuracy retained to replace these with their leading-order contributions $u_L = u$ and $v_L = v$. Further period-averaging then leads to

$$\langle u_L\rangle = \langle u\rangle + \left\langle\left(\int_0^t u dt\right)\frac{\partial u}{\partial x} + \left(\int_0^t v dt\right)\frac{\partial u}{\partial y}\right\rangle \tag{6.25}$$

Consistent with the arguments above, we may utilize the (second order) period-averaged Eulerian streaming velocity profile $\langle u\rangle$ from Eq. 6.19 in the first term, whereas we may utilize the leading-order contributions for u and v from Eqs. 6.2 and 6.5 in the second and third terms. Invoking these, after simplifying, results in the so-called mass transport (i.e. the period-averaged horizontal Lagrangian velocity) profile:

$$\frac{\langle u_L\rangle c}{U_{0m}^2} = \frac{5}{4} + \frac{3}{4}e^{-2\xi} - 2e^{-\xi}\cos\xi \tag{6.26}$$

This corresponds to the average velocity of individual fluid particles, and would hence correspond e.g. to the observed propagation velocity of a small patch of dye injected into the boundary layer at a particular $\xi = y/\delta_1$. At the top of the boundary layer (i.e. large ξ) the so-called *mass transport velocity* is

$$\frac{u_{L,\infty} c}{U_{0m}^2} = \frac{5}{4} \tag{6.27}$$

Hence, as might be expected the positive period averaged streaming velocity, Eq. 6.20, likewise corresponds to a positive mass transport velocity, as shown directly above. These would both be zero in the case of a horizontally uniform oscillatory boundary layer flow, but arise naturally in the present context where spatial non-uniformity (hence finite convective acceleration terms) has been taken into account.

Note that Eqs. 6.26 and 6.27 were originally derived by Longuet-Higgins (1953) and these results match those presented on his p. 567.

6.1.2 *Turbulent streaming beneath progressive waves*

Let us now turn attention to the related phenomena that will occur in wave boundary layers in the fully turbulent regime, again corresponding to either hydraulically rough beds ($k_s^+ = k_s U_f/\nu \geq 70$) and/or Reynolds numbers $Re = U_{0m} a/\nu \geq 1.5 \times 10^5$ where $a = U_{0m}/\omega$. We here adopt an Eulerian perspective.

It turns out that the expected streaming-induced period-averaged bed shear stress beneath progressive sinusoidal waves can be quite well approximated, based only on computed and/or measured results having spatial uniformity. This first requires a means of approximating the vertical velocity. Assuming a constant form wave propagation corresponds to the material derivative $d/dt = 0$ (this is quite reasonable on flat beds where the decay in wave height can be locally neglected), hence time and space derivatives are related via

$$\frac{d}{dt} = \frac{\partial}{\partial t} + c\frac{\partial}{\partial x} = 0 \qquad \text{or} \qquad \frac{\partial}{\partial x} = -\frac{1}{c}\frac{\partial}{\partial t} \tag{6.28}$$

After invoking this, the local continuity equation for the mean (phase-averaged) flow becomes

$$\frac{\partial \overline{u}}{\partial x} + \frac{\partial \overline{v}}{\partial y} = 0 \qquad \text{or} \qquad \frac{\partial \overline{v}}{\partial y} - \frac{1}{c}\frac{\partial \overline{u}}{\partial t} = 0 \tag{6.29}$$

Integrating to some vertical position δ taken as well above the boundary layer, we may then obtain

$$v_\infty = \frac{1}{c}\int_0^\delta \left(\frac{\partial \overline{u}}{\partial t}\right) dy \tag{6.30}$$

where to leading order (i.e. momentarily neglecting convective terms)

$$\frac{\partial \overline{u}}{\partial t} = -\frac{1}{\rho}\frac{\partial \overline{p}}{\partial x} + \frac{1}{\rho}\frac{\partial \tau}{\partial y} = \frac{\partial U_0}{\partial t} + \frac{1}{\rho}\frac{\partial \tau}{\partial y} \tag{6.31}$$

Note that seeking only the leading-order expression for $\overline{v}$ is justified, since this quantity only comes into play within (higher-order) convective terms in the horizontal Navier-Stokes equation. The vertical velocity is then, approximately

$$v_\infty = \frac{1}{c}\int_0^\delta \left(\frac{\partial U_0}{\partial t} + \frac{1}{\rho}\frac{\partial \tau}{\partial y}\right) dy = \frac{1}{c}\frac{\partial U_0}{\partial t}\delta - \frac{\tau_0}{\rho} \tag{6.32}$$

where zero shear stress far above the boundary layer has been assumed, as before. Multiplying both sides by U_0 and period-averaging (causing the time derivative to drop out), we finally obtain

$$\langle \tau_0 \rangle = -\rho \langle U_0 v_\infty \rangle = \frac{\langle U_0 \tau_0 \rangle}{c} \tag{6.33}$$

Hence, if the free stream velocity U_0 and bed shear stress τ_0 are known for a spatially-uniform wave boundary layer, they may be utilized directly in the right-hand-side of Eq. 6.33 to obtain an approximation of the period-averaged bed shear stress that will be induced by progressive wave streaming effects. Making this approximation, rather remarkably, requires zero knowledge of the actual period-averaged velocity profile. Note that the above considerations follow Brøker (1985) and are quite general; In the laminar case the process described above is an alternative way to obtain Eq. 6.15.

For turbulent wave boundary layers, determination of the required velocities and bed shear stresses generally requires either measurements or model (analytical or numerical) results, for both U_0 and especially the resulting τ_0. For the present purposes, results from oscillatory (i.e. sinusoidal free stream) turbulent wave boundary layer numerical simulations on both smooth and rough beds will be presented. These simulations utilize the one-dimensional vertical (1DV) Reynolds-averaged Navier-Stokes (RANS) equation model of Fuhrman et al. (2013) as described in Section 5.12, making use of the standard k-ω turbulence closure model (Wilcox, 2006, 2008). The results for the period-averaged streaming-induced bed shear stresses based on Eq. 6.33 are summarized in Figs. 6.3 and 6.4, for smooth-turbulent and rough-turbulent wave boundary layers, respectively. For smooth-turbulent oscillatory wave boundary layers, the streaming-induced period-averaged bed shear stress expectedly depends on the Reynolds number, and is well-approximated by

$$\frac{\langle \tau_0 \rangle c}{\rho U_{0m}^3} = 0.0119 Re^{-0.1795} \tag{6.34}$$

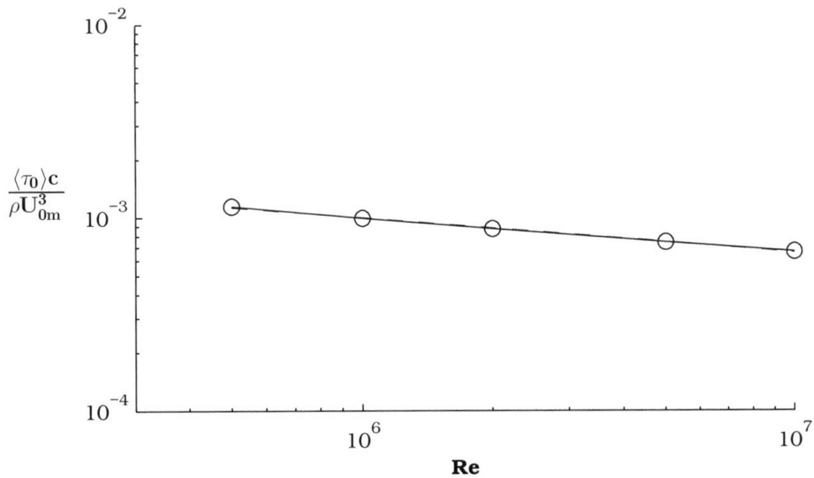

Fig. 6.3 Period-averaged bed shear stress beneath smooth-turbulent wave boundary layer. Full line with circles: computed with k-ω turbulence model; Dashed line: $0.0119(Re)^{-0.1795}$.

Conversely, in the rough-turbulent regime the non-dimensional period-averaged bed shear stress depends on the dimensionless roughness a/k_s, and is reasonably approximated by

$$\frac{\langle \tau_0 \rangle c}{\rho U_{0m}^3} = 0.013 \left(\frac{a}{k_s} \right)^{-0.29} \tag{6.35}$$

These expressions are depicted as the dashed lines in Figs. 6.3 and 6.4, which again match the exact numerical values (full lines with symbols) closely.

To determine the streaming velocities in turbulent wave boundary layer flows, Fuhrman et al. (2013) have also made long-term (spanning several hundred periods) simulations with the same model, where convective terms were approximated via Eq. 6.28, with vertical velocities obtained through subsequent numerical integration of the continuity equation Eq. 6.29, as also explained in Section 5.12. Their results for the period-averaged horizontal velocity (streaming) profiles above hydraulically smooth and rough beds are depicted in Figs. 6.5 and 6.6, respectively. As can be seen, the resulting (non-dimensional) streaming velocities are essentially constant. Based on this we may generalize Eq. 6.20 to

$$\frac{u_\infty c}{U_{0m}^2} = \Psi \tag{6.36}$$

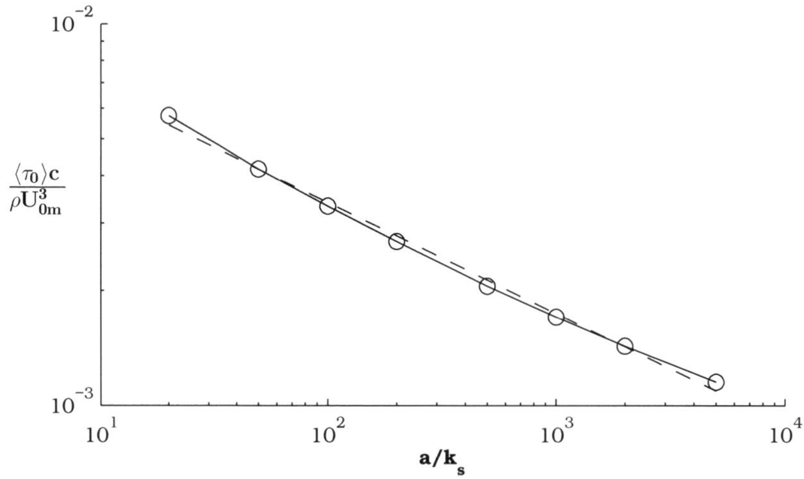

Fig. 6.4 Period-averaged bed shear stress beneath rough-turbulent wave boundary layer. Full line with circles: computed with k-ω turbulence model; Dashed line: $0.013(a/k_s)^{-0.29}$.

where $\Psi \approx 0.37$ for smooth-turbulent conditions and $\Psi \approx 0.34$ for rough-turbulent conditions. It can be noted that the latter value is in excellent agreement with the (very limited) range $0.3375 \leq \Psi =\leq 0.345$ found independently by Holmedal and Myrhaug (2009) using a similar RANS model coupled with k-ε turbulence closure. The essentially constant dimensionless streaming velocities under hydraulically rough conditions found using such two-equation turbulence closure models are in slight contrast to the very weak variation found by Brøker (1985) based on a momentum integral approach, see e.g. Fig. 2.25 of Fredsøe and Deigaard (1992). These differences can be considered as minor, however, and in all cases the predicted streaming velocities in rough-turbulent conditions are similar and considerably less than the $\Psi = 3/4 = 0.75$ predicted based on laminar theory, Eq. 6.20.

The reason for the reduced non-dimensional streaming in turbulent flows is directly linked to the significant reduction of the phase lead of the near bed flow within the boundary layer flows (i.e. from 45° near the bed in laminar boundary layers to say 5–10° in smooth-turbulent wave boundary layers, see Fig. 5.39b; see also Fig. 5.41 for rough bed results), which is again a natural consequence of increased vertical mixing of momentum once turbulent fluctuations are present.

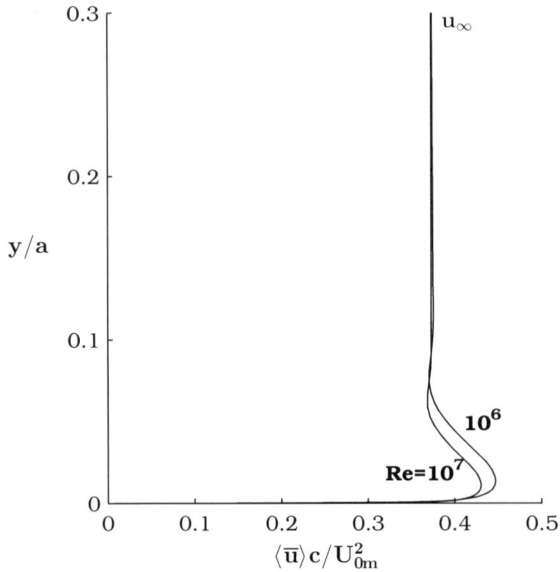

Fig. 6.5 Computed period-averaged velocity profiles in smooth turbulent streaming. Adapted from Fuhrman et al. (2013).

6.2 Streaming in converging-diverging oscillatory flow

In Section 6.1 the classical Longuet-Higgins type streaming phenomenon beneath progressive waves was considered in detail. This was clearly shown to arise from the convective acceleration terms, which became non-negligible after taking into account spatial variations along the direction of wave propagation. A related phenomenon occurs due to spatial non-uniformity associated with waves propagating on sloping beds, locally resulting in an oscillatory converging-diverging flow. As before in the previous section, let us first consider such effects based on laminar theory, which will in turn be followed by quantification of situations falling in the turbulent regime.

6.2.1 *Laminar theory*

Let us now consider the effects of a sloping bed on an oscillatory wave boundary layer e.g. as would occur in a converging-diverging flow in an oscillating tunnel, with small Reynolds number, $Re < 1.5 \times 10^5$, such that laminar theory applies. A convergent tunnel (with local half-depth h at $x = 0$ and slope $S = \tan\beta = -dh/dx$, see the definition sketch in Fig. 6.7)

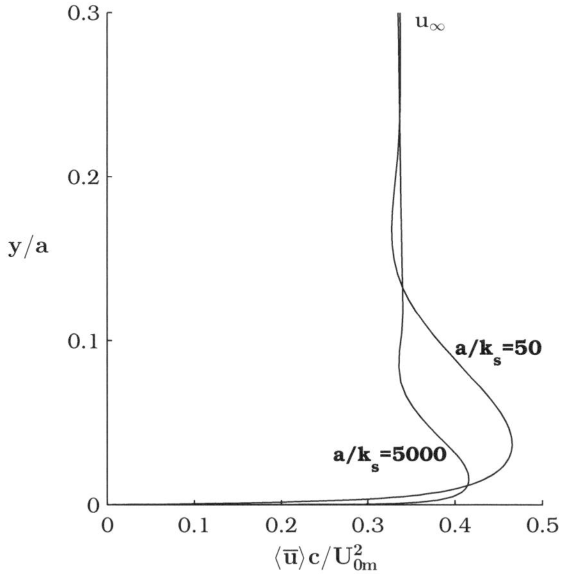

Fig. 6.6 Computed period-averaged velocity profiles in rough turbulent streaming. Adapted from Fuhrman et al. (2013).

leads to the depth and free stream velocity variation previously given in Eq. 5.152 and Eq. 5.153, respectively. Recall also that differentiating with respect to x is locally equivalent to multiplication by S/h, care of Eq. 5.154. To derive the laminar solution for the oscillatory flow on a mild slope, it is convenient to take the previously found oscillatory laminar solution Eq. 6.2 as the leading-order (zero slope) solution, now with $\theta = \omega t$, as we wish to here isolate the effects of the converging-diverging oscillatory flow:

$$u = U_{0m} \sin(\omega t) - U_{0m} e^{-\xi} \sin(\omega t - \xi) \tag{6.37}$$

where again $\xi = y/\delta_1$. Invoking this within Eq. 5.154 then directly gives the horizontal velocity gradient due to the sloping bed

$$\frac{\partial u}{\partial x} = \frac{S U_{0m}}{h} \big(\sin(\omega t) - e^{-\xi} \sin(\omega t - \xi)\big) \tag{6.38}$$

By integrating the local continuity equation Eq. 6.3, the corresponding vertical velocity may also be obtained:

$$\begin{aligned} v = \frac{U_{0m} \delta_1 S}{h} \bigg[& -\frac{1}{2} \cos(\omega t) + \frac{e^{-\xi}}{2} \cos(\xi - \omega t) \\ & + \frac{1}{2} \sin(\omega t) - \frac{\xi}{4} \sin(\omega t) + \frac{e^{-\xi}}{2} \sin(\xi - \omega t) \bigg] \end{aligned} \tag{6.39}$$

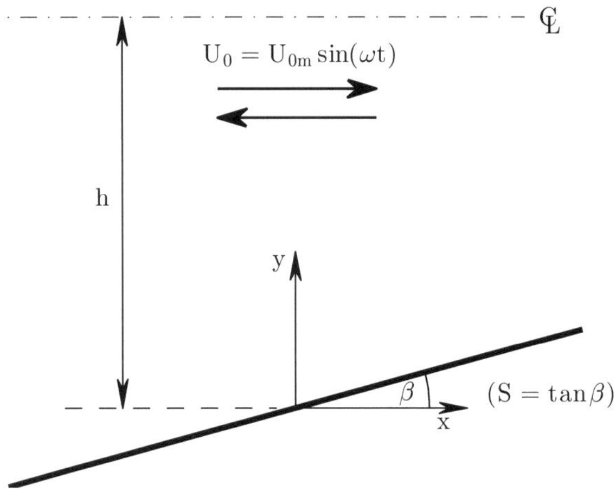

Fig. 6.7 Definition sketch for oscillatory converging-diverging flow. The lower half of a converging tunnel is depicted, where the top dashed-dotted line represents the tunnel centerline. The tunnel is assumed to be symmetric about this centerline.

Let us now consider the Navier-Stokes equation in the following form

$$\frac{\partial u}{\partial t} + u\frac{\partial u}{\partial x} + v\frac{\partial u}{\partial y} = -\frac{1}{\rho}\frac{\partial p}{\partial x} + \frac{1}{\rho}\frac{\partial \tau}{\partial y} \tag{6.40}$$

Upon period-averaging (assuming a periodic equilibrium condition has been reached), the time derivative will drop out, and this reduces to

$$\left\langle u\frac{\partial u}{\partial x}\right\rangle + \left\langle v\frac{\partial u}{\partial y}\right\rangle = -\left\langle \frac{1}{\rho}\frac{\partial p}{\partial x}\right\rangle + \left\langle \frac{1}{\rho}\frac{\partial \tau}{\partial y}\right\rangle \tag{6.41}$$

Outside the boundary layer, shear stresses (and therefore also $\partial u/\partial y$) may be neglected, leading to the period-averaged pressure gradient

$$\left\langle \frac{1}{\rho}\frac{\partial p}{\partial x}\right\rangle = -\left\langle U_0\frac{\partial U_0}{\partial x}\right\rangle = -\frac{SU_{0m}^2}{2h} \tag{6.42}$$

where the horizontal gradient operation Eq. 5.154 has been invoked on the leading-order free stream velocity Eq. 6.1. Notice that the (constant) pressure gradient induced by the converging-diverging flow is negative, that is, we have increasing pressures in the divergent direction (i.e. the direction of decreasing flow velocities), as should be expected based e.g. based on the Bernoulli equation. Making use of Eq. 6.37 and Eq. 6.39 likewise leads to the following period-averaged solutions for the horizontal and vertical convective terms

$$\left\langle u\frac{\partial u}{\partial x}\right\rangle = \frac{SU_{0m}^2}{2h}\left[1 + e^{-2\xi} - 2e^{-\xi}\cos\xi\right] \tag{6.43}$$

$$\left\langle v\frac{\partial u}{\partial y}\right\rangle = \frac{SU_{0m}^2}{2h}\left[e^{-\xi}\sin\xi - \xi e^{-\xi}\left(\cos\xi + \sin\xi\right)\right] \tag{6.44}$$

Invoking these within Eq. 6.41, and solving for the last term, leads to

$$\left\langle \frac{1}{\rho}\frac{\partial \tau}{\partial y}\right\rangle = \frac{SU_{0m}^2}{2h}\left[e^{-2\xi} - e^{-\xi}\left(2\cos\xi - \sin\xi\right) - \xi e^{-\xi}\left(\cos\xi + \sin\xi\right)\right] \tag{6.45}$$

Integrating once gives the period-averaged shear stress distribution

$$\frac{\langle\tau\rangle}{\rho} = \nu\frac{\partial u}{\partial y} = \frac{SU_{0m}^2\delta_1}{2h}\left[-\frac{e^{-2\xi}}{2} + e^{-\xi}(\cos\xi - 2\sin\xi) + \xi e^{-\xi}\cos\xi\right] \tag{6.46}$$

where the integration constant has been neglected, such that the shear stress vanishes for large ξ. Taking $\xi = 0$, the slope-induced period-averaged bed shear stress corresponds to

$$\frac{\langle\tau_0\rangle h}{\rho\delta_1 SU_{0m}^2} = \frac{1}{4} \tag{6.47}$$

As this is positive, it is directed in the convergent direction. The full solution for the period-averaged shear stress profile Eq. 6.46 is plotted in Fig. 6.8.

Integrating Eq. 6.46 once more, we obtain an expression for the period-averaged velocity profile

$$\frac{\langle u\rangle}{U_{0m}} = \frac{Sa}{h}\left[\frac{e^{-2\xi}}{4} + \frac{e^{-\xi}}{2}\cos\xi + 2e^{-\xi}\sin\xi + \frac{e^{-\xi}}{2}\xi(\sin\xi - \cos\xi) + C\right] \tag{6.48}$$

No-slip at the bed requires the integration constant to be

$$C = -\frac{3}{4} \tag{6.49}$$

Hence, the final solution for the period-averaged horizontal velocity is

$$\frac{\langle u\rangle}{U_{0m}} = \frac{Sa}{h}\left[\frac{e^{-2\xi}}{4} + \frac{e^{-\xi}}{2}\cos\xi + 2e^{-\xi}\sin\xi + \frac{e^{-\xi}}{2}\xi(\sin\xi - \cos\xi) - \frac{3}{4}\right] \tag{6.50}$$

For large ξ we find the streaming velocity

$$\frac{u_\infty}{U_{0m}} = -\frac{3}{4}\frac{Sa}{h} \tag{6.51}$$

The period-averaged velocity profile is plotted in Fig. 6.9. This solution is equivalent to that originally derived by Bijker et al. (1974), who instead presented his results in terms of $\partial a/\partial x$, which after invoking Eq. 5.154 is equivalent to

$$\frac{\partial a}{\partial x} = \frac{Sa}{h} \tag{6.52}$$

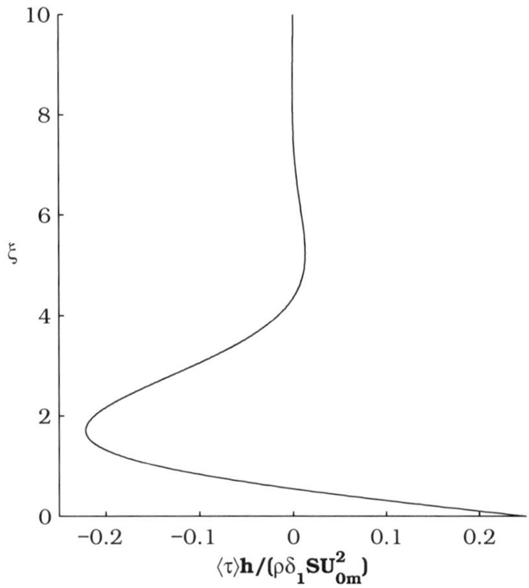

Fig. 6.8 Laminar solution for the period-averaged shear stress in an oscillatory converging-diverging flow.

where the quantity on the right hand side has been utilized above.

In line with the positive bed shear stress, even though the streaming velocity at the top of the boundary layer is technically negative, the laminar theory results in a near-bed region of positive streaming (likewise positive bed shear stress).

6.2.2 *Turbulent streaming in oscillatory converging-diverging flow*

Based on the preceding section, the analogous effects in turbulent oscillatory converging-diverging flow can be determined, using a k-ω turbulence closure model as before, where the local effects of a converging-diverging flow are accounted for as presented in Section 5.12.2. To maintain turbulent conditions that are analogous with the laminar theory considered in the preceding sub-section, it is assumed that there is zero shear outside the boundary layer. Hence no constant is added to the driving pressure gradient in Eq. 5.159 (doing otherwise would drive an additional current, which will in fact be demonstrated later in this sub-section, in connection

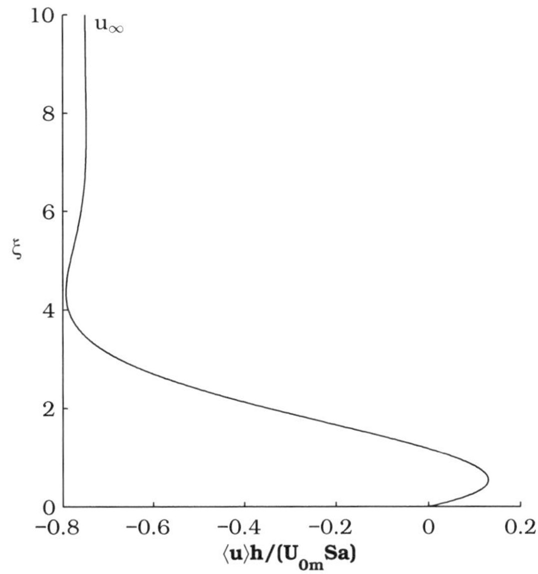

Fig. 6.9 Laminar wave boundary layer solution for the period-averaged horizontal velocity induced by a sloping bed.

with experimental studies in oscillating tunnels).

Utilizing the numerical model described above, several scenarios have been simulated to characterize the turbulent regimes of oscillatory converging-diverging flow, building further on the previous work of Fuhrman et al. (2009a). As examples, resulting period-averaged boundary layer streaming velocity profiles are depicted in Fig. 6.10 for a hydraulically smooth and rough bed, after equilibrium (negligible period-averaged acceleration) conditions have been reached. Qualitatively consistent with the laminar theory discussed previously (Fig. 6.9), both depict a thin region of positive streaming (directed towards the convergent end, see Fig. 6.7) near the bed, and negative streaming (directed towards the divergent end) further away from the bed. Roughness effects are seen to enhance the magnitude of the negative streaming velocities, as they will generally enhance turbulence and shear stresses. The larger boundary layer thickness in the rough bed cases is also clearly evident in these profiles.

The layer of positive near-bed streaming is important, and can be explained simply and physically based on considering the differences in the two half-cycles within an oscillatory converging-diverging flow: During the

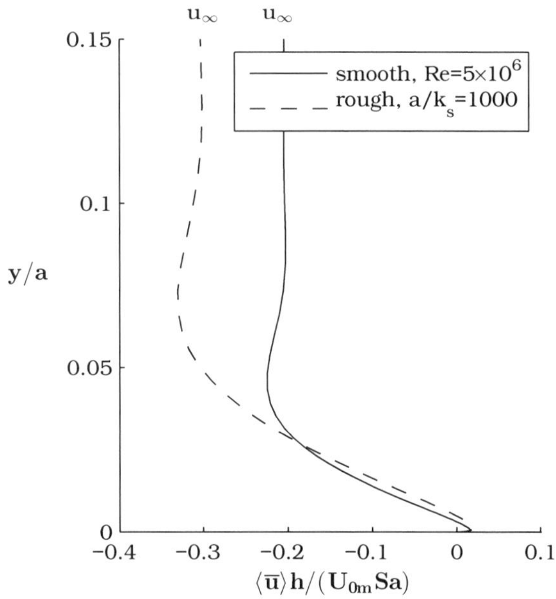

Fig. 6.10 Examples of wave boundary layer streaming due to converging-diverging bed slope effects. The positive direction is towards the convergent end, see the definition sketch in Fig. 6.7.

convergent half-cycle the flow will be subject to an external *favorable pressure gradient*, whereas during the divergent half-cycle the pressure gradient is *adverse*. The response to such conditions is well-known, and is illustrated conceptually in Fig. 6.14 (to be discussed in more detail later). Here it is seen that favorable (adverse) pressure gradients expectedly lead to increased (decreased) velocity gradients near the bed. Clearly, taking the average of these two scenarios, will lead to a net near-bed flux in the convergent direction, in line with the solutions depicted above.

The resulting period-averaged bed shear stresses and streaming velocities induced by the periodic converging-diverging flow will now be parameterized. On dimensional grounds, and based on Eq. 6.47 and Eq. 6.51, we must expect that extension from laminar to turbulent flow regimes will result in the following dependencies for the period-averaged bed shear stress and streaming velocities:

$$\frac{\langle \tau_0 \rangle}{\rho U_{0m}^2} \frac{h}{Sa} = f\left(Re, \frac{a}{k_s}\right) \tag{6.53}$$

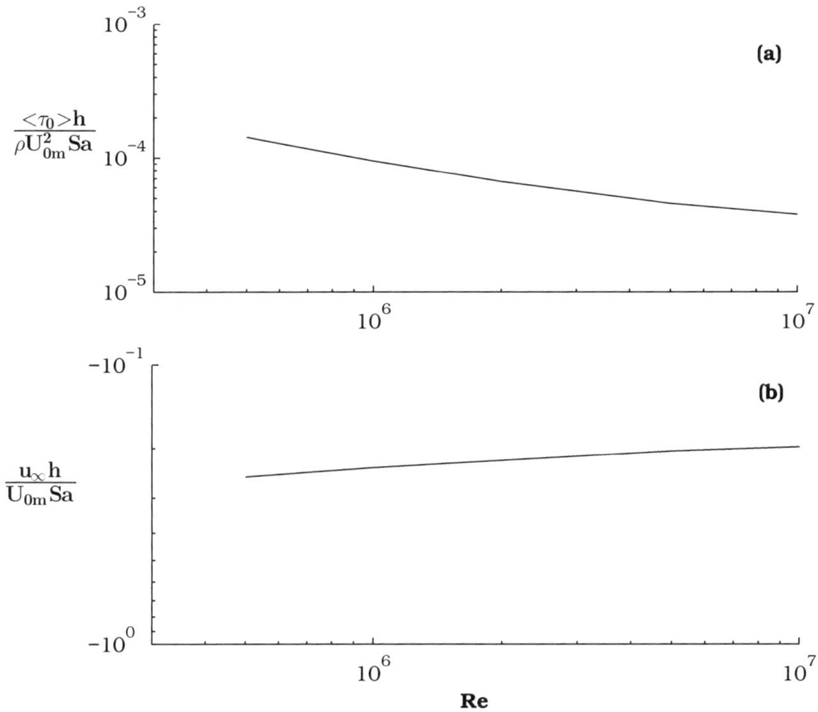

Fig. 6.11 Summary of computed (a) period-averaged bed shear stresses and (b) streaming velocities due to converging-diverging bed slope effects on smooth beds.

$$\frac{u_\infty}{\rho U_{0m}} \frac{h}{Sa} = f\left(Re, \frac{a}{k_s}\right) \tag{6.54}$$

i.e. where hydraulically smooth conditions will result in Reynolds number dependence and with hydraulically rough conditions alternatively yielding dependence on the dimensionless roughness a/k_s, where $a = U_{0m}/\omega$. This has been confirmed via systematic testing, as has the implied linear dependence on the convergence-divergence parameter Sa/h implied above. Systematic numerical study on smooth beds has resulted in the dependence shown in Fig. 6.11, and on rough beds that shown in Fig. 6.12. Consistent with expectations, these results allow quite general quantification of these basic parameters, while also generally showing the enhanced shear stresses and streaming when roughness effects are accounted for.

Turbulent oscillatory converging-diverging flows, similar to those considered above, have likewise been studied in an oscillating tunnel facility

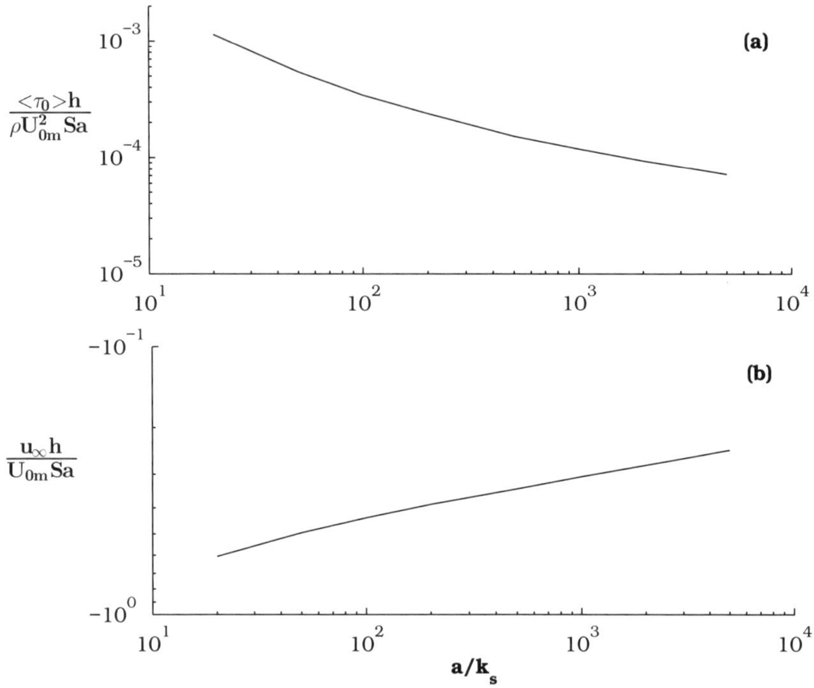

Fig. 6.12 Summary of computed (a) period-averaged bed shear stresses and (b) streaming velocities due to converging-diverging bed slope effects on rough beds.

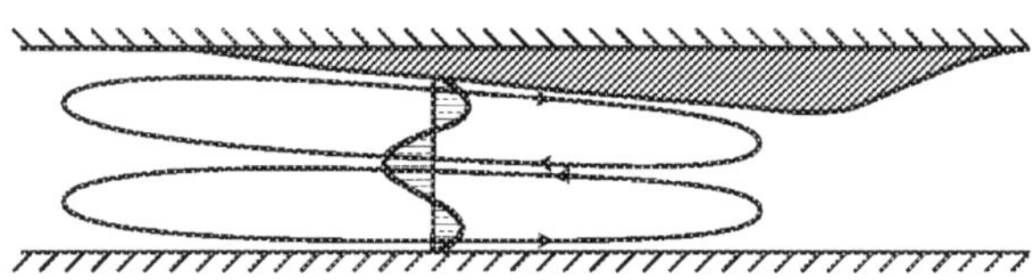

Fig. 6.13 Sketch of the oscillatory converging-diverging tunnel utilized by Sumer et al. (1993).

by Sumer et al. (1993) on smooth beds, where a convergent section was mounted on the top wall, as depicted in Fig. 6.13. Note that the study of Sumer et al. (1993) was previously discussed briefly in the preceding chapter, in Example 5.1. As an example of Sumer et al.'s (1993) results, comparison of their measured period-averaged streaming velocity profiles with those computed using a two-dimensional k-ω turbulence model (Fuhrman et al., 2009a) are shown in Fig. 6.15, for a set up where the top mounted

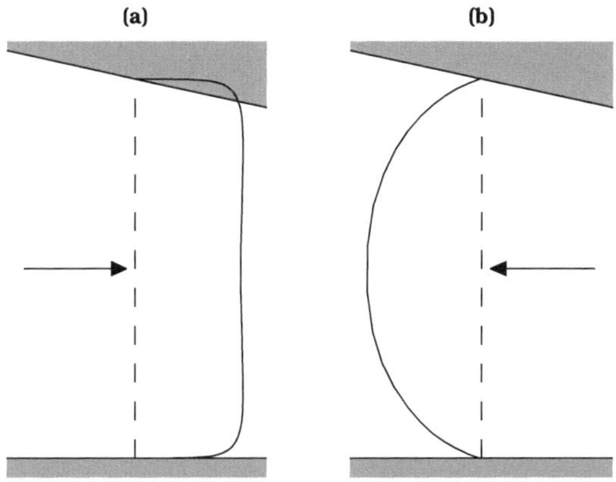

Fig. 6.14 Conceptual sketch illustrating flow subject to (a) a favorable external pressure gradient with flow in the convergent direction and (b) an adverse pressure gradient with flow in in the divergent direction. Adapted from Fuhrman et al. (2009a).

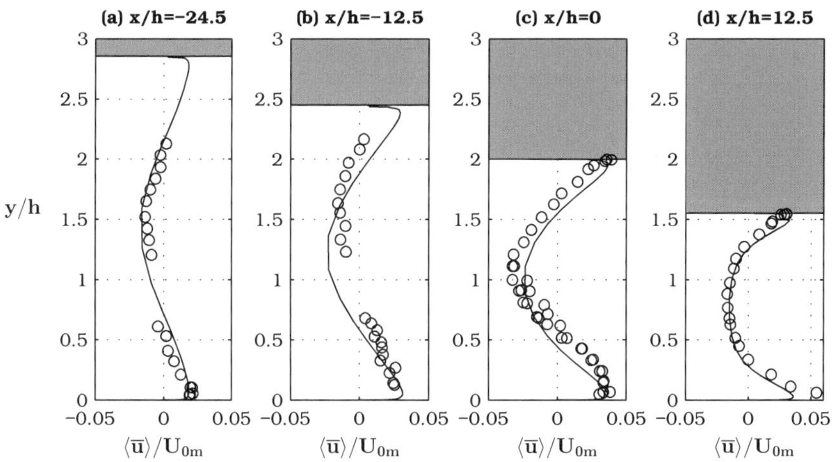

Fig. 6.15 Comparison of measured (circles, from test cases 14–17 of Sumer et al., 1993) and computed (full lines, from Fuhrman et al., 2009a) period-averaged velocity profiles with $Re \approx 6 \times 10^6$ at (a) $x/h = -12.5$, (b) $x/h = 0$, (c) $x/h = 12.5$, and (d) $x/h = 24.5$. Here x is the distance measured from the center of the converging section. Adapted from Fuhrman et al. (2009a).

convergent section had angle to the horizontal top of $2\beta \approx 2.062°$. The four cases considered (their tests 14–17) correspond to flows with $Re \approx 6 \times 10^6$ and $a/h \approx 31$ (taking h as their local tunnel half-heights in the middle of the converging section), corresponding physically to a bed slopes $S = \tan\beta \approx 0.018$, and hence $Sa/h \approx 0.56$. Note that the latter is quite large, and would likely exceed values encountered at typical wind wave scales, though may still be relevant at larger tsunami wave scales (Fuhrman et al., 2009a).

Notice that the negative streaming velocities in the divergent (leftward) direction (at $x = 0$ where h as defined above corresponds to the tunnel half-depth, in accordance with Fig. 6.15c) correspond to roughly $u_\infty/U_{0m} \approx -0.04$ i.e. to $u_\infty h/(U_{0m}Sa) \approx -0.07$. Conversely, their period-averaged bed shear stress from all of their cases can be shown to correspond to $\langle\tau_0\rangle h/(\rho U_{0m}^2 Sa) \approx 1.8\times10^{-4}$. For the given Reynolds number, these values (while being the correct order of magnitude) are both above (i.e. greater than) those expected from Fig. 6.11. The differences can be attributed due to additional currents that will be generated in such oscillating tunnel facilities, which though not considered before, would be physically necessary to ensure zero net flux over each vertical section. From consideration of the streaming profiles in the absence of such a current, e.g. those in Fig. 6.10, it is clear that this additional current must be directed in the convergent (rightward) direction, which would both elevate the period-averaged bed shear stress and reduce the magnitude of the (negative) streaming. Hence, this explains why the measured bed shear stresses and streaming velocities (at the tunnel centerline such that they can be compared with u_∞) are increased relative to the pure turbulent boundary layer considerations made previously, where zero shear stress outside the boundary layer was assumed.

Note that inclusion of such currents (driven e.g. simply by adding a constant to the driving pressure gradients in Eq. 5.159) has been shown to yield a good match with the experiments of Sumer et al. (1993), even within a 1DV modelling framework, as shown by Fuhrman et al. (2013). Essentially, their demonstrated results in the 1DV framework match very closely those in Fig. 6.15c.

For additional studies on the effects of a sloping bed on wave boundary layers, the interested reader is referred to e.g. Justesen (1988), Blondeaux et al. (2002), Zou and Hay (2003), Zou et al. (2003) and Lin et al. (2015).

6.3 Streaming due to changing bottom roughness

We will now examine streaming mechanisms in wave boundary layers that will occur locally due to another type of spatial non-uniformity: changing bed roughness. This is a situation of practical interest, as it is commonly encountered e.g. near stone protection layers around coastal structures, where there is a sudden transition from rip-rap stone protection to natural sediments. Moreover, it is likewise of fundamental interest e.g. in the understanding of potential natural sediment grain sorting mechanisms.

The results in this section will largely be based on the work of Fuhrman et al. (2011), who simulated the experimental conditions of Fredsøe et al. (1993) involving wave boundary layers over a sudden change in bottom roughness in an oscillating tunnel (these experiments were discussed briefly in the preceding chapter in Example 5.1), and subsequently performed a systematic (numerical) parametric study. This built on previous numerical work of Laursen et al. (1994), who utilized a one-dimensional Lagrangian description, wherein the bottom roughness was abruptly changed in time. For additional simulations of these experiments see Blondeaux and Vittori (2014). The presentation herein will be limited to the effects of a sudden change in bottom roughness, though it is noted that Fuhrman et al. (2011) likewise considered situations in which the roughness change is gradual.

The presentation in this section will be split into three parts: First, some relevant details on the (anisotropic) turbulence modeling approach utilized by Fuhrman et al. (2011) will be presented. Second, results involving oscillatory (sinusoidal) flow normal to a line of sudden roughness change (corresponding to $\alpha = 0°$, see the definition sketch in Fig. 6.16), will be considered. Third, results will be presented where the oscillatory flow is not normal to a line of sudden roughness change, corresponding to $0° < \alpha \leq 90°$. Note that for consistency with previous sections and chapters, $x_2 = y$ is maintained as the vertical coordinate, with $(x_1 = x, x_3 = z)$ defining the horizontal plane; See again Fig. 6.16.

6.3.1 *Anisotropic turbulence model of Fuhrman et al. (2011)*

For studying the effects of roughness change on oscillatory wave boundary layer flows, Fuhrman et al. (2011) utilized a RANS model coupled with k-ω turbulence closure, where the flow was driven by a body force B_i specified according to

$$(B_1, B_2, B_3) = (\cos\alpha, 0, \sin\alpha)U_{0m}\omega\cos(\omega t) \tag{6.55}$$

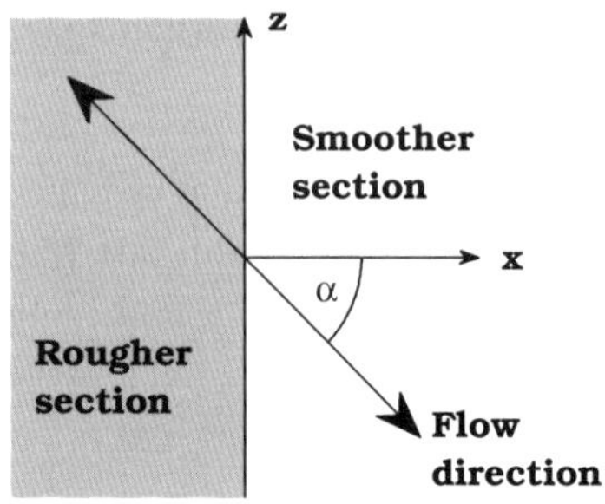

Fig. 6.16 Definition sketch (in plan) illustrating oscillatory flow with arbitrary flow direction α. Adapted from Fuhrman et al. (2011).

This was used to drive a sinusoidal horizontal free stream velocity of the form

$$(U_{01}, U_{02}, U_{03}) = (\cos\alpha, 0, \sin\alpha) U_{0m} \sin(\omega t) \tag{6.56}$$

having velocity magnitude U_{0m} and angular frequency $\omega = 2\pi/T$, where T is the wave period, as before. In what follows, without loss of generality, horizontal variations in bottom roughness will be considered in the x-direction i.e. with the transverse z-axis defining the line of bottom roughness transition. Hence the angle α defines the forced flow direction relative to the line of roughness transition, as again depicted in the conceptual sketch shown in Fig. 6.16. As the problem described above is uniform in z it conveniently remains two-dimensional in nature, though problems with $\alpha \neq 0°$ will obviously give rise to Reynolds-averaged velocity components in all three spatial directions.

In the turbulence model utilized by Fuhrman et al. (2011), the Reynolds stress tensor was defined according to the constitutive relation

$$\frac{\tau_{ij}}{\rho} = -\overline{u_i' u_j'} = 2\nu_t S_{ij} - a_{(i)} k \delta_{ij} \tag{6.57}$$

where δ_{ij} is the Kronecker delta, ν_t is the eddy viscosity,

$$k = \frac{1}{2}\overline{u_i' u_i'} \tag{6.58}$$

is the turbulent kinetic energy density, and the overbar denotes time averaging. Note that in Eq. 6.57 the subscript (i) indicates suppressed summation over this particular index, hence the last term in Eq. 6.57 only affects the turbulent normal stresses. The $a_{(i)}$ coefficients are defined as a function of the forced flow direction α according to

$$\left(a_{(1)}, a_{(2)}, a_{(3)}\right) = \left(\frac{8}{9}\cos^2\alpha + \frac{2}{3}\sin^2\alpha, \frac{4}{9}, \frac{2}{3}\cos^2\alpha + \frac{8}{9}\sin^2\alpha\right) \tag{6.59}$$

These relations were selected to guarantee fixed 4:3:2 ratios for the streamwise, transverse, and vertical fluctuating velocity components, respectively, at the limit of spatially uniform flow. As an example, for uniform flow with $\alpha = 0°$ (i.e. streamwise flow in the pure x-direction) their turbulence model predicts

$$-\frac{\tau_{11}}{\rho} = \overline{u'^2} = \frac{8}{9}k \tag{6.60}$$

$$-\frac{\tau_{22}}{\rho} = \overline{v'^2} = \frac{4}{9}k \tag{6.61}$$

$$-\frac{\tau_{33}}{\rho} = \overline{w'^2} = \frac{2}{3}k \tag{6.62}$$

Alternatively, with $\theta = 90°$ (i.e. streamwise flow in the transverse z-direction) it yields

$$-\frac{\tau_{11}}{\rho} = \overline{u'^2} = \frac{2}{3}k \tag{6.63}$$

$$-\frac{\tau_{22}}{\rho} = \overline{v'^2} = \frac{4}{9}k \tag{6.64}$$

$$-\frac{\tau_{33}}{\rho} = \overline{w'^2} = \frac{8}{9}k \tag{6.65}$$

These fixed ratios are known to be a good approximation throughout the log layer and much of the defect layer within flat-plate boundary layers (Wilcox, 2006, p. 309); See also Fig. 3.7. The 2:1 ratio for streamwise-to-vertical fluctuating components is likewise reasonably in line with steady uniform flow data over rough beds collected by Grass (1971); See again Fig. 3.15. Hence, by adopting Eq. 6.59 the model of Fuhrman et al. (2011), in a simple way, provides anisotropic turbulent normal stresses that are consistent with known uniform channel flow characteristics. This is in contrast to the traditionally-used standard Boussinesq approximation (Eq. 9.30 from Chapter 9), which utilizes constant $a_{(i)} = 2/3$. Note also that, as the coefficients in Eq. 6.59 sum to 2 for all α, the relation Eq. 6.58 is inherently satisfied, which can readily be seen by taking the trace (i.e. the sum of diagonal components) of Eq. 6.57 and invoking Eq. 6.3. The off-diagonal (shear) components of the Reynolds stress tensor, on the other hand, were left to be modeled in the standard way.

It should be stressed that the use of Eq. 6.59 assumes clear distinction between (known) streamwise, transverse, and vertical directions, with the vertical y-direction assumed normal to wall boundaries. Hence, it would not be valid e.g. where side-wall or corner effects are important. These would require nonlinear constitutive relations (e.g. Speziale, 1987; Wilcox, 2006, p. 308–311) e.g. for the prediction of secondary flows arising in non-circular pipes. This limitation was not an issue in the work of Fuhrman et al. (2011), which utilized periodic horizontal boundaries. The advantage of their anisotropic turbulence-modeling approach lies therefore merely in its simplicity for the idealized flow configurations considered, at the expense of greater generality offered by other, more complicated, approaches. As demonstrated by Fuhrman et al. (2011) via comparison with steady-state experimental results of Wang and Cheng (2006), despite its simplicity, the model is capable of predicting secondary flows induced by alternating rough and smooth bed patches, a phenomenon for which incorporating anisotropic turbulent normal stresses is essential, as will be explained.

6.3.2 *Normally-directed oscillatory flow*

Let us now consider the effects on an oscillatory (sinusoidal) wave boundary layer flow in the vicinity of a sudden change in roughness, when the flow is directly normal to the line of roughness change. This situation is depicted in elevation view in Fig. 6.17, which is based on the experimental set up used by Fredsøe et al. (1993), in addition to the plan view in Fig. 6.16 (with $\alpha = 0°$). In what follows k_s will denote Nikuradse's equivalent roughness for the rougher (left hand) section, whereas k_s' will denote the roughness of the smoother (right hand) section. Fredsøe et al. (1993) considered wave boundary layers in an oscillating tunnel facility (Fig. 5.21) involving both pebble-to-smooth roughness transition (k_s = 15 mm, k_s' = smooth, their Tests 1 and 2) as well as pebble-to-sand paper transitions (k_s = 15 mm, k_s' = 0.84 mm, their Test 3), and selected results from both scenarios will be considered as illustrations in what follows. We note that k_s here is Nikuradse's equivalent sand roughness, which is different from the actual roughness height, k, given in Fredsøe et al. (1993) as k = 4.8 mm and k' = 0.35 mm (Test 3), respectively.

The local effects of such a roughness change on a wave boundary layer can be explained, at least qualitatively, by considering the expected differences that will occur during positive and negative flow half cycles. These effects, it turns out, can be readily predicted based on the effects of roughness discussed previously in the preceding chapter, Section 5.4. During the positive half-cycle i.e. with the flow coming off the rougher section encounter-

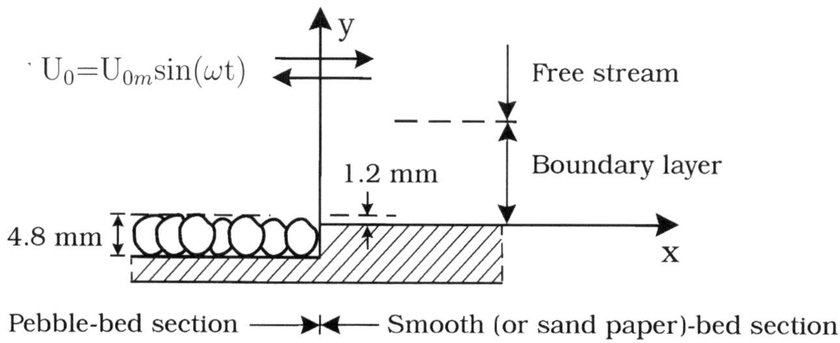

Fig. 6.17 Conceptual sketch of the roughness change within the oscillating tunnel experiments of Fredsøe et al. (1993). Adapted from Fuhrman et al. (2011).

ing the smoother section, the near-bed velocities (hence velocity gradients) will be relatively small, due to the retarding effect of the larger roughness. Conversely, during the negative half-cycle i.e. with the (leftward) flow exiting the smoother section, the near-bed flow will have unnaturally large velocity (hence velocity gradients) as it encounters the rougher section. Thus, when period-averaged, there will be near-bed period-averaged streaming directed towards the rougher section (leftward) in the vicinity of the roughness transition. This feature is clearly seen in Fig. 6.18, which depicts measured (Fredsøe et al., 1993, their Test 1) and computed (Fuhrman et al., 2011) period-averaged velocity profiles in the vicinity of a pebble-to-smooth roughness transition. The induced near-bed streaming is expectedly strongest in the immediate vicinity of the roughness change ($x = 0$) and is seen to gradually become weaker as the far field is approached. Blondeaux and Vittori (2014) also highlight that the streamwise variations of the horizontal velocity component induce a significant vertical component of the velocity (also clearly presented in the experimental measurements of Fredsøe et al., 1993), which should be taken into account if an accurate description of transport phenomena is to be obtained.

A change in bottom roughness likewise has a considerable effect on the turbulence in the vicinity of the transition. This can be seen in Fig. 6.19, which depicts a comparison of the root-mean-square of experimental and numerical horizontal turbulent fluctuating components ($u'_{rms} = \sqrt{\overline{u'^2}}$) at various cross-sections, both when the free stream flow is maximum ($\omega t = 90°$ i.e. flow directed rightward, top four sub-figures) as well as minimum ($\omega t = 270°$ i.e. flow directed leftward, bottom four sub-figures). Elevated

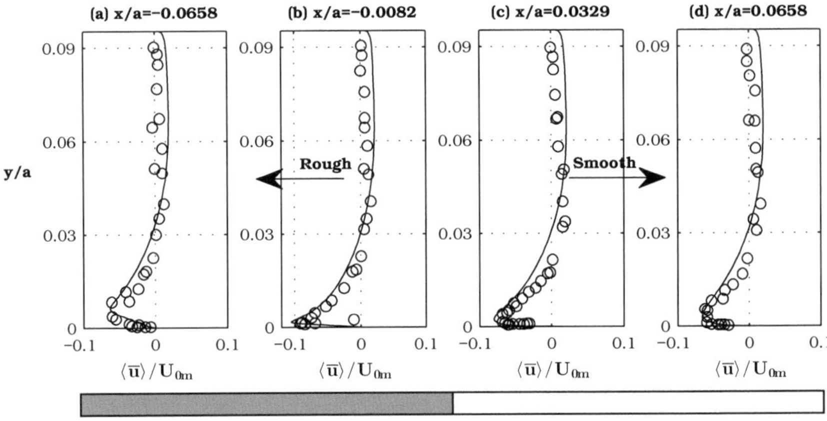

Fig. 6.18 Comparison of measured (circles, from Test 1 of Fredsøe et al., 1993) and computed (lines, from Fuhrman et al., 2011) period-averaged velocity profiles for a wave boundary layer ($T = 9.75$ s, $U_{0m} = 1.97$ m/s, $a = 3.04$ m, $Re = 6 \times 10^6$, $a/k_s = 200$) over a sudden change in bed roughness (smooth to pebble-bed section with $k_s = 0.015$ m). Adapted from Fuhrman et al. (2011).

turbulence levels are generally observed when the flow is positive i.e. coming off the rougher section. This is evident not only above the rougher section, but also over the smoother section, due to horizontal advection. When the flow is negative (leftward, bottom four sub-figures), the turbulence on the smoother section is comparatively small, though the production of near-bed turbulence is again significantly elevated over the rougher sections. This is especially apparent in the near-bed numerical results (full line) in Fig. 6.19f. Due to experimental constraints, measurements could not be made close enough to the bed to directly confirm this region, but its presence is consistent with the unnaturally larger velocity gradients that will be encountered very near the bed, as discussed previously. Similar results as above can also be found in the simulations of Blondeaux and Vittori (2014).

A final and important consequence of the unnaturally large near-bed velocities and velocity gradients experienced over the rougher section during the negative half-cycle is the (potentially significant) amplification of the bed shear stresses near the transition. This is illustrated in Fig. 6.20, which depicts the (measured and computed) maximum bed shear stress amplification α_τ, relative to the far field, experienced during Test 3 conditions of Fredsøe et al. (1993) (it is emphasized that the variable used

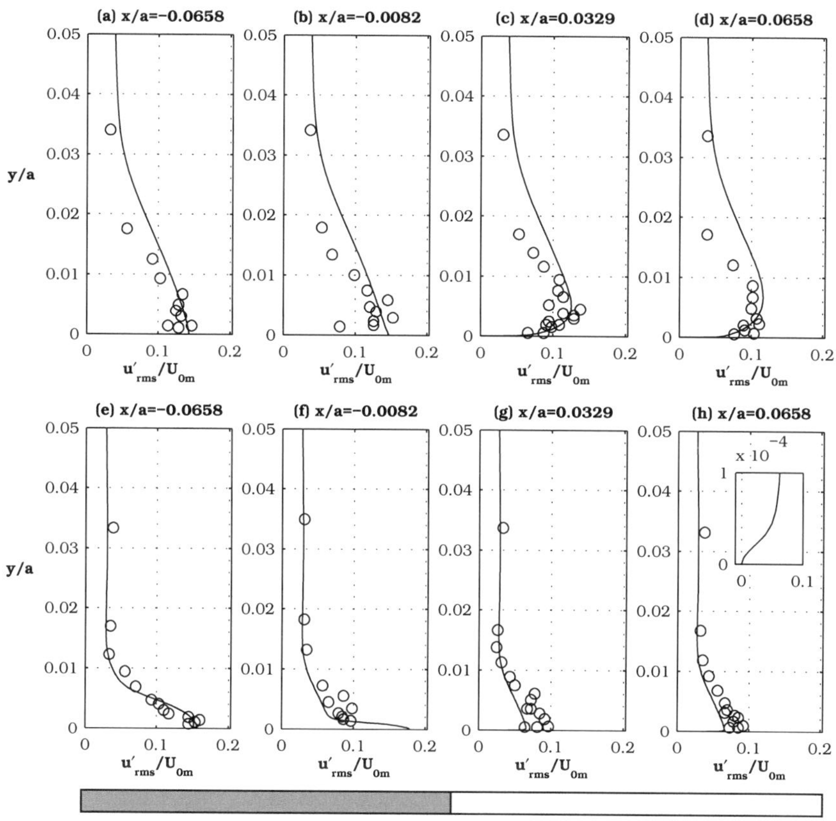

Fig. 6.19 Comparison of measured (circles, from Test 1 of Fredsøe et al., 1993) and computed (lines, from Fuhrman et al., 2011) profiles of the root-mean-squared horizontal turbulent fluctuations $u'_{rms} = \sqrt{\overline{u'^2}}$ over a sudden change in bed roughness (same case as in Fig. 6.18). Top panels: the free stream flow is maximum ($\omega t = 90°$). Bottom panels: free stream flow is minimum ($\omega t = 270°$). Adapted from Fuhrman et al. (2011).

for the bed shear stress amplification α_τ should not be confused with that used for the flow orientation angle α). Just to the left of the transition, for the conditions considered, a significant amplification of approximately a factor $\max(\alpha_\tau) = 1.8$ is observed. This is important to recognize, as it implies e.g. that stone protection layers designed solely based on the far field wave conditions (i.e. neglecting the transition) may not be sufficiently conservative to prevent mobilization of the larger protective grains.

Having successfully validated their numerical model against the experiments of Fredsøe et al. (1993), Fuhrman et al. (2011) subsequently per-

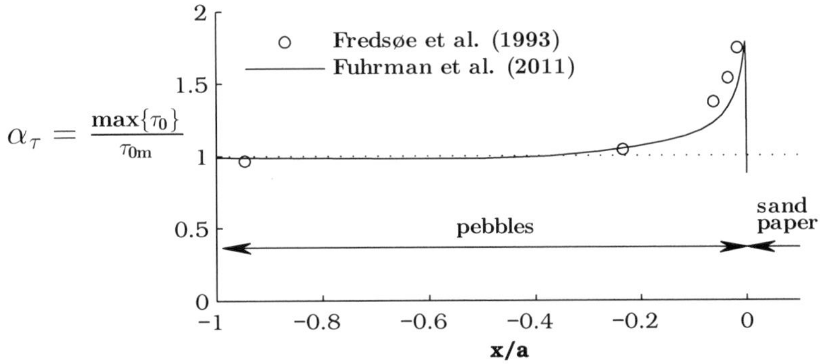

Fig. 6.20 Comparison of measured (circles, from Test 3 of Fredsøe et al., 1993) and computed (lines, from Fuhrman et al., 2011) bed shear stress amplification α over the rougher section ($x < 0$) for a wave boundary layer ($T = 9.75$ s, $U_{0m} = 2.05$ m/s, $a = 3.18$ m, $Re = 6.5 \times 10^6$, $a/k_s = 210$) over a sudden change in bed roughness (sand paper with $k_s' = 0.00084$ m to pebble-bed section with $k_s = 0.015$ m, $k_s/k_s' = 17.9$). Adapted from Fuhrman et al. (2011).

formed a systematic parametric study with their model (now on an effectively unconfined vertically domain) for the dimensionless maximum magnitude near-bed streaming velocity u_s/U_{0m} (with negative values directed towards the rougher section) as well as for the maximum bed shear stress amplification over the rougher section $\max\{\alpha_\tau\}$. Assuming fully rough conditions (hence results independent of Reynolds number), on dimensional grounds the following dependence for these quantities can be expected:

$$\left\{\frac{u_s}{U_{0m}}, \max\{\alpha_\tau\}\right\} = f\left(\frac{a}{k_s}, \frac{k_s}{k_s'}\right) \tag{6.66}$$

Fuhrman et al. (2011) considered the range $20 \leq a/k_s \leq 500$ and $1 \leq k_s/k_s' \leq 100$ and presented the design diagrams for these quantities shown in Fig. 6.21. Generally consistent with the comparisons against the experiments, over the parametric range considered, bed shear stress amplifications exceeding a factor of 2 were found, in conjunction with maximum streaming velocities u_s up to approximately 10% of the free stream magnitude U_{0m}. Both quantities increase naturally with k_s/k_s', as well as with a/k_s, the latter being due to e.g. increased flow velocities, with all other parameters held equal. The negative streaming velocities, u_s, parameterized above in Fig. 6.21b occurred near the bed, similar to those seen in Fig. 6.18.

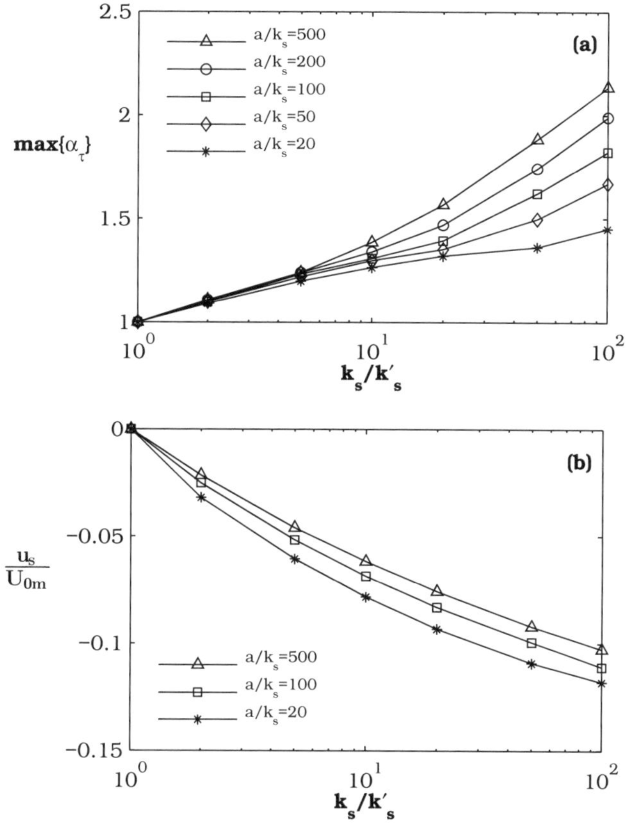

Fig. 6.21 Summary of maximum (a) bed shear stress amplification max$\{\alpha_r\}$ and (b) (in terms of magnitude) streaming velocity u_s due to a sudden change in bottom roughness. Adapted from Fuhrman et al. (2011).

6.3.3 *Obliquely-directed oscillatory flow*

Fuhrman et al. (2011) also considered the streaming induced by a roughness transition when the flow was obliquely directed i.e. $0° < \alpha \leq 90°$ (see Fig. 6.16), specifically with $k_s/k'_s = 100$. Some results showing the period-averaged streaming velocity profiles spanning $60° < \alpha \leq 90°$ are presented in Fig. 6.22. Based on Fig. 6.22a, for the conditions considered, with $\alpha \leq 60°$ the streaming will be directed leftwards i.e. towards the rougher section, for the same reasons as have been discussed previously with $\alpha = 0°$. However, for larger $\alpha \geq 70°$, Figs. 6.22b–d, the near bed streaming begins to change direction, eventually becoming directed towards the

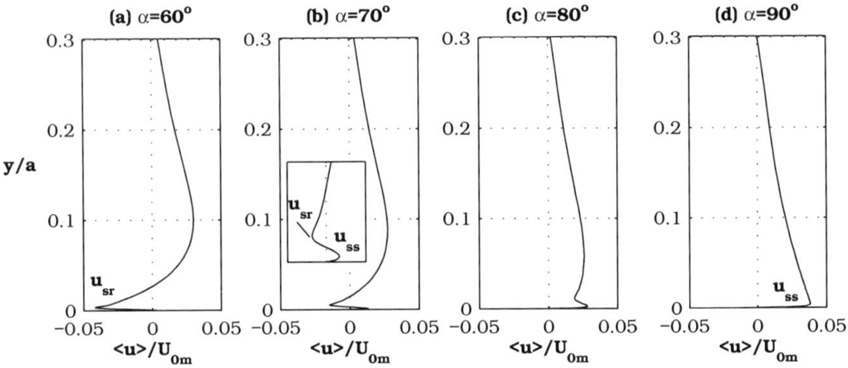

Fig. 6.22 Period-averaged streaming velocity profiles for various oscillatory flow angle α. Adapted from Fuhrman et al. (2011).

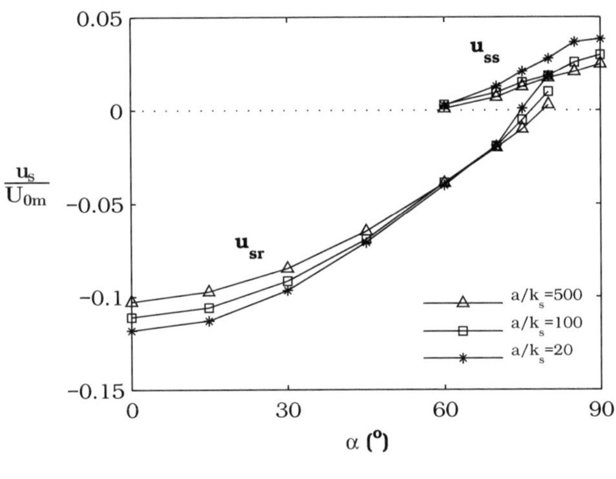

Fig. 6.23 Summary of largest computed period-averaged horizontal boundary layer streaming velocities u_s for simulations involving sudden roughness change ($k_s/k'_s = 100$) with varying oscillatory flow direction. Adapted from Fuhrman et al. (2011).

smoother section.

The peaks in magnitude for the period-averaged streaming velocities directed towards the rougher and smoother sections will, respectively, be denoted u_{sr} and u_{ss}, as defined in Fig. 6.22. Based on their results Fuhrman et al. (2011) quantified these versus α for cases with fixed $k_s/k'_s = 100$, as depicted in Fig. 6.23. As seen, the reversed streaming towards the smoother section for $\alpha \approx 90°$ can approach up to 5% of the free stream velocity magnitude U_{0m}, which is roughly half the streaming magnitude towards the rougher section found with $\alpha = 0°$.

The reason for the predicted secondary flows (i.e. transverse to the main flow direction) as α approaches 90° can be traced to anisotropy in the turbulent normal stresses τ_{xx} and τ_{yy}. This is consistent with the arguments of Speziale (1982), who (again, taking the transverse z-direction as streamwise) identified the importance of

$$\frac{\partial^2(\tau_{yy} - \tau_{xx})}{\partial x \partial y} \tag{6.67}$$

as an axial (streamwise) vorticity source term, while studying secondary flows within rectangular pipes. Because the turbulence generated over rougher sections will be larger than over smoother sections, large gradients in the turbulence quantities can develop in the plane perpendicular to the main flow with $\alpha = 90°$, hence activating Eq. 6.67, and promoting vorticity in the streamwise z direction, corresponding to the observed secondary flows in the x-y plane. Clearly, a turbulence model incorporating anisotropic normal Reynolds stresses, such as that presented in Section 6.3.1, is absolutely essential for the prediction of this phenomenon.

For the interested reader, the full derivation of the vector vorticity equation from the Reynolds-averaged Navier-Stokes (RANS) equations is detailed in Appendix 6.A. Based on this, the full derivation of the axial vorticity source term Eq. 6.67 is likewise provided in Appendix 6.B, such that the entirety of this chapter is self contained.

6.4 Streaming beneath non-sinusoidal waves

The sinusoidal wave-induced boundary layer flows considered previously yield a highly-useful basic understanding for the flow and features that develop beneath unsteady wave boundary layers. Strictly speaking, sinusoidal free stream flows correspond to those that can be physically expected according to the most basic *linear wave theory* (outlined in the Appendix to Chapter 5) which is valid when the wave heights are small. Sinusoidal flows can hence also be considered as the *leading-order* description, even in situations where surface nonlinearity is not negligible. However, when surface wave heights become large e.g. as waves shoal and steepen as they propagate towards the shoreline, nonlinearity will no longer be negligible and the waves will deviate from the leading-order sinusoidal description, becoming skewed (raised peaks and flattened troughs) and/or front-back asymmetric. From the wave boundary layer perspective, these non-sinusoidal effects will result, respectively, in what can un-ambiguously be termed: (1) free stream *velocity skewness* and (2) free stream *acceleration skewness*. In reality these may occur in combination, though for simplicity they will be considered individually in the forthcoming sub-sections.

6.4.1 *Streaming due to free-stream velocity skewness*

Let us first consider turbulent wave boundary layers that will be induced by purely skewed free stream velocities, which would be expected to develop beneath nonlinear progressive waves on flat beds (or those with gentle slopes). Experimental (e.g. Ribberink and Al-Salem, 1995; O'Donoghue and Wright, 2004a,b; Hassan and Ribberink, 2005) and numerical (Davies and Li, 1997; Hsu and Hanes, 2004; Holmedal and Myrhaug, 2006; Fuhrman et al., 2009a,b; Ruessink et al., 2009; Hassan and Ribberink, 2010; Fuhrman et al., 2013) studies of such flows have most commonly utilized a free stream velocity based on a Stokes second-order progressive wave, corresponding to

$$U_0 = U_{1m}\sin(\omega t) - U_{2m}\cos(2\omega t) \tag{6.68}$$

having time derivative

$$\frac{\partial U_0}{\partial t} = U_{1m}\omega\cos(\omega t) + 2U_{2m}\omega\sin(\omega t) \tag{6.69}$$

To isolate the effects of the non-sinusoidal wave shape, convective terms will here be ignored, corresponding to the horizontally-uniform situation that can be expected in a conventional oscillating tunnel environment (e.g. Fig. 5.21, without any modifications). As such, the flow may be assumed to be driven by a pressure gradient of the form

$$-\frac{1}{\rho}\frac{\partial \overline{p}}{\partial x} = \frac{\partial U_0}{\partial t} - P_x \tag{6.70}$$

where a constant pressure gradient term P_x is also included, which can drive additional steady currents; Unless otherwise noted we will consider $P_x = 0$. In the above U_{1m} and U_{2m} are the magnitude of the first- and second-harmonic velocity contributions, with skewness (as characterized by raised peaks and troughs) corresponding to $U_{2m} > 0$. Such a velocity-skewed signal is depicted in Fig. 6.24a, as an illustration, with the resulting bed shear stress predicted by a RANS model (using k-ω turbulence model) depicted in Fig. 6.24b. Note that for the case shown the parameters ($U_{1m} =$ 0.845 m/s, $U_{2m} = 0.27$ m/s, $\omega = 0.967$ s^{-1}, $k_s \approx 2.5d = 0.000525$ m;

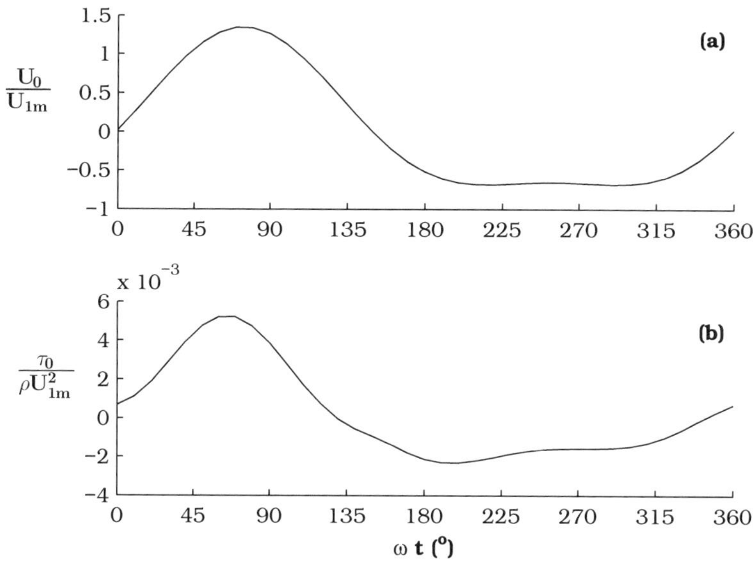

Fig. 6.24 Simulated (a) free stream velocity and (b) bed shear stress for the wave signal of Ribberink and Al-Salem (1995), corresponding to a velocity-skewed wave signal based on Eq. 6.68.

$Re \approx 8 \times 10^5$, $a/k_s = 1700$) have been selected to match those considered experimentally by Ribberink and Al-Salem (1995).

Let us now consider the boundary layer that will develop beneath a free stream signal that initially matches Eq. 6.68. As a first approximation for a wave boundary layer in the turbulent regime, the induced time variation of the bed shear stress can be approximated analytically by

$$\frac{\tau_0(t)}{\rho} = \frac{f_w}{2} U_0(t)|U_0(t)| \tag{6.71}$$

where the wave friction factor f_w can be calculated according to e.g. either Eq. 5.60 (smooth-turbulent regime) or Eq. 5.69 (rough-turbulent regime). (We note that Eq. 6.71 should not be confused with Eq. 5.56. Here $\tau_0(t)$ and $U_0(t)$ are the phase-resolved bed shear stress and free-stream velocity, respectively; The time dependence is added in these variables above to emphasize this point.) Based on this simple description, a good approximation for the period-averaged bed shear stress beneath a skewed free stream velocity of the form Eq. 6.68 may be obtained by dividing the two half-cycles as

$$\left\langle \frac{\tau_0}{\rho} \right\rangle = \frac{1}{2\pi}\int_0^{2\pi} \frac{f_w}{2} U_0|U_0| d\omega t \approx \frac{f_w}{4\pi}\left(\int_0^{\pi} U_0^2 d\omega t - \int_{\pi}^{2\pi} U_0^2 d\omega t\right) \quad (6.72)$$

leading to

$$\left\langle \frac{\tau_0}{\rho U_{1m}^2} \right\rangle \approx \frac{2 f_w}{3\pi}\frac{U_{2m}}{U_{1m}} \quad (6.73)$$

(For a more general method of predicting the time variation of the bed shear stress based on a given free stream velocity, see Section 5.11.3.)

From these considerations, it is now seen that a purely skewed free-stream velocity signal will, in the absence of external forces or a mean period-averaged flow, lead to a positive bed shear stress i.e. in the forward propagating direction. As convective terms have been neglected, and the driving period-averaged pressure gradient is zero, in this scenario the period-averaged Navier-Stokes equation (Eq. 6.8) would initially reduce to:

$$\left\langle \frac{\partial \overline{u}}{\partial t} \right\rangle = \left\langle \frac{1}{\rho}\frac{\partial \tau}{\partial y} \right\rangle \quad (6.74)$$

Note that even though the period-averaged bed shear stress will initially be *positive*, the right hand side will be *negative*, as the positive shear stress at the bed will decay to zero outside the boundary layer. Lacking a balancing force, the negative shear stress gradients will thus cause the period-averaged flow within the boundary layer to decelerate (i.e. $\partial\langle\overline{u}\rangle/\partial t < 0$ such that the flow becomes negative), until the shear stress becomes uniformly $\tau/\rho = 0$ throughout the column. Only at this stage will equilibrium be reached and the deceleration of the period-averaged flow cease. This, in the simplest of terms, illustrates why a skewed free stream velocity signal will promote period-average streaming in the negative direction.

An example period-averaged velocity profile, based on the signal depicted in Fig. 6.24, that has evolved precisely to this equilibrium situation is depicted in Fig. 6.25a. Here it is seen that the induced period-averaged streaming is indeed in the negative direction, as discussed above; Likewise, the period-averaged shear stress (full line in Fig. 6.25b) has evolved to become uniformly zero, as also expected from the above.

The existence of such wave boundary layer streaming induced by velocity-skewed free stream signals was first shown theoretically by Trowbridge and Madsen (1984a,b), who demonstrated the importance of a time varying eddy viscosity in promoting this phenomenon. Following e.g. Davies

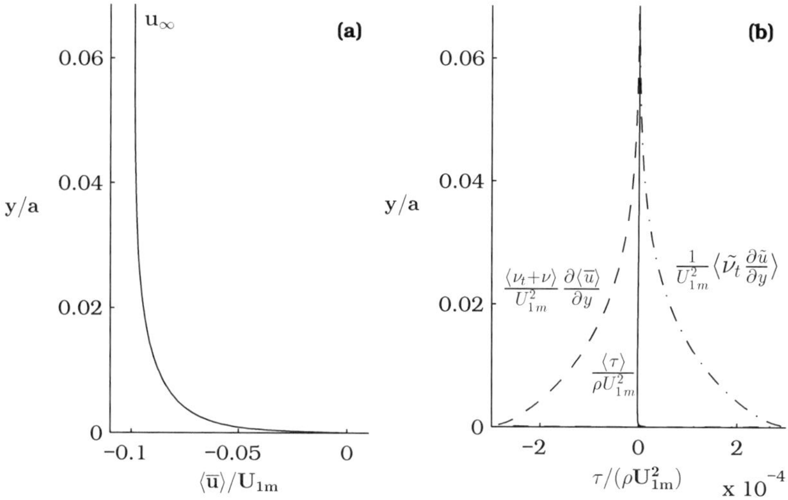

Fig. 6.25 Period-averaged (a) velocity profile and (b) shear stress distribution beneath a skewed wave signal of the form Eq. 6.68, with $U_{1m} = 0.845$ m/s, $U_{2m} = 0.27$ m/s, and $\omega = 0.967$ s^{-1} ($Re \approx 8 \times 10^5$, $a/k_s = 1700$, $U_{2m}/U_{1m} = 0.319$).

and Li (1997) and Holmedal and Myrhaug (2006), it is enlightening to further decompose the time-varying horizontal velocity and eddy viscosity into period-averaged and fluctuating ($\tilde{u}$ and $\tilde{\nu_t}$) components:

$$u(z,t) = \langle u(z,t) \rangle + \tilde{u}(z,t) \tag{6.75}$$

$$\nu_t(z,t) = \langle \nu_t(z,t) \rangle + \tilde{\nu_t}(z,t) \tag{6.76}$$

such that the period-averaged shear stress may be similarly decomposed as

$$\frac{\langle \tau \rangle}{\rho} = \left\langle (\nu + \nu_t) \frac{\partial u}{\partial y} \right\rangle = \langle \nu + \nu_t \rangle \frac{\partial \langle u \rangle}{\partial y} + \left\langle \tilde{\nu}_t \frac{\partial \tilde{u}}{\partial y} \right\rangle \tag{6.77}$$

where $\langle \tau \rangle = 0$ in the present situation. These components are likewise shown on Fig. 6.25b. Here the first term on the right hand side of Eq. 6.77 represents that part of the shear stress arising due to gradients in the period-averaged flow, whereas the second term represents the additional stresses induced by the fluctuations in the eddy viscosity, which arise due to asymmetric turbulence levels produced during the forward and negative half cycles of the flow. As we have begun our considerations from a situation where there is initially no period-averaged flow, it is the latter which initially dominates, driving the flow in the negative direction, as described above. As gradients in the period-averaged flow develop the first term then grows in strength, until it eventually balances the second, as is clearly seen in Fig. 6.25b, leading to the equilibrium situation with zero total period-averaged shear stress. Because this behavior arises due to the asymmetric turbulence that develops in the forward and negative directions, this boundary layer phenomenon beneath skewed free-stream wave signals is commonly referred to as *turbulence-induced streaming*, in contrast to the boundary layer streaming mechanisms due to either convective terms (progressive waves and converging-diverging flow) and roughness changes discussed previously.

While the previous considerations have explained the negative turbulence-induced streaming phenomenon that will occur beneath skewed waves, it is clear from Fig. 6.25a that the idealized zero-stress equilibrium situation will not match period-averaged velocity profiles that would be measured in an oscillating tunnel, in which an additional forward current must also arise to prevent a net flow over the cross-section (at least if this is assumed to be zero). This, slightly more general situation, now similarly driven by Eq. 6.70 but with $P_x = -2.28 \times 10^{-4} U_{1m}^2 / h_m < 0$, is illustrated in Fig. 6.26, which includes comparison with measurements made by Ribberink and Al-Salem (1995) (circles). Following Fuhrman et al. (2009a) the simulation is made using a model height of $h_m = 0.15$ m ($h_m/a \approx 0.18$), above which the measured period-averaged velocities are essentially uniform. (In a strictly 2D situation zero mass flux over the section must be expected; This is clearly not the case for the measurements shown, perhaps due to additional three-dimensional effects.) With the constant pressure gradient term now added, at equilibrium the period-averaged Navier-Stokes equation corresponds to

$$\left\langle \frac{1}{\rho} \frac{\partial \tau}{\partial y} \right\rangle - P_x = 0 \tag{6.78}$$

where the decomposition of the shear stress in Eq. 6.77 remains valid. Note that similar results as presented in Fig. 6.26 can likewise be found in Davies and Li (1997); Holmedal and Myrhaug (2006); Fuhrman et al. (2009a). Investigation of the decomposed shear stress components in Fig. 6.26b demonstrate that, while resembling those in Fig. 6.25a, they are now shifted positively (rightward), such that the vertical gradient in the total period-averaged shear stress is equal to the constant pressure gradient P_x, as implied by Eq. 6.78.

A commonly used measure with which to indicate the degree of free

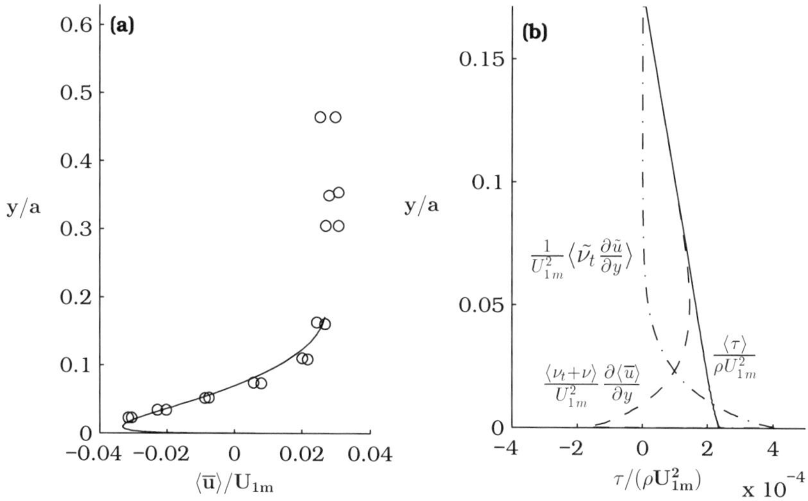

Fig. 6.26 Period-averaged (a) velocity profile and (b) shear stress distribution beneath a skewed wave signal of the form Eq. 6.68, with $U_{1m} = 0.845$ m/s, $U_{2m} = 0.27$ m/s, and $\omega = 0.967$ s^{-1} ($Re \approx 8 \times 10^5$, $a/k_s = 1700$, $U_{2m}/U_{1m} = 0.319$). The circles depict measurements from Ribberink and Al-Salem (1995) whereas the full lines correspond to simulations made using a k-ω model with domain height $h_m = 0.15$ m where a constant pressure gradient term ($P_x = -2.28 \times 10^{-4} U_{1m}^2/h_m$) has been added. (Compare with Fig. 6.25, where $P_x = 0$.)

stream velocity skewness is the so-called skewness coefficient R, defined as

$$R = \frac{U_{0,max}}{U_{0,max} - U_{0,min}} \tag{6.79}$$

with skewed velocity signals then corresponding to $R > 0.5$. For a free stream flow of the form Eq. 6.68 the maximum and minimum velocities correspond, respectively, to: $U_{0,max} = U_{1m} + U_{2m}$ and $U_{0,min} = U_{2m} - U_{1m}$ (provided that $U_{2m}/U_{1m} \leq 0.25$ which ensures no secondary peak will occur in the trough). For such a signal the skewness coefficient is then

$$R = \frac{U_{1m} + U_{2m}}{2U_{2m}} = \frac{1}{2}\left(1 + \frac{U_{2m}}{U_{1m}}\right) \tag{6.80}$$

To systematically quantify the boundary layer streaming and other properties induced by non-sinusoidal free stream signals, we will now consider a series of simulated cases based on the more general and highly-flexible wave shape signal proposed by Abreu et al. (2010):

$$U_0 = U_m \sqrt{1 - r^2} \cdot \frac{\sin(\omega t) + \frac{r \sin\phi}{1 + \sqrt{1 - r^2}}}{1 - r\cos(\omega t + \phi)} \tag{6.81}$$

where U_m defines the maximum velocity. These cases are driven by the pressure gradient Eq. 6.70, taking $P_x = 0$ (pure wave boundary layer cases), where the free-stream acceleration corresponds to:

$$\frac{\partial U_0}{\partial t} = U_m \omega \sqrt{1-r^2} \cdot \frac{\cos(\omega t) - r\cos\phi - \frac{r^2}{1+\sqrt{1-r^2}} \sin\phi \sin(\omega t + \phi)}{(1 - r\cos(\omega t + \phi))^2} \tag{6.82}$$

Note that with the parameter $r = 0$ the velocity signal Eq. 6.81 conveniently becomes sinusoidal, whereas with $\phi = -\pi/2$ and $r > 0$ it yields purely-skewed velocity signals similar in form to that shown previously in Fig. 6.24. (Alternatively, this signal with $\phi = 0$ and $r > 0$ yields acceleration-skewed flows, which are covered in the next sub-section.) The wave shape Eq. 6.81 likewise results in the following skewness coefficient

$$R = \frac{(1+r)\left(1 - r + \sqrt{1-r^2}\right)}{2\left(1 - r^2 + \sqrt{1-r^2}\right)} \tag{6.83}$$

having been previously defined in Eq. 6.79.

Results from the series of simulations involving velocity-skewed free-stream signals of the form Eq. 6.81 are summarized in Fig. 6.27, which depicts the variation versus R in: (1) the induced streaming velocity u_∞ obtained outside the boundary layer (see Fig. 6.25a), (2) the magnitude of the maximum and minimum bed shear stress ($\tau_{0,max}$ and $\tau_{0,min}$) relative that that in a sinusoidal flow (τ_{0m}), and (3) the maximum turbulent kinetic energy that occurs beneath crest and trough regions of the flow (k_{crest} and k_{trough}) relative to those beneath a (otherwise identical) sinusoidal flow (k_{0m}). Results for smooth beds (fixed $Re = 5 \times 10^6$) as well as a rough beds (fixed $a/k_s = 1000$) are shown. In all cases, the deviation in these quantities from unity with $R > 0.5$ is approximately linear. Due to the additional turbulence caused by roughness, the streaming in these cases is more dramatic, though on both smooth and rough beds u_∞/U_m can reach O(10%). The results for the amplification/reduction of the bed shear stress (Figs. 6.27b and 6.27c) are quite similar, which makes sense as these quantities scale similarly, i.e. $k \sim U_f^2$. The results demonstrate that the bed shear stress and turbulence generated beneath the crest of a highly-skewed wave can be nearly double that beneath a sinusoidal wave, whereas they can be halved in the trough, for sufficiently large R.

Reasonably similar simulations and parameterizations made by Fuhrman et al. (2009b) have shown that the relative effects of velocity skewness on such normalized wave boundary layer quantities in the turbulent regime are not altered significantly by changing either the Reynolds number

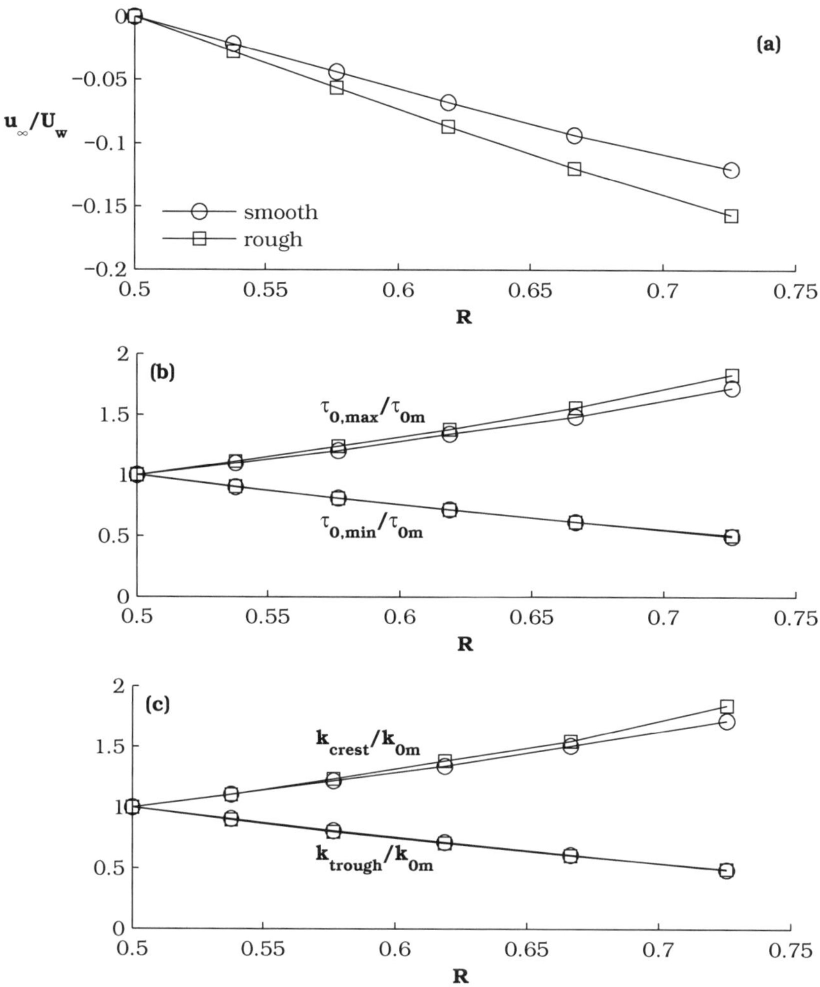

Fig. 6.27 Summary of (a) period-averaged streaming velocities, (b) bed shear stress amplification/reduction, and (c) turbulent kinetic energy amplification/reduction occurring during positive/negative half cycles beneath velocity skewed signals of the form Eq. 6.81. Smooth results correspond to $Re = aU_m/\nu = 5 \times 10^6$, with rough results having $a/k_s = 1000$ where $a = U_m/\omega$.

(in smooth-turbulent conditions) or the dimensionless roughness a/k_s (in rough-turbulent conditions). Hence, Fig. 6.27 can be taken as a reasonable approximation for the relative effects of velocity skewness for other values of Re and a/k_s than those specifically considered. As a demonstration, it can be noted that the case considered previously in Fig. 6.25 ($a/k_s = 1700$) corresponds to a skewness coefficient $R = 0.66$, approximated from Eq. 6.80. The observed streaming velocity from this case (Fig. 6.25a) corresponds to $u_\infty/U_{1m} \approx -0.10$, which is very close to what would be expected for a rough bed based on Fig. 6.27a.

For the interested reader it is finally noted that a study analyzing the accuracy of several different turbulence closure models based on RANS equations (relative to direct numerical simulation, DNS, results based on the model of Scandura, 2007) in predicting the wave boundary layer beneath velocity-skewed free stream signals can be found in Cavallaro et al. (2011). See also the DNS study of Ozdemir et al. (2019).

6.4.2 *Streaming due to free-stream acceleration skewness*

When nonlinear waves encounter a steep slope and/or result in breaking, rather than being purely skewed, their shape may alternatively become front-back asymmetric, resulting in so-called acceleration-skewed free-stream velocity signals. As an example, a time series representing a purely asymmetric free stream velocity signal is depicted in Fig. 6.28a, stemming from Eq. 6.81, with $\phi = 0$ and $r = 0.75$. The corresponding free-stream acceleration $\dot{U}_0 = \partial U_0/\partial t$ is depicted in Fig. 6.28b, which is clearly skewed, having elevated peaks and troughs, similar in form to the skewed velocity signal previously presented in Fig. 6.24a.

To isolate the effects of acceleration-skewness, let us consider the boundary layer that will develop beneath the acceleration-skewed signal depicted in Fig. 6.28a, initially free of any net currents. As before, convective terms will be neglected, and (unless noted otherwise) we will likewise assume the flow is free of external pressure gradients i.e. $P_x = 0$ (pure wave boundary layers in the absence of external currents). A numerical simulation of the turbulent wave boundary layer beneath this signal results in the bed shear stress time series depicted in Fig. 6.28c. Similar to Fig. 6.24b, it is seen that the acceleration-skewed free-stream flow has resulted in a larger bed shear stress magnitude beneath the peak than is observed in the trough. As a result, the period-averaged bed shear stress will again, therefore, initially be positive. This is qualitatively similar to what occurs beneath a velocity-skewed free stream flow.

The reason for the elevated positive bed shear stress in the previously-considered velocity-skewed flows was rather obvious, as the velocity itself was likewise elevated during the positive half cycle. The cause is less obvi-

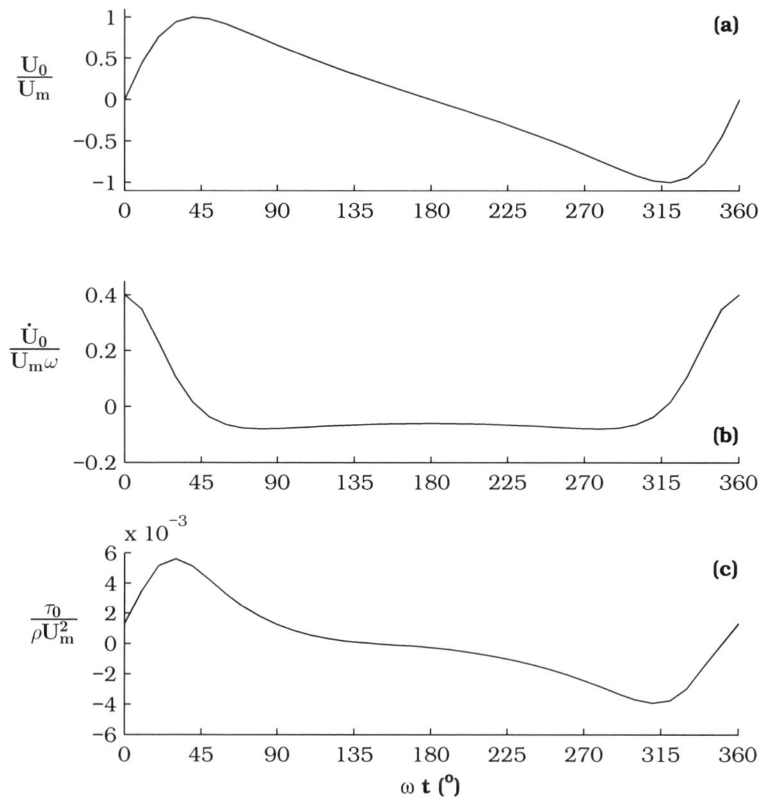

Fig. 6.28 The (a) free stream velocity, (b) acceleration, and (c) computed bed shear stress, based on an acceleration-skewed flow from Eq. 6.81 with $r = 0.75$, $\phi = 0$, $Re = 5 \times 10^6$, and $a/k_s = 1000$.

ous, however, for the present case involving an acceleration-skewed flow, as the velocity magnitude is, in fact, identical in peak (positive) and trough (negative) regions, see again Fig. 6.28a.

To better understand the cause of the elevated bed shear stresses beneath the crest of an acceleration-skewed flow relative to the trough, we must go a step further and instead focus on *velocity gradients.* For the sake of simplicity (neglecting phase differences within the boundary layer) one can envision the formation and growth of a new wave boundary layer upon each flow reversal. Comparing the positive and negative half-cycles for the free stream flow depicted in Fig. 6.28a, it is clear that the boundary layer will have comparatively little time to grow before reaching the crest when

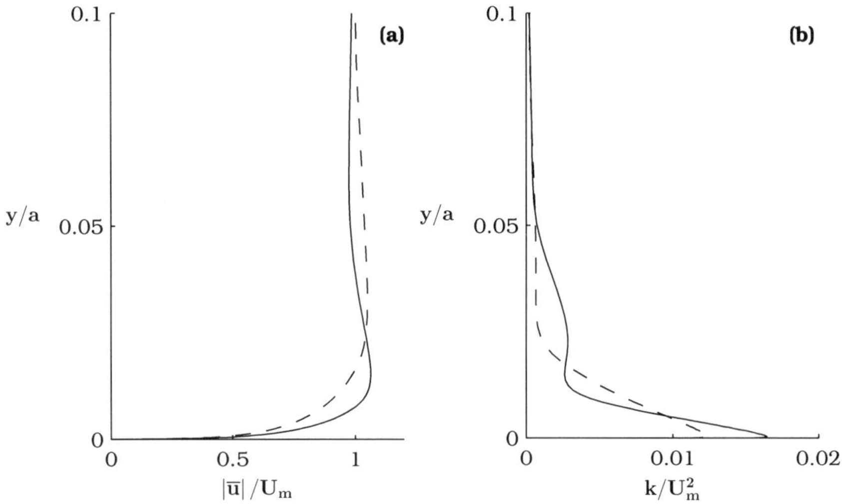

Fig. 6.29 Computed (a) velocity magnitude $|u|$ and (b) turbulent kinetic energy density k profiles at instants when the (acceleration-skewed) free stream velocity U_0 in Fig. 6.28a is at maximum (full lines) and minimum (dashed lines).

compared to the trough. That is, there will be a relatively small (large) boundary layer thickness when the free stream flow is at its maximum (minimum). As the peak and trough velocity are the same in magnitude, this therefore implies that there will be a relative increase in the magnitude of the velocity gradients during the peak flow (i.e. the velocity must grow from zero at the bed to maximum over a comparatively small vertical distance). The increased velocity gradients will likewise result in enhanced turbulence production in the boundary layer beneath the crest free stream flow.

These features can be clearly seen in Fig. 6.29, which depicts the simulated velocity magnitude (Fig. 6.29a) and turbulent kinetic energy density profiles (Fig. 6.29b), at instants when the free stream flow in Fig. 6.28a is maximum (full lines) and minimum (dashed lines). As both velocity gradients, as well as the turbulence, will be elevated beneath the free stream velocity crest, the reasons for the elevated positive bed shear stress magnitude seen in Fig. 6.28 are now clear. Figs. 6.28 and 6.29 can be taken as quantitative demonstrations of features discussed qualitatively by Nielsen (1992), p. 123.

As discussed previously for velocity-skewed flows, the positive period-averaged shear stress that would initially develop beneath a purely acceleration-skewed signal will initially be maximum at the bed, and reduce to zero in the free stream region far from the bed. From Eq. 6.74, this will therefore drive the period-averaged flow to become negative, until ultimately reaching equilibrium where the period-averaged shear stress is uniformly zero, and where the period-averaged and fluctuating components identified in Eq. 6.77 are in balance. Hence, for the same fundamental reasons, the *turbulence-induced* negative streaming phenomenon previously identified beneath skewed free stream velocity signals will likewise occur beneath acceleration-skewed free stream signals.

For the purposes of quantifying the boundary layer properties induced by acceleration skewness, a similar series of tests as considered in Fig. 6.27 have been simulated, for a wide range of acceleration-skewed free stream signals, again based on Eq. 6.81, but now with $\phi = 0$ and $r > 0$. Analogous to the skewness coefficient R defined in Eq. 6.79, we may generally define a so-called acceleration skewness coefficient according to

$$\beta = \frac{\dot{U}_{0,max}}{\dot{U}_{0,max} - \dot{U}_{0,min}} \tag{6.84}$$

where positive acceleration skewness corresponds to $\beta > 0.5$. For a signal specifically of the form Eq. 6.81 with $\phi = 0$ this corresponds to (Abreu et al., 2010):

$$\beta = \begin{cases} (1+r)/2, & r \leq 0.5 \\ 4r(1+r)/(1+2r)^2, & r > 0.5 \end{cases} \tag{6.85}$$

Similar to Fig. 6.27, the results versus β on the streaming velocity, amplification of the bed shear stress and turbulent kinetic energy during peaks and troughs (relative to those beneath an otherwise identical purely sinusoidal free stream velocity) are summarized in Fig. 6.30. At least for moderate β, the growth in each of these quantities is again approximately linear. As the streaming is again turbulence-induced, it is expectedly more pronounced on rough beds than on smooth beds. The relative amplification of the bed shear stress and turbulence beneath the crests are much stronger than are the reductions in the troughs, which more closely resemble those in a sinusoidal flow with equivalent magnitude and period. As a result, the magnitude of turbulence-induced streaming beneath acceleration-skewed flows, is, at most, say $u_\infty/U_m \approx -0.04$ for the range considered, which is comparatively mild relative to that found beneath velocity-skewed signals, see again Fig. 6.27. Note that typical values for coastal zones of this parameter are in the range $0.43 \leq \beta \leq 0.85$, which can be inferred from the large

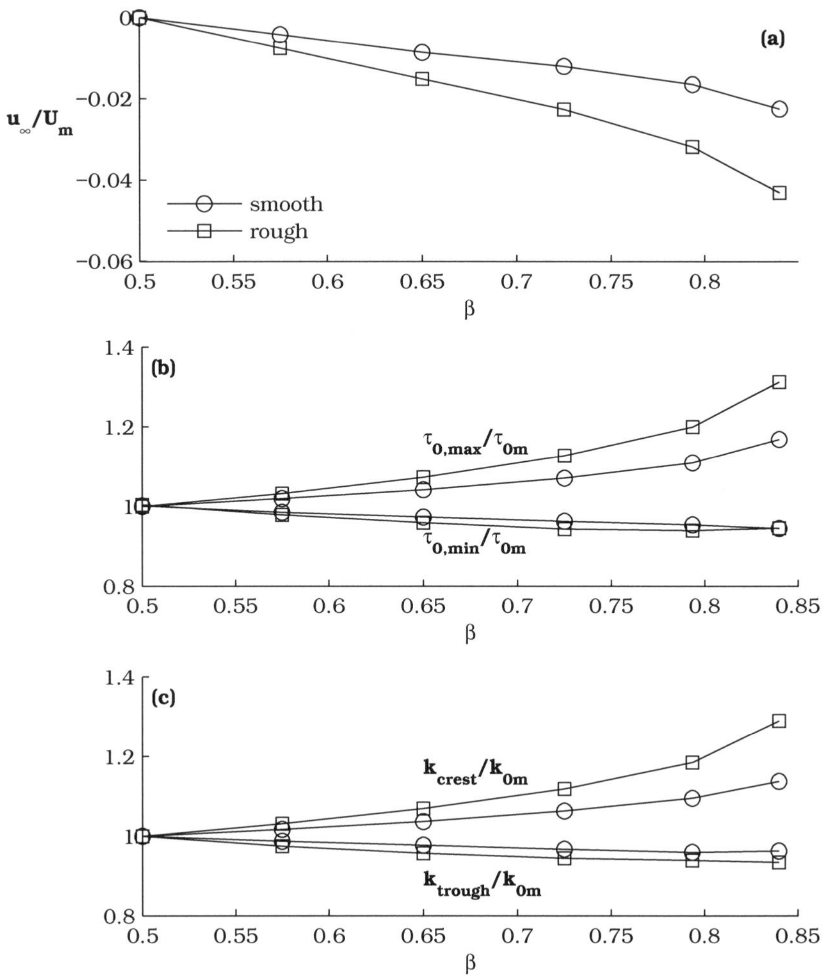

Fig. 6.30 Summary of (a) period-averaged streaming velocities, (b) bed shear stress amplification/reduction, and (c) turbulent kinetic energy amplification/reduction occurring during positive/negative half cycles beneath acceleration skewed signals of the form Eq. 6.81. Smooth results correspond to $Re = aU_m/\nu = 5 \times 10^6$, with rough results having $a/k_s = 1000$ where $a = U_m/\omega$.

collection of field measurements made by Elfrink et al. (2006). Hence, the range considered in Fig. 6.30 is realistic.

It can finally be noted that the negative boundary layer streaming predicted above to occur beneath acceleration-skewed flows is consistent with

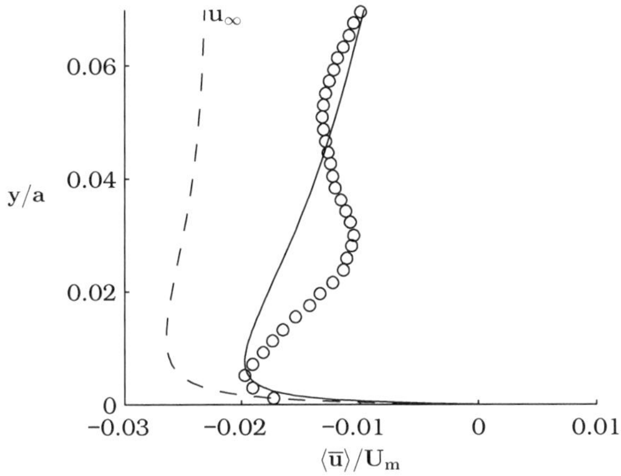

Fig. 6.31 Comparison of simulated ($T = 5$ s, $k_s = 1.09$ mm, $U_m = 0.905$ m/s; $\beta = 0.74$, $Re = 5.8\times10^5$, $a/k_s \approx 660$) and measured (circles van der A et al., 2011) period-averaged near-bed velocity profile beneath an acceleration skewed free stream signal. The full line is the result where a constant pressure gradient $P_x = -3.5 \times 10^{-5} U_m^2/h_m$ has been added, whereas the dashed line is the result with $P_x = 0$.

experimental findings made within an oscillatory tunnel facility by van der A et al. (2011). More specifically, for acceleration-skewed flows having $\beta \leq 0.75$ on both sand and gravel beds, they found negative streaming velocities up to approximately 2% of the free stream magnitude. This is very much in line with Fig. 6.30a for the considered range of β, as well as previously considered numerical results from Fuhrman et al. (2009b), which were the first to demonstrate this negative streaming phenomenon.

As demonstration that the numerical model utilized to produce Fig. 6.30 is capable of producing streaming profiles similar to those which have been measured, comparison of near-bed results based on a case from van der A et al. (2011) (sand bed with $\beta = 0.74$) is presented in Fig. 6.31, where other details are provided in the caption. Simulated results with $P_x = 0$ (dashed line) as well as with $P_x < 0$ (full line) are shown, with the latter clearly resulting in the better match. Thus, it is again clear that to match conditions within such oscillating tunnel facilities, it is necessary to add an additional constant forcing term to the pressure gradient as in Eq. 6.70, which drives a compensatory current in the opposite direction of the induced boundary layer streaming. In the present comparison, this constant P_x has been selected to yield a period-averaged velocity essentially matching the top measurement point in Fig. 6.31 at equilibrium.

For direct numerical simulation results involving the boundary layer beneath acceleration skewed flows, see Scandura et al. (2016).

6.5 Examples of streaming beneath real waves

Each of the previous sections has dealt with a single isolated mechanism that can promote period-averaged streaming phenomena in wave boundary layer flows, in one form or another. In reality, these processes will, of course, not necessarily occur in isolation; Rather, the various processes that have been covered may occur *simultaneously*. As a typical example, simply consider the ubiquitous case of nonlinear (finite amplitude) waves propagating on a sloping beach towards the coastline. In this situation it can be expected that, with the lone exception of the roughness-induced streaming mechanisms discussed in Section 6.3, all of the streaming mechanisms discussed previously would be active and competing against one another! Further complicating the situation will be the interaction with any net currents, which may arise due e.g. to tides, rip currents, return flows (compensating for positive mass fluxes that occur between the crest and trough of surface waves, see e.g. Figs. 5.59b, 5.96b and 6.32a), or undertow currents (offshore near-bed currents that occur in the surf zone, see e.g. Figs. 7.7b, 7.42 and 7.46 in the next chapter). In the surf zone there will likewise be additional turbulence generated by breaking waves, in addition to that generated in the boundary layer, which only complicates an already-complicated picture yet further!

The present section will by no means attempt to explain or investigate every conceivable combination resulting in boundary layer streaming, which would be exhaustive. (It can be noted that combination of several such effects have, however, been studied e.g. by Holmedal and Myrhaug (2009) and Kranenburg et al. (2012) who simulated several cases combining streaming effects due to progressive waves with those due to non-sinusoidal wave shapes; Holmedal et al. (2013) who simulated several scenarios involving progressive wave streaming coupled with co-linear currents; and Afzal et al. (2015) who simulated streaming beneath waves which were at an angle to currents.) Rather, in what follows some selected examples of observed boundary layer streaming (one at laboratory and one at field scale) will be considered, and an attempt at their interpretation will be made with consideration to the preceding sections.

As a first example of observed boundary layer streaming beneath real progressive waves, let us consider a case presented by Sumer and Fredsøe (2001) at laboratory scale, who measured the period-averaged velocity profile depicted in Fig. 6.32a. Note that this plot spans the entire vertical profile i.e. from the seabed to the wave crest, with the thin boundary layer streaming region indicated at the bottom. This example stems specifically from their case with wave period $T = 3.5$ s, wave height $H = 12$ cm, yielding free stream velocity magnitude $U_{0m} = 30.0$ cm/s (see Appendix 5.J, Eq. 5.213), with $U_{fm} = 2.01$ cm/s, measured on a water depth $h = 40$ cm on a bed with medium sediment diameter $d = 0.2$ mm (hence $k_s \approx 2.5d = 0.0005$ m). This yields the Reynolds numbers $Re = 5 \times 10^5$ and $k_s^+ = 10$, placing the wave boundary layer in the laminar regime, perhaps with a hint of roughness effects. It can be noted that the maximum streaming velocity in the boundary layer $u \approx 1.3$ cm/s is less than the value $u_\infty = 3.5$ cm/s predicted by Eq. 6.20, while matching more closely that predicted from the rough-turbulent expression Eq. 6.36 $u_\infty = 1.6$ cm/s. The reason the measured value is somewhat less than predicted by the laminar theory is thus likely partially due to the limited roughness effects, coupled with additional effects associated with the return current. Such a (negative) return current is clearly apparent in Fig. 6.32 for say $y < 34$ cm, and will commonly arise in closed wave flumes, to compensate primarily for the (positive) wave drift that is likewise readily apparent between the surface elevation crest and trough, 34 cm $\leq y \leq$ 48 cm (see Appendix 5.D). The presence of the return current will thus deviate from the idealized conditions considered theoretically e.g. in Section 6.1.1 by: (1) causing non-zero period-averaged shear stress at the top of the wave boundary layer, and (2) resulting in a more complicated wave-plus-current flow environment (see Section 5.6). Nevertheless, the observed streaming is in the propagating direction and is the correct order of magnitude as predicted theoretically, and is likewise shifted in the expected direction for an opposing current, as well as for the expected small, but finite, roughness effects.

As a second example, let us consider field measurements originally collected by Aagaard et al. (2012) at Pearl Beach on the Northwest shore of Broken Bay, New South Wales, Australia during the period June 12–24, 2011. During this time period data was collected outside the breaking point, during a storm, as well as pre-storm and post-storm, with observed boundary layer streaming being strongest during the pre-storm phase. The pre-storm conditions from this data set were subsequently re-analyzed by Wang et al. (2016), and two of their resulting near-bed averaged velocity profiles are depicted in Fig. 6.32b, corresponding to their "bursts" (to use

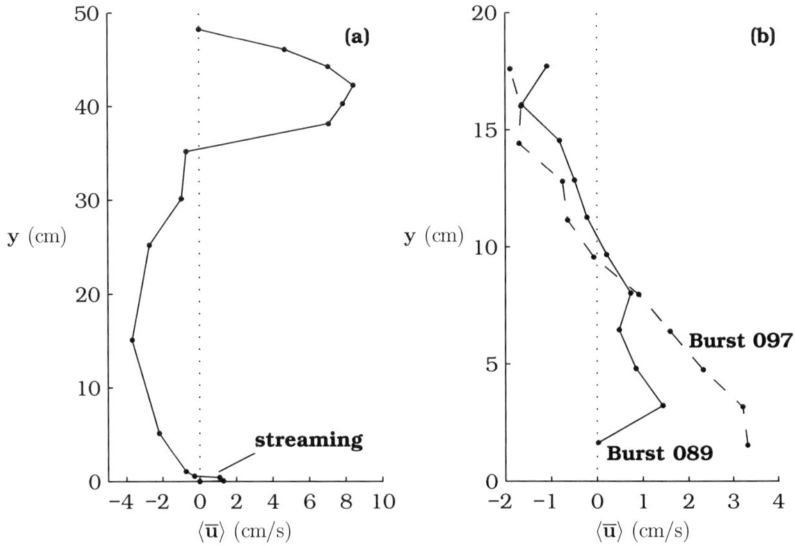

Fig. 6.32 (a) Example of a measured period-averaged velocity profile beneath a propagating wave train in the laboratory from Sumer and Fredsøe (2001). The boundary layer streaming can be seen in the thin region very near the bed. (b) Examples of measured near-bed streaming in field measurements, taken from Wang et al. (2016). In both cases the positive direction coincides with the direction of wave propagation.

their phrasing, not to be confused with the bursting process covered in Section 3.5) 089 and 097, both of which were measured at bed ripple midpoints (rather than at crests or troughs).

As the reported conditions for these bursts were not dramatically different, for the purposes of the present comparison we will simply average them, yielding: significant wave height $H_s = 0.75$ m, peak period $T_p = 8.7$ s, and water depth $h = 3.6$ m. Following Wang et al. (2016), we will likewise convert the significant wave height to an equivalent root-mean-squared wave height $H = H_s/\sqrt{2}$, for use in characterizing relevant surface wave and wave boundary layer properties based on linear wave theory. This yields: wave celerity $c = 5.8$ m/s, free stream velocity magnitude $U_m = 0.41$ m/s, and equivalent amplitude of orbital motion $a = 0.57$ m. From comparisons made by Wang et al. (2016), the roughness was best characterized based on the height of bed ripples which were present, hence we utilize their reported value corresponding to $k_s = 0.057$ m, which would clearly place these conditions well into the hydraulically rough regime. From Eq. 6.36, we may make a first estimate of the expected streaming velocity, corresponding to

$u_\infty \approx 1$ cm/s. This matches reasonably the positive streaming observed from burst 089, but is approximately a factor 3 lower than observed for burst 097 (similar results as above were in fact found using a k-ε model by Wang et al., 2016). Hence, based on this comparison the direction and order of magnitude of the observed streaming would seem to be correctly predicted, though it should be emphasized that the observed differences (even for rather similar conditions) highlights uncertainties in reliably predicting such quantities in field conditions.

Based on their full data set, even greater variability than might be taken from the two examples presented here was found, with 18 of the 24 analyzed series (corresponding to those at ripple troughs) demonstrating either no boundary layer drift or one that was offshore. This may be due e.g. to nonlinear wave shape effects (velocity or acceleration skewness) as covered in Section 6.4. Further complicating factors include the effects of irregular waves, interaction with any wave-driven currents, as well as interaction of non-sinusoidal wave shapes with ripples. See e.g. Ribberink et al. (2008) and Vittori and Blondeaux (2013) for more detailed discussions on the latter topic. For additional field observations of wave-induced streaming through submerged sea grass, see e.g. Luhar et al. (2013).

Note, finally, that other examples of wave boundary layer streaming in the present book can be observed in the period-averaged velocity profiles shown in Fig. 5.59. These figures show wave drift in the region between the trough and the crest, a negative return current in the main water column, as well as a small region of streaming in the forward direction very near the bed.

6.6 Importance of streaming and wave shape in coastal sediment transport

Wave boundary layer streaming in its various forms is of significant importance in coastal engineering, as such phenomena are important factors governing cross-shore (perpendicular to the shoreline) sediment transport and coastal profile morphology. Accurately predicting cross-shore sediment transport, and especially the morphology of the sea bed, beneath waves is notoriously difficult, primarily due to the large number of competing onshore and offshore driving factors simultaneously involved (many of which have been considered in the present chapter), see also the discussion and comparison made in Fuhrman et al. (2009b). While a complete account of this subject is well beyond the scope of the present chapter, some highlights of several selected studies from the literature will be presented in what follows, from which the interested reader may build.

Among the various streaming mechanisms considered through this chapter, the relative importance of conventional Longuet-Higgins-type progressive wave streaming (as covered in Section 6.1) and that due to non-sinusoidal wave shapes (as covered in Section 6.4), have perhaps received the most direct attention. The experimental study of progressive wave streaming naturally requires use of wave flumes, such that the spatial variations in the propagating direction are present. Dohmen-Janssen and Hanes (2002) presented detailed measurements of sediment concentrations and grain velocities inside the "sheet flow" layer under prototype surface gravity waves, in combination with measurements of suspensions processes above the sheet flow sediment transport layer. They compared their measured period-averaged sediment transport rates $\langle q_s \rangle$ versus the period-averaged cube of the free stream velocity signal $\langle U_0^3 \rangle$ for their cases with similar results from an oscillatory tunnel (hence no progressive wave streaming) from Ribberink and Al-Salem (1994). Both sets of results showed a roughly linear trend (a commonly used simple model for making sediment transport rate predictions), but found that the results from the progressive waves yielded net transport rates that were about a factor 2.5 times larger than in tunnels. This increase can at least partially be attributed to the effects of progressive wave streaming. Fuhrman et al. (2013) made a similar comparison numerically, and likewise found an increase in net sediment transport due to streaming, though less than the factor 2.5 mentioned above. While the exact quantitative effects of streaming in coastal sediment transport are still being actively researched, that such streaming beneath progressive waves would generally increase the onshore transports of sediment is certainly consistent with this phenomenon promoting both onshore-directed bed shear stresses and boundary layer streaming, as shown in Section 6.1.

The effects of non-sinusoidal wave shapes on sediment transport can be more readily studied in oscillating tunnel experiments. Such tunnel experiments involving velocity-skewed free stream signals include Ribberink and Al-Salem (1994, 1995), Ahmed and Sato (2003), O'Donoghue and Wright (2004a,b), Hassan and Ribberink (2005) and Silva et al. (2011), whereas the sediment transport beneath acceleration-skewed flows has been studied by van der A et al. (2010) and Silva et al. (2011). These have shown a general tendency for such nonlinear wave shapes to promote onshore (positively directed) transports of sediment. This is also as might be expected, given that the magnitude of the peak bed shear stress tends to be higher than that of the trough, see Figs. 6.24 and 6.28. However, such experiments have likewise shown that velocity-skewed signals can, somewhat non-intuitively, promote offshore (negatively) directed transports of fine sands (O'Donoghue and Wright, 2004a,b; Hassan and Ribberink, 2005). This is due to so-called phase-lag effects (Dohmen-Janssen et al., 2002), wherein fine sediments stirred up during the positive flow half-cycle do not have sufficient time to settle, and subsequently become transported backwards following flow reversal. The negative streaming associated with such flows (as shown in Section 6.4) will also play a role in the negative transport of suspended sediments in such situations. It can be noted that the RANS-based studies of Fuhrman et al. (2013) and Kranenburg et al. (2013) have, however, suggested that the transport reversal of fine sands mentioned above may, in fact, be an artifact of oscillating tunnel facilities, which obviously do not incorporate conventional progressive wave streaming (Section 6.1) or any related convective acceleration effects in the suspended sediment transport. When such effects were switched on in their numerical models, a "re-reversal" of the fine sand sediment transport to the positive (onshore) direction was predicted. For a detailed account of such modelling studies, see Fuhrman et al. (2013) and Kranenburg et al. (2013).

Methods for modelling the sediment transport beneath waves, including combined effects of streaming and wave shape, vary widely in their complexity. These range from relatively simple analytical (sometimes called engineering) approaches (e.g. Nielsen and Callaghan, 2003; Nielsen, 2006), to mixing length model-based approaches (e.g. Deigaard et al., 1999), to those based on two-equation turbulence models (Holmedal and Myrhaug, 2009; Blondeaux et al., 2012; Fuhrman et al., 2013; Kranenburg et al., 2013; Holmedal et al., 2013; Afzal et al., 2015; Caliskan and Fuhrman, 2017), to detailed two-phase (water and sediment models) (e.g. Hsu and Hanes, 2004; Bakhtyar et al., 2009; Yu et al., 2010; Kranenburg et al., 2014; Kim et al., 2018), just to name a few. Further development of such models is an area of active research.

6.7 References

(1) Aagaard, T., Hughes, M., Baldock, T., Greenwood, B., Kroon, A., and Power, H. (2012): Sediment transport processes and morphodynamics on a reflective beach under storm and non-storm conditions. Marine

Geology, vol. 326, 154–165.
(2) Abreu, T., Silva, P.A., Sancho, F. and Temperville, A. (2010): Analytical approximate wave form for asymmetric waves. Coast. Eng., vol. 57, 656–667.
(3) Afzal, M.S., Holmedal, L.E. and Myrhaug, D. (2015): Three-dimensional streaming in the seabed boundary layer beneath propagating waves with an angle of attack on the current. J. Geophys. Res., vol. 120, 4370–4391.
(4) Ahmed, A.S.M. and Sato, S. (2003): A sheetflow transport model for asymmetric oscillatory flows. Part I: Uniform grain size sediments. Coast. Eng. J., vol. 45, 321–337.
(5) Bakhtyar, R., Yeganeh-Bakhtiary, A., Barry, D.A. and Ghaheri, A. (2009): Two-phase hydrodynamic and sediment transport modeling of wave-generated sheet flow. Adv. Water Resources, vol. 32, 1267–1283.
(6) Bijker, E.W., Kalkwijk, J.P.T. and Pieter, T. (1974): Mass transport in gravity waves on a sloping bottom. In: Proc. 14th Int. Conf. Coast. Eng., pp. 447–465.
(7) Blondeaux, P., Brocchini, M. and Vittori, G. (2002): Sea waves and mass transport on a sloping beach. Proc. Roy. Soc. Lond. A, vol. 485, 2053–2082.
(8) Blondeaux, P., Vittori, G., Bruschi, A., Lalli, F., and Pesarino, V. (2012): Steady streaming and sediment transport at the bottom of sea waves. J. Fluid Mech., vol. 697, 115–149.
(9) Blondeaux, P., and Vittori, G. (2014): The flow over bedload sheets and sorted bedforms. Cont. Shelf Res., vol. 85, 9–20.
(10) Brøker, I. H. (1985): Wave generated ripples and resulting sediment transport in waves. Series paper no. 36, ISVA, Technical University of Denmark.
(11) Caliskan, U. and Fuhrman, D.R. (2017): RANS-based simulation of wave-induced sheet-flow transport of graded sediments. Coast. Eng., vol. 121, 90–102.
(12) Cavallaro, L., Scandura, P., and Foti, E. (2011): Turbulence-induced steady streaming in an oscillating boundary layer: On the reliability of turbulence closure models. Coast. Eng., vol. 58, 290–304.
(13) Davies, A.G. and Li, Z. (1997): Modelling sediment transport beneath regular symmetrical and asymmetrical waves above a plane bed. Cont. Shelf Res., vol. 17, 555–582.
(14) Deigaard, R., Jakobsen, J.B., and Fredsøe, J. (1999): Net sediment transport under wave groups and bound long waves. J. Geophys. Res.,

vol. 104(C6), 13559–13575.

(15) Dohmen-Janssen, C.M. and Hanes, D.M. (2002): Sheet flow dynamics under monochromatic nonbreaking waves. J. Geophys. Res., vol. 107, article no. C103149.

(16) Dohmen-Janssen, C.M., Kroekenstoel, D., Hassan, W.N. and Ribberink, J.S. (2002): Phase-lags in oscillatory sheet flow: experiments and bed load modelling. Coast. Eng., vol. 46, 61–87.

(17) Elfrink, B., Hanes, D.M. and Ruessink, B.G. (2006): Parameterization and simulation of near bed orbital velocities under irregular waves in shallow water. Coast. Eng., vol. 53, 915–927.

(18) Fredsøe, J. and Deigaard, R. (1992): Mechanics of Coastal Sediment Transport. World Scientific.

(19) Fredsøe, J., Sumer, B.M., Laursen, T. and Pedersen, C. (1993): Experimental investigation of wave boundary layers with a sudden change in roughness. J. Fluid Mech., vol. 252, 117–145.

(20) Fuhrman, D.R., Fredsøe, J. and Sumer, B.M. (2009a): Bed slope effects on turbulent wave boundary layers: 1. Model validation and quantification of rough-turbulent results. J. Geophys. Res., vol. 114, article no. C03024.

(21) Fuhrman, D.R., Fredsøe, J. and Sumer, B.M. (2009b): Bed slope effects on turbulent wave boundary layers: 2. Comparison with skewness, asymmetry, and other effects. J. Geophys. Res., vol. 114, article no. C03025.

(22) Fuhrman, D.R., Schløer, S. and Sterner, J. (2013): RANS-based simulation of turbulent wave boundary layer and sheet-flow sediment transport processes. Coast. Eng., vol. 73, 151–166.

(23) Fuhrman, D.R., Sumer, B.M. and Fredsøe, J. (2011): Roughness-induced streaming in turbulent wave boundary layers. J. Geophys. Res., vol. 116, article no. C10002.

(24) Grass, A.J. (1971): Structural features of turbulent flow over smooth and rough boundaries, J. Fluid Mech., vol. 50, 233–255.

(25) Hassan, W.N. and Ribberink, J.S. (2005): Transport processes of uniform and mixed sands in oscillatory sheet flow. Coast. Eng., vol. 52, 745–770.

(26) Hassan, W.N.M. and Ribberink, J. S. (2010): Modelling of sand transport under wave-generated sheet flows with a RANS diffusion model. Coast. Eng., vol. 57, 19–29.

(27) Holmedal, L.E., Johari, J. and Myrhaug, D. (2013): The seabed boundary layer beneath waves waves opposing and following a current. Cont.

Shelf Res., vol. 65, 27–44.
(28) Holmedal, L.E. and Myrhaug, D. (2006): Boundary layer flow and net sediment transport beneath asymmetrical waves. Cont. Shelf Res., vol. 26, 252–268.
(29) Holmedal, L.E. and Myrhaug, D. (2009): Wave-induced steady streaming, mass transport and net sediment transport in rough turbulent ocean bottom boundary layers. Cont. Shelf Res., vol. 29, 911–926.
(30) Hsu, T.-J. and Hanes, D.M. (2004): Effects of wave shape on sheet flow sediment transport. J. Geophys. Res., vol. 109, article no. C05025.
(31) Justesen, P. (1988): Turbulent Wave Boundary Layers. Series Paper 43, Institute of Hydrodynamic and Hydraulic Engineering, Technical University of Denmark.
(32) Kim, Y., Cheng, Z., Hsu, T.-J. and Chauchat, J. (2018): A numerical study of sheet flow under monochromatic nonbreaking waves using a free surface resolving Eulerian two-phase flow model. J. Geophys. Res.-Oceans, vol. 123, 4693–4719.
(33) Kranenburg, W.M., Hsu, T.-J., and Ribberink, J.S. (2014): Two-phase modeling of sheet-flow beneath waves and its dependence on grain size and streaming. Adv. Water Resources, vol. 72, 57–70.
(34) Kranenburg, W.M., Ribberink, J.S., Schretlen, J.J.L.M., and Uittenbogaard, R.E. (2013): Sand transport beneath waves: The role of progressive wave streaming and other free surface effects. J. Geophys. Res.-Earth Surface, vol. 118, 122–139.
(35) Kranenburg, W.M., Ribberink, J.S., Uittenbogaard, R.E., and Hulscher, S.J.M.H. (2012): Net currents in the wave bottom boundary layer: On waveshape streaming and progressive wave streaming. J. Geophys. Res.-Earth Surface, vol. 117, article no. F03005.
(36) Laursen, T.S., Fredsøe, J. and Sumer, B.M. (1994): Numerical prediction of wave boundary layer over a bed with a change in roughness. Coast. Eng., vol. 24, 81–96.
(37) Lin, C., Yeh, P.-H., Kao, M.-J., Yu, M.-H., Hsieh, S.-C., Chang, S.-C., Wu, T.-R. and Tsai, C.-P. (2015): Velocity fields in near-bottom and boundary layer flows in prebreaking zone of a solitary wave propagating over a 1:10 slope. J. Waterw. Port Coast. Ocean Eng., vol. 141, article no. 04014038.
(38) Longuet-Higgins, M.S. (1953): Mass transport in water waves. Phil. Trans. Roy. Soc. Lond. A, vol. 245, 535–581.
(39) Longuet-Higgins, M.S. (1958): The mechanics of the boundary-layer near the bottom in a progressive wave. In: Proc. 6th Int. Conf. Coast.

Eng., Miami, FL, pp. 184–193. (Appendix to Russell and Osorio, 1958).
(40) Luhar, M., Infantes, E., Orfila, A., Terrados, J. and Nepf, H.M. (2013): Field observations of wave-induced streaming through a submerged seagrass (Posidonia oceanica) meadow. J. Geophys. Res.-Oceans, vol. 118, 1955–1968.
(41) Nielsen, P. (1992): Coastal Bottom Boundary Layers and Sediment Transport. World Scientific.
(42) Nielsen, P. (2006): Sheet flow sediment transport under waves with acceleration skewness and boundary layer streaming. Coast. Eng., vol. 53, 749–758.
(43) Nielsen, P., and Callaghan, D.P. (2003): Shear stress and sediment transport calculations for sheet flow under waves. Coast. Eng., vol. 47, 347–354.
(44) O'Donoghue, T. and Wright, S. (2004a): Concentrations in oscillatory sheet flow for well sorted and graded sands. Coast. Eng., vol. 50, 117–138.
(45) O'Donoghue, T. and Wright, S. (2004b): Flow tunnel measurements of velocities and sand flux in oscillatory sheet flow for well-sorted and graded sands. Coast. Eng., vol. 51, 1163–1184.
(46) Ozdemir, C.E., Yu, X., Sororian, S., Zhu, L., Tyagi, M., Haddadian, S., Oliveira, D., Turnipseed, C.N. and Kefelegn, H. (2019): Direct numerical simulations of second-order Stokes wave driven smooth-walled oscillatory channel: investigation of net current formation. J. Turbulence, doi: 10.1080/14685248.2019.1649415
(47) Ribberink, J.S. and Al-Salem, A.A. (1994): Sediment transport in oscillatory boundary layers in cases of rippled beds and sheet flow. J. Geophys. Res., vol. 99, 12707–12727.
(48) Ribberink, J.S. and Al-Salem, A.A. (1995): Sheet flow and suspension of sand in oscillatory boundary layers. Coast. Eng., vol. 25, 205–225.
(49) Ribberink, J.S., van der Werf, J.J., O'Donoghue, T., and Hassan, W.N.M. (2008): Sand motion induced by oscillatory flows: Sheet flow and vortex ripples. J. Turbulence, vol. 9, 1–32.
(50) Ruessink, B.G., van den Berg, T.J.J. and van Rijn, L.C. (2009): Modeling sediment transport beneath skewed asymmetric waves above a plane bed. J. Geophys. Res., vol. 114, article no. C11021.
(51) Russell, R.C.H. and Osorio, J.D.C. (1958): An experimental investigation of drift profiles in a closed channel. In: Proc. 6th Int. Conf. Coast. Eng., Miami, FL, pp. 171–183.

(52) Scandura, P. (2007): Steady streaming in a turbulent oscillating boundary layer. J. Fluid Mech., vol. 571, 265–280.
(53) Scandura, P., Faraci, C., and Foti, E. (2016): A numerical investigation of acceleration-skewed oscillatory flows. J. Fluid Mech., vol. 808, 576–613.
(54) Silva, P.A., Abreu, T., van der A, D.A., Sancho, F., Ruessink, B.G., van der Werf, J. and Ribberink, J.S. (2011): Sediment transport in nonlinear skewed oscillatory flows: Transkew experiments. J. Hydraul. Res., vol. 49, 72–80.
(55) Speziale, C.G. (1982): On turbulent secondary flows in pipes of non-circular crosssection. Int. J. Eng. Sci., vol. 20, 863–872.
(56) Speziale, C.G. (1987): On nonlinear k-ℓ and k-ε models of turbulence. J. Fluid Mech., vol. 178, 459–475.
(57) Sumer, B.M. and Fredsøe, J. (2001): Wave scour around a large vertical circular cylinder, J. Waterw. Port Coast. Ocean Eng.-ASCE, vol. 127, 125–134.
(58) Sumer, B.M., Laursen, T.S. and Fredsøe, J. (1993): Wave boundary layer in a convergent tunnel. Coast. Eng., vol. 20, 317–342.
(59) Trowbridge, J. and Madsen, O.S. (1984a): Turbulent wave boundary layers 1. Model formation and first-order solution. J. Geophys. Res., vol. 89, 7989–7997.
(60) Trowbridge, J. and Madsen, O.S. (1984b): Turbulent wave boundary layers 2. Second-order theory and mass transport. J. Geophys. Res., vol. 89, 7999–8007.
(61) van der A, D.A., O'Donoghue, T., Davies, A.G. and Ribberink, J.S. (2011): Experimental study of the turbulent boundary layer in acceleration-skewed oscillatory flow. J. Fluid Mech., vol. 648, 251–283.
(62) van der A, D.A., O'Donoghue, T. and Ribberink, J.S. (2010): Measurements of sheet flow transport in acceleration-skewed oscillatory flow and comparison with practical formulations. Coast. Eng., vol. 57, 331–342.
(63) Vittori, G., and Blondeaux, P. (2013): Steady streaming induced by sea waves over rippled and rough beds. Cont. Shelf Res., vol. 65, 64–72.
(64) Wang, H., Aagaard, T., Holmedal, L.E. and Myrhaug, D. (2016): Analysis of observed streaming in field measurements, J. Waterw. Port Coast Ocean Eng.-ASCE, vol. 142, article no. 05015003.
(65) Wang, Z.-Q. and Cheng, N.-S. (2006): Time-mean structure of secondary flows in open channel with longitudinal bedforms. Adv. Water Resources, vol. 29, 1634–1649.

(66) Wilcox, D.C. (2006): Turbulence Modeling for CFD, 3rd edn. DCW Industries.

(67) Wilcox, D.C. (2008): Formulation of the k-ω turbulence model revisited. AIAA J., vol. 46, 2823–2838.

(68) Yu, X., Hsu, T.-J., and Hanes, D.M. (2010): Sediment transport under wave groups: Relative importance between nonlinear waveshape and nonlinear boundary layer streaming. J. Geophys. Res., vol. 115, article no. C02013.

(69) Zou, Q. and Hay, A.E. (2003): The vertical structure of the wave bottom boundary layer over a sloping bed: Theory and field measurements. J. Phys. Oceanogr., vol. 33, 1380–1400.

(70) Zou, Q., Hay, A.E. and Bowen, A.J. (2003): Vertical structure of surface gravity waves propagating over a sloping seabed: Theory and field measurements. J. Geophys. Res., vol. 108 (C8), article no. 3265, doi:10.1029/2002JC001432.

Appendix to Chapter 6: The vorticity equation

6.A Derivation of the vorticity equation

In this appendix we will derive the so-called (vector) vorticity equation. For this purpose, we will start from the incompressible Reynolds-averaged Navier-Stokes equations, in the following form:

$$\frac{\partial \overline{\mathbf{u}}}{\partial t} + (\overline{\mathbf{u}} \cdot \nabla)\overline{\mathbf{u}} = -\frac{1}{\rho}\nabla \overline{p} + \nu \nabla^2 \overline{\mathbf{u}} + \nabla \cdot \frac{\boldsymbol{\tau}}{\rho} \tag{6.86}$$

where $\overline{\mathbf{u}} = (\overline{u}, \overline{v}, \overline{w})$, $\boldsymbol{\tau}$ represents the turbulent Reynolds stresses, and

$$\nabla = \left(\frac{\partial}{\partial x}, \frac{\partial}{\partial y}, \frac{\partial}{\partial z}\right), \qquad \nabla^2 = \nabla \cdot \nabla \tag{6.87}$$

are, respectively, the gradient and Laplacian operators. Note that the convective term may be equivalently written as

$$(\overline{\mathbf{u}} \cdot \nabla)\overline{\mathbf{u}} = \nabla\left(\frac{1}{2}\overline{\mathbf{u}} \cdot \overline{\mathbf{u}}\right) - \overline{\mathbf{u}} \times \overline{\boldsymbol{\omega}} \tag{6.88}$$

where the (Reynolds-averaged) vorticity vector is defined according to

$$\overline{\boldsymbol{\omega}} = \nabla \times \overline{\mathbf{u}} = \left(\frac{\partial \overline{w}}{\partial y} - \frac{\partial \overline{v}}{\partial z}, \frac{\partial \overline{u}}{\partial z} - \frac{\partial \overline{w}}{\partial x}, \frac{\partial \overline{v}}{\partial x} - \frac{\partial \overline{u}}{\partial y}\right) \tag{6.89}$$

This has been previously defined in Eqs. 2.66 and 2.65 (there in terms of the instantaneous, rather than Reynolds-averaged, velocities). Taking the curl (vector product) of Eq. 6.86, after invoking Eq. 6.88, leads to

$$\frac{\partial \overline{\boldsymbol{\omega}}}{\partial t} - \nabla \times (\overline{\mathbf{u}} \times \boldsymbol{\omega}) = \nu \nabla^2 \overline{\boldsymbol{\omega}} + \nabla \times \left(\nabla \cdot \frac{\boldsymbol{\tau}}{\rho}\right) \tag{6.90}$$

where it has been utilized that

$$\nabla \times \nabla \phi = 0, \qquad \phi = \text{any scalar field} \tag{6.91}$$

Now note that

$$\nabla \times (\overline{\mathbf{u}} \times \overline{\boldsymbol{\omega}}) = -\overline{\boldsymbol{\omega}}(\nabla \cdot \overline{\mathbf{u}}) + (\overline{\boldsymbol{\omega}} \cdot \nabla)\overline{\mathbf{u}} - (\overline{\mathbf{u}} \cdot \nabla)\overline{\boldsymbol{\omega}} \tag{6.92}$$

where the first term on the right-hand-side may be eliminated due to local continuity:

$$\nabla \cdot \overline{\mathbf{u}} = 0 \tag{6.93}$$

Hence, we may now write Eq. 6.90 as

$$\frac{d\overline{\boldsymbol{\omega}}}{dt} = \frac{\partial \overline{\boldsymbol{\omega}}}{\partial t} + (\overline{\mathbf{u}} \cdot \nabla)\overline{\boldsymbol{\omega}} = (\overline{\boldsymbol{\omega}} \cdot \nabla)\overline{\mathbf{u}} + \nu \nabla^2 \overline{\boldsymbol{\omega}} + \nabla \times \left(\nabla \cdot \frac{\tau}{\rho}\right) \tag{6.94}$$

This is the (Reynolds-averaged) **vorticity equation**. In tensor notation, this is equivalently

$$\frac{d\overline{\omega}_i}{dt} = \frac{\partial \overline{\omega}_i}{\partial t} + \overline{u}_j \frac{\partial \overline{\omega}_i}{\partial x_j} = \overline{\omega}_j \frac{\partial \overline{u}_i}{\partial x_j} + \nu \frac{\partial^2 \overline{\omega}_i}{\partial x_j \partial x_j} + \frac{\epsilon_{ijk}}{\rho} \frac{\partial^2 \tau_{kl}}{\partial x_j \partial x_l} \tag{6.95}$$

where

$$\epsilon_{ijk} = \begin{cases} 1 & \text{if } (ijk) \text{ is an even (cyclic) permutation of (123) i.e.} \\ & \epsilon_{123} = \epsilon_{231} = \epsilon_{312} = 1, \\ -1 & \text{if } (ijk) \text{ is an odd (non-cyclic) permutation of (123) i.e.} \\ & \epsilon_{213} = \epsilon_{321} = \epsilon_{132} = -1, \\ 0 & \text{if two or more subscripts are the same e.g.} \\ & \epsilon_{111} = \epsilon_{112} = \epsilon_{313} = 0 \end{cases} \tag{6.96}$$

is the so-called Levi-Civita permutation tensor.

Note that, similar to the three velocity components, the three instantaneous vorticity components can generally be written as $\omega_i = \overline{\omega}_i + \omega_i'$ i.e. as the sum of averaged and (turbulent) fluctuating parts.

6.B Vorticity generation due to anisotropic normal Reynolds stresses

We will now demonstrate why gradients in anisotropic turbulent normal stresses result in a generation of streamwise vorticity, and hence secondary flows. For this purpose, consider an initially uni-directional turbulent flow that is uniform in the streamwise $x_3 = z$ direction. As secondary flows correspond to the streamwise (axial) component of the vorticity vector, it is natural then to consider Eq. 6.95 with $i = 3$, which reads

$$\frac{\partial \overline{\omega}_3}{\partial t} + \overline{u}_j \frac{\partial \overline{\omega}_3}{\partial x_j} = \overline{\omega}_j \frac{\partial \overline{u}_3}{\partial x_j} + \nu \frac{\partial^2 \overline{\omega}_3}{\partial x_j \partial x_j} + \frac{\epsilon_{3jk}}{\rho} \frac{\partial^2 \tau_{kl}}{\partial x_j \partial x_l} \tag{6.97}$$

Linearizing to retain only leading-order terms, and noting that initially $\overline{\omega}_3 = 0$, this simplifies further to

$$\frac{\partial \overline{\omega}_3}{\partial t} = \frac{\epsilon_{3jk}}{\rho} \frac{\partial^2 \tau_{kl}}{\partial x_j \partial x_l} \tag{6.98}$$

The right hand side again represents the (spatial gradients in) turbulent Reynolds stresses, which in full, corresponds to

$$\frac{\epsilon_{3jk}}{\rho} \frac{\partial^2 \tau_{kl}}{\partial x_j \partial x_l} = \frac{1}{\rho} \left[\frac{\partial}{\partial x_1} \left(\frac{\partial \tau_{21}}{\partial x_1} + \frac{\partial \tau_{22}}{\partial x_2} + \frac{\partial \tau_{23}}{\partial x_3} \right) - \frac{\partial}{\partial x_2} \left(\frac{\partial \tau_{11}}{\partial x_1} + \frac{\partial \tau_{12}}{\partial x_2} + \frac{\partial \tau_{13}}{\partial x_3} \right) \right] \tag{6.99}$$

From the assumed uniformity in x_3, however, this reduces to

$$\frac{\epsilon_{3jk}}{\rho} \frac{\partial^2 \tau_{kl}}{\partial x_j \partial x_l} = \frac{1}{\rho} \left[\frac{\partial^2 (\tau_{22} - \tau_{11})}{\partial x_1 \partial x_2} + \frac{\partial^2 (\tau_{21})}{\partial x_1^2} - \frac{\partial^2 (\tau_{12})}{\partial x_2^2} \right] \tag{6.100}$$

where the two normal stresses have now been grouped. Moreover, since it is again assumed that initially $\overline{u}_1 = \overline{u}_2 = 0$, it must be that

$$\tau_{12} = \tau_{21} = -\rho \overline{u_1' u_2'} = 0 \tag{6.101}$$

as there would be no reason for these turbulent fluctuations to be correlated (as $\overline{u}_1$ and $\overline{u}_2$ are both uniform their gradients are obviously zero!). Hence, for the conditions considered, Eq. 6.98 reduces simply to

$$\frac{\partial \overline{\omega}_3}{\partial t} = \frac{1}{\rho} \frac{\partial^2 (\tau_{22} - \tau_{11})}{\partial x_1 \partial x_2} \tag{6.102}$$

The right hand side is obviously a vorticity source term, and matches exactly that identified in Eq. 6.67.

From the above, it is now evident that the production of streamwise vorticity, and hence secondary flows, requires: (1) non-zero differences (anisotropy) in the two non-axial Reynolds stress components, and (2) non-zero gradients of these quantities in both non-axial directions. It is also immediately evident that any model predicting isotropic normal Reynolds stresses, e.g. those based on the standard constitutive relation, Eq. 9.30, will fail the first of these criteria, and would therefore not be capable of properly predicting turbulence-induced secondary currents.

Chapter 7

Flow and turbulence in breaking waves

A detailed understanding of the flow and sediment transport within the surf and swash zones is of major engineering importance. Here the surf zone is defined as the area extending in the cross shore direction from incipient wave breaking to the limit of runup, while the swash zone is the wet-dry portion of the beach ranging between the limit of drawdown and runup. The flow and turbulence induced by breaking waves make these regions of large importance within coastal sediment transport and morphology. In particular, surf and swash zone processes govern the exchange of sand between the emerged and submerged portions of the beach, and are therefore largely responsible for determining shoreline configurations, including e.g., beach accretion or recession.

We will, in this chapter, study flow (including turbulence) induced by breaking waves. We will first briefly introduce breaking waves and breaking regimes: spilling breaking waves, plunging breaking waves and surging breaking waves.

Detailed attention will first be concentrated on the flow beneath spilling breaking waves. In this context a simple conceptual model, namely the surface roller, will be introduced, which can be utilized to account for the additional fluxes of mass, energy, and momentum that will occur in spilling breaking waves. This will then be followed by the development of an analytical model for predicting the period-averaged turbulent shear stress and corresponding velocity distribution in the surf zone, where the latter gives rise to a well-known circulation current termed the *undertow*. Application of the developed undertow model will be exemplified through comparison with experimental measurements of the flow beneath breaking spilling waves.

Attention will then be concentrated on the flow (including turbulence) induced by more violent plunging breaking regular waves, as measured in experimental facilities. This will include details of the breaking process, including characterization of the bed shear stress throughout the surf and swash zones i.e. spanning from the pre-breaking region to the runup at the shoreline. We will also include a small section under plunging breaking waves for the implications of the described flow processes for sediment transport with special reference to the mechanism of sediment suspension under plunging breaking waves.

This section will be complemented by a smaller section on flow and turbulence processes in the surging-regular-wave regime.

Subsequently, we will turn our attention to plunging solitary waves. Similar to plunging regular waves, we will cover the subject extensively; the treatment will include a detailed description of the breaking process, the bed shear stress, and the flow resistance associated with the latter.

Finally, detailed attention will be paid to the computational fluid dynamics (CFD) simulation of breaking waves, with emphasis on applications using two-equation turbulence models. The background will be discussed, where historically such efforts have commonly resulted in the marked overproduction of turbulence, even prior to the onset of wave breaking. We will present results from recent analysis of this problem (Larsen and Fuhrman, 2018), which have demonstrated that: (1) this problem arises as a consequence of the unconditional instability of (seemingly all) widely-used two-equation turbulence models in the nearly potential flow region beneath surface waves, leading to unphysical exponential growth of the turbulent kinetic energy and the eddy viscosity; and (2) that such turbulence models can be easily *stabilized* with only minor modifications. The results of CFD simulations involving both spilling and plunging breaking waves (covering the wave evolution, undertow velocities and turbulence) will then be presented. Results will demonstrate clear advantage of utilizing stabilized turbulence closure models in such applications, as it is only these which enable the simulated breaking process to begin in an un-polluted environment.

Some familiarity with basic potential flow wave theory is assumed in the main text of this chapter. However, some essentials of linear potential flow wave theory are again described in the Appendix to Chapter 5. With the material presented there as background, the entirety of the present chapter should be accessible. Additionally, the Appendix to the present chapter derives the depth-integrated (and period-averaged) momentum equation, which is utilized in the development of the undertow model mentioned

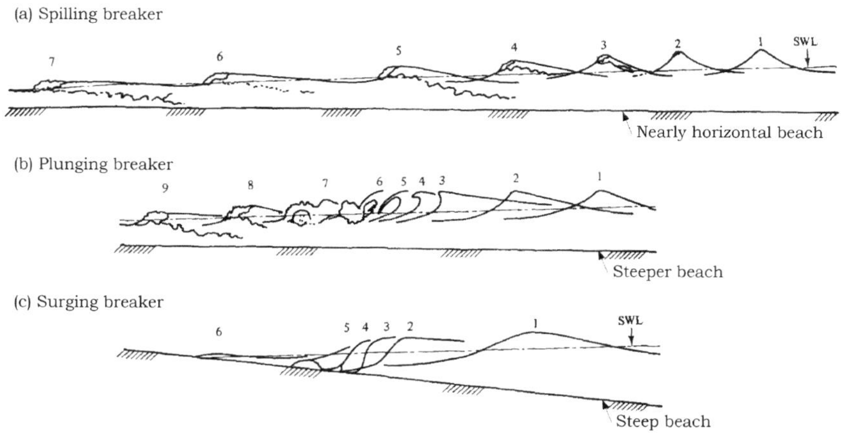

Fig. 7.1 Breaker types. Small numbers denote different stages of the breaking process. This figure has been adapted from Figs. 5.3.1, 5.3.3 and 5.3.4 of Svendsen (2006).

above (Section 7.2.2).

7.1 Wave breaking and breaker types

Waves approaching the shoreline undergo shoaling where the wave height H in the nearshore region generally increases until the point of breaking (see Appendix 5.I). Purely mathematical analysis of the shoaling process indicates that the wave height will approach infinity in very shallow waters. This is clearly unrealistic. Rather, at some depth, a wave of given characteristics will become unstable and break, corresponding to the point where the velocity at the wave crest exceeds the wave celerity c. Once this occurs, the wave will break, and dissipate energy in the form of turbulence (Dean and Dalrymple, 2014, p. 112).

The way in which waves break in shallow waters depends on the slope of the beach and the characteristics of the wave. Fig. 7.1 illustrates three types of wave breaking (namely spilling breaking, plunging breaking, and surging breaking), with the beach slope marked in the illustrations, namely nearly horizontal beach, steeper beach, and steep beach, respectively.

For nearly horizontal beaches (Fig. 7.1a), the waves tend to break in the spilling breaker regime; In this regime numerous breaking waves can be simultaneously present within the surf zone. For steeper beaches (Fig. 7.1b), the breaking will tend to occur in the plunging breaker regime, where the crest of the waves curl over forward and impinge onto part of the wave trough. For steep beaches (Fig. 7.1c), the waves break in the surging breaker regime with a narrow or nonexistent surf zone and characteristically high reflection (Dean and Dalrymple, 2004, p. 113).

The wave-breaker types are commonly related to the so-called surf similarity parameter (also known as the Iribarren number), ξ_0

$$\xi_0 = \frac{\tan(\beta)}{\sqrt{H_0/L_0}} \tag{7.1}$$

in which $\tan(\beta)$ is the beach slope, H_0 the deep-water wave height, and L_0 the deep-water wave length. Of the latter two quantities, the deep-water wave length is given by

$$L_0 = \frac{g}{2\pi}T^2 \tag{7.2}$$

in which T is the wave period, and g the acceleration due to gravity (the deep water wave length and other quantities are derived in Appendix 5.H). The deep-water wave height, on the other hand, can be calculated from known wave properties at any depth offshore of breaking (e.g. the toe depth in an experiment), based on linear wave theory and assumed conservation of energy flux:

$$H_0 = H\sqrt{\tanh(kh)\ [1+G]} \tag{7.3}$$

where

$$G = \frac{2kh}{\sinh(2kh)} \tag{7.4}$$

(see Appendix 5.I) in which H is the wave height and $k = 2\pi/L$ the wave number at some depth h prior to breaking (e.g. at the toe of a sloping region). Note that the wave number can be calculated from the linear dispersion relation (see Appendix 5.C)

$$\omega^2 = gk\tanh(kh) \tag{7.5}$$

in which $\omega = 2\pi/T$ is the angular frequency.

Based on the surf similarity parameter defined in Eq. 7.1, Galvin (1968) classified the wave breaker types as follows:

$$\xi_0 < 0.5: \text{ Spilling} \tag{7.6}$$

$$0.5 < \xi_0 < 3.3: \text{ Plunging} \tag{7.7}$$

$$\xi_0 > 3.3: \text{ Surging} \tag{7.8}$$

We note that Galvin (1968) has identified collapsing as a fourth type, which is a combination of plunging and surging.

We will, in the following sections, mainly study the flow (and turbulence) induced by both spilling and plunging breaking waves, and to a smaller extent, that by surging breaking waves.

7.2 Flow induced by spilling waves

Let us first consider the flow induced by spilling breaking waves. While perhaps less spectacular than their plunging counterparts, spilling breakers probably occur more frequently in nature, and are important contributors to turbulence, spray and bubble generation at the surface (Duncan, 2001), as well as in coastal sediment transport and morphology.

Previous experimental work on spilling breaking waves include the studies of Deigaard et al. (1991b), Ting and Kirby (1994, 1996), Huang et al. (2009, 2010), and Ting and Nelson (2011). Similarly, numerical studies of spilling breakers include the pioneering studies of Lin and Liu (1998), Christensen and Deigaard (2001) and Christensen (2006), as well as the more recent studies of Jacobsen et al. (2012), Brown et al. (2016), Lin et al. (2015), Chella et al. (2015, 2016), and Larsen and Fuhrman (2018), to name a few. For a comprehensive review on spilling breaking waves in general, see Duncan (2001).

The present section will focus primarily on the analytical development of a (reasonably) simple model for predicting the surf zone circulation current flow induced by spilling breaking waves, known as the undertow. (It is emphasized that undertow will also commonly develop in the surf zone initiated by plunging breakers as well, for fundamentally similar reasons.) We will first describe a simple model concept known as a surface roller, which is commonly applied in engineering practice to describe spilling breaking waves, in Section 7.2.1. This concept will then be utilized in the development of a theoretical model for predicting the shear stress and undertow velocity profiles in the surf zone in Section 7.2.2. The simple model to be developed builds especially on concepts previously introduced e.g. by Svendsen (1984a,b) (see also Section 12.2 of Svendsen, 2006) as well as Deigaard and Fredsøe (1989) and Fredsøe and Deigaard (1992). The developed model will be directly applied and tested against the experimental measurements of undertow velocities from Ting and Kirby (1994) in Section 7.2.3.

For a review of undertow models, see e.g. Section 3.1.6 of Christensen et al. (2002), as well as the more recent overview provided by Guannel and Ozkan-Haller (2014). For other studies on undertow the interested reader is referred e.g. to Deigaard (1993), Stive and de Vriend (1994), Garcez Faria et al. (2000), Reniers et al. (2004) and Newberger and Allen (2007), in addition to the other works mentioned above and below.

7.2.1 *The surface roller model*

The dynamics of breaking waves are inherently complicated, and for this reason often simple engineering type models have proven useful. A successful conceptual model particularly applicable to spilling breaking waves is the so-called surface roller, first introduced and used by Svendsen (1984b,a). The present description follows closely that from Fredsøe and Deigaard (1992). In this model the flow is divided into two parts. The first corresponds to the ordinary wave orbital motion which can, to a first approximation, be prescribed according to potential flow wave theory, as described in Appendix 5.B (or Appendix 5.G if shallow water approximations are appropriate, as will often be the case in nearshore regions). The second corresponds to the surface roller, which is treated simply as a volume of water riding on the front face of the broken wave at speed equal to the wave celerity c. This concept is depicted in Fig. 7.2, where the shallow water regime (hence uniform velocity profiles for the ordinary wave flow) is assumed. A photograph of an experimental spilling breaker with a clear roller region can additionally be seen in Fig. 7.3.

Let us denote the area of the surface roller (i.e. its volume per unit width) as A_r, as indicated in Fig. 7.2A. Based on this simple description, we can then account for the additional flux of mass (or volume), energy, or momentum induced by a spilling breaker in a lumped fashion. As A_r defines the volume (per unit width) of water associated with the passing of a single roller, the period-averaged discharge (per unit width) induced by the surface roller will be:

$$q_{roller} = \frac{A_r}{T} \tag{7.9}$$

This may be added to the wave drift contribution from Eq. 5.188, to estimate the total period-averaged discharge (per unit width) that would occur between the trough and crest region in a spilling breaking wave. In the shallow water regime, this may be approximated by

$$q_{drift} + q_{roller} = \frac{H^2\pi}{4Tkh} + \frac{A_r}{T} \tag{7.10}$$

where the expression for q_{drift} (first term on the right hand side) corresponds to the shallow water approximation of Eq. 5.188.

Similarly, the energy flux due to the passing of a single surface roller is

$$\rho A_r \frac{c^2}{2} \tag{7.11}$$

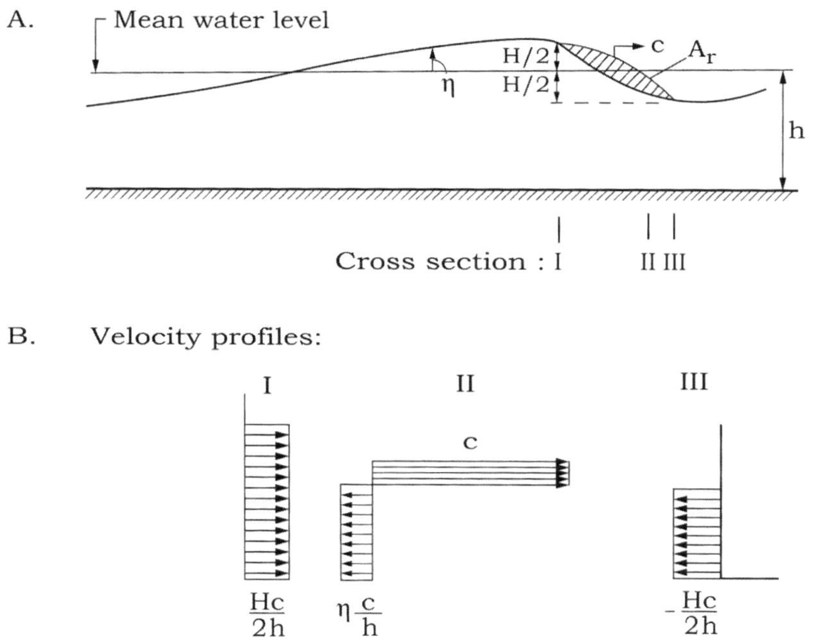

Fig. 7.2 Plots showing (a) the structure of a broken (shallow water) wave with a surface roller and (b) the velocity profiles at the three cross sections indicated. Adapted from Fredsøe and Deigaard (1992).

such that the period-averaged energy flux will correspond to

$$E_{f,roller} = \rho A_r \frac{c^2}{2T} \tag{7.12}$$

Note that the volume of surface rollers is sufficiently small that they will not contribute significantly to the pressure force (Fredsøe and Deigaard, 1992, p. 92), hence these are taken as negligible in the above and in what follows. To estimate the total period-averaged energy flux beneath a spilling wave, following Section 4.2 of Fredsøe and Deigaard (1992), Eq. 7.12 may thus be added to the energy flux due to the ordinary wave motion from Eq. 5.192 (or from Eq. 5.203 if shallow water approximations are valid i.e. $kh \lessapprox 0.3$). The total energy flux in a shallow water spilling breaking wave would thus correspond to:

$$E_f = Ec + \rho A_r \frac{c^2}{2T}, \qquad E = \frac{1}{8}\rho g H^2, \qquad c = \sqrt{gh} \tag{7.13}$$

where E is the energy density and c is taken as the linear shallow water wave celerity (see Appendix 5.G).

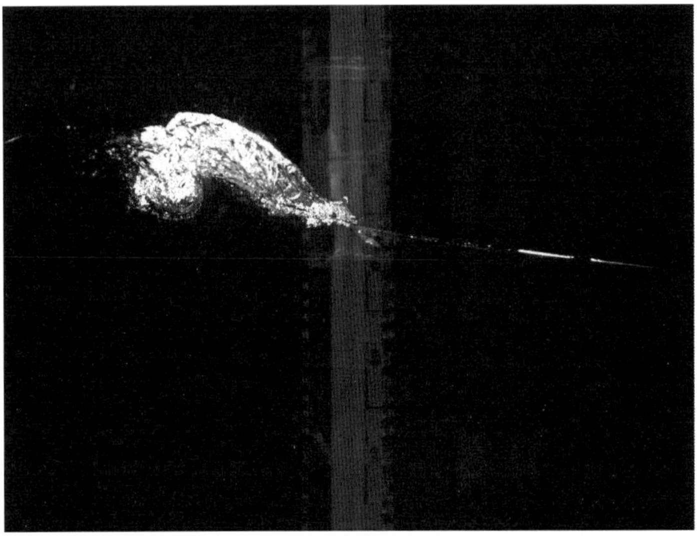

Fig. 7.3 Photograph of a spilling breaker with a clearly visible roller (white region, wave traveling from left to right, about to impact a monopile). This photo has been taken in wave tank measurements performed at the Technical University of Denmark. The roller is lit from below by a vertical laser sheet, which reflects the bubbles in the roller. [Photo provided courtesy of Dr. Malene H. Vested and Prof. Erik D. Christensen, from the study of Vested (2020).]

Finally, the momentum flux from the passing of a single roller will be

$$\rho A_r c \tag{7.14}$$

meaning that the period-averaged momentum flux will be

$$S_{xx,roller} = \rho \frac{A_r c}{T} \tag{7.15}$$

We may hence add this to the result from Eq. 5.197 (or Eq. 5.203 if the shallow water regime is assumed), to obtain an approximation for the total radiation stress associated with a spilling breaking wave. Utilizing the latter shallow water approximation, the total radiation stress for a shallow water spilling breaker can be approximated by

$$S_{xx} = \frac{3}{2}E + \rho \frac{A_r c}{T} \tag{7.16}$$

Of course, to utilize the expressions above in practice requires a means for approximating the roller area A_r. While more complicated options exist, the simplest model corresponds to

$$A_r = K_r H^2 \tag{7.17}$$

where Svendsen (1984b) suggests $K_r \approx 0.9$, based on measurements of breaking waves produced by a towed hydrofoil made by Duncan (1981).

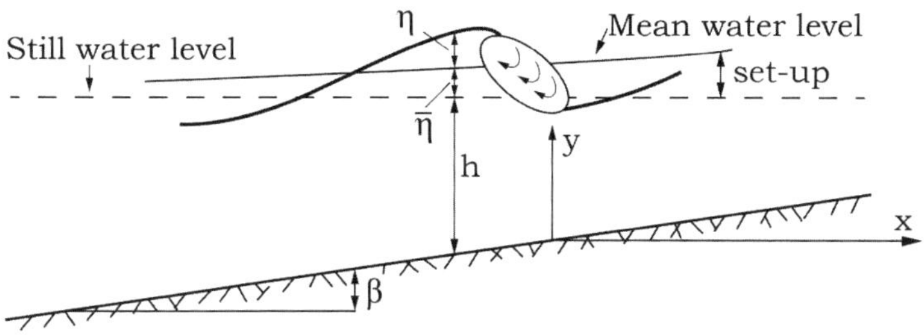

Fig. 7.4 Definition sketch. Adapted from Deigaard et al. (1991a).

7.2.2 *A simple model of undertow*

In what follows a simple theoretical model for predicting the circulating current, called the undertow, that arises in the surf zone will be developed. As will be seen, the theory will make use of the simple surface roller model covered above, and is hence especially applicable in the surf zone initiated by spilling breaking waves; It is emphasized, however, that it may also be valid as a first approximation for the undertow in the surf zone initiated by plunging waves, as these will typically also develop into a bore-like broken wave in the inner surf zone (i.e. the region near the shoreline). In what follows, explicit expressions for predicting the shear stress distribution and undertow velocity profile in the surf zone will be specifically developed and tested. All derivations in what follows will stem directly from governing flow equations. Please also see the definition sketch provided in Fig. 7.4.

7.2.2.1 *Governing equations*

Neglecting viscous terms in the Navier-Stokes equations, let us begin with the two-dimensional Euler equations, which may be written as

$$\frac{\partial u}{\partial t} + \frac{\partial (u^2)}{\partial x} + \frac{\partial (uv)}{\partial y} = -\frac{1}{\rho}\frac{\partial p}{\partial x} \tag{7.18}$$

$$\frac{\partial v}{\partial t} + \frac{\partial (uv)}{\partial x} + \frac{\partial (v^2)}{\partial y} = -\frac{1}{\rho}\frac{\partial p}{\partial y} - g \tag{7.19}$$

Let us decompose the velocities and pressure as follows:

$$u = \overline{u} + \tilde{u} + u', \qquad v = \tilde{v} + v', \qquad p = \overline{p} + \tilde{p} + p' \tag{7.20}$$

where overlined variables correspond to the mean (period-averaged) values, the tilde represents the periodic wave quantities, and the prime superscripts denote turbulent fluctuations, as usual.

Let us first consider the horizontal momentum equation Eq. 7.18. Inserting these and Reynolds (ensemble) averaging leads to:

$$\frac{\partial \tilde{u}}{\partial t} + \frac{\partial (\overline{u}^2)}{\partial x} + \frac{\partial (\tilde{u}^2)}{\partial x} + \frac{\partial (\overline{u'^2})}{\partial x} + \frac{\partial (\tilde{u}\tilde{v})}{\partial y} + \frac{\partial (\overline{u'v'})}{\partial y} = -\frac{1}{\rho}\frac{\partial (\overline{p}+\tilde{p})}{\partial x} \tag{7.21}$$

Let us now ascertain the importance of the various terms, under the assumption that the following quantities are small:

$$\frac{\overline{u}}{c}, \quad kh, \quad \frac{\tilde{v}}{c}, \quad \frac{\overline{u'^2}}{c^2}, \quad \frac{\overline{v'^2}}{c^2} \tag{7.22}$$

Physically, these correspond to assuming that: (1) the undertow velocity $\overline{u}$ is small relative to the wave-induced flow (here characterized by the wave celerity c), (2) the waves are in the shallow water regime (i.e. $kh \leq \pi/10 \approx 0.314$, see Appendix 5.G), and (3) turbulent normal stresses are small (Stive and Wind, 1982). Under these assumptions, the following scalings can be applied:

$$\begin{aligned} &x \to \frac{x^*}{k}, && y \to hy^*, && (\overline{p},\tilde{p}) \to \frac{\rho c^2}{\epsilon}(\overline{p}^*,\tilde{p}^*), \\ &\tilde{u} \to c\tilde{u}^*, && \overline{u'v'} \to c^2\overline{u'v'}^*, && (\overline{u'^2},\overline{v'^2}) \to \epsilon c^2(\overline{u'^2}^*,\overline{v'^2}^*), \\ &g \to \frac{c^2}{h\epsilon}g^*, && t \to \frac{h}{c}t^*, && (\overline{u},\tilde{v}) \to \epsilon c(\overline{u}^*,\tilde{v}^*) \end{aligned} \tag{7.23}$$

where ϵ is a marker representing any of the quantities assumed small above, and variables with an asterisk superscript are dimensionless counterparts. Substituting these leads, after simplification, to the non-dimensional equation

$$\frac{\partial \tilde{u}^*}{\partial t^*} + \epsilon^3\frac{\partial (\overline{u}^{*2})}{\partial x^*} + \epsilon\frac{\partial (\tilde{u}^{*2})}{\partial x^*} + \epsilon^2\frac{\partial (\overline{u'^2}^*)}{\partial x^*} + \epsilon\frac{\partial (\tilde{u}^*\tilde{v}^*)}{\partial y^*} + \frac{\partial (\overline{u'v'}^*)}{\partial y^*} = -\frac{\partial (\overline{p}^*+\tilde{p}^*)}{\partial x^*} \tag{7.24}$$

Retaining up to $O(\epsilon)$ terms and transforming back to dimensional variables, we are left with:

$$\frac{\partial \tilde{u}}{\partial t} + \frac{\partial (\tilde{u}^2)}{\partial x} + \frac{\partial (\tilde{u}\tilde{v})}{\partial y} + \frac{\partial (\overline{u'v'})}{\partial y} = -\frac{1}{\rho}\frac{\partial (\overline{p}+\tilde{p})}{\partial x} \tag{7.25}$$

We may express the Reynolds stress term according to the Boussinesq approximation, see Eq. 9.30, which (neglecting normal stresses) reads

$$-\overline{u'v'} = \frac{\overline{\tau}}{\rho} = \nu_t\frac{\partial \overline{u}}{\partial y} \tag{7.26}$$

Importantly, this brings the period-averaged current (undertow) velocity $\overline{u}$ (assumed small, but finite) back into the problem. We also have

$$\frac{\partial(\tilde{u}\tilde{v})}{\partial y} = \tilde{u}\frac{\partial \tilde{v}}{\partial y} + \tilde{v}\frac{\partial \tilde{u}}{\partial y} = -\tilde{u}\frac{\partial \tilde{u}}{\partial x} = -\frac{1}{2}\frac{\partial(\tilde{u}^2)}{\partial x} \tag{7.27}$$

where local continuity for the wave-induced flow, as well as assumed uniformity of the wave-induced horizontal velocity (in line with the shallow water assumption), corresponding respectively to:

$$\frac{\partial \tilde{u}}{\partial x} + \frac{\partial \tilde{v}}{\partial y} = 0, \qquad \frac{\partial \tilde{u}}{\partial y} = 0 \tag{7.28}$$

have been invoked. This is particularly convenient, since it enables the convective terms to be combined, while also eliminating the need to explicitly formulate the wave-induced vertical velocity $\tilde{v}$. Invoking these, the momentum equation becomes:

$$\frac{\partial \tilde{u}}{\partial t} + \frac{1}{2}\frac{\partial(\tilde{u}^2)}{\partial x} + \frac{1}{\rho}\frac{\partial(\overline{p} + \tilde{p})}{\partial x} = \frac{1}{\rho}\frac{\partial \overline{\tau}}{\partial y} \tag{7.29}$$

Period-averaging, such that the time derivative drops out, leads to:

$$\left\langle \frac{1}{2}\frac{\partial(\tilde{u}^2)}{\partial x} + \frac{1}{\rho}\frac{\partial \overline{p}}{\partial x} = \frac{1}{\rho}\frac{\partial \overline{\tau}}{\partial y} \right\rangle \tag{7.30}$$

Now let us similarly consider the vertical momentum equation, Eq. 7.19. Inserting the decompositions from Eq. 7.20 and then Reynolds and period averaging leads directly to:

$$\left\langle \frac{\partial(\tilde{u}\tilde{v})}{\partial x} + \frac{\partial(\overline{u'v'})}{\partial x} + \frac{\partial(\tilde{v}^2)}{\partial y} + \frac{\partial(\overline{v'^2})}{\partial y} = -\frac{1}{\rho}\frac{\partial \overline{p}}{\partial y} - g \right\rangle \tag{7.31}$$

Inserting Eq. 7.23 and simplifying, then yields the non-dimensional equation:

$$\left\langle \epsilon^3 \frac{\partial(\tilde{u}^*\tilde{v}^*)}{\partial x^*} + \epsilon^2 \frac{\partial(\overline{u'v'}^*)}{\partial x^*} + \epsilon^3 \frac{\partial(\tilde{v}^{*2})}{\partial y^*} + \epsilon^2 \frac{\partial(\overline{v'^2}^*)}{\partial y^*} = -\frac{\partial \overline{p}^*}{\partial y^*} - g^* \right\rangle \tag{7.32}$$

Keeping terms up to $O(\epsilon)$ as before, and transforming back to dimensional quantities thus leaves

$$\left\langle -\frac{1}{\rho}\frac{\partial \overline{p}}{\partial y} - g \right\rangle = 0 \tag{7.33}$$

Upon integration, choosing the integration constant such that there is zero pressure at the free surface, we obtain the hydrostatic pressure distribution induced by the mean water level $\overline{\eta}$ (corresponding to the so-called set up, as will be shown), as well as its x-derivative:

$$\frac{\overline{p}}{\rho} = g(\overline{\eta} - y) \qquad \Rightarrow \qquad \frac{1}{\rho}\frac{\partial \overline{p}}{\partial x} = g\frac{\partial \overline{\eta}}{\partial x} \tag{7.34}$$

Invoking Eqs. 7.34 into Eq. 7.30 and re-arranging finally yields the period-averaged horizontal momentum equation:

$$\frac{1}{\rho}\frac{\partial \overline{\tau}}{\partial y} = \left\langle \frac{1}{2}\frac{\partial(\tilde{u}^2)}{\partial x} \right\rangle + g\frac{\partial \overline{\eta}}{\partial x} \tag{7.35}$$

This equation will form the basis of the simple undertow model to be developed in what follows.

7.2.2.2 *Wave description*

To work out a model from the above description requires that the wave kinematics be specified. For this purpose, consider a local sinusoidal free surface wave variation

$$\eta = \frac{H}{2}\sin\theta, \qquad \theta = \omega t - \varphi(x), \qquad \varphi(x) = \int k dx \tag{7.36}$$

It is noted that the phase function θ above represents a generalized description for waves propagating over a variable bathymetry depth, compared to the constant depth expressions seen e.g. in Eq. 5.176 from Appendix 5.B.

Recognizing that the horizontal velocity $\tilde{u}$ only appears in the (higher-order) nonlinear term in Eq. 7.35, it is therefore sufficient to utilize the flat-bottom shallow-water solution for this quantity, Eq. 5.200, as a leading-order approximation. This corresponds to

$$\tilde{u} = \frac{H\omega}{2kh}\sin\theta + O\left(\frac{\partial H}{\partial x}\right) \tag{7.37}$$

where it is emphasized that here the wave height $H = H(x)$, wave number $k = k(x)$ and still water depth $h = h(x)$ are now all allowed to vary in x. (As justified above, an additional higher-order $O(\partial H/\partial x)$ contribution is neglected in Eq. 7.37. A complete local solution for $\tilde{u}$ and $\tilde{v}$ can if fact be obtained by satisfying the kinematic bottom and surface conditions, Eqs. 5.171 and 5.173, in addition to local continuity, Eq. 5.170; The above is sufficient for the derivation which follows, however.) It is again noted that the horizontal wave-induced velocity is uniform, as it is assumed that the waves are in the shallow water regime.

Let us now assume that the waves are propagating on a beach with local slope

$$\frac{\partial h}{\partial x} = -\tan\beta, \tag{7.38}$$

(this may be, but is not necessarily, constant in what follows). Additionally, we will also assume a simple breaking model, wherein the wave height H is always a constant fraction of the local water depth h:

$$\frac{H}{h} = K_b \tag{7.39}$$

where a value such as $K_b \approx 0.8$ is commonly used at incipient breaking and in the outer surf zone, whereas $K_b \approx 0.5$ is commonly used in the inner surf zone (see Fredsøe and Deigaard, 1992, pp. 86–87, and references therein). As these values are rather similar, we will simply assume that K_b is a constant. Under these assumptions, while also making use of the shallow water linear dispersion relation, Eq. 5.202, we find that

$$\frac{\partial H}{\partial x} = -K_b \tan\beta, \qquad \frac{\partial k}{\partial x} = \frac{k}{2h}\tan\beta \tag{7.40}$$

i.e. all x derivatives can conveniently be related to the local beach slope $\tan\beta$. These simplifications will be utilized extensively in what follows.

It is finally again assumed that the waves are breaking as spilling breakers, such that the surface roller model described in Section 7.2.1 applies. This conceptual model will be directly utilized in what follows.

7.2.2.3 *Shear stress distribution*

We will now derive the period-averaged shear stress distribution beneath the spilling breaking waves. Consider the period-averaged momentum equation, Eq. 7.35. Invoking the wave description for $\tilde{u}$ above, the first term on the right-hand side evaluates to:

$$\left\langle \frac{1}{2}\frac{\partial(\tilde{u}^2)}{\partial x} \right\rangle = -\frac{gK_b^2}{16}\tan\beta \tag{7.41}$$

Eq. 7.35 now becomes

$$\frac{1}{\rho}\frac{\partial \overline{\tau}}{\partial y} = -\frac{gK_b^2}{16}\tan\beta + g\frac{\partial \overline{\eta}}{\partial x} \tag{7.42}$$

Integrating once with respect to y gives

$$\frac{\overline{\tau}}{\rho} = \left[-\frac{gK_b^2}{16}\tan\beta + g\frac{\partial \overline{\eta}}{\partial x}\right] y + \frac{\overline{\tau}_0}{\rho} \tag{7.43}$$

where an integration constant, obviously corresponding to the (period-averaged) bed shear stress $\overline{\tau}_0/\rho$, has been added.

To find the bed shear stress we may utilize the depth-integrated and period-averaged momentum equation, which is derived in full in the present chapter's Appendix, Eq. 7.132. Solving for the period-averaged bed shear stress from Eq. 7.132, and assuming that the set up $\overline{\eta}$ is small relative to h, yields:

$$\frac{\overline{\tau}_0}{\rho} = -\frac{1}{\rho}\frac{\partial S_{xx}}{\partial x} - gh\frac{\partial \overline{\eta}}{\partial x} \tag{7.44}$$

For the radiation stress we utilize Eq. 7.16 coupled with Eq. 7.17, which accounts for both wave and surface roller contributions. The result, after simplifying, is:

$$\frac{\overline{\tau}_0}{\rho} = \frac{3ghK_b^2}{8}\tan\beta - gh\frac{\partial \overline{\eta}}{\partial x} + \frac{5gh^2K_b^2K_r\omega\tan\beta}{4c\pi} \tag{7.45}$$

Inserting this result back into Eq. 7.43 finally yields the period-averaged shear stress distribution

$$\frac{\overline{\tau}}{\rho} = \nu_t\frac{\partial \overline{u}}{\partial y} = \frac{gK_b^2\tan\beta}{16}(6h-y) - g\frac{\partial \overline{\eta}}{\partial x}(h-y) + \frac{5gh^2K_b^2K_r\omega\tan\beta}{4c\pi} \tag{7.46}$$

or in non-dimensional form:

$$\frac{\overline{\tau}}{\rho gh} = \underbrace{\frac{K_b^2\tan\beta}{16}\left(6-\frac{y}{h}\right)}_{\text{wave motion}} - \underbrace{\frac{\partial \overline{\eta}}{\partial x}\left(1-\frac{y}{h}\right)}_{\text{set up}} + \underbrace{\frac{5hK_b^2K_r\omega\tan\beta}{4c\pi}}_{\text{surface roller}} \tag{7.47}$$

These expressions for $\overline{\tau}$ are valid beneath the near-surface region (strictly speaking, from the sea bed up to the the wave trough level, though the latter may be reasonably approximated as the still water level given that the wave-induced flow has been approximated from linear wave theory).

The period-averaged shear stress $\overline{\tau}$ contains contributions from three quantities: (1) the wave motion, (2) the set up, and (3) the surface roller, as identified above. An example showing the typical variation of each component is provided in Fig. 7.5. Importantly, as these all have different y variation, there is no value of the mean surface slope $\partial\overline{\eta}/\partial x$ that will result in zero shear stress throughout the entirety of the water column. This, coupled with the expectation that there will be no net cross-shore flow at a given position, implies that there will therefore be a circulation current that forms within the surf zone, giving rise to the undertow! This is depicted conceptually in Fig. 7.6. The structure of this current will be formulated explicitly in the next sub-section.

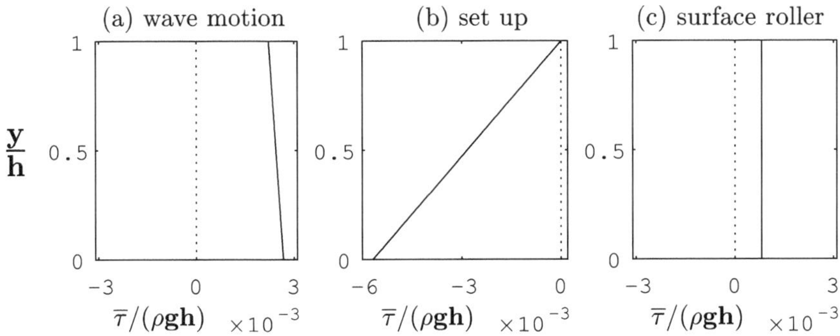

Fig. 7.5 Plot showing the individual contributions to the period-averaged shear stress distribution from the (a) wave motion, (b) set up and (c) surface roller. The wave conditions considered are the same as in Fig. 7.7, such that the three contributions shown individually here sum to that shown in Fig. 7.7a.

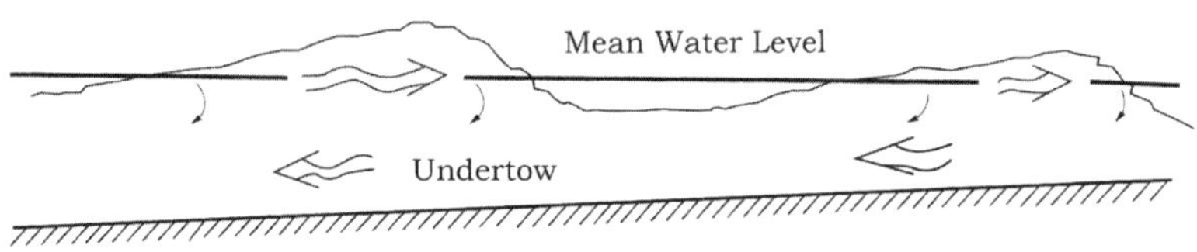

Fig. 7.6 The undertow circulation current. Adapted from Svendsen (1984a).

It is noted that the wave motion contribution in Eq. 7.47 differs from that found in Deigaard and Fredsøe (1989) (their Eq. 56) and Fredsøe and Deigaard (1992) (their Eq. 6.17 on their p. 176), which were derived from a (alternative and equally valid) control volume approach. The reason for this difference is simply because they assumed that the water depth h, and hence the wave number k, were constants, such that only the wave height H was varying in x, whereas the present model has assumed spatial variations in all three quantities. We have confirmed that the procedure above leads to results matching exactly those of Deigaard and Fredsøe (1989) and Fredsøe and Deigaard (1992) in the special case where h and k are assumed constant [DRF gratefully acknowledges helpful correspondence with Dr. Rolf Deigaard on this issue]. The resulting shear stress distribution in this case reads:

$$\frac{\overline{\tau}}{\rho} = \frac{gH}{8}\frac{\partial H}{\partial x}\left(\frac{y}{h} - 3\right) - g\frac{\partial \overline{\eta}}{\partial x}(h - y) - \frac{1}{T}\frac{\partial (cA_r)}{\partial x}, \qquad h = \text{constant} \quad (7.48)$$

This confirms that the derivation above is fully consistent with these earlier studies, provided that the underlying assumptions are kept consistent.

7.2.2.4 *Undertow velocity profile*

We may now integrate Eq. 7.46 to obtain the (period-averaged) undertow velocity profile, where for the sake of simplicity we assume the eddy viscosity ν_t to be constant in y (but not x):

$$\overline{u} = \frac{1}{\nu_t}\left[\frac{gK_b^2\tan\beta}{32}\left(12hy - y^2\right) + g\frac{\partial\overline{\eta}}{\partial x}\left(\frac{y^2}{2} - hy\right) + \frac{5gh^2K_b^2K_r\omega\tan\beta}{4c\pi}y\right] \tag{7.49}$$

Note that in the above the integration constant is taken as zero, such that there is no slip at the bed. Additional effects stemming from the wave boundary layer are neglected for simplicity. In non-dimensional form this is:

$$\frac{\overline{u}}{\sqrt{gh}} = \frac{1}{\nu_t}\left[\frac{chK_b^2\tan\beta}{32}\left(\frac{12y}{h} - \left(\frac{y}{h}\right)^2\right) + \frac{ch}{2}\frac{\partial\overline{\eta}}{\partial x}\left(\left(\frac{y}{h}\right)^2 - \frac{2y}{h}\right) + \frac{5h^2K_b^2K_r\omega\tan\beta}{4\pi}\frac{y}{h}\right] \tag{7.50}$$

The total period-averaged discharge (per unit width) in the bulk water column (i.e. $0 \le y \lessapprox h$) is then

$$q_{undertow} = \int_0^h \overline{u}dy = \frac{1}{\nu_t}\left[\frac{17gh^3K_b^2\tan\beta}{96} - \frac{gh^3}{3}\frac{\partial\overline{\eta}}{\partial x} + \frac{5gh^4K_b^2K_r\omega\tan\beta}{8c\pi}\right] \tag{7.51}$$

The period-averaged discharge (per unit width) due to wave drift will be

$$q_{drift} = \frac{H^2\pi}{4Tkh} = \frac{chK_b^2}{8} \tag{7.52}$$

This stems from the shallow water approximation of Eq. 5.188, which was also utilized in Eq. 7.10. Similarly, the period-averaged discharge (per unit width) due to the surface roller will, from Eq. 7.9, be

$$q_{roller} = \frac{A_r}{T} = \frac{h^2K_b^2K_r\omega}{2\pi} \tag{7.53}$$

Assuming that the beach is impermeable and that there is uniformity in the transverse (longshore) direction, the total net discharge (per unit width) should sum to zero:

$$q_{drift} + q_{roller} + q_{undertow} = 0 \tag{7.54}$$

As q_{drift} and q_{roller} are positive (i.e. in the propagating onshore direction), obviously $q_{undertow}$ should give a net negative (offshore) contribution, the strength of which is largely governed by the slope of the mean surface elevation $\partial\overline{\eta}/\partial x$, as is clear from Eq. 7.51. Invoking the three quantities above, we may solve explicitly for the value of $\partial\overline{\eta}/\partial x$ which satisfies the constraint Eq. 7.54, corresponding to:

$$\frac{\partial\overline{\eta}}{\partial x} = \underbrace{\frac{3cK_b^2\nu_t}{8gh^2}}_{\text{wave drift}} + \underbrace{\frac{3K_b^2K_r\nu_t\omega}{2\pi gh} + \frac{15hK_b^2K_r\omega\tan\beta}{8c\pi}}_{\text{surface roller}} + \underbrace{\frac{17K_b^2\tan\beta}{32}}_{\text{wave motion}} \quad (7.55)$$

where the net effect of the various contributions are as indicated above; Note that there are two contributions stemming from the surface roller, which arise from the accounting of its volume (the first of these) and momentum flux (the second of these). The resulting $\partial\overline{\eta}/\partial x$ will clearly be positive i.e. increasing mean water levels in the surf zone in the shoreward direction. This phenomenon is known as *wave set up*, see Fig. 7.4.

To complete the model above the eddy viscosity (again assumed vertically uniform for simplicity) must be specified. This can be formulated as proportional to the product of a local characteristic length scale (here taken as the still water depth h) and a characteristic velocity scale (here taken as the linear shallow water wave celerity $c = \sqrt{gh}$), leading to

$$\nu_t = K_t hc = K_t h\sqrt{gh} \quad (7.56)$$

where K_t is a constant. As both scales mentioned above can be expected to be relatively large compared to those of the turbulence, we should reasonably expect that K_t should be relatively small. Analysis of the eddy viscosity from CFD simulations of spilling breakers, e.g. Fig. 14 of Larsen and Fuhrman (2018), suggest that $K_t \approx 0.003$, though other authors suggest that this should be $O(0.01)$, e.g. Guannel and Ozkan-Haller (2014).

This completes the simple undertow model, which requires only assumed values of three coefficients K_b, K_r, and K_t (all of which are known with reasonable precision, as indicated above), in addition to relevant physical information, to be explicitly applied. An example application and comparison with experimental data is provided in the next sub-section.

7.2.3 *Comparison with measured undertow profile of Ting and Kirby (1994)*

We will now demonstrate an application of the simple undertow model derived above in this sub-section. For this purpose, we will consider the well-known spilling breaking experiment of Ting and Kirby (1994). In their experiment, the wave period was $T = 2$ s (hence $\omega = 2\pi/T = 3.14\ \mathrm{s}^{-1}$) and the beach slope was $\tan\beta = 1/35$. We will attempt comparison at their inner-most measurement position in the surf zone where the still water depth was $h = 0.097$ m. The gravitational acceleration $g = 9.81\ \mathrm{m/s^2}$ is used in all calculations below. Note that a more detailed account of these experiments and their measurements will be provided in Section 7.6.4, where results from numerical CFD simulations will be presented.

In what follows we will use the coefficients: $K_b = 0.5$ (typical of the inner surf zone), $K_r = 0.9$ (as suggested by Svendsen, 1984b) and $K_t = 0.0055$, which is in line with the range suggested above (this value has been calibrated to yield a good match with the measured undertow profile). The simple model of undertow derived above may then be applied as follows:

(1) From Eq. 5.202 the wave number is estimated as $k = 3.22\ \mathrm{m}^{-1}$ and the celerity as $c = \sqrt{gh} = 0.975$ m/s. Note that this yields $kh = 0.312$ i.e. this position is near the upper depth limit of the assumed shallow water wave regime, thus comparison is still reasonable.

(2) From Eq. 7.56, the eddy viscosity can be computed to be $\nu_t = 5.20 \times 10^{-4}\ \mathrm{m^2/s}$. While a small number, this is still much larger than the kinematic viscosity of water $\nu = 10^{-6}\ \mathrm{m^2/s}$ ($\nu_t/\nu = 520$ in this example; Compare this result with e.g. Fig. 7.39).

(3) From Eq. 7.55 the slope of the mean water level is calculated to be $\partial\overline{\eta}/\partial x = 0.00569$. This is positive, and hence corresponds to a set up, as should be expected.

(4) From Eq. 7.47, we can calculate the (dimensionless, period-averaged) shear stress distribution to be

$$\frac{\overline{\tau}}{\rho g h} = \frac{\overline{\tau}_0}{\rho g h} + 0.00525\frac{y}{h}, \qquad \frac{\overline{\tau}_0}{\rho g h} = -0.00222$$

where the bed shear stress $\overline{\tau}_0/\rho$ is as indicated above. Note that as this is negative, this result implies that there will indeed be an offshore directed current near the bed, the undertow! The linear shear stress distribution above is plotted in Fig. 7.7a, demonstrating a negative shear stress near the bed, which becomes positive further up in the water column.

(5) From Eq. 7.50 we calculate the (dimensionless, period-averaged) undertow velocity profile to be:

$$\frac{\overline{u}}{\sqrt{gh}} = -0.403\frac{y}{h} + 0.477\left(\frac{y}{h}\right)^2$$

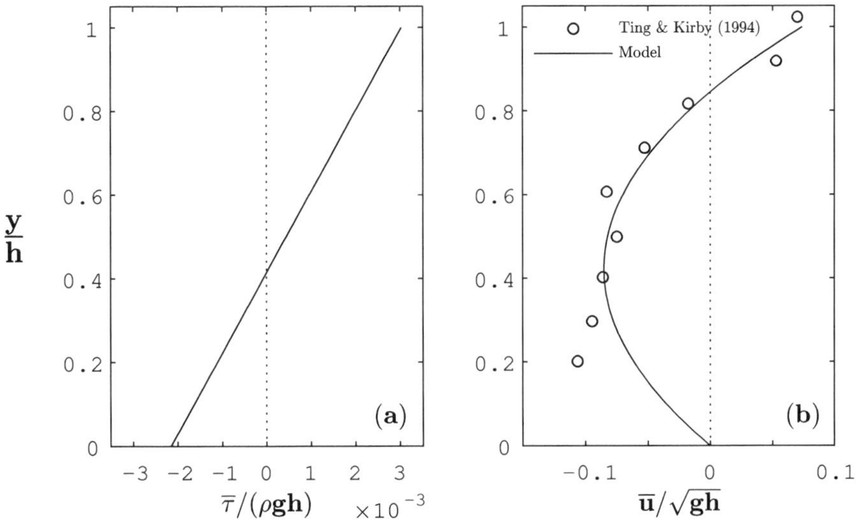

Fig. 7.7 Plot showing the distribution of (a) shear stress and (b) undertow velocity for the spilling breaking wave case of Ting and Kirby (1994) at their inner-most measurement position. In (b) their measured undertow velocities are also shown for comparison. Positive (negative) velocities are in the onshore (offshore) directions.

This is plotted in Fig. 7.7b, along with the experimental data from Ting and Kirby (1994). While there are deviations near the bed, the predicted undertow profile from the simple model formulated above matches the experimental data quite reasonably.

It is finally again noted that, while the present section has focused on the undertow specifically under spilling breaking waves, that undertow will also occur for fundamentally the same reasons in the surf zone initiated by plunging breakers. Comparison of more advanced computational fluid dynamics simulations of turbulence and undertow beneath spilling breakers will be presented in Section 7.6.4, and beneath plunging breakers in Section 7.6.5.

7.3 Flow induced by plunging waves

As mentioned already, surf and swash zone processes govern the exchange of sands between the emerged and submerged portions of the beach, and are therefore responsible for determining shoreline configurations, including e.g. beach accretion or recession. Reviews of various surf and swash zone processes can be found in the works of Battjes (1988), Butt and Russell (2000), Elfrink and Baldock (2002), Longo et al. (2002), Masselink and Puleo (2006), and Brocchini and Baldock (2008).

A large number of experimental studies over the years have greatly improved the present understanding of dynamics within the surf and swash zones. Several studies have taken detailed velocity measurements utilizing laser Doppler anemometry (LDA) (e.g. Stive 1984, 1985; Nadaoka et al. 1989; Ting and Kirby 1995, 1996; Cox and Kobayashi 2000; Stansby and Feng 2005; Lara et al. 2002, 2006; De Serio and Mossa 2006). Such measurements are effective, e.g., for determining time and ensemble-averaged properties of the turbulent flow field beneath waves at point locations, and are commonly utilized to validate numerical models of breaking waves (e.g. Lin and Liu 1998; Jacobsen et al. 2012). Several other experimental studies have alternatively utilized particle image velocimetry (PIV) (e.g., Chang and Liu 1998; Cox and Anderson 2001; Cowen et al. 2003; Kimmoun and Branger 2007; Huang et al. 2009, 2010; Sou et al. 2010; Ting and Nelson 2011; Ting 2013), which can provide instantaneous velocity fields beneath breaking waves. Other experimental works have focused on studying breaking wave processes with the aid of high speed video cameras (e.g., Chanson and Jaw-Fang 1997).

The above-mentioned works (and others), coupled with several advanced numerical modeling studies (e.g. Christensen and Deigaard 2001; Watanabe et al. 2005; Christensen 2006; Lakehal and Liovic 2011; Ma et al. 2011; Torres-Freyermuth et al. 2013), provide a detailed account e.g. of the flow velocities, turbulence characteristics, as well as other basic integral parameters (e.g. Buhr Hansen 1990), in the surf and swash zones. Understanding their role within sediment transport and morphology, however, remains a major challenge. Of fundamental importance in driving sediment transport and resultant morphology are the bed shear stresses, the interest of which has motivated a number of additional experimental works. In such studies, bed shear stresses in the laboratory or field have typically been obtained through logarithmic fitting of measured phase-averaged velocity profiles (e.g. Cox et al. 1996; Archetti and Brocchini 2002; O'Donoghue et al. 2010; Puleo et al. 2012), phase-averaged velocity gradients within the viscous sublayer or assumed quadratic forms (e.g. Huang et al. 2010), or via the use of shear plates (e.g. Barnes et al. 2009; Seelam et al. 2011). While giving valuable information about mean (ensemble averaged) values, such methods are unable to resolve important turbulent fluctuations in the bed shear stresses, which are likely considerable. Only a few researchers have conducted detailed bed shear stress measurements beneath breaking regular waves in the laboratory, utilizing hot-film probes mounted on smooth beds, Deigaard et al. (1991b), Yuksel et al. (1998) and Sumer et al. (2013). Earlier, Sumer et al. (2011) have also made detailed hot-film bed shear stress measurements beneath laboratory generated plunging solitary waves, which were recently used as numerical model validation by Adityawan et al. (2013), Li et al. (2019) and Larsen and Fuhrman (2019). Also warranting mention, Conley and Griffin (2004) have made direct hot-film field measurements of bed shear stresses within the swash zone.

While the above mentioned laboratory studies were predominantly conducted on rigid beds, many other laboratory studies have further investigated actual sediment transport and morphology of the surf and swash zones. Such studies include the works of Baldock et al. (2010, 2011), who investigated beach morphodynamics in the surf and swash zones at both small and large laboratory scales; as well as Alsina et al. (2012), who conducted large-scale experiments with special emphasis on studying the influence of swash conditions on the overall beach morphodynamics. Grasso et al. (2009, 2011) and Berni et al. (2013) have also studied beach morphological changes in the laboratory, making use of low density sediment material to achieve field similarities in both dimensionless Shields and Rouse parameters. Field morphological campaigns in recent years additionally include the works of e.g. Masselink et al. (2005), who measured depths, flow velocity and suspended sediment concentrations in the swash zone; Turner et al. (2008), who measured wave-by-wave bed levels in the swash zone; as well as Aagaard et al. (2012), who measured sediment transport and morphodynamics outside the breakpoint on a reflective beach.

Most of the previously mentioned laboratory studies focusing on surf and/or swash zone processes were conducted on either rigid or live sediment beds. An exception is the recent studies of Sumer et al. (2011 and 2013), who exploited a dual-use strategy, utilizing both rigid (smooth) and live sediment beds. This approach enabled detailed velocity and (hot film) bed shear stress measurements taken on rigid bed profiles to be directly related to near-bed pore-pressure measurements and subsequent morphological evolution observed on live sediment beds, under the same initial bed profile and wave conditions.

The presentation in the following subsections is mainly based on Sumer et al. (2013) while that in Section 7.5 dealing with flow and turbulence induced by plunging solitary waves is based on Sumer et al. (2011).

The motivation in the work of Sumer et al. (2013) was to provide an im-

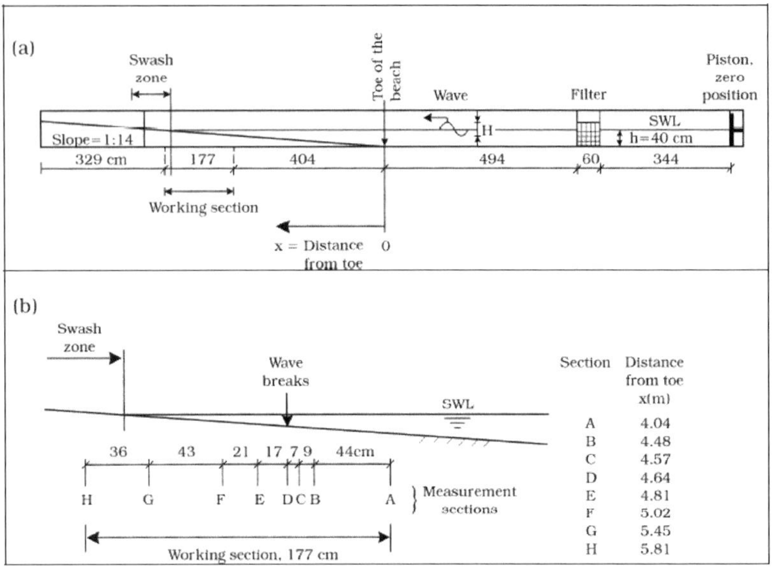

Fig. 7.8 Experimental setup of Sumer et al.'s (2013) rigid-bed experiments for plunging regular waves.

proved understanding of the entire breaking wave sequence and morphology in these conditions, with particular field relevance to plunging wave generated surf and swash zones. To this end, Sumer et al. (2013) carried out an extensive series of laboratory experiments where they utilized both rigid and live sediment beds, as mentioned previously. The next section will describe Sumer et al.'s (2013) experimental setup for the rigid-bed experiments. (Sumer et al.'s, 2013, sediment-bed experiments will be described later.)

7.3.1 *Experimental setup in Sumer et al. (2013) study*

The experiments were carried out in a wave flume, 28 m in length, 0.80 m in depth and 0.60 m in width, equipped with Active Wave Absorption Control System. The test setup is illustrated in Figs. 7.8a,b. The beach (with a slope of 1:14) was made of smooth-surface PVC plates. The junctions between the side walls of the flume and the PVC plates were sealed to ensure that no exchange of water takes place between the main body of the water and underneath the PVC plates.

Two kinds of measurements were made: bed shear stress measurements and water-surface elevation measurements. The bed shear stress was measured with a hot-film probe. A detailed description of the hot-film probe and its calibration was given in Sumer et al. (2011) in relation to Sumer et al.'s work on flow induced by a plunging solitary wave, and earlier, in Jensen et al. (1989), Lodahl et al. (1998), Carstensen et al. (2010) and Sumer et al. (2010) dealing with wave boundary layer flows, already discussed in Chapter 5.

The bed shear stress measurements were made at eight sections, with Sections A-C located at the offshore side of the wave breaking, and Sections D-H at the onshore side of the wave breaking (Fig. 7.8b). As the hot film does not sense the direction of the bed shear stress, synchronized velocity measurements were made at $y = 0.5$ mm (y being the vertical distance from the bed) just above the hot-film probe, in order to monitor the direction of the bed shear stress, using a Laser Doppler Anemometer (LDA) equipment.

The water-surface elevation measurements were made at three sections in each test, namely at the filter section (Fig. 7.8a), at the toe section, and at the section where the bed shear stress measurement was made. The bed shear stress measurements and the water-surface elevation measurements were synchronized.

In addition to the above mentioned measurements, synchronized flow visualizations were also performed, using a digital video recorder shooting from the side. Sumer et al. (2013) note that the flow visualization is limited to the side-view motion of wave breaking, and does not include the visualization of the so-called obliquely descending eddies observed by Nadaoka et al. (1989).

Finally, the location where the wave breaks is marked by a vertical arrow in Fig. 7.8b. The precise definition of the point of breaking adopted in the Sumer et al. (2013) study is that the wave breaks at the location where the wave overturns and touches down on the water surface. The swash zone, on the other hand, is marked in Fig. 7.8a, and the maximum point of runup lies at the onshore end of the swash zone, corresponding to an elevation of 7.3 cm from the still water level (SWL), as will be discussed in the following paragraphs.

The water depth at the toe of the beach in the test was $h = 40$ cm (Fig. 7.8a), the wave height $H = 5.9$ cm, and the wave period $T = 3.5$ s.

7.3.2 *Description of plunging breaking waves in Sumer et al. (2013) study*

7.3.2.1 *Breaking-wave and runup characteristics*

The maximum runup elevation from SWL was measured to be $R = 7.3$ cm, as mentioned in the preceding paragraphs. This compares well with a value $R = 7.6$ cm predicted by the empirical breaking wave runup expression of Hunt (1959), which is commonly expressed in the following form:

$$\frac{R}{H} = \xi \tag{7.57}$$

in which ξ is a surf similarity parameter, different from ξ_0 (Eq. 7.1), defined by

$$\xi = \frac{\tan(\beta)}{\sqrt{H/L_0}} \tag{7.58}$$

in which H is the wave height at the toe of the sloping region.

The wave breaking in the experiment, on the other hand, was observed to be in the plunging regime, and this was in good agreement with the breaker type defined by Eqs. 7.6–7.8, with the ξ_0 value being 1.42 (obtained from Eqs. 7.1–7.5), which is in the range $0.5 < \xi_0 < 3.3$ (Eq. 7.7).

7.3.2.2 *General description*

Fig. 7.9 illustrates the complete sequence of the plunging breaking wave over one wave period, observed in Sumer et al.'s (2013) experiment: (a) Shoaling and wave breaking (Fig. 7.9a); (b) Runup (Fig. 7.9a); and (c) Rundown (Fig. 7.9b). Numbers in the figure denote different stages of the breaking process. The pictures given in the figure are direct copies of the video frames. Bed shear stress measurement sections, Sections B-G, are also indicated in the figure for easy reference. (Sections A and H are outside the viewing area of Fig. 7.9.)

The wave breaking occurs with the wave curling over forward and impinging onto the water surface onshore with a small amount of air trapped inside the "tube" formed by the wave crest.

Fig. 7.10 illustrates the detailed picture of the breaking process between Stage 2 and Stage 4 depicted in Fig. 7.9.

The most prominent feature of the wave breaking is the generation of the vortex beneath the breaker, Vortex M (Fig. 7.10b). This is caused by the curling of the wave, in combination with the offshore-directed flow associated with the rundown stage of the wave breaking. Note that the flow

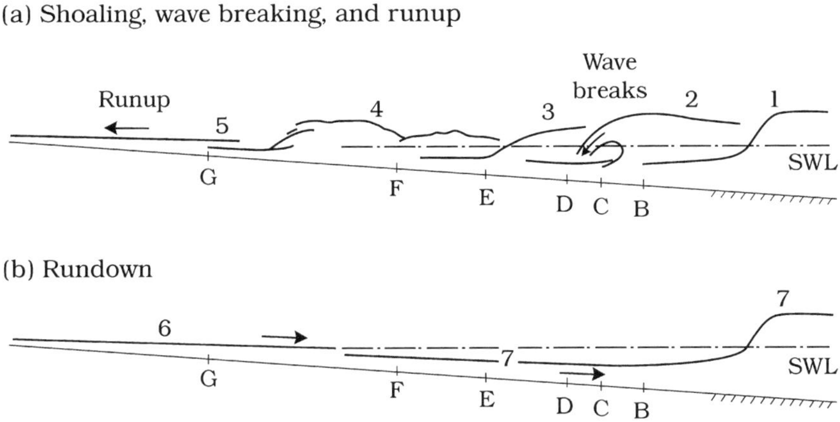

Fig. 7.9 Sequence of plunging wave breaking. Small numbers denote different stages. The times corresponding to the indicated stages: 1: $t + 0$ s; 2: $t + 0.2$ s; 3: $t + 0.4$ s; 4: $t + 0.734$ s; 5: $t + 2.23$ s; 6: $t + 2.267$ s; and 7: $t + 3.5$ s. Sumer et al. (2013).

direction in the sequence of flow pictures given in Fig. 7.10 was obtained from the LDA measurements described in the previous section. The generation of Vortex M is followed by that of two more vortices, Vortex N and Vortex P (Fig. 7.10e-f).

It can be noted that a reasonably similar pattern of vortices is also evident e.g. in the smooth-particle-hydrodynamics (SPH) numerical simulation of a weak plunger presented by Dalrymple and Rogers (2006, see their Fig. 2), and the overall behavior seems consistent with the "sequence of jet-splash motions" described previously by e.g. Miller (1976), Jansen (1986) and Battjes (1988).

Although the generation of these latter vortices is certainly associated with the wave splash as the consequence of the violent impact of the plunger onto the water surface, the sequential generation of the three vortices (Vortices M, N and P) may be viewed as a result of a shear-layer instability mechanism: The water between the wave crest and the wave trough has an onshore-directed velocity (with a magnitude approximately equal to the wave celerity, c) at the time of wave breaking, while the rest of the water at the breaking section has an offshore-directed velocity due to rundown. This type of flow is known to be subject to a shear-layer instability whereby the vorticity in the shear layer eventually concentrates in vortices (e.g., Sumer and Fredsøe, 2006, their p. 254 and Fig. 5.29). The present setting is dif-

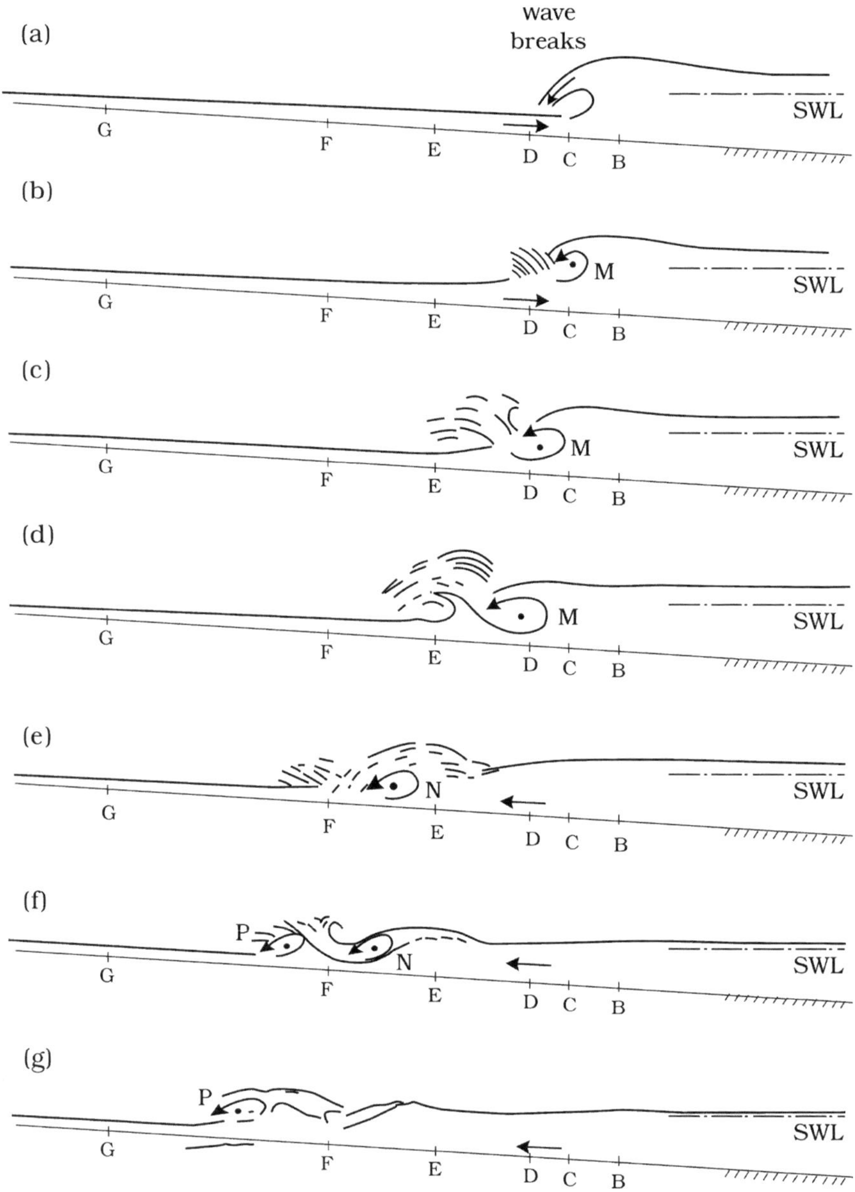

Fig. 7.10 Detailed sequence of plunging wave breaking between Stages 2 and 4 in the previous figure. The times shown are: (a) $t + 0.2$ s; (b) $t + 0.267$ s; (c) $t + 0.334$ s; (d) $t + 0.4$ s; (e) $t + 0.534$ s; (f) $t + 0.667$ s; and (g) $t + 0.734$ s. See Fig. 7.8 for the x coordinates of Sections B–G. Sumer et al. (2013).

ferent, however, from that in the classic shear-layer instability analysis; the flow in the present case has a free surface.

Christensen and Deigaard (2001), from their Large Eddy Simulations of spilling breakers, also refer to this kind of shear-layer instability, describing "... a shear layer and a surface roller are developing. From the shear layer, a number of eddies develop as the breaker propagates. ...".

We also note that Matsunaga and Honji (1980) described vortex generation for three kinds of breakers, what they call plunging, intermediate and surging, and discussed the role of the "backwash" vortex (corresponding to Vortex M of the present description) in the sediment transport.

Fig. 7.11 presents snapshots of spanwise-averaged vorticity, ω_z (see Eq. 6.89), along with spanwise-averaged longitudinal and vertical velocities while Fig. 7.12 presents those of 3D plots of *Lambda*2 contours, obtained by Kim and Hsu (2018, private communication), using a Large Eddy Simulation (LES) having the same setup of the rigid-bed experiments of Sumer et al. (2013) (Figs. 7.9 and 7.10). Here, *Lambda*2, usually adopted to qualitatively visualize the turbulent coherent structures, is defined as the eigenvalues of $S^2 + \Omega^2$ where S and Ω are, respectively, the symmetric and asymmetric parts of velocity gradient tensor (Jeong and Hussain, 1995). Fig. 7.12 shows *Lambda*2 $= -100$ plots, i.e., it represents spatially coherent contours at a given time which has an eigenvalue of *Lambda*2 $= -100$.

The authors (Kim and Hsu, 2018) give the following information for the details of their CFD simulation. The numerical model is implemented using the open-source CFD toolbox, OpenFOAM, with a minor modification of boundary conditions (e.g., Kim et al., 2017; Kim, 2018; Zhou et al., 2014, 2017). Among the various solvers within OpenFOAM, `interFoam` (Berberovic et al., 2009, and Klostermann et al., 2012) and a wave generation toolbox, called `waves2Foam` (Jacobsen et al., 2012, 2014), are adopted in this study. The numerical model solves the 3D filtered Navier-Stokes equations for two immiscible fluids (air and water phases). In `interFoam`, the interface between two immiscible fluids is tracked by the interface compression scheme, which follows the concept of volume of fluid (VOF) method (Hirt and Nichols, 1981). Some further details are provided in Section 7.6.3. The effects of sub-grid scale eddies are parameterized using the standard Smagorinsky closure (Smagorinsky, 1963). More details of the numerical model can be found elsewhere (e.g., Kim et al., 2017; Kim, 2018); See also Section 9.5 for a more in depth discussion of turbulence modelling with LES. The 3D LES numerical model domain is identical to that used in the laboratory experiment of Sumer et al. (2013). The model domain was com-

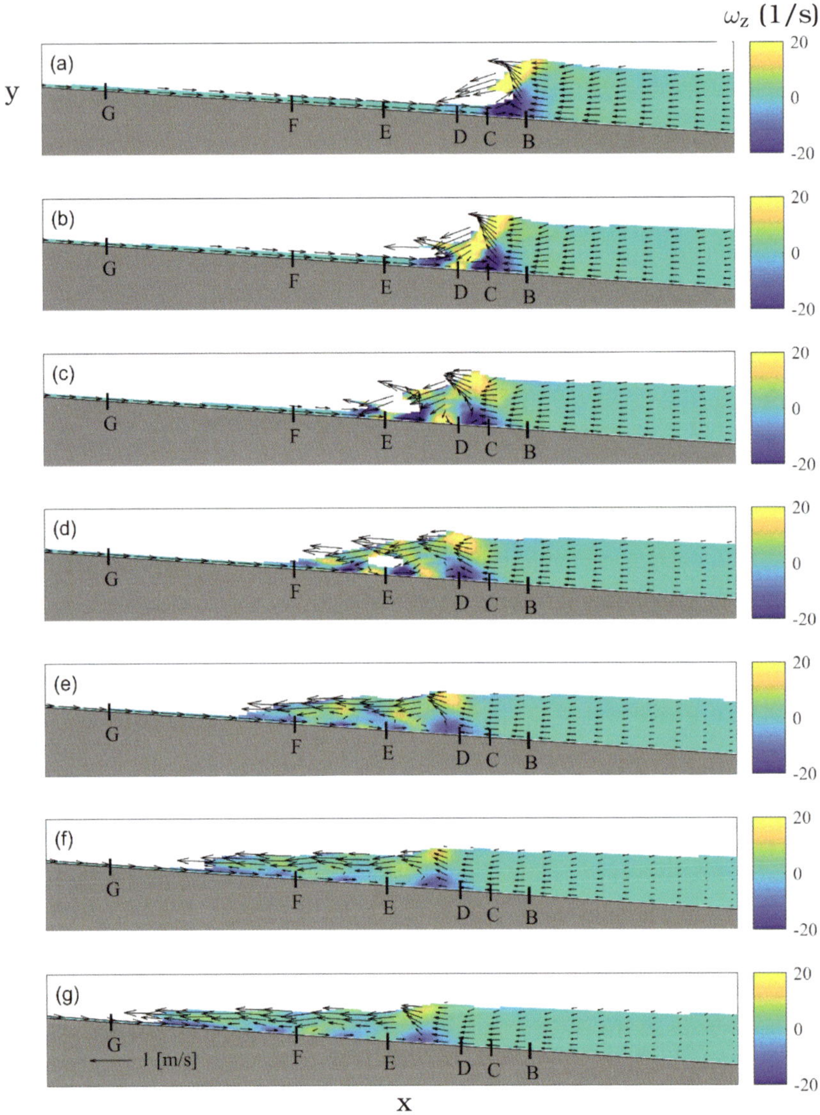

Fig. 7.11 LES simulation of Sumer et al. (2013) experiment (see previous figure), with spanwise-averaged vorticity (perpendicular to the picture plane), longitudinal and vertical velocities. The times: (a) $t + 0.2$ s; (b) $t + 0.3$ s; (c) $t + 0.4$ s; (d) $t + 0.5$ s; (e) $t + 0.6$ s; (f) $t + 0.7$ s; and (g) $t + 0.8$ s. Courtesy of Dr. Yeulwoo Kim and Prof. Tian-Jian Hsu.

Fig. 7.12 LES simulation of Sumer et al. (2013) experiment (see previous two figures), with 3D plots of *Lambda*2 contours. Generation of Vortices M, N and P in Fig. 7.10. The times shown from top to bottom are: $t+0$ s, $t+0.1$ s, $t+0.2$ s, ..., $t+0.6$ s. Courtesy of Dr. Yeulwoo Kim and Prof. Tian-Jian Hsu.

prised of 998, 60, 60 grid points in the streamwise (x), spanwise (z), and vertical (y) directions, respectively. The grid size in the vertical direction was non-uniform, with a minimum height of 3 mm at the onshore end of the slope. The grid sizes in the streamwise and spanwise direction was kept constant of 20 mm and 10 mm, respectively. Hence, a total number of 3.6 million computational grid points were used for this simulation. The 16th order stream function was used to generate the waves at the inlet boundary using `waves2Foam` (Jacobsen et al., 2012, 2014) to match the measured wave heights of Sumer et al. (2013) at the toe of the slope where the incident wave height (H), wave period (T), water depth (h) were set to be 5.9

cm, 3.5 s, and 40 cm, respectively.

As seen from Figs. 7.11 and 7.12, the model can reproduce the expected temporal and spatial evolutions of vortices caused by sequential wave breaking, consistent with Sumer et al. (2013). The velocity vectors clearly show that strong shear exist between backwash and main wave propagation being a main source of generating vortices. In general, Kim and Hsu's (2018) LES-simulation result is in good agreement with Sumer et al.'s observations, and it confirms remarkably well the generation of the three vortices (Vortices M, N and P, Fig. 7.10) observed in Sumer et al.'s (2013) experiment.

7.3.2.3 *Water-surface elevation*

Fig. 7.13 displays the time variation of the mean water-surface elevation at different measurement sections, Sections A, B, D, F and G, along with at the toe (the "reference signal"), while Fig. 7.14 displays the standard deviation of the water-surface elevation (or the r.m.s. value of the fluctuating component of the water-surface elevation), at the same sections. The time variations for Sections C, E and H are not included in the figures for reasons of space. The origin of time, $t = 0$, is taken such that it coincides with the zero up-crossing of the water-surface elevation at the toe (Fig. 7.13a). Here, the mean and the standard deviation of the water-surface elevation are calculated through ensemble averaging; see e.g., Eqs. 5.32 and 5.33.

The following observations can be made from Figs. 7.13 and 7.14:

1. The shoaling (Fig. 7.9a), which takes place until the wave breaks just before Section D, is clearly observed from Figs. 7.13a–c.

2. The runup and rundown stages of the process at the measurement sections are evident from the time series in Figs. 7.13b–f. Regarding the time variation recorded at Section G, as this section is rather close to the shoreline, the signal experiences a substantial amount of asymmetry (Fig. 7.13f).

3. The standard deviation of the surface elevation, σ_η, at the toe (Fig. 7.14b) is very small (smaller than about 3% of the wave height), indicating that the incoming wave form is practically constant.

4. However, σ_η begins to increase, already at Section A (Fig. 7.14c). This increase is associated with the wave crest (cf. Figs. 7.13b and 7.14c). The large value of σ_η during the passage of the wave crest (Fig. 7.13b) may be attributed to the steepening of waves due to shoaling. Furthermore, σ_η is largest at Section D (Fig. 7.14e), and this is obviously linked to the wave breaking.

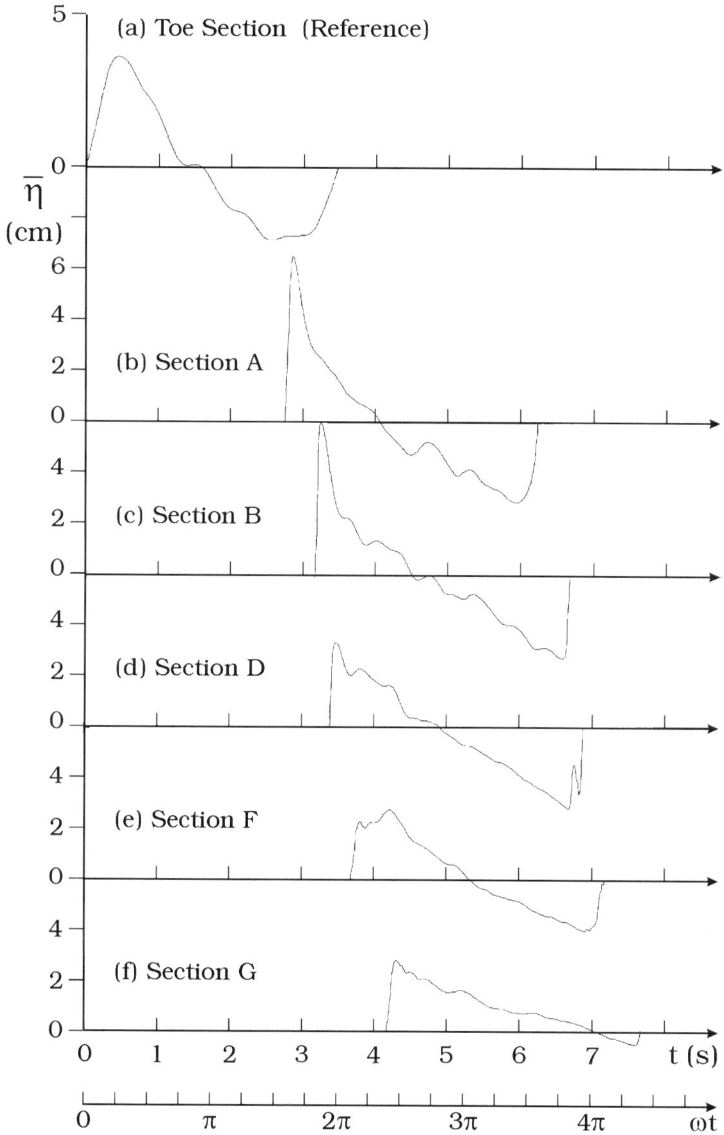

Fig. 7.13 Time variation of water-surface elevation at different sections. $t = 0$ coincides with the zero upcrossing of the surface elevation at the toe. See Fig. 7.8 for x coordinates of Sections A-G. Sumer et al. (2013).

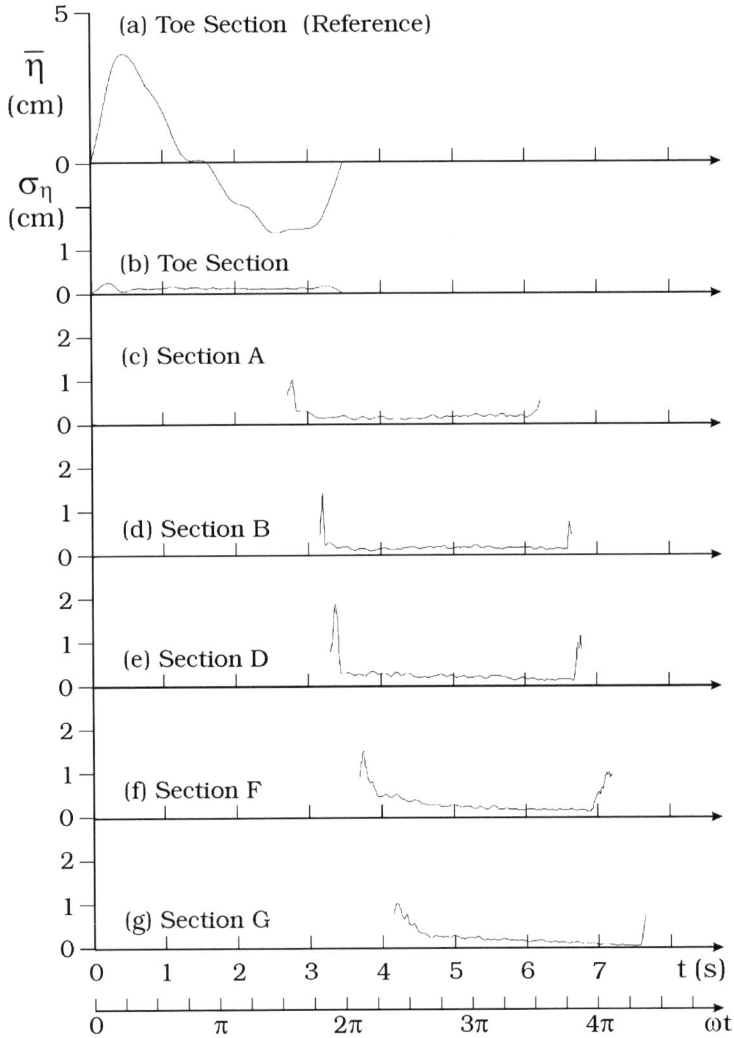

Fig. 7.14 Time variation of the r.m.s. value of the fluctuating component of the water-surface elevation at different section. See Fig. 7.8 for x coordinates of Sections A-G. Sumer et al. (2013).

7.3.3 *Bed shear stress in Sumer et al. (2013) study*

Fig. 7.15 presents the time variation of the mean bed shear stress, $\overline{\tau_0}$, while Fig. 7.16 presents that of the r.m.s. value of the fluctuating component of the bed shear stress, $\tau_0' = \tau_0 - \overline{\tau_0}$, at Sections A, B, D, F and G. The sign convention in Fig. 7.15 is such that the direction of the bed shear stress is positive in the onshore direction. As in Figs. 7.13 and 7.14, the time variations for Sections C, E and H are not included in Figs. 7.15 and 7.16 for reasons of space. Note that, for convenience, the bed shear stress data are plotted in terms of $\overline{\tau_0}/\rho$ and $(\overline{(\tau_0')^2})^{1/2}/\rho$ rather than $\overline{\tau_0}$ and $(\overline{(\tau_0')^2})^{1/2}$. Here ρ is the density of water. The mean value of the bed shear stress and the r.m.s. value of the fluctuating component of the bed shear stress are calculated through ensemble averaging (Eqs. 5.32 and 5.33).

The way in which the r.m.s. value is calculated (ensemble averaging) requires a very repeatable phase of the wave form. As pointed out in item 3 in the previous subsection, the incoming wave form measured at the toe section is practically constant. This ensures a repeatable wave form, and therefore that the measured finite r.m.s. values of the fluctuating bed shear stress are due to the hydrodynamic processes over the sloping bed, rather than a pseudo-turbulence induced by the fact that the average value is not repeated exactly with each wave. However, generation mechanisms of the bed shear stress fluctuations deserve an early discussion.

This issue has been discussed in greater detail in Sumer et al. (2011, Section 4.2.1) in the case of solitary plunging waves. Similar to Sumer et al. (2011): (1) the r.m.s. value measured at Sections A and B in Figs. 7.16b,c is associated with the wave-boundary-layer turbulence (before breaking), as will be detailed in the next subsection; (2) that measured at Sections D, F and G during the wave breaking and runup stage in Figs. 7.16d–f is associated with the wave breaking as well as the turbulence generated by the unsteady boundary layer developing during the runup (Fig. 7.9a); and (3) that measured at Sections D, F and G, corresponding to the rundown stage in Figs. 7.16d–f is associated with the turbulence generated by the unsteady boundary layer developing during this latter stage (Fig. 7.9b).

7.3.3.1 *Wave boundary layer prior to wave breaking*

Of the measurement sections displayed in Figs. 7.15 and 7.16, Sections A and B are two sections before the wave breaking occurs. The bed shear stress at these sections is governed by the wave boundary layer.

To give an idea, Sumer et al.'s (2013) measurements and analysis showed that, at Section A, the wave boundary layer is in the laminar regime in the trough half period (with a Reynolds number of $Re = 1.1 \times 10^5$), and it is

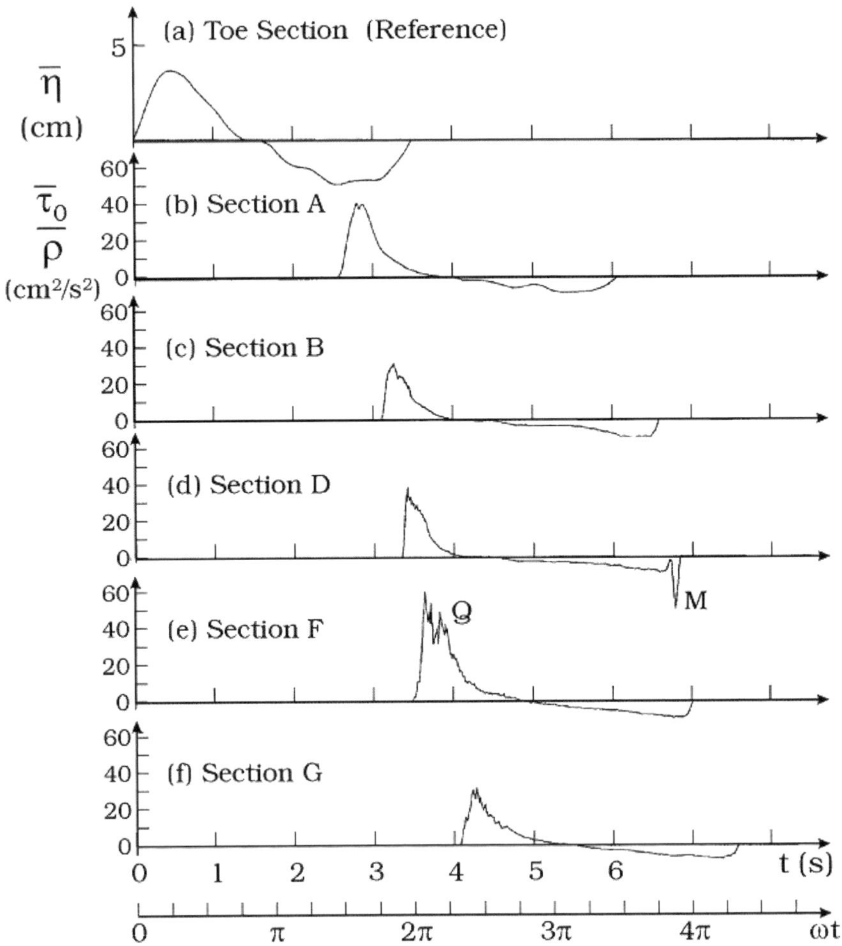

Fig. 7.15 Time variation of the mean bed shear stress. See Fig. 7.8 for x coordinates of Sections A-G. Sumer et al. (2013).

in the transitional regime in the crest half period ($Re = 1.4 \times 10^6$).

Sumer et al. (2013) compared the measured bed shear stress (the mean and the r.m.s. value of the fluctuations) with the existing information on wave boundary layers, and found that the measured values are in accordance with the corresponding data available in the literature. The reader is referred to the original publication for details.

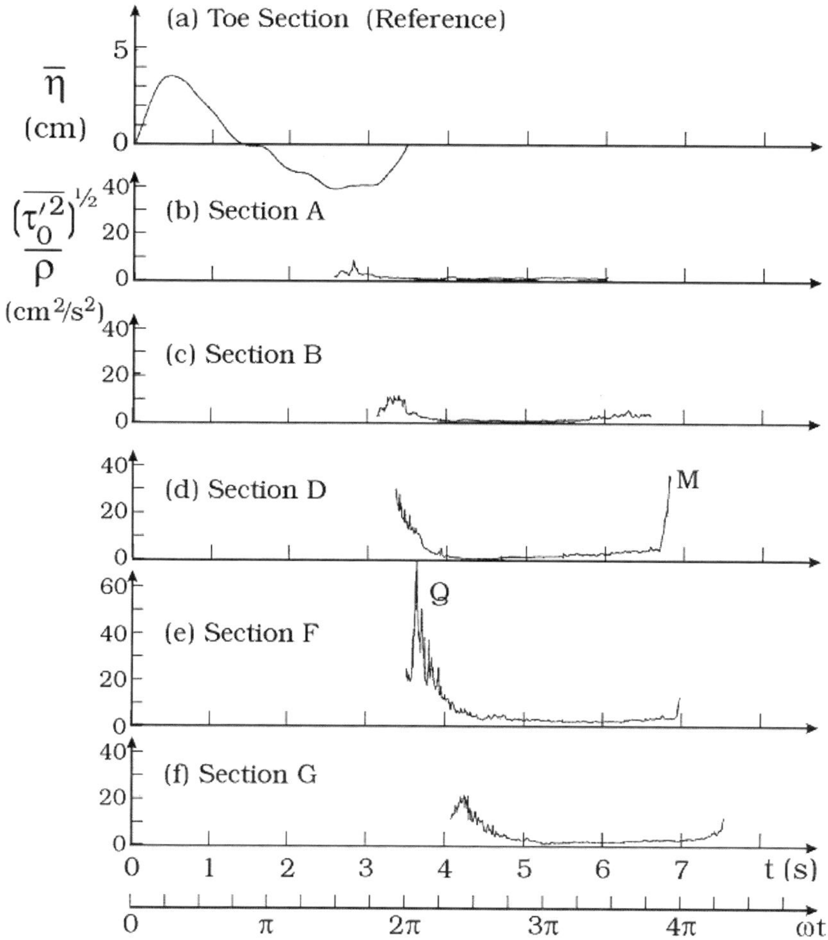

Fig. 7.16 Time variation of the r.m.s. value of the fluctuating component of the bed shear stress. See Fig. 7.8 for x coordinates of Sections A-G. Sumer et al. (2013).

7.3.3.2 *Wave breaking*

The way in which the mean bed shear stress varies with time at Section B is the same as in Section A where the bed shear stress is directed onshore (positive sign in the bed shear stress) during the crest half period while it is directed offshore (negative sign) during the trough half period (obviously with a phase lead with respect to the free-stream velocity).

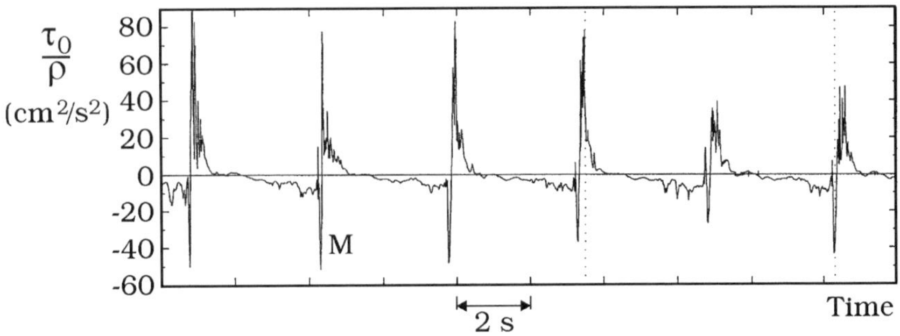

Fig. 7.17 Time series of the bed shear stress, measured at Section D. See Fig. 7.8 for the location of Section D. Sumer et al. (2013).

The way in which the bed shear stress varies at Section D, the breaking section, is similar to that at Sections A and B with the exception that the bed shear stress at Section D experiences a distinct, sudden, sharp dip in the trough half period, marked M in Fig. 7.15d. This dip precisely coincides with the time of the passage of Vortex M, described in conjunction with Fig. 7.10b–d. A similar sharp (but slightly smaller in magnitude) dip is also observed in the time variation of the bed shear stress at the neighboring Section C (not shown here). The time series from Section D plotted in Fig. 7.17 exhibits these sudden, sharp dips even more dramatically as they are not smoothed out with ensemble averaging. The time series from Section E (not shown here) also exhibited the same kind of pattern. It is to be noted that the peak in $(\overline{(\tau_0')^2})^{1/2}$ displayed in Fig. 7.16d (Section D), marked M, is associated with Vortex M.

Following the breaking, $\overline{\tau_0}$ experiences an increase during the runup stage (marked Q in Fig. 7.15e), which is due to the combined action of the breaking and the reduction of the flow depth as the flow continues onshore. Of particular interest is a substantial increase in turbulence in the bed shear stress $(\overline{(\tau_0')^2})^{1/2}$ following the wave breaking (Sections D–G, Fig. 7.16d–f). This increase has clearly significant implications for sediment transport. We note that the influence of externally generated turbulence on sediment transport was investigated by Sumer et al. (2003) in an open channel flow; Sumer et al.'s (2003) study showed that the externally generated turbulence caused a significant amount of increase in the sediment transport, up to a factor of 5, depending on the level of the external field of turbulence.

To illustrate the cross-shore variation of $\overline{\tau_0}$, the amplitude values of the

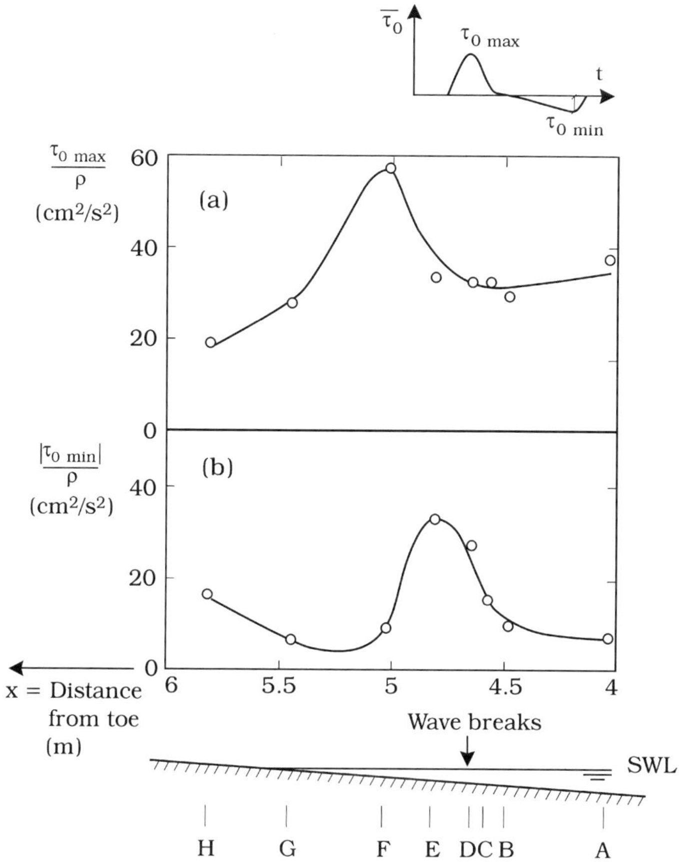

Fig. 7.18 Cross-shore variation of the maximum bed shear stress. (a): During the shoaling and runup stages; and (b): During the rundown stage. See Fig. 7.8 for locations of the sections A, B, etc. Sumer et al. (2013).

mean bed shear stress, $\tau_{0\,\mathrm{max}}$ and $\tau_{0\,\mathrm{min}}$ (see the definition sketch in the inset in Fig. 7.18), are picked up from Fig. 7.15, and plotted in Fig. 7.18. Data from Sections C, E and H are also included in Fig. 7.18 for completeness. Fig. 7.18a corresponds to the shoaling and the runup stages while Fig. 7.18b corresponds to the rundown stage. The variations of $\overline{\tau_0}$ as function of x can be interpreted as in the preceding paragraphs. The maximum bed shear stresses in Fig. 7.18 onshore of incipient breaking are increased significantly, by nearly a factor of 2 relative to the observed pre-breaking values.

Similar to Fig. 7.18, the maximum values of $(\overline{(\tau_0')^2})^{1/2}$ picked up from

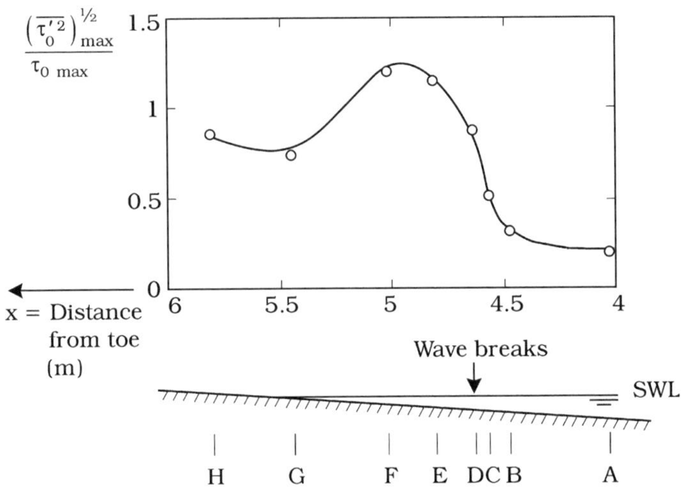

Fig. 7.19 Cross-shore variation of the normalized maximum value of the r.m.s. value of the fluctuating component of the bed shear stress. See Fig. 7.8 for locations of the sections A, B, etc. Sumer al. (2013).

Fig. 7.16 are plotted in Fig. 7.19, to illustrate the x variation of $(\overline{(\tau_0')^2})^{1/2}$. In order to facilitate comparison, the quantity $(\overline{(\tau_0')^2})^{1/2}_{\max}$ is plotted in normalized form $(\overline{(\tau_0')^2})^{1/2}_{\max}/\tau_{0\,\max}$, which enables us to reconcile with the existing turbulent wave boundary layer data. As seen, the quantity $(\overline{(\tau_0')^2})^{1/2}_{\max}/\tau_{0\,\max}$ approaches its "asymptotic" value $(\overline{(\tau_0')^2})^{1/2}_{\max}/\tau_{0\,\max} \simeq 0.2$, the wave-boundary-layer value, as x approaches Section A. (See the detailed discussion in Section 4.2.1 of Sumer et al., 2013.) This quantity increases, however, by a factor of 5–6 with x toward the breaking point, and takes its maximum value of 1.2 between Section E and Section F. Clearly, this increase with respect to the wave-boundary-layer level is caused by the breaking process.

Finally, Fig. 7.20 displays the cross-shore variation of the phase lead of the mean bed shear stress with respect to the mean water-surface elevation (at the section where the bed shear stress was measured). The phase lead plotted in the figure corresponds to the zero up-crossing points of the time variations of $\overline{\eta}$ and $\overline{\tau_0}$.

As seen from Fig. 7.20, the phase lead is significantly reduced over the Sections B–E from the wave-boundary-layer value, namely from 17.5^0 to almost nil. This is linked to very turbulent character of the flow; momentum-

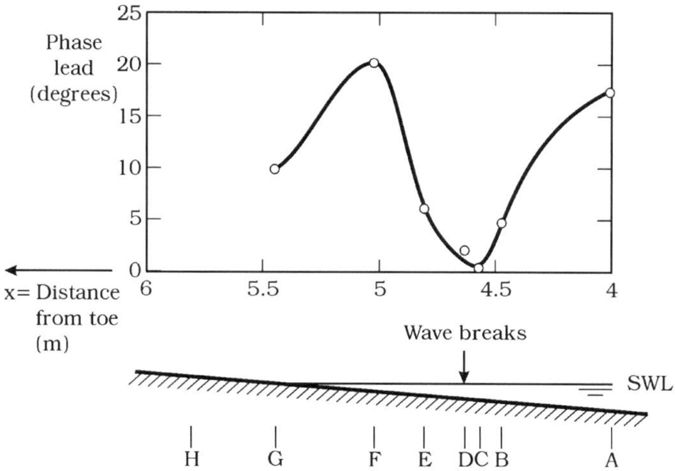

Fig. 7.20 Cross-shore variation of the phase lead of the bed shear stress with respect to that of the water-surface elevation, corresponding to the zero upcrossing points. Sumer et al. (2013).

rich fluid is brought to the near-bed region due to turbulence, and therefore the phase lead is greatly reduced. The phase lead increases, however, with increasing cross-shore distance.

7.3.3.3 *Flow resistance associated with breaking wave*

The maximum values of the friction velocity may be parameterized as follows. For this, deep-water quantities will be utilized, as the eventual wave breaking processes can most universally (and conveniently) be characterized by these conceptual offshore conditions prior to any shoaling (see the discussion in Appendices 5.H and 5.I). On purely dimensional grounds (neglecting viscous effects), it may be anticipated that the normalized maximum friction velocity may depend on two parameters, i.e., it will have a functional dependence of the form

$$\frac{U_{fm}}{\sqrt{gH_0}} = f\left(\frac{H_0}{L_0}, \tan\beta\right) \tag{7.59}$$

in which H_0 is the deep-water wave height, and L_0 the deep-water wave length, and U_{fm} the maximum value of the mean friction velocity caused by the breaking wave.

$$U_{fm} = \sqrt{\frac{\max(\tau_{0\,\max})}{\rho}} \tag{7.60}$$

Experience (e.g. Battjes 1974) has shown, however, that a number of surf zone processes can be adequately characterized by combining these two parameters into a single parameter, namely, the surf similarity parameter, as defined in Eq. 7.1. Following this convention, the bed shear stress is plotted in Fig. 7.21 in the non-dimensional form

$$\frac{U_{fm}}{\sqrt{gH_0}} = f(\xi_0) \tag{7.61}$$

Now, the friction velocity is governed by the wave breaking. The normalizing quantity $\sqrt{gH_0}$ can be taken, loosely, as a scale linked to the velocity of the plunging jet. The surf similarity parameter ξ_0, on the other hand, characterizes the type of the wave breaking, as discussed previously. Note that the spilling and plunging regimes from Eqs. 7.6 and 7.7, respectively, are indicated in Fig. 7.21.

Two kinds of data are plotted in Fig. 7.21: one from Sumer et al. (2013) (filled circle), and the rest from Deigaard et al. (1991b). The latter authors did similar experiments, using a hot-film probe to measure the bed shear stress on a slope of 1:30 with various wave heights and periods. Sumer et al. (2013, Table 3) summarized the test conditions for Deigaard et al.'s (1991b) data. The breaker type in Deigaard et al.'s (1991b) study was identified by visual observations. Evidently the data points corresponding to $\xi_0 = 0.365$ and 0.416 in Deigaard et al.'s study, although identified as plunging breakers, do not satisfy the plunging-breaker criterion in Eq. 7.7. In connection with this, Deigaard et al. (1991b) report that "... the plunging breakers were not very violent ...".

The tight clustering of the data in Fig. 7.21 suggests that the normalized maximum friction velocity is, in fact, very well parameterized by the surf similarity parameter ξ_0 alone. Fig. 7.21 illustrates a generally increasing trend (this is in fact evident even from Deigaard et al.'s, 1991b, data alone). The friction velocity increases with ξ_0 simply because the larger the value of ξ_0, the more "violent" the wave breaking. Hence, the increasing trend of the normalized friction velocity is consistent with physical expectations. Regarding the correlation exhibited in Fig. 7.21, it may be argued that the quantity $\sqrt{H_0}$ occurs on both axes of the plot, and therefore this can induce a false correlation. However, the range of $\sqrt{H_0}$ is quite narrow, namely $\sqrt{H_0} = 2.2$–$3.3\ \mathrm{cm}^{1/2}$, the lower bound corresponding to the filled circle,

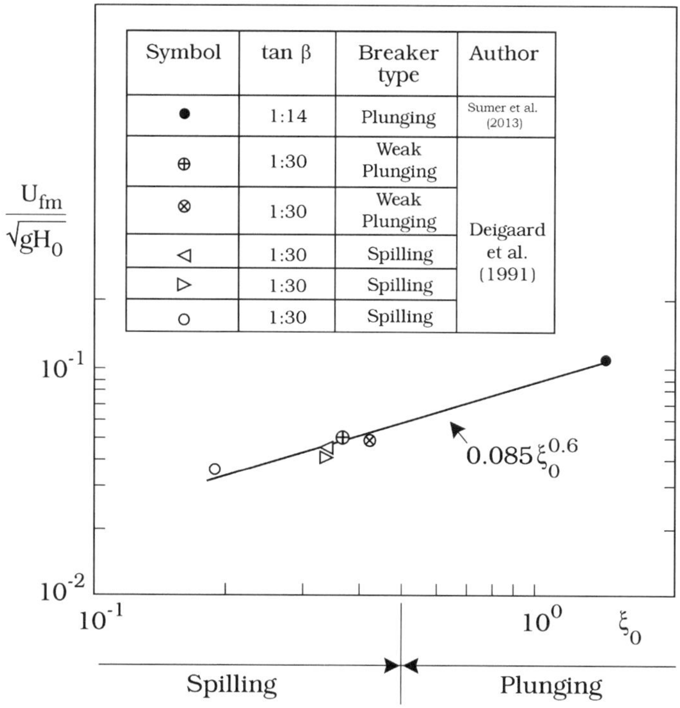

Fig. 7.21 Normalized friction velocity vs. surf-similarity parameter over a smooth (or hydraulically smooth) bed. Sumer et al. (2013). In the figure legend Deigaard et al. (1991) corresponds to the reference Deigaard et al. (1991b).

and the upper bound to the open circle in Fig. 7.21, and clearly this cannot explain a factor of 3 change in the value of $U_{fm}/\sqrt{gH_0}$ corresponding to this range.

The data in Fig. 7.21 are well represented by the following empirical expression:

$$\frac{U_{fm}}{\sqrt{gH_0}} = 0.085\ \xi_0^{0.6}, \qquad 0.19 < \xi_0 < 1.42 \tag{7.62}$$

Caution must be observed when using the above equation outside the parameter range indicated in the above equation. However, the range of the surf similarity parameter is sufficiently broad to cover, for the most part, most practical situations, namely spilling and plunging breaking conditions not only in the laboratory but also in the field. We also note that this equation is valid inside the surf zone (see Fig. 7.18), and the data were obtained from rigid-bed experiments with smooth, plane sloping boundaries (the Sumer et al., 2013, study and the study of Deigaard et al., 1991b). However, clearly, the shape of the morphology is also important for the maximum value of the friction velocity, and therefore this aspect of the problem also needs to be borne in mind when using Eq. 7.62 in practice.

7.3.4 *Mechanism of sediment suspension by plunging waves*

As mentioned previously, Sumer et al. (2013), in their study of flow and sediment transport induced by plunging waves, exploited a dual-use strategy, utilizing both rigid (smooth) and live sediment beds in the experiments, which enabled detailed velocity and (hot film) bed shear stress measurements taken on a rigid bed to be directly related to sediment-bed measurements and subsequent morphological evolution observed on live sediment bed, under the same initial bed profile and wave conditions. Sumer et al.'s (2013) rigid-bed results are highlighted in the previous sections. The next subsections will discuss Sumer et al.'s (2013) sediment-bed experiments. The section will end with a detailed discussion on mechanism of sediment suspension by plunging waves. The bed morphology was also a major focus area of Sumer et al.'s (2013) study. Sumer et al.'s (2013) results regarding the bed morphology will not be covered here, as it is outside the subject matter of this book. The interested reader is referred to the original publication, Sumer et al. (2013, Section 5.3).

7.3.4.1 *Sediment-bed experiments in Sumer et al. (2013) study*

The sediment-bed experiments in Sumer et al. (2013) were carried out in the same flume. The beach (with exactly the same slope as in the rigid-bed experiments, namely 1:14) was formed with quartz sand, the grain size being $d_{50} = 0.18$ mm. The sand bed thickness was practically nil at the toe while it was 65 cm at the end of the beach, 9.10 m from the toe. The wave conditions were exactly the same as in the rigid-bed experiments, as already mentioned. Sumer et al. (2013) discussed that the bed shear stress in the case of the sediment bed can, to a first approximation, be approximated by that measured in the rigid-bed experiments, considering the fact that the sediment bed, for the most part, acted in their experiments as a hydraulically smooth boundary, or as a transitional boundary at most.

Three kinds of measurements were made in the sediment-bed experiments: pore-water pressure measurements (synchronized with the video recording from the side), water-surface elevation measurements, and bed profile measurements. Specifics of Sumer et al.'s sediment-bed measurements are described in greater detail in the original publication (Sumer et al., 2013). We will focus herein on the mechanism of sediment suspension induced by plunging breaking waves the hydrodynamics of which we have discussed in the preceding paragraphs in great detail.

7.3.4.2 *Pore-water pressure gradient measurements*

The pore-water pressure measurements in Sumer et al. (2013) were made at eight sections, not exactly the same sections in the rigid-bed experiments, but slightly shifted (8 cm) offshore. (We note that this was unintentional. See the details in Sumer et al., 2013). The pore-water pressure was measured at seven depths, $y = -0.3, -2.85, -5.0, -7.25, -9.8, -12.9$ and -16.4 cm, at all eight sections (Sections 1–8), in the middle of the flume in which y is the distance from the bed surface (y axis directed upwards).

Sumer et al. (2013) present data related to the measured pore-water pressure. Of the latter, the pore-water pressure measurements at the "surface layer" prove to be most interesting. The difference between the two uppermost measurement points, in terms of pressure gradient $-\partial p/\partial y$, is actually a vertical force on the surface sediment. It can be readily seen that this force is upward directed if $-\partial p/\partial y > 0$, and the opposite if $-\partial p/\partial y < 0$.

7.3.4.3 *Impulsive lift force on sediment during the passage of plunger-induced vortex*

Fig. 7.22 displays the time variation of the mean pressure gradient at the surface layer of the bed at the measurement Section 5. This section is very close to the breaking section. The pressure gradient in Fig. 7.22 is plotted in the non-dimensional form $-\partial(p/\gamma)/\partial y$ for convenience where γ is the specific weight of water. As stressed above, $-\partial(p/\gamma)/\partial y > 0$ represent a "lift" force on the surface sediment, directed upward.

The figure shows that the pressure gradient experiences a spike, marked M, during the rundown stage, Sumer et al. (2013). This spike coincides precisely with the passage over the pressure rack at Section 5 of Vortex M, Fig. 7.23. This corresponds to an impulsive, upward-directed pressure gradient force, with a magnitude of $(\overline{\frac{-\partial(p/\gamma)}{\partial y}})_{\max} \simeq 1.4$ (Fig. 7.22), which is 1.7 times larger than the submerged weight of the sediment $(s-1)(1-n) = O(0.9)$ in which s is the specific gravity of sediment grains, and n the porosity of the bed sediment. This causes an impulsive "suction" of the bed

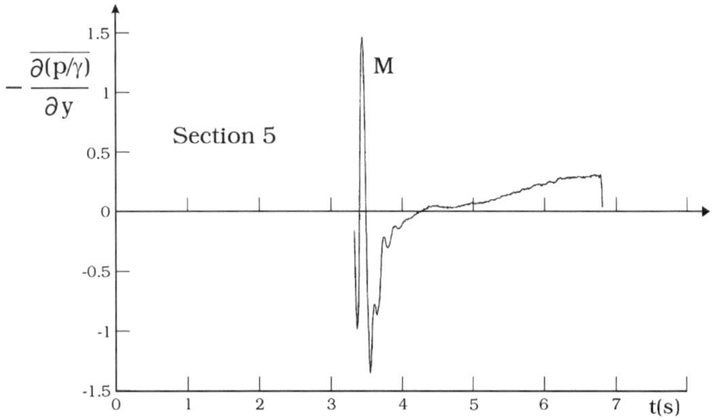

Fig. 7.22 Time variation of the mean pressure gradient in the soil at the surface layer of the bed. Positive pressure gradient corresponds to an upward-directed pressure-gradient force on the sediment. Section 5 is very close to the wave-breaking section. Sumer et al. (2013).

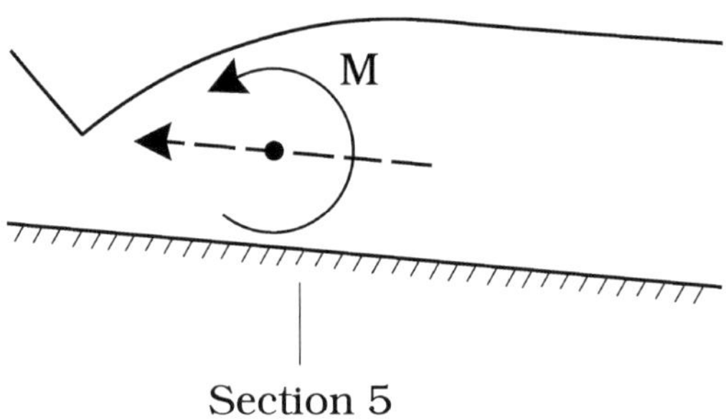

Fig. 7.23 Vortex M passing over Section 5, generated by the combined effect of the wave curling and the rundown.

sediment into the main body of the water, as revealed by the synchronized flow visualization, as will be discussed shortly. We note that Baldock et al. (2001) presented field measurements of pressure gradients in the surface layers of a sand beach. They found very large upward-directed pressure gradients ($O(0.5-1)$ and larger) frequently observed in the top 1.5 cm of the beach material. A striking similarity between the Sumer et al.'s (2013) laboratory result (Fig. 7.22) and Baldock et al.'s (2001) field results is to be noted.

The above-mentioned upward-directed pressure gradient in the sediment bed is attributed to the generation of a pressure minimum as a result of the anticlockwise vortex moving in the onshore direction (Fig. 7.23). This is because the flow velocity just above the bed at the bottom of the vortex is rather small (since the translational motion and the rotational motion of the vortex at this location are opposite to each other), and the opposite effect at the top of the vortex. This will result in a pressure minimum in excess of the static situation, pressures being larger at the bottom of the vortex, and the opposite effect at the top.

7.3.4.4 *Mechanism of sediment suspension*

Fig. 7.24 displays a sequence of video frames with accompanying sketches, illustrating the suspension of sediment from the bed. The video is from Test 2 of Sumer et al.'s (2013) experiments where the pore-water pressure was measured, synchronized with the video recording. Fig. 7.22 clearly shows that the suspension of the bed sediment is caused by Vortex M, generated by the wave breaking (Fig. 7.10b–d). This is elaborated further in the following paragraphs.

The value of the Shields parameter corresponding to this stage (Vortex M) is given in Sumer et al. (2013) as follows

$$\theta = \frac{|\tau_{0\,\mathrm{min}}/\rho|}{g(s-1)d_{50}} = 0.9 \tag{7.63}$$

in which $\tau_{0\,\mathrm{min}}$ (see the inset in Fig. 7.18 for the definition) is the bed shear stress corresponding to the dip in the time variation of the bed shear stress depicted in Fig. 7.15d at the instant where Vortex M is passing overhead (Fig. 7.10c). The latter dip is seen even more strongly in the time series displayed in Fig. 7.17, as mentioned previously. Note that the flow at this instant is directed offshore. Likewise, the value of the upward-directed pressure gradient at the bed corresponding to the same instant is

$$\left(-\frac{\partial(p/\gamma)}{\partial y}\right)_{\mathrm{max}} = 1.4 \tag{7.64}$$

as discussed in the previous subsection. (Note that, in the above prediction, the value of $\tau_{0\,\mathrm{min}}/\rho$ is picked up from Fig. 7.15d, while the value of the upward-directed pressure gradient, $\left(-\frac{\partial(p/\gamma)}{\partial y}\right)_{\mathrm{max}}$, is picked up from Fig. 7.22.)

The Shields parameter corresponding to the passage of Vortex M, $\theta = 0.9$ (Eq. 7.63), is larger than the threshold for the sheet-flow sediment transport, $\theta = 0.5$–0.6 (Sumer et al., 1996). On the other hand, the

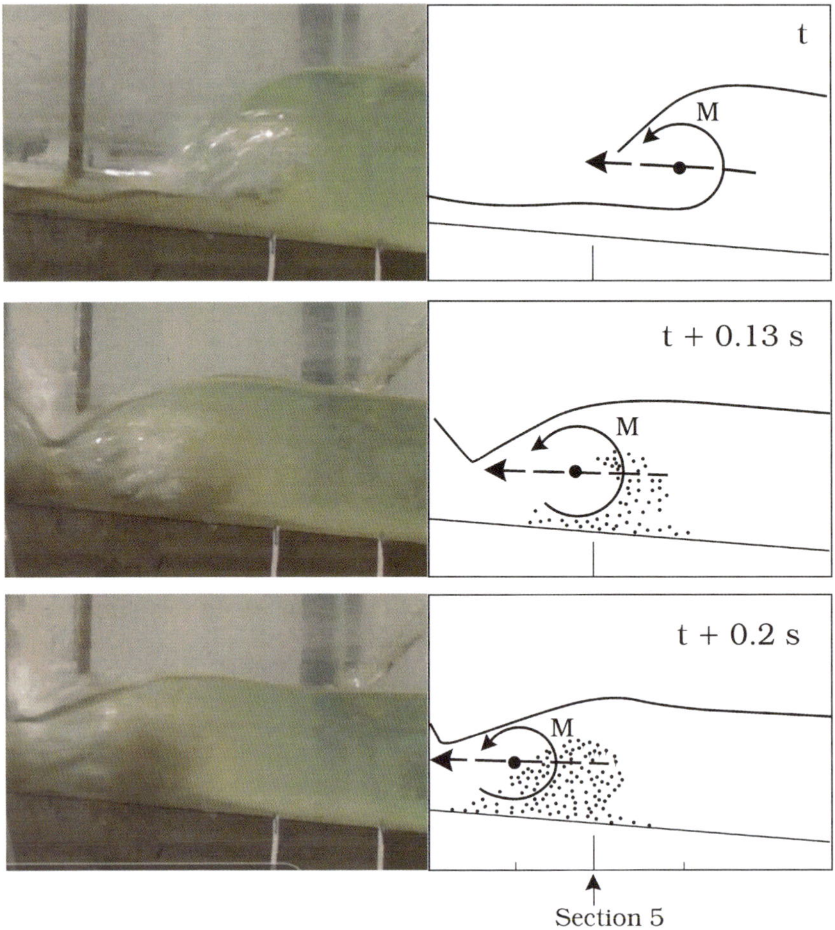

Fig. 7.24 Sequence of pictures illustrating the suspension of sediment from the bed. Vortex M creates a pressure minimum as described in the text. Sumer et al. (2013).

fall-velocity-to-friction-velocity ratio for the sediment used in the test is $w_s/(|\tau_{0\,\mathrm{min}}/\rho|)^{1/2} \simeq 2.2/27^{1/2} \simeq 0.4$, and this value is smaller than unity, and therefore the sheet-flow sediment transport occurs in the suspension mode (Sumer et al., 1996).

On the other hand, the upward-directed pressure gradient at the bed corresponding to the passage of Vortex M is 1.4 (Eq. 7.64), and as pointed out earlier, this value is significantly larger than the submerged weight of the sediment $(s-1)(1-n) = O(0.9)$.

Now, the above analysis implies the following. With the formation of Vortex M and its induced sheet-flow, suspension-mode sediment transport, the bed sediment will be swept in the sheet-flow regime in the offshore direction as Vortex M passes overhead. At the same time, the bed sediment will undergo a very large upward-directed pressure gradient force through the passage over the bed of Vortex M. As a result of the combined action of these two effects, the sediment driven offshore will be "sucked" vigorously into the main body of the water, forming a sediment "plume", in the way illustrated in Fig. 7.24.

Sumer et al. (2013) note that this analysis is based on the mean values. Clearly, both effects are subject to considerable amount of turbulence (see Fig. 7.19 above, and also Fig. 19 in Sumer et al., 2013, for the turbulence in the pore-water pressure gradient), and therefore the agitating forces (the bed shear stress and the upward-directed pressure gradient, corresponding to the passage of Vortex M) will be even more pronounced, resulting in more dramatic occurrence of suspension of the bed sediment at times.

Formation of such sediment plumes has been captured spectacularly by Clark Little on a natural beach (`www.clarklittlephotography.com`), reproduced here in Fig. 7.25.

It may be noted that the key role of Vortex M in the formation of sediment plumes was also noted by Matsunaga and Honji (1980), based on their flow visualization study (see the discussion in Matsunaga and Honji in conjunction with their Fig. 4). Considering the fact that the experimental conditions of Sumer et al.'s (2013) study and the study of Matsugana and Honji are similar, with the shallow, offshore flow interacting strongly with the breaking wave, similarities between the results of these two studies are indeed not unexpected.

Fig. 7.26 displays a snapshot from Test 3 of Sumer et al. (2013) with its accompanying sketch. The picture corresponds to the stage of the breaking process illustrated in Fig. 7.10g. The locations of Sections C–G are also marked in the figure for easy reference. Fig. 7.26 clearly shows that there are three distinct sediment plumes, Plumes m, n and p, in which sediment is put into suspension from the bed in a sequential order; Plume m is caused by Vortex M, as described in the preceding paragraphs. This is followed by the sediment suspension, resulting in Plume n, and subsequently, that resulting in Plume p, generated by Vortices N and P (Fig. 7.10 e–g), respectively.

Fig. 7.25 Suspension of sediment from the bed. Formation of sediment plume. Reproduced, with Clark Little Photography License, from: www.clarklittlephotography.com.

7.4 Flow induced by surging waves

We will, in this section, describe the flow induced by surging breaking waves with reference to the work of Jensen et al. (2014). The purpose of Jensen et al.'s study was to investigate the destabilizing effect of two hydrodynamic quantities on the armour layer installed on the slope of a rubble-mound breakwater, these two hydrodynamic quantities being the lift force induced by pore-pressure gradients, and the wall shear stress. To this end, Jensen et al. (2014) carried out experiments in the laboratory. The following section will give a brief description of these experiments.

7.4.1 *Experiments of Jensen et al. (2014)*

Fig. 7.27 illustrates the experimental setup in Jensen et al.'s (2014) study. The wave flume was 25 m in length, 0.6 m in width, and 0.8 m in depth. The water depth at the toe was maintained at $h = 0.4$ m. The front and rear slope of the breakwater was arranged with a slope of 1:1.5. The core material of the breakwater model was made of gravel the size $d_{50} = 1.8$

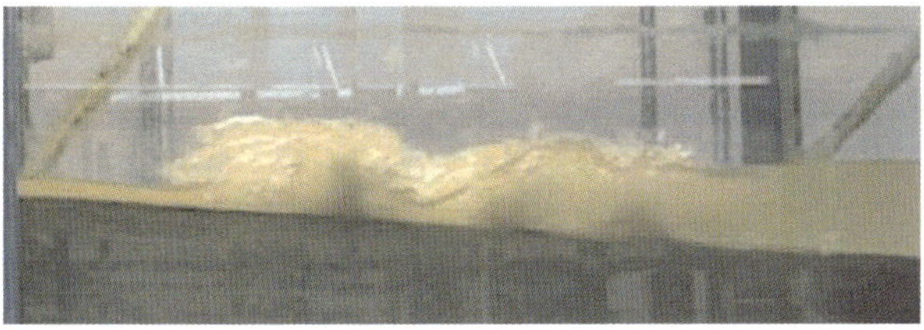

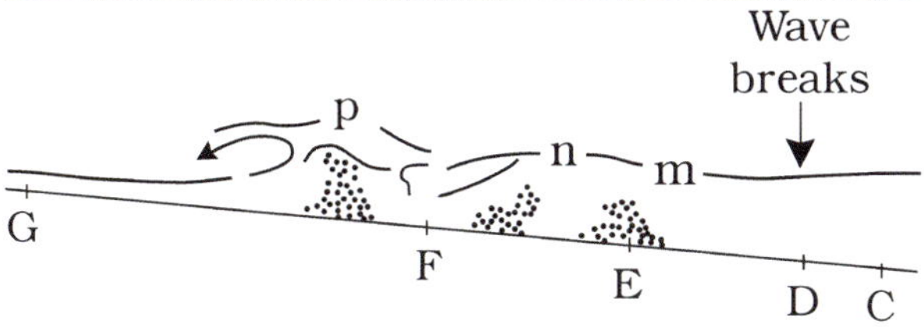

Fig. 7.26 Suspension of sediment from the bed. The time corresponds to that in Fig. 7.10g. Sediment Plumes m, n and p are caused by Vortices M, N and P, respectively. Sumer et al. (2013).

cm. In order to maintain a stable and reproducible structure, a perforated steel plate was attached on the front slope of the breakwater on top of the core material. The plate had a thickness of 2 mm. The perforations were made as quadratic voids with dimension of 1 cm and a void-to-plate ratio (porosity) of 0.41. This corresponded approximately to the porosity of the underlying core material of $n = 0.4$. The armour layer was made out of spherical plastic elements with a diameter of $d = 3.8$ cm in one series of the experiments, and crushed stones with a size of $d_{50} = 5.0$ cm in another. The spherical plastic elements were glued to the perforated steel plate in a 90° arrangement. The crushed stones were also glued to the perforated steel plate.

Three kinds of measurements were performed in Jensen et al.'s (2014) study: pressure measurements, velocity measurements, and surface elevation measurements, similar to Sumer et al. (2013). Profiles of the two above-mentioned hydrodynamic quantities (pressure and velocity) were measured normal to the breakwater slope at 12 sections equally spaced with 5 cm. For all 12 sections, the pore-water pressure was measured at six depths in the core material, $y = [0; -2.0; -3.0; -5.0; -7.0; -10.0]$ cm in which y is the distance from the core-material surface (under the armour stones) with the y-axis directed outward. Detailed description of the measurements is given in the original publication (Jensen et al., 2014).

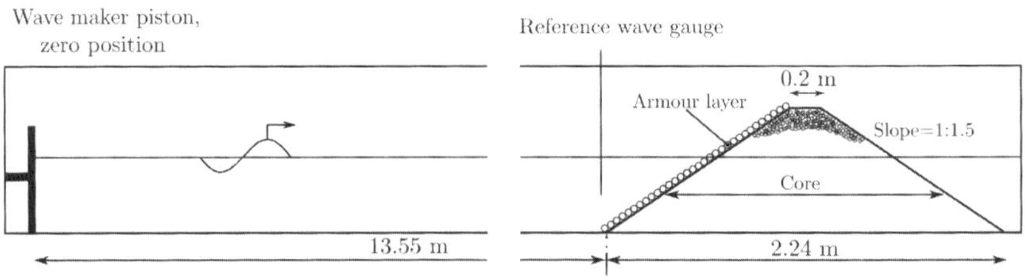

Fig. 7.27 Experimental setup of Jensen et al. (2014) for plunging and surging regular waves.

Table 7.1 Breaking waves in Jensen et al.'s (2014) experiments.

H_I (cm)	T (s)	ξ_0	Breaker type
11.7	1.0	1.76	Plunging
11.6	1.5	2.64	Plunging
11.4	1.7	3.05	Plunging
11.9	3.0	5.79	Surging

Jensen et al. (2014) experimented with four kinds of breaking waves over both types of the armour layer (plastic-spherical elements, and the crushed stones). Of the four types of breaking waves, the first three types were plunging-breaker waves while the fourth one was surging-breaker waves, Table 7.1.

The following section will describe (1) the flow induced by the plunging wave with $T = 1$ s (first row in Table 7.1), and (2) that induced by the surging wave (last row in Table 7.1) of Jensen et al. (2014). We include the plunging wave case to relate Jensen et al.'s study to the previous section, and more importantly to describe the flow induced by the surging wave of Jensen et al. with reference to that induced by their plunging waves. The quantity H_I in Table 7.1 is the incident wave height, T the period, and ξ_0 the surf similarity parameter (Eq. 7.1).

7.4.2 *Description of plunging and surging breaking waves in Jensen et al. (2014) study*

7.4.2.1 *Plunging breaking wave*

Fig. 7.28 presents the runup and rundown stages of the plunging breaker with a wave period of $T = 1.0$ s and $\xi_0 = 1.76$. The measuring sections for pore-water pressure are indicated in the figure with measuring Section

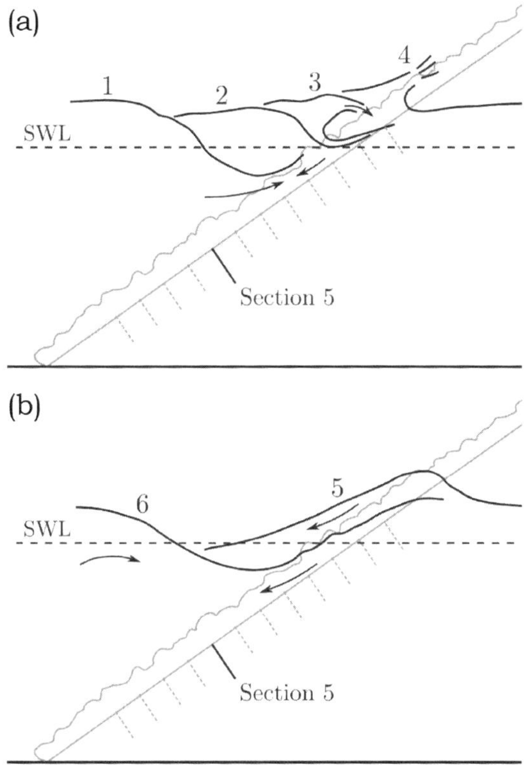

Fig. 7.28 Plunging wave of Jensen et al. (2014) study. The times shown are: 1: $t = 0.2$ s; 2: $t = 0.4$ s; 3: $t = 0.5$ s; 4: $t = 0.7$ s; 5: $t = 0.8$ s; and 6: $t = 1.1$ s, where $t = 0$ s corresponds to the zero-up crossing of the water-surface elevation at the toe of the breakwater slope.

5 highlighted for reference in the original publication.

The sequence of the wave breaking in Fig. 7.28 is, for the most part, similar to that of Sumer et al. (2013), Fig. 7.9, although there are a number of differences in the two experimental setups. Two major differences are that, in Sumer et al. (2013), the bed acted as a hydraulically smooth wall, while, in Jensen et al. (2014), it was rough; and, likewise, in Sumer et al. (2013), the slope of the beach was very mild (1:14), while, in Jensen et al. (2014), it was very steep (1:1.5). Nevertheless, the flow induced by the wave breaking in the two cases is, not surprisingly, quite similar, and this is simply because the breaking regime in both experiments was the same, namely the plunging breaking with the value of the surf similarity

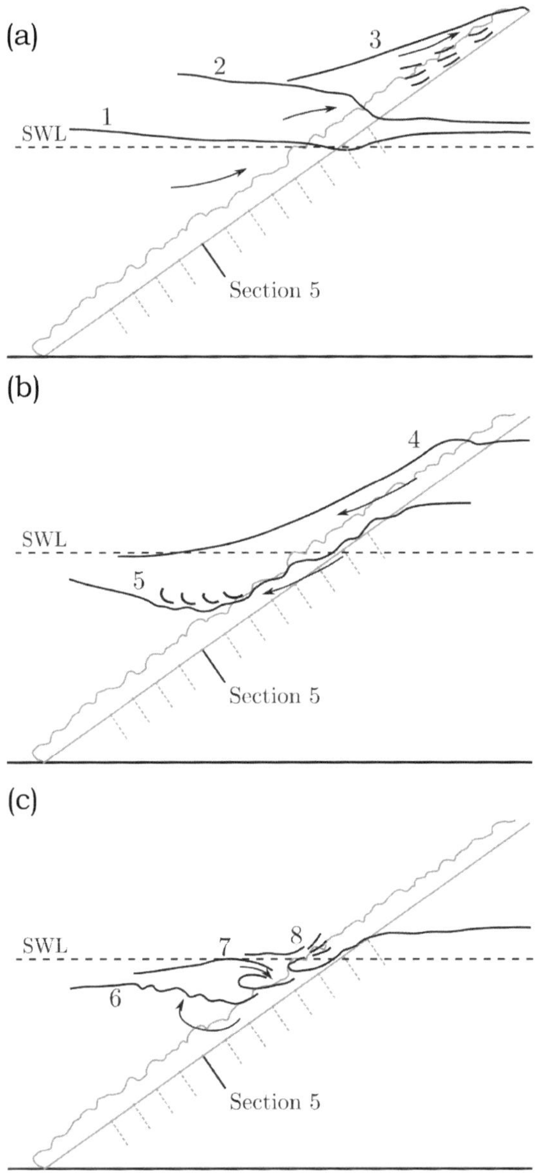

Fig. 7.29 Surging wave of Jensen et al. (2014) study. The times shown are: 1: $t = 0.1$ s; 2: $t = 0.5$ s; 3: $t = 1.0$ s; 4: $t = 1.2$s; 5: $t = 1.8$ s; 6: $t = 1.9$ s; 7: $t = 2.2$ s; and 8: $t = 2.4$ s, where $t = 0$ s corresponds to the zero-up crossing of the water-surface elevation at the toe of the breakwater slope.

parameter being $\xi_0 = 1.42$ in Sumer et al. (2013) and $\xi_0 = 1.76$ in Jensen et al. (2014), both in the range $0.5 < \xi_0 < 3.3$ (Eq. 7.7). (The detailed description of the runup and rundown processes in Jensen et al. is given in the original publication.)

We note, however, that no vortex formation (Vortex M in Fig. 7.10) has been observed in Jensen et al.'s (2014) plunging wave experiments, revealed by the video recording of Jensen et al. (2014). This is in contrast to Sumer et al. (2014), Fig. 7.10. This no-vortex regime may be attributed to the complete absence (or very weak presence if any) of a shear layer in Jensen et al.'s (2014) breaking process. This may be because the wave in Jensen et al.'s (2014) setting is not a progressive wave, but rather a standing wave (due to strong reflection at the beach), and therefore no apparent onshore-directed velocity ($\sim c$) at the time of wave breaking exists in the Jensen et al. setting, an essential component of the shear-layer instability to generate vortices Vortices M, N and P (see Section 7.3.2). So, there exists no shear-layer instability, and therefore no vortex formation. Clearly, this may have implications for sediment transport.

7.4.2.2 *Surging breaking wave*

Fig. 7.29 presents the runup and rundown stages for the surging breaker with a wave period of $T = 3.0$ s and $\xi_0 = 5.79$.

Comparison of Figs. 7.28 and 7.29 indicate that the flow induced by the surging breaker is vastly different from that of the plunging breaker. The runup phase was initiated with the water surface being close to the still water level (stage 1). In the runup, the water surface does not approach the breakwater front slope with a steep angle as it has been seen for the plunging wave. Without any "severe" wave breaking, the runup is allowed to move along the breakwater slope to its maximum runup level (stage 2 and 3).

Fig. 7.29b presents the rundown stage where a large downward directed velocity is reached at the armour layer (stage 4). The rundown water impacts the main body of water which, at this instant, has not been affected by the successive incoming wave (stage 5). At the maximum rundown level, a stirring zone occurs with some air entrainment. Immediately after the maximum rundown level is reached, a secondary runup begins to emerge, as shown in Fig. 7.29c (stage 6). This forms a breaking bore or hydraulic jump which was also described in Pedersen and Gjevik (1983) for surging waves (stage 7). The secondary runup creates some air entrainment during the breaking process (stage 7 and 8). Further details of the surging breaking experienced in Jensen et al. (2014) are given in the original publication.

7.4.2.3 *Wall-normal pore-pressure gradient*

Recall that the wall-normal pore-pressure gradient at the surface of the sediment slope represents a lift force on the surface sediment (Section 7.3.4). Jensen et al. (2014) obtained this quantity from their pore pressure measurements, and plotted its variation along the breakwater slope. The variations for Jensen et al.'s both cases (the plunging breaking wave case, Fig. 7.28, and the surging breaking wave case, Fig. 7.29) appear to be similar to that of Sumer et al.'s (2013) plunging wave case. However, the maximum wall-normal pore-pressure gradient in Jensen et al.'s cases both occurred practically at the maximum rundown levels while it occurred in Sumer et al.'s (2013) case at the location where Vortex M (Fig. 7.10b–d) passed overhead.

Another difference between Jensen et al. (2014) and Sumer et al. (2013) in this context is that the maximum value of the pressure gradient experienced in Jensen et al.'s cases, $\left(-\frac{\partial(p/\gamma)}{\partial y}\right)_{\max}$, appears to be a factor of 2 smaller.

7.5 Flow induced by plunging solitary waves

We have discussed various surf and swash zone processes in the previous sections, Sections 7.1–7.4. A canonical problem that has often been studied in the literature, having relevance to these processes, is the runup of breaking solitary waves on a sloping beach. (We note that while wave flume/basin studies involving solitary waves have often traditionally been motivated by the study of tsunamis, Madsen et al. (2008) have relatively recently pointed out that, when scaled up, such experiments are in fact more likely relevant to wind wave scales. See the relevant discussion in Section 5.11.) In terms of specific boundary layer quantities, we add that non-dimensional time scales relevant to tsunamis may in fact be reached in oscillating tunnel facilities, e.g. as discussed in Fuhrman et al., 2009, p. 15, if the tunnel half-depth is taken as analogous to the water depth.)

The hydrodynamics within the evolution, runup, and rundown of solitary waves on sloping beds have been quite extensively studied in previous laboratory experiments (e.g. Synolakis, 1987; Synolakis and Skjelbreia,

1993; Lin et al., 1999; Li and Raichlen, 2001, 2002; Hsiao et al., 2008; Higuera et al., 2018). While many of these have focused largely on the runup and rundown processes, a number of relatively recent papers have presented more detailed experimental measurements of additional features. For example, the specific role of the jet in the runup of a plunging solitary wave has been investigated by Li and Raichlen (2003). Velocity kinematics have been measured by Jensen et al. (2003) and Heller et al. (2005), while turbulence properties under broken solitary waves have been measured by Ting (2006, 2008), extending the author's previous work published in a series of papers on turbulence under broken ordinary oscillatory waves (see e.g. Ting and Kirby, 1995), as well as by Baldock et al. (2009). Chang et al. (2009) also present detailed measurements of swash flow velocities and pressures during the runup process. Additionally, Seelam and Baldock (2009) have measured the phase-resolved total bed stress beneath non-breaking solitary waves as well as breaking solitary bores on slopes using a shear cell, although with this method turbulent fluctuations are not resolved. Direct (shear cell) bed shear stress measurements have also been measured in bore-driven swash by Barnes et al. (2009). Also meriting mention, Lara et al. (2011a) investigated (numerically) the evolution of breaking solitary waves over a porous underwater step.

Studies involving sediment transport mechanisms within the runup of solitary waves are comparatively rare, and appear to be limited to the work of Tonkin et al. (2003), who considered scour from solitary waves around a circular cylinder; Kobayashi and Lawrence (2004), who investigated the cross-shore sediment transport and beach morphology under breaking (both positive and negative) solitary waves; and Young et al. (2010), who investigated the erosion and deposition patterns under breaking positive solitary waves, with additional modeling performed by Xiao et al. (2010) as well as recently by Li et al. (2019). In related work, Alsina et al. (2009) have presented measurements involving the advection of suspended sediment in the swash zone stemming from solitary broken bores.

The previously mentioned laboratory studies focusing on surf and/or swash zone processes were conducted for the most part on either rigid or live sediment beds. An exception is the recent studies of Sumer et al. (2011 and 2013), who exploited a dual-use strategy, utilizing both rigid (smooth) and live sediment beds. As already mentioned in Section 7.3, this approach enabled detailed velocity and (hot film) bed shear stress measurements taken on rigid bed profiles to be directly related to near-bed pore-pressure measurements and subsequent morphological evolution observed on live sediment beds, under the same initial bed profile and wave conditions. The focus of Sumer et al.'s (2013) study was flow and sediment transport induced by breaking plunging regular waves, which has been discussed quite extensively in the preceding paragraphs, Section 7.3, while that of Sumer et al. (2011) was flow and sediment transport induced by plunging solitary waves. The latter study will be discussed in greater detail in the following paragraphs.

The motivation of Sumer et al.'s (2011) work was to add to the previously mentioned studies, by providing an improved understanding of the entire sequence of wave breaking within the context of solitary waves (shoaling, wave breaking, runup, rundown, hydraulic jump, and trailing waves), and especially on the resulting bed shear stresses, sediment transport patterns, and related bed morphology.

For the above purpose, results of two parallel experiments involving the evolution and runup of plunging solitary waves on a sloping bed have been presented in the original publication, which are regarded as a simple, idealized means of studying dynamics within isolated breaking wave events (although two processes in ordinary regular waves are missing in the aforementioned idealized case, namely the process controlling the wave setup, and that controlling the water table in the porous seabed of the beach).

The first of the two parallel experiments in Sumer et al. (2011) has utilized a rigid bed, allowing e.g. direct (hot film) measurements of bed shear stresses, including their turbulent fluctuations, in exactly the same fashion as in Sumer et al. (2013), already discussed in Section 7.3. The second has utilized an erodible bed consisting of fine sand, allowing for the measurement of pore-water pressures, pressure gradient forces, as well as for observation of the resulting morphological changes. The two experimental conditions have been otherwise maintained as similar as possible, both in terms of the generated waves and initial bottom configuration, making observations from both sets of experiments easily relatable.

7.5.1 *Experimental setup in Sumer et al. (2011) study*

The wave flume used for the rigid-bed experiments in Sumer et al. (2011) was the same as that described in Sumer et al. (2013), discussed in Section 7.3.1 earlier. Fig. 7.30 displays the longitudinal section of the wave flume, and also the measurement sections, reproduced from Sumer et al. (2011). The bed shear stress was also measured in Sumer et al. (2011), using the hot film technique discussed in Section 7.3.1 previously in conjunction with

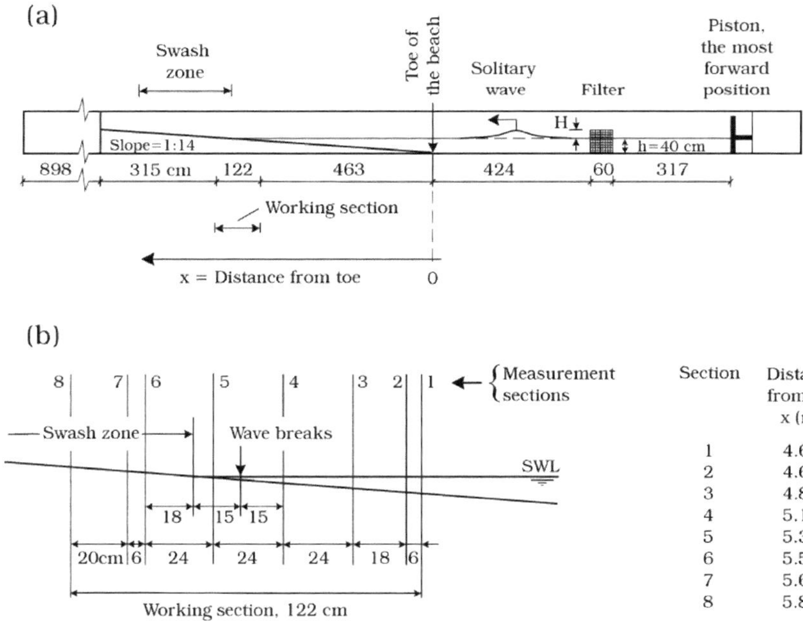

Section	Distance from toe x (m)
1	4.63
2	4.69
3	4.87
4	5.11
5	5.35
6	5.59
7	5.65
8	5.85

Fig. 7.30 Experimental setup of Sumer et al.'s (2011) rigid-bed experiments for plunging solitary waves.

Sumer et al.'s (2013) study of flow induced by plunging regular waves.

The solitary wave traveling onshore is illustrated in Fig. 7.30. Its characteristics, namely the wave height, H, and the water depth, h, are indicated in the figure. (See also the definition sketch of a solitary wave given earlier in Fig. 5.64 in conjunction with solitary wave boundary layers.) The offshore water depth in Sumer et al.'s (2011) test was $h = 40$ cm, and the height of the solitary wave (measured from the still water level, SWL, Fig. 7.30) was $H = 7.1$ cm, with the offshore surface elevation given by the small amplitude solitary wave theory,

$$\eta = H\text{sech}^2(\omega t) \tag{7.65}$$

in which ω is

$$\omega = \sqrt{\frac{3}{4}gH}\frac{1}{h} \tag{7.66}$$

where t is time and g the acceleration due to gravity. Similar to sinusoidal waves, a time scale can be defined by

$$T = \frac{2\pi}{\omega} = 2\pi\sqrt{\frac{4}{3gH}}h \tag{7.67}$$

which can be interpreted as the time scale characterizing the width of the surface-elevation time series, sometimes also referred to as an "effective period". This quantity in Sumer et al.'s (2011) experiments was $T = 3.48$ s.

The waves in the experiments were reproducible. Sumer et al. (2011) note that the surface elevation time series of twenty individual waves at the toe of the slope, when plotted on top of each other, collapsed on a single curve, revealing the reproducibility of the waves.

7.5.2 *Description of plunging solitary wave in Sumer et al. (2011) study*

7.5.2.1 *Breaking-wave and runup characteristics*

The maximum runup elevation from SWL in Sumer et al.'s (2011) plunging solitary wave experiment was measured to be $R = 18$ cm. This compares well with a value $R = 18.9$ cm predicted by the empirical expression (for breaking solitary wave runup) of Fuhrman and Madsen (2008):

$$\frac{R}{H} = 3.9\xi_s^{0.42} \tag{7.68}$$

with ξ_s

$$\xi_s = \frac{\tan(\beta)}{H/h} \tag{7.69}$$

in which $\tan(\beta)$ is the beach slope and ξ_s is the so-called solitary wave surf similarity parameter, which was reduced to the elementary form above based on the concept originally presented by Kobayashi and Karjadi (1994).

It is stressed that the parameter ξ_s is analogous to the parameter ξ for regular waves (Eq. 7.58), which is commonly used to parameterize their runup. It would not, therefore, necessarily be expected to predict wave breaker types, which are typically characterized based on a surf similarity parameter defined in terms of the deep-water wave height H_0 (ξ_0, see Eq. 7.1). Alternatively, breaker types for solitary waves have been directly characterized by Grilli et al. (1997) using a so-called slope parameter:

$$S_0 = 1.521\frac{\tan(\beta)}{\sqrt{H/h}} \tag{7.70}$$

Sumer et al. (2011) report that their S_0 value is 0.26, which is within Grilli et al.'s suggested plunging breaker regime ($0.025 < S_0 < 0.30$). Incidentally, the solitary wave breaking types are characterized in Grilli et al. (1997) as

$$S_0 < 0.025: \quad \text{Spilling} \tag{7.71}$$

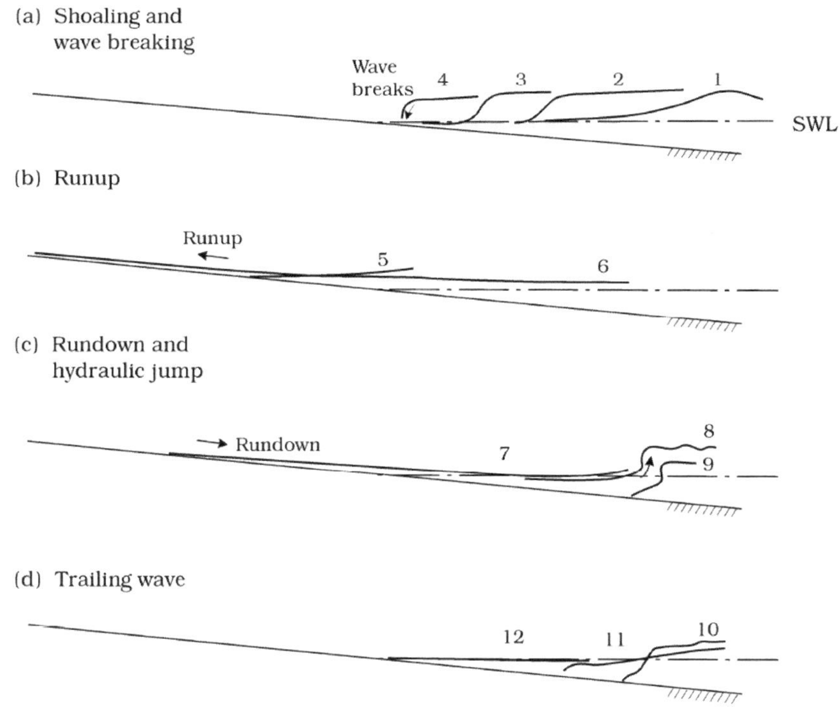

Fig. 7.31 Sequence of solitary wave breaking. Small numbers indicate different stages of the process. Sumer et al. (2011).

$$0.025 < S_0 < 0.30 : \quad \text{Plunging} \tag{7.72}$$

$$0.30 < S_0 < 0.37 : \quad \text{Surging} \tag{7.73}$$

Now, from the above considerations, Sumer et al. (2011) conclude that their observations are reasonably in line with previous work, in terms of both the runup and the breaker type.

7.5.2.2 *General description*

Fig. 7.31 schematically illustrates the complete sequence of the plunging solitary wave observed in the Sumer et al. (2011) experiments: (a) Shoaling and wave breaking; (b) Runup; (c) Rundown and hydraulic jump; and (d) Trailing wave. (The numbers in the figure denote different stages of the breaking process.)

In the stage "shoaling and wave breaking", the plunging breaking occurs with the wave curling over forward and impinging onto the water surface onshore with a small amount of air trapped inside the "tube" formed by the wave crest, in the same manner as in the plunging regular waves, Section 7.3.

Sumer et al.'s (2011) solitary plunging-wave study conducted somewhat earlier than the authors' plunging-regular-wave study, reported in Sumer et al. (2013), Section 7.3. As mentioned in Section 7.3.2, the most prominent feature of the plunging regular waves is the generation of the vortex beneath the breaker, Vortex M (Fig. 7.10b), and this is followed by the generation of two more vortices, Vortex N and Vortex P (Fig. 7.10e–f), Section 7.3.2. This feature had been overlooked in Sumer et al.'s (2011) solitary-plunging-wave work. The videos of the latter study have been re-examined to observe if the aforementioned vortex system exists in plunging solitary waves, also noted in Sumer et al. (2013, Section 6). From the videos, it appears that a well-defined Vortex M (although noticeably weak) is generated. However, the generation of Vortex M is not followed by that of the smaller vortices, Vortex N and Vortex P, in contrast to the plunging regular waves. This is attributed to the absence of the rundown from the "previous cycle" of the solitary motion, resulting in a very weak shear-layer instability (if any), and therefore the absence of vortices Vortex N and Vortex P. See Section 7.3.2 for the discussion about the shear layer instability in the generation of the vortex system observed in plunging regular waves.

With respect to Vortex M, Sumer et al. (2013, p. 6179) commented that the strong, sudden dip in the trace of the bed shear stress (Fig. 7.17), and the strong, sharp peak in the trace of the pressure gradient (Fig. 7.22), both caused by Vortex M, are not observed in the case of the solitary wave because of the non-existence of the offshore-directed flow (rundown) in the solitary-wave case. We note, however, that this may be because the measurement sections were not placed precisely at or very near the wave-breaking section in Sumer et al.'s (2011) plunging-solitary-wave experiment, and therefore these effects were not captured in the measurements. It appears that this aspect is worth exploring further.

Finally, we note that, in the "rundown and hydraulic jump" stage, the hydraulic jump involves (1) initially almost a vertical face, which eventually curls over toward the shore, impinging on the fast incoming run-down flow, and (2) a system of vortices adjacent to the bed on the shoreward side of the hydraulic jump. This sequence of events has been captured by Higuera et al. (2018) in their experiment with their solitary wave over a slope of 1:3

with a value of the slope parameter (Eq. 7.70) of $S_0 = 0.84$; See Fig. 7.32, with (a): Hydraulic jump, (b): Curling of water surface over toward the shore, and (b) and (c): Formation of Vortices A, B and C. The solitary wave in Higuera et al.'s (2018) experiments was a non-breaking wave (also revealed by the value of $S_0 = 0.84$). Although, in Sumer et al.'s (2011) experiments, the aforementioned events, namely the the curling over toward the shore of the water surface, and the formation of the vortex system beneath the hydraulic jump, could not be resolved, it is expected, however, that the same sequence of events should also occur in the Sumer et al. (2011) study. This is because the hydrodynamics of the process leading to these events should remain unchanged, regardless of the incoming solitary wave (breaking, the Sumer et al. (2011) wave, or non-breaking, the Higuera et al. (2018) wave).

7.5.2.3 *Water-surface elevation*

Fig. 7.33 displays the time series of the surface elevation at different measurement sections, Sections 1, 3, 5 and 8, along with that at the toe of the beach (see Fig. 7.30 for the locations of the measurement sections). The time series for Sections 2, 4, 6 and 7 are not included in the figure for reasons of space. The origin of time, $t = 0$, is taken as the instant where the wave crest passes the toe section (Fig. 7.30a). In order to picture the various stages of the wave breaking represented in Fig. 7.33, and relate them to Fig. 7.31, we recall (see Fig. 7.30b) that Sections 1 and 3 are located at the offshore side, and Sections 5 and 8 at the onshore side of the breaking point.

The following observations can be made from Fig. 7.33:

1. The shoaling (Fig. 7.31a), which takes place until the wave breaks just before Section 5, is clearly observed from Figs. 7.33a–c.

2. The signal corresponding to Section 1 (Fig. 7.33b) begins to oscillate, starting with time $t \simeq 7$ s (J in Fig. 7.33b). Sumer et al. (2011) report that the synchronized video recordings reveal that these oscillations are due to the hydraulic jump following the rundown stage (Fig. 7.31c). The wave generated by the hydraulic jump travels offshore, and the crest of this wave arrives at the toe of the beach at time $t = 9.5$ s (marked K in Fig. 7.33a).

3. The runup and rundown stages of the process at the measurement sections are evident from the time series in Figs. 7.33b–e. Regarding the time series recorded at Section 8, as this section is in the intermittently wet-dry area, the signal is cut for small and large times, namely for $t < 3$

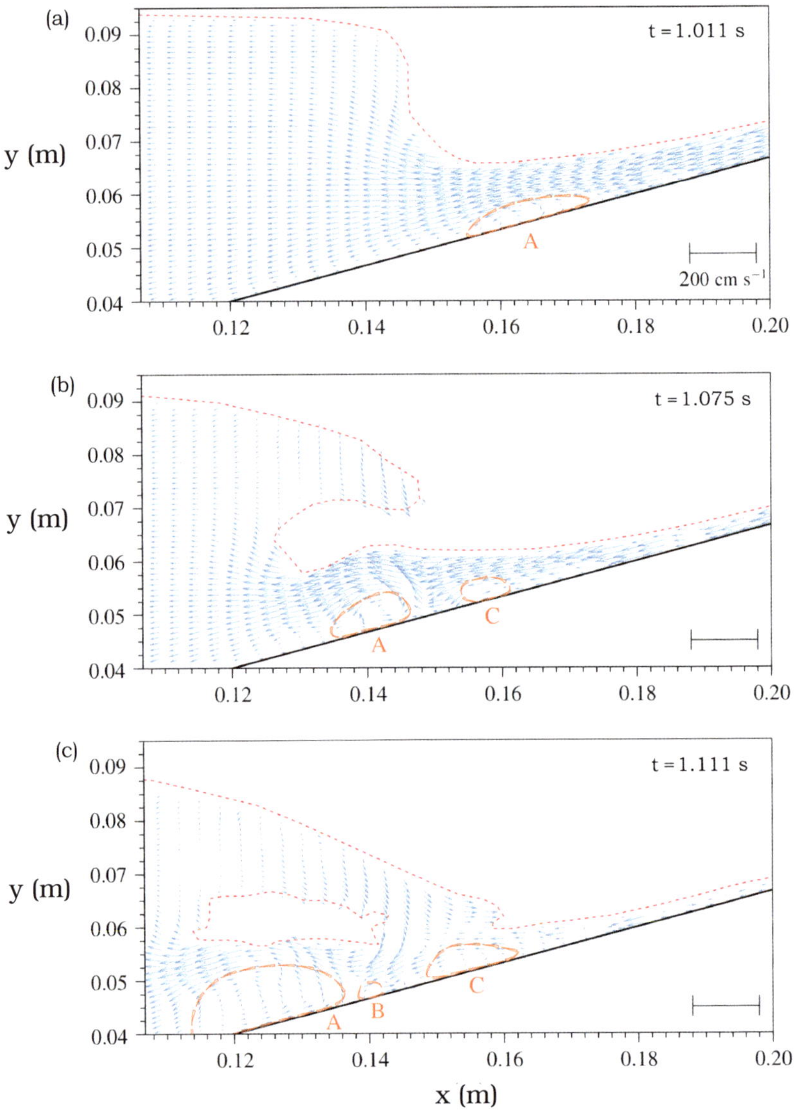

Fig. 7.32 Measured velocity fields during "Rundown and Hydraulic Jump" stage of the non-breaking solitary wave of Higuera et al. (2018). (a): Hydraulic jump. (b): Curling of water surface. (b) and (c): Formation of Vortices A, B and C.

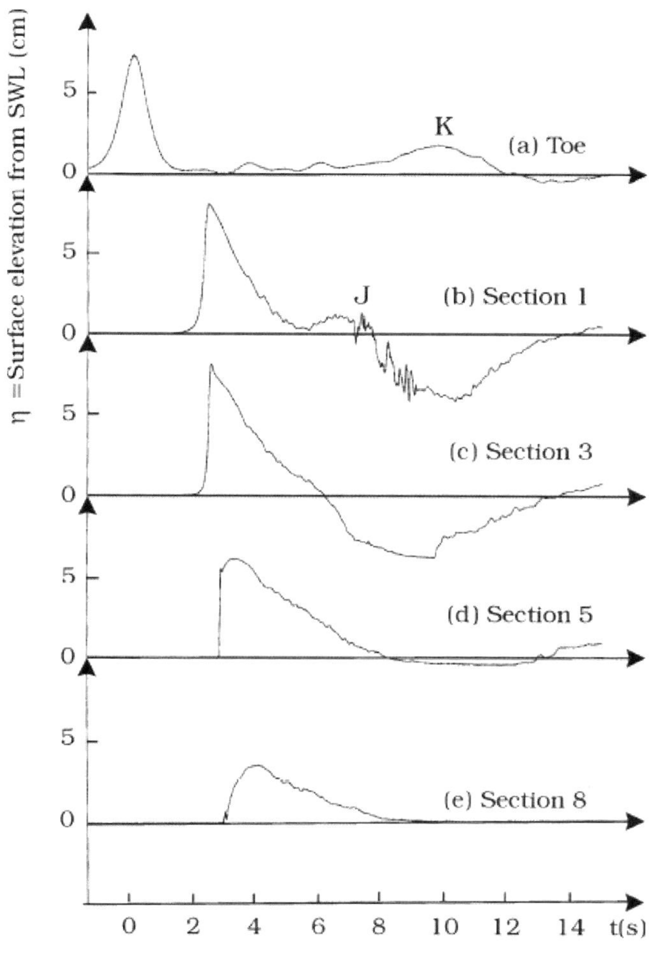

Fig. 7.33 Time series of surface elevation at different section. See Fig. 7.30 for the locations of different sections. Sumer et al. (2011).

s and for $t > 8$ s.

7.5.3 *Bed shear stress in Sumer et al. (2011) study*

Fig. 7.34 presents the time variation of the mean bed shear stress while Fig. 7.35 presents that of the r.m.s. value of the fluctuating component of the bed shear stress $\tau_0' = \tau_0 - \overline{\tau_0}$ at Sections 1, 2, 5, 6 and 8. Recall (see Fig.

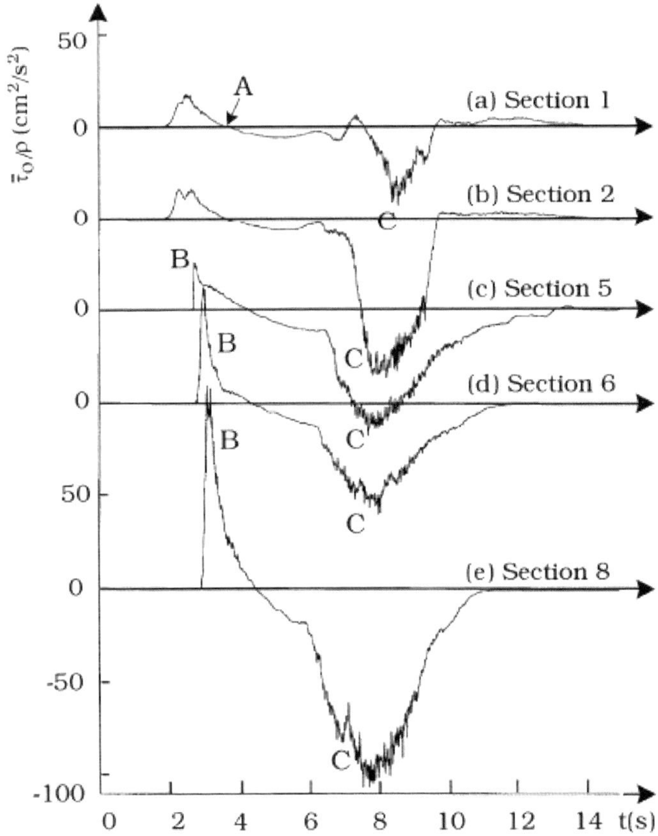

Fig. 7.34 Time series of bed shear stress at different section. See Fig. 7.30 for the locations of different sections. $t = 0$ corresponds to the instant the wave crest passes the toe section. Sumer et al. (2011).

7.30b) that Sections 1 and 2 are located at the offshore side, and Sections 5, 6 and 8 at the onshore side of the breaking point. Note that the bed shear stress data are plotted in Figs. 7.34 and 7.35 in terms of $\overline{\tau_0}/\rho$ and $(\overline{(\tau_0')^2})^{1/2}/\rho$ rather than $\overline{\tau_0}$ and $(\overline{(\tau_0')^2})^{1/2}$ for convenience, as in the case of plunging regular waves (Section 7.3.3).

The mean value of the bed shear stress is calculated through ensemble averaging according to

$$\overline{\tau_0}(t) = \frac{1}{N}\sum_{j=1}^{N}[\tau_0(t)]_j \tag{7.74}$$

in which $[\tau_0(t)]_j$ is the time series of τ_0 in run j, and N is the number of runs (i.e., the sample size) at that section. Likewise, $(\overline{(\tau_0')^2})^{1/2}$

$$(\overline{(\tau_0')^2})^{1/2}(t) = \left\{\frac{1}{N-1}\sum_{j=1}^{N}\{[\tau_0(t)]_j - \overline{\tau_0}(t)\}^2\right\}^{1/2} \tag{7.75}$$

It is important to emphasize that the way in which the "turbulence" is calculated according to Eq. 7.75 requires a very repeatable phase of the wave form. This was checked in Sumer et al. (2011) by plotting twenty, arbitrarily selected time series of the surface elevation (at the toe of the beach) on top of each other; it was observed that these time series collapsed on a single curve with r.m.s. differences being nil, as mentioned previously.

Now, the r.m.s. value of the fluctuating component of the bed shear stress exhibited in Fig. 7.35 calls for an early discussion about the generation of these fluctuations in the bed shear stress in the Sumer et al. (2011) experimental setting.

7.5.3.1 *Generation mechanisms of bed shear stress fluctuations, turbulence in bed shear stress*

We start off with the section far off on the onshore side, Section 8 (Fig. 7.35e). There are two peaks in the time variation of $(\overline{(\tau_0')^2})^{1/2}$. The first peak is associated with the runup stage while the second peak is associated with the rundown stage. For the runup stage, the bed shear stress fluctuations are consistent with a dual generation for turbulence by the wave breaking prior to the runup, and by the turbulence generated by the unsteady boundary layer that develops during the runup. For the rundown stage, on the other hand, they are generated solely by the turbulence generated in the unsteady boundary layer of the downrush flow (Fig. 7.31c). The previously described picture is qualitatively the same for all sections at the onshore side of the breaking section, namely Sections 5-8 (Fig. 7.35c–e).

Regarding the offshore side of the breaking point (Sections 1 and 2 in Figs. 7.35a,b, respectively), here, too, there are two peaks in the time variation of $(\overline{(\tau_0')^2})^{1/2}$. The first peak (D in Figs. 7.35a,b) is associated with the wave boundary layer experienced at these sections while the second peak is associated with the rundown stage and the hydraulic jump (C in

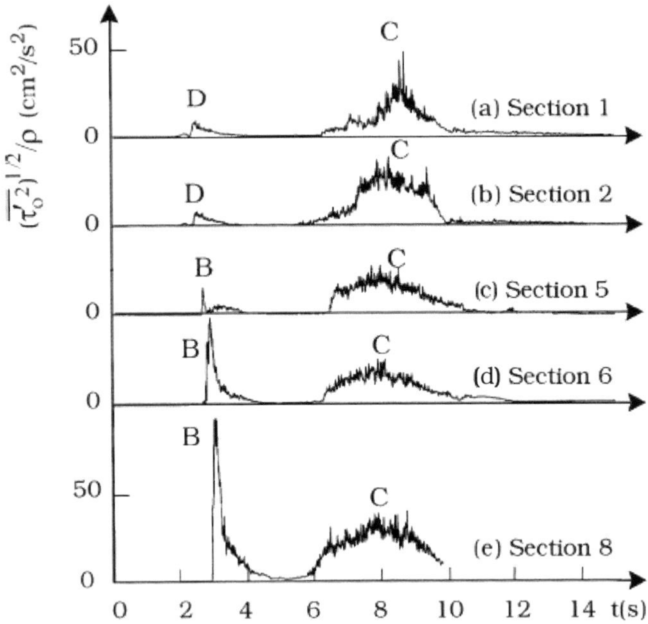

Fig. 7.35 Time variation of the r.m.s. value of the fluctuating component of the bed shear stress. $t = 0$ corresponds to the instant the wave crest passes the toe section. Sumer et al. (2011).

Figs. 7.35a,b). Regarding the first peak, Sumer et al. (2011) discussed that the first peak is generated by the wave boundary layer, which is revealed by the presence of turbulence in the time series of the bed shear stress in the deceleration period (see Sumer et al., 2011, Fig. 7). Sumer et al. (2011) note that the estimated value of the Reynolds number of the boundary layer is 3×10^5, a value slightly lower than the critical Reynolds number for laminar-to-turbulence transition, 5×10^5 (see Chapter 5). Sumer et al. (2011) discuss this issue, namely early transition to turbulence in terms of the Reynolds number, in detail, and the reader is referred to Sumer et al. (2011, their Section 4.2.1).

7.5.3.2 *Mean bed shear stress*

The behaviour of the mean bed shear stress at Section 1 (Fig. 7.34a) for times 2 s $\lesssim t \lesssim$ 6 s is typical for a solitary wave boundary layer, see Chapter 5 (Section 5.10), and also Sumer et al. (2010). $\overline{\tau_0}$ changes direction (marked A in Fig. 7.34a) during deceleration stage of the solitary wave boundary layer as the wave continues to propagate in the onshore direction. Sumer et al. (2011) discuss extensively the mean bed shear stress prior to the wave breaking, and its relation to the boundary layer in the original publication.

Following the breaking, $\overline{\tau_0}$ experiences a tremendous increase during the runup stage (marked B in Figs. 7.34c–e), which is due to the combined action of the breaking and the reduction of the flow depth as the flow continues onshore. Of particular interest is the amplification in the swash zone (Section 8, Fig. 7.34e). The increase in $\overline{\tau_0}$ with respect to that of the approaching boundary layer flow is like O(8). This clearly has a large implication for sediment transport in the swash zone.

Similarly, $\overline{\tau_0}$ experiences very large values during the rundown stage, as seen from Figs. 7.34a–e, marked C in the figure. In the swash zone, the maximum value attained during the rundown is as large as that measured during the runup (Fig. 7.34e). Here, too, very large values of the bed shear stress imply that the sediment transport during the rundown is expected to increase tremendously.

To illustrate the x-variation of the bed shear stress, Sumer et al. (2011) plotted the maximum values of the mean bed shear stress picked up from Fig. 7.34, and also data from Sections 3, 4 and 7 as well; the end result is plotted in Fig. 7.36. Fig. 7.36a corresponds to the shoaling and the runup stages while Fig. 7.36b corresponds to the rundown and the hydraulic jump stages.

7.5.3.3 *R.m.s. value of fluctuating bed shear stress*

Now, returning to the r.m.s. value of the fluctuating component of the bed shear stress, $(\overline{(\tau_0')^2})^{1/2}$, Fig. 7.35, it is seen that it increases massively in the swash zone (measurement sections, Section 6 and 8) during runup (Figs. 7.35d and e, marked B in the figure), and during rundown and hydraulic-jump stages (Figs. 7.35a–e, marked C). The generation mechanisms of the fluctuations in the bed shear stress have already been discussed in the preceding paragraphs. We will continue the discussion to address the question of how the present data compares with the data from classic boundary layer flows.

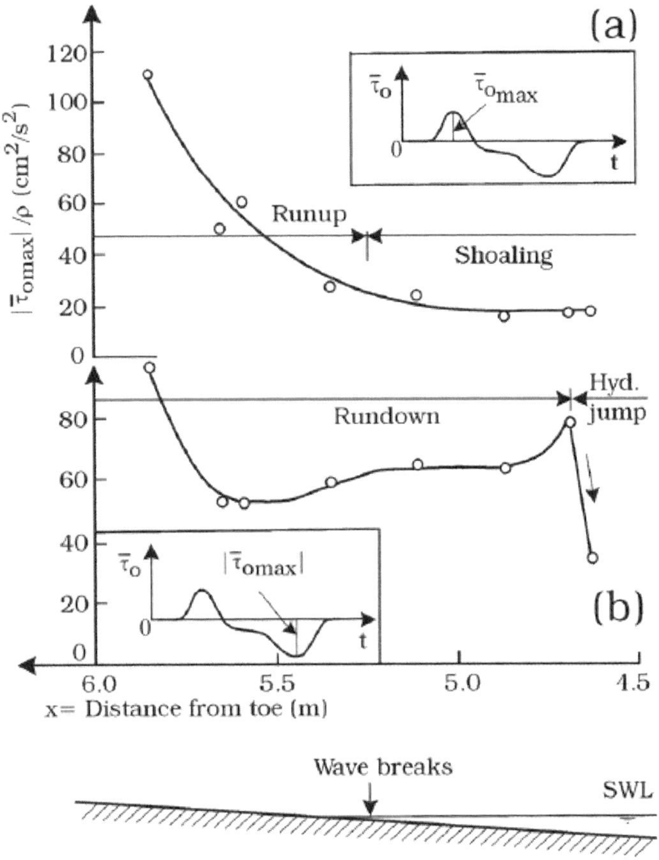

Fig. 7.36 Measured x-variation of the maximum value of the mean bed shear stress. (a) During the shoaling and runup stages. (b) During the rundown and hydraulic jump stages. Sumer et al. (2011).

In order to compare the present breaking solitary wave data with the corresponding data from boundary layers, Sumer et al. (2011) recast the latter data in terms of the maximum value of $(\overline{(\tau_0')^2})^{1/2}$, normalized by the corresponding maximum value of $\overline{\tau_0}$, and plotted as function of x, Fig. 7.37 (for the shoaling and runup stages, Fig. 7.37a, and, for the rundown and hydraulic jump stages, Fig. 7.37b).

First of all, the normalized quantity $(\overline{(\tau_0')^2})^{1/2}_{\max}/\overline{\tau_0}_{\max}$ takes values in the range from 0.2 to 0.4 during shoaling, for $x < 5.25$ m. The range 0.2–0.4 is not radically different from the range $(\overline{(\tau_0')^2})^{1/2}/\overline{\tau_0}$= 0.25–0.40

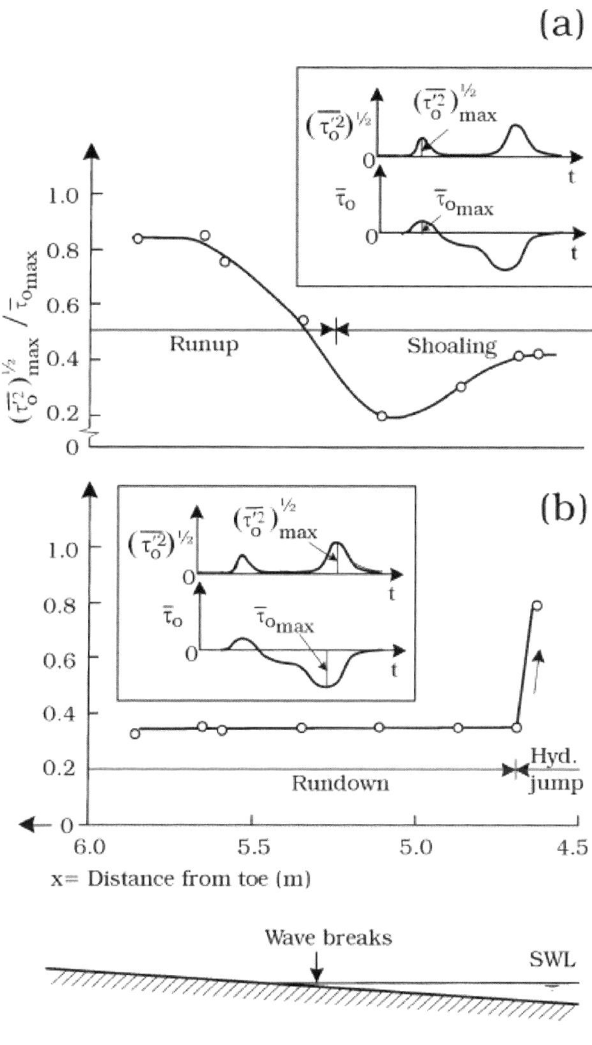

Fig. 7.37 Measured x-variation of the normalized maximum value of the r.m.s. value of the fluctuating component of the bed shear stress. (a) During the shoaling and runup stages. (b) During the rundown and hydraulic jump stages. Sumer et al. (2011).

reported in conjunction with the current-boundary-layer research (Alfredsson et al., 1988, Eckelmann, 1974, Mitchell and Hanratty, 1966, among others), and that $(\overline{(\tau_0')^2})^{1/2}_{\max}/\overline{\tau_0}_{\max} = 0.25$–$0.30$ reported in conjunction with the oscillatory-boundary-layer research (Lodahl et al., 1998). However, it is evident from Fig. 7.37a that, with the wave breaking, followed by the runup, $(\overline{(\tau_0')^2})^{1/2}_{\max}/\overline{\tau_0}_{\max}$ increases quite significantly, and reaches values as high as 0.85. Clearly this is the "external" field of turbulence introduced by the wave breaking process.

Secondly, it is seen from Fig. 7.37b that $(\overline{(\tau_0')^2})^{1/2}_{\max}/\overline{\tau_0}_{\max}$ is maintained constant, at the value $(\overline{(\tau_0')^2})^{1/2}_{\max}/\overline{\tau_0}_{\max}= 0.35$, during rundown all the way down to $x = 4.7$ m where the flow experiences a hydraulic jump. The value $(\overline{(\tau_0')^2})^{1/2}_{\max}/\overline{\tau_0}_{\max} = 0.35$ is in strikingly good agreement with the previously cited range of turbulence for current boundary layers, namely $(\overline{(\tau_0')^2})^{1/2}/\overline{\tau_0}= 0.25$–$0.40$. This supports the argument put forward earlier that the fluctuating shear stress is generated by the turbulent boundary layer process of the downrush flow during this stage. However, $(\overline{(\tau_0')^2})^{1/2}_{\max}/\overline{\tau_0}_{\max}$ experiences a sudden jump when the downrush flow eventually undergoes the hydraulic jump at the section $x = 4.7$ m. (Sumer et al., 2011, note: "Unfortunately no measurements were available from the present study for x-sections at the offshore side of the hydraulic jump to resolve the turbulence beyond this section ...".)

7.5.3.4 *Flow resistance associated with plunging solitary wave*

With regard to this issue, Sumer et al. (2013) have, in Section 6 "Comparison with Plunging Solitary Waves" of their paper, developed the following analysis, utilizing the flow resistance formulation discussed in Sumer et al. (2013) for plunging regular waves, which is summarized in Section 7.3.3.

Given the similarity in conditions (apart from e.g. relatively small differences in the incoming wave height), the maximum bed shear stresses from the solitary wave experiments of Sumer et al. (2011) corresponding to $U_{fm} = 10.5$ cm/s, based on their Fig. 9a, and the plunging regular wave case of Sumer et al. (2013) ($U_{fm} = 7.6$ cm/s; see Table 3 in Sumer et al., 2013) are reasonably similar.

To compare these more systematically, an "equivalent" surf similarity parameter for Sumer et al.'s (2011) solitary wave experiment may be formulated. This may be calculated, to a first approximation, using Eq. 7.1 in which the deep-water wave length L_0 is calculated from Eq. 7.2, T is taken from Eq. 7.67, and H_0 is calculated according to Eqs. 7.3 and 7.4. Furthermore, invoking linear shallow water theory at the toe of the slope, whereby energy flux conservation (Eqs. 7.3 and 7.4) simplifies to

$$H_0 = H\sqrt{2kh} \qquad (7.76)$$

and the dispersion relation Eq. 7.5 to

$$\omega^2 = gk^2h \tag{7.77}$$

leads to the following simple approximation for the equivalent surf similarity parameter ξ_0:

$$\xi_0 \approx \frac{\sqrt{8\pi}}{3^{5/8}} \frac{\tan(\beta)}{(H/h)^{9/8}} \approx 2.52 \frac{\tan(\beta)}{(H/h)^{9/8}} \tag{7.78}$$

This is quite similar in form (neglecting constants, and taking 9/8 as approximately unity) to the simplified parameter (Eq. 7.69):

$$\xi_s = \frac{\tan(\beta)}{H/h} \tag{7.79}$$

proposed by Fuhrman and Madsen (2008) for predicting solitary wave runup, Section 7.5.2. Direct use of Eq. 7.78 yields $\xi_0 = 1.26$, which is expectedly similar to a value $\xi_0 = 1.28$ found from Eq. 7.1, based on the preceding discussion, but without invoking the shallow water approximations at the toe.

Now, using the value $U_{fm} = 10.5$ cm/s from Sumer et al. (2011) yields a non-dimensional maximum friction velocity $U_{fm}/\sqrt{gH_0} = 0.138$, whereas direct application of Eq. 7.62 predicts 0.0986. Hence, the friction velocities representing Sumer et al.'s (2011) solitary wave experiment would lie above the best-fit line in Fig. 7.21, by roughly 40%; in other words, Eq. 7.62, when applied to plunging solitary waves, underestimates quite substantially the flow resistance. Sumer et al. (2013) argue that (1) these differences are likely due in large part to the previously alluded to differences in the breaking processes. They also add that it should also be mentioned that there are likewise uncertainties in quantitatively approximating a representative solitary wave period for use in defining ξ_0, and not to mention obvious questions regarding the validity of invoking deep-water quantities for solitary waves, which will only maintain a constant form on the particular depth they were generated. That said, nevertheless, this was done for the sake of consistency.

7.5.4 *Sediment-bed experiments of Sumer et al. (2011) study*

The sediment-bed experiments were carried out in another flume in Sumer et al.'s (2011) study, 28 m in length, 1 m in depth and 4.0 m in width. Similar to the rigid-bed experiments, solitary waves were produced by a piston type wave generator. The beach (with exactly the same slope as in the rigid-bed experiments, namely 1:14) was formed with sand, the grain size being $d_{50} = 0.18$ mm. Three kinds of measurements were made, similar to Sumer et al. (2013): pore-water pressure measurements, surface elevation measurements, and bed profile measurements.

Sumer et al. (2013) note that, in the light of the knowledge of the mechanism of sediment suspension unveiled in the Sumer et al. study (2013), they have re-examined the videos shot in the sediment bed tests in Sumer et al.'s (2011) experiments, and that these videos are not clear to reveal the way in which the bed sediment is entrained into the flow in the case of the plunging solitary waves. Whether or not the impulsive, vigorous suspension of the sediment from the bed of the Sumer et al. (2013) experiments (caused by Vortex M, Figs. 7.23 and 7.24) also occurs in the case of the solitary wave is unknown.

Sumer et al.'s (2011) pore-water pressure measurements showed, however, that the sediment at (or near) the surface of the bed experiences upward-directed pressure gradient forces during the rundown and hydraulic-jump stages. This force is caused by the delay in the pore pressure in responding to the fluid loading (infiltration and exfiltration).

The aforementioned measurements indicated that the magnitude of this force can reach values as much as 30% of the submerged weight of the sediment, which has clearly significant implications for sediment transport.

As in the study of Sumer et al. (2013), the bed morphology was also a major focus area of Sumer et al.'s (2011) study of plunging solitary waves. Sumer et al.'s (2011) results regarding the bed morphology will not be covered here, as it is outside our main interest in this book. The interested reader is referred to the original publication, Sumer et al. (2011, Section 5.2).

7.6 Numerical simulation of breaking waves

7.6.1 *Background*

With the continual increase in computational power, breaking waves have become more and more commonly studied numerically through computational fluid dynamics (CFD). Some of the earliest studies involving the computational study of breaking waves include Sakai et al. (1986), who used the marker-and-cell (MAC) method developed by Harlow and Welch (1965), and Lemos (1992), who was among the first to apply the volume of fluid method (VOF), originally developed by Hirt and Nichols (1981). Such studies include large eddy simulation (LES) models (e.g. Watanabe and Saeki, 1999; Hieu et al., 2004; Losada et al., 2005; Christensen, 2006), smooth particle hydrodynamics (SPH) models (e.g. Dalrymple and Rogers, 2006; Shao, 2006; Makris et al., 2016; De Padova et al., 2018), as well as those based on Reynolds-averaged Navier-Stokes (RANS) equations coupled with various two-equation turbulence closure models. Such studies include those utilizing closures within both the k-ε (e.g. Lin and Liu, 1998; Bradford, 2000; Hsu et al., 2002; Lara et al., 2011b; Xie, 2013; Hsu et al., 2015; Brown et al., 2016) and k-ω (e.g. Mayer and Madsen, 2000; Jacobsen et al., 2012, 2014; Brown et al., 2016; Chella et al., 2016; Devolder et al., 2018) families, the latter of which is presented in detail in Section 9.4.

In many of the RANS model studies of breaking waves mentioned above, there has been a marked and collective tendency to predict turbulence levels that are much higher than have been measured. Alarmingly, such over-predicted turbulence has often been most apparent prior to breaking, where (in contrast to experimental findings and intuition) computed turbulent kinetic energy levels can be the same order of magnitude as within the surf zone (see e.g. Lin and Liu, 1998; Bradford, 2000; Hsu et al., 2015; Brown et al., 2016). These effects can cause pre-mature wave decay, and such pollution can carry over well into the outer surf zone, thus affecting the computed sub-surface kinematics and subsequent surf zone dynamics as a whole. This thus represents a significant fundamental problem, and one that has only recently been fully analyzed and arguably solved.

The fundamental cause of the widespread over-prediction of turbulence in RANS models of surface waves was first explained theoretically by Mayer and Madsen (2000), who formally proved that the standard k-ω model of Wilcox (1988) is *conditionally unstable* when applied to a region of potential flow beneath surface waves, i.e. that the model can lead to exponential growth of both the turbulent kinetic energy and eddy viscosity. Inspired by their work, Larsen and Fuhrman (2018) re-visited this problem. They demonstrated that seemingly all (at least all that they analyzed) two-equation turbulence closures in wide use are, in fact, *unconditionally unstable* in such regions. Moreover, they have demonstrated how such closure models can be simply and elegantly re-formulated, such that they are formally stable in the nearly-potential flow region beneath surface waves, while defaulting to a desired standard closure in sheared flow regions of primary interest (i.e. the bottom boundary layer and surf zones), thus solving this long-standing problem in a fundamentally sound way.

The remainder of the present section presents a synopsis of their work, including both analytical as well as numerical results. The results from their stability analysis will be covered in detail in Section 7.6.2. A description of their resulting two-phase flow computational fluid dynamics model will be provided in Section 7.6.3, which utilizes their stabilized turbulence model to provide closure. Simulated results involving spilling breaking waves will then be presented in Section 7.6.4, and plunging breaking waves in Section 7.6.5.

7.6.2 *Stability analysis of two-equation closure models*

7.6.2.1 *Generalized k-ω model of Larsen and Fuhrman (2018)*

Larsen and Fuhrman (2018) adopted the following generalized version of the widely-used (Wilcox, 2006) k-ω model, comprised of a transport equation for the turbulent kinetic energy density $k = 1/2(\overline{u_i'u_i'})$:

$$\frac{\partial \overline{\rho} k}{\partial t} + \overline{u}_j \frac{\partial \overline{\rho} k}{\partial x_j} = \overline{\rho} P_k - \overline{\rho} P_b - \overline{\rho}\beta^* k\omega + \frac{\partial}{\partial x_j}\left[\left(\mu + \overline{\rho}\sigma^* \frac{k}{\omega}\right)\frac{\partial k}{\partial x_j}\right] \tag{7.80}$$

and the specific dissipation rate ω:

$$\frac{\partial \overline{\rho}\omega}{\partial t} + \overline{u}_j \frac{\partial \overline{\rho}\omega}{\partial x_j} = \overline{\rho} P_\omega - \overline{\rho}\beta\omega^2 + \overline{\rho}\frac{\sigma_d}{\omega}\frac{\partial k}{\partial x_j}\frac{\partial \omega}{\partial x_j} + \frac{\partial}{\partial x_j}\left[\left(\mu + \overline{\rho}\sigma \frac{k}{\omega}\right)\frac{\partial \omega}{\partial x_j}\right] \tag{7.81}$$

The shear production term for k was defined in the standard way

$$P_k = \tau_{ij}\frac{\partial \overline{u}_i}{\partial x_j} = p_0 \nu_t, \qquad p_0 = 2S_{ij}S_{ij} \tag{7.82}$$

where

$$\tau_{ij} = -\overline{u_i'u_j'} = 2\nu_t S_{ij} - \frac{2}{3}k\delta_{ij} \tag{7.83}$$

is the Reynolds stress tensor, expressed in accordance with the Boussinesq approximation,

$$S_{ij} = \frac{1}{2}\left(\frac{\partial \overline{u}_i}{\partial x_j} + \frac{\partial \overline{u}_j}{\partial x_i}\right) \tag{7.84}$$

is the mean strain rate tensor, and ν_t is the eddy viscosity. Additionally, a buoyancy production term for k was included, formulated as

$$P_b = -\frac{g_i}{\overline{\rho}}\overline{u_i'\rho'} = p_b \nu_t, \qquad p_b = \alpha_b^* N^2, \qquad N^2 = \frac{g_i}{\overline{\rho}}\frac{\partial \overline{\rho}}{\partial x_i} \tag{7.85}$$

where $g_i = (0, -g, 0)$ is gravitational acceleration and N is the so-called Brunt-Vaisala frequency. This term accounts for production (or suppression) of turbulent kinetic energy, due to correlations in the fluctuating velocity and fluid density which may occur e.g. in two-phase (e.g. air and water) flows. The eddy viscosity was defined according to

$$\nu_t = \frac{k}{\tilde{\omega}} \tag{7.86}$$

The ω production term was formulated as

$$P_\omega = \alpha \frac{\omega}{k} \frac{\tilde{\omega}}{\tilde{\tilde{\omega}}} P_k = \alpha \frac{\omega}{\tilde{\tilde{\omega}}} p_0 \tag{7.87}$$

where it is seen that the eddy viscosity has been effectively eliminated from this term.

The closure model defined above corresponds to a generalized form of the Wilcox (2006) turbulence closure model (for two dimensional flows), which is again presented in detail in Section 9.4. (For a broader overview of turbulence modelling, see the entirety of Chapter 9.) In addition to Wilcox's basic model, the form above: (1) retains the (air-water) mixture density $\overline{\rho}$ inside several derivative terms, (2) includes the previously defined buoyancy production term P_b in Eq. 7.80, and (3) incorporates two (rather than the more usual one) limited versions of ω (i.e. $\tilde{\omega}$ and $\tilde{\tilde{\omega}}$) within Eqs. 7.86 and 7.87, for reasons which will be made clear in what follows.

With suitable selections of $\tilde{\omega}$, $\tilde{\tilde{\omega}}$, and closure coefficients the generalized model above can conveniently be reduced to a number of common variations of the k-ω model in the literature. Following Larsen and Fuhrman (2018), the standard closure coefficients of Wilcox (2006) are adopted, as defined in Section 9.4, Eqs. 9.39–9.41. Additionally, we adopt the value $\alpha_b^* = 1.36$, which was formally derived by Larsen and Fuhrman (2018).

7.6.2.2 *Turbulence production in potential flow beneath waves*

Before proceeding further, let us first briefly review the production of turbulence beneath potential flow progressive waves, as originally described by Mayer and Madsen (2000). Following their work, and for the sake of simplicity, consider the velocity fields given by Stokes first-order wave theory (the leading-order description for the potential flow region beneath small amplitude waves), as given in Eqs. 5.178 and 5.179. Mayer and Madsen (2000) showed that if the velocity profiles stemming from such a linear potential flow wave solution were substituted into Eq. 7.82, after period averaging, this results in a non-zero turbulence production term (Eq. 7.82) with:

$$\langle p_O \rangle = \frac{k^2 H^2 \omega^2}{2} \frac{\cosh(2ky)}{\sinh^2(kh)} \tag{7.88}$$

After further depth-averaging, this is:

$$\langle\langle p_O \rangle\rangle = \frac{k^2 H^2 \omega^2}{2kh \tanh{(kh)}} \tag{7.89}$$

where it is emphasized that k is the wave number and ω is the angular frequency in the above, $\langle\cdot\rangle$ denotes period-averaging, and $\langle\langle\cdot\rangle\rangle$ denotes depth- and period-averaging.

The above demonstrates that, while the theoretical potential-flow region beneath progressive surface waves is of course irrotational, it is not free of strain. This result is quite important, as it means that *there will be a non-zero production of turbulent kinetic energy in the potential flow region beneath surface waves, provided that the eddy viscosity is finite*, see Eq. 7.82. As will be shown, if there is not sufficient dissipation, this can (and in fact, will) cause standard, widely-employed two-equation turbulence closures to become unstable, corresponding to non-physical exponential growth of turbulent kinetic energy in such regions.

7.6.2.3 *Analysis of standard k-ω models*

To see the potential ramifications of the above let us consider a region of nearly-potential flow, such that

$$p_\Omega = 2\Omega_{ij}\Omega_{ij} \ll p_O, \qquad \Omega_{ij} = \frac{1}{2}\left(\frac{\partial \overline{u_i}}{\partial x_j} - \frac{\partial \overline{u_j}}{\partial x_i}\right) \tag{7.90}$$

in a fluid of constant density (hence $P_b = p_b = 0$), where Ω_{ij} is the mean rotation rate tensor. For the sake of analytical simplicity, let us assume that p_O is fixed at some finite value. Following Mayer and Madsen (2000), diffusive and convective terms will be neglected in the analysis, which is justifiable in the region above the bottom boundary layer. In this case the governing generalized turbulence model Eqs. 7.80 and 7.81 reduce to the following ordinary differential equations

$$\frac{\partial k}{\partial t} = \frac{k}{\tilde{\omega}} p_O - \beta^* \omega k \tag{7.91}$$

$$\frac{\partial \omega}{\partial t} = \alpha \frac{\omega}{\tilde{\tilde{\omega}}} p_O - \beta \omega^2 \tag{7.92}$$

where k and ω in the above are again, respectively, the turbulent kinetic energy density and the specific dissipation rate.

Let us now conduct an asymptotic stability analysis of the Wilcox (2006) k-ω (see Section 9.4) model under the conditions stated above, which contains the standard Wilcox (1988) k-ω model as a special case. In the context of the generalized model presented in Section 7.6.2.1 this model corresponds to setting

$$\tilde{\omega} = \tilde{\tilde{\omega}} = \max\left[\omega, \lambda_1\sqrt{\frac{p_0}{\beta^*}}\right] \tag{7.93}$$

with $\lambda_1 = 7/8$. If the first term in Eq. 7.93 is active, then this corresponds to setting $\omega = \tilde{\omega} = \tilde{\tilde{\omega}}$ in the above, leading to the further reduced equations

$$\frac{\partial k}{\partial t} = \frac{k}{\omega}p_0 - \beta^*\omega k \tag{7.94}$$

$$\frac{\partial \omega}{\partial t} = \alpha p_0 - \beta\omega^2 \tag{7.95}$$

This corresponds to the standard Wilcox (1988) closure model, and is the same form as originally considered by Mayer and Madsen (2000), who formally proved it to be *conditionally unstable*.

As pointed out by Larsen and Fuhrman (2018), from inspection of Eq. 7.95, it is seen that this equation is de-coupled from the k equation Eq. 7.94. Hence, regardless of its initial value, ω will ultimately evolve to a constant such that $\partial\omega/\partial t = 0$, corresponding in this case to

$$\omega_\infty = \sqrt{\frac{p_0\alpha}{\beta}} \approx 2.71\sqrt{p_0} \tag{7.96}$$

Moreover, once the specific dissipation rate reaches its asymptotic value $\omega = \omega_\infty$, the k equation becomes linearized, ultimately leading to simple ordinary differential equations of the form

$$\frac{1}{k}\frac{\partial k}{\partial t} = \frac{1}{\nu_t}\frac{\partial \nu_t}{\partial t} = \Gamma_\infty \tag{7.97}$$

These have solutions such that

$$k \sim \nu_t \sim e^{\Gamma_\infty t} \tag{7.98}$$

where Γ_∞ is the asymptotic unstable growth rate (with instability corresponding to $\Gamma_\infty > 0$ i.e. exponential growth). For the standard Wilcox (1988) k-ω model under present consideration this works out to be

$$\Gamma_\infty = (\beta - \alpha\beta^*)\sqrt{\frac{p_0}{\alpha\beta}} \approx 0.125\sqrt{p_0} \tag{7.99}$$

Hence, this means that the Wilcox (1988) k-ω model is *unconditionally unstable* i.e. under the conditions assumed above, it will inevitably lead to exponential growth of the turbulent kinetic energy density k and eddy viscosity ν_t. Such exponential growth very near that predicted analytically above was verified directly through CFD simulation of a periodic wave train by Larsen and Fuhrman (2018), confirming the analysis.

Let us now extend the analysis to cover the Wilcox (2006) k-ω model. For this purpose, consider the situation where the second (stress limiting) argument in Eq. 7.93 dominates, then Eq. 7.91 and Eq. 7.92 become

$$\frac{\partial k}{\partial t} = \frac{k}{\lambda_1}\sqrt{\beta^* p_0} - \beta^* \omega k \tag{7.100}$$

$$\frac{\partial \omega}{\partial t} = \frac{\alpha\omega}{\lambda_1}\sqrt{\beta^* p_0} - \beta\omega^2 \tag{7.101}$$

As before, setting the right-hand-side of Eq. 7.101 equal to zero, we now find that ω will tend to the asymptotic value

$$\omega_\infty = \frac{\alpha}{\lambda_1\beta}\sqrt{\beta^* p_0} \approx 2.52\sqrt{p_0} \tag{7.102}$$

Inserting this value for ω back into Eq. 7.100 then leads directly to linearized expressions of the form Eq. 7.97, where the unstable growth rate is

$$\Gamma_\infty = \frac{(\beta - \alpha\beta^*)}{\lambda_1\beta}\sqrt{\beta^* p_0} \approx 0.116\sqrt{p_0} \tag{7.103}$$

Hence, the Wilcox (2006) k-ω model is likewise *unconditionally unstable* in nearly potential flow regions having finite strain. Notably, it is seen that the stress limiter reduces the unstable growth rate slightly relative to the standard Wilcox (1988) k-ω model.

7.6.2.4 *Stability analysis of other turbulence closure models*

Making use of similar techniques as described above, Larsen and Fuhrman (2018) likewise analyzed several other popular two-equation turbulence closure models for stability, including those in both k-ω and k-ε closure families. They found that the instability discussed above is wide spread, and that all of the models they analyzed were similarly unconditionally unstable in nearly potential flow regions having finite strain. This notably includes the ad-hoc modification used by Mayer and Madsen (2000), who changed their production terms P_k and P_ω such that they were based on the vorticity rather than the strain rate (equivalent to setting $\tilde{\omega} = \tilde{\tilde{\omega}} = \omega p_0/p_\Omega$ exclusively in these terms). The resulting asymptotic values for ω_∞ and

Table 7.2 Summary of asymptotic ω_∞ values and unstable exponential growth rates Γ_∞ for existing two-equation turbulence closures analyzed by Larsen and Fuhrman (2018).

Model	Reference	$\approx \omega_\infty$	$\approx \Gamma_\infty$
Standard k-ω	Wilcox (1988)	$2.71\sqrt{p_0}$	$0.125\sqrt{p_0}$
k-ω	Mayer and Madsen (2000)	$2.71\sqrt{p_\Omega}$	$0.125\sqrt{p_\Omega}$
k-ω	Wilcox (2006)	$2.52\sqrt{p_0}$	$0.116\sqrt{p_0}$
k-ω SST	Menter (1994)	$2.72\sqrt{p_0}$	$0.066\sqrt{p_0}$
Standard k-ε	Launder and Sharma (1974)	$2.31\sqrt{p_0}$	$0.226\sqrt{p_0}$
RNG k-ε	Yakhot et al. (1991)	$2.70\sqrt{p_0}$	$0.142\sqrt{p_0}$

the unstable growth rates Γ_∞ for each of the closure models analyzed by Larsen and Fuhrman (2018) are summarized in Table 7.2. These findings are generally consistent with the over-predicted turbulence levels found in CFD modeling in the nearly-potential flow region which exist prior to waves breaking e.g. by Lin and Liu (1998), Bradford (2000), Hsu et al. (2015), and Brown et al. (2016).

Animations demonstrating the instability of the turbulent kinetic energy using several of the two-equation turbulence closure models listed in Table 7.2 (involving both wave train propagation as well as spilling breaking waves) can be found at:

`https://doi.org/10.11583/DTU.8180708`

Comparisons with these standard closure models with their respective "stabilized" versions (as described in the next section) can also be seen in the animations above.

7.6.2.5 *Analysis of the Larsen and Fuhrman (2018) k-ω model*

As seen above, the introduction of the stress limiter in the Wilcox (2006) model reduced the unstable growth rate relative to the standard Wilcox (1988) model which contains no limiter, see Table 7.2. Recognizing this, Larsen and Fuhrman (2018) proposed the following modifications to the stress limiting features. First, for the sake of full consistency with Eq. 7.80 in two-phase (air and water) flow applications, as commonly used for studying breaking waves, they proposed to generalize $\tilde{\tilde{\omega}}$ slightly such that the buoyancy production was also included:

$$\tilde{\tilde{\omega}} = \max\left[\omega, \lambda_1\sqrt{\frac{p_0 - p_b}{\beta^*}}\right] \tag{7.104}$$

This modification has no bearing on the formal stability of the model; If p_b is omitted in the above, then this is equivalent to the Wilcox (2006) stress limiter seen in Eq. 7.93, which may alternatively be utilized. Much more importantly in the present context, to formally stabilize the resulting closure model in a region of potential flow with finite strain (as considered at length above) they proposed to modify $\tilde{\omega}$ to:

$$\tilde{\omega} = \max\left[\tilde{\tilde{\omega}}, \lambda_2 \frac{\beta}{\beta^* \alpha} \frac{p_0}{p_\Omega} \omega\right] \tag{7.105}$$

where $\lambda_2 \ll 1$ is an additional stress limiter coefficient. Note that the new addition to the stress limiter in Eq. 7.105, by design, will only be active in a region of nearly-potential flow with finite strain i.e. where $p_0 \gg p_\Omega$, precisely the conditions where such standard models are formally unstable. Moreover, note that (for single-phase incompressible flows) if $\lambda_2 = 0$ the model becomes equivalent to the Wilcox (2006) model, whereas if $\lambda_1 = \lambda_2 = 0$ this model becomes equivalent to the standard k-ω model of Wilcox (1988). It is hence fully compatible with the traditional closures considered above.

To demonstrate the formal asymptotic stability of the model proposed by Larsen and Fuhrman (2018), let us first repeat the above analysis of the standard Wilcox (1988) k-ω model (momentarily setting $\lambda_1 = 0$), but now with the new limiter active in $\tilde{\omega}$. In this case, the k equation Eq. 7.91 becomes

$$\frac{\partial k}{\partial t} = \frac{\alpha \beta^*}{\lambda_2 \beta} \frac{k p_\Omega}{\omega} - \beta^* \omega k \tag{7.106}$$

whereas the ω equation from Eq. 7.92 remains equivalent to Eq. 7.95, and hence ω will ultimately evolve to ω_∞ from Eq. 7.96. Inserting this value into Eq. 7.106 leads to linearized expressions of the form Eq. 7.97, where the exponential growth rate is

$$\Gamma_\infty = \sqrt{\frac{p_0 \alpha}{\beta}} \beta^* \left(\frac{p_\Omega}{\lambda_2 p_0} - 1\right) \tag{7.107}$$

From this it can be seen that the modified model of Larsen and Fuhrman (2018) will be formally *stable* (i.e. $\Gamma_\infty \leq 0$) in a fluid region where

$$\frac{p_\Omega}{p_0} \leq \lambda_2 \tag{7.108}$$

This hence provides a clear physical meaning for the added stress limiter coefficient λ_2, which defines the threshold of p_Ω / p_0 identifying a region as effectively potential flow.

Let us now retain λ_1 and repeat the analysis of the Wilcox (2006) model subject to the modifications proposed by Larsen and Fuhrman (2018). In this case the k equation will remain equivalent to Eq. 7.106, whereas the ω equation will be equivalent to Eq. 7.101, and hence ω will tend towards ω_∞ in Eq. 7.102. Inserting this value as a constant within Eq. 7.106 then similarly leads to linearized expressions of the form Eq. 7.97, where the unstable growth rate is

$$\Gamma_\infty = \sqrt{p_0}\beta^* \left(\frac{p_\Omega \lambda_1}{p_0 \lambda_2 \sqrt{\beta^*}} - \frac{\alpha\sqrt{\beta^*}}{\lambda_1 \beta} \right) \tag{7.109}$$

This will thus be stable provided that

$$\frac{p_\Omega}{p_0} \leq \lambda_2 \frac{\alpha\beta^*}{\lambda_1^2 \beta} \approx 0.863\lambda_2 \tag{7.110}$$

where the standard coefficient values have been inserted. As the lead coefficient is near unity, it is seen that this further generalization does not greatly affect the previously mentioned physical interpretation of λ_2. Note also that once $\omega = \omega_\infty$ if $\lambda_1 < \sqrt{\alpha\beta^*/\beta} \approx 0.813$ then the first argument will inevitably dominate in Eq. 7.104, and the threshold for stability will again be given by Eq. 7.108.

Hence, the modification to the stress limiting feature proposed by Larsen and Fuhrman (2018) formally solves the longstanding instability problem considered at length above in a rather simple and elegant way, requiring only minimal changes to existing closure models. In uniform boundary layer flows $p_0 = p_\Omega$, and thus the new limiter will be completely inactive. Moreover, in other more complicated sheared flow regions p_Ω and p_0 are typically the same order of magnitude and the new addition to the limiter will similarly remain inactive. Hence, the modified closure will, by design, effectively reduce to a standard closure, except in regions of nearly-potential flow having finite strain (clearly defined by the choice of λ_2), where it would otherwise be unconditionally unstable, as shown above.

Larsen and Fuhrman (2018) likewise demonstrated that all of the other widely-utilized, but formally unconditionally unstable, closure models mentioned in Table 7.2 can also be formally stabilized via essentially similar simple modifications to their stress limiting features. For full details on the modifications required for these models the interested reader is referred there.

To avoid the potential for excessive and un-physical production of turbulence in nearly-potential flow regions, as has consistently been observed in numerous prior simulations of surface waves, it is recommended that these (or otherwise formally stabilized and un-intrusive) modified closure models be utilized in any future CFD simulations of surface waves with two-equation RANS closure models. The fundamental benefits of doing so will be demonstrated in the remainder of this chapter.

7.6.3 *Model description for breaking wave simulations*

Following Larsen and Fuhrman (2018), CFD simulation of breaking waves will now be considered in the context of two-phase (air and water) flow. For this purpose, the turbulence model defined by Eqs. 7.80 and 7.81 will be utilized to provide closure to a CFD model solving the incompressible Reynolds-averaged Navier-Stokes (RANS) equations:

$$\frac{\partial \overline{\rho} \overline{u}_i}{\partial t} + \overline{u}_j \frac{\partial \overline{\rho} \overline{u}_i}{\partial x_j} = -\frac{\partial \overline{p}^*}{\partial x_i} - g_j x_j \frac{\partial \overline{\rho}}{\partial x_i} + \frac{\partial}{\partial x_j}(2\mu S_{ji} + \tau_{ij}) \tag{7.111}$$

where the fluid mixture density $\overline{\rho}$ is retained inside the derivatives. This was coupled with the local continuity equation

$$\frac{\partial \overline{u}_i}{\partial x_i} = 0 \tag{7.112}$$

In the above $\overline{p}^*$ is the pressure in excess of hydrostatic. A scalar field γ is used to track the two fluids, where $\gamma = 0$ represents pure air and $\gamma = 1$ pure water, with any intermediate value representing a mixture. The distribution of γ is governed by an advection equation

$$\frac{\partial \gamma}{\partial t} + \frac{\partial(\overline{u}_i \gamma)}{\partial x_i} + \frac{\partial(\overline{u}_i^r \gamma(1-\gamma))}{\partial x_i} = 0 \tag{7.113}$$

where $\overline{u}_i^r$ is a relative velocity used to compress the interface, as documented in Berberovic et al. (2009). Any fluid property Φ in the flow is assumed to be given by

$$\Phi = \gamma \Phi_{water} + (1-\gamma)\Phi_{air} \tag{7.114}$$

i.e. properties are weighted linearly based on the local value of γ.

The governing equations have been solved within the open-source OpenFOAM environment (version `foam-extend 3.1`), making use of the `waves2FOAM` toolbox developed by Jacobsen et al. (2012) for wave generation. This toolbox is built upon the widely used `interFoam` solver, which utilizes the volume of fluid (VOF) method; For further details on this solver the interested reader is referred to Deshpande et al. (2012). In cases presented in the present section the time step has been adjusted such that a maximum Courant number $Co = |\overline{u}_i|\Delta t/\Delta x_i = 0.05$ is maintained at

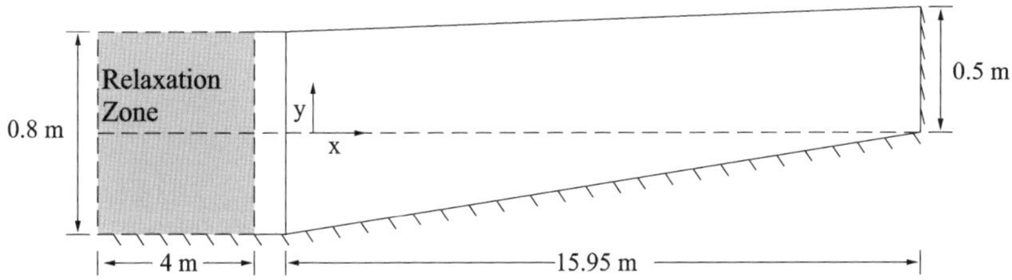

Fig. 7.38 Layout of the computational domain for the simulation of the Ting and Kirby (1994) experiments. Adapted from Larsen and Fuhrman (2018).

all times. This ensures accurate velocity kinematics and enables nearly constant-form propagation of wave trains for long durations, as shown in laminar (i.e. no turbulence closure) simulations made by Larsen et al. (2019). At the bottom, the generalized wall functions from Fuhrman et al. (2014) have been utilized, as described in Section 9.4.2. For full details on the initial conditions and mesh resolution utilized, the interested reader may consult Larsen and Fuhrman (2018).

7.6.4 *Simulation of spilling breaking waves of Ting and Kirby (1994)*

Let us now consider CFD simulations of the well-known spilling breaking wave experiment of Ting and Kirby (1994) based on the model presented in Section 7.6.3. The model domain for these simulations consists of a flat region having water depth $h = 0.4$ m, connected to a region having constant $\tan\beta = 1/35$ slope. For these simulations numerically exact nonlinear wave (potential flow) solutions of Fenton (1988) (with period $T = 2$ s and wave height $H = 0.125$ m, yielding $kh = 0.664$ and $kH = 0.207$, where here k is the wave number) have been generated on the horizontal bed. Note that this case yields a surf similarity parameter of

$$\xi_0 = \frac{\tan\beta}{\sqrt{H_0/L_0}} = 0.20 \tag{7.115}$$

which is well within the spilling breaker regime according to Eq. 7.6. At the left inlet boundary a relaxation zone of 4 m was used, which serves to absorb any waves reflected by the slope. In Fig. 7.38 a layout of the computational domain is seen. For the purposes of consistent comparison, the origin is positioned at the same depth ($h = 0.38$ m) as in the experiments. The bed was made of plywood, and therefore the model bed is assigned a Nikuradse

Table 7.3 Summary of parameters used in the five different simulations of the Ting and Kirby (1994) spilling breaking wave experiment. All models include the buoyancy production term.

Case	k-ω Model	λ_1	λ_2	$\tilde{\tilde{\omega}}$	$\tilde{\omega}$
1	Wilcox (1988)	0	0	$=\omega$	$=\omega$
2	Wilcox (2006)	0.875	0	$=\tilde{\omega}$	Eq. 7.93
3	Larsen and Fuhrman (2018)	0	0.05	$=\omega$	Eq. 7.105
4	Larsen and Fuhrman (2018)	0.875	0.05	Eq. 7.104	Eq. 7.105
5	Larsen and Fuhrman (2018)	0.2	0.05	Eq. 7.104	Eq. 7.105

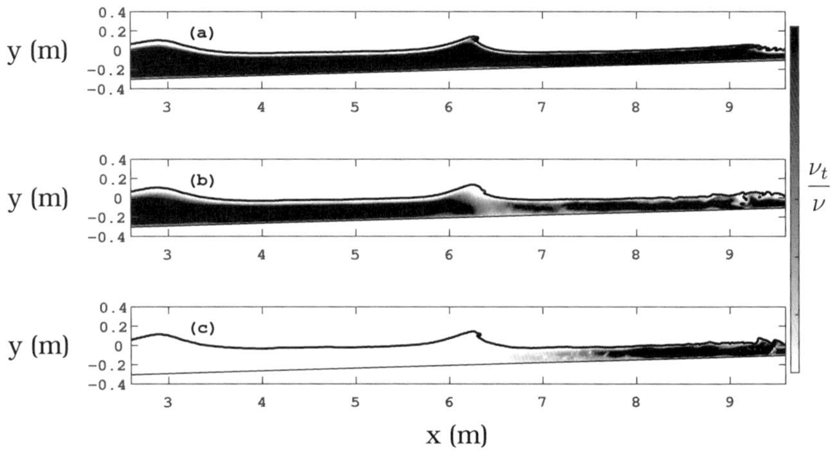

Fig. 7.39 Computed ν_t/ν at $t/T = 100$ with the wave right at breaking (spilling breaking case) computed with the (a) standard Wilcox (1988) k-ω model (Case 1), (b) standard Wilcox (2006) k-ω model (Case 2), and (c) the stabilized Larsen and Fuhrman (2018) k-ω model (exemplified with Case 5). The free surface and bottom are marked by a solid line. Adapted from Larsen and Fuhrman (2018).

equivalent sand roughness $k_s = 10^{-4}$ m.

For the sake of systematic comparison, Larsen and Fuhrman (2018) considered five different turbulence model variants based on the generalized k-ω closure presented in Section 7.6.2.1, as indicated in Table 7.3. Cases 1 and 2, respectively, correspond to the traditional k-ω closures (with buoyancy production also added) of Wilcox (1988) and Wilcox (2006), whereas Cases 3–5 correspond to stabilized variations of the Larsen and Fuhrman (2018) model, as presented in Section 7.6.2.5. In what follows an abbreviated account of their full results will be presented, with results referred to by their case number from Table 7.3.

To clearly demonstrate the marked differences in the simulation of surface waves with traditional (unstable) versus stabilized k-ω closure models, Fig. 7.39 shows computed snapshots depicting the surface and non-dimensional eddy viscosity ν_t/ν from Case 1 (standard k-ω model of Wilcox, 1988), Case 2 (standard k-ω model of Wilcox, 2006), and Case 5 (example of a stabilized k-ω model of Larsen and Fuhrman, 2018), following a long simulation time of 100 periods, such that steady (repeatable) conditions were reached. Each plot is frozen at an instant where the wave is very close to breaking, such that the location of the surf zone is evident, beginning at around $x_b = 6.3$ m. As can be seen both standard closures (Fig. 7.39a,b) result in eddy viscosities ν_t that are several orders of magnitude larger than ν, even before the waves break. Such large values are not physical: in real waves significant turbulence (hence eddy viscosity) in the potential flow core region beneath surface waves should not be expected prior to breaking, see e.g. the measured turbulence levels from Ting and Kirby (1994) and van der A et al. (2017), the PIV experiments of Chang and Liu (1998), Kimmoun and Branger (2007), Belden and Techet (2011), and Grue and Jensen (2006) or the dye experiment of Duncan et al. (1999). Rather, these artificially high eddy viscosities are a direct consequence of the instability of these models in the nearly-potential flow region leading up to the surf zone, as described in Section 7.6.2.3.

It is again emphasized that the buoyancy production term in Eq. 7.80, as suggested in this context e.g. by Devolder et al. (2017, 2018), is active in Figs. 7.39a,b (as well as in Figs. 7.39c). While this feature creates a local sink near the air-water interface, it does little to prevent the over-production of turbulence throughout the bulk of the water column beneath the waves (see Figs. 7.39a,b). This is consistent with the previous analysis, which has shown that *it is the entirety of the nearly potential flow region that is unstable*, and that this is not merely a problem confined to the near-surface or air-water interface region.

Conversely, the stabilized closure model of Larsen and Fuhrman (2018), exemplified in Fig. 7.39c (showing Case 5), avoids the unphysical production of pre-breaking turbulence entirely: Significant eddy viscosity with the stabilized model is rightly confined to the near bed boundary layer and surf zone, and this was found to be the case regardless of the specific λ_1 value utilized (i.e. Cases 3 and 4 resulted in qualitatively similar plots, see Larsen and Fuhrman, 2018). To reiterate: Only the formally stable closure models leads to an un-polluted wave field at the point where the wave breaking process (the very phenomenon of computational interest!) begins.

To provide further insight regarding the various simulations, Fig. 7.40 shows comparison for the same three models of the computed and experimental surface elevation envelopes as well as the mean water levels for the same three models mentioned above (Cases 1, 2 and 5). The solid lines represent the mean (ensemble averaged) values, whereas the shaded area represents plus or minus one standard deviation, to give an indication of wave-to-wave variability. It is seen that the models considered, in general, capture the evolution of the mean surface elevations well (though it can be noted that the unstable models would lead to unphysical wave decay prior to breaking if greater propagation distance were allotted, as directly shown by Larsen and Fuhrman, 2018). The results demonstrate initially increasing wave heights during the shoaling process, followed by decay after breaking. The most notable difference in the results shown in Fig. 7.40 is that the model with the Wilcox (2006) stress limiter off (i.e. Case 1 with $\lambda_1 = 0$, shown in Fig. 7.40a) demonstrates essentially no wave-to-wave variability in the surf zone. As demonstrated by Larsen and Fuhrman (2018), this is because the waves, in fact, do not break (spill) properly with the Wilcox (2006) stress limiter switched off i.e. with $\lambda_1 = 0$. The observed decay of the wave envelope in Fig. 7.40a can be attributed to turbulent diffusion of the wave due to the excessively large eddy viscosity in the crest region (see Fig. 7.39a), rather than a convincing sequence of spilling breaking. The cases with $\lambda_1 > 0$ (e.g. Cases 2 and 5 shown in Fig. 7.40b,c) both lead to proper spilling breakers, and as a result demonstrate a much more dynamic surf zone with more considerable wave-to-wave variability. This is in line with the experimental observations, see e.g. Fig. 3b of Ting and Kirby (1994). Hence these results demonstrate the importance of the Wilcox (2006) stress limiter in leading to properly spilling waves and hence correct description of the breaking process as a whole.

As further comparison, Figs. 7.41 and 7.42 respectively compare computed (Larsen and Fuhrman, 2018) and measured (Ting and Kirby, 1994) averaged turbulent kinetic energy density k and horizontal undertow velocity profiles at a total of eight positions. [Note that Fig. 7.42 is a corrected version of Fig. 13 of Larsen and Fuhrman (2018), where in subplots b–h the dimensional rather than dimensionless experimental values were inadvertently plotted; This does not greatly affect the visual results or any conclusions made, since $\sqrt{gh} \approx 1$ m/s at these positions (minor visual differences can be seen in subplots b,c,d).] The plots in these figures correspond to two pre-breaking positions (subplots a and b), as well as six in the surf zone (subplots c–h). In the experiments only one or two components

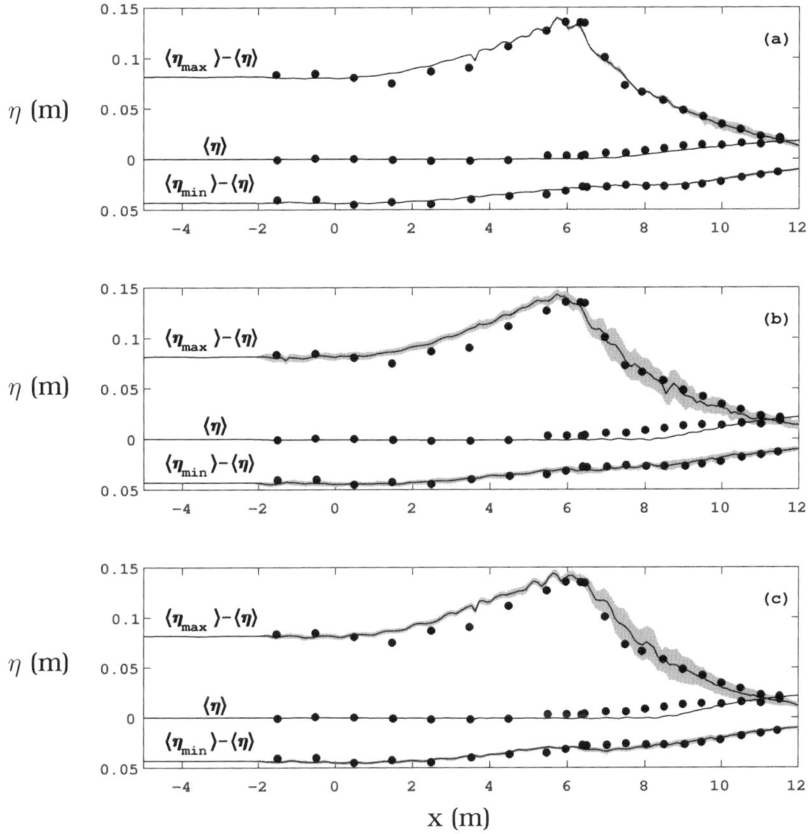

Fig. 7.40 Comparison of modelled (lines) and measured (circles, from Ting and Kirby, 1994) surface elevation envelopes (top and bottom lines) and mean water levels (middle lines) for the spilling breaking case. Results in (a), (b) and (c) correspond to Cases 1, 2 and 5, respectively. Solid lines represents the mean surface elevation and the shaded area represents plus and minus one standard deviation. Adapted from Larsen and Fuhrman (2018).

of the velocity were measured, so k has been approximated by either

$$k \approx 0.75(\overline{u'^2} + \overline{v'^2}) \tag{7.116}$$

at locations where two components were measured or

$$k \approx 1.25\overline{u'^2} \tag{7.117}$$

where only a single component was measured. These approximations are based on surf-zone measurements of Scott et al. (2005).

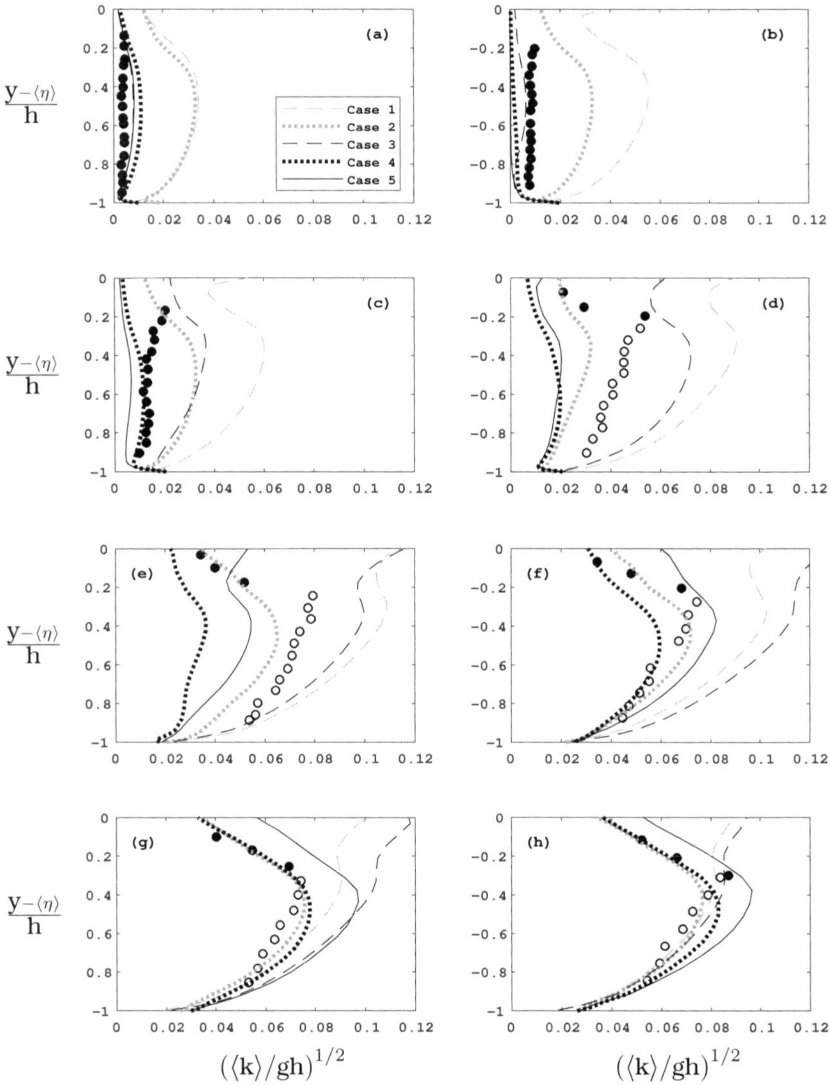

Fig. 7.41 Comparison of modelled (lines) and measured (open circles: from Eq. 7.116, filled circles: from Eq. 7.117) turbulent kinetic energy k profiles (spilling breaking case) at (a) $x = -1.265$ m ($\tilde{x} = (x - x_b)/h_b = -38.518$), (b) $x = 5.945$ m ($\tilde{x} = -2.286$), (c) $x = 6.665$ m ($\tilde{x} = 1.332$), (d) $x = 7.275$ m ($\tilde{x} = 4.397$), (e) $x = 7.885$ m ($\tilde{x} = 7.462$), (f) $x = 8.495$ m ($\tilde{x} = 10.528$), (g) $x = 9.11$ m ($\tilde{x} = 13.618$) and (h) $x = 9.725$ m ($\tilde{x} = 16.709$). Adapted from Larsen and Fuhrman (2018).

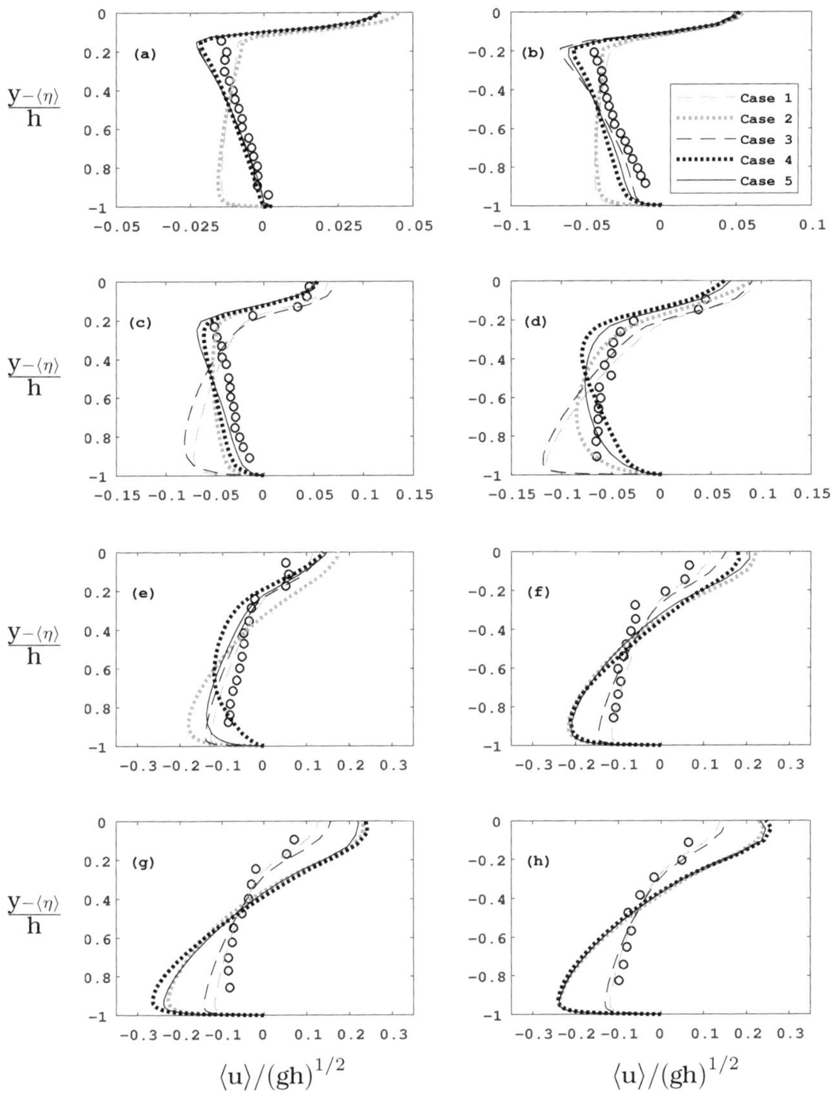

Fig. 7.42 Comparison of modelled (lines) and measured (circles) undertow velocity profiles (spilling breaking case) at (a) $x = -1.265$ m ($\tilde{x} = (x - x_b)/h_b = -38.518$), (b) $x = 5.945$ m ($\tilde{x} = -2.286$), (c) $x = 6.665$ m ($\tilde{x} = 1.332$), (d) $x = 7.275$ m ($\tilde{x} = 4.397$), (e) $x = 7.885$ m ($\tilde{x} = 7.462$), (f) $x = 8.495$ m ($\tilde{x} = 10.528$), (g) $x = 9.11$ m ($\tilde{x} = 13.618$) and (h) $x = 9.725$ m ($\tilde{x} = 16.709$).

Careful inspection of Figs. 7.41 and 7.42 reveals the following:

(1) Consistent with the preceding discussion and stability analysis, only results computed with stabilized closure models ($\lambda_2 = 0.05$, Cases 3–5) avoid pronounced un-physical over-production of turbulent kinetic energy prior to breaking, e.g. Fig. 7.41a,b;
(2) The over-production of turbulence prior to breaking causes unphysically large flow resistance in the upper part of the water column, resulting in qualitatively erroneous structure of the undertow profile prior to breaking (e.g. Cases 1 and 2 in Fig. 7.42a,b). This structure is largely corrected by the stabilized models, Cases 3–5;
(3) The over-prediction of pre-breaking turbulence can likewise carry over well into the outer surf zone, e.g. remnants are clearly visible in Fig. 7.41c–f;
(4) The Wilcox (2006) stress limiter (i.e. $\lambda_1 > 0$, Cases 2, 4, and 5) also helps avoid over-production of turbulent kinetic energy in the outer surf zone, e.g. Fig. 7.41c,d;
(5) As a consequence of the above, only the stabilized models can be said to result in the qualitatively correct evolution of the undertow structure from outside to the outer surf zone, e.g. Fig. 7.42a–e. The best result overall is arguably Case 5, corresponding to a stabilized model ($\lambda_2 = 0.05$) coupled with an intermediate value for the Wilcox (2006) stress limiter coefficient $\lambda_1 = 0.2$.
(6) Results from otherwise-identical standard or stabilized closure models become essentially similar in the inner surf zone, e.g. Fig. 7.41g,h and Fig. 7.42g–h. This is expected, as the stabilized model of Larsen and Fuhrman (2018), by design, defaults to a desired standard closure in sheared regions (such as the surf zone).

It is also finally emphasized that the locally superior match with the experimental undertow profiles in the inner surf zone achieved with those models having $\lambda_1 = 0$ (Cases 1 and 3, Fig. 7.42g,h) should be interpreted as fortuitous, since the wave breaking process, turbulence production, and undertow profiles leading up to these positions with these models were clearly erroneous. Larsen and Fuhrman (2018) demonstrated that these local differences were due to increased eddy viscosity in the upper column at these inner-most positions, which were more a remnant of the their over-production of turbulence earlier in the breaking process, than a correct description of the breaking process as a whole. Indeed, the turbulent kinetic energy profiles computed with Cases 1 and 3 (Fig. 7.41g,h) deviate significantly from the measurements in the upper part of the column at these same positions i.e. even if the computed eddy viscosity may be "right" at these positions, it is apparently for "wrong" reasons. Care must hence be taken not to interpret results at any one position in a vacuum; they are all part of a simulated breaking process, and results at one position can be affected by those further offshore.

Most comparable models in the literature have similarly predicted exaggerated undertow profiles in the inner surf zone, similar to Cases 2, 4, and 5 in Fig. 7.42g,h (e.g. Brown et al., 2016). (Notably, results from the simple undertow model developed previously in this chapter, Fig. 7.7b, do not exhibit such exaggeration, though it is emphasized the eddy viscosity was again tuned to achieve the result shown there.) How to achieve a CFD model based on RANS equations which yields quantitatively correct undertow profiles in the inner surf zone, while also correctly describing the breaking process leading up to these positions, remains open at the time of this writing, to the best of the authors' knowledge.

7.6.5 *Simulation of plunging breaking waves of Ting and Kirby (1994)*

We will now further demonstrate results from CFD simulations of plunging breaking waves, now considering well-known plunging case experiments of Ting and Kirby (1994). These have likewise been widely compared against in the literature. Similar to the spilling case considered in the previous sub-section, these experiments were performed in a small scale wave flume with depth $h = 0.4$ m, again connected to a region having $\tan\beta = 1/35$ slope.

For these simulations the waves are again generated with the `waves2FOAM` toolbox developed by Jacobsen et al. (2012), as in the preceding sub-section, now with period $T = 5$ s and wave height $H = 0.128$ m (such that $kh = 0.24$ and $H/h = 0.32$, where here k is the wave number). Note that this case yields a surf similarity parameter

$$\xi_0 = \frac{\tan\beta}{\sqrt{H_0/L_0}} = 0.60 \tag{7.118}$$

which is consistent with the plunging regime indicated in Eq. 7.7.

The results presented in what follows are simulated using the turbulence model from Larsen and Fuhrman (2018), as presented in Larsen (2018), and for full details the interested reader is directed there. As a first demonstration, Fig. 7.43 depicts the computed ν_t/ν after nearly 50 periods, using both

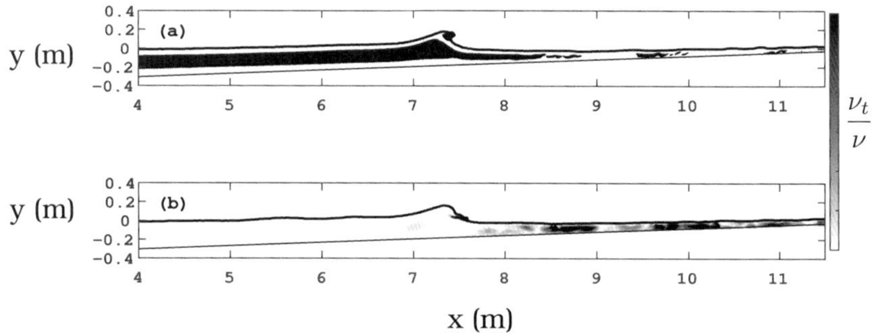

Fig. 7.43 Computed ν_t/ν at $t/T = 47.8$ with the wave right at breaking (plunging breaking case) computed with the (a) standard Wilcox (1988) k-ω model ($\lambda_1 = \lambda_2 = 0$ such that $\omega = \tilde{\omega} = \tilde{\tilde{\omega}}$) and (b) the stabilized Larsen and Fuhrman (2018) k-ω model with $\lambda_1 = 0.2$ and $\lambda_2 = 0.05$. The free surface and bottom are marked by a solid line. Larsen (2018).

a standard (Fig. 7.43a, settings corresponding to Case 1 from Table 7.3) and stabilized (Larsen and Fuhrman, 2018) (Fig. 7.43b, settings corresponding to Case 5 from Table 7.3) k-ω turbulence model. Similar to Fig. 7.39 for the spilling case, this demonstrates how standard turbulence closures, due to their inherent unconditional instability in the nearly-potential flow core region beneath surface waves (as established in Section 7.6.2), can once again lead to severely polluted results prior to the onset of wave breaking.

It is also again emphasized that the buoyancy production term in Eq. 7.80, as suggested in this context e.g. by Devolder et al. (2017, 2018), is active in both Fig. 7.43a,b. While this feature again creates a local sink in the turbulent kinetic energy (hence eddy viscosity) near the air-water interface, it again does rather little to prevent the non-physical over-production of turbulence throughout the bulk of the water column beneath the surface wave (see Fig. 7.43a, the region prior to breaking i.e. say $x < 7.3$ m in the model results). This problem is removed entirely by incorporating the stabilization described by Larsen and Fuhrman (2018), as depicted in Fig. 7.43b. This simulation results in a clear plunging breaking wave, largely free of turbulence in the core region at incipient breaking, in accordance with the experimental observations as will be shown, whereas turbulence is allowed to grow naturally inside the surf zone and in the bottom boundary layer, as should be expected on physical grounds.

Having once again demonstrated the qualitative differences between standard and stabilized closure models above (Fig. 7.43, and in the pre-

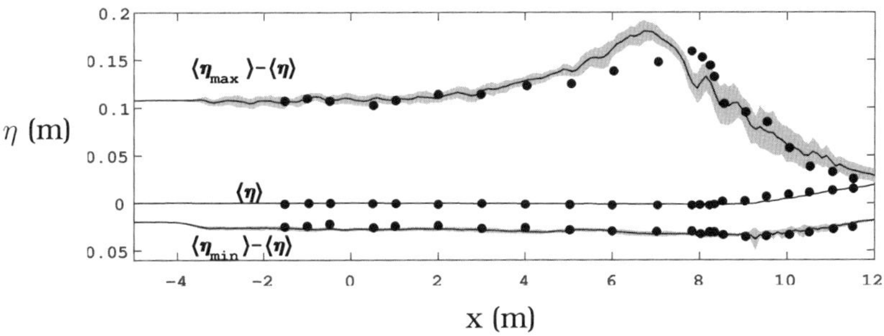

Fig. 7.44 Comparison of modelled (lines) and measured (circles, from Ting and Kirby, 1994) surface elevation envelopes (top and bottom lines) and mean water levels (middle lines) for the plunging breaking case. Results in (a), (b) and (c) correspond to Cases 3, 4 and 5, respectively. Solid lines represents the mean surface elevation and the shaded area represents plus and minus one standard deviation. Larsen (2018).

ceding sub-section for spilling breaking waves), we will focus exclusively on results utilizing the stabilized k-ω model of Larsen and Fuhrman (2018) in the remainder of what follows, as only this model avoids creating a polluted turbulence environment prior to breaking. Specifically, we will consider three different set-ups corresponding to the settings from Cases 3–5 in Table 7.3, which will again be referred to by their respective case numbers for simplicity and consistency.

Similar to before, the resulting evolution of the free surface envelope, as well as the mean water level, is presented in Fig. 7.44, here exemplified with Case 5 ($\lambda_1 = 0.2$, $\lambda_2 = 0.05$). This corresponds to the recommended settings from the spilling breaking case. The shaded region again represents plus or minus one standard deviation, thus indicating the wave-to-wave variability. The model results in a more rapid steepening of the waves due to shoaling (i.e. prior to breaking) than was measured. This is not related to the formal stabilization of the models, as similar results have been found using standard (non-stabilized) RANS models e.g. in Brown et al. (2016) as well as in the LES model of Christensen (2006). Results using other stabilized settings (Cases 3 and 4) are quite similar, and are hence not shown for the sake of brevity.

The resulting averaged turbulent kinetic energy density, $k = 1/2\overline{u_i'u_i'}$, and undertow velocity profiles are respectively presented at seven cross-shore positions in Figs. 7.45 and 7.46. These correspond to one position prior to breaking (sub-plot a), one at approximately the break point (sub-plot b) and five positions in the surf zone (sub-plots c–g). Careful evaluation

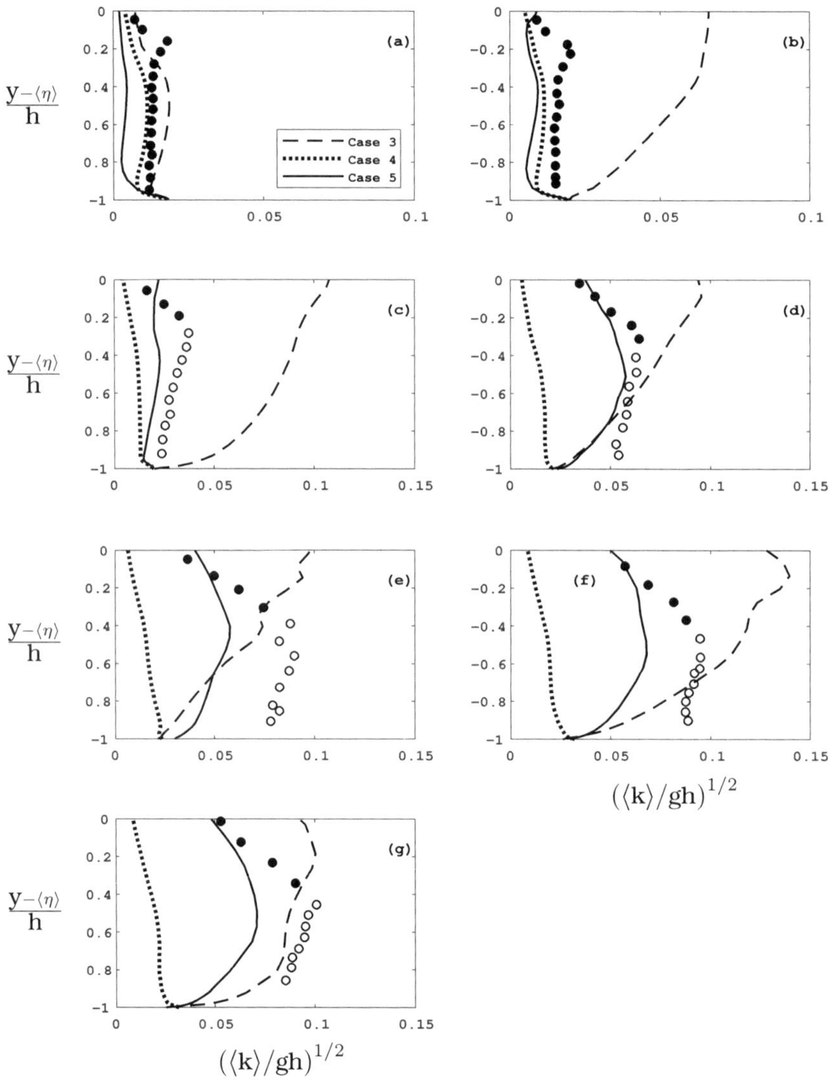

Fig. 7.45 Comparison of modelled (lines) and measured (open circles: from Eq. 7.116, filled circles: from Eq. 7.117) turbulent kinetic energy k profiles (plunging breaking case) at (a) $x = 7.95$ m ($\tilde{x} = (x - x_b)/h_b = -3.247$, $x_b = 7.795$ m), (b) $x = 7.795$ m ($\tilde{x} = 0$), (c) $x = 8.345$ m ($\tilde{x} = 3.571$), (d) $x = 8.795$ m ($\tilde{x} = 6.494$), (e) $x = 9.295$ m ($\tilde{x} = 9.740$), (f) $x = 9.795$ m ($\tilde{x} = 12.987$), (g) $x = 10.395$ m ($\tilde{x} = 16.883$). Larsen (2018).

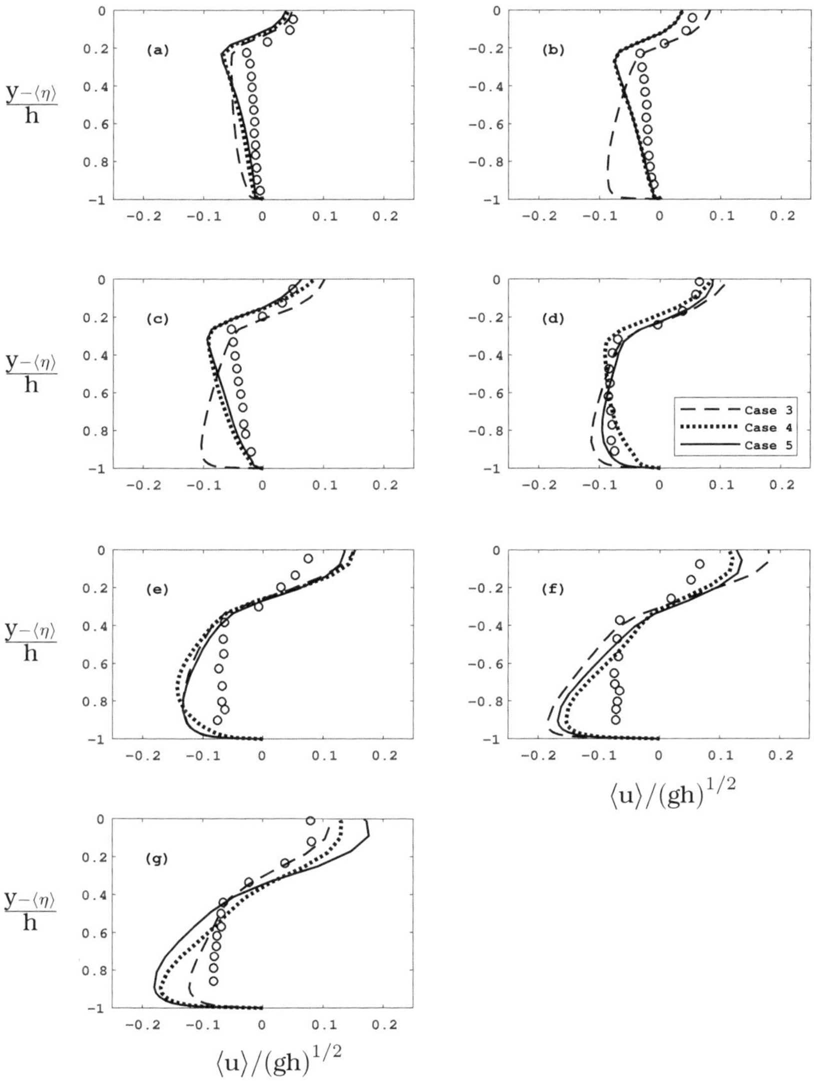

Fig. 7.46 Comparison of modelled (lines) and measured (circles) undertow velocity profiles (plunging breaking case) at (a) $x = 7.95$ m ($\tilde{x} = (x - x_b)/h_b = -3.247$), (b) $x = 7.795$ m ($\tilde{x} = 0$), (c) $x = 8.345$ m ($\tilde{x} = 3.571$), (d) $x = 8.795$ m ($\tilde{x} = 6.494$), (e) $x = 9.295$ m ($\tilde{x} = 9.740$), (f) $x = 9.795$ m ($\tilde{x} = 12.987$), (g) $x = 10.395$ m ($\tilde{x} = 16.883$). Larsen (2018).

of these plots reveals the following:

(1) The result with the Wilcox (2006) stress limiter off (Case 3 with $\lambda_1 = 0$, dashed lines) severely over-predicts turbulence levels in the outer surf zone, whereas the default value suggested by Wilcox (2006) (Case 4, $\lambda_1 = 0.875$, dotted lines) damps turbulence too much in the inner surf zone. While not perfect, Case 5 (with the intermediate value $\lambda_1 = 0.2$, solid lines) produces the best overall evolution of the turbulent kinetic energy throughout the surf zone, as seen in Fig. 7.45;
(2) The over-predicted turbulent kinetic energy for Case 3 leads to qualitatively erroneous structure of the undertow profile in the outer surf zone, Fig. 7.46b,c;
(3) All of the variations presented over-estimate the magnitude of the undertow in the inner surf zone by roughly a factor of two. This is similar to the results achieved with the spilling breakers, Fig. 7.42g,h;
(4) On balance, the best overall result tested seemingly corresponds to Case 5 (with $\lambda_1 = 0.2$ and $\lambda_2 = 0.05$). Note that these were also the recommended values based on the overall assessment of the spilling breaking results.

The observations discussed above based on the CFD simulation of plunging breaking waves are very much in line with expectations based on the previously-considered simulations of the spilling breaking case in Section 7.6.4. That the setup used for Case 5 ($\lambda_1 = 0.2$, $\lambda_2 = 0.05$) arguably gives the best overall results for both the considered spilling and plunging cases seems to support the use of the intermediate value $\lambda_1 = 0.2$ for the Wilcox (2006) stress-limiter coefficient in the general CFD simulation of breaking waves, within the stabilized k-ω model of Larsen and Fuhrman (2018).

The present sub-section has considered simulation of and comparison against plunging breaking wave experiments at relatively small laboratory scales (Ting and Kirby, 1994). It can additionally be noted that extensive experimental campaigns considering large-scale plunging breaking waves over barred cross-shore profiles have been recently conducted by van der Zanden et al. (2016, 2017, 2018, 2019), van der A et al. (2017) and Fromant et al. (2019). These additional large-scale data sets are likely to be of considerable importance in the further validation of CFD models in the future. The book cover, in fact, depicts snapshots from CFD simulations of plunging breaking waves based on the bichromatic wave experiments of van der Zanden et al. (2019) (the book cover plots are provided courtesy of Dr. Bjarke Eltard Larsen).

7.6.6 *Summary and concluding remarks*

The present section on the numerical simulation of breaking waves has systematically analyzed several widely-used two-equation turbulence closure models for applications involving surface waves. Building on the prior analysis of Mayer and Madsen (2000) who proved conditional instability, the analysis of Larsen and Fuhrman (2018) presented above demonstrates that such models are in fact *unconditionally unstable* in the nearly-potential flow region beneath waves, leading to exponential growth of the turbulent kinetic energy and eddy viscosity. This explains the marked over-production of turbulence even prior to wave breaking, which has been commonly observed in simulations of breaking waves over the past two decades. Local surface modifications such as adding buoyancy-modification, while creating a sink of turbulent kinetic energy near the air-water interface, do little to avoid this problem as a whole (see again Figs. 7.39a,b and 7.43a).

The analysis of Larsen and Fuhrman (2018) likewise shows how such models can be easily stabilized via simple modification to the eddy viscosity, requiring minimal changes to existing codes. The results shown above (e.g. Fig. 7.39c for spilling and Fig. 7.43b for plunging breaking waves) demonstrate that only such stabilized models avoid creating a computational environment polluted by unphysical generation of turbulence prior to incipient wave breaking, commonly the very phenomenon of interest in such simulations. Given the potentially adverse effects of doing otherwise, apparent from several previous studies as well as demonstrated herein, it is recommended that formally stabilized turbulence models be utilized in any future CFD studies involving surface waves (non-breaking or breaking), in which RANS equations are coupled with two-equation closure models.

7.7 References

(1) Aagaard, T., Hughes, M., Baldock, T., Greenwood, B., Kroon, A. and Power, H. (2012): Sediment transport processes and morphodynamics on a reflective beach under storm and non-storm conditions. Marine Geology, vol. 326–328, 154–165.

(2) Adityawan, M.B., Tanaka, H. and Lin, P. (2013): Boundary layer approach in the modeling of breaking solitary wave runup. Coast. Eng., vol. 73, 167–177.

(3) Alfredsson, P.H., Johansson, A.V., Haritonidis, J.H. and Eckelmann, H. (1988): The fluctuating wall shear stress and the velocity field in

the viscous sublayer. Phys. Fluids, vol. 31, 1026–1033.
(4) Alsina, J.M., Caceres, I., Brocchini, M. and Baldock, T.E. (2012): An experimental study of sediment transport and bed evolution under different swash zone morphodynamics. Coast. Eng., vol. 68, 31–43.
(5) Alsina, J.M., Falchetti, S. and Baldock, T.E. (2009): Measurements and modeling of the advection of suspended sediment in the swash zone by solitary waves. Coast. Eng., vol. 56, 621–631.
(6) Archetti, R. and Brocchini, M. (2002): An integral swash zone model with friction: an experimental and numerical investigation. Coast. Eng., vol. 45, 89–110.
(7) Baldock, T.E., Alsina, J.A., Caceres, I., Vicinanza, D., Contestabile, P., Power, H. and Sanchez-Arcilla, A. (2011): Large-scale experiments on beach profile evolution and surf and swash zone sediment transport induced by long waves, wave groups and random waves. Coast. Eng., vol. 58, 214–217.
(8) Baldock, T.E., Baird, A.J., Horn, D.P. and Mason, T. (2001): Measurements and modeling of swash-induced pressure gradients in the surface layers of a sand beach. J. Geophys. Res., vol. 106, No. C2, 2653–2666.
(9) Baldock, T.E., Cox, D., Maddux, T., Killian, J. and Fayler, L. (2009): Kinematics of breaking tsunami wavefronts: A data set from large scale laboratory experiments. Coast. Eng., vol. 56, 506–516.
(10) Baldock, T.E., Manoonvoravong, P., and Pham, K.S. (2010): Sediment transport and beach morphodynamics induced by free long waves, bound long waves and wave groups. Coast. Eng., vol. 57, 898–916.
(11) Barnes, M.P., O'Donoghue, T., Alsina, J.M. and Baldock, T.E. (2009): Direct bed shear stress measurements in bore-driven swash. Coast. Eng., vol. 56, 853–867.
(12) Battjes, J.A. (1974): Surf similarity. Proc. 14th Int. Conf. Coast. Eng., vol. 1, ASCE, New York, pp. 466–480.
(13) Battjes, J.A. (1988): Surf-zone dynamics. Ann. Rev. Fluid Mech., vol. 20, 257–293.
(14) Belden, J. and Techet, A.H. (2011): Simultaneous quantitative flow measurement using PIV on both sides of the air-water interface for breaking waves, Exp. Fluids, vol. 50, 149–161.
(15) Berberovic, E., van Hinsberg, N.P., Jakirlic, S.I., Roisman, V. and Tropea, C. (2009): Drop impact onto a liquid layer of finite thickness: Dynamics of the cavity evolution, Phys. Rev., E79, article no. 036306, doi:10.1103/PhysRevE.79.036306.
(16) Berni, C., Barthelemy, E. and Michallet, H. (2013): Surf zone cross-

shore boundary layer velocity asymmetry and skewness: an experimental study on a mobile bed. J. Geophys. Res., vol. 118, doi:10.1002/jgrcc.20125.
(17) Bradford, S.F. (2000): Numerical simulation of surf zone dynamics, J. Waterw. Port Coast. Ocean. Eng.-ASCE, vol. 126, 1–13.
(18) Brocchini, M. and Baldock, T.E. (2008): Recent advances in modeling swash zone dynamics: influence of surf-swash interaction on nearshore hydrodynamics and morphodynamics. Rev. Geophys., vol. 46, RG3003.
(19) Brown, S.A., Greaves, D.M., Magar, V. and Conley, D.C. (2016): Evaluation of turbulence closure models under spilling and plunging breakers in the surf zone. Coast. Eng., vol. 114, 177–193.
(20) Buhr Hansen, J. (1990): Periodic waves in the surf zone: Analysis of experimental data. Coast. Eng., vol. 14, 19–41.
(21) Butt, T. and Russell, P. (2000): Hydrodynamics and cross-shore sediment transport in the swash zone of natural beaches: a review. J. Coast. Res., vol. 16, 255–268.
(22) Carstensen, S., Sumer, B.M. and Fredsøe, J. (2010): Coherent structures in wave boundary layers. Part 1. Oscillatory motion. J. Fluid Mech., vol. 646, 169–206.
(23) Chang, K.-A. and Liu, P.L.-F. (1998): Velocity, acceleration and vorticity under a breaking wave. Phys. Fluids, vol. 10, 327–329.
(24) Chang, Y.-H., Hwang, K.-S. and Hwung, H.-H. (2009): Large-scale laboratory measurements of solitary wave inundation on a 1:20 slope. Coast. Eng., vol. 56, 1022–1034.
(25) Chanson, H. and Jaw-Fang, L. (1997): Plunging jet characteristics of plunging breakers. Coast. Eng., vol. 31, 125–141.
(26) Chella, M.A., Bihs, H., Myrhaug, D., and Muskulus, M. (2015): Breaking characteristics and geometric properties of spilling breakers over slopes. Coast. Eng., vol. 95, 4–19.
(27) Chella, M.A., Bihs, H., Myrhaug, D. and Muskulus, M. (2016): Hydrodynamic characteristics and geometric properties of plunging and spilling breakers over impermeable slopes. Ocean Modelling, vol. 103, 53–72.
(28) Christensen, E.D. (2006): Large eddy simulation of spilling and plunging breakers. Coast. Eng., vol. 53, 463–485.
(29) Christensen, E.D. and Deigaard, R. (2001): Large eddy simulation of breaking waves. Coast. Eng., vol. 42, 53–86.
(30) Christensen, E.D., Walstra, D.-J. and Emerat, N. (2002): Vertical vari-

ation of the flow across the surf zone. Coast. Eng., vol. 45, 169–198.
(31) Conley, D.C. and Griffin, J.G. (2004): Direct measurements of bed shear stress under swash in the field. J. Geophys. Res., vol. 109, article no. C03050, doi:10.1029/2003JC001899.
(32) Cowen, E.A., Sou, I.M., Liu, P.L.-F. and Raubenheimer, B. (2003): Particle image velocimetry measurements within laboratory-generated swash zone. J. Eng. Mech., vol. 129, 119–1129.
(33) Cox, D.T. and Anderson, S.L. (2001): Statistics of intermittent surf zone turbulence and observations of large eddies using PIV. Coast. Eng. J., vol. 43, 121–131.
(34) Cox, D.T. and Kobayashi, N. (2000): Identification of intense, intermittent coherent motions under shoaling and breaking waves. J. Geophys. Res., vol. 105 (C6), 14223–14236.
(35) Cox, D.T., Kobayashi, N. and Okayasu, A. (1996): Bottom shear stress in the surf zone. J. Geophys. Res., vol. 101 (C6), 14337–14348.
(36) Dalrymple, R.A. and Rogers, B.D. (2006): Numerical modeling of water waves with the SPH method. Coast. Eng., vol. 53, 141–147.
(37) De Padova, D., Brocchini, M., Buriani, F., Corvaro, S., De Serio, F., Mossa, M. and Sibilla, S. (2018): Experimental and numerical investigation of pre-breaking and breaking vorticity within a plunging breaker. Water, vol. 10, article no. 387, doi:10.3390/w10040387.
(38) De Serio, F. and Mossa, M. (2006): Experimental study on the hydrodynamics of regular breaking waves. Coast. Eng., vol. 53, 99–113.
(39) Dean, R.G. and Dalrymple, R.A. (1984): Water Wave Mechanics for Engineers and Scientists. Prentice Hall, Englewood Cliffs, New Jersey 07632.
(40) Deigaard, R. (1993): A note on three-dimensional shear stress distribution in a surf zone. Coast. Eng., vol. 20, 157–171.
(41) Deigaard, R. and Fredsøe, J. (1989): Shear stress distribution in dissipative water waves. Coast. Eng., vol. 13, 357–378.
(42) Deigaard, R., Justesen, P. and Fredsøe, J. (1991a): Modelling of undertow by a one-equation turbulence model. Coast. Eng., vol. 15, 431–458.
(43) Deigaard, R., Mikkelsen, M.B. and Fredsøe, J. (1991b): Measurements of the bed shear stress in a surf zone. Progress Report No. 73, pp. 21–30, December 1991. Inst. Hydrodyn. and Hydraulic Eng., Tech. Univ. Denmark.
(44) Deshpande, S.S., Anumolu, L. and Trujillo, M.F. (2012): Evaluating the performance of the two-phase flow solver interfoam, Comput. Sci.

Discov., vol. 5, article no. 014016.

(45) Devolder, B., Rauwoens, P. and Troch, P. (2017): Application of a buoyancy-modified k-ω SST turbulence model to simulate wave runup around a monopile subjected to regular waves using OpenFOAM. Coast. Eng. 125, 81–94.

(46) Devolder, B., Troch, P. and Rauwoens, P. (2018): Performance of a buoyancy-modified k-ω and k-ω SST turbulence model for simulating wave breaking under regular waves using OpenFOAM. Coast. Eng., vol. 138, 49–65.

(47) Duncan, J.H. (1981): An experimental investigation of breaking waves produced by a towed hydrofoil. Proc. Roy. Soc. Lond. A, vol. 377, 331–348.

(48) Duncan, J.H. (2001): Spilling breakers. Ann. Rev. Fluid Mech., vol. 33, 519–547.

(49) Duncan, J.H., Qiao, H.B., Philomin, V. and Wenz, A. (1999): Gentle spilling breakers: crest profile evolution, J. Fluid Mech., vol. 379, 191–222.

(50) Eckelmann, H. (1974): The structure of the viscous sublayer and the adjacent wall region in a turbulent channel flow. J. Fluid Mech., vol. 65, 439–459.

(51) Elfrink, B. and Baldock, T. (2002): Hydrodynamics and sediment transport in the swash zone: a review and perspectives. Coast. Eng., vol. 45, 149–167.

(52) Fenton, J.D. (1988): The numerical solution of steady water wave problems. Comput. Geosci., vol. 14, 357–368.

(53) Fredsøe, J. and Deigaard, R. (1992): Mechanics of Coastal Sediment Transport. World Scientific.

(54) Fromant, G., Hurther, D., van der Zanden, J., van der A, D.A., Caceres, I., O'Donoghue, T. and Ribberink, J.S. (2019): Wave boundary layer hydrodynamics and sheet flow properties under large-scale plunging-type breaking waves. J. Geophys. Res.-Oceans, vol. 124, 75–98.

(55) Fuhrman, D.R., Baykal, C., Sumer, B. M., Jacobsen, N.G. and Fredsøe, J. (2014): Numerical simulation of wave-induced scour and backfilling processes beneath submarine pipelines. Coast. Eng., vol. 94, 10–22.

(56) Fuhrman, D.R., Fredsøe, J. and Sumer, B. M. (2009): Bed slope effects on turbulent wave boundary layers: 1. Model validation and quantification of rough-turbulent results, J. Geophys. Res., vol. 114, article no. C03024, doi:10.1029/2008JC005045.

(57) Fuhrman, D.R. and Madsen, P.A. (2008): Surf similarity and solitary

wave runup. J. Waterw. Port Coast. Ocean Eng.-ASCE, vol. 134, 195–198.
(58) Galvin, C.J. (1968): Breaker type classification on three laboratory beaches. J. Geophys. Res., vol. 73, 3651–3659.
(59) Garcez Faria, A.F., Thornton, E.B., Lippmann, T.C. and Stanton, T.P. (2000): Undertow over a barred beach. J. Geophys. Res., vol. 105 (C7), 16999–17010.
(60) Grasso, F., Michallet, H., Barthelemy, E. and Certain, R. (2009): Physical modeling of intermediate cross-shore beach morphology: Transients and equilibrium states. J. Geophys. Res., vol. 114, article no. C09001, doi:10.1029/2009JC005308.
(61) Grasso, F., Michallet, H. and Barthelemy, E. (2011): Sediment transport associated with morphological beach changes forced by irregular asymmetric, skewed waves. J. Geophys. Res., vol. 116, article no. C03020, doi:10.1029/2010JC006550.
(62) Grilli, S.T., Svendsen, I.A. and Subramanya, R. (1997): Breaking criterion and characteristics for solitary waves on slopes. J. Waterw. Port Coast. Ocean Eng.-ASCE, vol. 123, 102–112.
(63) Grue, J. and Jensen, A. (2006): Experimental velocities and accelerations in very steep wave events in deep water, Eur. J. Mech. B/Fluids, vol. 25, pp. 554–564.
(64) Guannel, G. and Ozkan-Haller, H.T. (2014): Formulation of the undertow using linear wave theory. Phys. Fluids, vol. 26, article no. 056604.
(65) Harlow, F.H. and Welch, J.E. (1965): Numerical calculation of time-dependent viscous incompressible flow of fluid with free surface. Phys. Fluids, vol. 8, 2182–2189.
(66) Heller, V., Unger, J., and Hager, W. (2005): Tsunami runup - A hydraulic perspective. J. Hydraul. Eng., vol. 131, 743–747.
(67) Hieu, P.D., Katsutohi, T. and Ca, V.T. (2004): Numerical simulation of breaking waves using a two-phase flow model. App. Math. Mod., vol. 28, 983–1005.
(68) Higuera, P., Liu, P.L.-F., Lin, C., Wong, W.-Y. and Kao, M.-J. (2018): Laboratory-scale swash flows generated by a non-breaking solitary wave on a steep slope. J. Fluid Mech., vol. 847, 186–227.
(69) Hirt, C.W., and Nichols, B.D. (1981): Volume of fluid (VOF) method for the dynamics of free boundaries, J. Computational Phys., vol. 39, 201–225, doi:10.1016/0021-9991(81)90145-5.
(70) Hsiao, S.-C., Hsu, T.W., Lin, T.-C. and Chang, Y.-H. (2008): On the evolution and runup of breaking solitary waves on a mild sloping beach.

Coast. Eng., vol. 55, 975–988.

(71) Hsu, T.W., Hsieh, C.M., Tsai, C.Y. and Ou, S.H. (2015): Coupling VOF / PLIC and embedding method for simulating wave breaking on a sloping beach, J. Marine Sc. Tech.-Taiwan, vol. 23, 498–507.

(72) Hsu, T.-J., Sakakiyama, T. and Liu, P.L.-F. (2002): A numerical model for wave motions and turbulence flows in front of a composite breakwater, Coast. Eng., vol. 46, 25–50.

(73) Huang, Z.-C., Hsiao, S.-C., Hwung, H.-H. and Chang, K.-A. (2009): Turbulence and energy dissipations of surf-zone spilling breakers. Coast. Eng., vol. 56, 733–746.

(74) Huang, Z.-C., Hwung, H.-H., Hsiao, S.-C. and Chang, K.-A. (2010): Laboratory observation of boundary layer flow under spilling breakers in surf zone using particle image velocimetry. Coast. Eng., vol. 57, 343–357.

(75) Hunt, I.A. (1959): Design of seawalls and breakwaters. J. Waterw. Harbors Div. ASCE, vol. 85 (WW3), 123–152.

(76) Jacobsen, N.G., Fredsøe, J., Jensen, J.H. (2014): Formation and development of a breaker bar under regular waves. Part 1: Model description and hydrodynamics, Coast. Eng., vol. 88, 182–193.

(77) Jacobsen, N.G., Fuhrman, D.R. and Fredsøe, J. (2012): A wave generation toolbox for the open-source CFD library: OpenFoam®. Int. J. Numer. Meth. Fluids, vol. 70, 1073–1088.

(78) Jansen, P.C.M. (1986): Laboratory observations of the kinematics in the aerated region of breaking waves. Coast. Eng., vol. 9, 453–477.

(79) Jensen, A., Pedersen, G.K. and Wood, D.J. (2003): An experimental study of wave runup at a steep beach. J. Fluid Mech., vol. 486, 161–188.

(80) Jensen, B., Christensen, E.D. and Sumer, B.M. (2014): Pressure-induced forces and shear stresses on rubble mound breakwater armour layers in regular waves. Coast. Eng., vol. 91, 60–75.

(81) Jensen, B.L., Sumer, B. M. and Fredsøe, J. (1989): Turbulent oscillatory boundary layers at high Reynolds numbers. J. Fluid Mech., vol. 206, 265–297.

(82) Jeong, J. and Hussain, F. (1995): On the definition of a vortex. J. Fluid Mech., vol. 285, 69–94.

(83) Kim, Y. (2018): A Numerical Investigation of Turbulent Coherent Structures and Sediment Transport under Waves in Shallow Coastal Zone. PhD Thesis, Univ. of Delaware, Newark, DE, USA.

(84) Kim, Y. and Hsu, T.-J. (2018, private communication): Large eddy

simulation of plunging breaking waves. Center for Appl. Coastal Research, University of Delaware, Newark, DE 19716, USA.
(85) Kim, Y., Zhou, Z., Hsu, T.-J. and Puleo, J.A. (2017): Large eddy simulation of dam-break-driven swash on a rough-planar beach. J. Geophys. Res.-Oceans, vol. 122(2), 1274–1296, https://doi.org/10.1002/2016JC012366.
(86) Kimmoun, O. and Branger, H. (2007): A particle image velocimetry investigation on laboratory surf-zone breaking waves over a sloping beach. J. Fluid Mech., vol. 588, 353–397.
(87) Klostermann, J., Schaake, K. and Schwarze, R. (2012): Numerical simulation of a single rising bubble by VOF with surface compression, Int. J. Numer. Meth. Fluids, vol. 71(8), 960–982, doi:10.1002/fld.3692.
(88) Kobayashi, N. and Karjadi, E.A. (1994): Surf-similarity parameter for breaking solitary-wave runup. J. Waterw. Port Coast. Ocean Eng.-ASCE, vol. 120, 645–650.
(89) Kobayashi, N. and Lawrence, A.R. (2004): Cross-shore sediment transport under breaking solitary waves. J. Geophys. Res., vol. 109, article no. C03047, doi:10.1029/2003JC002084.
(90) Lakehal, D. and Liovic, P. (2011): Turbulence structure and interaction with steep breaking waves. J. Fluid Mech., vol. 674, 522–577.
(91) Lara, J.L., Losada, I.J. and Cowen, E.A. (2002): Large-scale turbulence structures over an immobile gravel-bed inside the surf zone. In: Proc. 28th Int. Conf. Coast. Eng., vol. 1, pp. 1050–1061.
(92) Lara, J.L., Losada, I.J. and Liu, P.L.-F. (2006): Breaking waves over a mild gravel slope: Experimental and numerical analysis. J. Geophys. Res., vol. 111, article no. C11019, doi:10.1029/2005JC003374.
(93) Lara, J.L., Losada, I.J., Maza, M. and Guanche, R. (2011a): Breaking solitary wave evolution over a porous underwater step. Coast. Eng., vol. 58, 837–850.
(94) Lara, J.L., Ruju, A. and Losada, I.J. (2011b): Reynolds averaged Navier-Stokes modelling of long waves induced by a transient wave group on a beach. Proc. Roy. Soc. A., vol. 467, 1215–1242.
(95) Larsen, B.E. (2018): Tsunami-Seabed Interactions. PhD Thesis (Principal Supervisor: D.R. Fuhrman). Technical University of Denmark.
(96) Larsen, B.E. and Fuhrman, D.R. (2018): On the over-production of turbulence beneath surface waves in Reynolds-averaged Navier-Stokes models. J. Fluid Mech., vol. 853, 419–460.
(97) Larsen, B.E. and Fuhrman, D.R. (2019): Full-scale CFD simulation of tsunamis. Part 1: Model validation and runup. Coast. Eng., vol. 151,

22–41.
(98) Larsen, B.E., Fuhrman, D.R. and Roenby, J. (2019): Performance of interFoam on the simulation of progressive waves. Coast. Eng. J., vol. 61, 380–400.
(99) Launder, B.E. and Sharma, B.I. (1974): Application of the energy-dissipation model or turbulence to the calculation of flow near a spinning disc. Lett. Heat Mass Transfer, vol. 1, 131–138.
(100) Lemos, C.M. (1992): Wave breaking: A numerical study. In: Lecture Notes in Engineering, vol. 71, Springer, Berlin.
(101) Li, J., Qi, M. and Fuhrman, D.R. (2019): Numerical modeling of flow and morphology induced by a solitary wave on a sloping beach. Appl. Ocean Res., vol. 82, 259–273.
(102) Li, Y. and Raichlen, F. (2001): Solitary wave runup on plane slopes. J. Waterw. Port Coast. Ocean Eng.-ASCE, vol. 127, 33–44.
(103) Li, Y. and Raichlen, F. (2002): Non-breaking and breaking solitary wave runup. J. Fluid Mech., vol. 465, 295–318.
(104) Li, Y. and Raichlen, F. (2003): Energy balance model for breaking solitary wave runup. J. Waterw. Port Coast. Ocean Eng.-ASCE, vol. 129, 47–59.
(105) Lin, P., Chang, K.-A. and Liu, P.L.-F. (1999): Runup and rundown of solitary waves on sloping beaches. J. Waterw., Port Coastal Ocean Eng.-ASCE, vol. 125, 247–255.
(106) Lin, P. and Liu, P.L.F. (1998): A numerical study of breaking waves in the surf zone. J. Fluid Mech., vol. 359, 239–264.
(107) Lin, T.-C., Wu, H.-L., Lo, S.-M., Hsiao, S.-C. and Hwung, H.-H. (2015): Flow characteristics of spilling and plunging breakers in the surf zone. J. Marine Sci. Tech., vol. 23, 638–648.
(108) Lodahl, C.R., Sumer, B.M. and Fredsøe, J. (1998): Turbulent combined oscillatory flow and current in a pipe. J. Fluid Mech., vol. 373, 313–348.
(109) Longo, S., Petti, M. and Losada, I.J. (2002): Turbulence in the swash and surf zones: a review. Coast. Eng., vol. 45, 129–147.
(110) Losada, I.J., Lara, J.L., Christensen, E.D. and Garcia, N. (2005): Modelling of velocity and turbulence fields around and within low-crested rubble-mound breakwaters. Coast. Eng., vol. 52, 887–913.
(111) Ma, G., Shi, F. and Kirby, J.T. (2011): A polydisperse two-fluid model for surf zone bubble simulation. J. Geophys. Res., vol. 116, article no. C05010, doi:10.1029/2010JC006667.
(112) Madsen, P.A., Fuhrman, D.R. and Schäffer, H.A. (2008): On the solitary wave paradigm for tsunamis. J. Geophys. Res., vol. 113, article

no. C12012, doi:10.1029/2008JC004932.
(113) Makris, C.V., Memos, C.D. and Krestenitis, Y.N. (2016): Numerical modeling of surf zone dynamics under weakly plunging breakers with SPH method. Ocean Model., vol. 98, 12–35.
(114) Masselink, G., Evans, D., Hughes, M.G., Russel, P. (2005): Suspended sediment transport in the swash zone of a dissipative beach. Marine Geology, vol. 216, 169–189.
(115) Masselink, G. and Puleo, J.A. (2006): Swash-zone morphodynamics. Cont. Shelf Res., vol. 26, 661–680.
(116) Matsunaga, N. and Honji, H. (1980): The backwash vortex. J. Fluid Mech., vol. 99, 813–815.
(117) Mayer, S. and Madsen, P.A. (2000): Simulations of breaking waves in the surf zone using a Navier-Stokes solver. Proc. 27th Int. Conf. Coast. Eng. (Sydney, Australia), pp. 928–941.
(118) Menter, F.R. (1994): Two-equation eddy-viscosity turbulence models for engineering applications. AIAA J., vol. 32, 1598–1605.
(119) Miller, R.L. (1976): Role of vortices in surf zone prediction: sedimentation and wave forces. In Beach and Nearshore Sedimentation, Soc. Econ. Paleontol. Mineral. Spec. Publ. 24, ed. R.A. Davis, R.L. Ethington, 92–114.
(120) Mitchell, J.E. and Hanratty, T.J. (1966): A study of turbulence at a wall using an electrochemical wall shear-stress meter. J. Fluid Mech., vol. 26, 199–221.
(121) Nadaoka, K., Hino, M. and Koyano, Y. (1989): Structure of the turbulent flow field under breaking waves in the surf zone. J. Fluid Mech., vol. 204, 359–387.
(122) Newberger, P.A. and Allen, J.S. (2007): Forcing a three-dimensional, hydrostatic, primitive-equation model for application in the surf zone: 1. Formulation. J. Geophys. Res., vol. 112, article no. C08018.
(123) O'Donoghue, T., Pokrajac, D. and Hondebrink, L. (2010): Laboratory and numerical study of dambreak-generated swash on impermeable slopes. Coast. Eng., vol. 57, 513–530.
(124) Pedersen, G. and Gjevik, B. (1983): Runup of solitary waves. J. Fluid Mech., vol. 135, 283–299.
(125) Puleo, J.A., Lanckriet, T. and Wang, P. (2012): Near bed cross-shore velocity profiles, bed shear stresses and friction on the foreshore of a microtidal beach. Coast. Eng., vol. 68, 6–16.
(126) Reniers, A.J.H.M., Thornton, E.B., Stanton, T.P. and Roelvink, J.A. (2004): Vertical flow structure during Sandy Duck: Observations and

modeling. Coast. Eng., vol. 51, 237–260.
(127) Sakai, T., Mizutani, T., Tanaka, H. and Tada, Y. (1986): Vortex formation in plunging breakers. In: Proc. 20th Conf. Coast. Eng., Vol. 1 (Taipei, Taiwan), pp. 272–285.
(128) Scott, C.P., Cox, D.T., Maddux, T.B. and Long, J.W. (2005): Large-scale laboratory observations of turbulence on a fixed barred beach. Meas. Sci. Technol. 16, 1903–1912.
(129) Seelam, J.K. and Baldock, T.E. (2009): Direct bed stress measurement under solitary tsunami-type waves and breaking tsunami wave fronts. In: Coastal Dynamics 2009, Paper No. 49.
(130) Seelam, J.K., Guard, P.A. and Baldock, T.E. (2011): Measurement and modeling of bed shear stress under solitary waves. Coast. Eng., vol. 58, 937–947.
(131) Shao, S. (2006): Simulation of breaking wave by SPH method coupled with k-ε model. J. Hydraul. Res., vol. 44, 338–349.
(132) Smagorinsky, J. (1963): General circulation experiments with the primitive equations: I. The basic experiment. Monthly Weather Rev., vol. 91, 99–164.
(133) Sou, I.M., Cowen, E.A. and Liu, P.L.-F. (2010): Evolution of the turbulence structure in the surf and swash zones. J. Fluid Mech., vol. 644, 193–216.
(134) Stansby, P.K. and Feng, T. (2005): Kinematics and depth-integrated terms in surf zone waves from laboratory measurements. J. Fluid Mech., vol. 529, 279–310.
(135) Stive, M.J.F. (1984): Energy dissipation in waves breaking on gentle slopes. Coast. Eng., vol. 8, 99–127.
(136) Stive, M.J.F. (1985): A scale comparison of waves breaking on a beach. Coast. Eng. 9, 151–158.
(137) Stive, M.J.F. and de Vriend, H.J. (1994): Shear stresses and mean flow in shoaling and breaking waves. In: Proc. 24th Int. Conf. Coast. Eng., pp. 594–608.
(138) Stive, M.J.F. and Wind, H.G. (1982): A study of radiation stress and set-up in the nearshore region. Coast. Eng. 6, 1–25.
(139) Sumer, B.M., Chua, L.H.C., Cheng, N.-S. and Fredsøe, J. (2003): The influence of turbulence on bedload sediment transport. J. Hydraul. Eng.-ASCE, vol. 129, 585–596.
(140) Sumer, B.M. and Fredsøe, J. (2006): Hydrodynamics Around Cylindrical Structures. World Scientific. Revised edition.
(141) Sumer, B.M., Guner, H.A.A., Hansen, N.M., Fuhrman, D.R. and

Fredsøe, J. (2013): Laboratory observations of flow and sediment transport induced by plunging regular waves. J. Geophys. Res.-Oceans, vol. 118, 6161–6182.

(142) Sumer, B.M., Jensen, P.M., Sørensen, L.B., Fredsøe, J., Liu, P.L.-F. and Carstensen, S. (2010): Coherent structures in wave boundary layers. Part 2. Solitary motion. J. Fluid Mech., vol. 646, 207–231.

(143) Sumer, B.M., Kozakiewicz, A., Fredsøe, J. and Deigaard, R. (1996): Velocity and concentration profiles in the sheet flow layer of movable bed. J. Hydraul. Eng.-ASCE, vol. 122, 549–558.

(144) Sumer, B.M., Sen, M.B., Karagali, I., Ceren, B., Fredsøe, J., Sottile, M., Zilioli, L. and Fuhrman, D.R. (2011): Flow and sediment transport induced by a plunging solitary wave. J. Geophys. Res., vol. 116, article no. C01008, doi: 10.1029/2010JC006435.

(145) Svendsen, I.A. (1984a): Mass flux and undertow in a surf zone. Coast. Eng., vol. 8, 347–365.

(146) Svendsen, I.A. (1984b): Wave heights and set-up in a surf zone. Coast. Eng., vol. 8, 303–329.

(147) Svendsen, I.A. (2006): Introduction to Nearshore Hydrodynamics. World Scientific.

(148) Synolakis, C. (1987): The runup of solitary waves. J. Fluid Mech., vol. 185, 523–545.

(149) Synolakis, C. and Skjelbreia, J.E. (1993): Evolution of maximum amplitude of solitary waves on plane beaches. J. Waterw. Port Coast Ocean Eng.-ASCE, vol. 119, 323–342.

(150) Ting, F.C.K. (2006): Large-scale turbulence under a solitary wave. Coast. Eng., vol. 53, 441–462.

(151) Ting, F.C.K. (2008): Large-scale turbulence under a solitary wave: Part 2 Forms and evolution of coherent structures. Coast. Eng., vol. 55, 522–536.

(152) Ting, F.C.K. (2013): Laboratory measurements of large-scale near-bed turbulent flow structures under plunging regular waves. Coast. Eng., vol. 77, 120–139.

(153) Ting, F.C.K. and Kirby, J.T. (1994): Observations of undertow and turbulence in a laboratory surf zone. Coast. Eng., vol. 24, 51–80.

(154) Ting, F.C.K. and Kirby, J.T. (1995): Dynamics of surf-zone turbulence in a strong plunging breaker. Coast. Eng., vol. 24, 177–204.

(155) Ting, F.C.K. and Kirby, J.T. (1996): Dynamics of surf-zone turbulence in a spilling breaker. Coast. Eng. 27, 131–160.

(156) Ting, F.C.K. and Nelson, J.R. (2011): Laboratory measurements of

large-scale near-bed turbulent flow structures under spilling regular waves. Coast. Eng., vol. 58, 151–172.

(157) Tonkin, S., Yeh, H., Kato, F. and Sato, S. (2003): Tsunami scour around a cylinder. J. Fluid Mech., vol. 496, 165–192.

(158) Torres-Freyermuth, A., Puleo, J.A. and Pokrajac, D. (2013): Modeling swash-zone hydrodynamics and shear stresses on planar slopes using Reynolds-Averaged Navier-Stokes equations. J. Geophys. Res.-Oceans, vol. 118, 1019–1033, doi:10.1002/jgrc.20074.

(159) Turner, I.L., Russell, P.E. and Butt, T. (2008): Measurement of wave-by-wave bed-levels in the swash zone. Coast. Eng., vol. 55, 1237–1242.

(160) van der A, D.A., van der Zanden, J., ODonoghue, T., Hurther, D., Caceres, I., McLelland, S.J. and Ribberink, J.S. (2017): Large-scale laboratory study of breaking wave hydrodynamics over a fixed bar. J. Geophys. Res., vol. 122, 3287–3310.

(161) van der Zanden, J., van der A, D.A., Caceres, I., Hurther, D., McLelland, S.J., Ribberink, J.S. and O'Donoghue, T. (2018): Near-bed turbulent kinetic energy budget under a large-scale plunging breaking wave over a fixed bar. J. Geophys. Res.-Oceans, vol. 123, 1429–1456.

(162) van der Zanden, J., van der A, D.A., Caceres, I., Larsen, B.E., Fromant, G., Petrotta, C., Scandura, P. and Li, M. (2019): Spatial and temporal distributions of turbulence under bichromatic breaking waves. Coast. Eng., vol. 146, 65–80.

(163) van der Zanden, J., van der A, D.A., Hurther, D., Caceres, I., O'Donoghue, T. and Ribberink, J.S. (2016): Near-bed hydrodynamics and turbulence below a large-scale plunging breaking wave over a mobile barred bed profile. J. Geophys. Res.-Oceans, vol. 121, 6482–6506.

(164) van der Zanden, J., van der A, D.A., Hurther, D., Caceres, I., O'Donoghue, T. and Ribberink, J.S. (2017): Suspended sediment transport around a large-scale laboratory breaker bar. Coast. Eng., vol. 125, 51–69.

(165) Vested, M.H. (2020): Extreme Wave Interaction and Wave Loads on Offshore Wind Turbine Structures. PhD thesis, Technical University of Denmark.

(166) Watanabe, Y. and Saeki, H. (1999): Three-dimensional large eddy simulation of breaking waves. Coast. Eng. J., vol. 41, 281–301.

(167) Watanabe, Y., Saeki, H. and Hosking, R.J. (2005): Three-dimensional vortex structures under breaking waves. J. Fluid Mech., vol. 545, 291–328.

(168) Wilcox, D.C. (1988): Reassessment of the scale-determining equation for advanced turbulence models. AIAA J., vol. 26, 1299–1310.

(169) Wilcox, D.C. (2006): Turbulence Modeling for CFD, 3rd edn. DCW Industries.

(170) Xiao, H., Young, Y.L. and Prevost, J.H. (2010): Hydro- and morphodynamic modeling of breaking solitary waves over a fine sand beach. Part II: Numerical simulation. Marine Geology, vol. 269, 119–131.

(171) Xie, Z. (2013): Two-phase flow modelling of spilling and plunging breaking waves. App. Math. Model., vol. 37, 3698–3713.

(172) Yakhot, V., Thangam, S., Speziale, C.G., Orszag, S.A. and Gatski, T.B. (1991): Development of turbulence models for shear flows by a double expansion technique. Inst. Computer Application Sci. Eng., NAS1-18605.

(173) Young, Y.L., Xiao, H. and Maddux, T. (2010): Hydro- and morphodynamic modeling of breaking solitary waves over a fine sand beach. Part I: Experimental study. Marine Geology, vol. 269, 107–118.

(174) Yuksel, Y., Cevik, E.O. and Kapdasli, S. (1998): Bed shear stress distribution over beach profiles. J. Coast. Res., vol. 14, 1044–1053.

(175) Zhou, Z., Hsu, T.-J., Cox, D.T. and Liu, X. (2017): Large-eddy simulation of wave-breaking induced turbulent coherent structures and suspended sediment transport on a barred beach, J. Geophys. Res.-Oceans, vol. 122, 207–235, doi:10.1002/2016JC011884.

(176) Zhou, Z., Sangermano, J., Hsu, T.-J. and Ting, F.C.K.(2014): A numerical investigation of wave-breaking induced turbulent coherent structure under a solitary wave, J. Geophys. Res.-Oceans, vol. 119, 6952–6973, doi:10.1002/2014JC009854.

Appendix to Chapter 7: The depth-integrated momentum equation

In this appendix we will derive the depth-integrated momentum equation, which was utilized in the development of the simple undertow model in Section 7.2.2. For this purpose, consider the horizontal momentum (Navier-Stokes) equation

$$\frac{\partial u}{\partial t} + \frac{\partial (u^2)}{\partial x} + \frac{\partial (uv)}{\partial y} + \frac{1}{\rho}\frac{\partial p}{\partial x} = \frac{1}{\rho}\frac{\partial \tau}{\partial z} \tag{7.119}$$

In what immediately follows, for convenience we will place the origin at the mean water level (MWL), and integrate each term from the bed at $y = -h$ to the free surface (above MWL) at $y = \eta$. We will make extensive use of the Leibniz rule, which reads:

$$\int_{a(x)}^{b(x)} \frac{\partial f}{\partial x} dy = \frac{\partial}{\partial x}\int_{a(x)}^{b(x)} f dy + \frac{\partial a}{\partial x} f(y=a) - \frac{\partial b}{\partial x} f(y=b) \tag{7.120}$$

Integrating term by term yields the following:

$$\int_{-h}^{\eta} \frac{\partial u}{\partial t} dy = \frac{\partial}{\partial t}\int_{-h}^{\eta} u dy - \frac{\partial \eta}{\partial t} u_s \tag{7.121}$$

$$\int_{-h}^{\eta} \frac{\partial (u^2)}{\partial x} dy = \frac{\partial}{\partial x}\int_{-h}^{\eta} u^2 dy - \frac{\partial h}{\partial x} u_0^2 - \frac{\partial (\eta)}{\partial x} u_s^2 \tag{7.122}$$

$$\int_{-h}^{\eta} \frac{\partial (uv)}{\partial y} dy = u_s v_s - u_0 v_0 \tag{7.123}$$

$$\frac{1}{\rho}\int_{-h}^{\eta} \frac{\partial p}{\partial x} dy = \frac{1}{\rho}\frac{\partial}{\partial x}\int_{-h}^{\eta} p dy - \frac{1}{\rho}\frac{\partial h}{\partial x} p_0 \tag{7.124}$$

$$\frac{1}{\rho}\int_{-h}^{\eta} \frac{\partial \tau}{\partial y} dy = -\frac{\tau_0}{\rho} \tag{7.125}$$

where $u_s = u(y = \eta)$, $v_s = v(y = \eta)$, $u_0 = u(y = -h)$, $v_0 = v(y = -h)$, $p_0 = p(y = -h)$, and the pressure and shear stress at the surface are assumed to be zero. Collecting the above results in:

$$\begin{aligned}\frac{\partial}{\partial t}\int_{-h}^{\eta} u dy + \frac{\partial}{\partial x}\int_{-h}^{\eta} u^2 dy + \frac{1}{\rho}\frac{\partial}{\partial x}\int_{-h}^{\eta} p dy - \frac{1}{\rho}\frac{\partial h}{\partial x}p_0 \\ - u_0\underbrace{\left(v_0 + \frac{\partial h}{\partial x}u_0\right)}_{=0} - u_s\underbrace{\left(\frac{\partial \eta}{\partial t} + \frac{\partial \eta}{\partial x}u_s - v_s\right)}_{=0} = -\frac{\tau_0}{\rho}\end{aligned} \tag{7.126}$$

where the two underbraced quantities can be taken as zero according to the kinematic bottom condition (Eq. 5.171) and the kinematic surface condition (Eq. 5.169), respectively. The resulting depth-integrated momentum equation thus reads:

$$\frac{\partial}{\partial t}\int_{-h}^{\eta} u dy + \frac{\partial}{\partial x}\int_{-h}^{\eta} u^2 dy + \frac{1}{\rho}\frac{\partial}{\partial x}\int_{-h}^{\eta} p dy - \frac{1}{\rho}\frac{\partial h}{\partial x}p_0 = -\frac{\tau_0}{\rho} \tag{7.127}$$

Upon period-averaging and multiplying by ρ we obtain:

$$\left\langle \frac{\partial}{\partial x}\int_{-h}^{\eta} \left(\rho u^2 + p\right) dy - \frac{\partial h}{\partial x}p_0 = -\tau_0 \right\rangle \tag{7.128}$$

Let us now translate our origin such that the sea bottom is locally at $y = 0$ (the usual convention utilized in the present book) for the sake of familiarity, such that the still water level (SWL) is now locally at $y = h$. Additionally, let us decompose the free surface elevation above SWL into a mean water level ($\overline{\eta}$, now representing the set-up) as well as an oscillatory component (now taken as η). These correspond to substituting $-h \to 0$ and $\eta \to h + \overline{\eta} + \eta$, respectively, into the lower and upper limits of the definite integral in Eq. 7.128. The period-averaged pressure at the bottom is then $p_0 = \rho g(h + \overline{\eta})$, resulting in:

$$\frac{\partial}{\partial x}\left\langle \int_{0}^{h+\overline{\eta}+\eta} \left(\rho u^2 + p\right) dy \right\rangle - \rho g(h + \overline{\eta})\frac{\partial h}{\partial x} = -\left\langle \tau_0 \right\rangle \tag{7.129}$$

Since

$$\begin{aligned}-\rho g(h + \overline{\eta})\frac{\partial h}{\partial x} &= -\rho g(h + \overline{\eta})\frac{\partial (h + \overline{\eta})}{\partial x} + \rho g(h + \overline{\eta})\frac{\partial \overline{\eta}}{\partial x} \\ &= -\rho g\frac{1}{2}\frac{\partial (h + \overline{\eta})^2}{\partial x} + \rho g(h + \overline{\eta})\frac{\partial \overline{\eta}}{\partial x}\end{aligned} \tag{7.130}$$

Eq. 7.129 may be equivalently expressed as:

$$\frac{\partial}{\partial x}\left\{\left\langle \int_{0}^{h+\overline{\eta}+\eta} \left(\rho u^2 + p\right) dy \right\rangle - \frac{\rho g(h + \overline{\eta})^2}{2}\right\} + \rho g(h + \overline{\eta})\frac{\partial \overline{\eta}}{\partial x} = -\left\langle \tau_0 \right\rangle \tag{7.131}$$

or even more compactly as:

$$\frac{\partial S_{xx}}{\partial x} + \rho g(h+\overline{\eta})\frac{\partial \overline{\eta}}{\partial x} = -\langle \tau_0 \rangle \tag{7.132}$$

where

$$S_{xx} = \left\langle \int_0^{h+\overline{\eta}+\eta} \left(\rho u^2 + p\right) dy \right\rangle - \frac{\rho g(h+\overline{\eta})^2}{2} \tag{7.133}$$

is a generalized definition of the radiation stress (with the water depth h in Eq. 5.193 replaced by $h+\overline{\eta}$ in the above i.e. Eq. 7.133 accounts for the more general case where $\overline{\eta} \neq 0$).

A slightly simplified form of Eq. 7.132 is utilized in the main chapter text, appearing as Eq. 7.44 (where it was additionally assumed that $\overline{\eta} \ll h$).

Chapter 8

Diffusion and dispersion

Diffusion/dispersion is another important property of turbulent flows. Some examples are: (1) Waste products released into a recipient environment, such as a stream or the ocean, undergo rapid diffusion; (2) Smoke (or exhaust gases) released into the atmosphere diffuses rapidly when there is strong wind; (3) Likewise, dredged material (sand, silt, etc.), when dumped into the sea, is subjected to large diffusion.

The objective of this chapter is to describe the diffusion and dispersion processes in turbulent flows. First, we will study the diffusion process in a turbulent flow field (a general analysis), using a so-called single particle (Lagrangian) analysis. The latter analysis will lead to the concept of flux due to turbulent diffusion, a concept known from molecular diffusion. The latter will enable us to make an Eulerian formulation of the turbulent diffusion process.

We will subsequently turn our attention to longitudinal dispersion, a process caused by the combined action of the turbulent diffusion and the spatial velocity variation in the flow. The theory, based on the aforementioned single-particle analysis, will be applied to describe the process of longitudinal dispersion in open channels. This will be developed for both neutrally-buoyant dispersants as well as for sediment. The analysis includes two more application areas, namely longitudinal dispersion in a river, and that in an oscillating tunnel, the latter being closely associated with diffusion/dispersion in wave boundary layers. The dispersion in the surf zone is also included.

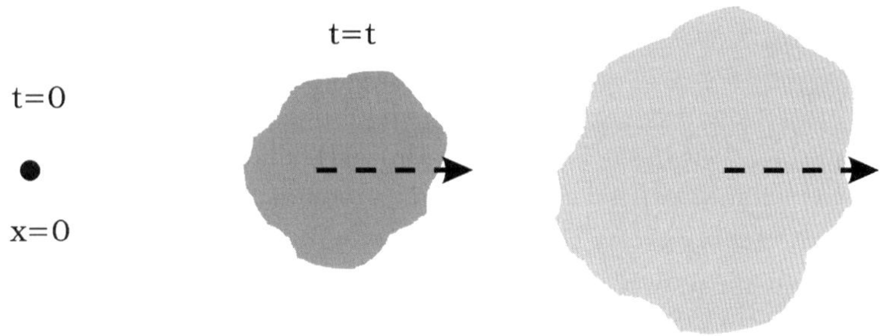

Fig. 8.1 Definition sketch.

8.1 One-particle analysis

Release a lump of dispersant into the flow in the form of an *instantaneous point source*. The point of release is $\mathbf{x} = 0$, and time $t = 0$ (Fig. 8.1). This dispersant will undergo diffusion as it is convected downstream, as sketched in Fig. 8.1. Note that we will use the term *dispersant* to refer to any kind of passive quantity such as waste products, suspended sediment, heat, etc. Also, we will assume that there is no interaction between the dispersant and the flow; the dispersant acts as a "passive" quantity.

The diffusion of the dispersant is caused by the following mechanism: The instantaneous point source is actually formed by many dispersant particles. Since the flow is turbulent, each particle will follow a random path. This will cause the particles to get farther and farther apart as they travel (Fig. 8.2). The end result will be the continuous spreading (diffusion) of the dispersant, as sketched in Fig. 8.1.

The diffusion of a passive quantity can be studied by adopting one of the following two approaches: (1) Release the dispersant in the form of a lump at $\mathbf{x} = 0$ and at time $t = 0$; and follow the cloud of the dispersant as it "grows" and is convected downstream; or (2) Release one particle at $\mathbf{x} = 0$ and at time $t = 0$, and follow its path $\mathbf{x} = \mathbf{x}(t)$; Repeat this exercise many times, and determine the statistical properties of $\mathbf{x}(t)$. This approach is called a *one-particle analysis*. In the following paragraphs, the latter approach will be adopted.

For simplicity, we shall study the process in *one dimension*. The particles are released at $x = 0$, at time $t = 0$. They follow random paths $x(t)$. The objective of the analysis is to determine the statistical properties of $x(t)$. These statistical properties are (Fig. 8.3):

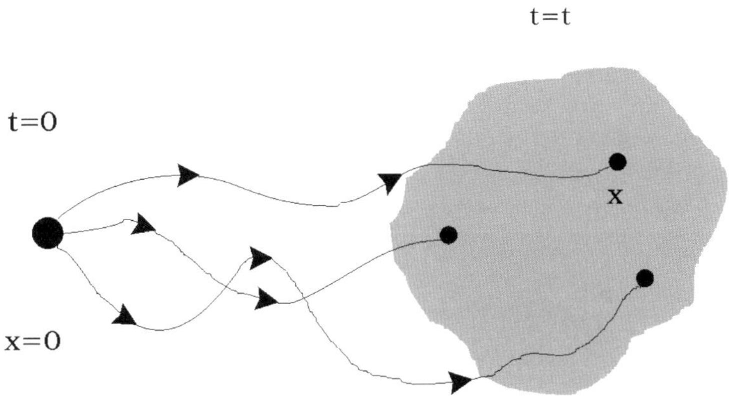

Fig. 8.2 Random paths of particles.

1. The mean value of x, namely $\overline{x}$; this will give the mean position of the dispersing cloud as it is convected downstream.
2. The variance of x, namely $\overline{(x-\overline{x})^2}$; this will give a measure of the degree of dispersion of the cloud.
3. Other statistical moments; and
4. The probability density function (p.d.f.) of the particle position, namely $p(x)$. This quantity is proportional to the concentration, as will be demonstrated later in the chapter.

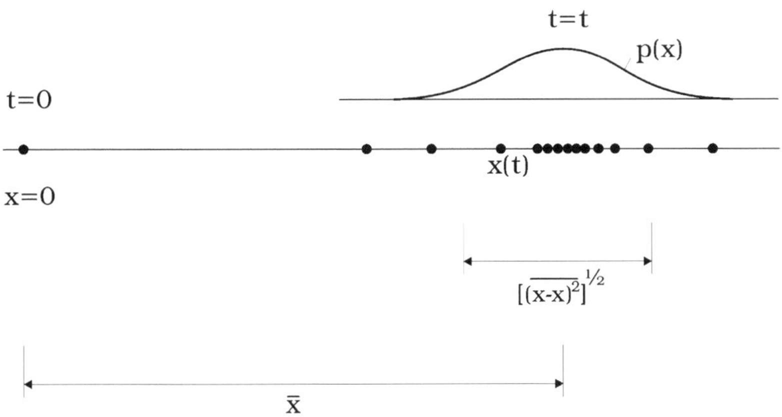

Fig. 8.3 One-dimensional diffusion.

The particle velocity (the Lagrangian velocity) is:

$$V = \frac{dx}{dt} \tag{8.1}$$

Integrating gives

$$x(t) = \int_0^t V(t')\, dt' \tag{8.2}$$

Now, ensemble averaging gives the *mean* particle position (the position of the centroid of the particle cloud):

$$\overline{x(t)} = \overline{\int_0^t V(t')\, dt'} = \int_0^t \overline{V(t')}\, dt' \tag{8.3}$$

Here, the ensemble averaging is

$$\overline{\phi} = \frac{\sum_{n=1}^{N} \phi_n}{N} \tag{8.4}$$

in which ϕ is any hydrodynamic quantity, ϕ_n the value of ϕ in one realization, and N the total number of realizations. (In the present context, one realization is one release of a particle).

The *variance*, on the other hand, can be calculated in the following way. Let

$$y(t) = x(t) - \overline{x(t)} \tag{8.5}$$

Then, from Eq. 8.2

$$\begin{aligned} y(t) = x(t) - \overline{x(t)} &= \int_0^t V(t')\, dt' - \int_0^t \overline{V(t')}\, dt' \\ &= \int_0^t [V(t') - \overline{V(t')}]\, dt' \end{aligned} \tag{8.6}$$

The integrand here is the fluctuating part of the particle velocity

$$V' = V - \overline{V} \tag{8.7}$$

Hence,

$$y(t) = \int_0^t V'(t')\, dt' \tag{8.8}$$

Differentiating $\overline{y^2(t)}$ with respect to t

$$\frac{d}{dt}\overline{y^2(t)} = 2\, \overline{y \frac{dy}{dt}} \tag{8.9}$$

and from Eq. 8.8

$$\frac{d}{dt}\,\overline{y^2(t)} = 2\,\overline{\left[\int_0^t V'(t')\,dt'\right]\,V'(t)} \tag{8.10}$$
$$= 2\int_0^t \overline{V'(t')\,V'(t)}\,dt'$$

The integrand in the preceding equation, $\overline{V'(t')\,V'(t)}$, recalling the Lagrangian correlation coefficient $R_L(t-t')$ (Eq. 4.46), may be written as:

$$\overline{V'(t')\,V'(t)} = \overline{V'^2}\,\frac{\overline{V'(t')V'(t)}}{\overline{V'^2}} = R_L(t-t')\,\overline{V'^2} \tag{8.11}$$

Then,

$$\frac{d}{dt}\,\overline{y^2(t)} = 2\,\overline{V'^2}\int_0^t R_L(t-t')\,dt' \tag{8.12}$$

To summarize at this point, (1) the mean particle position is given by Eq. 8.3; and (2) the variance of the particle position is given by Eq. 8.12.

Now, we will continue with the analysis for the following two extreme cases: (1) For small times ($t \ll T_L$); and (2) For large times ($t \gg T_L$) in which T_L is the Lagrangian macro time scale defined by

$$T_L = \int_0^\infty R_L(t)\,dt \tag{8.13}$$

(Eqs. 4.46 and 4.47, here replacing the separation time τ with t for convenience). Recall that Lagrangian macro time scale, T_L, may be interpreted as the time during which a flow structure keeps its coherence (Section 4.2.2).

8.1.1 *Diffusion for small times*

For small times ($t \ll T_L$), the particle velocity will be nearly the same as that at the point where the particle starts its travel, namely at $x = 0$ at time $t = 0$. This velocity is the Eulerian velocity at $x = 0$, namely $u(x = 0)$:

$$V(t) \simeq V(t = 0) \equiv u(x = 0) \tag{8.14}$$

The mean particle position, from Eq. 8.3 is:

$$\overline{x(t)} = \int_0^t \overline{V(t')}\,dt' = \int_0^t \overline{u(x=0)}\,dt' = \overline{u(x=0)}\int_0^t dt' = \overline{u(x=0)}\,t \tag{8.15}$$

Regarding the variance, for small times ($t \ll T_L$), first of all, $\overline{V'^2} \simeq \overline{u'^2(x=0)}$, and secondly, $R_L(t-t') \simeq 1$. Therefore, Eq. 8.12 becomes

$$\frac{d}{dt}\,\overline{y^2(t)} = 2\,\overline{u'^2(x=0)}\,t$$

and integrating gives

$$\overline{y^2(t)} = \overline{u'^2(x=0)}\; t^2 \tag{8.16}$$

Now, assume that the flow/turbulence is *homogeneous*. Then the statistical properties of velocity should be invariant with respect to the position:

$$\overline{u(x=0)} \equiv \overline{u(x)} = \overline{u} = \text{ constant} \tag{8.17}$$

and

$$\overline{u'^2(x=0)} \equiv \overline{u'^2(x)} = \overline{u'^2} = \text{ constant} \tag{8.18}$$

Thus, the mean particle position and the variance will be (from Eqs. 8.15, 8.16, 8.17, 8.18)

$$\overline{x(t)} = \overline{u}\; t \tag{8.19}$$

and

$$\overline{y^2(t)} = \overline{u'^2}\; t^2 \tag{8.20}$$

The p.d.f. of the particle position, on the other hand, will be the same as that of the Eulerian velocity provided that u should be replaced with x/t.

8.1.2 *Diffusion for large times*

Mean and variance of particle position

For large times ($t \gg T_L$), the analysis will be developed further for the following special case. Namely, the particle velocity is a stationary random function (s.r.f.) of time.

(The particle velocity is a s.r.f. of time if the probability density function of particle velocity is invariant with respect to time, i.e.,

$$p[V(t)] \equiv p[V(t+\tau)] \tag{8.21}$$

meaning that the statistical properties of the particle velocity such as the mean $\overline{V}$ and the variance $\overline{(V-\overline{V})^2} = \overline{V'^2}$ are time invariant. Homogeneity in space is analogous to being stationary in time.)

First, consider the mean particle position. From Eq. 8.3, taking $\overline{V(t')}$ outside the integral because it is constant,

$$\overline{x(t)} = \int_0^t \overline{V(t')}\, dt' = \overline{V}\; t \tag{8.22}$$

The mean particle velocity should be equal to the Eulerian mean velocity, $\overline{u}$. The fact that the Lagrangian velocity is a stationary random function of time requires the homogeneity in the Eulerian sense; therefore, $\overline{V} \equiv \overline{u}$. Hence,

$$\overline{x(t)} = \overline{u}\ t \tag{8.23}$$

Next, consider the variance of the particle position. From Eq. 8.12:

$$\frac{d}{dt}\overline{y^2(t)} = 2\ \overline{V'^2} \int_0^t R_L(t-t')\ dt' \tag{8.24}$$

Introducing the notation

$$\xi = t - t' \tag{8.25}$$

then Eq. 8.24 will be

$$\frac{d}{dt}\overline{y^2(t)} = 2\ \overline{V'^2} \int_{\xi=0}^t R_L(\xi)\ d\xi \tag{8.26}$$

Integrating by parts (see Eq. 4.222) gives

$$\overline{y^2(t)} = 2\ \overline{V'^2} \left(\int_{\xi=0}^t R_L(\xi)\ d\xi \right)\ t - 2\overline{V'^2} \int_{\xi=0}^t \xi R_L(\xi)\ d\xi \tag{8.27}$$

For large times ($t \gg T_L$), t is so large that the integrals in the preceding equation converge (provided, of course, that the form of $R_L(\xi)$ is such that the integrals do converge). Furthermore, the second term on the right hand side is small compared with the first term. Hence, for large times,

$$\overline{y^2(t)} = 2\ \overline{V'^2} \left(\int_{\xi=0}^t R_L(\xi)\ d\xi \right)\ t \tag{8.28}$$

The integral for large times can be approximated as

$$\int_{\xi=0}^t R_L(\xi)\ d\xi \simeq \int_{\xi=0}^\infty R_L(\xi)\ d\xi = T_L \tag{8.29}$$

(from Eq. 8.13). Inserting the latter equation in Eq. 8.28, and also considering that $\overline{V'^2}$ is identically equal to $\overline{u'^2}$ (due to the fact that $V(t)$ is a stationary random function of time), the variance is obtained as

$$\overline{y^2(t)} = 2\ \overline{u'^2} T_L\ t \tag{8.30}$$

To summarize, the mean particle position and the variance are given by Eqs. 8.23 and 8.30, respectively.

As seen from Eqs. 8.20 and 8.30, the variance is

$$\overline{y^2(t)} \sim t^2 \text{ for small times } (t \ll T_L) \tag{8.31}$$

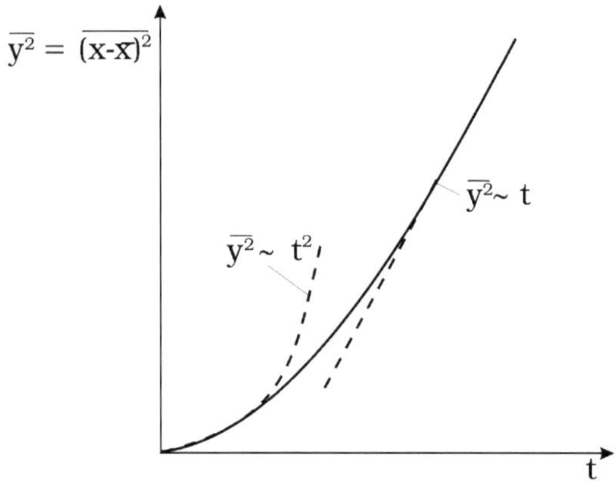

Fig. 8.4 Variance of particle position.

and

$$\overline{y^2(t)} \sim t \text{ for large times } (t \gg T_L) \tag{8.32}$$

This implies that the spreading occurs very fast in the beginning of the process ($\overline{y^2(t)} \sim t^2$), and then the rate at which the spreading occurs reaches an equilibrium stage where $\overline{y^2(t)} \sim t$, as sketched in Fig. 8.4.

Probability density function (p.d.f.) of particle position

The particle displacement (Eq. 8.2)

$$x(t) = \int_0^t V(t')\, dt'$$

can be considered as the sum of a series integral over a fixed interval, say τ,

$$x(t) = \int_0^t V(t')\, dt' = \sum_n \int_{n\tau}^{(n+1)\tau} V(t')\, dt' \tag{8.33}$$

such that each integral has only a partial statistical connection with each other. For times $t \gg T_L$, the number of such integrals becomes sufficiently large so that we have a situation like that covered by the central limit theorem (Batchelor and Townsend, 1956):

The central limit theorem states that *for a series of statistically independent terms, the p.d.f. of the sum tends to the Gaussian distribution (normal distribution), as the number of terms approaches infinity, irrespective of the probability distributions of each term, which form the sum.*

Therefore, for large times $t \gg T_L$, the p.d.f. of $x(t)$ will be

$$p(x) = \frac{1}{\sqrt{2\pi}\,\sqrt{\overline{(x-\overline{x})^2}}} \exp\left(-\frac{(x-\overline{x})^2}{2\overline{(x-\overline{x})^2}}\right) \tag{8.34}$$

i.e. the Gaussian distribution. Changing the notation to $y = x - \overline{x}$ (for convenience), we will have

$$p(y) = \frac{1}{\sqrt{2\pi}\,\sqrt{\overline{y^2}}} \exp\left(-\frac{y^2}{2\,\overline{y^2}}\right) \tag{8.35}$$

Inserting Eq. 8.30 in the preceding equation,

$$p(y) = \frac{1}{\sqrt{2\pi}\,\sqrt{2\,\overline{u'^2}T_L\,t}} \exp\left(-\frac{y^2}{2\,(2\,\overline{u'^2}T_L\,t)}\right) \tag{8.36}$$

We shall return to this equation later in the section.

Now, consider the diffusion equation

$$\frac{\partial c}{\partial t} = D\frac{\partial^2 c}{\partial y^2} \tag{8.37}$$

in which D is the diffusion coefficient, y the space coordinate, and t the time. Consider the initial condition

$$t = 0: \qquad c = \delta(y) \tag{8.38}$$

in which $\delta(y)$ is the Dirac delta function, which is defined by

$$\delta(y) = 0 \text{ for } y \neq 0$$
$$\delta(y) = \infty \text{ for } y = 0$$

and

$$\int_{-\infty}^{\infty} \delta(y)dy = 1$$

The solution to the diffusion equation with the above initial condition is

$$c(y) = \frac{1}{\sqrt{2\pi}\,\sqrt{2Dt}} \exp\left(-\frac{y^2}{2\,(2\,D\,t)}\right) \tag{8.39}$$

Now, compare the above equation with Eq. 8.36. From this comparison, the following deductions can be made:

1. The p.d.f. of the particle position for large times ($t \gg T_L$) satisfies the diffusion equation (cf., 8.37)

$$\frac{\partial p}{\partial t} = D\frac{\partial^2 p}{\partial y^2} \tag{8.40}$$

2. The diffusion coefficient of the diffusion process is

$$D = \overline{u'^2}T_L \tag{8.41}$$

which can, from Eq. 8.30, also be written as

$$D = \overline{u'^2}T_L = \frac{1}{2}\frac{d\overline{y^2(t)}}{dt} \tag{8.42}$$

D is called the *turbulent diffusion coefficient.*

The theory summarized in the preceding paragraphs was developed by Taylor (1921). It was later extended to the three-dimensional case by Batchelor (1949). For further reading, consult Batchelor and Townsend (1956) and Monin and Yaglom (1973).

Flux due to turbulent diffusion

The probability that a particle is present in the interval $(y, y + dy)$ is

$$\Pr[y < Y < y + dy] = \frac{m}{M} \tag{8.43}$$

where M is the total number of released particles, and m is the number of particles occupying the interval $(y, y + dy)$.

This probability can also be written as

$$\Pr[y < Y < y + dy] = p(y)\ dy \tag{8.44}$$

From the preceding two equations, one gets

$$p(y) = \frac{1}{M}\frac{m}{dy} \tag{8.45}$$

Here, m/dy, the number of particles per unit length, is the concentration $\bar{c}$ in the one-dimensional case. So,

$$p(y) = \frac{\bar{c}(y)}{M} \tag{8.46}$$

The probability that a particle is present in the half space $(-\infty, y)$ is

$$\Pr[-\infty < Y < y] = \frac{m_1}{M} \tag{8.47}$$

in which m_1 is the number of particles occupying the interval $(-\infty, y)$. Likewise, this probability can be written as

$$\Pr[-\infty < Y < y] = \int_{-\infty}^{y} p(y)\ dy \tag{8.48}$$

Then from Eqs. 8.47 and 8.48

$$m_1 = M \int_{-\infty}^{y} p(y)\ dy \tag{8.49}$$

Differentiating the above equation with respect to time gives

$$-\frac{\partial m_1}{\partial t} = -M \frac{\partial}{\partial t} \int_{-\infty}^{y} p(y)\ dy \tag{8.50}$$

Here $-\partial m_1/\partial t$ is the rate of change in the number of the particles which occupy the region $(-\infty, y)$, i.e., the decrease in the number of particles in the region $(-\infty, y)$ per unit time. This means that there is a continuous supply of particles from the region $(-\infty, y)$ to the region $(y, +\infty)$ at a rate $-\partial m_1/\partial t$, i.e., the number of particles passing through the section y per unit time is $-\partial m_1/\partial t$.

The expression for $p(y)$ is given in Eq. 8.35. Substituting this equation in Eq. 8.50, and integrating and then differentiating with respect to time gives

$$-\frac{\partial m_1}{\partial t} = M\ \frac{y}{2t}\ p(y) \tag{8.51}$$

Differentiating Eq. 8.35 with respect to y, on the other hand, gives

$$\frac{\partial p}{\partial y} = -\frac{y}{2Dt}\ p(y) \tag{8.52}$$

From the preceding two equations,

$$-\frac{\partial m_1}{\partial t} = -MD\frac{\partial p}{\partial y}$$

or

$$-\frac{\partial m_1}{\partial t} = -D\frac{\partial (Mp)}{\partial y} \tag{8.53}$$

From Eq. 8.46, $\overline{c} = Mp$. Inserting this in the above equation gives

$$-\frac{\partial m_1}{\partial t} = -D\frac{\partial \overline{c}}{\partial y} \tag{8.54}$$

Changing notation to

$$\Gamma = -\frac{\partial m_1}{\partial t}$$

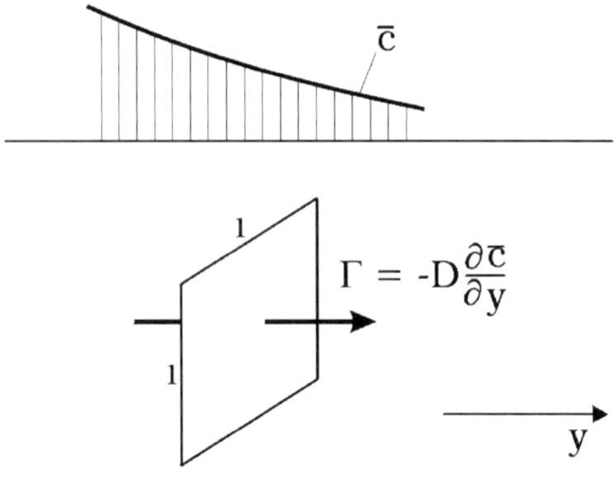

Fig. 8.5 Flux due to turbulent diffusion.

then Eq. 8.54 becomes

$$\Gamma = -D\frac{\partial \bar{c}}{\partial y} \tag{8.55}$$

The quantity Γ, the number of particles passing through the section y per unit time (Fig. 8.5), is called the *flux* (the *flux due to turbulent diffusion*). Notice the similarity between the flux due to turbulent diffusion and that due to molecular diffusion, namely $\Gamma_{mol} = -D_{mol}\frac{\partial c}{\partial y}$ (Fick's law) in which D_{mol} is the molecular diffusion coefficient.

8.2 Conservation of mass: Eulerian analysis

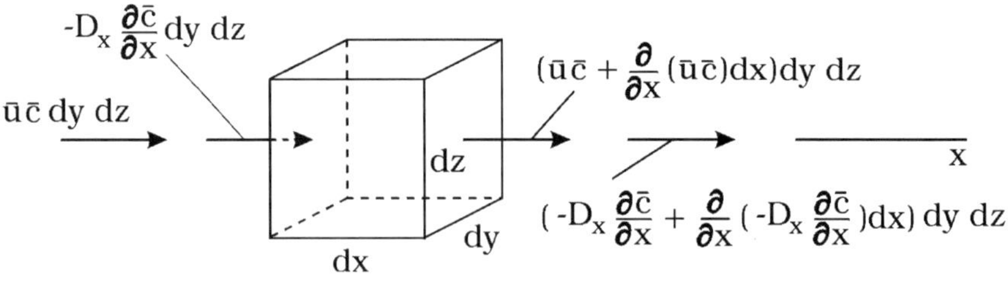

Fig. 8.6 Definition sketch.

Consider the fluid element $dx\ dy\ dz$ in Fig. 8.6. The influx of mass in the x-direction is

$$\overline{u}\,\overline{c}\,dy\,dz \qquad \text{(due to convection)} \tag{8.56}$$

and

$$-D_x\,\frac{\partial \overline{c}}{\partial x}\,dy\,dz \qquad \text{(due to turbulent diffusion)} \tag{8.57}$$

(from Eq. 8.55).

The efflux in the same direction, on the other hand, is

$$\left[\overline{u}\,\overline{c} + \frac{\partial}{\partial x}(\overline{u}\,\overline{c})\,dx\right]dy\,dz \qquad \text{(due to convection)} \tag{8.58}$$

and

$$\left[-D_x\,\frac{\partial \overline{c}}{\partial x} + \frac{\partial}{\partial x}\left(-D_x\,\frac{\partial \overline{c}}{\partial x}\right)dx\right]dy\,dz \qquad \text{(due to turbulent diffusion)} \tag{8.59}$$

Thus the net influx in the x-direction is

$$-\frac{\partial}{\partial x}(\overline{u}\,\overline{c})\,dx\,dy\,dz - \frac{\partial}{\partial x}\left(-D_x\,\frac{\partial \overline{c}}{\partial x}\right)dx\,dy\,dz \tag{8.60}$$

Similar expressions for the y- and z-directions are:

$$-\frac{\partial}{\partial y}(\overline{v}\,\overline{c})\,dx\,dy\,dz - \frac{\partial}{\partial y}\left(-D_y\,\frac{\partial \overline{c}}{\partial y}\right)dx\,dy\,dz \tag{8.61}$$

$$-\frac{\partial}{\partial z}(\overline{w}\,\overline{c})\,dx\,dy\,dz - \frac{\partial}{\partial z}\left(-D_z\,\frac{\partial \overline{c}}{\partial z}\right)dx\,dy\,dz \tag{8.62}$$

On the other hand, the rate of change of mass in the volume $dx\,dy\,dz$ is

$$\frac{\partial}{\partial t}(\overline{c}\,dx\,dy\,dz) \tag{8.63}$$

The conservation of mass requires that the rate of change of mass should be equal to the net influx. Hence, from Eqs. 8.60–8.63, and using the continuity equation (Eq. 2.6), namely

$$\frac{\partial \overline{u}}{\partial x} + \frac{\partial \overline{v}}{\partial y} + \frac{\partial \overline{w}}{\partial z} = 0 \tag{8.64}$$

one gets

$$\frac{\partial \overline{c}}{\partial t} + \overline{u}\frac{\partial \overline{c}}{\partial x} + \overline{v}\frac{\partial \overline{c}}{\partial y} + \overline{w}\frac{\partial \overline{c}}{\partial z} = \frac{\partial}{\partial x}\left(D_x\,\frac{\partial \overline{c}}{\partial x}\right) + \frac{\partial}{\partial y}\left(D_y\,\frac{\partial \overline{c}}{\partial y}\right) + \frac{\partial}{\partial z}\left(D_z\,\frac{\partial \overline{c}}{\partial z}\right) \tag{8.65}$$

or, in index notation

$$\frac{\partial \overline{c}}{\partial t} + \overline{u_i}\frac{\partial \overline{c}}{\partial x_i} = \frac{\partial}{\partial x_i}\left(D_{\underline{i}}\,\frac{\partial \overline{c}}{\partial x_i}\right) \tag{8.66}$$

and including the molecular diffusion:

$$\frac{\partial \overline{c}}{\partial t} + \overline{u_i}\frac{\partial \overline{c}}{\partial x_i} = \frac{\partial}{\partial x_i}\left((D_{\underline{i}} + D_{mol})\,\frac{\partial \overline{c}}{\partial x_i}\right) \tag{8.67}$$

However, note that, normally, $D_{mol} \ll D_{\underline{i}}$. The above equation is commonly known as the turbulent diffusion equation.

In the case of heavy particles (sediment in suspension, for example), the experimental evidence shows that reasonable agreement between theory and experiment can be achieved in most cases by assuming that the downward flux of particles is equal to $\overline{c}w_s$ (provided that the particle fall velocities are small, possibly in the Stokes range), and the turbulent diffusion coefficient of heavy particles is equal to that of fluid particles. Here, w_s is the fall velocity of heavy particles in otherwise still water. Methods for its practical calculation are outlined in the chapter Appendix. The turbulent diffusion equation in the case of heavy particles will then read

$$\frac{\partial \overline{c}}{\partial t}+\overline{u}\frac{\partial \overline{c}}{\partial x}+\overline{v}\frac{\partial \overline{c}}{\partial y}+(\overline{w}-w_s)\frac{\partial \overline{c}}{\partial z} = \frac{\partial}{\partial x}\left(D_x\frac{\partial \overline{c}}{\partial x}\right)+\frac{\partial}{\partial y}\left(D_y\frac{\partial \overline{c}}{\partial y}\right)+\frac{\partial}{\partial z}\left(D_z\frac{\partial \overline{c}}{\partial z}\right) \tag{8.68}$$

8.3 Longitudinal dispersion

8.3.1 *Mechanism of longitudinal dispersion*

The mechanism of longitudinal dispersion may be best described by reference to an open-channel flow, Fig. 8.7. Release a lump of dispersant in the form of a point source at section $x = 0$ and $t = 0$. Consider that the released dispersant is formed by many particles. Now, one such particle is subject to two effects:

1. The turbulent motion in the y-direction; Under this effect, the particle travels up and down in a random fashion.

2. The streamwise motion in the x-direction. The particle takes the velocity $\overline{u}(y)$ in which y is the vertical position of the particle, not to be confused with $y = x - \overline{x}$ in the previous section (Eq. 8.5). The particle actually takes a velocity equal to $\overline{u}(y) + u'$ in which u' is the fluctuating component of the streamwise velocity present at the instant when the particle is at the position y. However, the latter velocity is small compared with $\overline{u}(y)$, and therefore we will neglect u' in the analysis.

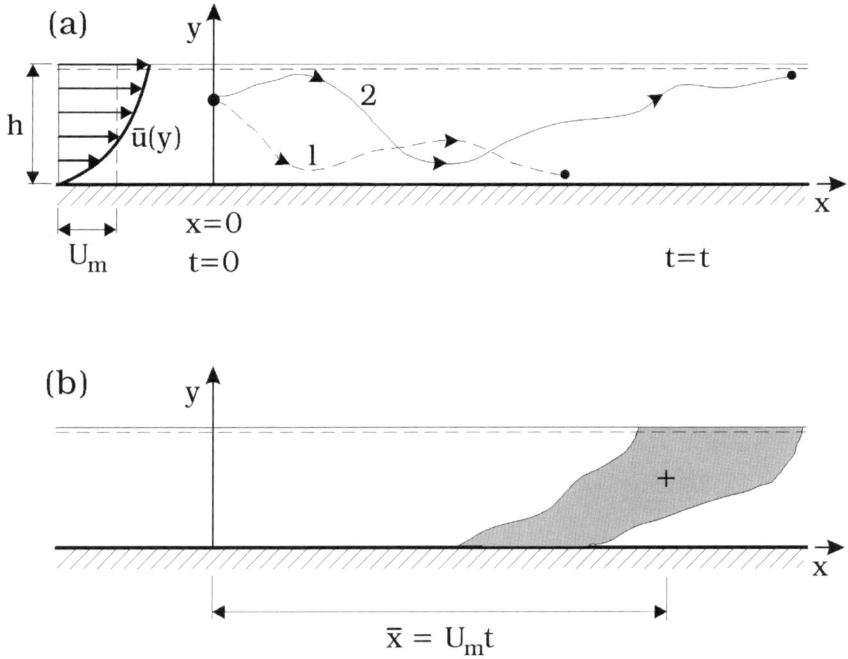

Fig. 8.7 Longitudinal dispersion.

The dispersant particles near the bottom will experience relatively smaller velocities $\overline{u}(y)$, and therefore will be moved to short distances (Path 1 in Fig. 8.7a) while those near the free surface will experience relatively larger velocities $\overline{u}(y)$, and therefore will be moved to long distances (Path 2 in Fig. 8.7a). This implies that the particles will get farther and farther apart in the longitudinal direction as they travel. Thus, the end result will be a progressive dispersion of the dispersant in the longitudinal direction, as sketched in Fig. 8.7b. This process is called the *longitudinal dispersion.*

When compared with the ordinary diffusion process studied in the previous section (cf., Fig. 8.1 and 8.2), the key element here is that the dispersion is caused mainly by the velocity variation $\overline{u}(y)$ across the depth, whereas the dispersion in the ordinary diffusion process is caused by the random paths of particles; no spatial variation in the mean velocity exists in the latter process. For the above reason, the present longitudinal diffusion process is sometimes called the mechanical dispersion.

From the preceding considerations, it can easily be seen that:

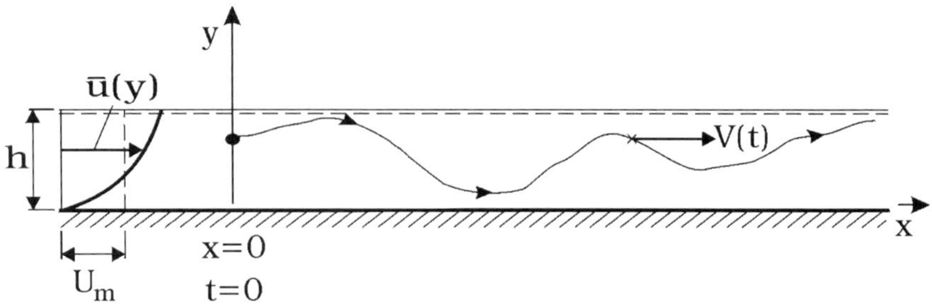

Fig. 8.8 Particle velocity V sampled from the velocity profile $\overline{u}(y)$.

1. There will be no longitudinal dispersion, when the velocity distribution is uniform across the depth; and

2. Likewise, there will be no longitudinal dispersion when there is no turbulent motion across the depth (i.e., when there is no turbulent diffusion across the depth).

8.3.2 *Application of one-particle analysis*

Fig. 8.8 shows the path of a particle. Here, $V(t)$ is the streamwise velocity of the particle. $V(t)$ is actually the sum of two velocities: (1) The velocity $\overline{u}(y)$ at the position where the particle is at that particular time t, and; (2) The x-component of the fluctuating velocity u', which occurs at that position, and at that particular time, t.

The fluctuating velocity is small compared with $\overline{u}(y)$, and therefore can be neglected, as already stated in the preceding paragraphs. Hence, the particle velocity can be taken as $V(t) = \overline{u}(y(t))$.

Now, release many particles at time $t = 0$, and determine their velocities, $V(t)$, at time t. Let the probability density function (p.d.f.) of these velocities be $p(V(t))$. Next, do the same exercise, but this time determine their velocities at time $t + \tau$. Let the p.d.f. of the latter velocities be $p(V(t + \tau))$. If t is sufficiently large, then the two probability density functions will be identical

$$p(V(t)) \equiv p(V(t + \tau)) \tag{8.69}$$

This implies that the velocity $V(t)$ is a stationary random function of time (Section 8.1.2).

The immediate implication of this is that the theory given in the previous section in conjunction with the ordinary diffusion process for large times (Section 8.1.2) can be applied to the present process because, as in the previous case, the particle velocity is a stationary random function of time. By application of the one-particle analysis, the following conclusions can be drawn:

1. The mean particle position (or the position of the centroid of the dispersant) will be at (Fig. 8.7b)

$$\overline{x}(t) = U_m t \tag{8.70}$$

(from Eq. 8.23). Here, U_m is the mean Eulerian velocity, and equal to the cross-sectional average velocity (Fig. 8.8)

$$U_m = \frac{1}{h}\int_0^h \overline{u}(y)\, dy \tag{8.71}$$

2. The variance will be

$$\overline{(x-\overline{x})^2} = 2\,\overline{V'^2}T_{L1}\, t \tag{8.72}$$

(from Eq. 8.30). Here T_{L1} is the Lagrangian time scale defined by

$$T_{L1} = \int_0^\infty R_{L1}(t)\, dt \tag{8.73}$$

in which $R_{L1}(t)$ is the Lagrangian correlation coefficient

$$R_{L1}(t) = \frac{\overline{V'(\tau)V'(\tau+t)}}{\overline{V'^2}} \tag{8.74}$$

in which

$$V' = V - U_m \tag{8.75}$$

Here, the sub-index 1 is used to distinguish the present application from the ordinary diffusion.

3. The p.d.f. of the particle position tends to the Gaussian (normal) distribution for large times ($t \gg T_{L1}$):

$$p(x) = \frac{1}{\sqrt{2\pi}\,\sqrt{\overline{(x-\overline{x})^2}}}\exp\left(-\frac{(x-\overline{x})^2}{2\overline{(x-\overline{x})^2}}\right) \tag{8.76}$$

(from Eq. 8.34).

4. The diffusion coefficient of the process is

$$D_1 = \overline{V'^2}T_{L1} = \frac{1}{2}\frac{d\overline{(x-\overline{x})^2}}{dt} \tag{8.77}$$

(from Eq. 8.42). Here, D_1 is called the *longitudinal dispersion coefficient.*

5. The p.d.f. of particle position satisfies the diffusion equation

$$\frac{dp}{dt} = D_1 \frac{\partial^2 p}{\partial (x - \overline{x})^2} \tag{8.78}$$

(from Eq. 8.40). Here, d/dt is the time derivative when considering the moving coordinate $x - \overline{x}(t)$. Changing to the fixed coordinate x,

$$\frac{d}{dt} = \frac{\partial}{\partial t} + \frac{\partial}{\partial x}\frac{\partial x}{\partial t} = \frac{\partial}{\partial t} + \frac{\partial \overline{x}(t)}{\partial t}\frac{\partial}{\partial x} = \frac{\partial}{\partial t} + U_m \frac{\partial}{\partial x} \tag{8.79}$$

Then, Eq. 8.78 becomes

$$\frac{\partial p}{\partial t} + U_m \frac{\partial p}{\partial x} = D_1 \frac{\partial^2 p}{\partial x^2} \tag{8.80}$$

6. From Eqs. 8.80 and 8.46

$$\frac{\partial \langle c \rangle}{\partial t} + U_m \frac{\partial \langle c \rangle}{\partial x} = D_1 \frac{\partial^2 \langle c \rangle}{\partial x^2} \tag{8.81}$$

in which $\langle c \rangle$ is the cross-sectional average concentration:

$$\langle c \rangle = \frac{1}{h} \int_0^h \overline{c}\, dy \tag{8.82}$$

Eq. 8.81 is called the *dispersion equation.*

The above analysis shows that the longitudinal dispersion process is represented by a one-dimensional diffusion equation with two constant coefficients, U_m and D_1 (Eq. 8.81).

The longitudinal dispersion process can also be described by the general diffusion equation (Eq. 8.65):

$$\frac{\partial \overline{c}}{\partial t} + \overline{u}\frac{\partial \overline{c}}{\partial x} + \overline{v}\frac{\partial \overline{c}}{\partial y} = \frac{\partial}{\partial x}\left(D_x \frac{\partial \overline{c}}{\partial x}\right) + \frac{\partial}{\partial y}\left(D_y \frac{\partial \overline{c}}{\partial y}\right) \tag{8.83}$$

Clearly, the one-dimensional equation (Eq. 8.81) is much simpler than the two-dimensional representation (Eq. 8.83). This is, however, at the expense of the calculated concentration. In the one-dimensional representation, the cross-sectional average concentration is calculated, whereas, in the two-dimensional representation, the point-concentration is calculated. However, this is not a problem because the calculation with the cross-sectional average concentration is normally good enough for longitudinal-dispersion problems.

For the application of the previously mentioned one-dimensional dispersion equation, two quantities need to be known: the cross-sectional average velocity, U_m, and the longitudinal dispersion coefficient, D_1.

The velocity U_m is obtained simply by integrating the velocity distribution over the cross-section, and dividing it by the cross-sectional area.

This is done if the velocity distribution is known theoretically (e.g. in the form of the logarithmic law). Otherwise, the velocity distribution over the cross-section is measured, for example, as in the case of a natural stream (Fig. 8.14), and the mean velocity calculation is based on the measured velocity.

The longitudinal dispersion coefficient D_1, on the other hand, can be calculated theoretically (see the next section), or numerically. The latter is based on the measured velocity distribution over the cross section, as will be detailed in Section 8.4. The longitudinal dispersion coefficient can also be determined experimentally in the laboratory, or in the field, by injecting a finite amount of dispersant (salt solution, for example, in a laboratory test), and subsequently tracking the concentration versus time variation at a measurement section downstream.

Taylor (1953, 1954) was the first to introduce the concept of longitudinal dispersion. In his pioneering 1954 paper, he notes: "It was pointed out to me in private conversation with G.K. Batchelor, that the case of flow at constant speed through a long uniform pipe is one of the cases to which this analysis can legitimately be applied, ..." (Note by the authors: Taylor refers here to the one-particle analysis introduced in Section 8.1.) Taylor's (1954) theoretical work and his experiments showed that the longitudinal dispersion coefficient in a turbulent pipe flow is $D_1/(RU_f) = 10$ in which R is the radius of the pipe, and U_f is the friction velocity. Taylor's work laid the groundwork for the later research on dispersion in natural streams, estuaries, industrial flows such as flow through a pipe, etc.

8.4 Longitudinal dispersion in an open channel

In this section, we shall calculate the longitudinal dispersion coefficient for an open channel flow (Fig. 8.7). We shall do this exercise for suspended heavy particles (for example, sand, or silt in water). This feature will be represented by the fall velocity of the particles, w_s. However, the case of the neutrally-buoyant particles/dispersant will be captured by setting w_s equal to zero in the end solutions.

The analysis presented in the following paragraphs is mainly based on the work of Sumer (1974). It may be noted that the method used in Sumer's (1974) analysis was first introduced by Aris (1956) and also employed by Sayre (1968).

8.4.1 *Formulation*

The conservation of mass (Fig. 8.6) from Eq. 8.68 is:

$$\frac{\partial \overline{c}}{\partial t} + \overline{u}(y)\frac{\partial \overline{c}}{\partial x} - w_s\frac{\partial \overline{c}}{\partial y} = \frac{\partial}{\partial x}\left(\epsilon(y)\frac{\partial \overline{c}}{\partial x}\right) + \frac{\partial}{\partial y}\left(\epsilon(y)\frac{\partial \overline{c}}{\partial y}\right) \tag{8.84}$$

in which w_s is the fall velocity of the particles, x is the streamwise distance and y is the vertical distance, measured from the bottom of the channel, and $\epsilon(y)$ is the turbulent diffusion coefficient (for convenience, we use the same notation for this quantity as in Sumer, 1974). It is assumed that $\epsilon(y)$ is the same for both x- and y-directions.

The boundary conditions at the bottom ($y = 0$), and at the free surface ($y = h$), see Fig. 8.7, are:

$$y = 0 \text{ and } h: \qquad \epsilon(y)\frac{\partial \overline{c}}{\partial y} + w_s\overline{c} = 0 \tag{8.85}$$

implying that there is no net transport across these boundaries.

The initial condition is (for a uniformly distributed plane source)

$$t = 0: \qquad \overline{c}(x, y, 0) = \frac{M}{h}\delta(x) \tag{8.86}$$

in which M is the total number of particles released into the flow and $\delta(x)$ is the Dirac delta function.

Furthermore, we write

$$\overline{u}(y) = U_m(1 + \chi(y)), \text{ and } \epsilon(y) = D_m\psi(y) \tag{8.87}$$

in which U_m and D_m are the cross-sectional average velocity and the cross-sectional average diffusion coefficient, respectively. We shall return to these quantities and the non-dimensional functions $\chi(y)$ and $\psi(y)$ later in the section.

Define the pth moment of concentration

$$C_p = \int_{-\infty}^{\infty} \xi^p C d\xi \tag{8.88}$$

in which ξ is the normalized streamwise distance defined by

$$\xi = \frac{x - U_s t}{h} \tag{8.89}$$

in which U_s is the mean particle velocity, and C is the normalized concentration defined by

$$C = \frac{\overline{c}}{M/h^2} \tag{8.90}$$

Multiplying the two sides of the above equations (Eqs. 8.84–8.86) by ξ^p, and integrating the equations from $-\infty$ to $+\infty$ yields:

$$\frac{\partial C_p}{\partial \tau} - (\mu - \mu_s + \mu\chi)pC_{p-1} - \nu_s \frac{\partial C_p}{\partial \eta} = p(p-1)\psi C_{p-2} + \frac{\partial}{\partial \eta}\left(\psi \frac{\partial C_p}{\partial \eta}\right) \quad (8.91)$$

$$\eta = 0 \text{ and } 1: \qquad \psi \frac{\partial C_p}{\partial \eta} + \nu_s C_p = 0 \quad (8.92)$$

and

$$\tau = 0: \qquad C_p(\eta, 0) = 1 \text{ and } 0, \text{ respectively, for } p = 0 \text{ and } p > 0 \quad (8.93)$$

Here,

$$\eta = \frac{y}{h}, \qquad \tau = \frac{D_m t}{h^2} \quad (8.94)$$

are the normalized distance from the bottom of the channel and the normalized time, respectively. Also, μ, μ_s and ν_s are the normalized mean flow velocity, the normalized mean particle velocity and the normalized mean fall velocity, respectively:

$$\mu = \frac{U_m h}{D_m}, \qquad \mu_s = \frac{U_s h}{D_m}, \qquad \nu_s = \frac{w_s h}{D_m} \quad (8.95)$$

Regarding the flow velocity, $\overline{u}(y)$, and turbulent diffusion coefficient, $\epsilon(y)$, first of all, from Eq. 3.96,

$$\overline{u} - U_m = \frac{1}{\kappa} U_f \left[\ln\left(\frac{y}{h}\right) + 1\right] \quad (8.96)$$

(the logarithmic law), U_f being the bed friction velocity, and κ the von Karman constant. The distribution of the turbulent diffusion coefficient, on the other hand, is found from the analogy between the momentum flux (Eq. 3.100) and the flux due to turbulent diffusion

$$\epsilon = \frac{-\Gamma}{\frac{d\overline{c}}{dy}} = \frac{-\rho\overline{u'v'}}{\rho\frac{d\overline{u}}{dy}} \quad (8.97)$$

Taking $-\rho\overline{u'v'} \simeq \rho U_f^2(1 - \frac{y}{h})$ (see Eq. 3.21), and using the above analogy (Eq. 8.97), the diffusion coefficient

$$\epsilon(y) = \kappa U_f \, y \left(1 - \frac{y}{h}\right) \quad (8.98)$$

and D_m is

$$D_m = \frac{1}{h}\int_0^h \epsilon(y)\, dy = \frac{1}{6}\kappa h U_f \quad (8.99)$$

The functions $\chi(y)$ and $\psi(y)$ in Eqs. 8.87 are then

$$\mu\chi = \frac{6}{\kappa^2}(1 + \ln\eta), \text{ and } \psi = 6\eta(1 - \eta) \quad (8.100)$$

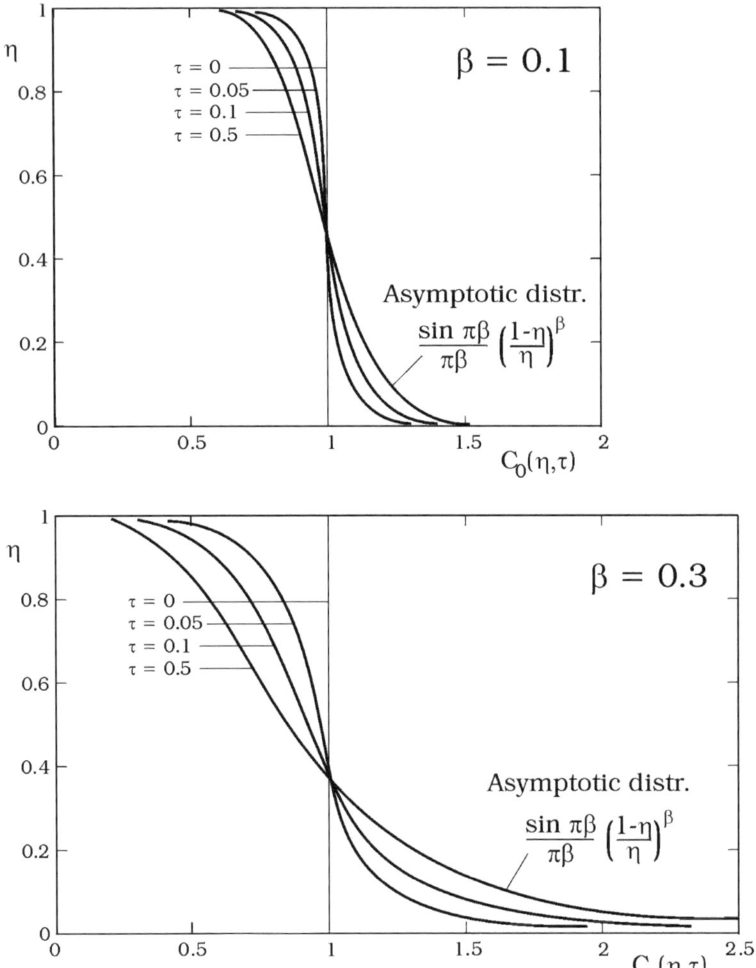

Fig. 8.9 $C_0(\eta, \tau)$ for two different sediments with (a): $\beta = \frac{w_s}{\kappa U_f} = 0.1$ and (b): $\beta = \frac{w_s}{\kappa U_f} = 0.3$. Sumer (1974).

8.4.2 *Zeroth moment of concentration*

For $p = 0$, the solution to the Eqs. 8.91–8.93 is found to be as follows

$$\begin{aligned} C_0 &= \frac{\sin(\pi\beta)}{\pi\beta}\left(\frac{1-\eta}{\eta}\right)^{\beta} \\ &\quad + \sum_{K=1}^{\infty} a_K \left(\frac{1-\eta}{\eta}\right)^{\beta} F(-K, 1+K; 1-\beta; \eta)\exp\{-6(K^2+K)\tau\} \end{aligned} \tag{8.101}$$

in which a_K

$$a_K = \frac{(2K+1)\,\Gamma(K+1-\beta)}{\Gamma^2(1-\beta)\,\Gamma(K+1+\beta)}\;{}_3F_2(-K, K+1, 1; 1-\beta, 2; 1) \qquad (8.102)$$

F is a hypergeometric function, ${}_3F_2$ is a generalized hypergeometric series, and Γ is the Gamma function (Gradshteyn and Ryzhik, 1965). The quantity β here is the fall velocity parameter (also known as the Rouse parameter) defined by

$$\beta = \frac{\nu_s}{6} = \frac{w_s}{\kappa U_f} \qquad (8.103)$$

and the solution is found for $\beta \neq 0$, $\beta < 1$.

Fig. 8.9 displays $C_0(\eta, \tau)$ for two values of β. Note that the physical meaning of C_0 is that C_0 is identical to the p.d.f. of the particle position across the depth. As seen from the figure, the larger the value of β, the more weighted the distributions towards the bottom, as expected.

From Eq. 8.101

$$C_0(\eta) = \frac{\sin(\pi\beta)}{\pi\beta}\left(\frac{1-\eta}{\eta}\right)^{\beta} \qquad \text{as} \qquad \tau \to \infty \qquad (8.104)$$

On physical grounds, Sumer (1974) argues that, for the case of neutrally buoyant particles ($\beta = 0$), C_0 should go to unity, $C_0 \to 1$. Indeed the above equation reveals this.

8.4.3 *Mean particle velocity*

For this, we define

$$m_p = \int_0^1 C_p \, d\eta \qquad (8.105)$$

Now, integrate Eqs. 8.91 and 8.93 from $\eta = 0$ to 1, and using the boundary conditions (Eq. 8.92):

$$\frac{dm_p}{d\tau} - p\int_0^1 (\mu - \mu_s + \mu\chi)C_{p-1}\, d\eta - p(p-1)\int_0^1 \psi C_{p-2}\, d\eta = 0 \quad (8.106)$$

$$\tau = 0: \qquad m_p(0) = 1 \text{ and } 0\text{, respectively, for } p = 0 \text{ and } p > 0 \qquad (8.107)$$

The preceding two equations for $p = 1$

$$\frac{dm_1}{d\tau} - \int_0^1 (\mu - \mu_s + \mu\chi)C_0\, d\eta = 0 \qquad (8.108)$$

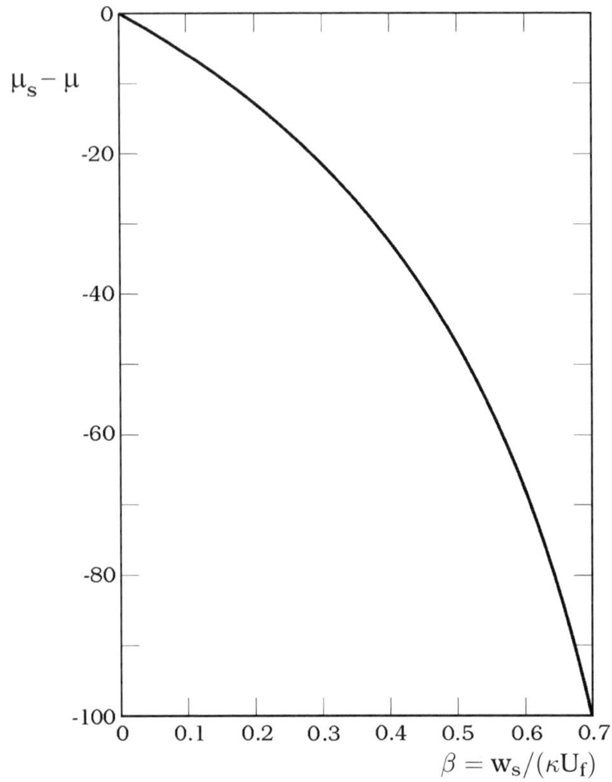

Fig. 8.10 Mean velocity of heavy particles relative to the mean flow velocity. Sumer (1974).

and

$$m_1(0) = 0 \text{ at } \tau = 0 \tag{8.109}$$

Note that m_1 represents the mean particle displacement relative to the ξ axis, and $dm_1/d\tau$ represents the mean particle velocity relative to this axis. Since this axis moves with the mean particle velocity (Eq. 8.89), then $dm_1/d\tau \to 0$ as time $\tau \to \infty$. Hence, from Eq. 8.108, for large times

$$-\int_0^1 (\mu - \mu_s + \mu\chi) C_0(\eta, \infty)\, d\eta = 0$$

or, from Eq. 8.104,

$$\mu_s - \mu = \int_0^1 \mu\chi\, C_0(\eta, \infty)\, d\eta = \int_0^1 \mu\chi \frac{\sin(\pi\beta)}{\pi\beta} \left(\frac{1-\eta}{\eta}\right)^\beta d\eta \tag{8.110}$$

This is the mean particle velocity relative to the mean flow velocity (Eq. 8.95). Fig. 8.10 illustrates how the mean particle velocity varies with the fall velocity parameter. Note that for $\beta = 0$, the case of neutrally-buoyant particles, obviously this velocity should be equal to the mean flow velocity, meaning that $\mu_s - \mu = 0$, as revealed by the figure.

8.4.4 *Longitudinal dispersion coefficient*

Before we continue with the analysis to calculate the longitudinal dispersion coefficient, some dimensional considerations would prove useful. The longitudinal dispersion coefficient for neutrally-buoyant particles, D_1, should depend solely on the following two quantities:

$$D_1 = f(h, U_f) \tag{8.111}$$

in which h is the flow depth, and U_f the bed friction velocity. This is because the longitudinal dispersion is governed by (1) the streamwise velocity that the particles experience relative to the mean flow velocity, namely, $\overline{u}(y) - U_m$, and (2) the turbulent diffusion in the vertical direction, i.e., $\epsilon(y)$. The latter quantities, $\overline{u}(y) - U_m$ and $\epsilon(y)$, are both dependent on the parameters h, and U_f (see Eqs. 8.96 and 8.98, respectively). Using dimensional analysis, the above dimensional expression, Eq. 8.111, can readily be turned into the following non-dimensional expression:

$$\frac{D_1}{hU_f} = \text{constant} \tag{8.112}$$

As seen, provided that it is normalized by the product of hU_f, the longitudinal dispersion coefficient is not a function of other flow parameters such as the Reynolds number.

In the case of heavy particles, there will be one additional parameter, namely the fall-velocity parameter β (Eq. 8.103). Therefore, the longitudinal dispersion coefficient from Eq. 8.112 is:

$$\frac{D_1}{hU_f} = f(\beta) \tag{8.113}$$

The objective of this section is to determine the constant value in Eq. 8.112 (neutrally-buoyant particles) and the function $f(\beta)$ in Eq. 8.113 (heavy particles).

Now, returning back to the analysis, the longitudinal dispersion coefficient can, from Eqs. 8.88 and 8.105, be written in the following normalized form as

$$\frac{D_1}{hU_f} = \frac{\kappa}{6}\left(\frac{1}{2}\frac{dm_2}{d\tau}\right)_{\tau\to\infty} \tag{8.114}$$

From Eq. 8.106,

$$\frac{1}{2}\frac{dm_2}{d\tau} = 6\int_0^1 \eta(1-\eta)C_0\, d\eta + \int_0^1 (\mu - \mu_s + \mu\chi)C_1\, d\eta \tag{8.115}$$

in which C_1 is found from Eqs. 8.91–8.93, setting $p = 1$

$$C_1 = \phi(\eta) + \sum_{K=1}^{\infty} a_K f_K(\eta)\exp\{-6(K^2+K)\tau\} \tag{8.116}$$
$$+ \sum_{K=1}^{\infty} d_K\left(\frac{1-\eta}{\eta}\right)^{\beta} F(-K, 1+K; 1-\beta; \eta)\exp\{-6(K^2+K)\tau\}$$

in which

$$\phi = A_\phi\left(\frac{1-\eta}{\eta}\right)^{\beta} - \frac{\sin(\pi\beta)}{\pi\beta}\left(\frac{1-\eta}{\eta}\right)^{\beta}\int_0^{\eta}\left(\frac{1-\eta}{\eta}\right)^{-\beta}\frac{1}{6\eta(1-\eta)}d\eta \tag{8.117}$$
$$\times \int_0^{\eta}(\mu - \mu_s + \mu\chi)\left(\frac{1-\eta}{\eta}\right)^{\beta} d\eta$$

in which $\beta < 1$. It may be noted that $C_1 \to \phi$ as $\tau \to \infty$.

From Eqs. 8.115, 8.117 and 8.104, the dispersion coefficient is obtained as

$$\frac{D_1}{hU_f} = \frac{\kappa}{6}(1-\beta^2) - \frac{\kappa}{36}\frac{\sin(\pi\beta)}{\pi\beta}\int_0^1[(\mu-\mu_s)+\mu\chi]\left(\frac{1-\eta}{\eta}\right)^{\beta} d\eta \tag{8.118}$$
$$\times \int_0^{\eta}\left(\frac{1-\eta}{\eta}\right)^{-\beta}\frac{1}{\eta(1-\eta)}d\eta \int_0^{\eta}[(\mu-\mu_s)+\mu\chi]\left(\frac{1-\eta}{\eta}\right)^{\beta} d\eta$$

in which $\beta < 1$.

For the case of *neutrally buoyant particles/dispersant*, putting $\beta = 0$, one gets

$$\frac{D_1}{hU_f} = \frac{\kappa}{6} - \frac{\kappa}{36}\int_0^1 \mu\chi d\eta \int_0^{\eta}\frac{1}{\eta(1-\eta)}d\eta \int_0^{\eta}\mu\chi d\eta \tag{8.119}$$

The expression in Eq. 8.119 was first obtained by Elder (1959). The integral can be calculated numerically. The result is (taking κ, the von Karman constant, 0.41)

$$\frac{D_1}{hU_f} = 0.07 + 5.86 = 5.93 \tag{8.120}$$

As seen from Eq. 8.119, the dispersion coefficient consists of two terms; the first term on the right hand side of the equation, $\kappa/6$, (the numerical value of which is 0.07, Eq. 8.120) represents the effect of the turbulent diffusion

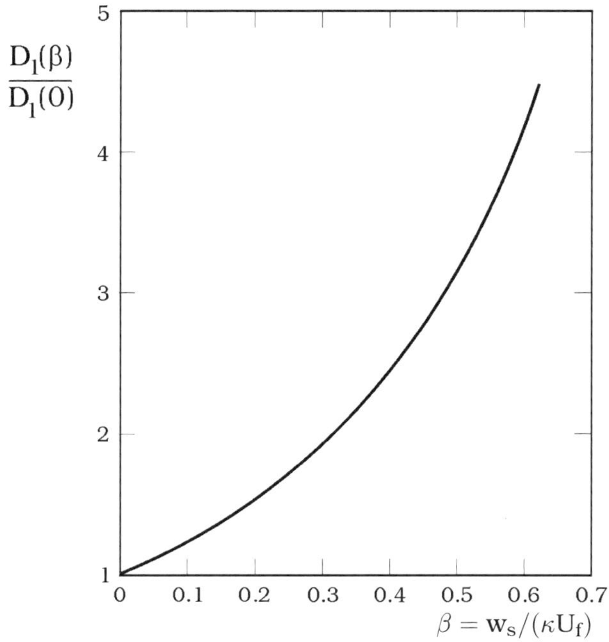

Fig. 8.11 Ratio of the longitudinal dispersion coefficient of heavy particles to that of neutrally-buoyant particles. Sumer (1974).

in the streamwise direction; it stems from the first term on the right hand side of Eq. 8.84. The second term, the main portion of the longitudinal dispersion coefficient (the numerical value of which is 5.86, Eq. 8.120) is caused by the combined action of the turbulent diffusion across the depth y, and the variation of the flow velocity in the y-direction. Incidentally, the above analysis confirms the dimensional analysis discussed in conjunction with Eq. 8.112.

Returning to the case of *heavy particles*, the dispersion coefficient from Eq. 8.118 is plotted in Fig. 8.11. The figure shows that the dispersion coefficient increases with increasing β. This can be explained as follows. A heavy particle spends most of the time near the bottom where the velocity gradient is largest. On the other hand, the larger the velocity gradient, the larger the dispersion coefficient. Therefore, a heavier particle (i.e., with larger β) should have a larger dispersion coefficient.

We note that, similar to the case of neutrally-buoyant particles, the above analysis reveals the dimensional considerations discussed in conjunction with Eq. 8.112, i.e., the fact that the longitudinal dispersion coefficient is a function of the fall-velocity parameter β alone.

It may be noted that diffusion and dispersion of particles in turbulent flows can also be studied by numerical simulations, using particle methods. Sullivan (1971) and Sumer (1973) were the first to use a random-walk model for the purpose of simulating the longitudinal dispersion processes in a turbulent open-channel flow, Sullivan (1971) for neutrally-buoyant particles, and Sumer (1973) for heavy particles. In simulations, typically the path of an initially marked particle position is calculated as the particle migrates through the statistical field variables. Many such paths are used to compile the probability distribution by means of an ensemble average.

See Chapter 10, Exercise 2, for a detailed "hands on" exercise involving the numerical simulation of longitudinal dispersion of particles in a channel flow using a random-walk model.

Example 8.1. Settlement of sediment in an open-channel. Application to settling basins.

We will give another example to illustrate the use of turbulent diffusion equation for heavy particles in a problem relevant to hydraulic engineering, namely, the settlement of solid particles in an open channel.

The setting is exactly the same as in the previous problem studied in Section 8.4.1; We have a turbulent flow in an open channel with a depth of h (Fig. 8.7), and with an initial, uniformly distributed plane source of sediment particles. Recall that, in the previous problem, there was no net transport at the bottom of the channel, and therefore the particles released in the channel were maintained in the body of the flow all the time. We now consider that the particles coming to the bottom deposit there, with no re-entrainment. This is essentially the process encountered in settling canals or settling basins in civil engineering practice, as will be discussed later.

We adopt the same set of equations as in the previous problem (Eqs. 8.84–8.87) with the exception that the bottom boundary condition at $y = 0$ will change to that describing the condition that the particles deposit on the bottom with no re-entrainment.

The turbulent diffusion equation (Eq. 8.84) is:

$$\frac{\partial \overline{c}}{\partial t} + \overline{u}(y)\frac{\partial \overline{c}}{\partial x} - w\frac{\partial \overline{c}}{\partial y} = \frac{\partial}{\partial x}\left(\epsilon(y)\frac{\partial \overline{c}}{\partial x}\right) + \frac{\partial}{\partial y}\left(\epsilon(y)\frac{\partial \overline{c}}{\partial y}\right), \qquad (8.121)$$

the boundary condition at the free surface ($y = h$), Eq. 8.85, is:

$$y = h: \qquad \epsilon(y)\frac{\partial \bar{c}}{\partial y} + w\bar{c} = 0 \tag{8.122}$$

expressing that there is no net transport across this boundary, and the initial condition for a uniformly distributed plane source at $x = 0$ (Eq. 8.87) is:

$$t = 0: \qquad \bar{c}(x, y, 0) = \frac{M}{h}\delta(x) \tag{8.123}$$

in which M is the total number of particles released into the flow, and $\delta(x)$ is the Dirac delta function, Section 8.4.1.

Regarding the bottom boundary condition, this condition requires that, at the bottom, the total flux in the vertical direction for sediment (upward directed), $-\epsilon(y)\frac{\partial \bar{c}}{\partial y} - w\bar{c}$, should be equal to the downward flux of sediment, i.e., the total amount of sediment passing through a unit area of the bottom per unit time, $-w\bar{c}$. Therefore, at the bottom:

$$\left[-\epsilon(y)\frac{\partial \bar{c}}{\partial y} - w\bar{c}\right] = -w\bar{c}, \tag{8.124}$$

i.e., the boundary condition at the bottom

$$y = 0: \qquad \epsilon(y)\frac{\partial \bar{c}}{\partial y} = 0 \tag{8.125}$$

Sumer (1977) developed analytical solutions for the above set of equations, using Aris' moment method, described in Section 8.4.1 in conjunction with the problem of longitudinal dispersion process.

Fig. 8.12 displays vertical distributions of sediment particles for $\beta = 3$ for various times, obtained by Sumer (1977) (see Eq. 8.103 for the definition of the fall velocity parameter β). The horizontal axis in the figure is

$$\frac{C_0(\eta, \tau)}{m_0(\tau)} \tag{8.126}$$

in which $C_0(\eta, \tau)$ is defined by

$$C_0 = \int_{-\infty}^{\infty} C \, d\xi \tag{8.127}$$

with

$$\xi = \frac{x - U_m t}{h} \tag{8.128}$$

and

$$m_0(\tau) = \int_0^1 C_0(\eta, \tau) \, d\eta \tag{8.129}$$

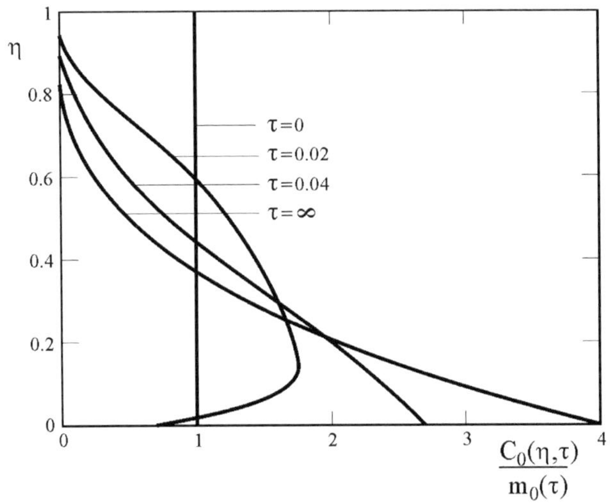

Fig. 8.12 Vertical distribution of suspended particles when particles coming to the bottom settle there. $\beta = 3$. Sumer (1977).

Notice that ξ is, in the present context, defined slightly different from that in Eq. 8.89. When compared with the vertical distributions in Fig. 8.9, it is seen that the vertical distribution of particles experiences a sharp decrease in particle concentration in the upper layers of the flow, presumably due to the very large value of the fall velocity parameter in the present case, namely $\beta = 3$.

One of the important parameters in the present problem is the mean rate at which particles settle out of suspension. This quantity is denoted by λ. Sumer (1977, p. 1329) was able to obtain an analytical expression for this quantity, utilizing his solution for the above set of equations (Eqs. 8.121–8.125):

$$\lambda = \left(\sum_{K=1}^{\infty} \frac{\beta a_K}{6(K+\beta)^2(K-1+\beta)^2} \right)^{-1} \tag{8.130}$$

in which a_K

$$a_K = \frac{(2K+2\beta-1)\,\Gamma(K+2\beta)}{\Gamma(1+\beta)\,\Gamma(K+\beta+1)} P_{K-1}^{(\beta+1,\,\beta-1)}(-1) \tag{8.131}$$

in which $\Gamma(x)$ is the Gamma function (Abramowitz and Stegun, 1968, p. 255), and $P_n^{(a,\,b)}(x)$ is the Jacobi polynomial (the same reference, p. 793). Sumer's (1977) solution for λ (Fig. 5 in Sumer, 1977) is given in Fig. 8.13.

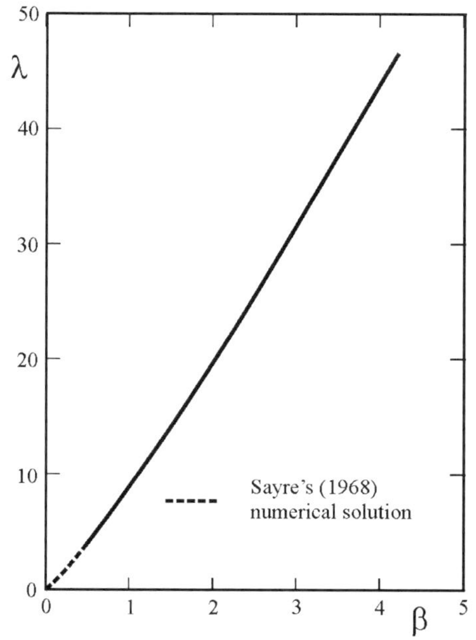

Fig. 8.13 Mean rate at which particles settle out of suspension (inverse of mean retention time) as function of the fall-velocity (Rouse) parameter. Sumer (1977).

The variation in Fig. 8.13 can be approximated by the following simple expression for $\beta \gtrsim 1$:

$$\lambda = 12.2\,\beta - 4.5, \qquad \beta \gtrsim 1 \tag{8.132}$$

Application to settling basins. In practice, the accumulation of sediment in irrigation, hydropower, and municipal canals interfere with their operations. This is prevented by various sediment exclusion techniques. Of these, settling basins are an effective method for removal of suspended sediment from canals; the underlying principle is to provide a section wide and long enough so that the reduced flow velocity will make it possible for the sediment to settle out of suspension due to gravity. The sediment accumulated on the bottom of the settling basin is then drawn off occasionally either by mechanical means, or by flushing downstream by opening of the downstream gates, the flushing being achieved by the steep slope (like a few percent) of the basin floor.

Let r denote the removal ratio, namely, the percentage of sediment that settles out of suspension. The removal ratio can be related to the mean rate at which particles settle out of suspension, λ, by the following equation (Sumer, 1977, p. 1332):

$$r = 1 - \exp\left(-\frac{\lambda}{\mu}\frac{x}{h}\right) \quad (8.133)$$

in which x is the settling length, μ is the non-dimensional mean flow velocity defined in Eq. 8.95:

$$\mu = \frac{U_m h}{D_m} \quad (8.134)$$

and D_m, the mean diffusion coefficient (Eq. 8.99) is:

$$D_m = \frac{1}{6}\kappa h U_f \quad (8.135)$$

From Eqs. 8.133–8.135, the design settling length, corresponding to a certain removal ratio, r, is obtained as

$$L = -\frac{6h}{\kappa\lambda}\frac{U_m}{U_f}\ln(1-r) \quad (8.136)$$

Numerical example. Given that the mean flow velocity is $U_m = 40$ cm/s, the settling velocity of sediment particles the size $d \cong 0.1$ mm is $w_s \cong 1$ cm/s, and the water depth $h = 1.5$ m, find the length of a settling basin required to achieve $r = 90\%$ removal.

To calculate the length of the settling basin from Eq. 8.136, we need the bottom friction velocity, U_f. For this, we use the flow resistance equation from Eq. 3.130, considering that the sediment bottom is hydraulically smooth:

$$\frac{1}{\sqrt{8}}\frac{U_m}{U_f} = 2.0\log\left(\frac{U_m(4r_h)}{\nu}\sqrt{8}\frac{U_f}{U_m}\right) - 0.8 \quad (8.137)$$

in which the diameter D in Eq. 3.130 is replaced by $4r_h$ with r_h being the hydraulic radius (see Section 3.4). Taking $r_h \cong h$, the latter equation gives $U_f = 1.4$ cm/s.

We check if the bottom is hydraulically smooth by calculating the grain Reynolds number, $dU_f/\nu = 0.01 \times 1.4/0.01 = 1.4$, which indicates that the bottom acts as a hydraulically smooth boundary, and therefore, Eq. 8.137 can be used to predict the bottom friction velocity.

The fall velocity parameter (Rouse parameter) will then be

$$\beta = \frac{w_s}{\kappa U_f} = \frac{1}{0.4 \times 1.4} = 1.8 \quad (8.138)$$

By inserting $\beta = 1.8$ in Eq. 8.132, λ is found to be $\lambda = 17$, and from Eq. 8.136, the length of the settling basin is found to be $L = 86$ m for the removal ratio $r = 90\%$.

An important point here is that the no re-entrainment condition has to be checked. For this, the Shields parameter is:

$$\theta = \frac{U_f^2}{g(s-1)d} = \frac{1.4^2}{981 \times (2.65 - 1) \times 0.01} = 0.12 \qquad (8.139)$$

Considering that the grain Reynolds number is $dU_f/\nu = 1.4$, the above value of the Shields parameter, $\theta = 0.12$, corresponds to a very weak sediment motion at the bottom (see, the Shields diagram, for example, in Sumer and Fredsøe, 2002, Fig. 1.2). This value of the Shields parameter also ensures that no re-suspension of sediment particles from the bed will occur (see, the critical curve for the initiation of suspension from the bed, Fig. 3.49, Chapter 3). Also, we note that, when plotted on the diagram in Fig. 3.51, the point $(dU_f/\nu, \theta) = (1.4, 0.12)$ lies rather close to the critical curve $w_s/U_f = 1$. This implies that the sediment coming close to the bed from upper elevations is unlikely to be maintained in suspension. So, collectively, it can be considered that the no re-entrainment condition is essentially satisfied.

When the above exercise was repeated for a flow velocity of $U_m =$ 20 cm/s, the corresponding quantities will be found as $U_f = 0.75$ cm/s, $dU_f/\nu = 0.75$, $\theta = 0.035$, $\beta = 3.3$, and $\lambda = 36$. With these latter values, the required length of the settling basin for the same removal ratio, $r = 90\%$, will be $L = 38$ m.

8.5 Longitudinal dispersion in rivers

In the previous section, the longitudinal dispersion coefficient has been calculated for an open-channel flow for neutrally-buoyant particles/dispersant, and found to be $D_1/(hU_f) \simeq 6$. Field experiments in rivers show, however, that $D_1/(hU_f)$ takes values much larger than 6. Values reported from field studies can be as large as $D_1/(hU_f) = O(7000)$; For example, $D_1/(hU_f) \simeq 7500$ in the Missouri river, as reported by Fischer et al. (1979). This implies that the value obtained for the case of the open channel, namely $D_1/(hU_f) \simeq 6$, cannot apply to real streams. This issue was studied by Fischer (1966, 1967). It was shown that, in the case of real streams, although the velocity distribution satisfies the logarithmic law in the vertical direction, the variation of the velocity across the river width apparently plays the key role.

A quantitative estimate of the dispersion coefficient in a real stream can be obtained by neglecting the vertical variation of the velocity distribution,

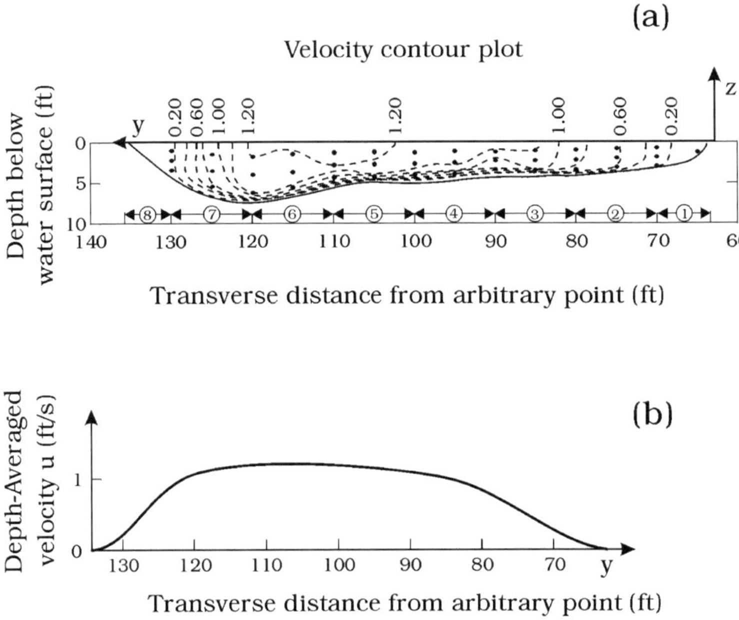

Fig. 8.14 (a): The cross-sectional velocity distribution at Renton Junction in the Green-Duwamish River, Washington. Contours are lines of constant velocity, in increment of 0.2 ft/s; (b): The transverse profile of the depth-averaged velocity corresponding to the velocity distribution shown in (a). Adapted from Fischer et al. (1979).

but instead, by considering the transverse velocity profile, i.e., the velocity profile across the river width. (See, for example, the velocity distribution in Fig. 8.14. The figure illustrates the transverse velocity variation in the Green-Duwamish River in Washington, and the corresponding transverse velocity profile.) Considering the transverse velocity variation in a river, Fisher (1966, 1967) obtained the following expression for the longitudinal dispersion coefficient

$$D_1 = -\frac{1}{A}\int_0^W (u(y) - U_m)\, d(y)\, dy \int_0^y \frac{1}{\epsilon_t d(y)}\, dy \int_0^y (u(y) - U_m)\, d(y)\, dy \tag{8.140}$$

in which A is the cross-sectional area of the river, W the width at the surface, U_m the mean flow velocity, y the transverse distance measured from one bank of the river (Fig. 8.14) (not to be confused with the vertical distance used in the previous section), $d(y)$ the depth (the depth is a function of y, Fig. 8.14), ϵ_t is the turbulent diffusion coefficient in the transverse direction (it may or may not be a function of y), and $u(y)$ is the depth-averaged velocity defined by (Fig. 8.14):

$$u(y) = \frac{1}{d(y)} \int_{-d(y)}^{0} \overline{u}(y,z)\, dz \tag{8.141}$$

Returning to the expression obtained for the open-channel flow in the previous section, Eq. 8.119, (omitting the first term on the right hand side of the equation for simplicity)

$$\frac{D_1}{hU_f} = -\frac{\kappa}{36} \int_0^1 \mu\chi d\eta \int_0^\eta \frac{1}{\eta(1-\eta)} d\eta \int_0^\eta \mu\chi d\eta$$

or, in terms of dimensional quantities:

$$\frac{D_1}{hU_f} = -\frac{1}{h} \int_0^h (\overline{u}(y) - U_m) dy \int_0^y \frac{1}{\epsilon(y)} dy \int_0^y (\overline{u}(y) - U_m) dy$$

or alternatively

$$\frac{D_1}{hU_f} = -\frac{1}{h^2} \int_0^h (\overline{u}(y) - U_m) h dy \int_0^y \frac{1}{\epsilon(y) h} dy \int_0^y (\overline{u}(y) - U_m) h dy \tag{8.142}$$

As seen, this expression is quite similar to that obtained by Fischer (1966, 1967), Eq. 8.140. The only difference is that, in the case of the river dispersion, y is the transverse distance, while, in the case of the open-channel dispersion, y is the vertical distance. This comparison clearly reveals that, in the case of rivers, the transverse variation of the velocity is the main cause of the longitudinal dispersion.

Fischer et al. (1979) give a numerical example to illustrate how to use Eq. 8.140 in a field application.

Finally, it may be noted that real streams may have irregularities such as bends, sandbars, side pockets, pools, riffles, bridge piers, man-made revetments, and even occasional junked car, or sunken barges. Every irregularity contributes to longitudinal dispersion (Fischer et al., 1979, and Pedersen, 1977). Empirical expressions have been given for the longitudinal dispersion coefficient in terms of gross flow characteristics of the river (Fischer et al., 1979, p. 136). These can be used to make first assessments of the dispersion coefficient.

8.6 Longitudinal dispersion in an oscillating tunnel

An oscillating tunnel is typically a facility where fluid is moved back and forth. There are several types of oscillating tunnels. Fig. 8.15 displays a

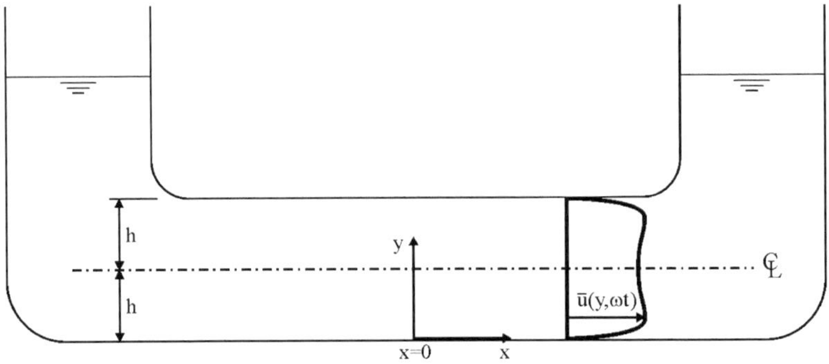

Fig. 8.15 U-shaped oscillating tunnel.

sketch of a U-shaped oscillating water tunnel. Oscillating tunnels facilitate, among other things, the study of wave boundary layers, as mentioned in Chapter 5. Large-scale oscillating tunnels (dimensions being like $O(10\text{ m})$ in length, $O(0.3\text{–}1\text{ m})$ in the vertical and spanwise directions, and periods being like $O(5\text{–}10\text{ s})$) enable researchers to obtain fully-developed turbulent wave boundary layers similar to those encountered in the field under storm conditions.

Description of the longitudinal dispersion process

Consider the oscillating tunnel in Fig. 8.15. Release a dispersant at $x = 0$ and $t = 0$ in the form of an instantaneous plane source, uniformly distributed across the entire depth of the tunnel, $0 \le y \le 2h$. Consider that the free stream velocity at the time of release is at its maximum, with $U_0 = U_{0m}\cos(\omega t)$, a sinusoidal variation. (We take the latter variation as $\cos(\omega t)$ rather than $\sin(\omega t)$ for convenience.) Here, t is the time, T is the period of the oscillatory motion, and $\omega = 2\pi/T$ is the angular frequency of the motion.

Fig. 8.16 displays a sequence of pictures, illustrating schematically the spreading of the dispersant over slightly more than one period of the oscillatory motion from the instant of release (over the time $0 \le t \le 5T/4$). The dispersant undergoes longitudinal dispersion as the cloud is moved forward and backward under the oscillatory motion of the flow. This longitudinal dispersion is caused by the combined action of the shear across the depth (namely, $u(y, \omega t) - U_m(\omega t)$), and the turbulent diffusion in the vertical direction (namely, $\epsilon(y, \omega t)$), in exactly the same manner as in the case of

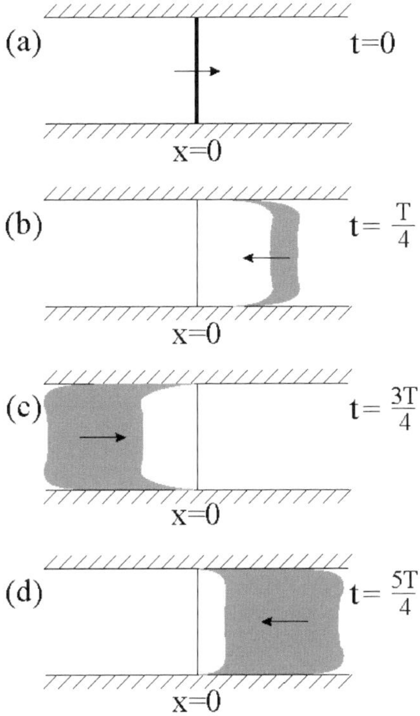

Fig. 8.16 Dispersant undergoes longitudinal dispersion as the dispersant cloud is moved forth and back in the oscillating tunnel.

the steady current, discussed in Sections 8.3 and 8.4. Here, $U_m(\omega t)$ is the depth-averaged velocity at phase ωt.

Note that if there is no shear across the depth, no longitudinal dispersion will occur (save the longitudinal turbulent diffusion), and therefore the dispersant will be moved forth and back, maintaining its planar form across the depth all the time.

Fig. 8.17 displays the dispersant cloud at time t. The displacement of any particle (marked A in the figure) $x(\omega t)$ is

$$x(\omega t) = \overline{x(\omega t)} + x'(\omega t) \tag{8.143}$$

(see Fig. 8.17) in which $\overline{x(\omega t)}$ is the mean position of particles corresponding to the time ωt, and $x'(\omega t)$ is the displacement of particles relative to the mean. The mean particle displacement can formally be written as

$$\overline{x(\omega t)} = a_1 \sin(\omega t) \tag{8.144}$$

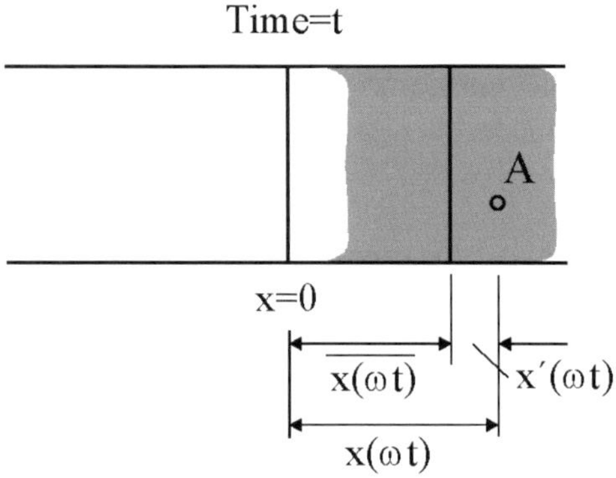

Fig. 8.17 Dispersant cloud at time t (or alternatively, at time ωt). Particle position x' relative to the mean.

in which a_1 is the "amplitude" associated with the motion of the centroid of the cloud. (The amplitude a_1 is always smaller than the amplitude of the motion in the free stream region, a. Clearly, in the absence of the longitudinal dispersion, the amplitude a_1 will reduce to a.)

Now, the longitudinal dispersion of particles is characterized by the quantity $\overline{[x'(\omega t)]^2} = \overline{\left[x(\omega t) - \overline{x(\omega t)}\right]^2}$, the variance of the particle displacement at time ωt. Two important features of this quantity are as follows:

1. $\overline{[x'(\omega t)]^2}$ increases continuously period after period, as implied by Fig. 8.16; and

2. $\overline{[x'(\omega t)]^2}$ has a periodicity equal to $T/2$, the half period of the motion. (Take $x'(\omega t)$ like $\sin(\omega t)$, and see that $[x'(\omega t)]^2 \sim \sin^2(\omega t)$ will have a periodicity of $T/2$.) The latter conclusion can also be drawn from the fact that $[x'(\omega t)]^2$ cannot "sense" as to whether the motion is from left to right, or from right to left, and therefore the periodicity in $[x'(\omega t)]^2$ should be half of the period of the motion.

From the above considerations, the variation of $\overline{[x'(\omega t)]^2}$ with time should be like that sketched in Fig. 8.18. We shall return to Fig. 8.18 shortly.

Numerical simulation: Kirca et al. (2016) study

Kirca et al. (2016) studied the longitudinal dispersion in an oscillating tunnel with special reference to wave boundary layers. In Kirca et al.'s (2016) study, the longitudinal dispersion process was investigated numerically, us-

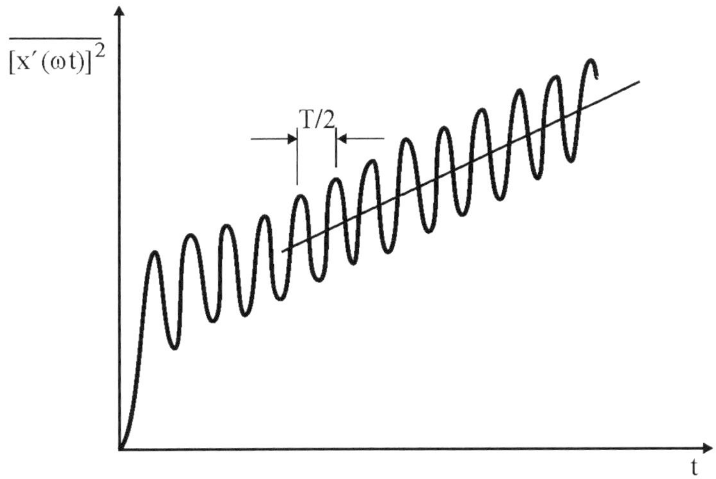

Fig. 8.18 Typical variance of particle position in oscillating flow, with T being the period of the oscillating motion.

ing a random-walk model. Kirca et al. (2016) picked up the input data for the mean and turbulent characteristics of the flow from the solution of a transitional (low Reynolds number) two-equation k-ω Reynolds-averaged Navier-Stokes (RANS) model, and plugged it in the random-walk dispersion model. The flow model, adapted from Fuhrman et al. (2013), was validated against the experiments of Jensen et al. (1989). (This same model is described in detail in Section 5.12). The simulations were carried out for both neutrally-buoyant particles, and heavy particles (sediment). The results of the numerical simulations were compared with experiments conducted in the same study (Kirca et al., 2016), utilizing the oscillating tunnel at the Hydraulics Laboratory, Technical University of Denmark, the same tunnel as that used in Jensen et al.'s (1989) flow measurements. The results from the numerical simulations were found to be in general agreement with the experiments.

Fig. 8.19 presents the variance of particle displacement versus time, obtained in one of Kirca et al.'s (2016) numerical simulations. The quantity δ in the figure is the boundary layer thickness (see Fig. 5.4 for the definition) of the oscillatory flow in the tunnel, and U_{fm} the maximum value of the bed friction velocity. Kirca et al. (2016) note that the period of the flow in the simulation was $TU_{fm}/\delta = 16.3$. Fig. 8.19 indicates that the time variation of $\overline{[x'(\omega t)]^2}$ has a periodicity of $tU_{fm}/\delta = 8.15$. As seen, this is

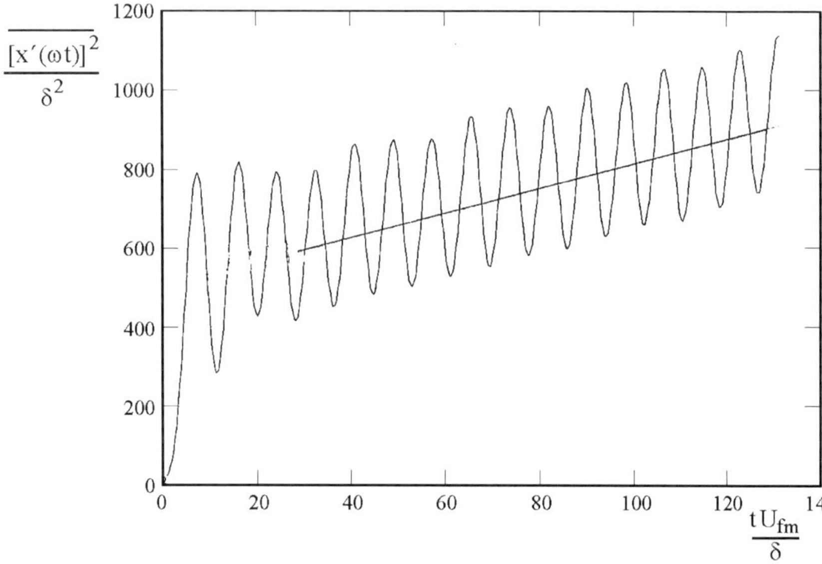

Fig. 8.19 Variance vs. time in one of oscillating-tunnel tests in Kirca et al. (2016) numerical simulations. Neutrally-buoyant particles. $a/h = 21.4$ and $Re_w = 6.2 \times 10^6$.

precisely half of the period of the oscillatory motion, confirming that the period of the time variation of $\overline{[x'(\omega t)]^2}$ is $T/2$.

Fig. 8.19 indicates that, for large times ($t \gtrsim 2T$), the period-averaged variation of $\overline{[x'(\omega t)]^2}$ (the solid line in the figure) varies with time linearly, as expected, and therefore can be utilized to determine the longitudinal dispersion coefficient:

$$D_1 = \frac{1}{2}\frac{d}{dt}\left\langle \overline{[x'(\omega t)]^2} \right\rangle \tag{8.145}$$

in which the brackets indicate period averaging.

Longitudinal dispersion coefficient

The longitudinal dispersion coefficient, D_1, should depend on the following quantities:

$$D_1 = f(h, U_{fm}, U_{0m}, a, \nu, \delta) \tag{8.146}$$

From dimensional analysis:

$$\frac{D_1}{\delta U_{fm}} = f\left(Re_w, \frac{a}{h}\right) \tag{8.147}$$

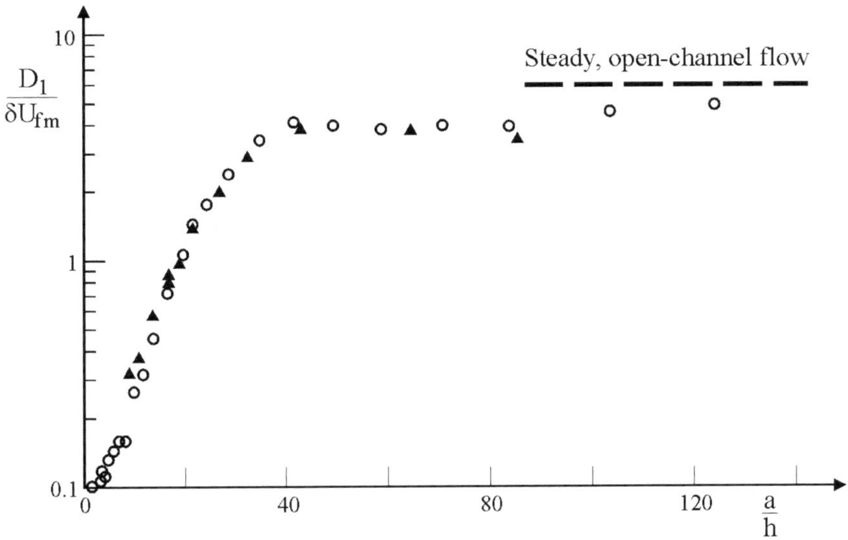

Fig. 8.20 Longitudinal dispersion coefficient for neutrally-buoyant particles vs. non-dimensional amplitude. Oscillating tunnel. Circles: $Re_w = 6.2 \times 10^6$. Triangles: variable Re_w in the range $Re_w = 2.7 \times 10^5$–9.9×10^7. Kirca et al. (2016).

in which Re_w is the Reynolds number defined by

$$Re_w = \frac{aU_{0m}}{\nu} \tag{8.148}$$

(It may be noted that, although the present flow is not exactly the same as in Lodahl et al., 1998, the flow in an oscillatory motion of fluid in a circular pipe is similarly described by two parameters, the Reynolds number, and the radius-to-the-Stokes-layer-thickness, Lodahl et al., 1998. The latter parameter, in the present flow, plays the same role as the inverse of a/h.)

Recall that the analysis carried out for the longitudinal dispersion in an open channel flow (Section 8.4.4) has indicated that the longitudinal dispersion coefficient is not dependent on the Reynolds number. By analogy to the open-channel case, the Reynolds number in Eq. 8.147 can be omitted, and therefore, the longitudinal dispersion coefficient can be written as

$$\frac{D_1}{\delta U_{fm}} = f\left(\frac{a}{h}\right) \tag{8.149}$$

Fig. 8.20 displays the variation of the longitudinal dispersion coefficient as function of a/h, obtained by Kirca et al. (2016) from their numerical simulations. Two sets of data are plotted in Fig. 8.20, one corresponding to

$Re_w = 6.2 \times 10^6$ (circles), and the other (triangles) to a variety of Reynolds number with $Re_w = 2.7 \times 10^5$–9.9×10^7. The following conclusions can be drawn from Fig. 8.20:

1. As seen clearly, the previously mentioned two sets of data practically collapse, revealing the assumption made in conjunction with Eqs. 8.147 and 8.149 that the Reynolds-number dependence in Eq. 8.147 can be omitted.

2. The dispersion coefficient increases with increasing a/h. This is because the larger the ratio a/h, the more developed the boundary layer, and therefore the larger the thickness of the boundary layers over the bed and the top of the tunnel. Clearly, the thicker the boundary layer, the larger the dispersion coefficient, as discussed in the preceding sections. Hence the dispersion coefficient should increase with increasing values of a/h.

3. As a/h increases, the two boundary layers at the bed and at the top of the tunnel will cover larger and larger depths. With $a/h \to \infty$, the boundary layers will cover the entire depth of the tunnel, and therefore the dispersion coefficient should be expected to approach an asymptotic value $D_1/(\delta U_{fm}) \to$ Constant as $a/h \to \infty$. The figure shows that $D_1/(\delta U_{fm}) \simeq 6$ for $a/h \gtrsim 40$. The latter asymptotic case actually corresponds to the case of an open-channel flow, studied in Section 8.4. As seen, the present result, $D_1/(\delta U_{fm}) \simeq 6$, reconciles with the value calculated analytically in Section 8.4.4, $D_1/(hU_f) \simeq 5.9$, Eq. 8.120.

As mentioned previously, Kirca et al. (2016) did numerical simulation tests also for heavy particles. Fig. 8.21 displays the results of these tests for the fall-velocity-parameter values $\beta = 0.1$ and 0.3 along with the result for the neutrally-buoyant particles ($\beta = 0$), reproduced from the previous figure. Here, the fall-velocity (Rouse) parameter β is defined by

$$\beta = \frac{w_s}{\kappa U_{fm}} \tag{8.150}$$

(similar to the steady flow case, Eq. 8.103). Fig. 8.21 shows the same trend as in the previous figure with respect to the variation of a/h. The figure further indicates that the dispersion coefficient increases with increasing β. This is consistent with the results obtained for the open-channel-flow case (see Fig. 8.11). A close inspection of the numerical values of the longitudinal dispersion coefficient for heavy particles for the asymptotic case $a/h \to \infty$ in Fig. 8.21 indicates that the results of Kirca et al.'s (2016) numerical simulations agree well also quantitatively with the results of the analytical solution plotted in Fig. 8.11.

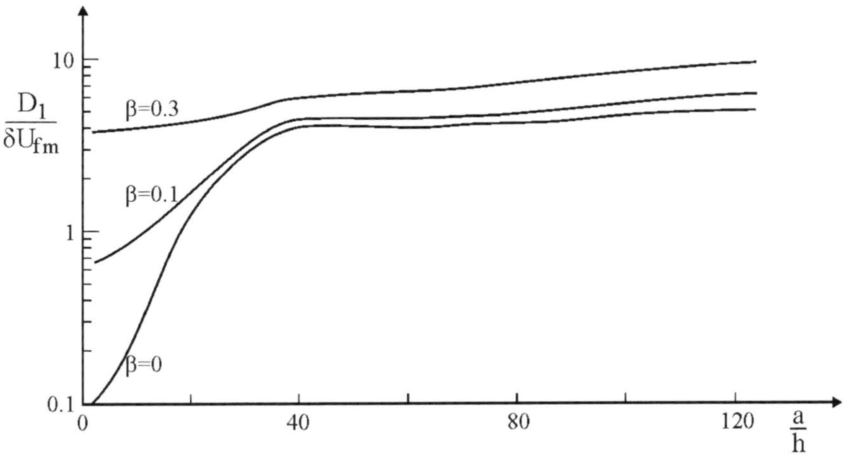

Fig. 8.21 Longitudinal dispersion coefficient vs. the parameter a/h as function of particle fall-velocity parameter. $\beta = 0$ corresponds to the neutrally-buoyant particles, reproduced from the previous figure. Kirca et al. (2016).

Example 8.2. Longitudinal dispersion of dredged sediment in wave boundary layers

In this section, a practical example from Kirca et al. (2016) will be given to illustrate how the presented numerical model can be used to simulate the longitudinal dispersion of sediment in wave boundary layers. The following case is considered:

Excavated material (fine sand with $d_{50} = 0.10$ mm) from a dredging operation is disposed to a coastal area at a depth of 10 m. The fall velocity of the material is $w_s = 0.65$ cm/s. Some time after the disposal, the material disperses under the action of storm waves with a root-mean-square wave height of $H_{rms} = 2.6$ m and zero-crossing period of $T_z = 8$ s. The objective is to simulate this process, using the presented numerical model.

First, the flow model is run for the determination of the mean flow and turbulence characteristics developing under the wave boundary layer. The free-stream velocity is taken as the horizontal orbital velocity at the bed found from the linear wave theory, $U_{m0} = 1.01$ m/s. The model upper boundary (i.e., frictionless lid) is set to $h = 1$ m. It can be shown by the linear wave theory that the orbital velocity at the top (at the model upper boundary) is practically equal to that at the bed ($U_{m0}(y = 1 \text{ m}) = 1.02$ m/s). Thus, the oscillatory flow model results are directly applicable for

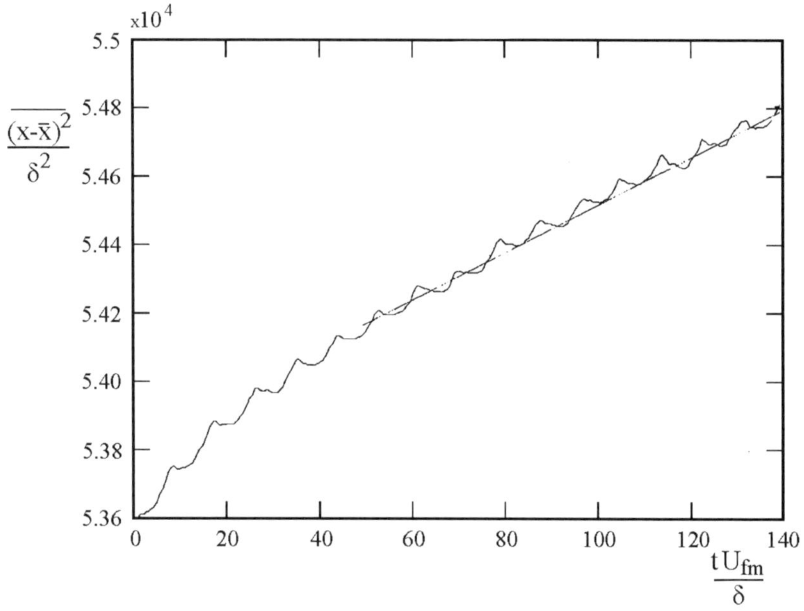

Fig. 8.22 Non-dimensional variance of particle position vs. non-dimensional time for the first eight waves (solid line: phase-averaged variance) for the wave-boundary-layer case. Water depth: 10 m, $\beta = 0.3$, $H_{rms} = 2.6$ m, $T_z = 8$ s. Kirca et al. (2016).

this wave boundary layer case.

The flow model yielded a friction velocity of $U_{fm} = 5.5$ cm/s, and a boundary layer thickness of $\delta = 2.5$ cm, compatible with the ones that can be found from the classic wave boundary layer formulae (e.g., Fredsøe and Deigaard, 1992, Chapter 2). For these conditions, the non-dimensional settling velocity of the particles corresponds to $\beta = w_s/(\kappa U_{fm}) = 0.3$. With the output of the flow model, the random-walk dispersion model is run with $N = 10000$ particles. The particles are released from the bed. Kirca et al. (2016) note that none of the particles were seen to travel higher than 0.32 m from the bed. Fig. 8.22 displays the variance of particle position versus time for the first eight waves of the storm. From the variance-versus-time data, the dispersion coefficient is calculated as $D_1/(\delta U_{fm}) = 3.5$, or $D_1 = 4.8 \times 10^{-3}$ m^2/s.

In order to illustrate the spreading of the dumped sediment as a result of the longitudinal dispersion caused by the storm waves, it is assumed that the initial layer of the dumped sediment (before the storm breaks)

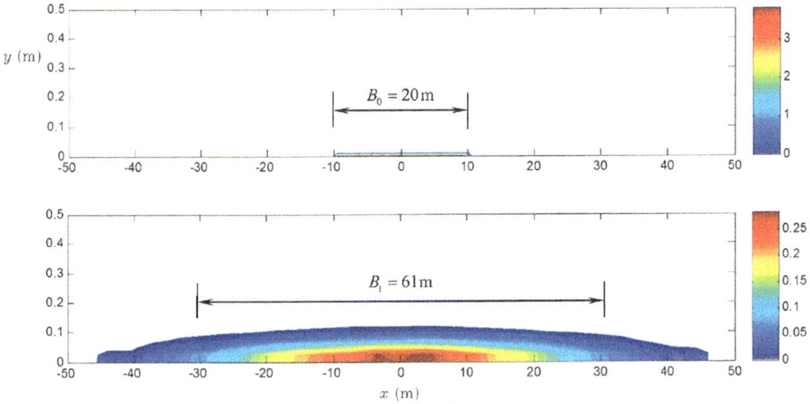

Fig. 8.23 Initial (top panel) and final (bottom panel) concentration distribution (probability density distribution of particles) of the disposed sediment. Water depth: 10 m, $\beta = 0.3$, $H_{rms} = 2.6$ m, $T_z = 8$ s, $t_{storm} = 8$ hours. Kirca et al. (2016).

covered a width of $B_0 = 20$ m in a uniform manner with a small thickness compared to depth. The storm lasts eight hours. Find the area over which the sediment is dispersed during the course of the entire storm.

The increase of the variance of the sediment-particle position in the longitudinal direction can be calculated by use of Eq. 8.145:

$$\left\langle \overline{[x(\omega t)]^2} \right\rangle - \left\langle \overline{[x(\omega t)]^2} \right\rangle_0 = 2D_1 t_{storm} \tag{8.151}$$

in which the initial variance $\left\langle \overline{[x(\omega t)]^2} \right\rangle_0$:

$$\left\langle \overline{[x(\omega t)]^2} \right\rangle_0 = \frac{1}{N} \sum_{n=1}^{N} \left(n\frac{B_0}{N} - \frac{B_0}{2} \right)^2 \simeq 33 \text{ m}^2 \tag{8.152}$$

With this,

$$\left\langle \overline{[x(\omega t)]^2} \right\rangle = \left\langle \overline{[x(\omega t)]^2} \right\rangle_0 + 2D_1 t_{storm}$$

$$\left\langle \overline{[x(\omega t)]^2} \right\rangle = 33 + 2 \times (4.8 \times 10^{-3}) \times (8 \times 3600) = 277 \text{ m}^2 \tag{8.153}$$

With a back calculation

$$\left\langle \overline{[x(\omega t)]^2} \right\rangle = \frac{1}{N} \sum_{n=1}^{N} \left(n\frac{B}{N} - \frac{B}{2} \right)^2 \simeq 277 \text{ m}^2 \tag{8.154}$$

(c.f. Eq. 8.152), the equivalent width of the sediment layer after the storm can be found as $B \simeq 61$ m. This example is illustrated in Fig. 8.23, obtained by the model simulation, which shows the initial and final concentration distributions, in terms of the probability density function of the sediment particle position.

8.7 Dispersion in the surf zone

Spreading of contaminants in the surf zone is also a concern in practice. The surf zone is the area extending in the cross-shore direction from wave breaking to the limit of runup, Chapter 7. It is viewed as a narrow strip (comparable to the characteristic wave length) where several important processes take place, such as: (1) longshore current caused by radiation stresses (Fredsøe and Deigaard, 1992, p. 129); (2) wave drift (see Appendix 5.D, as well as e.g. Fredsøe and Deigaard, 1992, p. 6), both in the longshore direction and in the cross-shore direction; (3) undertow (see Section 7.2.2); and (4) very strong turbulence caused by wave breaking (see Chapter 7).

A contaminant released in the surf zone will undergo dispersion as it is convected in the longshore and cross-shore directions. Convection of the contaminant in the longshore direction is due to the longshore current (and, to a smaller degree, due to the wave drift in the longshore direction), and that in the cross shore direction is due to the wave drift in the cross-shore direction, mentioned in the preceding paragraph.

The spreading, on the other hand, is due to the dispersion mechanisms existing in the surf zone. The longshore spreading is due the longitudinal dispersion mechanism, similar to that in a river (Section 8.5), and essentially caused by the combined action of the flow turbulence (mainly, generated by the wave breaking, and also by the shear near the bed) and the variation of the longshore velocity across the section normal to the longshore direction. In contrast, the cross-shore spreading is due to the cross-shore dispersion caused by the combined action of the turbulence and the variation of the cross-shore velocity across the section normal to the cross-shore direction.

Winckler et al. (2013) have recently studied this problem. Assuming small wave amplitude, they have adopted the following linearized form of turbulent diffusion/dispersion equation, to describe the convection and dispersion process in the surf zone:

$$\frac{\partial C}{\partial t} + u\frac{\partial C}{\partial x} + v\frac{\partial C}{\partial y} = \frac{1}{\overline{h}}\left[\frac{\partial}{\partial x}\left(\overline{h}K_x\frac{\partial C}{\partial x}\right) + \frac{\partial}{\partial y}\left(\overline{h}K_y\frac{\partial C}{\partial y}\right)\right] + \frac{q}{\overline{h}} \quad (8.155)$$

in which C is the depth-averaged concentration, x the coordinate axis in the cross-shore direction, y the coordinate axis in the longshore direction, $u = u(x, y, t)$ and $v = v(x, y, t)$ the cross-shore and longshore velocities, respectively, $\overline{h}$ the time-averaged water depth over a period, K_x and K_y the cross-shore and longshore diffusivities, respectively, reflecting the cross-shore and longshore dispersion processes described in the preceding paragraph, and q the rate of contaminant discharge per unit area.

In Winckler et al.'s (2013) study, the wave field in the shoaling zone is modelled by the linear theory for monochromatic waves (see the Appendix to Chapter 5), and further it is assumed that the breaking wave height is proportional to the local depth in the surf zone (this is essentially the same type of breaking model as used in Section 7.2.2, Eq. 7.39). The simplest scenario of longshore current on a straight beach of constant slope is considered. For the latter, the longshore current velocity is predicted by a (slightly) modified version of Longuet-Higgins' theory. Considering that the time scale of horizontal diffusion is large compared with the wave period, the perturbation method of multiple scales is applied to derive the transport equation from Eq. 8.155 for the convection and dispersion of a contaminant for the present problem. We note that Winckler et al. remark that no attempts are made to calibrate the diffusivities K_x and K_y because of lack of reliable data. Instead, they used an expression put forward by Longuet-Higgins (1970) for the eddy viscosity (Winckler et al. 2013).

From the study, Winckler et al. (2013) obtained the following key results:

1. The total convection velocity is found to be the sum of the steady current caused by radiation stresses, and a relatively small contribution from wave drift.

2. The dispersant is found to drift shoreward in addition to the expected transport along the shore, as described in the preceding paragraphs.

The reader can consult the original paper by Winckler et al. (2013) where the authors presented computed examples and also comparisons with available laboratory experiments were discussed in greater detail. The paper also includes an extensive review of the subject.

8.8 References

(1) Abramowitz, M. and Stegun, I.A. (1968): Handbook of Mathematical Functions. Dover Publications, Inc., New York, N.Y.
(2) Aris, R. (1956): On the dispersion of a solute in a fluid flowing through a tube. Proc. Roy. Soc. Lond. A, vol. 235, 67–77.
(3) Batchelor, G.K. (1949): Diffusion in a field of homogeneous turbulence. Australian J. Sci. Res. A, vol. 2, 473–450.
(4) Batchelor, G.K. and Townsend, A.A. (1956): Turbulent diffusion. In: Surveys in Mechanics (ed. G.K. Batchelor and R.M. Davies), Cam-

bridge University Press, pp. 352–399.
(5) Elder, J.W. (1959): The dispersion of marked fluid in turbulent shear flow. J. Fluid Mech., vol. 5, 544–560.
(6) Fischer, H.B. (1966): Longitudinal Dispersion in Laboratory and Natural Streams. Tech. Report KH-R-12, California Institute of Technology, Pasadena, California.
(7) Fischer, H.B. (1967): The mechanics of dispersion in natural streams. J. Hydraul. Div.-ASCE, vol. 93, 187–216.
(8) Fischer, H.B., List, E.J., Koh, R.C.Y., Imberger, J. and Brooks, N.H. (1979): Mixing in Inland and Coastal Waters. Academic Press, Inc.
(9) Fredsøe, J. and Deigaard, R. (1992): Mechanics of Coastal Sediment Transport. World Scientific.
(10) Fuhrman, D.R., Schløer, S. and Sterner, J. (2013): RANS-based simulation of turbulent wave boundary layer and sheet-flow sediment transport processes. Coast. Eng., vol. 73, 151–166.
(11) Gradshteyn, I.S. and Ryzhik, I.M. (1965): Table of Integrals, Series and Products. Academic Press.
(12) Jensen, B.L., Sumer, B.M. and Fredsøe, J. (1989): Turbulent oscillatory boundary layers at high Reynolds numbers. J. Fluid Mech., vol. 206, 265–297.
(13) Kirca, V.S.O., Sumer, B.M., Steffensen, M., Jensen, K.L. and Fuhrman, D.R. (2016): Longitudinal dispersion of heavy particles in an oscillating tunnel and application to wave boundary layers. J. Ocean Eng. Mar. Energy., vol. 2, 59-83.
(14) Lodahl, C.R., Sumer, B.M. and Fredsøe, J. (1998): Turbulent combined oscillatory flow and current in a pipe. J. Fluid Mech., vol. 373, 313–348.
(15) Longuet-Higgins, M.S. (1970): Longshore currents generated by obliquely incident sea waves: 2. J. Geophys. Res., vol. 75, 6790–6801.
(16) Monin, A.S. and Yaglom, A.M. (1973): Statistical Fluid Mechanics: Mechanics of Turbulence, vol. 1, MIT Press. Cambridge, Mass.
(17) Pedersen, F.B. (1977): Prediction of Longitudinal Dispersion in Natural Streams. Series Paper No. 14, Technical University of Denmark, Institute of Hydrodyn. and Hyd. Eng. (ISVA).
(18) Sayre, W.W. (1968): Dispersion of Mass in Open-Channel Flow. Hydraulic Papers, No. 3, February 1968, Colorado State University, Fort Collins.
(19) Sullivan, P.J. (1971): Longitudinal dispersion within a two-dimensional shear flow. J. Fluid Mech., vol. 49, 551–576.
(20) Sumer, B.M. (1973): Simulation of dispersion of suspended particles.

J. Hydraul. Div.-ASCE, vol. 99, 1705–1726.

(21) Sumer, B.M. (1974): Mean velocity and longitudinal dispersion of heavy particles in turbulent open-channel flow. J. Fluid Mech., vol. 65, 11–28.

(22) Sumer, B.M. (1977): Settlement of solid particles in open-channel flow. J. Hydraul. Div.-ASCE, vol. 103, 1323–1337.

(23) Sumer, B.M. and Fredsøe, J. (2002): The Mechanics of Scour in the Marine Environment. World Scientific.

(24) Taylor, G.I. (1921): Diffusion by continuous movements. Proc. Lond. Math. Soc., vol. 20, 196–212.

(25) Taylor, G.I. (1953): Dispersion of soluble matter in solvent flowing slowly through a tube. Proc. Roy. Soc. Lond. A, vol. 219, 186–203.

(26) Taylor, G.I. (1954): The dispersion of matter in turbulent flow through a pipe. Proc. Roy. Soc. Lond. A, vol. 223, 446–468.

(27) Winckler, P., Liu, P.L.-F. and Mei, C.C. (2013): Advective diffusion of contaminants in the surf zone. J. Waterw. Port Coast. Ocean Eng.-ASCE, vol. 139, 437–454.

Appendix to Chapter 8: Calculation of the settling velocity

The *fall* or *settling velocity*, w_s, of a grain is defined as the terminal velocity attained when the grain is settling in an extended fluid under the action of gravity (Fredsøe and Deigaard, 1992, p. 198). In this appendix some simple methods for its practical calculation/estimation are presented.

8.A General formulation

Consider an idealized spherical grain having diameter d and density ρ_p falling at velocity w_s in a fluid with density ρ and kinematic viscosity ν. The combined effects of gravity and buoyancy result in the downward force

$$F_s = g(\rho_p - \rho)\frac{\pi}{6}d^3 \tag{8.156}$$

whereas the (upward) drag force on the falling particle may be generally formulated in terms of a drag coefficient c_D according to:

$$F_D = \frac{1}{2}\rho c_D w_s^2 \frac{\pi}{4}d^2 \tag{8.157}$$

At terminal velocity these will be equal i.e. $F_s = F_D$, leading to the following general expression for the fall velocity:

$$w_s = \sqrt{\frac{4(s-1)gd}{3c_D}} \tag{8.158}$$

where $s = \rho_p/\rho$ is the specific gravity (relative density) of the sediment grain. Note that for natural (quartz based) sediment $s \approx 2.65$.

8.B Small Reynolds number: Stokes law

The drag coefficient c_D generally depends on the settling Reynolds number

$$Re_s = \frac{w_s d}{\nu} \tag{8.159}$$

For small $Re_s \leq 1$, then the flow around the particle is said to be within the Stokes (or creeping flow) regime i.e. where advective (inertial) forces are small relative to viscous forces. In this regime the drag coefficient can be shown to theoretically be

$$c_D = \frac{24}{Re_s} \tag{8.160}$$

which when substituted into Eq. 8.158 leads to the settling velocity

$$w_s = \frac{(s-1)gd^2}{18\nu} \tag{8.161}$$

Both Eqs. 8.160 and 8.161, and other variants thereof, are commonly known as *Stokes law* (Stokes, 1851).

It is again emphasized that this is only valid for $Re_s \leq 1$. Taking $g = 9.81$ m/s^2, $\nu = 10^{-6}$ m^2/s, and $s = 2.65$, this corresponds to sediment grain diameters of approximately $d \leq 0.104$ mm (with corresponding settling velocities $w_s \leq 0.00965$ m/s).

8.C Larger Reynolds numbers

For larger settling Reynolds numbers $Re_s > 1$, the drag coefficient can be formulated e.g. according to the empirical expression of Fredsøe and Deigaard (1992), their Eq. 7.12:

$$c_D = 1.4 + \frac{36}{Re_s} \tag{8.162}$$

This expression was shown to match experimental measurements quite well.

Utilizing this expression, the settling velocity w_s may be solved for explicitly, as follows. As shown by Fuhrman et al. (2013), inserting Eq. 8.162 into Eq. 8.158, squaring both sides, and re-organizing leads to the following quadratic polynomial in w_s:

$$a_s w_s^2 + b_s w_s + c_s = 0 \tag{8.163}$$

where

$$a_s = 4.2, \qquad b_s = 108\frac{\nu}{d}, \qquad c_s = -4gd(s-1) \tag{8.164}$$

The settling velocity is thus

$$w_s = \frac{-b_s + \sqrt{b_s^2 - 4a_s c_s}}{2a_s} \tag{8.165}$$

corresponding to the largest root of the quadratic equation Eq. 8.163.

As an example for the interested reader, taking the grain size as $d = 0.3$ mm and again maintaining $g = 9.81$ m/s^2, $\nu = 10^{-6}$ m^2/s and $s = 2.65$, the above yields the settling velocity $w_s = 0.0375$ m/s (corresponding to a settling Reynolds number $Re_s = 11$).

It is finally noted that at high concentrations the settling velocity of sediment particles becomes *hindered* i.e. smaller than would be predicted from the above. For methods of estimating the settling velocity in such high concentration conditions, see e.g. the classic work of Richardson and Zaki (1954), as well as the more recent work of Baldock et al. (2004).

8.D References

(1) Baldock, T.E., Tomkins, M.R., Nielsen, P. and Hughes, M.G. (2004): Settling velocity of sediments at high concentrations. Coast. Eng., vol. 51, 91–100.

(2) Fredsøe, J. and Deigaard, R. (1992): Mechanics of Coastal Sediment Transport. World Scientific.

(3) Fuhrman, D.R., Schløer, S. and Sterner, J. (2013): RANS-based simulation of turbulent wave boundary layer and sheet-flow sediment transport processes. Coast. Eng., vol. 73, 151–166.

(4) Richardson, J.F. and Zaki, W.N. (1954): Sedimentation and fluidisation, Part 1. Trans. Inst. Chem. Eng., vol. 32, 35–53.

(5) Stokes, G.G. (1851): On the effect of internal friction of fluids on the motion of pendulums. Trans. Cambridge Phil. Soc., vol. 9, 8–106. (The formula for the terminal velocity appears on p. 52, Eq. 127.)

Chapter 9

Mathematical modelling of turbulence

The present chapter describes the mathematical modelling of turbulence, also colloquially known simply as "turbulence modelling". Turbulence modelling in various forms has already been presented and/or employed in several of the previous chapters, hence this chapter is intended to supplement much of the preceding book. In what follows the closure problem of turbulence is first covered, establishing the necessity of resorting to turbulence modelling in the first place. The remainder of the chapter is then dedicated to describing some selected models for establishing turbulence closure in practice. In particular one closure from each of the three main model types are presented. These correspond to the mixing length model (an algebraic model), the k-ω model (a turbulence energy equation model), and a large eddy simulation (LES) model (a direct simulation model). Simple examples employing each of the specific models described are also presented. The chapter ends with a brief discussion of how computational costs associated with various methods (direct numerical simulation - DNS, LES, and turbulence energy equation models) scale with respect to increases in the Reynolds number.

As indicated above, the present chapter will only present a synopsis of some selected popular methods for the practical modelling turbulence. It is not intended as a detailed literature survey. The excellent book of Wilcox (2006) is dedicated solely to the topic of turbulence modelling, and can be consulted for a more detailed account of turbulence models in general. A comprehensive historical review of turbulence modelling specifically within the field of hydraulics has also been published by Rodi (2017). The present chapter will focus exclusively on turbulence modelling for single phase flows of an incompressible fluid. For an overview of turbulence modelling in porous media flow see Losada et al. (2016).

9.1 The closure problem

We have previously derived the basic equations of turbulent flows in Chapter 2. These equations governing the mean flow are the continuity equation (Eq. 2.6):

$$\frac{\partial \overline{u}_i}{\partial x_i} = 0 \tag{9.1}$$

and the Reynolds-averaged Navier-Stokes (RANS) equations (Eq. 2.14):

$$\rho\left(\frac{\partial \overline{u}_i}{\partial t} + \overline{u}_j \frac{\partial \overline{u}_i}{\partial x_j}\right) = \rho \overline{g}_i + \frac{\partial}{\partial x_j}(\overline{\sigma}_{ij} - \rho\overline{u'_i u'_j}) \tag{9.2}$$

where

$$\overline{\sigma}_{ij} = -\overline{p}\delta_{ij} + \mu\left(\frac{\partial \overline{u}_i}{\partial x_j} + \frac{\partial \overline{u}_j}{\partial x_i}\right) \tag{9.3}$$

is the stress on the fluid. The Reynolds stress tensor is

$$-\rho\overline{u'_i u'_j} = \begin{bmatrix} -\rho\overline{u'_1 u'_1} & -\rho\overline{u'_1 u'_2} & -\rho\overline{u'_1 u'_3} \\ -\rho\overline{u'_2 u'_1} & -\rho\overline{u'_2 u'_2} & -\rho\overline{u'_2 u'_3} \\ -\rho\overline{u'_3 u'_1} & -\rho\overline{u'_3 u'_2} & -\rho\overline{u'_3 u'_3} \end{bmatrix} \tag{9.4}$$

in which there are, due to symmetry, six independent Reynolds stresses (three diagonal and three off-diagonal components).

As has already been discussed (Chapter 2), there are four equations for the mean flow (i.e. the continuity equation Eq. 9.1 and three RANS equations from Eq. 9.2), whereas there are ten unknowns (namely, three components of the velocity $\overline{u}_i$, the pressure $\overline{p}$, as well as six components of the Reynolds stress $-\rho\overline{u'_i u'_j}$). There being more unknowns than equations, the system is not closed, and this is known as the *closure problem of turbulence*. To close the system, we must somehow devise a sufficient number of equations describing the effects of turbulence on the mean flow such that the unknowns may be solved for. This gives rise to the science (and art!) of *turbulence modelling*.

For comparison, note that in the much simpler case of laminar flow there are four unknowns (namely, three components of the velocity, u_i, and the pressure, p), and four equations, i.e. the continuity equation (Eq. 2.4) and three N.-S. equations (Eqs. 2.8 and 2.9). Therefore, the system, in this case, is closed. The same accounting would likewise hold in a so-called direct numerical simulation (DNS) of turbulent flow (to be discussed in what follows), which is also based on the primitive Navier-Stokes equations. This thus demonstrates that the closure problem for turbulence arises as a consequence of resorting to a Reynolds-averaged description of turbulent flows.

9.2 Types of turbulence models

There are several kinds of turbulence models in modern use. The most important of these are:

(1) Algebraic models;
(2) Turbulence energy equation models; and
(3) Simulation models.

Algebraic models, also known as the zero-equation models, are the simplest of all turbulence models. Here (1) the Reynolds stress is expressed as the product of a turbulence viscosity (eddy viscosity) and the mean strain rate via the so-called Boussinesq eddy-viscosity approximation, and (2) the eddy viscosity is computed or somehow specified, often in terms of a so-called mixing length. Indeed, of the algebraic models, it is the mixing length model that is the simplest and the most "popular" one, and this model will therefore be described in Section 9.3.

Turbulence energy equation models are more advanced. Here, too, the Reynolds stress is expressed as the product of a turbulence viscosity (eddy viscosity) and the mean strain rate. However, the turbulence viscosity is not computed in terms of a mixing length (or otherwise specified directly), but rather it is formulated in terms of the turbulence kinetic energy (per unit mass), $k = 1/2(\overline{u_i' u_i'})$. There are two principal types of turbulence energy equation models: one-equation models, and two-equations models. The latter are the most popular in modern applications, and as an example the so-called k-ω model will be described in Section 9.4.

Simulation models are the most advanced for describing turbulent flows, and include methods known as Direct Numerical Simulation (DNS), Large Eddy Simulation (LES), and Detached Eddy Simulation (DES). These models directly simulate the turbulence process itself, and therefore they do not involve the aforementioned closure problem. In a DNS study, a complete time-dependent solution of the primitive Navier-Stokes equations and the continuity equation is sought, and turbulent fluctuations (in space and time at all scales) are resolved directly rather than being parameterized. Therefore the solution furnishes the time series of (1) the three components of the velocity and (2) the pressure. Up until now it is only possible to simulate relatively low Reynolds number-flows with the DNS method due to their large computational expense. (Examples stemming from DNS simulations can be seen e.g. in Figs. 4.19, 5.7 and 5.9.) For larger Reynolds numbers, the scales of the dissipative part of turbulence motion are so small that this kind of small scale motion cannot be resolved in a numerical calculation (the overall computational costs required to resolve such motion increases approximately with Re^3, where Re is the Reynolds number, see Section 9.6). However, the upper limit of the Reynolds number range is pushed to higher and higher values as more and more computer power becomes available. Rogallo and Moin (1984) and Moin and Mahesh (1998) give reviews of the DNS method where the principal/early publications are cited. The LES model offers something of a compromise, the basic idea being to resolve the large turbulent eddies (large-scale motions), while still modelling the additional effects of the smallest eddies at the smaller sub-grid scales. The LES model is described in Section 9.5 along with an application example.

9.3 Mixing length model of Prandtl (1925)

The mixing length model is an algebraic model, originally formulated by Prandtl (1925). In what follows, we will describe this model with reference to the horizontally uniform steady boundary-layer flow close to a smooth or rough wall (i.e. in the constant stress layer, see Section 3.1), as is depicted in Fig. 9.1.

9.3.1 *The eddy viscosity*

As mentioned above, to begin formulation of the mixing length turbulence model we may draw an analogy to Newton's friction law

$$\tau = \mu \frac{du}{dy} \qquad \text{or} \qquad \frac{\tau}{\rho} = \nu \frac{du}{dy} \tag{9.5}$$

and formulate the turbulence shear stress (as was first suggested by Boussinesq, 1877) according to:

$$-\rho\overline{u'v'} = \mu_t \frac{d\overline{u}}{dy} \qquad \text{or} \qquad -\overline{u'v'} = \nu_t \frac{d\overline{u}}{dy} \tag{9.6}$$

in which μ_t and $\nu_t = \mu_t/\rho$ are respectively the dynamic and kinematic turbulence viscosity (or eddy viscosity). The eddy viscosity may essentially be thought of as a "turbulence exchange coefficient for momentum" (Tennekes and Lumley, 1972, p. 43). The eddy viscosity is distinctly different than the fluid viscosity, however, in that it is a property of the turbulence and flow field rather than the fluid!

Use of an eddy viscosity is of fundamental importance in the practical modelling of turbulent flows. Establishing a reasonable physical justifica-

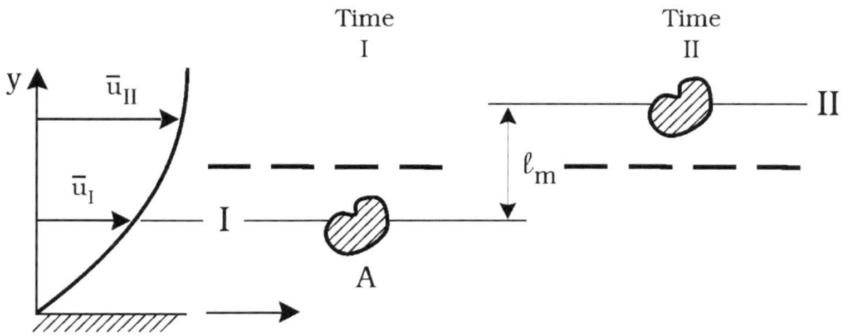

Fig. 9.1 A fluid parcel traveling from Level I to Level II.

tion is hence important, and will be attempted in what follows. For the shear component of the Reynolds stress $-\rho\overline{u'v'}$ (describing the mean vertical exchange of horizontal momentum due to turbulence) to be non-zero clearly requires that the two fluctuating components (u' and v') be *correlated* (see Section 4.2). In other words, even though $\overline{u'}$ and $\overline{v'}$ are, by definition, zero, the average of their product apparently may not be, and Eq. 9.6 implies that the velocity gradient in the mean flow ($d\overline{u}/dy$) apparently plays a key role in making this so. Why this is can be explained relatively simply from inspection e.g. of the mean velocity profile $\overline{u}(y)$ in Fig. 9.1. In such a flow, at any vertical position y within the boundary layer, we must expect a tendency for positive horizontal fluctuations ($u' > 0$, i.e. instantaneous horizontal flow greater than the mean) to occur in conjunction with $v' < 0$, as such higher (than average) streamwise velocity flow would most likely have come from above i.e. transported downward. Similarly, we should expect a tendency for negative horizontal fluctuations ($u' < 0$) to occur when $v' > 0$, since such lower (than average) streamwise velocity fluid would most likely have come from below. These simple considerations explain why u' and v' are always, in fact, *negatively correlated* in simple steady uniform boundary layer flows, and that the vertical gradient in the mean flow $d\overline{u}/dy$ is essential to causing this correlation. Indeed, if $d\overline{u}/dy = 0$ there would be no reason to expect correlation of any kind between the turbulent fluctuations u' and v' and the shear component of the Reynolds stress tensor would be zero. These considerations are directly reflected in Eq. 9.6 and provide some physical justification of its use, at least as a first approximation.

9.3.2 *Prandtl's mixing length hypothesis*

Obviously, in order to utilize Eq. 9.6 in practice, requires a means of determining the previously-introduced eddy viscosity. Prandtl formulated this as follows:

$$\mu_t = \rho \ell_m V_t \qquad \text{or} \qquad \nu_t = \ell_m V_t \tag{9.7}$$

where V_t is a velocity scale characterizing the turbulent fluctuations and ℓ_m is the so-called *mixing length.* This quantity conceptually defines the characteristic distance a fluid parcel is transported across the flow by turbulent fluctuations before becoming mixed with the surrounding fluid (hence taking on the flow properties at the new level). The mixing length is something akin to the Lagrangian macro length scale presented in Section 4.2.2. The mixing length concept is illustrated in Fig. 9.1, which depicts a parcel of fluid traveling between two levels (I and II, separated by the mixing length ℓ_m) having different mean velocity $\overline{u}$ for the steady boundary layer flow under present consideration. Eq. 9.7 is known as *Prandtl's mixing length hypothesis.*

In his formulation of the eddy viscosity Prandtl was originally inspired by the kinetic theory of gases, which predicts the fluid viscosity in terms of the mean free path length of molecules and their r.m.s. speed (see e.g. Davidson, 2015, p. 111). Basing the eddy viscosity in terms of the product of such characteristic turbulence length and velocity scales is also justifiable on simple dimensional grounds.

(It may be noted that, in a more recent publication, Nielsen and Teakle, 2004, develop a theory of turbulent diffusion of momentum and suspended particles, a finite-mixing-length theory, in which the turbulence viscosity is expressed as $\mu_t = \rho \ell_m V_t (1 + O(\ell_m/L)^2)$. Here L is the scale of the velocity distribution $\overline{u}(y)$. It may also be noted that different diffusivities exhibited by different "species", like density, heat and salinity, in turbulent flows may be qualitatively explained in terms of Nielsen and Teakle's theory. See also Absi, 2005, in relation to the latter publication.)

9.3.3 *Prandtl's second hypothesis*

To achieve closure with the simple mixing length model outlined above requires a means of determining the characteristic turbulent fluctuating velocity V_t. Prandtl formulated this as equivalent to the absolute difference in the streamwise velocity between level I and II i.e. conceptually as $V_t \approx |\overline{u}_{II} - \overline{u}_I|$. Assuming that the mixing length ℓ_m is small, this can be expressed in terms of the gradient of the mean flow as:

$$V_t = \ell_m \left| \frac{d\overline{u}}{dy} \right| \tag{9.8}$$

The expression above is known as *Prandtl's second hypothesis*. The physical justification of the above is straightforward after incorporating the definition of the mixing length above within the preceding qualitative justification of Eq. 9.6. As previously discussed, at a given position y, positive/negative turbulent fluctuations in u' will tend to correspond to flow parcels coming from above/below. In accordance with its definition above, taking the mixing length ℓ_m as the characteristic vertical distance from which such parcels come leads directly to the characteristic fluctuating velocity V_t in Eq. 9.8.

It is emphasized that V_t is just a scale. Any missing constants will be effectively lumped into the mixing length ℓ_m (to be determined below), and we may hence freely use the equal sign, rather than a weaker proportionality, in Eq. 9.8. In this spirit, it likewise makes no difference if V_t characterizes e.g. horizontal or vertical fluctuations, as these are the same order of magnitude (see e.g. Figs. 3.7, 3.14 and 3.15).

Combining Eqs. 9.7 and 9.8 leads to the eddy viscosity finally formulated as:

$$\mu_t = \rho \ell_m^2 \left| \frac{d\overline{u}}{dy} \right| \qquad \text{or} \qquad \nu_t = \ell_m^2 \left| \frac{d\overline{u}}{dy} \right| \tag{9.9}$$

As is clear, use of the absolute value of the velocity gradient ensures that this quantity is a positive scalar.

9.3.4 *The mixing length and solution for flow above a smooth wall*

The final step to achieve closure is to determine the mixing length in the preceding equation, which must be determined or calibrated individually for a given problem of interest. Until now the analysis is given for a general case involving a steady horizontally uniform boundary layer flow, and therefore it can be implemented for any distance y from the wall. However, from now on, we will further restrict our attention to the region close to the wall in the so-called constant stress layer (see Chapter 3). That is, in what follows it will be assumed that y is so small that $\tau \simeq \tau_0$, requiring that

$$y \ll h \tag{9.10}$$

in which h is the flow depth in the case of an open channel flow. (This may be replaced with the pipe radius in the case of a pipe flow or the boundary layer thickness in the case of a boundary layer over a wall. See the discussion in Section 3.1.1.)

On the other hand, let us also assume that y is still sufficiently large such that the transport of momentum is indeed due to turbulence i.e. viscosity will be neglected in the transport of momentum such that $\tau \simeq \tau_0 \simeq -\rho\overline{u'v'}$. We require that the transport of momentum is due to turbulence because the previously described mixing length (Section 9.3.2) is essentially based on the turbulent transport of momentum. For flow near a smooth wall this requires that

$$y^+ > 30 - 70 \tag{9.11}$$

in which $y^+ = yU_f/\nu$, $U_f = \sqrt{\tau_0/\rho}$ being the friction velocity (Chapter 3). Combining Eqs. 9.10 and 9.11 above leads to the following constraint for the logarithmic region:

$$30 - 70 < y^+ \ll Re_\tau, \qquad Re_\tau = \frac{hU_f}{\nu} \tag{9.12}$$

showing that the upper limit of the log layer is dependent on the Reynolds number, which must obviously be sufficiently large for a logarithmic layer to exist.

Now, we should expect that the mixing length (for y^+ values in the range stated in Eq. 9.12) increases with increasing distance from the wall, and the simplest form of this dependence is

$$\ell_m = \kappa y \tag{9.13}$$

in which κ is a constant. The resulting model equations then correspond to Eqs. 9.6, 9.9 and 9.13. Taking $\tau_0/\rho = U_f^2 = -\overline{u'v'}$ and assuming the velocity gradient to be positive, these equations lead to

$$\frac{1}{U_f}\frac{d\overline{u}}{dy} = \frac{1}{\kappa y} \qquad \text{or} \qquad \frac{d(\overline{u}/U_f)}{dy^+} = \frac{1}{\kappa y^+} \tag{9.14}$$

After integration, we then obtain the velocity profile

$$\frac{\overline{u}}{U_f} = \frac{1}{\kappa}\ln y^+ + B \tag{9.15}$$

from which it is seen that κ turns out to be the von Karman constant ($\kappa = 1/A = 0.4$) introduced in Chapter 3, and where experiments suggest $B = 5.1$ (see e.g. Eq. 3.44). To reiterate: Eqs. 9.13 and 9.15 are only valid in the y^+ range indicated in Eq. 9.12.

van Driest (1956) revised Eq. 9.13 to accommodate the influence of viscosity in the transport of momentum. This led to

$$\ell_m = \kappa y\left[1 - \exp\left(-\frac{y^+}{A_d}\right)\right] \tag{9.16}$$

in which $A_d = 25$ is called the damping coefficient. This equation for the mixing length is valid for the range $0 \leq y \ll h$. From Eq. 9.16, it is seen that

$$\ell_m \to \kappa y, \text{ for large values of } y^+ \tag{9.17}$$

while

$$\ell_m \to 0, \text{ as } y^+ \to 0 \tag{9.18}$$

The latter implies that there will be no turbulent transport of momentum as the wall is approached. Note that for flow above a rough wall (covered in the next sub-section) the mixing length will inevitably change from the above.

Eqs. 9.6, 9.9 and 9.16 are the resulting model equations. For the problem under present consideration, i.e. the steady boundary-layer flow close to a smooth wall (in the constant stress layer, see again Section 3.1), there are the three model equations identified just above and one Reynolds equation from Eq. 9.2, where for the present case only the final two terms would be active. Simplifying this for the present case (steady boundary layer in the constant stress region), after integration with respect to y, results in:

$$\tau_0 = \rho U_f^2 = \mu \frac{d\overline{u}}{dy} - \rho\overline{u'v'} \qquad \text{or} \qquad \frac{\tau_0}{\rho} = U_f^2 = \nu \frac{d\overline{u}}{dy} - \overline{u'v'} \tag{9.19}$$

Hence, we have altogether four equations. There are likewise four unknowns, namely $\overline{u}$, $-\rho\overline{u'v'}$, μ_t (or ν_t), and ℓ_m. The system is therefore closed and the unknowns can be determined.

Of particular interest is the velocity, $\overline{u}$, which can be found after inserting the model equations into Eq. 9.19. This leads to the solution previously obtained in Section 3.3.1 (Eq. 3.108):

$$\overline{u} = 2U_f \int_0^{y^+} \frac{dy^+}{1 + \left\{1 + 4\kappa^2 y^{+2}\left[1 - \exp\left(-\frac{y^+}{A_d}\right)\right]^2\right\}^{1/2}} \tag{9.20}$$

This is the celebrated *van Driest* velocity distribution, depicted in Fig. 3.20. This equation is again valid above a smooth wall in the constant stress layer, namely for the range $0 \leq y \ll h$ (equivalently $0 \leq y^+ \ll Re_\tau$).

For small values of y^+, the van Driest profile reduces to the linear velocity distribution in the viscous sublayer (Eq. 3.35),

$$\frac{\overline{u}}{U_f} = y^+ \tag{9.21}$$

whereas for large values of y^+ it reduces to the logarithmic velocity distribution in the logarithmic layer, Eq. 9.15. It is seen from Fig. 3.20 that the van Driest profile agrees remarkably well with experimental data.

9.3.5 *The mixing length and solution for flow above transitional and rough walls*

We will now consider further generalization of the mixing length to also incorporate effects of bottom roughness. In this case, the mixing length may be given as (Cebeci and Chang, 1978)

$$\ell_m = \kappa(y + \Delta y)\left[1 - \exp\left(-\frac{(y^+ + \Delta y^+)}{A_d}\right)\right] \tag{9.22}$$

y in the above equation is measured from the level where the velocity $\overline{u}$ is zero. This level lies almost at the top of the roughness elements, as discussed in Section 3.3.1 (Here y should not be confused with the distance measured from the theoretical wall, which corresponds to $y + \Delta y$; See the detailed discussion in Section 3.3.1.)

In Eq. 9.22 Δy is the so-called coordinate displacement, or the coordinate shift (see Example 3.3 in Section 3.3.1). This was given by Cebeci and Chang (1978) as

$$\Delta y^+ = 0.9\left[\sqrt{k_s^+} - k_s^+ \exp\left(-\frac{k_s^+}{6}\right)\right]; \qquad 5 < k_s^+ < 2000 \tag{9.23}$$

in which k_s^+ is the roughness Reynolds number (k_s being Nikuradse's equivalent sand roughness):

$$k_s^+ = \frac{k_s U_f}{\nu} \tag{9.24}$$

Here the distance from the wall y should be in the interval $0 \leq y \ll h$ where

the upper boundary ensures that the y levels lie in the constant stress layer.

The solution for $\overline{u}$ for the set of the model equations along with the Reynolds equation is obtained in Section 3.3.1 (Eq. 3.113):

$$\overline{u} = 2U_f \int_0^{y^+} \frac{dy^+}{1 + \left\{1 + 4\kappa^2 (y^+ + \Delta y^+)^2 \left[1 - \exp\left(-\frac{(y^+ + \Delta y^+)}{A_d}\right)\right]^2\right\}^{1/2}} \tag{9.25}$$

Fig. 3.21 displays the velocity profiles obtained from the preceding equation including the smooth-wall profile i.e. the solution in Eq. 9.25 is generally valid for the flow near smooth, transitional, and rough walls. In Fig. 3.21, a represents the linear profile in the viscous sublayer (Eq. 9.21), b the logarithmic profile (Eq. 9.15) in the case of the hydraulically smooth wall, and

c the logarithmic profile in the case of the completely rough wall (Eq. 3.73), namely, by putting $\tilde{y} = y + \Delta y$,

$$\overline{u} = AU_f \ln\left(\frac{30\tilde{y}}{k_s}\right), \qquad A = \frac{1}{\kappa} = 2.5 \tag{9.26}$$

with $\tilde{y}$ corresponding to the vertical distance above the theoretical wall.

9.3.6 *Turbulence quantities*

Due to assumed statistical uniformity in the transverse $x_3 = z$ direction, the components $-\rho\overline{u_1'u_3'}$ and $-\rho\overline{u_2'u_3'}$ in the Reynolds stress tensor Eq. 9.4 would be zero in the present example involving a simple steady boundary layer flow. This leaves the remaining turbulence quantities $-\rho\overline{u'^2}$, $-\rho\overline{v'^2}$, $-\rho\overline{w'^2}$ and $-\rho\overline{u'v'}$, as the unknowns. As is hopefully clear from the above, the Reynolds shear stress, $-\rho\overline{u'v'}$, can be determined from Eq. 9.6, once the mean flow velocity $\overline{u}$ is computed. The mixing length model is "designed" only to be able to handle the velocity $\overline{u}$, and from this the Reynolds shear stress $-\rho\overline{u'v'}$, however. It does not directly offer solutions for the other three unknowns, i.e. the turbulence quantities $-\rho\overline{u'^2}$, $-\rho\overline{v'^2}$ and $-\rho\overline{w'^2}$, corresponding to the normal (diagonal) components of the Reynolds stress tensor. In the case considered (a steady uniform boundary layer flow) these quantities would be uniform in the streamwise direction, and would hence not directly influence the mean flow.

9.4 The k-ω model of Wilcox (2006)

As an example of a turbulence energy equation model, we will now consider the widely used two-equation k-ω turbulence closure model. The k-ω model was originally suggested (in an early form with little formal development) by Kolmogorov (1942), and later independently by Saffman (1970), with further developments made e.g. by Wilcox (1988) and Wilcox (2006), among others. In what follows we will consider the version of the model in the form originally presented by Wilcox (2006, 2008). While there are many other two-equation turbulence closure models in use (e.g. the k-ω SST (shear stress transport) model of Menter (1994) or the k-ε model of Launder and Sharma (1974)), this variant of the k-ω model has been specifically selected because it:

- can simulate a full variety of turbulent flows (steady and unsteady

boundary layers, free shear flows, separation, etc.);
- accurately describes flows from outer regions all the way down to a wall (including the viscous sublayer above smooth walls);
- performs well in flows involving adverse pressure gradients;
- can naturally simulate flows over both smooth and rough surfaces without modification;
- can easily be modified to a "low Reynolds number" variant (giving improved accuracy of turbulence in the buffer layer, and even enabling, to some extent, simulation of laminar-to-turbulent transition in terms of the mean flow); and
- remains rather elegant in its formulation (requiring only six closure coefficients and avoiding use of so-called blending functions).

In the authors' experience, this model performs well in steady current flows (to be detailed later in Section 9.4.4), unsteady wave boundary flows (detailed e.g. in Chapter 5, Section 5.12 as well as in Chapter 6), and even (with slight modifications) for modelling more complicated problems such as breaking waves (Chapter 7, Section 7.6). Indeed, much of the turbulence modelling results included throughout this book have been based on this specific model, or close variations thereof.

9.4.1 *Model equations*

9.4.1.1 *Standard model formulation*

In an attempt to de-mystify the origin of the model equations, as well as help keep clear what is theoretically based, and what is not, in the development of turbulence energy equation models, we will start our presentation from the basic equations presented in Chapter 2. Recall that the kinetic energy of turbulence per unit mass is defined as:

$$k = \frac{K_t}{\rho} = \frac{1}{2}\overline{u_i' u_i'} = \frac{1}{2}(\overline{u'^2} + \overline{v'^2} + \overline{w'^2}) \tag{9.27}$$

After dividing Eq. 2.31 by ρ, neglecting fluctuations in the body force i.e. setting $g' = 0$, performing some further (but straight-forward) manipulation of the viscous terms, while also invoking local continuity, we may transform Eq. 2.31 to the following exact transport equation for k:

$$\underbrace{\frac{\partial k}{\partial t}}_{(1)} + \underbrace{\overline{u}_j \frac{\partial k}{\partial x_j}}_{(2)} = \underbrace{-\overline{u_i' u_j'} \frac{\partial \overline{u}_i}{\partial x_j}}_{(3)} - \underbrace{\nu \overline{\frac{\partial u_i'}{\partial x_k} \frac{\partial u_i'}{\partial x_k}}}_{(4)} + \frac{\partial}{\partial x_j}\left[\underbrace{\nu \frac{\partial k}{\partial x_j}}_{(5)} - \underbrace{\frac{1}{2}\overline{u_i' u_i' u_j'}}_{(6)} - \underbrace{\frac{1}{\rho}\overline{p' u_j'}}_{(7)}\right] \tag{9.28}$$

The derivation, as just described, of the exact k equation above is presented in full in Appendix 9.A.

The individual numbered terms in Eq. 9.28 account for: (1) **unsteadiness**, (2) **convection**, (3) **production**, (4) **dissipation** (per unit mass i.e. the rate at which turbulent kinetic energy is converted into thermal energy), (5) **molecular diffusion**, (6) **turbulent transport** (the rate of turbulent kinetic energy transport due to turbulent fluctuations themselves), and (7) **pressure diffusion** (accounting for turbulent transport resulting from correlation of pressure and velocity fluctuations). Note that the dissipation term (4) is in principal, the isotropic version of ε from Eq. 4.82, as proved in Appendix 4.G; Nevertheless these quantities are commonly termed the same, and throughout the present chapter the convention

$$\varepsilon = \nu \overline{\frac{\partial u_i'}{\partial x_k} \frac{\partial u_i'}{\partial x_k}} \tag{9.29}$$

will be maintained.

As seen, Eq. 9.28 contains several terms with unknown correlations, hence making it of little direct use in the practical modelling of turbulence. The "art" of turbulence modelling in the present framework therefore lies in our ability to approximate these unknown quantities as simply as possible, while of course maintaining sufficient accuracy! In the k-ω model, the Reynolds stress tensor $(-\rho\overline{u_i' u_j'})$ in the production term (which it is emphasized also appears in the RANS equations, Eq. 9.2) is most commonly formulated in terms of an eddy viscosity ν_t based on the so-called Boussinesq approximation:

$$\frac{\tau_{ij}}{\rho} = -\overline{u_i' u_j'} = 2\nu_t S_{ij} - \frac{2}{3} k \delta_{ij} \tag{9.30}$$

where

$$S_{ij} = \frac{1}{2}\left(\frac{\partial \overline{u}_i}{\partial x_j} + \frac{\partial \overline{u}_j}{\partial x_i}\right) \tag{9.31}$$

is the mean strain-rate tensor. The form above can be considered as a generalization of that presented previously in Eq. 9.6 in the formulation of the mixing length model, there for the special case of a steady and uniform boundary layer flow. Relating the Reynolds shear stresses to gradients in the mean flow, as done via the Boussinesq approximation, again has quite reasonable physical basis (see the preceding discussion in Section 9.3.1), at least as a first approximation.

Notice that, unlike the simplified form in Eq. 9.6, the Boussinesq approximation above, in principal, enables determination of both turbulent shear and normal stresses. The latter may not be particularly accurate for most real flows, however. Consider e.g. that in a simple uniform boundary layer flow Eq. 9.30 predicts isotropic normal stresses i.e. $\overline{u'^2} = \overline{v'^2} = \overline{w'^2} = 2/3k$, which clearly differ from measurements e.g. Figs. 3.7, 3.14 and 3.15. Fundamentally important, nevertheless, it is seen that Eq. 9.30 satisfies the definition of k from Eq. 9.27 i.e. that $\overline{u'_i u'_i} = 2k$.

To enable practical evaluation of the other unknown quantities in Eq. 9.28 it is convenient to adopt a second model variable for describing the turbulence. While there are several alternatives (one being ε itself, thus leading to the previously mentioned k-ε model), a proven alternative is to utilize the so-called specific dissipation rate ω, thus leading to the k-ω model. This quantity is formally defined by its relationship to the dissipation:

$$\varepsilon = \nu \overline{\frac{\partial u'_i}{\partial x_k} \frac{\partial u'_i}{\partial x_k}} = \beta^* k\omega \tag{9.32}$$

where β^* is a closure coefficient, to be defined below.

Physically speaking, with the adopted approach, we essentially attempt to describe the turbulence with only two characteristic scalar parameters: k (having velocity squared units) and ω (having inverse time units). Hence, in the simplest of terms, we may interpret $\sqrt{k}$ as the scale characterizing the magnitude of the turbulent velocity fluctuations, whereas ω can be interpreted as a characteristic velocity gradient (also having inverse time units) of the turbulence, such that this variable in a sense characterizes how rapidly the turbulence fluctuates in space (as such, ω is sometimes described as a characteristic vorticity of the turbulence). Note that on dimensional grounds we may now also define a model length scale as

$$\ell = \frac{\sqrt{k}}{\omega} \tag{9.33}$$

This is proportional (but not equal) to the previously-defined mixing length ℓ_m. (This will be shown in a forthcoming discussion of wall functions, leading to Eq. 9.56.)

Similarly, the turbulent transport and (usually relatively small) pressure diffusion (sixth and seventh terms in Eq. 9.28) are commonly grouped together, and assumed to be reasonably described as a gradient diffusion process, leading to a term similar in form to the molecular diffusion (term 5 in Eq. 9.28, which will itself be retained in exact form). In the molecular diffusion term the lead coefficient is simply the kinematic fluid viscosity ν. However, as we are here attempting to account for turbulent diffusion, the leading part is instead taken as proportional to k/ω (closely linked to the kinematic eddy viscosity, to be presented below, see Eq. 9.37). This leads to:

$$-\frac{1}{2}\overline{u_i'u_i'u_j'} - \frac{1}{\rho}\overline{p'u_j'} = \sigma^* \frac{k}{\omega}\frac{\partial k}{\partial x_j} \tag{9.34}$$

where σ^* is another model closure coefficient.

After invoking the model approximations above (note that while these have reasonable physical basis, they have certainly not been rigorously derived!), we may then transform the exact k equation from Eq. 9.28 to the following *model* k equation:

$$\frac{\partial k}{\partial t} + \overline{u}_j \frac{\partial k}{\partial x_j} = \frac{\tau_{ij}}{\rho}\frac{\partial \overline{u}_i}{\partial x_j} - \beta^* k\omega + \frac{\partial}{\partial x_j}\left[\left(\nu + \sigma^*\frac{k}{\omega}\right)\frac{\partial k}{\partial x_j}\right] \tag{9.35}$$

Similarly, the specific dissipation rate ω (as with most any physical quantity) can be assumed to undergo a similar combination of processes (convection, production, dissipation, and molecular/turbulent diffusion). In the Wilcox (2006) model, the second model ω equation is thus given as:

$$\frac{\partial \omega}{\partial t} + \overline{u}_j \frac{\partial \omega}{\partial x_j} = \alpha\frac{\omega}{k}\tau_{ij}\frac{\partial \overline{u}_i}{\partial x_j} - \beta\omega^2 + \frac{\sigma_d}{\omega}\frac{\partial k}{\partial x_j}\frac{\partial \omega}{\partial x_j} + \frac{\partial}{\partial x_j}\left[\left(\nu + \sigma\frac{k}{\omega}\right)\frac{\partial \omega}{\partial x_j}\right] \tag{9.36}$$

Most of the terms above are directly analogous to those in Eq. 9.35, the lone exception being the term proportional to σ_d. This term is known as **cross diffusion**, and was added by Wilcox (2006) to mitigate the model's sensitivity to the free stream value of ω, as originally suggested by Speziale et al. (1992). It is again important to recognize that, like many of the terms in the model k equation (Eq. 9.35), our arrival at the ω equation above is essentially founded on dimensional analysis coupled with experience, rather than based on rigorous derivation. This equation is essentially empirical in nature, and hence requires inclusion of several added closure coefficients (α, β, σ_d, and σ), as seen in the right hand side of Eq. 9.36.

To achieve closure the eddy viscosity must finally be defined. In the Wilcox (2006) model this can be generally defined according to:

$$\nu_t = \alpha^* \frac{k}{\tilde{\omega}} \tag{9.37}$$

where

$$\tilde{\omega} = \max\left[\omega, \lambda_1 \sqrt{\frac{2S_{ij}S_{ij}}{\beta_0^*/\alpha^*}}\right] \tag{9.38}$$

The max function in the above introduces a **stress-limiter** modification, serving to limit the value of the eddy viscosity in regions where turbulence production exceeds dissipation. This feature was first introduced by Coakley (1983), and has been shown by Huang (1999) to yield larger separation bubbles and improve incompressible and transonic-flow predictions. More relevant in the present context, Larsen and Fuhrman (2018) have also found that this feature helps limit the over-production of turbulence near the tip of breaking waves and in the outer surf zone (see Section 7.6; They also further extended the limiter concept to produce seemingly the first formally stable model in the potential flow region beneath surface waves, see Section 7.6.2.3). Wilcox (2006) suggests the value $\lambda_1 = 7/8$ for the stress limiter coefficient. Setting $\lambda_1 = 0$ can also be used to switch this feature off entirely, upon which the eddy viscosity simplifies to the more traditional relation $\nu_t = k/\omega$ (provided that $\alpha^* = 1$).

The closure coefficients for the standard Wilcox (2006) k-ω model are:

$$\alpha^* = 1, \quad \alpha = 0.52, \quad \beta = \beta_0 f_\beta, \quad \beta_0^* = \beta^* = 0.09$$
$$\sigma = 0.5, \quad \sigma^* = 0.6, \quad \sigma_{do} = 0.125 \tag{9.39}$$

$$\sigma_d = H\left\{\frac{\partial k}{\partial x_j}\frac{\partial \omega}{\partial x_j}\right\}\sigma_{do} \tag{9.40}$$

$$\beta_0 = 0.0708, \qquad f_\beta = \frac{1+85\chi_\omega}{1+100\chi_\omega}, \qquad \chi_\omega = \left|\frac{\Omega_{ij}\Omega_{jk}S_{ki}}{(\beta^*\omega)^3}\right| \tag{9.41}$$

(Note that because α^* takes a value of unity, it can simply be omitted in the standard closure model; We include it in the above for the sake of generality, such that the "standard" and "low Reynolds number" model variations elegantly share a common set of governing equations.)

These closure coefficients have been carefully and systematically tuned such that the model results are in agreement with well-known experimental observations in canonical flow situations (for full details see Wilcox, 2006,

2008). In short: the ratio $\beta^*/\beta_0 \approx 1.27$ is required to ensure correct decay of homogeneous, isotropic turbulence. The values for α and β^* can be established by examining the logarithmic layer in steady boundary layer flows. The optimum choice for σ is based on detailed analysis of the defect and viscous sublayers. Finally, the values for σ^* and β_0 are based on analysis of free shear flows (e.g. jets and plane wakes). In the above $H\{\cdot\}$ is the Heaviside step function, which takes a value of unity if the argument is positive and zero otherwise. Note that in two-dimensional flows $\chi_\omega = 0$, whence $f_\beta = 1$ and $\beta = \beta_0$. The dependence of β on χ_ω is based on the work of Pope (1978), and has a significant effect for round and radial jets (Wilcox, 2006, p. 127).

9.4.1.2 *Low Reynolds number model formulation*

A nice feature of the k-ω turbulence model is that it can be relatively easily modified to yield a "low Reynolds number" variant, requiring only that certain closure coefficients be re-defined. This version can improve accuracy of the turbulent kinetic energy e.g. in the buffer layer region, and can even be utilized to simulate, to some extent, the mean flow and turbulence in laminar-to-turbulence transition. (It is emphasized, however, that as this model is still working within the confines of RANS equations, it cannot describe the finer details associated with transition e.g. formation and development of turbulent spots, etc. Simulation of such features would require much higher resolution with DNS models.)

As an alternative to the standard "high Reynolds number" k-ω model described above, Wilcox (2006) also presents such a "low Reynolds number" variant. This was utilized e.g. by Williams and Fuhrman (2016) for the simulation of wave boundary layer flows over a wide range of scales, ranging from laboratory to full-scale tsunami wave scales. In this model, rather than being constants, selected closure coefficients are instead made to depend on a so-called turbulence Reynolds number:

$$Re_T = \frac{k}{\omega\nu} \tag{9.42}$$

Selected other closure coefficients are then re-defined as follows:

$$\sigma^* = 0.6\alpha^*, \qquad \sigma = 0.5\alpha^* \tag{9.43}$$

$$\alpha = \frac{13}{25}\frac{\alpha_0 + Re_T/R_\omega}{1 + Re_T/R_\omega}\left(\alpha^*\right)^{-1} \tag{9.44}$$

$$\beta^* = \beta_0^* \frac{100\,\beta_0/27 + (Re_T/R_\beta)^4}{1 + (Re_T/R_\beta)^4} \tag{9.45}$$

where:

$$\alpha^* = \frac{\alpha_0^* + Re_T/R_k}{1 + Re_T/R_k} \tag{9.46}$$

and $R_k = 6$, $R_\omega = 2.61$, $R_\beta = 8$, $\alpha_0^* = \beta_0/3$, $\alpha_0 = 1/9$, and $\beta_0^* = 0.09$ (the value of β^* for fully-turbulent flow). These coefficients have been tuned to yield transition to turbulence in accordance with experimental observations involving initially laminar flow above a flat plate, as discussed in detail in the Appendix to Chapter 1.

The three functions (α^*, α, and β^*) are plotted versus Re_T in Fig. 9.2. From this figure it is seen that, as Re_T is increased these coefficients tend asymptotically to their "high Reynolds number" (constant) values (indicated as the dashed lines) prescribed within the standard model. It is also seen by inspection that for low Re_T the added coefficient α^* becomes small. As this factor is applied directly on the eddy viscosity, Eq. 9.37, this enables the turbulence model to effectively de-couple from the RANS equations, Eq. 9.2. Hence, this single model can be utilized to simulate flows both in the laminar or turbulent regimes, including transition from one regime to the other (again, with the obvious limitations mentioned previously; as this model is still a closure to RANS equations, it does not directly resolve the turbulence as in DNS, or even LES, simulations).

To gain first hand experience with such a model, the interested reader is encouraged to complete hands-on Exercise 4 in Chapter 10.

9.4.2 *Wall boundary conditions*

With the k-ω model described above there are two primary strategies for imposing boundary conditions at frictional wall boundaries. These are: (1) to resolve the flow all the way down to the wall, thus requiring boundary conditions directly at the wall surface, or (2) to only resolve the flow beginning some distance from the wall (typically at the cell center or grid point nearest the wall), thus requiring what are known as *wall functions.* These two strategies are detailed below.

9.4.2.1 *Surface wall boundary conditions*

A rather remarkable feature of the k-ω closure model is that it can, without modification, quite accurately describe flow all the way down to a wall

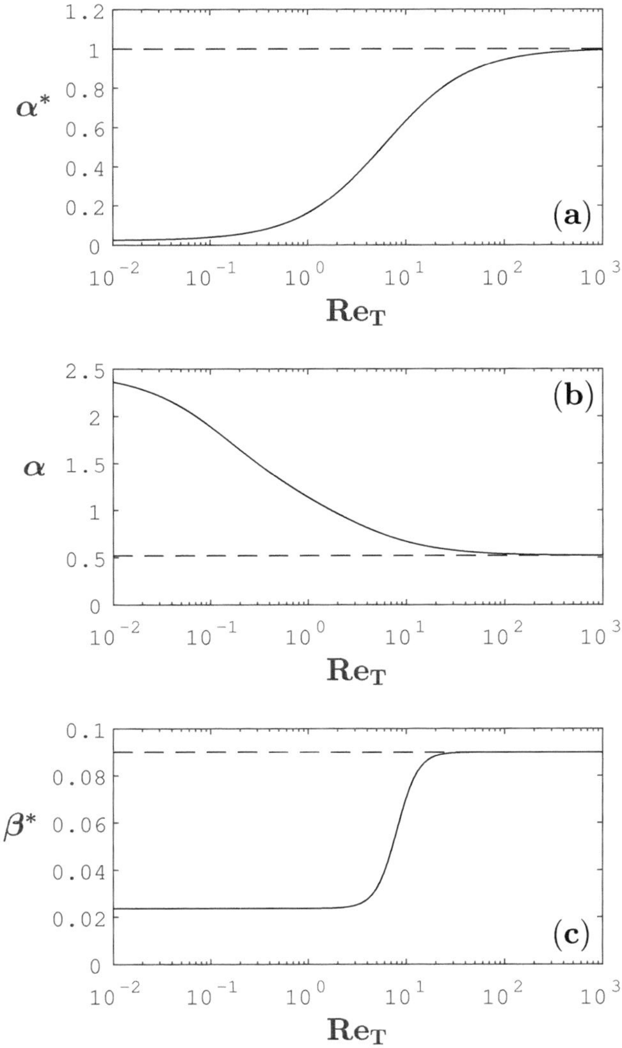

Fig. 9.2 Variation of the low Reynolds number k-ω turbulence model closure coefficients with turbulence Reynolds number $Re_T = k/(\omega\nu)$. The horizontal dashed lines represent the fully turbulent values.

surface. If such resolution is desired, then boundary conditions must be prescribed directly at the wall surface boundary.

For prescribing the value of the specific dissipation rate directly on a

Table 9.1 Table of recommended settings for the Wilcox (2006) k-ω turbulence model, regarding the stress limiter coefficient λ_1 and the rough wall coefficient K_r in Eq. 9.48.

Turbulence closure:	Standard	Standard	Low *Re*
Wall boundary condition:	$k = 0$	$\partial k/\partial n = 0$	$k = 0$
λ_1 (smooth wall)	7/8	7/8	7/8
λ_1 (rough wall)	0	7/8	0
K_r	80	180	80

wall surface boundary, Wilcox (2006) suggests:

$$\omega = \frac{U_f^2}{\nu} S_R, \qquad y = 0 \tag{9.47}$$

where S_R depends on the roughness Reynolds number $k_s^+ = k_s U_f/\nu$, where $U_f = \sqrt{|\tau_0|/\rho}$ is the instantaneous friction velocity, according to

$$S_R = \begin{cases} \left(\frac{200}{k_s^+}\right)^2, & k_s^+ \leq 5 \\ \frac{K_r}{k_s^+} + \left[\left(\frac{200}{k_s^+}\right)^2 - \frac{K_r}{k_s^+}\right] e^{5-k_s^+}, & k_s^+ > 5 \end{cases} \tag{9.48}$$

For rough walls k_s should be specified as the proper physical value. For large k_s^+ the above leads to $\omega = K_r U_f/k_s$ i.e. the boundary condition becomes independent of the viscosity, as it should for a fully rough wall. At the smooth wall limit Eq. 9.48 leads to $\omega = 200^2\nu/k_s^2$ corresponding to a "slightly-rough" boundary condition. In such cases k_s should be specified such that the flow is in the hydraulically smooth regime i.e. $k_s^+ < 5$. In these circumstances, we generally recommend e.g. using say $0.1 \leq k_s^+ \leq 2$.

The value of the rough-wall coefficient K_r depends on the specific closure model and k boundary condition that are utilized. Three combinations that work well in practice are indicated in Table 9.1. The values in Table 9.1 are based on the model validations and experience presented in Fuhrman et al. (2010) as well as subsequent experience.

As indicated in Table 9.1, a $k = 0$ Dirichlet wall boundary condition may be generally utilized in conjunction with the k-ω model, corresponding to a no slip condition applied to the turbulent fluctuations. If this option is utilized, a viscous sublayer region will develop regardless of wall type (i.e. smooth or rough) and must be resolved. This is quite natural if the wall is smooth, and a good rule of thumb for the near-wall cell size in such circumstances is

$$\Delta y^+ = \frac{\Delta y U_f}{\nu} \leq 1 \tag{9.49}$$

which ensures that there will be several cells or grid points in the viscous sublayer region, spanning $y^+ \leq 5$.

Forcing a viscous sublayer on a hydraulically rough wall, however, is somewhat contrary to the physics within the hydraulically rough regime, which is founded on the notion that the viscous sublayer is disrupted entirely. Indeed, satisfying Eq. 9.49 in the hydraulically rough regime requires a mesh that is Reynolds number dependent, even though the physics above the wall largely are not! Note also that if the $k = 0$ wall boundary condition is utilized on a rough wall (with either standard or low Reynolds number closure variants) we recommend switching off the stress limiting feature by setting $\lambda_1 = 0$, as it can pollute results on rough walls (though it may be freely left on when smooth walls are considered), as discussed by Fuhrman et al. (2010). This is also indicated in Table 9.1.

Alternatively, as discussed by Fuhrman et al. (2010) and Fuhrman et al. (2011), utilizing a Neumann $\partial k/\partial n = 0$ boundary condition (corresponding generally to a zero diffusive flux of turbulent kinetic energy through the wall), where n is the direction normal to the wall, is more physically consistent with the rough wall physics, while also remaining applicable on smooth walls.

Let us first consider the smooth wall case. On a hydraulically smooth wall where the zero-gradient condition is imposed, k will still naturally tend to zero (in accordance with the no slip condition discussed above), and both the zero-value and zero-gradient conditions will be satisfied. That both $k = 0$ and $\partial k/\partial n = 0$ conditions hold on a smooth wall is supported by both experimental (e.g. Eckelmann, 1974, his Figs. 5 and 13) as well as theoretical arguments. To demonstrate the theoretical case, consider the boundary layer flow above a smooth wall at $y = 0$. Following Wilcox (2006), p. 177, we may Taylor expand the near-wall fluctuating velocities according to:

$$u' = a(x, z, t)y + O(y^2) \tag{9.50}$$

$$v' = b(x, z, t)y^2 + O(y^3) \tag{9.51}$$

$$w' = c(x, z, t)y + O(y^2) \tag{9.52}$$

Each of the above satisfies the no-slip requirement at $y = 0$, whereas $v' \sim y^2$ (rather than $v' \sim y$) is required such that the fluctuating velocity field may satisfy local continuity, $\partial u'_i/\partial x_i = 0$. The above implies that the near-wall turbulent kinetic energy density will vary as

$$k = \frac{1}{2}\overline{u'_i u'_i} = \frac{1}{2}\left(\overline{u'^2} + \overline{v'^2} + \overline{w'^2}\right) = \frac{1}{2}\overline{(a^2 + c^2)}y^2 + O(y^3) \tag{9.53}$$

or more simply, that near the wall $k \sim y^2$, and hence $\partial k/\partial y = 0$ at $y = 0$. Thus, both $k = 0$ and $\partial k/\partial n = 0$ boundary conditions are applicable on a hydraulically smooth wall, though there would not appear to be any great advantage to imposing the latter over the former.

On hydraulically rough walls, however, the zero gradient condition can be advantageous, as it allows finite k (hence ν_t) at the location of the theoretical wall, and hence prevents creation of a "fictitious" viscous sublayer. This is in line with the physics near the "theoretical wall" which typically lies a distance (0.15–0.3)d below the top of the roughness elements, where d is the physical roughness height, corresponding e.g. to the grain size on a plane sediment bed (see Section 3.1.3, as well as Bayazit (1976, 1983); Fredsøe et al. (1999); Sumer et al. (2001)), where pore space thus exists between individual grains. The zero gradient condition on rough walls is directly supported by experimental observations (Nezu, 1977; Sumer et al., 2003; Fuhrman et al., 2010).

Fuhrman et al. (2010) showed that, as a result of the added physical consistency discussed above with the zero-gradient condition, the near-wall grid spacing on rough beds could, more naturally, be based on k_s (rather than a viscous length scale ν/U_f as in Eq. 9.49). In line with their suggestions, we suggest the following rule of thumb for selecting a near-wall cell size above a rough wall:

$$\frac{\Delta y}{k_s} \leq 0.02 \tag{9.54}$$

if a zero gradient $\partial k/\partial n = 0$ wall boundary condition is imposed. In practice, in intermediate roughness conditions (i.e. $5 \leq k_s^+ \leq 70$) taking the near-wall cell size Δy as the maximum from Eq. 9.49 or 9.54 will typically be sufficiently conservative, provided that the zero-gradient boundary condition for k is utilized. Notice that Δy based on Eqs. 9.49 and 9.54 become equivalent when $k_s^+ = 50$. Hence, the latter (zero gradient) constraint offers potential computational savings (larger near-wall grid spacing that is independent of the Reynolds number) for flows where k_s^+ exceeds this value.

9.4.2.2 *Wall functions*

If resolving the flow all the way down to the wall surface is too computationally expensive, or otherwise not desired, an alternative to utilizing the wall surface boundary conditions is to use so-called *wall functions*. The main idea is to assume a velocity profile e.g. from a turbulent flow stemming either from Eq. 9.15 or Eq. 9.26, depending on the flow regime. From the assumed profile applied at the nearest cell center or grid point, the friction velocity U_f can then be determined, and used to calculate reasonable values for k and ω at this location. The most commonly utilized wall functions for these variables correspond to:

$$k = \frac{U_f^2}{\sqrt{\beta^*}}, \qquad \omega = \frac{U_f}{\sqrt{\beta^*}\kappa y_c} \tag{9.55}$$

where here y_c represents the normal distance from the wall to where the wall functions are to be applied (e.g. the nearest cell center or grid point).

The values in Eqs. 9.55 are only valid in the logarithmic layer, hence their proper application assumes that y_c lies in this region (ideally say $y^+ \geq 30$ for smooth walls or $y_c/k_s \geq 0.2$ for rough walls). This presents obvious limitations to both mesh refinement testing as well as validity of a given mesh in unsteady flow situation. (Note that the standard wall functions above, after invoking Eq. 9.33, yield the length scale $\ell = \sqrt{k}/\omega = \beta^{*1/4}\kappa y$. Through comparison with Eq. 9.13 it is thus seen that the length scale as defined in the k-ω model can be related to the mixing length according to

$$\ell = \beta^{*1/4} l_m \tag{9.56}$$

i.e. $\ell \approx 0.55 l_m$ after inserting the standard (fully turbulent) value for the closure coefficient $\beta^* = 0.09$.)

To get around the restriction that the cells nearest the wall must necessarily lie within the logarithmic layer (the location of which will be varying for non-uniform and/or unsteady flow problems) several authors (e.g. Roulund et al., 2005; Dixen et al., 2013; Fuhrman et al., 2014) have utilized a generalized wall function approach. We will here specifically describe the method utilized by Fuhrman et al. (2014). Rather than assuming a logarithmic velocity distribution explicitly, they instead (following similar methodology of Roulund et al., 2005; Dixen et al., 2013) based the determination of the friction velocity on the velocity profile of Cebeci and Chang (1978), as indicated in Eqs. 9.22, 9.23 and 9.25, applied at $y = y_c$ (hence $y^+ = y_c^+$, where here y_c is typically e.g. the cell center or grid point spacing nearest the wall). Similarly, the generalized wall functions for k and ω were taken as

$$\frac{k}{U_f^2} = \min\left\{A_w y_c^{+2}, \frac{1}{\sqrt{\beta^*}}\right\} \tag{9.57}$$

$$\frac{\omega\nu}{U_f^2} = \max\left\{\frac{B_w}{y_c^{+2}}, \frac{1}{\sqrt{\beta^*}\kappa y_c^+}\right\} \tag{9.58}$$

The first arguments in these functions ensure that these variables follow their proper scaling $k \sim y^2$ and $\omega \sim 1/y^2$ in the viscous sublayer region (see Eq. 9.53). Fuhrman et al. (2014) utilized the values $A_w = 1/(\delta^{+2}\sqrt{\beta^*}) = 0.02466$ and $B_w = \delta^+/(\sqrt{\beta^*}\kappa) = 96.885$, ensuring a continuous transition to the (fully-turbulent logarithmic region) second arguments at $y^+ = \delta_v^+$, where $\delta_v^+ = 11.626$ was taken as the viscous sublayer thickness (in dimensionless wall coordinates; This threshold is based on the intersection of extrapolated linear (Eq. 9.21) and logarithmic (Eq. 9.15) velocity profiles commonly assumed to characterize the viscous sublayer and log-layer regions above smooth walls).

It is emphasized that the first arguments introduced in Eqs. 9.57 and 9.58 only provide asymptotic consistency for k and ω near a smooth wall, and are not applicable very near the theoretical wall in the hydraulically rough regime. This would require further generalization. (To the best of the authors' knowledge, a completely generalized strategy for imposing wall functions for any combination of near-wall y_c, k_s, and flow has not been published.) Eqs. 9.57 and 9.58 may still be utilized on rough walls, but it is then important that the positions where they are applied lie in the logarithmic flow region, i.e. ideally that $y_c/k_s \geq 0.2$.

Regardless of the form utilized, at the location where the wall functions are enforced the eddy viscosity must be re-defined locally at the point/cell where the wall functions are imposed such that

$$\frac{\tau_0}{\rho} = U_f^2 = (\nu + \nu_t)\frac{\partial \overline{u}}{\partial y}\bigg|_{y=y_c} \tag{9.59}$$

is satisfied. This ensures overall consistency in the calculation of shear stresses from numerically computed velocity gradients.

The set of generalized wall functions above have been utilized within several CFD studies involving scour (e.g. by Fuhrman et al., 2014; Baykal et al., 2015, 2017; Larsen et al., 2016, 2017), as well as breaking waves by Larsen and Fuhrman (2018). A summary of the latter results is presented in Section 7.6. See also the reasonably similar implementation of generalized wall functions of Roulund et al. (2005) and Dixen et al. (2013).

9.4.3 *Numerical computation*

To sum up, the set of equations which generally needs to be solved when employing the k-ω turbulence model consists of:

(1) The continuity equation, Eq. 9.1;

(2) The (three) RANS equations, Eqs. 9.2–9.4;
(3) The k and ω model equations, Eqs. 9.35–9.36 (two equations total); and
(4) The Boussinesq approximation, Eq. 9.30 (six equations total i.e. three turbulent normal stresses and three shear stresses).

The system indicated above is comprised of 12 equations. This is the same as the number of variables i.e. the 10 variables indicated in Section 9.1 (three velocities, the pressure, and six Reynolds stresses) in addition to the two introduced model variables (k and ω). The system defined above is hence closed! Note that as the eddy viscosity ν_t in Eq. 9.37 is defined in terms of k and ω (as well as the mean flow via the stress limiter) it need not be counted as a separate independent variable.

The system of equations above may be utilized in conjunction with either the surface wall boundary conditions for k and ω or wall functions, as described above (see Section 9.4.2). Usually a zero normal gradient is imposed for the pressure boundary condition at solid walls in combination with no slip conditions for the three velocity variables, though their precise specifications may vary and inevitably depend somewhat on the problem at hand.

Notice that two-equation models, like the k-ω model formulated above, are regarded as *complete* turbulence models i.e. they can be employed to predict properties of a turbulent flow without requiring a priori knowledge of the turbulence structure. This is in contrast to e.g. the simpler mixing length model presented in Section 9.3, which requires specification of the mixing length specifically tailored to the problem being considered.

Also unlike the mixing length model, this set of equations defined above is too complex for one to typically seek analytical solutions. Instead, the equations are generally solved numerically. To this end, one resorts to a CFD (Computational Fluid Dynamics) model wherein RANS equations are coupled with a turbulence model (e.g. the k-ω model in the present case). CFD codes may also offer various turbulence-modelling options. In these codes, the momentum, continuity and turbulence-model equations may e.g. be transformed into curvilinear coordinates, and they may be linearized and decoupled. The equations are discretized, using a computational mesh (e.g., the latter can be based on a multi-block structure with each block containing N^3 computational cells where N is the number of cells in each spatial direction.)

Widely used modern CFD software packages commonly utilized in hydraulic and coastal engineering include several commercial models, as well as numerous research/specialized codes. Examples of the latter include those developed and/or applied by e.g.: Troch and De Rouck (1999), Mayer and Madsen (2002), Christensen (2006), Ong et al. (2009), del Jesus et al. (2012), Lara et al. (2012) and Bihs et al. (2016), just to mention a few. Open source software such as OpenFOAM, has exploded in popularity in coastal and hydraulic engineering applications in recent years. Developments within this framework include those of Jacobsen et al. (2012), Higuera et al. (2013), Larsen and Fuhrman (2018), Jacobsen et al. (2014), Chauchat et al. (2018), Kim et al. (2018), Jasak et al. (2019) and Roenby et al. (2016). For additional examples demonstrating turbulence modelling in hydraulic/coastal applications within the OpenFOAM framework, see e.g. Liu and Garcia (2008), Fuhrman et al. (2010, 2011, 2014), Jacobsen and Fredsøe (2014), Baykal et al. (2015, 2017), Maza et al. (2015), Vukcevic et al. (2016a,b), Brown et al. (2016), Zhou et al. (2017), Devolder et al. (2017, 2018), Larsen et al. (2019), Liu et al. (2019a,b), and Kim et al. (2019), to mention a few.

It is beyond the scope of the present book to get into the details of the various numerical methods and schemes that are typically employed for temporal and spatial discretization, the potential combinations of which are immense. For an introduction to the numerical methods commonly used, the interested reader may consult several other books dedicated exclusively to such topics e.g. those of Ferziger and Peric (2002) and Hirsch (2007).

9.4.4 *An application example*

As an application example we will present results where the Wilcox (2006) k-ω model has been employed to simulate steady uniform boundary layer flows, to demonstrate its performance in what can be considered as among the most elementary of turbulent boundary layer flows. In what follows, we will compare computed results using the Wilcox (2006) k-ω model (as implemented by Fuhrman et al., 2013; Williams and Fuhrman, 2016) against experimental measurements of Fuhrman et al. (2010) on both smooth and rough beds. The experimental measurements were originally made by Dr. Martin Dixen in the tilting flume in the DTU Mechanical Engineering Hydraulics Laboratory, based on LDA measurements of all three velocity components.

In the results that follow the flow has been driven by a specified pressure gradient (used as a body force), selected to yield the proper friction velocity once equilibrium conditions are reached i.e. by setting $P_x = U_f^2/h$ in Eq. 5.159. The top boundary has been modeled as a slip wall (i.e. scalar quantities and tangential velocities have zero gradient, whereas the normal velocity has zero value). In accordance with our previous discussion, results with the standard Wilcox (2006) k-ω model utilize a $\partial k/\partial y = 0$ boundary condition at the bed, whereas those using the low Reynolds number variation of the model use a $k = 0$ bottom boundary condition. Both use the surface ω boundary condition from Eq. 9.47, while invoking S_R from Eq. 9.48, with the K_r and λ_1 settings as indicated in Table 9.1.

Let us first consider results for the hydraulically smooth steady uniform boundary layer case of Fuhrman et al. (2010), characterized by the Reynolds number $Re = hV/\nu = 1.9 \times 10^4$, V being the depth-averaged velocity. Figs. 9.3 and 9.4 compare the modeled and measured velocity and turbulent kinetic energy density k profiles computed, respectively, utilizing the standard and low Reynolds number variants of the Wilcox (2006) k-ω model presented previously. The velocity distributions are presented in semi-logarithmic (sub-figures a) and linear (sub-figures b) plots. The agreement between the model and the measured velocity profiles is very good, and both model variants produce nearly identical flow profiles which are in line with the logarithmic velocity distribution (dashed lines) once the constant stress region is reached.

Comparison of Figs. 9.3c and 9.4c demonstrate improved prediction in the near-wall turbulent kinetic energy when utilizing the "low Reynolds number" k-ω model. The standard closure under-predicts the turbulent kinetic energy by about 30% near the bed, whereas this region is captured much more accurately with the low Reynolds number model. This improvement makes sense, as this corresponds to the buffer layer region where the flow is "transitioning" from a region dominated by viscous stresses to one dominated by (turbulent) Reynolds stresses further from the wall. Similar findings as depicted here can be seen in simulated channel/pipe flows in the book of Wilcox (2006); See e.g. his Figs. 4.28–4.29 and 4.38–4.39. Note finally the insets in the upper right hand corners of Figs. 9.3c and 9.4c, which depict zoomed-in regions near the wall. Figs. 9.3c demonstrates e.g. how even though the $\partial k/\partial y = 0$ bottom boundary condition has been formally imposed in this case, the model naturally still predicts $k = 0$ at the smooth wall, consistent with the preceding discussion.

Let us next consider the hydraulically rough steady uniform boundary layer case of Fuhrman et al. (2010), characterized by the flow Reynolds number $Re = hV/\nu = 1.4 \times 10^4$ and the roughness Reynolds number $k_s^+ =$

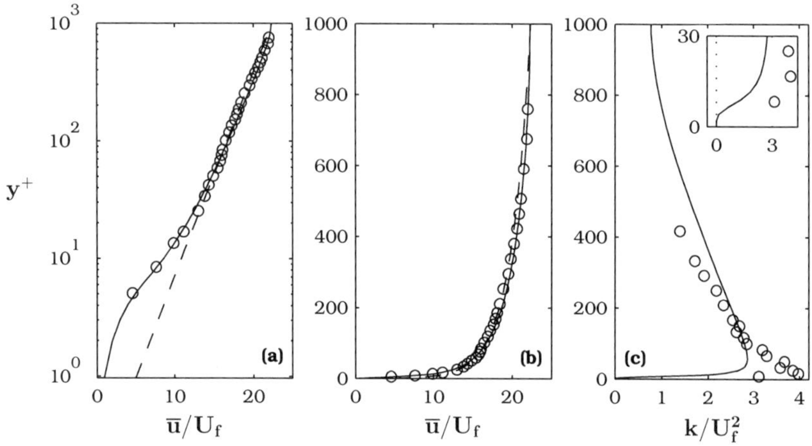

Fig. 9.3 Comparison of computed (full lines, using standard k-ω model with $k_s^+ = 2$), logarithmic (dashed lines, $\overline{u}/U_f = 1/\kappa \ln(y^+) + 5.1$, $\kappa = 0.4$), and measured (circles, from Fuhrman et al., 2010, with $Re = 1.9 \times 10^4$) (a and b) velocity and (c) turbulent kinetic energy density profiles for a steady boundary layer flow on a hydraulically smooth wall. The inset in (c) shows k-profiles very near the wall.

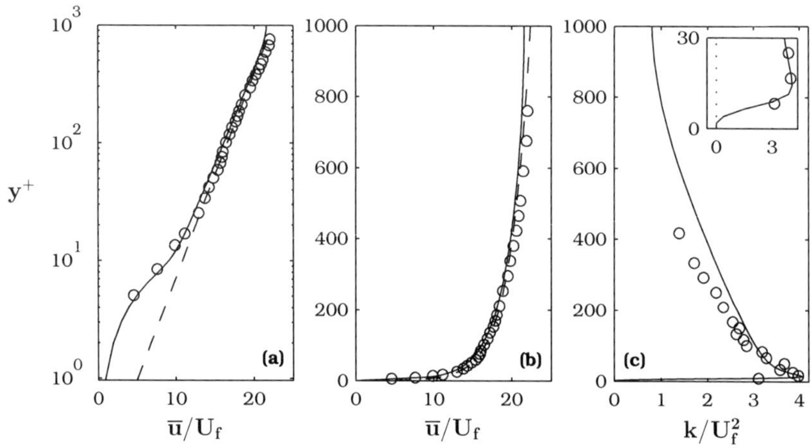

Fig. 9.4 As in Fig. 9.3, but with computed results (full lines) from the low Reynolds number k-ω model.

$k_s U_f/\nu = 200$. Similar to the smooth-bed case above, Figs. 9.5 and 9.6 compare the computed and measured velocity and turbulent kinetic energy

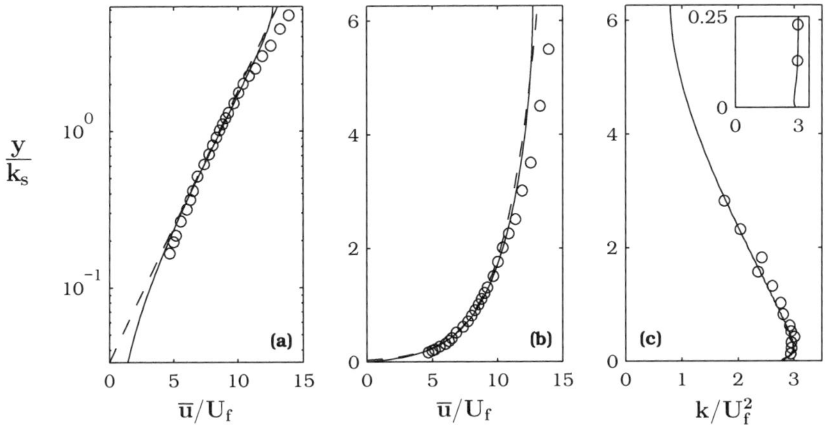

Fig. 9.5 Comparison of computed (full lines, using the standard k-ω model), logarithmic (dashed lines, $\overline{u}/U_f = 1/\kappa \ln(30y/k_s)$, $\kappa = 0.4$), and measured (from Fuhrman et al., 2010, with $k_s^+ = 200$, $Re = 1.4 \times 10^4$) (a and b) velocity and (c) turbulent kinetic energy density profiles for a steady boundary layer flow on a hydraulically rough wall. The inset in (c) shows k-profiles very near the wall.

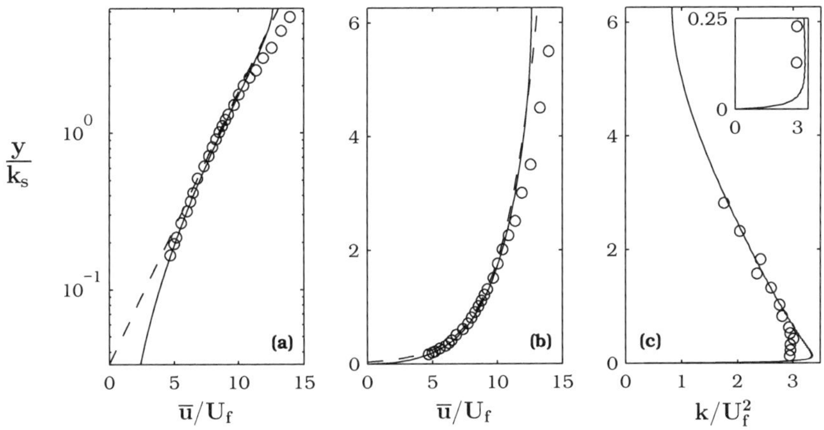

Fig. 9.6 As in Fig. 9.5, but with computed results (full lines) from the low Reynolds number k-ω model.

density distributions, with the standard and low Reynolds number k-ω model variants. Regarding the flow velocities, the agreement between the model results and the experiment is generally good, although both models under-predict the velocity by slightly less than 20% near the free surface. This may be due to the present use of a frictionless rigid lid at this boundary, rather than directly simulating a free surface. The models capture the velocity quite well in the logarithmic region and in the buffer zone.

Regarding the turbulent kinetic energy density, both model results are quite good, with the standard model (in conjunction with the $\partial k/\partial y = 0$ bottom boundary condition) predicting better results near the wall. Comparison of the insets in Figs. 9.5c and 9.6c demonstrate how the zero-gradient boundary condition employed with the standard closure model (Figs. 9.5) allows finite k at the wall, thus avoiding the viscous sublayer region near the wall that is forced by the $k = 0$ boundary condition utilized with the low Reynolds number model (Fig. 9.6). Based on these results, though reasonable accuracy can be achieved with either model, there would not appear to be any great advantage in utilizing the low Reynolds number model over the standard closure on rough beds.

The present section establishes the performance of the Wilcox (2006) k-ω model in simulating steady boundary layer flows on smooth and rough beds. Other applications with the standard k-ω model presented above (or slight variations thereof) are prevalent throughout this book. These include several applications involving turbulent wave boundary layers in Chapters 5 and 6, as well as CFD simulations of wave breaking in Chapter 7. Generally, the standard "high Reynolds number" closure model should be sufficient for many engineering applications involving turbulent flows, especially those where the flow and/or bed shear stress is of utmost importance or e.g. for flows above rough beds (often the focus of coastal and civil engineering applications). The "low Reynolds number" variant is, nevertheless, a straight-forward extension that can be utilized to study situations where e.g. the mean flow and turbulence in laminar-to-turbulence transition are of special interest. An additional application of the low Reynolds number closure on an unsteady flow problem can be seen in Figs. 5.94 and 5.95.

To gain first hand experience in turbulence modelling of some simple (unsteady) wave boundary layer flows, the interested reader is again encouraged to complete Exercise 4 in Chapter 10.

9.5 Large eddy simulation (LES)

To demonstrate an example of a simulation model, we will finally describe the formulation and application of an LES model. The following is based largely on the work of Lohmann et al. (2006). Note that a detailed account of turbulence models for LES has been given in a review paper by Meneveau and Katz (2000).

9.5.1 *Model equations*

The principal idea in the method of LES is to average the Navier-Stokes equations over a small volume of space. (Recall that the Navier-Stokes equations are averaged over time to obtain the Reynolds equations, i.e. Eqs. 9.1 and 9.2). Normally, the size of the averaging volume is chosen such that the small-scale component of turbulence is, for convenience, not resolved. The aforementioned space-averaging gives the following set of equations: (1) the continuity equation,

$$\frac{\partial u_i}{\partial x_i} = 0 \tag{9.60}$$

and (2) the equation of motion

$$\frac{\partial u_i}{\partial t} + u_j \frac{\partial u_i}{\partial x_j} = -\frac{1}{\rho}\frac{\partial p}{\partial x_i} + \nu \frac{\partial}{\partial x_j}\left(\frac{\partial u_i}{\partial x_j} + \frac{\partial u_j}{\partial x_i}\right) - \frac{\partial \overline{u_i' u_j'}}{\partial x_j} \tag{9.61}$$

in which u_i are the space-averaged (resolved) component of the velocity, p is the space-averaged (resolved) component of the pressure, and u_i' are the fluctuating components of the velocity over the averaging volume (called the subgrid-scale component of the velocity, to differentiate it from the term "fluctuation" in the conventional Reynolds decomposition), and $\overline{u_i' u_j'}$ are the subgrid-scale stresses, analogous to the Reynolds stresses in the case of the Reynolds decomposition.

By analogy to the Boussinesq approximation, the shear components (off diagonal components of the Reynolds stress tensor) of the subgrid-scale stresses (SGS) may be taken as

$$\overline{u_i' u_j'} = \nu_t \left(\frac{\partial u_i}{\partial x_j} + \frac{\partial u_j}{\partial x_i}\right), \qquad i \neq j \tag{9.62}$$

in which ν_t is the eddy viscosity associated specifically with the unresolved component of turbulence. On the other hand, the corresponding normal stresses (the diagonal components of the Reynolds stress tensor) can be taken as automatically lumped in the pressure term. After invoking the above, Eq. 9.61 hence becomes

$$\frac{\partial u_i}{\partial t} + u_j \frac{\partial u_i}{\partial x_j} = -\frac{1}{\rho}\frac{\partial p}{\partial x_i} + \frac{\partial}{\partial x_j}\left[(\nu + \nu_t)\left(\frac{\partial u_i}{\partial x_j} + \frac{\partial u_j}{\partial x_i}\right)\right] \tag{9.63}$$

(It is emphasized that the eddy viscosity ν_t here is different from that in the k-ω model defined in the previous section. As stressed above, in the present definition, Eq. 9.62, ν_t is associated only with the unresolved small-scale turbulence averaged over the "averaging" volume, whereas ν_t defined in conjunction with the k-ω model is associated with the turbulence over the entire spectrum).

On dimensional grounds, the turbulence viscosity may be written as

$$\nu_t = C_s^2 \Delta^2 s \tag{9.64}$$

in which Δ is a length scale and C_s is a non-dimensional constant. (By making an analogy to Prandtl's mixing length hypothesis, it may be seen that s must be a quantity like $\partial u/\partial y$ in the case of a uniform boundary layer flow). We shall return to this quantity below.

Clearly, the resolved quantities, $u_i(x_i, t)$, $p(x_i, t)$ and $\nu_t(x_i, t)$ in Eqs. 9.62–9.64 should be dependent on the size of the averaging volume. The smaller the size of this volume, the better the resolution (both in space and time), and therefore the more information on space and time variations of $u_i(x_i, t)$, $p(x_i, t)$ and $\nu_t(x_i, t)$, implying that these quantities are dependent on the size of the averaging volume. This feature in LES is handled through ν_t as follows.

As seen in Eq. 9.64, ν_t involves a length scale, Δ. This quantity is taken as the length scale representing the size of the averaging volume

$$\Delta = (\Delta x_1 \Delta x_2 \Delta x_3)^{1/3} \tag{9.65}$$

in which Δx_1, Δx_2, Δx_3 are the grid sizes in the x_1-, x_2-, and x_3-directions, respectively. (The averaging volume being considered to be $\Delta x_1 \Delta x_2 \Delta x_3$). Furthermore, the quantity s in Eq. 9.64 is taken as

$$s = (2S_{ij}S_{ij})^{1/2} \tag{9.66}$$

where

$$S_{ij} = \frac{1}{2}\left(\frac{\partial u_i}{\partial x_j} + \frac{\partial u_j}{\partial x_i}\right) \tag{9.67}$$

is the strain-rate tensor corresponding to the resolved part of the flow. It is easy to see that s can also be written in the following form

$$s = \left[\left(\frac{\partial u_i}{\partial x_j} + \frac{\partial u_j}{\partial x_i}\right)\frac{\partial u_i}{\partial x_j}\right]^{1/2} \tag{9.68}$$

The model described above (Eqs. 9.60–9.66) is known as the Smagorinsky (1963) model for LES. As seen, only one "closure" coefficient is required, namely the Smagorinsky constant C_s. The value of the Smagorinsky constant is determined from experiments, where experience has shown e.g. that $C_s = 0.2$ in isotropic turbulence whereas $C_s = 0.1$ in channel flow (Lohmann et al., 2006).

To sum up, the solution of the above set of equations, namely Eqs. 9.60–9.62, and Eqs. 9.64–9.67, basically gives the time series of the hydrodynamic quantities (the velocity components and the pressure), similar to the time series obtained in a laboratory experiment using, for example, the Laser Doppler Anemometer (LDA), or the Particle Image Velocimetry (PIV) techniques. Once we obtain this raw data, then we process the data in terms of statistical quantities such as the mean, the variance, other statistical moments, and even correlations and spectra, in exactly the same fashion as we process data from experiments. (It should be added that, clearly, the time series obtained through the LES "experiments" are resolved with a resolution equal to the grid size selected in the simulation, just as the experimentally obtained time series are resolved with a resolution equal to the size of the measurement volume of the measuring instrument, e.g. to the size of the measurement volume of the LDA probe.)

9.5.2 *An application example*

We will now present the results of a study where the LES model described above was used to simulate the wave boundary layer in a U-shaped oscillating tunnel (Fig. 9.7); The study stems from a PhD study conducted at the Technical University of Denmark, published by Lohmann et al. (2006). The CFD code employed in the study is called NS3, a three-dimensional numerical flow solver, which was made available by DHI.

The LES simulation covers the unsteady boundary layer that develops beneath an oscillatory free stream velocity signal given by

$$U_0 = U_{0m} \sin(\omega t)$$

in which $\omega = 2\pi/T$ is here the angular frequency, with T being the wave period. The case being considered corresponds to test 10 from the experiments of Jensen et al. (1989), with $T = 9.72$ s and $U_{0m} = 2.0$ m/s (such that $Re = aU_{0m}/\nu = 6 \times 10^6$, where $a = U_{0m}/\omega \approx 3$ m is the amplitude of the free stream orbital motion). (See Chapter 5 for a much more detailed discussion of wave boundary layers, as well as a greater discussion of the

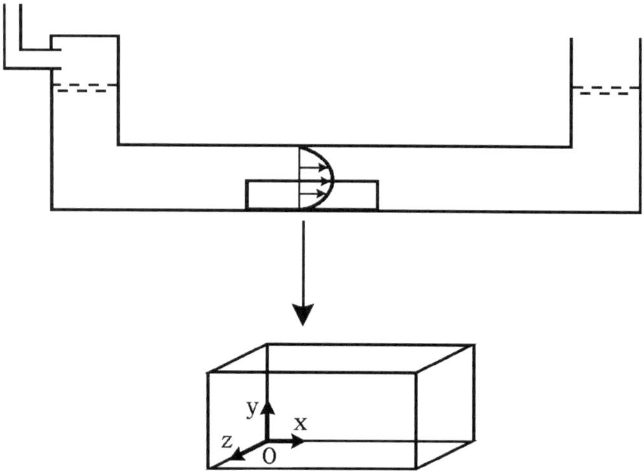

Fig. 9.7 A conceptual sketch depicting a U-shaped oscillating tunnel, from Lohmann et al. (2006).

Jensen et al. (1989) experiments.)

The computational domain utilized is three-dimensional, with height 0.145 m (corresponding to the tunnel half-height, length 0.9 m and width 0.4 m (see Section 5.3.1). The simulations utilize a no slip condition at the bed coupled with a symmetry condition at the top boundary. Periodic conditions are imposed at lateral boundaries. The domain is discretized into $96 \times 48 \times 80$ computational cells, with horizontal mesh size corresponding to $\Delta x = 9.375$ mm (streamwise direction) and $\Delta z = 5$ mm (transverse direction). In the vertical direction $\Delta y = 0.057$ mm at the bed and increases exponentially to 4.07 mm at the top. Jensen et al. (1989) found a maximum friction velocity $U_f \approx 0.08$ m/s for this case, such that in wall units the smallest grids correspond approximately to $\Delta x^+ = 750$, $\Delta y^+ = 4.56$ and $\Delta z^+ = 400$ (Lohmann et al., 2006). The time step utilized is 0.001 s.

The van Driest damping function, $1 - \exp\left(-\frac{y^+}{A_d}\right)$, as seen e.g. in Eq. 9.16, is generally used as a standard part of the Smagorinsky subgrid-scale formulation near walls. In the present work this was formulated as in Ferziger and Peric (2002), p. 280, such that C_s is modified near the wall according to:

$$C_s = C_{s0}\left[1 - \exp\left(-\frac{y^+}{A_d}\right)\right]^2 \tag{9.69}$$

where $A_d = 25$. The base value $C_{s0} = 0.1$ was utilized, corresponding to

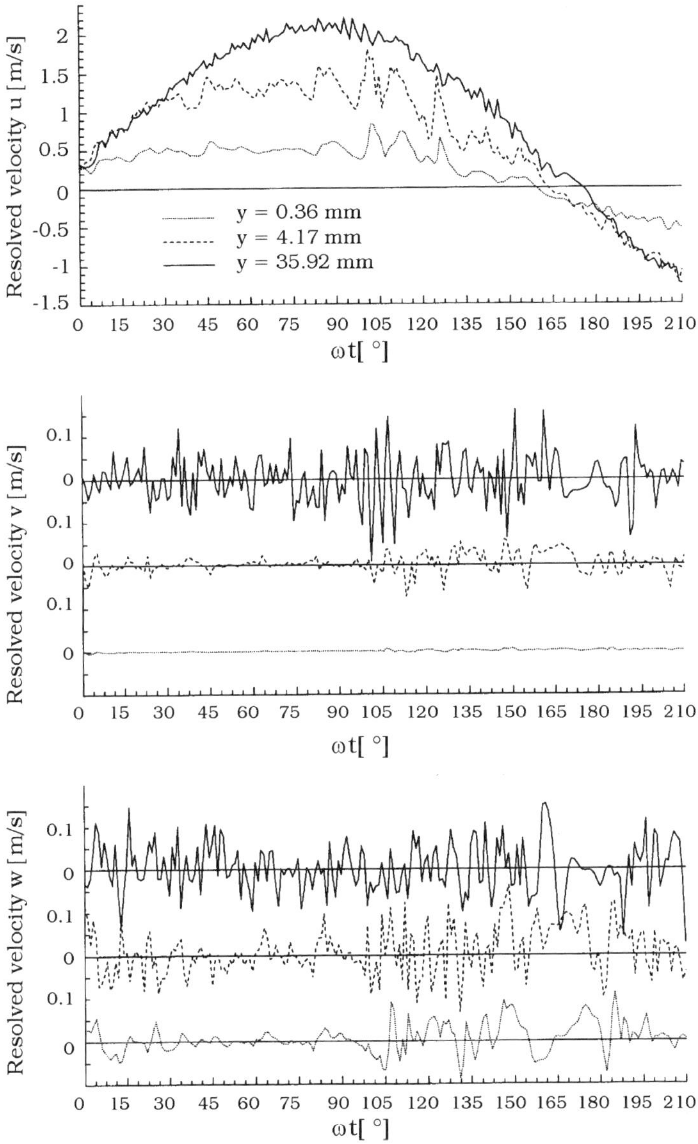

Fig. 9.8 Computed time series of velocity components using the LES model, from Lohmann et al. (2006).

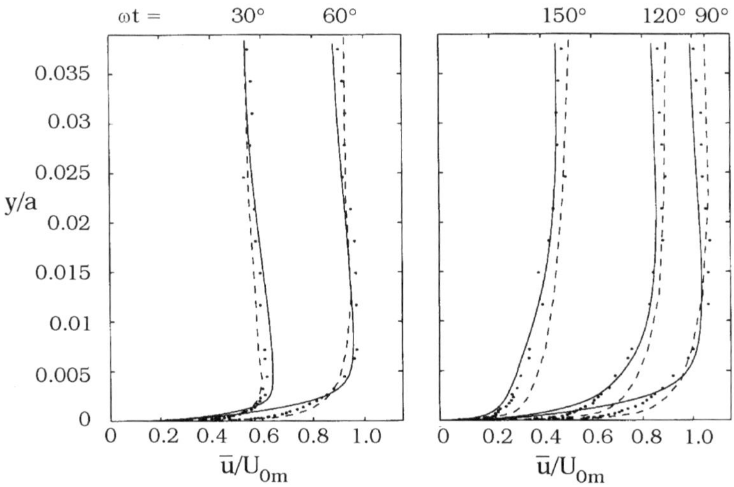

Fig. 9.9 Sequence of velocity profiles over half period of the motion. Full line: LES results (Lohmann et al., 2006), dashed line: k-ε model results (Justesen, 1988), dots: Experimental results (Jensen et al., 1989).

the proper value in channel flow, and was found to be most appropriate based on comparison with experimental data by Lohmann et al. (2006). Similar (though not identical) damping has been used in LES of breaking waves, see e.g. Christensen (2006).

Fig. 9.8 presents the raw time series of the three components of the resolved velocity at three different values of the distance y from the bed. As seen from Fig. 9.8, the time series are qualitatively no different from those obtained experimentally using LDA (c.f. Fig. 5.22 in Chapter 5). Ensemble averaging has been used to calculate the mean value of the resolved velocity, Eqs. 5.32 and 5.33. Lohmann et al. (2006) found that, for the chosen phase values, averaging first over the x-z plane at a given y-location, and then over 3 periods was sufficient to achieve reliable ensemble averages.

Fig. 9.9 gives a sequence of ensemble-averaged velocity profiles at different phase values ωt over one half period of the motion. The distance from the wall is normalized by the amplitude of the free-stream motion a. The figure also gives two sets of other data: The first is from the numerical study of Justesen (1988) where a turbulence energy equation model, namely the k-ε model, was used (this model is similar in form to the two-equation k-ω model presented in Section 9.4, but with the dissipation ε, defined in

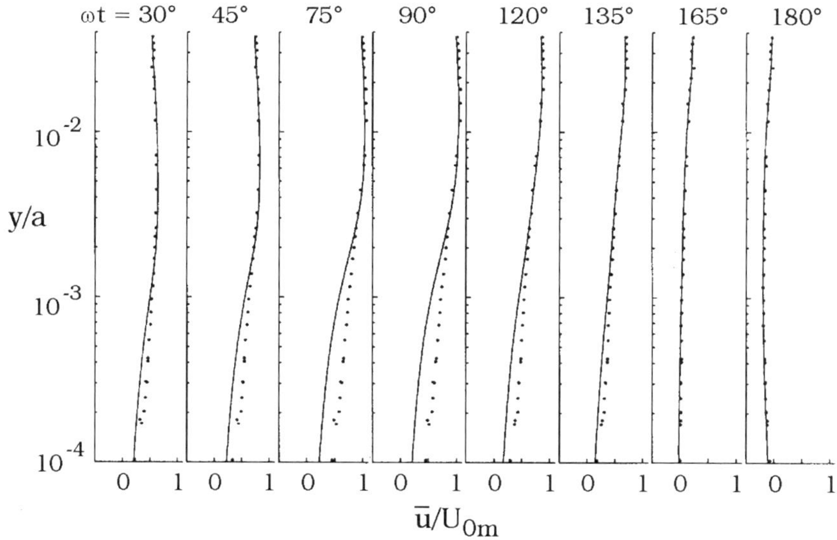

Fig. 9.10 Sequence of ensemble-averaged velocity profiles over half period of the motion. Results are plotted on a semi-logarithmic scale, from Lohmann et al. (2006). Dots: Experimental results from Jensen et al. (1989).

Eq. 9.32, retained as the quantity modeled in the second equation); The second data set is directly from the experimental measurements of Jensen et al. (1989). Fig. 9.9 shows that the LES model does a better job than the k-ε model. In Fig. 9.10, the velocity information is plotted on a semi-logarithmic graph. The figure illustrates how the LES solution behaves for small values of the distance from the bed in relation to the experimental data. From the figure, although superior to the k-ε model, the LES model under-predicts the velocities for small y values. (Note that two-equation k-ω model results from this same test case can be seen in Figs. 5.86–5.89.)

Figs. 9.11 (normal stresses) and 9.12 (shear stresses) present the vertical distribution of turbulence quantities obtained by the LES model along with the experimental data. Although there are discrepancies between the model prediction and the experiment, the model predicts the turbulence fairly well. Notably, unlike what would be obtained from a turbulence energy equation model employing the conventional Boussinesq approximation, Eq. 9.30, the computed normal stresses from the LES model in Fig. 9.11 are not isotropic. Yet, one must note that the model near the bed fails to some extent to

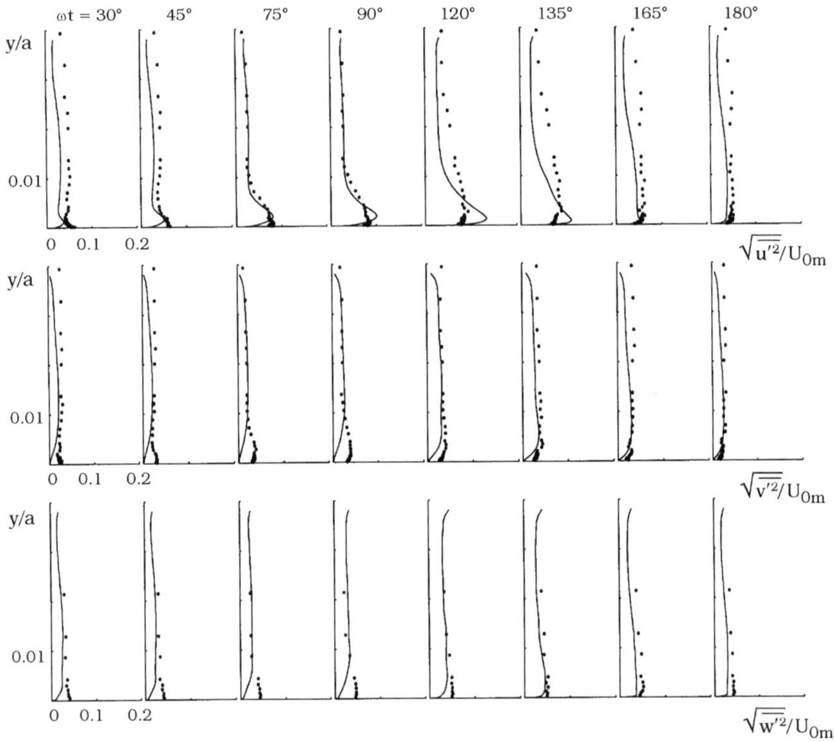

Fig. 9.11 Sequence of turbulence profiles, from Lohmann et al. (2006). Dots: Experimental results from Jensen et al. (1989).

predict the turbulence accurately. This may be a major shortcoming when processes (to be studied by the model) have significant elements related to the turbulence near the bed, particularly related to the y- and z-components of turbulence.

The above demonstrates LES applications for cases involving turbulent oscillatory wave boundary layer flow. Other demonstrations of LES simulations presented in this book include simulation of plunging breaking waves by Kim and Hsu (2018), e.g. Figs. 7.11 and 7.12.

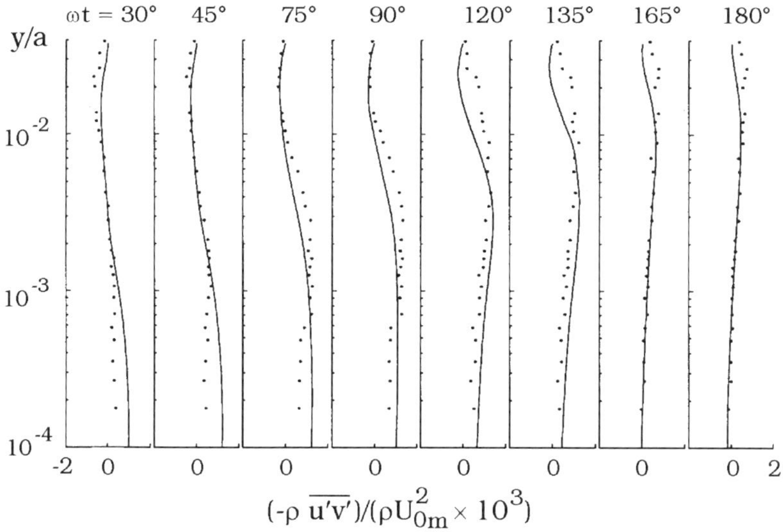

Fig. 9.12 Shear stress profiles from Lohmann et al. (2006). Dots: Experimental results from Jensen et al. (1989).

9.6 Scaling of computational costs in the modelling of turbulent flows

We will finally end this chapter with a discussion on how the computational costs can be expected to scale with respect to the Reynolds number for various turbulence modelling alternatives. Specifically, we will consider those associated with DNS, LES, and turbulence energy equation models, as these are the most popular approaches used in modern engineering applications and research purposes.

9.6.1 *Direct numerical simulation (DNS)*

DNS simulations, in principal, require resolution of all scales i.e. ranging from the large scale motion down to the smallest scales of turbulence, the Kolmogorov scales, introduced previously in Chapter 4. Let us take the length scale $\mathcal{L}$ defined on dimensional grounds in Eq. 4.91 as characteristic of the larger scale eddies. From Eq. 4.92 the ratio of this and the

Kolmogorov length scale will vary as

$$\frac{\mathcal{L}}{\eta_K} \sim Re_{\mathcal{L}}^{3/4}, \qquad Re_{\mathcal{L}} = \frac{\mathcal{L}\sigma_u}{\nu} \tag{9.70}$$

where the Reynolds number $Re_{\mathcal{L}}$ is as defined in Eq. 4.91, with σ_u being the standard deviation of the turbulence fluctuations. Turbulence being inherently three dimensional, we can hence expect the number of points required in space N_{space} to scale like

$$N_{space} \sim \left(\frac{\mathcal{L}}{\eta_K}\right)^3 \sim Re_{\mathcal{L}}^{9/4} \tag{9.71}$$

Let us further define an analogous time scale $\mathcal{T} \sim \mathcal{L}/\sigma_u$ characterizing larger scale fluctuations (similar to the macro time scale). Again requiring resolution of all relevant (large down to Kolmogorov) scales will require the temporal resolution to vary as

$$N_{time} \sim \frac{\mathcal{T}}{\tau_K} \sim \frac{\mathcal{L}\varepsilon^{1/2}}{\sigma_u \nu^{1/2}} = \left(\frac{\mathcal{L}\sigma_u}{\nu}\right)^{1/2} \frac{\mathcal{L}^{1/2}\varepsilon^{1/2}}{\sigma_u^{3/2}} = Re_{\mathcal{L}}^{1/2} \tag{9.72}$$

where the definition of the Kolmogorov time scale from Eq. 4.87 has been invoked, as has the left-most relationship in Eq. 4.91 (yielding $\varepsilon^{1/2} = \sigma_u^{3/2}/\mathcal{L}^{1/2}$). Hence, we find that the total computational cost will scale like

$$N_{total} \sim N_{space} \times N_{time} \sim Re_{\mathcal{L}}^{11/4}, \qquad \frac{11}{4} = 2.75 \tag{9.73}$$

i.e. very nearly proportional to $Re_{\mathcal{L}}^3$. Such scaling is prohibitively expensive, except for relatively small Reynolds numbers. For example, merely doubling the Reynolds number for a given flow will increase the required computational costs by a factor $2^{2.75} \approx 7$ whereas increasing the Reynolds number by an order of magnitude (much more interesting) inflates computational costs by the factor $10^{2.75} \approx 600$! (Only single digit accuracy is maintained in the estimations above.)

To obtain some idea of the historical evolution of the Reynolds numbers that have been achieved, it can be noted that the Reynolds numbers in DNS simulations of channel flow have ranged e.g. from $Re_\tau = U_f h/\nu \approx 180$ in 1987 (Kim et al., 1987) up to $Re_\tau \approx 5200$ in 2015 (Lee and Moser, 2015).

9.6.2 *Large eddy simulation (LES)*

Next, let us similarly consider the scaling of computational costs in LES modelling as the Reynolds number is increased. The principal idea of LES modelling is to resolve the large eddies while modelling the smaller eddies having sub-grid scales. Ideally such resolution would extend into the inertial subrange (upon which the nature of the sub-grid turbulence would be locally isotropic and part of the universal equilibrium range, see Section 4.3.3). The inertial subrange can be reasonably characterized by the Taylor micro scales as introduced and discussed in Chapter 4. From Eq. 4.93, the ratio of the large scale eddies to those characterized by the Taylor micro scales will vary like

$$\frac{\mathcal{L}}{\lambda_f} \sim Re_{\mathcal{L}}^{1/2} \tag{9.74}$$

meaning that the required number of computational cells in space will scale like

$$N_{space} \sim \left(\frac{\mathcal{L}}{\lambda_f}\right)^3 \sim Re_{\mathcal{L}}^{3/2} \tag{9.75}$$

As the corresponding time scales can both be reasonably estimated from the length scales utilizing Taylor's frozen turbulence approximation, Section 4.2.3, we may similarly estimate from Eq. 9.74 that the required temporal resolution will analogously scale like

$$N_{time} \sim \frac{\mathcal{T}}{\tau_E} \sim Re_{\mathcal{L}}^{1/2} \tag{9.76}$$

Hence, the total computational costs in LES can be expected to scale approximately as

$$N_{total} \sim N_{space} \times N_{time} \sim Re_{\mathcal{L}}^2 \tag{9.77}$$

This is still quite expensive, but much less prohibitive than found above for DNS. Using the same examples as before, doubling the Reynolds number for a given flow will increase the required computational costs by a factor $2^2 = 4$ whereas increasing the Reynolds number by an order of magnitude will increase computational costs by a factor of "only" approximately $10^2 = 100$.

9.6.3 *Turbulence energy equation models*

Let us finally consider the scaling that can be expected with turbulence energy equation models, e.g. the k-ω model presented in Section 9.4. For the sake of simplicity, let us consider the demands that can be expected to model a simple steady boundary layer flow on a flow depth $h \sim \mathcal{L}$ subject to changes in the Reynolds number. As the turbulent fluctuations are parameterized in this approach, rather than resolved (either fully as in DNS or partially as in LES), the problem is steady and hence there is nothing requiring resolution in time (at least from an accuracy or physical stand point; note that for simplicity we do not consider numerical stability issues in the present discussion, which can further complicate such matters). Formally, in the same terms as in the preceding discussions, we can then state that $N_{time} \sim 1$. (More generally, this can be interpreted as temporal resolution being governed by the unsteady properties of the mean flow, rather than on the turbulence.)

The scaling of computational costs in terms of the spatial resolution required with this model, on the other hand, depend somewhat on the strategy adopted, especially regarding the treatment of the near wall flow. If conventional wall functions are adopted, then the viscous sublayer is simply not modelled, and the near wall cells with size Δy need to be within the logarithmic region of the constant stress layer, which extends to say $y \approx 0.1h$. Hence the required near-wall cells would scale with the larger scales of the flow, and we may state that

$$N_{total} \sim N_{space} \times N_{time} \sim \frac{h}{\Delta y} \sim \frac{\mathcal{L}}{\mathcal{L}} = 1 \tag{9.78}$$

i.e. no Reynolds number dependence in the computational costs. Similar findings can be even more generally demonstrated above a rough wall, where the smallest cells required would either scale like $\Delta y \sim h$ (as above if wall functions are used) or as $\Delta y \sim k_s$ (if wall surface boundary conditions are utilized, see Eq. 9.54), which in either case does not involve the viscosity (hence Reynolds number).

Conversely, if a viscous sublayer is to be resolved above a smooth wall (which the k-ω model is perfectly capable of doing if this is desired), then the required cells near the wall would be like $\Delta y = \nu/U_f \sim \nu/\sigma_u$ (corresponding to $\Delta y^+ = \Delta y U_f/\nu = 1$). This hence yields

$$N_{total} \sim N_{space} \times N_{time} \sim \frac{h}{\Delta y} \sim \frac{\mathcal{L}\sigma_u}{\nu} = Re_{\mathcal{L}} \tag{9.79}$$

i.e. a linear increase in total costs with Reynolds number.

Hence, turbulence energy equations models can simulate turbulent flows with computational costs that at best (at least solely from the viewpoint of computational costs) are independent of the Reynolds number and at worst scale something like $N_{total} \sim Re_{\mathcal{L}}$. This is much less expensive than with either DNS or LES, discussed above. The governing equations can also, in principal, be solved in either one-, two-, or three-dimensions depending only on the nature of the mean flow properties defining a given problem (in contrast to DNS/LES simulations, which require full/partial resolution of turbulent fluctuations in three dimensions, even if the mean flow is only one- or two-dimensional). It is hence easy to see why turbulence energy equation models, such as the k-ω model, remain popular with engineers, scientists, and researchers, as they offer a balance between detailed accuracy and computational costs. Such popularity can be expected to continue for the foreseeable future.

9.7 References

(1) Absi, R. (2005) Comment on "Turbulent diffusion of momentum and suspended particles: A finite-mixing-length theory" [Phys. Fluids 16, 2342 (2004)]. Phys. Fluids, vol. 17, article no. 079101.
(2) Bayazit, M. (1976): Free surface flow in a channel of large relative roughness. J. Hydraul. Res., vol. 14, 115–126.
(3) Bayazit, M. (1983). Flow structure and sediment transport mechanics in steep channels. In: B.M. Sumer and A. Muller (eds.), Euromech 156: Mechanics of Sediment Transport, Balkema, Rotterdam, pp. 197–206.
(4) Baykal, C., Sumer, B.M., Fuhrman, D.R., Jacobsen, N.G. and Fredsøe, J. (2015): Numerical investigation of flow and scour around a vertical circular cylinder. Phil. Trans. Roy. Soc. A, vol. 373, article no. 20140104.
(5) Baykal, C., Sumer, B.M., Fuhrman, D.R., Jacobsen, N.G. and Fredsøe, J. (2017): Numerical simulation of scour and backfilling processes around a circular pile in waves. Coast. Eng., vol. 122, 87–107.
(6) Bihs, H., Kamath, A., Chella, M.A., Aggarwal, A. and Arntsen, Ø.A. (2016): A new level set numerical wave tank with improved density interpolation for complex wave hydrodynamics. Comput. Fluids, vol. 140, 191–208.
(7) Boussinesq, J. (1877): Essai sur la théorie des eaux courantes. Mémoires présentés par divers savants à l'Académie des Sciences, vol. 23, 1–680.
(8) Brown, S.A., Greaves, D.M., Magar, V. and Conley, D.C. (2016): Evaluation of turbulence closure models under spilling and plunging breakers in the surf zone. Coast. Eng., vol. 114, 177–193.
(9) Cebeci, T. and Chang, K.C. (1978): Calculation of incompressible rough-wall boundary-layer flows. AIAA J., vol. 16, 730–735.
(10) Chauchat, J., Cheng, Z., Nagel, T., Bonamy, C. and Hsu, T.-J. (2017):

SedFoam-2.0: A 3-D two-phase flow numerical model for sediment transport. Geoscientific Model Development, vol. 10, 4367–4392.
(11) Christensen, E.D. (2006): Large eddy simulation of spilling and plunging breakers. Coast. Eng., vol. 53, 463–485.
(12) Coakley, T.J. (1983): Turbulence modeling methods for the compressible Navier-Stokes equations. AIAA Paper 83–1693.
(13) Davidson, P.A. (2015): Turbulence: An Introduction for Scientists and Engineers, 2nd edn. Oxford University Press.
(14) del Jesus, M., Lara, J.L. and Losada, I.J. (2012): Three-dimensional interaction of waves and porous coastal structures. Part I: Numerical model formulation. Coast. Eng., vol. 64, 57–72.
(15) Devolder, B., Rauwoens, P. and Troch, P. (2017): Application of a buoyancy-modified k-ω SST turbulence model to simulate wave runup around a monopile subjected to regular waves using OpenFOAM. Coast. Eng. 125, 81–94.
(16) Devolder, B., Troch, P. and Rauwoens, P. (2018): Performance of a buoyancy-modified k-ω and k-ω SST turbulence model for simulating wave breaking under regular waves using OpenFOAM. Coast. Eng., vol. 138, 49–65.
(17) Dixen, M., Sumer, B.M. and Fredsøe, J. (2013): Numerical and experimental investigation of flow and scour around a half-buried sphere. Coast. Eng., vol. 73, 84–105.
(18) Eckelmann, H. (1974): The structure of the viscous sublayer and the adjacent wall region in a turbulent channel flow. J. Fluid Mech., vol. 65, 439–459.
(19) Ferziger, J.H. and Peric, M. (2002): Computational Methods for Fluid Dynamics, 3rd edn. Springer.
(20) Fredsøe, J., Andersen, K.H. and Sumer, B.M. (1999): Wave plus current over a ripple-covered bed. Coast. Eng., vol. 38, 177–221.
(21) Fuhrman, D.R., Baykal, C., Sumer, B.M., Jacobsen, N.G. and Fredsøe, J. (2014): Numerical simulation of wave-induced scour and backfilling processes beneath submarine pipelines. Coast. Eng., vol. 94, 10–22.
(22) Fuhrman, D.R., Dixen, M. and Jacobsen, N.G. (2010): Physically-consistent wall boundary conditions for the k-ω turbulence model. J. Hydraul. Res., vol. 48, 793–800.
(23) Fuhrman, D.R., Schløer, S. and Sterner, J. (2013): RANS-based simulation of turbulent wave boundary layer and sheet-flow sediment transport processes. Coast. Eng., vol. 73, 151–166.
(24) Fuhrman, D.R., Sumer, B.M. and Fredsøe, J. (2011): Roughness-

induced streaming in turbulent wave boundary layers. J. Geophys. Res., vol. 116, article no. C10002.
(25) Higuera, P. and Lara, J.L. and Losada, I.J. (2013): Simulating coastal engineering processes with OpenFOAM. Coast. Eng., vol. 71, 119–134.
(26) Hirsch, C. (2007): Numerical Computation of Internal and External Flows: The Fundamentals of Computational Fluid Dynamics, 2nd edn. Butterworth-Heinemann.
(27) Huang, P.G. (1999): Physics and computations of flows with adverse pressure gradients. In E.A.M. Salas (ed.), Modeling Complex Turbulent Flows, Kluwer, pp. 245–258.
(28) Jacobsen, N. G. and Fredsøe, J. (2014): Cross-shore redistribution of nourished sand near a breaker bar. J. Waterw. Port Coast. Ocean Eng.-ASCE, vol. 140, 125–134.
(29) Jacobsen, N.G., Fuhrman, D.R. and Fredsøe, J. (2012): A wave generation toolbox for the open-source CFD library: OpenFoam®. Int. J. Numer. Meth. Fluids, vol. 70, 1073–1088.
(30) Jasak, H., Vukcevic, V., Gatin, I. and Lalovic, I. (2019): CFD validation and grid sensitivity studies of full scale ship self propulsion. Int. J. Naval Arch. Ocean Eng., vol. 11, 33–43.
(31) Jensen, B.L., Sumer, B.M. and Fredsøe, J. (1989): Turbulent oscillatory boundary layers at high Reynolds numbers. J. Fluid Mech., vol. 206, 265–297.
(32) Justesen, P. (1988): Turbulent Wave Boundary Layers. Series paper no. 43, ISVA, Technical University of Denmark (A PhD study supervised by J. Fredsøe and B. M. Sumer).
(33) Kim, J., Moin, P. and Moser, R. (1987): Turbulence statistics in fully developed channel flow at low Reynolds number. J. Fluid Mech., vol. 177, 133–166.
(34) Kim, Y., Cheng, Z., Hsu, T.-J. and Chauchat, J. (2018): A numerical study of sheet flow under monochromatic nonbreaking waves using a free surface resolving Eulerian two-phase flow model. J. Geophys. Res.-Oceans, vol. 123, 4693–4719.
(35) Kim, Y. and Hsu, T.-J. (2018, private communication): Large eddy simulation of plunging breaking waves. Center for Appl. Coastal Research, University of Delaware, Newark, DE 19716, USA.
(36) Kim, Y., Mieras, R.S., Cheng, Z., Anderson, D., Hsu, T.-J., Puleo, J.A. and Cox, D. (2019): A numerical study of sheet flow driven by velocity and acceleration skewed near-breaking waves on a sandbar using SedWaveFoam. Coast. Eng., vol. 152, article no. 103526,

doi:10.1016/j.coastaleng.2019.103526.
(37) Kolmogorov, A.N. (1942): Equations of turbulent motion of an incompressible fluid. Izvestia Academy of Sciences, USSR; Physics, vol. 6, 56–58.
(38) Lara, J.L., del Jesus, M. and Losada, I.J. (2012): Three-dimensional interaction of waves and porous coastal structures. Part II: Experimental validation. Coast. Eng., vol. 64, 26–46.
(39) Larsen, B.E. and Fuhrman, D.R. (2018): On the over-production of turbulence beneath surface waves in Reynolds-averaged Navier-Stokes models. J. Fluid Mech., vol. 853, 419–460.
(40) Larsen, B.E., Fuhrman, D.R., Baykal, C. and Sumer, B.M. (2017): Tsunami-induced scour around monopile foundations. Coast. Eng., vol. 129, 36–49.
(41) Larsen, B.E., Fuhrman, D.R. and Roenby, J. (2019): Performance of interFoam on the simulation of progressive waves. Coast. Eng. J., vol. 61, 380–400.
(42) Larsen, B.E., Fuhrman, D.R. and Sumer, B.M. (2016): Simulation of wave-plus-current scour beneath submarine pipelines. J. Waterw. Port Coast. Ocean Eng.-ASCE, vol. 142, article no. 04016003.
(43) Launder, B.E. and Sharma, B.I. (1974): Application of the energy-dissipation model or turbulence to the calculation of flow near a spinning disc. Lett. Heat Mass Transfer, vol. 1, 131–138.
(44) Lee, M. and Moser, R.D. (2015): Direct numerical simulation of turbulent channel flow up to $Re_\tau \approx 5200$. J. Fluid Mech., vol. 774, 395–415.
(45) Liu, X. and Garcia, M.H. (2008): Three-dimensional numerical model with free water surface and mesh deformation for local sediment scour. J. Waterw. Port Coast. Ocean Eng.-ASCE, vol. 134, 203–217.
(46) Liu, S., Ong, M.C. and Obhrai, C. (2019a): Numerical simulations of breaking waves and steep waves past a vertical cylinder at different Keulegan-Carpenter numbers. J. Offshore Mech. Arctic Eng., vol. 141, article no. 041806, doi:10.1115/1.4043278.
(47) Liu, S., Jose, J., Ong, M.C. and Gudmestad, O.T. (2019b): Characteristics of higher-harmonic breaking wave forces and secondary load cycles on a single vertical circular cylinder at different Froude numbers. Marine Structures, vol. 64, 54–77.
(48) Lohmann, I.P., Fredsøe, J., Sumer, B.M. and Christensen, E.D. (2006): Large eddy simulation of the ventilated wave boundary layer. J. Geophys. Res., vol. 111, article no. C06036.
(49) Losada, I.J., Lara, J.L. and del Jesus, M. (2016): Modeling the inter-

action of water waves with porous coastal structures. J. Waterw. Port Coast. Ocean Eng.-ASCE, vol. 142, article no. 03116003.
(50) Mayer, S. and Madsen, P.A. (2000): Simulations of breaking waves in the surf zone using a Navier-Stokes solver. Proc. 27th Int. Conf. Coast. Eng. (Sydney, Australia), pp. 928–941.
(51) Maza, M., Lara, J.L. and Losada, I.J. (2015): Tsunami wave interaction with mangrove forests: A 3-D numerical approach. Coast. Eng., vol. 98, 33–54.
(52) Meneveau, C. and Katz, J. (2000): Scale-invariance and turbulence models for large-eddy simulation. Annu. Rev. Fluid Mech., vol. 32, 1–32.
(53) Menter, F.R. (1994): Two-equation eddy-viscosity turbulence models for engineering applications. AIAA J., vol. 32, 1598–1605.
(54) Moin, P. and Mahesh, K. (1998): Direct numerical simulation: A tool in turbulence research. Annu. Rev. Fluid Mech., vol. 30, 539–578.
(55) Nezu, I. (1977): Turbulent Structure in Open Channel Flows. PhD thesis, Kyoto University, Japan.
(56) Nielsen, P. and Teakle, I.A.L. (2004): Turbulent diffusion of momentum and suspended particles: A finite-mixing-length theory. Phys. Fluids, vol. 16, 2342–2348.
(57) Ong, M.C., Utnes, T., Holmedal, L.E., Myrhaug, D. and Pettersen, B. (2009): Numerical simulation of flow around a smooth circular cylinder at very high Reynolds numbers. Marine Structures, vol. 22, 142–153.
(58) Pope, S.B. (1978): An explanation of the turbulent round-jet/plane-jet anomaly. AIAA J., vol. 72, 331–340.
(59) Prandtl, L. (1925): Uber die ausgebildete turbulenz. ZAMM, vol. 5, 136–139.
(60) Rodi, W. (2017): Turbulence modeling and simulation in hydraulics: A historical review. J. Hydraul. Eng.-ASCE, vol. 143, article no. 03117001.
(61) Roenby, J., Bredmose, H. and Jasak, H. (2016): A computational method for sharp interface advection. Roy. Soc. Open Sci., vol. 3, article no. 160405, doi: 10.1098/rsos.160405.
(62) Rogallo, R. and Moin, P. (1984): Numerical simulation of turbulent flows. Annu. Rev. Fluid Mech., vol. 16, 99–137.
(63) Roulund, A., Sumer, B.M., Fredsøe, J. and Michelsen, J. (2005): Numerical and experimental investigation of flow and scour around a circular pile. J. Fluid Mech., vol. 534, 351–401.
(64) Saffman, P.G. (1970): A model for inhomogenous turbulent flow. Proc.

Roy. Soc. Lond. A, vol. 317, 417–433.
(65) Smagorinsky, J. (1963): General circulation experiments with the primitive equations: 1. The basic experiment. Monthly Weather Rev., vol. 91, 99–164.
(66) Speziale, C.G., Abid, R. and Anderson, E.C. (1992). A critical evaluation of two-equation models for near wall turbulence. AIAA J., vol. 30, 324–331.
(67) Sumer, B.M., Chua, L.H.C., Cheng, N.-S. and Fredsøe, J. (2003): Influence of turbulence on bed load sediment transport. J. Hydraul. Eng.-ASCE, vol. 129, 585–596.
(68) Sumer, B.M., Cokgor, S. and Fredsøe, J. (2001): Suction removal of sediment from between armour blocks. J. Hydraul. Eng.-ASCE, vol. 127, 293–306.
(69) Tennekes, H. and Lumley, J.L. (1972): A First Course in Turbulence. MIT Press.
(70) Troch, P. and De Rouck, J. (1999): An active wave generating-absorbing boundary condition for VOF type numerical model. Coast. Eng., vol. 38, 223–247.
(71) van Driest, E.R. (1956): On turbulent flow near a wall. J. Aeronautical Sciences, vol. 23, 1007–1011, 1036.
(72) Vukcevic, V., Jasak, H. and Malenica, S. (2016a): Decomposition model for naval hydrodynamic applications, Part I: Computational method. Ocean Eng., vol. 121, 37–46.
(73) Vukcevic, V., Jasak, H. and Malenica, S. (2016b): Decomposition model for naval hydrodynamic applications, Part I: Computational method. Ocean Eng., vol. 121, 76–88.
(74) Wilcox, D.C. (1988): Reassessment of the scale-determining equation for advanced turbulence models. AIAA J., vol. 26, 1299–1310.
(75) Wilcox, D.C. (2006): Turbulence Modeling for CFD, 3rd edn. DCW Industries.
(76) Wilcox, D.C. (2008): Formulation of the k-ω turbulence model revisited. AIAA J., vol. 46, 2823–2838.
(77) Williams, I.A. and Fuhrman, D.R. (2016): Numerical simulation of tsunami-scale wave boundary layers. Coast. Eng., vol. 110, 17–31.
(78) Zhou, Z., Hsu, T.-J., Cox, D., and Liu, X. (2017): Large-eddy simulation of wave-breaking induced turbulent coherent structures and suspended sediment transport on a barred beach. J. Geophys. Res.-Oceans, 122, 207–235.

Appendix to Chapter 9: The exact k equation

9.A Derivation of the exact k equation

From Eqs. 2.31 we have the energy equation for the fluctuating flow (i.e. turbulence):

$$\frac{\partial K_t}{\partial t}+\frac{\partial}{\partial x_\alpha}\left[K_t\overline{u}_\alpha+\frac{1}{2}\rho\overline{u'_iu'_iu'_\alpha}+\overline{p'u'_\alpha}-\overline{u'_i\mu\left(\frac{\partial u'_i}{\partial x_\alpha}+\frac{\partial u'_\alpha}{\partial x_i}\right)}\right]$$
$$=\rho\overline{u'_\alpha g'_\alpha}-\Phi_t+\left(-\rho\overline{u'_iu'_\alpha}\right)\frac{\partial\overline{u}_i}{\partial x_\alpha} \tag{9.80}$$

where from Eq. 2.33

$$\Phi_t=\mu\overline{\left(\frac{\partial u'_i}{\partial x_\alpha}+\frac{\partial u'_\alpha}{\partial x_i}\right)\frac{\partial u'_i}{\partial x_\alpha}} \tag{9.81}$$

is the viscous dissipation of the turbulent energy per unit volume of fluid and per unit time. Recall that the turbulent kinetic energy per unit mass is:

$$k=\frac{K_t}{\rho}=\frac{1}{2}\overline{u'_iu'_i} \tag{9.82}$$

Inserting the expression for Φ_t into Eq. 9.80, dividing by the density ρ (noting that $\nu=\mu/\rho$), neglecting fluctuations in the body force i.e. setting $g'=0$, and re-arranging some terms, this may be written as

$$\frac{\partial k}{\partial t}+\frac{\partial}{\partial x_\alpha}[k\overline{u}_\alpha]=-\overline{u'_iu'_\alpha}\frac{\partial\overline{u}_i}{\partial x_\alpha}+\frac{\partial}{\partial x_\alpha}\left[-\frac{1}{2}\overline{u'_iu'_iu'_\alpha}-\frac{1}{\rho}\overline{p'u'_\alpha}\right]$$
$$-\nu\overline{\left(\frac{\partial u'_i}{\partial x_\alpha}+\frac{\partial u'_\alpha}{\partial x_i}\right)\frac{\partial u'_i}{\partial x_\alpha}}+\frac{\partial}{\partial x_\alpha}\left[\overline{u'_i\nu\left(\frac{\partial u'_i}{\partial x_\alpha}+\frac{\partial u'_\alpha}{\partial x_i}\right)}\right] \tag{9.83}$$

Making use of local continuity

$$\frac{\partial u_\alpha}{\partial x_\alpha} = 0 \tag{9.84}$$

the viscous terms may be expanded and then simplified as follows:

$$\begin{aligned}
&-\nu\overline{\left(\frac{\partial u_i'}{\partial x_\alpha} + \frac{\partial u_\alpha'}{\partial x_i}\right)\frac{\partial u_i'}{\partial x_\alpha}} + \frac{\partial}{\partial x_\alpha}\left[\overline{u_i'\nu\left(\frac{\partial u_i'}{\partial x_\alpha} + \frac{\partial u_\alpha'}{\partial x_i}\right)}\right] \\
&= -\nu\overline{\frac{\partial u_i'}{\partial x_\alpha}\frac{\partial u_i'}{\partial x_\alpha}} - \nu\overline{\frac{\partial u_\alpha'}{\partial x_i}\frac{\partial u_i'}{\partial x_\alpha}} + \nu\frac{\partial}{\partial x_\alpha}\left[\overline{u_i'\frac{\partial u_i'}{\partial x_\alpha}}\right] + \nu\overline{\frac{\partial u_\alpha'}{\partial x_i}\frac{\partial u_i'}{\partial x_\alpha}} \\
&= -\nu\overline{\frac{\partial u_i'}{\partial x_\alpha}\frac{\partial u_i'}{\partial x_\alpha}} + \nu\frac{\partial}{\partial x_\alpha}\left[\frac{1}{2}\frac{\partial \overline{u_i'u_i'}}{\partial x_\alpha}\right] \\
&= -\nu\overline{\frac{\partial u_i'}{\partial x_\alpha}\frac{\partial u_i'}{\partial x_\alpha}} + \nu\frac{\partial}{\partial x_\alpha}\left[\frac{\partial k}{\partial x_\alpha}\right]
\end{aligned} \tag{9.85}$$

where the definition of k from Eq. 9.82 has been invoked at the end. Inserting this result back into Eq. 9.83 then yields

$$\begin{aligned}
\frac{\partial k}{\partial t} + \overline{u}_\alpha\frac{\partial k}{\partial x_\alpha} = &-\overline{u_i'u_\alpha'}\frac{\partial \overline{u}_i}{\partial x_\alpha} + \frac{\partial}{\partial x_\alpha}\left[-\frac{1}{2}\overline{u_i'u_i'u_\alpha'} - \frac{1}{\rho}\overline{p'u_\alpha'}\right] \\
&- \nu\overline{\frac{\partial u_i'}{\partial x_\alpha}\frac{\partial u_i'}{\partial x_\alpha}} + \nu\frac{\partial}{\partial x_\alpha}\left[\frac{\partial k}{\partial x_\alpha}\right]
\end{aligned} \tag{9.86}$$

where local continuity has again been invoked on the convective term (second term on the left hand side). Re-arranging slightly, this is equivalently

$$\frac{\partial k}{\partial t} + \overline{u}_\alpha\frac{\partial k}{\partial x_\alpha} = -\overline{u_i'u_\alpha'}\frac{\partial \overline{u}_i}{\partial x_\alpha} - \nu\overline{\frac{\partial u_i'}{\partial x_\alpha}\frac{\partial u_i'}{\partial x_\alpha}} + \frac{\partial}{\partial x_\alpha}\left[\nu\frac{\partial k}{\partial x_\alpha} - \frac{1}{2}\overline{u_i'u_i'u_\alpha'} - \frac{1}{\rho}\overline{p'u_\alpha'}\right] \tag{9.87}$$

This matches the exact k (turbulent kinetic energy) equation presented in Eq. 9.28, as well as the result found e.g. in the book of Wilcox (2006), p. 108.

9.B Reference

(1) Wilcox, D.C. (2006): Turbulence Modeling for CFD, 3rd edn. DCW Industries.

Chapter 10

Hands-on exercises

10.1 Introduction

This final chapter contains four hands-on exercises. The exercises are designed to cover a wide variety of turbulent flows (in most cases using actual real data sets!), analysis methods, as well as modelling techniques, as covered in the previous chapters. All exercises are designed to be solved in Matlab, and have been developed over many years as report assignments within the Turbulence Theory/Turbulent Flows course taught at the Technical University of Denmark (DTU).

Exercise 1 considers analysis of the flow and turbulence properties stemming from measurements made in an **open channel flow**, and is hence of direct relevance to material covered in Chapters 2 and 3. Exercise 2 considers the development and application of a simple model of **turbulent dispersion**, with obvious relevance to Chapter 8. Exercise 3 considers the statistical, correlation, and spectral analysis of measured velocities in a **turbulent air jet**. It is therefore especially relevant to the concepts introduced in Chapter 4. Finally, Exercise 4 focuses directly on **turbulence modelling** of unsteady (oscillatory) **wave boundary layer** flows, including comparison with data measured in an oscillating tunnel flow on both **smooth** and **rough beds**. It is hence especially relevant to material from Chapters 5, 6 and 9.

The authors sincerely hope that these exercises give the interested reader a true "hands-on" experience in the analysis and modelling of turbulent flows. In working through the exercises, it is likewise intended that students experience in the joys (and sometimes challenges) of working with and interpreting actual data measured in laboratory settings.

10.2 Data sets

As discussed above, three of the four exercises involve analysis of and/or comparison against actual real life data sets, which have been measured in laboratories at the Technical University of Denmark (DTU). Details on the contents of each data set will be described directly in the relevant exercise statements that follow; see the forthcoming Tables 10.2, 10.3 and 10.5. The individual data sets correspond to:

- Exercise 1: Steady turbulent open channel flow (filename: `Exercise1.mat`);
- Exercise 3: Turbulent air jet flow (filename: `Exercise3.mat`);
- Exercise 4: Oscillatory wave boundary layer flow (Jensen et al., 1989, filename: `Exercise4.mat`).

All of the digital material intended to accompany the present book (the three data sets above, the MatRANS model described in the next section, as well as the Matlab examples from Chapter 4) can be found in the data collection available online at:

`https://doi.org/10.11583/DTU.c.4508648`

10.3 MatRANS k-ω turbulence model in Matlab

Several of the exercises also involve applications of an actual turbulence closure model, and for this purpose the authors provide the MatRANS model (a simple one-dimensional vertical, 1DV, model solving Reynolds-averaged Navier-Stokes equations, coupled with a k-ω model in Matlab). This model is as originally described by Fuhrman et al. (2013), with subsequent developments to the turbulence modelling capabilities made by Williams and Fuhrman (2016) and Kirca et al. (2016), who extended the model to include the "low Reynolds number" version.

This model has been described in detail in the present book in Chapter 5, Section 5.12, with the final 1DV formulation corresponding to the framed equations in that section (primarily Eqs. 5.147, 5.162 and 5.163, combined

Table 10.1 Description of contents from MatRANS output file specified in the `OutFileName` variable. When loaded this will contain the single structure array `MatRANS` described below. Note that the period T and free stream velocity magnitude U_{0m} are only relevant for wave boundary layer simulations.

Structure array	Field	Book notation	Size	Description
`MatRANS`			1×1	Selected output from MatRANS simulation
	`n_t`	n_t	1×1	Number of saved time levels (s)
	`n_y`	n_y	1×1	Number of y grid points (s)
	`t`	t	$n_t \times 1$	Saved time levels (s)
	`y`	y	$1 \times n_y$	Model y positions (m)
	`u`	$\overline{u}(t, y)$	$n_t \times n_y$	Horizontal velocities (m/s)
	`k`	$k(t, y)$	$n_t \times n_y$	Turbulent kinetic energy density (m^2/s^2)
	`omega`	$\omega(t, y)$	$n_t \times n_y$	Specific dissipation rate (1/s)
	`nu_t`	$\nu_t(t, y)$	$n_t \times n_y$	Kinematic eddy viscosity (m^2/s)
	`tau0`	$\overline{\tau}_0(t)$	$n_t \times 1$	Bed shear stress (Pa)
	`nu`	ν	1×1	Fluid kinematic viscosity (m^2/s)
	`rho`	ρ	1×1	Fluid density (kg/m^3)
	`k_s`	k_s	1×1	Bottom roughness (m)
	`h_m`	h_m	1×1	Model domain height (m)
	`T`	T	1×1	Wave period (s)
	`U0m`	U_{0m}	1×1	Free stream velocity magnitude (m)

with the eddy viscosity defined in Eq. 5.164). The model can be downloaded (filename: `MatRANS.zip`) from the data collection link provided at the end of Section 10.2.

Examples to be used in the following exercises are provided already set up, and can be found in the `MatRANS/TurbulenceBook/` folder, which is then sub-divided into `Exercise1` and `Exercise4` folders. Running the model is straightforward, accomplished by simply executing the `runMatRANS.m` script! Any alterations required to the input will be explained directly in the relevant exercise statements (namely, Exercise 4).

In all cases, upon completion selected results from the provided book examples will be exported to a single output file, with name corresponding to the `OutFileName` input string variable (e.g. `out_MatRANS.mat`). This output file consists of a single Matlab data structure array called `MatRANS`, with fields described in detail in Table 10.1. All field names have been chosen to closely coincide with the book notation. With this, it should be straightforward for users familiar with Matlab to make all desired animations or plots requested in the following exercises. Code to make some example plots of the flow/turbulence profiles are provided at the end of the `runMatRANS.m` input execution scripts to help facilitate this.

10.4 Exercise 1: Analysis of a turbulent boundary layer in an open channel

Experiments

For this exercise you will consider a series of velocity measurements made in a steady and (streamwise) uniform boundary layer flow in the tilting flume at the Hydraulic Engineering Laboratory at the Technical University of Denmark (DTU). A two-component Laser Doppler Velocimeter (LDV), also known as Laser Doppler Anemometer (LDA), was used to measure the u and v components of the water velocity, where x is in the streamwise direction and y in the vertical direction. The flow depth in the experiments was maintained at $h = 7$ cm, and the flow velocity at the water surface at $U_0 \approx 30$ cm/s. The cross section of the flume is rectangular, with width $b = 30$ cm. The u and v velocities were measured at the following 23 y (cm) locations: 0.02, 0.03, 0.05, 0.07, 0.1, 0.14, 0.18, 0.2, 0.25, 0.3, 0.4, 0.5, 0.6, 0.7, 0.8, 1.0, 1.25, 1.5, 2.0, 2.5, 3.0, 4.0, 5.0. The sampling length was approximately 180 s for each point, and the sampling interval was set automatically by the LDA equipment (but will be in all circumstances smaller than O(20 ms)). For the purposes of this exercise, you may assume that the kinematic water viscosity is $\nu = 10^{-6}$ m^2/s.

Data input, processing and analysis

All data processing is intended to be done in Matlab. The data set to be analyzed is provided in the workspace file `Exercise1.mat` from the steady turbulent open channel flow data set. This can be obtained as presented in Section 10.2. This file can be easily imported into Matlab using the `load` command. Once loaded, the data is contained entirely in the single Matlab data structure variable `Channel`, the contents of which are described in full detail in Table 10.2.

The naming of the various fields comprising the `Channel` variable is seen to closely coincide with the book notation; Data can hence hopefully be easily and intuitively accessed. For example: `Channel(5).u` will return the raw u values from the fifth measurement point, etc. Other data can be similarly accessed by replacing the field `u` with any of those indicated in Table 10.2. Similarly, data from other measurement positions can be accessed simply by replacing the `5` with any other integer ranging from `1` to `23`.

Table 10.2 Description of contents for Matlab data set contained in `Exercise1.mat`. The number of measurements (sample length) N for each position varies.

Structure array	Field	Book notation	Size	Description
`Channel`			1×23	Data set for 23 y measurement positions
	`y`	y	1×1	Measurement position (m)
	`u`	u	N	Streamwise velocities (m/s)
	`v`	v	N	Vertical velocities (m/s)
	`t`	t	N	Measurement times (s)
	`tt`	t_t	N	Travel times (s)
	`h`	h	1×1	Channel depth (m)
	`nu`	ν	1×1	Kinematic water viscosity (m^2/s)

Since the data set includes irregularly sampled data, some form of weighting technique must be used to properly estimate statistical quantities (mean, variance, etc.). For example, weighting based simply on the the transit time t_t (corresponding to the time taken for a measured flow particle to pass through the measurement volume) will correspond (e.g. for the streamwise component) to using for the mean:

$$\overline{u} = \frac{\sum_{i=1}^{N} u_i t_{t,i}}{\sum_{i=1}^{N} t_{t,i}} \tag{10.1}$$

and for the r.m.s. of the turbulent fluctuations:

$$\sqrt{\overline{u'^2}} = \left(\frac{\sum_{i=1}^{N} u'^2 t_{t,i}}{\sum_{i=1}^{N} t_{t,i}} \right)^{1/2} \tag{10.2}$$

where N is the sample length. Analogous expressions may be used for other quantities (e.g. $\sqrt{\overline{v'^2}}$ and $\sqrt{-\overline{u'v'}}$; For the latter case e.g. u'^2 should be replaced with $-u'v'$ within the numerator of Eq. 10.2).

Action items

(1) Plot the mean velocity $\overline{u}$ as function of y using linearly scaled axes. In addition to the raw data, you may include the points $u(y = 0) = 0$ (no slip at the bed) and $u(y = h) = U_0 = 0.30$ m/s for completeness.

(2) From the velocity profile $\overline{u}(y)$ in item 1 calculate the depth-averaged velocity V via

$$V = \frac{1}{h} \int_0^h \overline{u} dy \tag{10.3}$$

(Hint: the Matlab function `trapz` may be used to compute the integral.)

(3) Estimate the friction velocity U_f from the Darcy-Weisbach equation

$$U_f = \sqrt{\frac{f}{2}}V \tag{10.4}$$

where f is the friction coefficient, calculated from the Blausius formula

$$f = \frac{0.0557}{Re^{0.25}} \tag{10.5}$$

where $Re = r_h V/\nu$ is the Reynolds number, with $r_h = A/P$ the hydraulic radius (where $A = hb$ is the cross-sectional flow area, and $P = 2h + b$ is the wetted perimeter). Eq. 10.5 is a good approximation to the more general flow-resistance equation, Eq. 3.130, for $Re < 0.25 \times 10^5$ (Schlichting, 1979, p. 597 and Fig. 20.1).

(4) Plot $\overline{u}$ versus y on a semi-log scale (i.e. logarithmic scale for y and linear scale for $\overline{u}$). Identify the approximate lower and upper bounds of the logarithmic region. The upper bound should be taken as the top of the constant stress region ($y \leq 0.1h$ or equivalently $y^+ \leq 0.1Re_\tau$, where $y^+ = yU_f/\nu$ and $Re_\tau = hU_f/\nu$) whereas the lower bound should be taken as where $y^+ = 30$, where U_f is estimated from item 3 above (this bound should be re-checked afterwards!). By fitting a straight line to the data in the logarithmic-layer portion of the velocity profile (i.e. of the form $\overline{u} = m\log(y) + b$), and comparing the result with Eq. 3.42, calculate U_f directly from the data (Hint: use Matlab's `polyfit` function to determine the coefficients m and b above). Compare this U_f with the estimation made in item 3.

(5) Using the U_f from the log-fitting exercise (item 4) above re-plot the velocity profile in a dimensionless fashion i.e. as $\overline{u}/U_f$ vs. y^+ as well as vs. y/h. Identify various regions of the flow e.g. the viscous sublayer, the buffer layer, the logarithmic layer, and the outer region.

(6) Plot the van Driest velocity distribution from Eq. 3.108 on the same diagram and compare the experimentally determined velocity profile with the van Driest profile. (Hint: Calculate the required cumulative integral numerically in Matlab using `cumtrapz`.)

(7) Plot the turbulence data in wall units, namely $\sqrt{\overline{u'^2}}/U_f$, $\sqrt{\overline{v'^2}}/U_f$ and $\sqrt{-\overline{u'v'}}/U_f$ as a function of y^+ over the range $0 \leq y^+ \leq 100$. Can you see that these turbulence quantities become approximately constant once the logarithmic region is reached?

(8) Plot the turbulence data in terms of outer-flow parameters, namely $\sqrt{\overline{u'^2}}/U_f$, $\sqrt{\overline{v'^2}}/U_f$ and $\sqrt{-\overline{u'v'}}/U_f$ as a function of y/h, now spanning the entire flow depth i.e. $0 \leq y/h \leq 1$.

(9) Also, do the same for k/U_f^2, where $k = 1/2\left(\overline{u'^2} + \overline{v'^2} + \overline{w'^2}\right)$ is the turbulent kinetic energy (per unit mass), as a function of y/h. For this purpose, you may use the approximation

$$\overline{w'^2} = 1.8\overline{v'^2} \tag{10.6}$$

for the missing transverse component of the fluctuating velocity.

(10) Calculate the Reynolds stress through

$$-\rho\overline{u'v'} = \overline{\tau} - \mu\frac{\partial\overline{u}}{\partial y} \tag{10.7}$$

where $\overline{\tau}$ is given by

$$\overline{\tau} = \rho U_f^2\left(1 - \frac{y}{h}\right) \tag{10.8}$$

(Hint: For calculating $\partial\overline{u}/\partial y$ use Matlab's `gradient` function.) Plot the non-dimensional Reynolds stress $-\rho\overline{u'v'}/U_f^2$ as a function of y/h. Compare the latter with that obtained directly from the measurements in item 7.

(11) For the steady and streamwise uniform boundary layer under consideration, write out the simplified form of the energy equation for turbulence, Eq. 2.31. This should lead to (among other things) a turbulence energy production term corresponding to $-\rho\overline{u'v'}\partial\overline{u}/\partial y$. Plot this quantity as a function of y/h.

(12) Run the MatRANS k-ω model for this case i.e. simply execute the `MatRANS/TurbulenceBook/Exercise1/runMatRANS.m` script in Matlab, which is already appropriately set up for this case. See again Table 10.1 for a complete description of the model output. Compare the experimental data from above with the turbulence model results for $\overline{u}/U_f$ and k/U_f^2 vs. y/h.

10.5 Exercise 2: A simple numerical model of dispersion in a turbulent boundary layer flow

For this exercise consider the following: A certain amount of dispersant is dumped into a river or channel. The time when the dumping takes place is $t = 0$. It is assumed that the dispersant is distributed uniformly over the cross-section at $x = 0$ at the instant of dumping. The dispersant will be convected in the (streamwise) x-direction. At the same time, it will be subject to dispersion in the x-direction due to the combined effect of turbulent diffusion and the velocity gradient in the y-direction. As a result, the cloud will become more and more dispersed as it is convected downstream.

Flow model

In this exercise, you will study this problem by a numerical simulation of the dispersion process in a channel flow on the computer. For this purpose the same channel flow as in Exercise 1 will be considered, with flow velocity and turbulence properties taken from the end state of the MatRANS turbulence model results used previously (Exercise 1, Action item 12; Note that this simulation is set up such that the flow depth is $h = 0.07$ m and friction velocity is $U_f = 0.0134$ m/s once the steady state is reached). Hence, it is assumed that the `runMatRANS.m` script mentioned there has already been executed, and that the resulting workspace file containing the `MatRANS` structure array (with contents again described in full detail in Table 10.1) has been created. This structure array thus contains the computed horizontal mean flow velocity $\overline{u}$, the turbulent kinetic energy density k, and the specific dissipation rate ω at discrete y positions.

From the basic flow model results the variance of the vertical turbulent fluctuations can be assumed to follow

$$\overline{v'^2} = \sigma_v^2 = \frac{1}{3}k \tag{10.9}$$

Additionally, the mixing length ℓ_m can be taken from Eqs. 9.56 and 9.33 to be

$$\ell_m = \frac{1}{\beta^{*1/4}}\frac{\sqrt{k}}{\omega} \tag{10.10}$$

where the fully turbulent value $\beta^* = 0.09$ (Eq. 9.39) for the closure coefficient can be utilized. This should be valid in the logarithmic region and above.

From the direct model output above physical quantities can be obtained directly at the discrete y positions contained in `MatRANS.y`. Relevant quantities at other y positions can then easily be interpolated as needed (Hint: Matlab's `interp1` function can be utilized, but use of `griddedInterpolant` is much faster and recommended). As it is only the end (steady state) flow conditions that are to be used, it may be convenient to assign the end state flow results as simple vector variables e.g. `u=MatRANS.u(end,:)`, and similarly for `k` and `omega`, though this is not strictly necessary.

Random walk model

With the basic flow properties essentially given, the primary aim of the present exercise is to develop and apply a so-called random walk model for simulating the longitudinal dispersion process in a channel flow. For this purpose consider the motion of a single particle having position $(x_p^{(n)}, y_p^{(n)})$ at time t, where the superscript (n) refers to an integer time index i.e. not raising to a power. The evolution of this particle to a new position $(x_p^{(n+1)}, y_p^{(n+1)})$ at $t + \Delta t$ can be obtained from the following algorithm, which is reasonably similar to that used in Kirca et al. (2016):

- The small time increment Δt can be locally calculated from the mixing length ℓ_m and the standard deviation of the vertical turbulent fluctuations σ_v (both based on position $y = y_p^{(n)}$) according to

$$\Delta t = \frac{\ell_m}{\sigma_v} \tag{10.11}$$

- The instantaneous vertical fluctuation can be obtained by

$$v_p = \alpha_r \sigma_v - w_s \tag{10.12}$$

  where α_r is a normally distributed random number having unit standard deviation (Hint: Use Matlab's `randn` function; You may wish to set a seed at the beginning using Matlab's `rng` function, such that your results are repeatable). In the above w_s is the settling velocity of the particles (hence the special case where $w_s = 0$ corresponds to neutrally buoyant particles.)
- The new vertical particle position at time $t+\Delta t$ can then be computed via

$$y_p^{(n+1)} = y_p^{(n)} + v_p \cdot \Delta t \tag{10.13}$$

  If the obtained $y_p^{(n+1)} > h$ or $y_p^{(n+1)} < y_{bot}$ then the particle should be simply mirrored back into the domain via reflection about the exceeded upper or lower boundary, such that $y_{bot} \le y_p^{(n+1)} \le h$. Here y_{bot} defines the lower y limit for the dispersion model domain. For the present purposes this can be taken as where $y_{bot}^+ = y_{bot} U_f/\nu = 70$ i.e. such that y_{bot} defines the beginning of the logarithmic layer.
- The average horizontal (translation) velocity of the particle u_p over the time increment Δt can then be estimated as

$$u_p = \frac{\overline{u}\left(y = y_p^{(n)}\right) + \overline{u}\left(y = y_p^{(n+1)}\right)}{2} \tag{10.14}$$

- Finally, the new horizontal position of the particle can be determined from

$$x_p^{(n+1)} = x_p^{(n)} + u_p \cdot \Delta t \tag{10.15}$$

after which the time level index can be updated from (n) to $(n+1)$, and the process described above repeated.

The algorithm above describes a model for a single particle. However, as many particles as desired can be considered in succession, and the results then combined to demonstrate the dispersion process of a cloud. For the purposes of the present exercise, you should construct your model such that the dispersant will be modelled with N particles released one-by-one, with initial positions distributed uniformly over the modelled flow depth $y_{bot} < y < h$ at $x = 0$ at time $t = 0$.

In the presentation of all results, please use the following non-dimensional quantities:

$$Y = \frac{y}{h}, \qquad X = \frac{x}{h}, \qquad T = \frac{tU_f}{h} \tag{10.16}$$

Released particles should be modelled until the dimensionless time $T_{end} = 25$ is reached. It is noted that the time step Δt (hence ΔT) will be irregularly spaced and vary for each particle modelled. Therefore, once the end time is reached for a given particle, it is suggested to re-sample the particle X and Y positions at regular time intervals, such that the particle positions representing the dispersant cloud can be re-assembled at common times. (Hint: This can again be easily achieved using either the `interp1` or `griddedInterpolant` function.)

Action items

(1) Load the workspace from the MatRANS simulation from Exercise 1. Plot the resulting velocity profile ($\overline{u}/U_f$ vs. y/h), the turbulent kinetic energy density profile (k/U_f^2 vs. y/h), and the resulting mixing length profile (ℓ_m/h vs. y/h) at the end state. As a check, compare the results for the first two of these plots with those presented in Fig. 9.4.

(2) From the velocity profile compute the dimensionless depth-averaged velocity V/U_f, where

$$V = \frac{1}{h}\int_0^h \overline{u}dy \tag{10.17}$$

(Hint: Use `trapz` to compute the integral.)

(3) Consider first cases involving neutrally buoyant particles, i.e. with settling velocity $w_s = 0$. Run the model for some small values of N and show the resulting particle paths of a few particles.

(4) Perform simulations where $N = 100, 200, 500, 1000, 2000, 5000$ (or more!) uniformly distributed particles are released. Plot the mean particle position $\overline{X}$ and the variance $\overline{(X - \overline{X})^2}$ at time $T = 25$ as a function of the number of particles N, and see how many particles are needed for $\overline{X}$ and $\overline{(X - \overline{X})^2}$ to tend to practically constant (converged) values with increasing N.

(5) For the largest N achieved above (in reasonable computational time), calculate $\overline{X}$, $\overline{Y}$ and $\overline{(X - \overline{X})^2}$ at times $T = 5$, 10, 15, 20, and 25. Plot these quantities as a function of time T.

(6) Calculate the (dimensionless) ultimate mean velocity of the cloud from

$$U_m^* = \frac{U_m}{U_f} = \frac{d\overline{X}}{dT} \tag{10.18}$$

How does U_m^* compare with V/U_f from item 2?

(7) Make a probability density function for the end particle positions X and Y (Hint: Consider for simplicity using Matlab's `histogram` function). Compare the probability density functions for X and Y, respectively, with a Gaussian (normal) and uniform distribution.

(8) From the plot of $\overline{(X - \overline{X})^2}$ vs. T above, calculate the non-dimensional longitudinal dispersion coefficient from

$$D_1^* = \frac{D_1}{hU_f} = \frac{1}{2}\frac{d}{dT}\overline{(X - \overline{X})^2} \tag{10.19}$$

(Hint: consider using `polyfit`.) How does D_1^* compare with expectations based on Eq. 8.120. (Hint: If the model is implemented correctly, they should be rather close!)

(9) Now consider a released dispersant that is not neutrally buoyant, but rather one composed of slightly heavy particles having settling velocity $w_s = 0.15U_f$. Repeat item 7 above and make probability density functions for X and Y at the end state for the heavy particles. Compare these with those from the neutrally buoyant particles, and explain the differences.

(10) Finally, calculate the dimensionless ultimate mean velocity (U_m^*) and the longitudinal dispersion coefficient (D_1^*) of the heavy dispersant. Compare these with those of the neutrally-buoyant dispersant and physically explain the reasons for the differences observed.

Table 10.3 Description of contents in Matlab data contained in `Exercise3.mat`.

Structure array	Field	Book notation	Size	Description
`Jet`			1×12	Data set for 12 exit velocities
	`u`	u	500000×1	Flow velocities (m/s)
	`t`	t	500000×1	Measurement times (s)
	`V`	V	1×1	Exit velocity (m/s)
	`D`	D	1×1	Exit diameter (m)
	`nu`	ν	1×1	Kinematic air viscosity (m^2/s)
	`L`	L	1×1	Measurement position (m)

10.6 Exercise 3: Statistical, correlation, and spectral analysis of turbulent air jet flow

Introduction

For this exercise, you will consider a turbulent air velocity signal measured in a free round jet flow in the DTU Mechanical Engineering Fluid Mechanics Laboratory. The experiment was conducted using a jet with exit diameter $D = 0.03$ m with exit velocity V. Velocities were measured utilizing a single hot wire (Constant Temperature Anenometry, CTA, Dantec Dynamics probe type 55P11) sensor, at a fixed distance of $L = 0.6$ m ($20D$) away from the exit, with a sampling frequency of 50,000 Hz. For the purposes of this exercise, you may assume a kinematic fluid (air) viscosity of $\nu = 1.51 \times 10^{-5}$ m^2/s.

The full data set is comprised of 12 tests with various fixed values for V (varied over the range 16.0 m/s $\leq V \leq$ 38.4 m/s). It is provided in the `Exercise3.mat` Matlab workspace file, which can be obtained as presented in Section 10.2. The data is contained in a Matlab structure array variable called `Jet`, with fields fully detailed in Table 10.3. It can be accessed e.g. for Test 12 as: `Jet(12).u` (for the velocities), and similarly for the other fields. All data are in SI units.

For the purposes of this exercise you are to consider primarily Test 12, which had an exit velocity $V = 38.4$ m/s. (Hint: For help in performing many of the action items below, it may be useful to consult the various Matlab examples provided in Chapter 4.)

Action items

(1) Load the data and subtract the mean value from the raw velocity signal for Test 12. Plot the resultant turbulent fluctuating velocity

$u' = u(t) - \overline{u}$ time series. Plot both the full signal, as well as zoomed-in segments, such that you can clearly identify the passing of some individual eddies or coherent structures in the flow.

(2) Report basic statistics for the signal i.e. the mean (of the raw signal) $\overline{u}$, variance σ_u^2, skewness S_u, and kurtosis F_u in a table (see Section 4.1). Also report the turbulence intensity $\sigma_u/\overline{u}$, see e.g. Eq. 4.53.

(3) Plot the probability density function (p.d.f.) $p(u')$ for the turbulent signal, and compare this with the normal distribution from Eq. 4.3.

(4) Plot the time correlation function $R_E(\tau)$ (Eq. 4.38) for the measured turbulent signal. (For the purposes of this exercise it is suggested to compute $R_E(\tau)$ until the first zero crossing, see Program 3 in Chapter 4.) From this function determine the Eulerian macro time scale T_E and the Eulerian micro time scale τ_E. Can you identify regions of passing coherent structures corresponding approximately to these time scales in the zoomed-in time series plotted in item 1 above?

(5) Making use of Taylor's frozen turbulence approximation, e.g. Eq. 4.55, work out approximations for the corresponding macro (Λ_f) and micro (λ_f) length scales of turbulence. Discuss the applicability of this approximation, based on the turbulence intensity $\sigma_u/\overline{u}$ computed in item 2 above.

(6) Assuming isotropic turbulence as an approximation, can you also estimate the Kolmogorov time scale τ_K, the dissipation rate (per unit mass) ε, and the Kolmogorov length scale η_K? (Hint: See Section 4.3.3.)

(7) Make a log-log plot of the raw (using fast Fourier transform FFT) energy density spectra $S(f)$ in the frequency domain. Be sure to check that the integral of the spectrum yields the variance! (Hint: Please make sure you are analyzing u' and not u!) Also add a smoothed out version of this spectrum (e.g. using Matlab's `pwelch` function).

(8) Assuming Taylor's frozen turbulence approximation applies such that wave numbers can be determined by $k = 2\pi f/\overline{u}$, convert the frequency domain spectrum $S(f)$ above to a one-dimensional wave number spectrum $F(k)$, according to Eq. 4.125. Be sure to check that the integral of this spectrum now yields half the variance, in accordance with the convention from Eq. 4.124. Compare the resulting wave number spectrum with the von Karman model spectrum given in Eq. 4.131.

(9) On the wave number spectrum, identify the energy containing range, the universal equilibrium range, the inertial subrange, and the dissipation subrange.

(10) Convert the characteristic length scales found above to characteristic wave numbers by taking their inverse i.e. using $k_{macro} \simeq \Lambda_f^{-1}$, $k_{micro} \simeq \lambda_f^{-1}$ and $k_{Kolmogorov} \simeq \eta_K^{-1}$. Identify where these are located on your wave number spectrum $F(k)$. Do these fall into the expected ranges identified above?

(11) Repeat the analysis above considering all of the other measurement signals having other values for the exit velocity V (Test 1–12). For each test record the three characteristic length scales Λ_f, λ_f and η_K as well as the Reynolds number $Re = VD/\nu$. Make scatter plots (log-log scaled) of the resulting length scale ratios Λ_f/η_K and Λ_f/λ_f versus Re. Also add trend lines such that $\Lambda_f/\eta_K \sim Re^{3/4}$ and $\Lambda_f/\lambda_f \sim Re^{1/2}$, as suggested by Eqs. 4.92 and 4.93. Do the measured scale ratios follow the expected Reynolds number dependence?

10.7 Exercise 4: Turbulence modelling of oscillatory wave boundary layer flows

Introduction

In this exercise you will perform a numerical modelling study of oscillatory wave boundary layers for various Reynolds numbers, as well as on both smooth and rough beds. In doing so, you will obtain first-hand experience in turbulence modelling, by properly setting up and running the provided MatRANS k-ω model in Matlab (Fuhrman et al., 2013). You will likewise compare your model results with either theory or laboratory measurements made in DTU's oscillating tunnel environment (Jensen et al., 1989).

Model description

The model to be used in this assignment is the so-called MatRANS model, originally developed by Fuhrman et al. (2013). The model solves one-dimensional vertical (1DV) Reynolds-averaged Navier-Stokes (RANS) equations, coupled with two-equation (Wilcox, 2006) k-ω turbulence closure ($k = 1/2\overline{u_i' u_i'}$ being the turbulent kinetic energy density, ω being the specific dissipation rate), in the Matlab environment. The model is as originally described in Fuhrman et al. (2013), with extension to include the "low Reynolds number" version, as described by Williams and Fuhrman (2016) and Kirca et al. (2016). The resulting model equations are described in full detail in Section 5.12, where the final governing 1DV equations correspond to the framed (boxed) equations in that section (primarily Eqs. 5.147, 5.162 and 5.163, combined with the eddy viscosity defined in Eq. 5.164). The model is again available for download at the link provided at the end of Section 10.2.

Note that while the full model includes a method for approximating convective terms, as described in Section 5.12 (underlined terms), these are to be neglected in the present exercise (achieved by keeping `streaming=0` in the model input). This is appropriate for situations where the flow is statistically uniform in the streamwise direction, as in a tunnel environment.

The driving pressure gradient is as described in Eqs. 5.159 and 5.161, again with all underlined terms neglected and with $P_x = 0$, such that a free stream velocity of the form

$$U_0 = U_{0m} \sin\left(\frac{2\pi}{T} t\right) \tag{10.20}$$

will be driven, where T is the wave period. As boundary conditions, a frictionless rigid lid is imposed at the top boundary, whereby vertical derivatives of $\overline{u}$, k, and ω are set to zero. Alternatively, the bottom boundary is considered a friction wall, and a no-slip boundary condition is imposed, i.e. velocity variables are set to $\overline{u} = 0$, as well as $k = 0$. The bottom boundary condition for the specific dissipation rate ω is given according to Eq. 9.47.

For the purposes of this exercise you are to utilize the transitional (low Reynolds number) version of the model (achieved by keeping `turb=2` in the model input, as provided), wherein selected closure coefficients are flow dependent as specified in Eq. 9.43–9.46, and all others are specified according to Eqs. 9.39–9.41 (with $\beta = \beta_0$ as for two-dimensional flows). This single model should be capable of simulating both laminar and turbulent regimes, including transition from one to another, without modification.

Model input and output

All model input variables are set in the `runMatRANS.m` file, a working example of which is provided in the:

```
MatRANS/TurbulenceBook/Exercise4/Case1
```

directory, which is hopefully self-explanatory. The only input variables that should require changing to complete this exercise are: `U_1m` (corresponding to the free stream velocity magnitude U_{0m}), `dy0` (corresponding to the near wall grid size Δy), and in a single case `k_s` (corresponding to Nikuradse's

Table 10.4 Summary of wave boundary layer conditions to be considered. The corresponding test from the experimental work of Jensen et al. (1989) is also indicated. All calculations below use the reported kinematic viscosity $\nu = 1.14 \times 10^{-6}$ m^2/s.

Case	Test no. from Jensen et al. (1989)	T (s)	U_{0m} (m/s)	k_s (m)	Re	a/k_s
1	3	9.72	0.23	smooth	7.2×10^4	–
2	7	9.72	0.68	smooth	6.3×10^5	–
3	10	9.72	2.0	smooth	5.4×10^6	–
4	13	9.72	2.0	0.00084	5.4×10^6	3700

equivalent sand grain roughness k_s). All physical variables in the model have SI units.

Following simulation of each case, a single output file will be created as explained in full detail in Section 10.3, and organized as indicated in Table 10.1. Following each simulation, it is strongly encouraged that you view an animation of the results e.g. by executing the provided `animateMatRANS.m` script, which also serves as an example on how to access and plot the model results.

Turbulence modelling cases

For the present exercise you will consider a series of four cases summarized in Table 10.4, where the Reynolds number Re is defined according to Eq. 5.1. The model is to be set up to mimic wave boundary layers in DTU's oscillating water tunnel facility, see Fig. 5.21. For the purposes of this exercise, you are to use a model height corresponding to half the physical tunnel height (0.29 m) i.e. $h_m = 0.145$ m, hence it is only the bottom boundary layer that will be simulated (in reality, wave boundary layers will develop at both the top and bottom boundaries in a tunnel environment!). For each case, you are to simulate five wave periods, utilizing only the results from the last period for your analysis. This duration is sufficient to effectively achieve cyclic behavior in the model domain. (Please note that it is implied that the phase starts over at $\omega t = 0°$ at the beginning of each period.)

Data set

Note that each of the selected cases corresponds to experimental conditions considered e.g. in Jensen et al. (1989), hence selected (processed) data from these experiments will be utilized for comparison. The data set is provided in the Matlab workspace file `Exercise4.mat`. This can be obtained as pre-

Table 10.5 Description of contents in Matlab data contained in `Exercise4.mat`. The number of measurement positions for each case n_y varies for each variable and from case to case.

Structure array	Field	Book notation	Size	Description
`WBL`			1×4	Selected data for 4 experiments
	`test`	–	1×1	Test no. from Jensen et al. (1989)
	`T`	T	1×1	Period (s)
	`U0m`	U_{0m}	1×1	Free stream velocity magnitude (m/s)
	`k_s`	k_s	1×1	Bottom roughness (m) (or `smooth`)
	`omegat`	ωt	1×13	Phase angles (°) for velocities
	`y_u`	y	$n_y \times 13$	y positions (m) for $\overline{u}$
	`u`	$\overline{u}(y, \omega t)$	$n_y \times 13$	Ensemble averaged velocities (m/s)
	`y_uuvv`	y	$n_y \times 13$	y positions (m) for $\overline{u'u'}$, $\overline{v'v'}$
	`uu`	$\overline{u'u'}(y, \omega t)$	$n_y \times 13$	Reynolds normal stresses (m^2/s^2)
	`vv`	$\overline{v'v'}(y, \omega t)$	$n_y \times 13$	Reynolds normal stresses (m^2/s^2)
	`y_uv`	y	$n_y \times 13$	y positions (m) for $\overline{u'v'}$
	`uv`	$\overline{u'v'}(y, \omega t)$	$n_y \times 13$	Reynolds shear stresses (m^2/s^2)
	`omegat_tau0`	ωt	varies	Phase angles (°) for τ_0
	`tau0`	τ_0	varies	Bed shear stress (Pa)
	`nu`	ν	1×1	Kinematic water viscosity (m^2/s)
	`rho`	ρ	1×1	Water density (kg/m^3)

sented in Section 10.2. Once loaded, the data is contained in the single Matlab structure array `WBL`, with fields as indicated in Table 10.5, allowing (hopefully) easy and intuitive access for analyzing and plotting. For example, the following Matlab commands:

```
c=3; n=7; WBL(c).omegat(n),
plot(WBL(c).u(:,n),WBL(c).y_u,'o')
```

will: (1) output the phase being considered to the screen (in this example $\omega t = 90°$, since `n=7` and the velocity data is provided at 15° increments starting at 0°) for Case 3 (since `c=3`) and (2) plot the ensemble averaged velocity profile $\overline{u}(y)$ for this case at this phase. Additionally, the time series of the bed shear stress $\overline{\tau_0}(\omega t)$ can be obtained by plotting the field `tau0` versus `omegat_tau0`. All fields included in the `WBL` structure array are intended to be self explanatory and are consistent with the book notation.

Action items

(1) Based on your knowledge of wave boundary layers from Chapter 5, for each of the four cases make an *a priori* estimate of the wave friction factor f_w and the expected maximum friction velocity U_{fm}. For this

purpose it is imperative that you utilize an expression for f_w that is appropriate for the expected flow regime (i.e. laminar, Eq. 5.59; smooth-turbulent, Eq. 5.60; or rough-turbulent, Eq. 5.69). [Note that if a case falls into a transitional regime, e.g. Case 2, then a conservative choice of f_w should be used i.e. one resulting in the largest U_{fm}.]

(2) Use the estimated U_{fm} values to determine the required near-wall grid spacing Δy for each case, such that Eq. 9.49 is satisfied. For Cases 1–3 (smooth beds) also determine an appropriate roughness k_s (model variable `k_s`) such that $k_s^+ = k_s U_{fm}/\nu = 0.1$ is satisfied. Based on your *a priori* calculations, what is the expected maximum roughness Reynolds number k_s^+ for Case 4? Is this case in the hydraulically rough regime? Please list all of the values found for items 1 and 2 in a table.

(3) **Case 1: Laminar wave boundary layer,** $Re = 7.2 \times 10^4$.
In the provided `Case1` directory verify that the model input for Case 1 in the `runMatRANS.m` script is correct and in accordance with your calculations above. Run the simulation simply by executing the `runMatRANS.m` script. Simulation results will be saved into the output workspace file `out_MatRANS.mat`, as designated by the `OutFileName` model input variable. When the simulation is complete be sure to animate the results to make sure everything looks reasonable.

Make plots comparing the computed and theoretical laminar velocity profiles (u/U_{0m} vs. y/a) at four phases: $\omega t = 0°$, $45°$, $90°$, and $135°$, where now $\omega = 2\pi/T$ is the wave angular frequency. Additionally, make a plot comparing the computed bed shear stress time series ($\tau_0/(\rho U_{0m}^2)$ versus ωt) with the laminar theory.

(4) **Case 2: Transitional wave boundary layer,** $Re = 6.3 \times 10^5$.
Now copy your `Case1` directory to a new directory called `Case2`, adjust the necessary model input, and run the simulation for Case 2. After animating the results, make a plot again comparing the computed bed shear stress time series over the final period with the laminar theory, as well as with the provided data of Jensen et al. (1989) (their test 7). Similarly, based on the modeled/measured bed shear stresses, calculate the time evolution of f_w^* defined in Eq. 5.22 for $0° \leq \omega t \leq 130°$. Compare the model results with the constant f_w from laminar theory (Eq. 5.59) as well as against f_w^* based on the measurement data. From these plots, at which phase ωt do you see the laminar-to-turbulent transition begin? How does this compare with expectations based on e.g. Fig. 5.17.

(5) **Case 3: Smooth-turbulent wave boundary layer,** $Re = 5.4 \times 10^6$.
Similarly, create a new `Case3` directory, adjust the necessary model input, and run/animate the simulation for Case 3. Compare the computed velocity ($\overline{u}/U_{0m}$ vs. y/a) and turbulent kinetic energy (k/U_{0m}^2 vs. y/a) profiles at the four phases: $\omega t = 0°$, $45°$, $90°$, and $135°$ with the provided oscillatory tunnel measurements from Jensen et al. (1989) (their test 10). Note that as only two components of velocity were measured, you may use the approximation (Justesen, 1991)

$$k = 0.65\left(\overline{u'^2} + \overline{v'^2}\right) \tag{10.21}$$

to obtain the turbulent kinetic energy (per unit mass) k. Similarly, compare the modelled/measured profiles of the dimensionless Reynolds stress $-\overline{u'v'}/U_{0m}^2$ at the same four phases as above. (Note that this can be computed from the model results as: $-\overline{u'v'} = \nu_t \partial\overline{u}/\partial y$; Hint: use Matlab's `gradient` function to compute the velocity gradient). Finally, make a plot comparing the computed and measured bed shear stress time series ($\tau_0/(\rho U_{0m}^2)$ versus ωt). At approximately what phase ωt do you see laminar-to-turbulent transition occur now? (Hint: It should occur almost immediately!)

(6) **Case 4: Rough-turbulent wave boundary layer,** $Re = 5.4 \times 10^6$.
Now copy your `Case3` directory to a new `Case4` directory, adjust the near-wall grid size and wall roughness k_s, and run/animate the simulation for Case 4. Compare the computed velocity, turbulent kinetic energy, and Reynolds stress profiles at the same four phases as before with the provided oscillatory tunnel measurements from Jensen et al. (1989) (now their test 13). Similarly, compare the computed and measured bed shear stress time series, as before.

(7) Finally, for the specific phase $\omega t = 90°$ create new plots directly comparing the computed velocity and turbulent kinetic energy profiles from Cases 3 and 4. Can you observe/explain the physical effects of the bottom roughness on the flow and turbulence?

As indicated above, all of the preceding simulations are intended to be made utilizing the implemented "low Reynolds number" (transitional) version of the k-ω model (controlled by setting the input variable `turb=2`). Readers interested in additionally comparing against the "standard" (high Reynolds number version i.e. with constant closure coefficients in accordance with Eq. 9.39–9.41) are welcome/encouraged to repeat the simulations of e.g. Cases 3 and 4 with the input simply changed to utilize `turb=1`.

Similarly, the turbulence model can be turned off entirely by setting `turb=0`, which should thus yield very nearly the same results as above for Case 1. (If the models are otherwise properly set up, this should be the only change required!)

10.8 References

(1) Fuhrman, D.R., Schløer, S. and Sterner, J. (2013): RANS-based simulation of turbulent wave boundary layer and sheet-flow sediment transport processes. Coast. Eng., vol. 73, 151–166.
(2) Jensen, B.L., Sumer, B.M. and Fredsøe, J. (1989): Turbulent oscillatory boundary layers at high Reynolds numbers. J. Fluid Mech., vol. 206, 265–297.
(3) Justesen, P. (1991): A note on turbulence calculations in the wave boundary layer. J. Hydraul. Res., vol. 29, 699–711.
(4) Kirca, V.S.O., Sumer, B.M., Steffensen, M., Jensen, K.L. and Fuhrman, D.R. (2016): Longitudinal dispersion of heavy particles in an oscillating tunnel and application to wave boundary layers. J. Ocean Eng. Mar. Energy., vol. 2, 59–83.
(5) Schlichting, H. (1979): Boundary Layer Theory. McGraw-Hill.
(6) Wilcox, D.C. (2006): Turbulence Modeling for CFD, 3rd edn. DCW Industries.
(7) Williams, I.A. and Fuhrman, D.R. (2016): Numerical simulation of tsunami-scale wave boundary layers. Coast. Eng., vol. 110, 17–31.

Author index

Aagaard, T., 461, 465, 470, 497, 573
Abarbanel, H.D.I., 218, 219
Abid, R., 692
Abramowitz, M., 620, 637
Abreu, T., 451, 457, 466, 470
Absi, R., 650
Adityawan, M.B., 497, 573
Adrian, R.J., 13
Afzal, M.S., 460, 465, 466
Aggarwal, A., 687
Ahmed, A.S.M., 464, 466
Akhavan, R, 255, 388
Al-Salem, A.A., 446, 447, 450, 451, 464, 469
Alanso, C.V., 106, 107, 150
Alfredsson, P.H., 250, 394, 545, 573
Allen, J.S., 481, 582
Alsina, J.M., 497, 531, 574
Andersen, K.H., 151, 390, 688
Anderson, D., 689
Anderson, E.C., 692
Anderson, S.L., 496, 576
Anumolu, L., 576
Arbøll, L.K., 393
Archetti, R., 496, 574
Aris, R., 609, 619, 637
Armenio, V., 396
Arnskov, M.M., 296, 306, 308, 388, 398
Arntsen, Ø.A., 687
Asano, T., 296, 388
Aydin, M., 121, 122, 150

Bagnold, R.A., 147, 150, 290, 292, 294, 389
Baird, A.J., 574
Bakhtyar, R., 465, 466
Bakker, W.T., 296, 297, 389, 399
Balachandar, S., 395, 396
Baldock, T.(E.), 465, 496, 497, 520, 531, 573–575, 577, 583, 643
Baldwin, B.S., 262, 277, 278, 289, 290, 398
Baltzer, J.R., 13
Barnes, M.P., 496, 531, 574
Barry, D.A., 466
Barthelemy, E., 574, 578
Batchelor, G.K., 57, 63, 147, 150, 223, 228, 235, 241, 389, 598, 600, 609, 637
Battjes, J.A., 496, 501, 516, 574
Bayazit, M., 85, 86, 90, 150, 666, 687
Baykal, C., 394, 577, 668, 670, 687, 688, 690
Beach, R.A., 390
Belden, J., 561, 574
Berberovic, E., 503, 558, 574
Berni, C., 497, 574
Bihs, H., 575, 670, 687
Bijker, E.W., 296, 389, 427, 466
Binder, G., 399
Blackwelder, R.F., 40, 391, 399

Blondeaux, P., 264, 334, 337, 340, 341, 343–345, 354, 389, 390, 392, 395, 399, 434, 435, 439, 440, 463, 465, 466, 470
Bonamy, C., 687
Bose, S.K., 145, 150
Boussinesq, J., 648, 687
Bowen, A.J., 471
Bradford, S.F., 549, 555, 575
Branger, H., 496, 561, 580
Bredmose, H., 691
Brocchini, M., 466, 496, 574–576
Brodkey, R.S., 118–120, 124, 125, 150, 152–154
Brooks, N.H., 638
Browand, F., 21, 39, 250, 389
Brown, S.A., 481, 549, 555, 567, 569, 575, 670, 687
Bruschi, A., 466
Brøker, I.H., 376, 389, 421, 423, 466
Buchhave, P., 181, 219
Buhr Hansen, J., 496, 575
Buriani, F., 576
Buscombe, D., 393
Butt, T., 496, 575, 585

Ca, V.T., 578
Cacchione, D.A., 391
Caceres, I., 574, 577, 585
Caliskan, U., 380, 389, 465, 466
Callaghan, D.P., 465, 469
Callander, R.A., 187, 220, 225, 232, 234, 235
Cantwell, B.(J.), 17–19, 21–23, 39, 110, 150
Carlsen, N.A., 294, 392
Carstensen, S., 9, 12, 26, 39, 123, 150, 245, 246, 248–256, 259–263, 289, 334–340, 344, 351, 389, 393, 398, 499, 575, 584
Cavallaro, L., 454, 466
Cebeci, T., 100, 105, 106, 150, 654, 667, 687
Ceren, B., 584
Certain, R., 578
Cevik, E.O., 586
Chambers, F.W., 21, 39, 250, 389
Champagne, F.H., 206, 207, 220
Chan, I.-C., 359, 362, 365, 389
Chang, K.-A., 496, 561, 575, 579, 581
Chang, K.C., 100, 105, 106, 150, 654, 667, 687
Chang, S.-C., 468
Chang, Y.-H., 531, 575, 578
Chanson, H., 496, 575
Chauchat, J., 468, 670, 687, 689
Chella, M.A., 481, 549, 575, 687
Cheng, N.-S., 147, 150, 438, 470, 583, 692
Cheng, Z., 468, 687, 689
Christensen, E.D., 394, 481, 484, 496, 503, 549, 569, 575, 579, 581, 680, 688, 690
Christiansen, N., 398
Christoffersen, J.B., 297, 390
Chua, L.H.C., 391, 583, 692
Clamond, D., 401, 410
Coakley, T.J., 660, 688
Cokgor, S., 153, 692
Coleman, N.L., 106, 107, 150
Coles, D., 39, 271, 390
Combs, G.D., 298, 303, 304, 391
Conley, D.C., 497, 575, 576, 687
Constantinescu, G., vi
Contestabile, P., 574
Corino, E.R., 124, 150
Corvaro, S., 576
Costamagna, P., 255, 264, 345, 390
Cote, R., 220
Cowen, E.A., 394, 496, 576, 580, 583
Cox, D.(T.), 496, 574, 576, 583, 586, 689, 692
Crise, A., 396

Dalrymple, R.A., 479, 480, 501, 549, 576
Darrigol, O., 63
Dash, S.M., 369, 400
Davidson, P.A., 650, 688
Davies, A.G., 297, 313, 318, 390, 397, 399, 446, 448, 450, 466, 470

De Padova, D., 549, 576
De Rouck, J., 670, 692
De Serio, F., 496, 576
de Vriend, H.J., 481, 583
Dean, R.G., 401, 411, 479, 480, 576
Defina, A., 122, 123, 151
Deigaard, R., v, 124–131, 133–135, 137, 139, 140, 153, 178, 220, 282, 287, 290, 294, 321, 330, 376, 390, 391, 398, 423, 465–467, 481–483, 485, 489, 491, 496, 497, 503, 516–518, 575–577, 584, 634, 636, 638, 641–643
del Jesus, M., 670, 688, 690
Deshpande, S.S., 558, 576
Detert, D., 122, 151
Devolder, B., 549, 561, 568, 577, 670, 688
Dey, S., 145, 150
Didenkulova, I., 396
Dimotakis, P., 39
Dixen, M., 264, 290–295, 390, 667, 668, 670, 688
Dohmen-Janssen, C.M., 464, 465, 467
Drake, D.E., 391
Dryden, H.R., 203, 205, 220
Duncan, J.H., 481, 484, 561, 577
Dutykh, D., 401, 410

Eckelman, L.D., 152
Eckelmann, H., 110, 111, 150, 151, 154, 163–166, 220, 545, 573, 577, 665, 688
Eckmann, D.M., 302, 390
Elder, J.W., 17, 39, 616, 638
Elfrink, B., 458, 467, 496, 577
Emerat, N., 575
Emmons, H.W., 15–17, 21, 22, 39, 246, 390
Engelund, F., 85, 130, 137, 151
Evans, D., 582

Falchetti, S., 574
Faraci, C., 396, 470
Fayler, L., 574
Feng, T., 496, 583
Fenton, J.D., 401, 410, 559, 577
Ferziger, J.H., 670, 678, 688
Finnicum, D.S., 298, 390
Fischer, H.B., 623–625, 638
Fornarelli, F, 264, 390
Foster, D.L., 255, 390
Foti, E., 396, 466, 470
Fredsøe, J., v, 10–13, 39, 85, 137, 144, 150, 151, 153, 262, 264, 282, 287, 289, 290, 292–294, 297, 311–313, 318, 320–325, 328, 329, 345, 388–392, 394, 398, 423, 435, 438–442, 461, 462, 466–468, 470, 481–483, 489, 491, 501, 575–577, 579, 581, 583, 584, 623, 634, 636, 638, 639, 641–643, 666, 670, 687–692, 714
Frigaard, S., 393
Fromant, G., 572, 577, 585
Fuhrman, D.R., 294, 357, 359–363, 366–370, 372, 380, 383, 384, 386, 387, 389, 391, 393, 394, 400, 401, 411, 421, 422, 424, 425, 429, 432–444, 446, 450, 452, 459, 463–467, 478, 481, 493, 497, 530, 534, 547, 549–551, 553–564, 566–569, 572, 573, 577, 579–581, 583, 584, 629, 638, 642, 643, 660, 661, 664–668, 670–673, 687–690, 692, 696, 708, 714
Fyfe, A.J., 395

Gad-el-Hak, M., 17–21, 40, 250, 391
Galvin, C.J., 480, 578
Garcez Faria, A.F., 481, 578
Garcia, M.H., 125, 135, 152, 330, 395, 670, 690
Garcia, N., 581
Gatin, I., 689
Gatski, T.B., 586
Ghaheri, A., 466
Gilbrech, D.A., 298, 303, 304, 391
Gjevik, B., 529, 582
Glenn, S.M., 391
Goseberg, N., 356, 391

Gradshteyn, I.S., 613, 638
Grant, W.D., 264, 296, 297, 311, 313, 318, 391
Grass, A.J., 72, 83, 84, 91, 110, 114–117, 120–124, 130, 135, 151, 163, 164, 220, 330, 391, 397, 437, 467
Grasso, F., 497, 578
Gravesen, H., 130, 151
Greaves, D.M., 575, 687
Green, A.E., 198, 221
Greenwood, B., 465, 573
Griffin, J.G., 497, 576
Grilli, S.T., 534, 578
Grotberg, J.B., 302, 390
Grue, J., 561, 578
Guanche, R., 580
Guannel, G., 481, 493, 578
Guard, P.A., 583
Gudmestad, O.T., 690
Guner, H.A.A., 583

Haddadian, S., 469
Hager, W., 578
Hanes, D.M., 446, 464, 465, 467, 468, 471
Hanratty, T.J., 152, 298, 390, 394, 545, 582
Hansen, E., 85, 151
Hansen, N.M., 583
Hara, T., 391
Haritonidis, J.H., 573
Harlow, F.H., 548, 578
Hassan, W.N.(M.), 446, 464, 465, 467, 469
Hatipoglu, F., 390
Hay, A.E., 434, 471
Hazen, D.G., 297, 391
Heller, V., 531, 578
Hickey, J.-P., 41, 154
Hieu, P.D., 549, 578
Higuera, P., 531, 536–538, 578, 670, 689
Hino, M., 255, 262, 264, 265, 277, 289, 290, 295, 326–328, 391, 582
Hinze, J.O., 110, 151, 174, 187, 188, 220, 227, 229, 230, 235
Hirsch, C., 670, 689
Hirt, C.W., 503, 549, 578
Holman, R.A., 390
Holmedal, L.E., 376, 391, 423, 446, 449, 450, 460, 465–468, 470, 691
Hondebrink, L., 582
Honji, H., 503, 523, 582
Horn, D.P., 574
Hosking, R.J., 585
Howarth, L., 174, 221, 225, 235
Hsiao, S.-C., 531, 578, 579, 581
Hsieh, C.M., 579
Hsieh, S.-C., 468
Hsu, T.-J., 395, 396, 446, 465, 468, 471, 503–506, 549, 579, 580, 586, 682, 687, 689, 692
Hsu, T.W., 549, 555, 578, 579
Huang, P.G., 660, 689
Huang, Z.-C., 481, 496, 579
Hughes, M.(G.), 465, 573, 582, 643
Hulscher, S.J.M.H., 468
Humphery, J.D., 297, 398
Hunt, I.A., 500, 579
Hunt, J.C.R., 199, 220
Huntley, D.A., 297, 391
Hurther, D., 577, 585
Hussain, A.K.M.F., 298, 398
Hussain, F., 503, 579
Huynh-Thanh, S., 297, 318, 392
Hwang, K.-S., 575
Hwung, H.-H., 575, 579, 581

Imberger, J., 638
Infantes, E., 469
Iwagaki, Y., 388
Izumi, Y., 220

Jørgensen, F.E., 398
Jackson, R.G., 125, 130, 151
Jacobsen, N.G., 481, 496, 503, 505, 549, 558, 567, 577, 579, 670, 687–689
Jakirlic, S.I., 574
Jakobsen, J.B., 390, 466
Jansen, P.C.M., 501, 579
Jasak, H., 670, 689, 691, 692

Jaw-Fang, L., 496, 575
Jensen, A., 531, 561, 578, 579
Jensen, B., 524–530, 579
Jensen, B.L., 254–259, 261–268, 270, 271, 273–278, 282–290, 294, 298, 320, 328, 339, 347–349, 351, 356, 369, 380–385, 392, 398, 499, 579, 629, 638, 677, 678, 680–683, 689, 696, 708, 710–714
Jensen, J.H., 579
Jensen, J.J., 392
Jensen, K.L., 140, 151, 245, 247, 276–281, 330, 392, 393, 638, 714
Jensen, P.M., 398, 584
Jeong, J., 503, 579
Jimenez, J., 195, 220
Jirka, G.H., 151
Johansson, A.V., 573
Johari, J., 467
Jonsson, I.G., 264, 265, 287, 290, 292, 294, 295, 297, 390, 392, 394
Jordinson, R., 32, 33, 40
Jose, J., 690
Justesen, P., 264, 295–297, 376, 380, 382, 392, 393, 434, 468, 576, 680, 689, 713, 714

Kaimal, J.C., 203, 220
Kajiura, K., 292, 393
Kalkwijk, J.P.T., 466
Kamath, A., 687
Kamm, R.D., 388
Kamphuis, J.W., 85, 152, 262, 289, 290, 293–295, 393
Kao, M.-J., 468, 578
Kapdasli, S., 586
Karagali, I., 584
Karjadi, E.A., 534, 580
Kashiwayanagi, M., 391
Kato, F., 585
Katsutohi, T., 578
Katz, J., 675, 691
Kawamura, I., 399
Kefelegn, H., 469
Kemp, P.H., 264, 296, 311, 393
Keulegan, G.H., 331, 393
Killian, J., 574
Kim, J., 684, 689
Kim, Y., 465, 468, 503–506, 579, 580, 670, 682, 689
Kimmoun, O., 496, 561, 580
King, H.L., 390
Kirby, J.T., 481, 493–496, 531, 559–563, 567, 569, 572, 581, 584
Kirca, V.S.O., 380, 393, 628–635, 638, 696, 703, 708, 714
Kit, E., 397
Klebanoff, P.S., 17, 40, 78, 152
Kline, S.J., 38, 40, 112, 114, 116, 125, 152, 153
Klopman, G., 324, 325, 393, 397
Klostermann, J., 503, 580
Kobayashi, N., 496, 531, 534, 576, 580
Koh, R.C.Y., 638
Kolmogorov, A.N., 193, 220, 655, 690
Koyano, Y., 582
Kozakiewicz, A., 153, 391, 584
Kranenburg, W.M., 460, 465, 468
Kreplin, H.P., 164, 166, 220
Krestenitis, Y.N., 582
Kroekenstoel, D., 467
Kroon, A., 465, 573
Kyriacou, A., 397

Lacy, J.R., 370–372, 393
Lakehal, D., 496, 580
Lalli, F., 466
Lalovic, I., 689
Lanckriet, T., 582
Lara, J.L., 496, 531, 549, 580, 581, 670, 688–691
Larsen, B.E., 360, 361, 367, 369, 393, 394, 478, 481, 493, 497, 549–551, 553–564, 566–573, 580, 581, 585, 660, 668, 670, 690
Laufer, J., 78, 80, 94, 110, 117, 152
Launder, B.E., 7, 12, 555, 581, 655, 690
Laursen, T.S., 391, 398, 435, 467, 468, 470

Lawrence, A.R., 531, 580
Lee, M., 684, 690
Lee, M.K., 124, 152
Lemos, C.M., 548, 581
Leutheusser, H.J., 121, 122, 150
Li, J., 497, 531, 581
Li, M., 585
Li, Y., 531, 581
Li, Z., 446, 449, 450, 466
Lim, T., 396
Lin, C., 434, 468, 578
Lin, P., 481, 496, 531, 549, 555, 573, 581
Lin, T.-C., 481, 578, 581
Liovic, P., 496, 580
Lippmann, T.C., 578
List, E.J., 638
Liu, P.L.-F., 244, 331, 334, 339, 347–352, 359, 362, 365, 387–389, 394, 396, 398, 481, 496, 549, 555, 561, 575, 576, 578–581, 583, 584, 639
Liu, S., 670, 690
Liu, X., 586, 670, 690, 692
Lo, S.-M., 581
Lodahl, C.(R.), 9, 12, 254, 255, 264, 297–299, 301–305, 307, 309–316, 323, 339, 345, 394, 399, 499, 546, 581, 631, 638
Lohmann, I.P., 264, 394, 675, 677–683, 690
Long, J.W., 583
Longo, S., 496, 581
Longuet-Higgins, M.S., 406, 411, 417, 420, 468, 637, 638
Losada, I.J., 549, 580, 581, 645, 688–691
Lozano-Duran, A., 41, 154
Luhar, M., 463, 469
Lumley, J.L., 133, 152, 194, 196, 221, 648, 692
Lundgren, H., 264, 265, 311, 394, 406, 411

Ma, G., 496, 581
Maddux, T.(B.), 574, 583, 586
Madsen, O.S., 264, 296, 311, 313, 318, 376, 391, 399, 448, 470
Madsen, P.A., 357, 359, 362, 367, 394, 401, 411, 530, 534, 547, 549, 551–555, 573, 577, 581, 582, 670, 691
Magar, V., 575, 687
Mahesh, K., 648, 691
Makris, C.V., 549, 582
Malenica, S., 692
Manoonvoravong, P., 574
Mansour-Tehrani, M., 151, 397
Mantz, P.A., 144, 145, 152
Mao, Z., 298, 394
Mason, T., 574
Masselink, G., 496, 497, 582
Matsubara, M., 250, 394
Matsunaga, N., 503, 523, 582
Mayer, S., 549, 551–555, 573, 582, 670, 691
Maza, M., 580, 670, 691
Mazzuoli, M., 246, 248, 250, 252, 256, 260, 395
McLelland, S.J., 585
Mei, C.C., 331, 395, 639
Memos, C.D., 582
Meneveau, C., 675, 691
Menter, F.R., 555, 582, 655, 691
Michallet, H., 574, 578
Michelsen, J., 691
Mieras, R.S., 689
Mikkelsen, M.B., 576
Miller, R.L., 501, 582
Mitchell, J.E., 545, 582
Mizutani, T., 583
Moin, P., 13, 40, 41, 154, 220, 648, 689, 691
Monin, A.S., 7, 12, 23–25, 40, 74, 76, 78, 84, 85, 94, 95, 152, 600, 638
Moser, R., 684, 689, 690
Mossa, M., 496, 576
Muskulus, M., 575
Myrhaug, D., 295–297, 376, 391, 395, 397, 423, 446, 449, 450, 460, 465–468, 470, 575, 691

Nadaoka, K., 496, 499, 582
Nagel, T., 687
Nakagawa, H., v, 110, 152
Nakagawa, M., 388
Nakayama, A., 391
Navier, C.L.M.H., 63
Nelson, J.R., 481, 496, 584
Nepf, H.M., 469
Newberger, P.A., 481, 582
Nezu, I., v, 99, 110, 152, 666, 691
Nichols, B.D., 503, 549, 578
Nielsen, P., v, 286, 287, 294, 324, 370, 395, 456, 465, 469, 643, 650, 691
Nikora, V., 151
Nikuradse, J., 82–85, 102, 108, 152
Nino, Y., 125, 135, 152, 330, 395
Nychas, S.G., 124, 125, 152

O'Donoghue, T., 399, 446, 464, 465, 469, 470, 496, 574, 577, 582, 585
Obhrai, C., 690
Offen, G.R., 38, 40, 112, 125, 152, 153
Oguz, B., 126–132, 135, 142, 153, 178, 220, 330, 398
Okayasu, A., 576
Oliveira, D., 469
Ong, M.C., 670, 690, 691
Orfila, A., 331, 394, 469
Orszag, S.A., 586
Osorio, J.D.C., 469
Ou, S.H., 579
Ozdemir, C.E., 252, 340, 341, 395, 396, 454, 469
Ozkan-Haller, H.T., 481, 493, 578

Park, G.I., 39–41
Park, Y.S., 331, 394, 396
Pedersen, C., 391, 467
Pedersen, F.B., 625, 638
Pedersen, G.(K.), 396, 529, 579, 582
Peric, M., 670, 678, 688
Perry, A., 250, 396
Pesarino, V., 466
Petrotta, C., 585
Pettersen, B., 691
Petti, M., 581
Pham, K.S., 574
Philomin, V., 577
Pieter, T., 466
Plocher, D.A., 21, 39, 250, 389
Pokrajac, D., 582, 585
Pope, S.B., 206, 220, 661, 691
Power, H., 465, 573, 574
Pralitis, J., 389
Prandtl, L., 7, 12, 648, 650, 651, 676, 691
Praturi, A.K., 125, 153
Prevost, J.H., 586
Puleo, J.A., 496, 580, 582, 585, 689

Qi, M., 581
Qiao, H.B., 577

Rabinovich, A.B., 362, 396
Raichlen, F., 206, 207, 220, 531, 581
Ramaprian, B.R., 298, 304, 396, 399
Raubenheimer, B., 576
Raudkivi, A.J., 187, 220, 225, 232, 234, 235
Rauwoens, P., 577, 688
Reed, K., 395
Reniers, A.J.H.M., 481, 582
Reynolds, O., 1, 2, 7, 8, 12, 13, 44, 52
Reynolds, W.C., 40, 152
Ribberink, J.S., 399, 446, 447, 450, 451, 463–465, 467–470, 577, 585
Richardson, J.F., 643
Richardson, L.F., 190, 220
Riley, J.J., 40, 391
Rodi, W., v, vi, 99, 152, 645, 691
Roelvink, J.A., 582
Roenby, J., 581, 670, 690, 691
Rogallo, R., 648, 691
Rogers, B.D., 501, 549, 576
Roisman, V., 574
Rotta, J.C., 102, 104–106, 153
Roulund, A., 667, 668, 691
Rubin, D.M., 393
Ruessink, B.G., 446, 467, 469, 470

Ruju, A., 580
Runstadler, P.W., 40, 152
Russell, P.(E.), 496, 575, 582, 585
Russell, R.C.H., 469
Ryzhik, I.M., 613, 638

Sørensen, L.B., 398, 584
Sørensen, T., 264, 265, 394
Saeki, H., 549, 585
Saffman, P.G., 655, 691
Sakai, T., 548, 583
Sakakiyama, T., 579
Saleh, W.M., 397
Salon, S., 264, 396
Sana, A., 399
Sanchez-Arcilla, A., 574
Sancho, F., 466, 470
Sangermano, J., 586
Sarpkaya, T., 262, 298, 303, 304, 396
Sato, S., 464, 466, 585
Sawamoto, M., 391
Sayre, W.W., 609, 638
Scandura, P., 250, 252, 340, 341, 396, 454, 460, 466, 470, 585
Schaake, K., 580
Schimmels, S., 356, 362, 364, 366, 368, 396
Schlichting, H., 7, 8, 13, 23–25, 27, 29–36, 40, 56, 61, 63, 82, 107–109, 153, 260, 302, 313, 314, 396, 700, 714
Schlurmann, T., 391
Schløer, S., 391, 467, 638, 643, 688, 714
Schraub, F.A., 40, 152
Schretlen, J.J.L.M., 468
Schubauer, G.B., 17, 34, 40
Schwarze, R., 580
Schäffer, H.A., 357, 394, 581
Scott, C.P., 563, 583
Seelam, J.K., 496, 531, 583
Seifert, A., 250, 396
Sen, M.B., 584
Sergeev, S.I., 302, 397
Shao, S., 549, 583
Shapiro, A.H., 388
Sharma, B.I., 555, 581, 655, 690
Sharma, J.N., 401, 411
Shemer, L., 298, 397
Shi, F., 581
Shields, A., 144, 145, 153
Sibilla, S., 576
Silva, P.A., 464, 466, 470
Simons, R.(R.), 264, 290, 292, 293, 296, 311, 327, 328, 393, 397, 400
Skarda, J., 41, 154
Skjelbreia, J.E., 530, 584
Skramstad, H.K., 34, 40
Slaattelid, O.H., 297, 395
Sleath, J.F.A., v, 254, 262, 264, 265, 268, 275, 286, 287, 289, 290, 293, 294, 296, 297, 397
Smagorinsky, J., 503, 583, 677, 692
Sororian, S., 469
Sottile, M., 584
Sou, I.M., 496, 576, 583
Soulsby, R.L., 264, 297, 311, 316–318, 371, 390, 397, 398
Spalart, P.R., 262, 277, 278, 289, 290, 398
Speziale, C.G., 438, 445, 470, 586, 659, 692
Squire, H.B., 30, 40
Sriram, V., 396
Stansby, P.K., 496, 583
Stanton, T.P., 578, 582
Steffensen, M., 393, 638, 714
Stegun, I.A., 620, 637
Sterner, J., 391, 467, 638, 643, 688, 714
Stettler, J.C., 298, 398
Stewart, R.W., 406, 411
Stive, M.J.F., 481, 486, 496, 583
Stoesser, T., vi
Stokes, G.G., 61, 63, 642, 643
Stuart, R.J., 151
Subramanya, R., 578
Sullivan, P.J., 618, 638
Sumer, B.M., 10–13, 39, 80, 86, 90, 124–140, 142–145, 150, 151, 153, 178, 220, 246, 264, 271, 272, 276, 280, 283, 290, 292–295, 320, 328–

330, 334–356, 384, 386–394, 398, 399, 432–434, 461, 462, 467, 468, 470, 497–523, 525, 527, 529–537, 539–548, 575, 577, 579, 581, 583, 584, 609, 610, 612–614, 617–623, 638, 639, 666, 687–692, 714
Sutherland, A.J., 130, 153
Svendsen, I.A., 406, 407, 411, 479, 481, 482, 484, 491, 494, 578, 584
Swart, D.H., 294, 398
Synolakis, C., 530, 584

Tada, Y., 583
Takasu, S., 391
Tanaka, H., 244, 331–333, 339, 362, 399, 573, 583
Tardu, S.F., 298, 399
Taylor, G.I., 175, 178, 198, 220, 221, 225, 234, 235, 600, 609, 639
Teakle, I.A.L., 650, 691
Techet, A.H., 561, 574
Teh, E., 396
Tehrani, M.M., 397
Temperville, A., 297, 318, 392, 466
Tennekes, H., 194, 196, 221, 648, 692
Terrados, J., 469
Thangam, S., 586
Thomas, A.S.W., 21, 39, 250, 389
Thomas, G.P., 398
Thomson, R.E., 362, 396
Thornton, E.B., 578, 582
Ting, F.C.K., 481, 493–496, 531, 559–563, 567, 569, 572, 584, 586
Tollmien, W., 31–33, 35, 40
Tomkins, M.R., 643
Tonkin, S., 531, 585
Townsend, A.A., 598, 600, 637
Troch, P., 577, 670, 688, 692
Tromans, P.S., 254, 298, 301, 303, 399
Tropea, C., 574
Trowbridge, J., 376, 399, 448, 470
Trujillo, M.F., 576
Tsai, C.-P., 468
Tsai, C.Y., 579
Tu, S.W., 298, 304, 396, 399
Turner, I.L., 497, 585
Turnipseed, C.N., 469
Tuzi, R., 345, 399
Tyagi, M., 469

Uittenbogaard, R.E., 468
Unger, J., 578
Utnes, T., 691

van den Berg, T.J.J., 469
van der A, D.A., 369, 399, 459, 464, 470, 561, 572, 577, 585
van der Werf, J.J., 469, 470
van der Zanden, J., 572, 577, 585
van Doorn, T., 296, 389
van Driest, E.R., 99, 154, 269, 353, 399, 652, 692
van Dyke, M., 32, 33, 40
van Hinsberg, N.P., 574
van Kesteren, W.G.M., 297, 399
van Rijn, L.C., 469
Velte, C.M., 181, 219
Verschaeve, J., 396
Verzicco, R., 255, 264, 399
Vested, M.H., 484, 585
Vicinanza, D., 574
Vittori, G., 252, 255, 260, 264, 334, 337, 340, 341, 343–345, 354, 389, 390, 392, 395, 399, 435, 439, 440, 463, 466, 470
von Karman, T., 174, 203, 221, 225, 235
Vukcevic, V., 670, 689, 692

Wallace, J.M., 39–41, 118, 150, 154
Walstra, D.-J., 575
Wang, H., 461–463, 470
Wang, P., 582
Wang, Z.-Q., 438, 470
Watanabe, Y., 496, 549, 585
Wei, T., 96, 97, 154
Welch, J.E., 548, 578
Wenz, A., 577
Werle, H., 33, 41
Westfall, P.H., 162, 221

Wilcox, D.C., 195, 221, 357, 378, 380, 383, 385, 400, 421, 437, 438, 471, 549–551, 553–557, 560–562, 566, 568, 572, 586, 645, 655, 659–661, 664, 665, 670, 671, 674, 692, 694, 708, 714
Williams, A.J., 391
Williams, I.A., 357, 359, 362, 363, 366, 368, 370, 372, 380, 383, 384, 386, 387, 400, 661, 670, 692, 696, 708, 714
Willmarth, W.W., 96, 97, 154
Winckler, P., 636, 637, 639
Wind, H.G., 486, 583
Wong, W.-Y., 578
Wood, D.J., 579
Wray, A.A., 220
Wright, S., 446, 464, 465, 469
Wu, H.-L., 581
Wu, T.-R., 468
Wu, X., 8, 13, 35–41, 110, 111, 154
Wurpts, A., 391
Wygnanski, I.(J.), 250, 396, 397
Wyngaard, J.C., 220

Xiao, H., 531, 586
Xie, Z., 549, 586

Yaglom, A.M., 7, 12, 23–25, 40, 74, 76, 78, 84, 85, 94, 95, 152, 600, 638
Yakhot, V., 555, 586
Yamaji, H., 399
Yeganeh-Bakhtiary, A., 466
Yeh, H., 585
Yeh, P.-H., 468
Young, Y.L., 531, 586
Yu, M.-H., 468
Yu, X., 465, 469, 471
Yuan, J., 369, 400
Yuksel, Y., 497, 586

Zaki, W.N., 643
Zhang, X., 327, 328, 400
Zhou, Z., 503, 580, 586, 670, 692
Zhu, L., 469
Zilioli, L., 584
Zou, Q., 434, 471

Subject index

adverse pressure gradient, 107, 243, 259, 264, 271, 333, 339, 344, 346, 383, 430, 433, 656
aliasing, 206, 209
Aris' moment method (in analysis of turbulent dispersion), 619

basic equations (of turbulent flows), 43, 66, 646, 656
bed shear stress, 6, 69, 674, 697
  in breaking waves, 478, 490, 494, 496, 497, 499, 500, 508–514, 516, 518, 521, 523, 531, 532, 536, 539–546
  in tsunamis, 359, 360, 363–372
  in wave boundary layers, xxv, 244, 247, 248, 251, 255–259, 262, 263, 265–267, 269, 273, 279, 283, 288, 292–294, 297, 306, 311, 315–318, 323, 328–331, 333–337, 339–347, 349, 350, 352, 354, 383, 387, 388, 417, 418, 420–423, 427, 428, 430–432, 434, 440–443, 446–448, 452–458, 464, 465, 711–713
boundary conditions, 30, 31, 67, 69, 70, 93, 241, 379, 402, 503, 610, 613, 618, 619, 662–666, 669, 671, 674, 686, 709
Boussinesq approximation, 373, 437, 486, 550, 657, 658, 669, 675, 681
breaking waves, 2, 3, 11, 12, 460, 477–482, 484, 485, 489, 495, 496, 500, 519, 524, 526, 548–550, 555, 558, 559, 567, 569, 572, 573, 656, 660, 668, 680, 682
  plunging, 12, 477–481, 485, 495, 497–503, 509, 516–519, 526, 527, 529–534, 536, 540, 546–548, 550, 567–573, 682
  sediment transport and morphology, 477, 481, 496, 497, 531
  spilling, 12, 477–485, 489, 493–495, 503, 516, 517, 534, 550, 555, 559, 560, 562–565, 567–569, 572, 573
  surging, 12, 477–480, 503, 524, 526, 528–530, 535
buffer layer, 38, 75, 79, 92, 97, 99, 100, 103, 656, 661, 671, 700
buoyancy production, 550, 551, 555, 560, 561, 568
bursting process, 65, 109–112, 114, 116, 117, 120, 121, 123, 126, 127, 131, 132, 134, 140, 146, 159, 164, 462
  implications for sediment transport, 126
  implications for sediment transport, 126, 131, 134, 140, 146

in wave boundary layers, 281, 326, 327
over rough wall, 120
over rough walls, 120, 121, 123

changing bottom roughness, 434
closure problem of turbulence, 47, 645, 646
combined wave and current boundary layer, 296, 316
computational fluid dynamics (CFD), v, xxvi, 367, 478, 493–495, 503, 548, 550, 554, 555, 558, 559, 567, 572, 573, 668, 669, 674, 677
constant stress layer (or region), 69–71, 73, 75, 89, 92, 97, 141, 146, 149, 271, 272, 648, 651, 653, 654, 671, 686, 700
constitutive equation (for an incompressible Newtonian fluid), 46, 61–63
continuity equation, 29, 43, 45–47, 53, 54, 61, 66, 240, 376, 402, 414, 420, 422, 425, 558, 603, 646, 647, 668, 675
converging-diverging oscillatory flow, 413, 424, 425
correlation
correlation functions, 159, 172–175, 177, 178, 180–182, 187, 202–206, 215, 218, 219, 223, 224, 226, 231, 707
two-point double correlation, 172, 173, 184
critical Reynolds number, 23, 26, 30, 33, 34, 255, 260, 262, 330, 348, 542

deep water approximations, 408
defect layer, 437, 661
diffusion, xxi, xxiv, 79, 355, 562, 591–593, 595, 596, 599, 600, 602–608, 610, 611, 615–618, 622, 624, 626, 627, 636, 637, 639, 650, 657, 659, 701
one-particle analysis, 592
dimensional analysis, 67, 68, 73, 77, 82, 90, 92, 93, 155–157, 192, 615, 617, 630, 659
Dirac delta function, xxiv, 208, 599, 610, 619
direct numerical simulation (DNS), xxvi, 5, 8, 35, 37, 39, 110, 198, 199, 245–248, 250, 252, 255, 256, 260, 262, 277, 278, 280, 289, 290, 327, 334, 341, 345, 354, 388, 454, 460, 645–648, 661, 662, 683–687
dispersion
longitudinal dispersion, xxii, 591, 604–609, 615, 617–619, 631, 703, 705
longitudinal dispersion in an oscillating tunnel or waves, 625–628, 630–634, 636
longitudinal dispersion in rivers, 623–625
one-particle analysis, 606, 607, 609
dissipation range, 203, 205, 206, 707
drag coefficient, xxi, 147, 641, 642
drawdown (or rundown), 477, 500, 501, 506, 509, 513, 519, 520, 526, 529–532, 535–538, 541, 543–546, 548
Dryden spectrum, 182, 203–205, 215

eddy viscosity, xxiv, 98, 313, 355, 378, 379, 383, 384, 436, 448, 449, 478, 492–494, 549–552, 554, 561, 562, 566–568, 573, 637, 647, 648, 650, 651, 657, 659, 660, 662, 668, 669, 675, 676, 697, 709
Einstein's summation (convention), 45, 54, 62, 63
energy cascade, 190, 195, 197, 198
energy equation, 43, 47–50, 78, 645, 647, 655, 656, 680, 681, 683, 685–687, 693, 694, 701
energy flux, 50, 406, 409, 480, 482, 483, 546
ensemble averaging, xxv, 43, 44, 267, 300, 305, 335, 388, 486, 506,

509, 512, 594, 680
equation(s) of motion, 43, 45, 46, 48, 55, 56, 63, 136, 140, 240, 675
Eulerian (frame of reference), xxiii, xxv, 176–178, 181, 182, 210, 330, 404, 414, 417, 419, 420, 591, 595–597, 602, 607, 707

fast Fourier transform (FFT), xxvi, 210, 211, 214, 218, 707
favourable pressure gradient, 107, 271, 345, 430, 433
flow around a circular cylinder, 3, 9
flow over a flat plate, xxi, xxiii, 3, 4, 7, 25, 31, 35, 39, 65
flow resistance, 25, 369, 622, 700
  in breaking waves, 478, 515, 546, 547, 566
  in combined wave and current boundary layers, 316
  in solitary wave boundary layer s, 347
  in solitary wave boundary layers, 347, 354
  in steady boundary layer flows and in pipes and channels, 107
  in wave boundary layers, 6, 237, 261, 262, 283, 288–290, 293, 314
Fourier transform, xxvi, 184, 187, 209, 210, 218, 219, 707
Froude number, xxii, 404
frozen turbulence approximation, 178–180, 184, 201, 215, 685, 707

homogeneous turbulence, 36, 174, 223, 225, 226, 232, 234, 596, 661
hydraulic jump, 529, 532, 535–538, 541, 543–546
hydraulically rough, 82, 282, 359, 420, 423, 431, 462, 665, 666, 668, 671, 673, 712
hydraulically smooth, 82, 101, 121, 141, 282, 321, 356, 359, 414, 422, 429, 431, 517, 518, 527, 622, 654, 664–666, 671, 672
hydrodynamic stability theory, 5, 7, 8, 27, 31, 32, 34, 35

inertial subrange, 176, 191, 193, 195, 198, 201, 203, 205, 206, 218, 685, 707
initiation of suspension, xxiv, 65, 126, 141, 142, 144, 145, 148, 623
isotropic turbulence, 36, 159, 173–176, 184–188, 191, 192, 200, 201, 203, 204, 206, 223–226, 228, 230–234, 661, 677, 685, 707

Kaimal spectrum, 203, 215
Kolmogorov scale(s), 193, 195, 198, 203, 205, 683
  Kolmogorov length scale, xxiv, 192, 194, 684, 707
  Kolmogorov time scale, xxv, 192, 684, 707
  Kolmogorov velocity scale, xxv, 192
Kolmogorov's $-5/3$ law, 193, 201, 203, 205, 206, 218
Kronecker delta, xxiv, 45, 62, 436
kurtosis, 162–164, 166, 168, 707

Lagrangian (frame of reference), xxiii, 176–178, 181, 419, 435, 591, 594, 595, 597, 607, 650
large eddy simulation (LES), vi, xxvi, 503–506, 549, 569, 645, 647, 648, 662, 675–677, 679–686
law of the wall, 68, 70–72, 75, 157
lift coefficient, xxi, 139
logarithmic layer (or region), 73, 75, 89, 92, 93, 97, 100, 103, 266, 270, 271, 275, 276, 284, 383, 652, 653, 661, 667, 668, 674, 686, 700, 702, 703

macro scale(s), xxiv, 176, 184, 204
  macro length scale, 176, 178, 179, 181, 650
  macro time scale, 165, 177, 178,

184, 595, 684, 707
mass transport, 322, 419, 420
mass transport velocity, 419, 420
mathematical modelling, vii, 373, 645
Matlab, 159, 165, 166, 168, 170, 181, 182, 209–211, 213, 214, 217–219, 380, 695–703, 705–708, 710, 711, 713
MatRANS, viii, 380, 696, 697, 701, 702, 704, 708–710, 712
mean value, 18, 43, 140, 267, 268, 280, 299, 300, 335, 509, 523, 540, 593, 680, 706
mechanism of (sediment or particle) suspension
in open channels, rivers, etc., 65, 126, 130, 131, 133–136, 144
in plunging waves, 478, 518, 519, 521, 548
micro scale(s), xxiv, 174, 176, 180, 191, 194, 195, 204, 233, 234, 685
micro length scale, 175, 225
micro time scale, 177, 181, 182, 184, 707
mixing length, xxii, 97–100, 465, 645, 647, 648, 650–655, 658, 667, 669, 676, 702–704

Navier-Stokes (N.-S.) equations, 27, 28, 45–47, 53, 55, 63, 180, 187, 190, 193, 198, 226, 239, 250, 277, 415, 416, 421, 426, 448, 450, 485, 503, 587, 646, 647, 675
neutral stability curve, 8, 9, 31–33, 35
Nikuradse's equivalent (sand grain) roughness, xxii, 26, 82, 85–87, 91, 101, 239, 260, 283, 284, 314, 323, 356, 359, 438, 654, 709, 710
non-sinusoidal waves, 445
acceleration skewness, 446, 454, 457, 463
velocity skewness, 446, 451–453
normal (or Gaussian) distribution, 161–163, 168, 598, 599, 607, 707

one-dimensional spectrum, 186, 200–202, 204–206, 208, 209, 214, 219, 231
one-dimensional vertical (1DV), 239, 373–375, 377–379, 387, 388, 421, 434, 696, 708
one-equation (turbulence closure) model, 295, 647
one-particle analysis, 592, 606, 607, 609
origin of turbulence, 1, 2
Orr-Sommerfeld equation, 27, 30, 31
oscillatory flow in pipes, 8, 300
over-production of turbulence, 478, 561, 566, 568, 573, 660

pipe flow, xxii, 3, 7, 8, 70, 75, 92, 93, 95, 106, 109, 181, 302, 609, 651, 671
plunging breaking waves, 477, 478, 500, 550, 567, 572, 573, 682
flow and turbulence, 477, 480, 495
mechanism of sediment suspension, 478, 518, 519
plunging solitary waves, 478, 497, 530, 532, 533, 535, 536, 546, 547
sediment transport and morphology, 532, 548
potential flow, 15, 37, 239, 367, 401, 405, 410, 415, 417, 478, 482, 549, 551, 552, 554–557, 559, 561, 568, 573, 660
Prandtl's mixing length hypothesis, 98, 650, 676
Prandtl's second hypothesis, 98, 650, 651
probability density function (p.d.f.), xxii, xxvi, 159–165, 168, 170, 593, 596, 598–600, 606–608, 613, 635, 705, 707

radiation stress, xxiii, 406, 407, 484, 490, 589, 636, 637

re-laminarization, 9, 239, 259, 303–305, 310
Reynolds conditions, 44
Reynolds decomposition, 43, 44, 46, 675
Reynolds equation(s), 43, 46, 48, 66, 68, 70, 97, 99, 653, 654, 675
Reynolds stress tensor, 46, 76, 436, 437, 550, 646, 649, 655, 657, 675
Reynolds' tube experiment, 2
Reynolds-averaged Navier-Stokes (RANS) equations, 46, 239, 373, 374, 377, 388, 421, 445, 454, 473, 549, 558, 567, 573, 629, 646, 657, 661, 662, 669, 696, 708
rough wall, 5, 26, 65–67, 79, 81–85, 90, 91, 95, 100, 101, 103, 104, 116, 117, 120–125, 139, 143, 163, 259, 260, 282, 369, 383–385, 648, 653–655, 664–668, 673, 686
Rouse parameter, xxiv, 497, 613, 621, 622, 632
runup, xxii, 367, 370, 477, 478, 499, 500, 506, 509, 512, 513, 526, 529–532, 534, 535, 537, 541, 543–547, 636

secondary flows (or currents), 76, 297, 414, 438, 445, 474, 475
sediment transport, v, vii, 65, 85, 86, 126, 237, 280, 307, 334, 356, 370, 414, 463–465, 477, 478, 481, 496, 503, 512, 518, 521, 529, 531, 532, 543, 548
  lift force (on sediment particles), xxii, 65, 126, 136–140, 143–145, 276, 519, 524, 530
  particle motions (near the bottom), 65, 126, 134, 135, 330
  suspension, xxiv, 65, 86, 126, 127, 129–142, 144–149, 276, 464, 478, 518, 519, 521–525, 548, 604, 620–623
separation (of flow), 11, 656, 660
settling (or fall) velocity, xxiii, 86, 127, 133, 134, 147, 522, 604, 609–611, 613, 615, 618–622, 632–634, 641–643, 703, 705
settling basin(s), 618, 621–623
shallow water approximations, 407, 482–484, 492, 547
shear-layer instability, 251, 255, 339, 501, 503, 529, 536
Shields parameter, xxiv, 87, 128, 141, 144, 145, 147–149, 497, 521, 623
shoaling, 11, 357, 409, 479, 500, 506, 513, 515, 532, 535–537, 544, 545, 562, 569, 637
skewness, 161–166, 168, 707
smooth wall, 7, 26, 65–67, 71, 73, 76, 78–82, 85, 90–92, 97, 101–103, 116, 117, 120, 121, 123, 124, 260, 315, 328, 381, 382, 384, 527, 651–654, 656, 664–668, 671, 672, 686
solitary wave boundary layers, 3, 6, 7, 244, 246, 331, 334, 336–338, 340, 344–348, 350–352, 354, 386, 533, 543
spectral analysis, 159, 181, 184, 219, 695, 706
stability diagram, 31
standard deviation, 138, 140, 161, 162, 280, 281, 506, 562, 563, 569, 684, 703
steady turbulent boundary layers, 65, 78
Stokes law, 642
Stokes length, xxiv, 5, 8, 242, 243, 278, 279, 287, 298, 300, 301, 345, 414
streaming, 375, 413, 414, 460–465
  beneath propagating waves, 319, 322, 413, 414, 417–425
  due to changing bottom roughness, 329, 414, 434, 439, 442–445
  due to converging-diverging oscillatory flow, 329, 413, 424, 427–432, 434
  due to non-sinusoidal wave

shape, 414, 445, 446, 448, 450–454, 457–459
streaming velocity, 417, 419, 420, 422, 423, 427–432, 434, 442–444, 452–454, 457–459, 461, 462
surf similarity parameter, 12, 409, 480, 500, 516, 517, 526, 529, 534, 546, 559, 567
surf zone, vi, vii, 401, 460, 477, 479–481, 485, 489, 490, 493–495, 516, 517, 549, 561, 562, 566–569, 572, 591, 636, 637, 660
surface roller, 477, 481–483, 485, 489–493, 503
surging breaking waves, 477, 480, 524, 526
flow and turbulence, 478, 524
suspension criterion (or criteria), 127, 140, 144, 147–149
swash zone, vi, 329, 477, 478, 495–499, 530, 531, 543

theoretical wall, 86, 87, 89, 90, 100–102, 105, 124, 125, 655, 666, 668
three-dimensional spectrum, 186, 200–203, 206, 208, 209
time averaging, 43, 44, 118, 172, 227, 300
Tollmien-Schlichting waves, 32, 33
transition to turbulence, 5, 6, 11, 15, 23, 26, 35, 239, 245, 254, 255, 259, 260, 263, 301–303, 327, 330, 345, 346, 351, 388, 542, 662
transitional wall, 82, 84, 97, 100, 109, 116
tsunami, 238, 319, 356–367, 369–372, 434, 530, 661
tsunami wave boundary layers, 238, 356, 357, 359–361, 370, 371
turbulence energy equation models, 645, 647, 655, 656, 680, 681, 683, 685, 687
turbulence intensity, 36, 96, 97, 180, 707
turbulence modelling, vi, xxiv, xxv, 77, 96, 97, 100, 106, 239, 295, 374, 503, 551, 645, 646, 656, 657, 669, 674, 683, 695, 696, 708
turbulence quantities, 76–78, 90, 91, 95, 96, 117, 233, 273, 276, 278, 285, 330, 445, 655, 681, 700
turbulence-induced streaming, 450, 457
turbulent diffusion, xxi, xxiv, 79, 355, 562, 591, 600, 602–604, 606, 610, 611, 615, 616, 618, 624, 627, 636, 650, 659, 701
turbulent diffusion equation, 604, 618, 636
turbulent dispersion, 695
turbulent dispersion equation, 608, 636
turbulent spots, 4–6, 9, 15–17, 19, 21, 23, 24, 26, 35, 37–39, 123, 255, 388, 661
in solitary wave boundary layers, 337, 342–345, 350
in wave boundary layers, 9, 123, 245, 246, 248, 250, 252, 254–257, 259–262, 264, 276, 280, 326
two-equation (turbulence closure) model, 239, 378, 423, 465, 478, 549, 550, 552, 554, 555, 558, 573, 629, 647, 655, 669, 680, 681, 708
stability analysis of, 550, 553, 554, 566

undertow, 460, 477, 478, 481, 485–488, 490–495, 562, 565–567, 569, 571, 572, 587, 636
universal equilibrium range, 191, 193, 195, 685, 707

van Driest (velocity) profile, 100, 101, 269, 271, 353, 653, 700
variance, 161, 162, 165, 168, 200, 201, 214, 215, 219, 225, 228, 229, 232, 593–598, 607, 628, 629, 634, 635, 677, 699, 702, 705, 707
velocity defect, xxi, 286
velocity-defect law, 93–95

viscous sublayer, xxiv, 71, 73, 75, 79, 82, 92, 97, 100, 101, 103, 109–111, 120, 124, 163, 283, 496, 653, 654, 656, 661, 665, 666, 668, 674, 686, 700
von Karman constant, xxi, xxiv, 75, 99, 143, 269, 355, 611, 616, 652
von Karman spectrum, 203, 215, 217
vortex stretching, 195, 196, 198
vortex tubes, 198, 250–254, 337, 339–342, 344, 347, 349, 388
vorticity, xxiv, xxv, 39, 59, 195–199, 414, 445, 473–475, 501, 503, 504, 554, 658
vorticity equation, 196, 414, 445, 473, 474

wall functions, 559, 658, 662, 666–669, 686
wave boundary layers, vii, xxiii, xxv, 2, 3, 5, 6, 8, 9, 12, 26, 44, 123, 237–239, 244, 264, 282, 283, 288–290, 293, 295, 300, 313, 314, 319, 320, 325, 326, 328, 329, 333, 347, 356, 373, 375–377, 382, 410, 413, 414, 434, 435, 438, 440, 442, 445, 446, 452, 454, 460–462, 492, 499, 509, 510, 514, 541, 591, 626, 628, 633, 634, 661, 674, 677, 695–697, 708, 710–712
  apparent roughness, 314, 315, 372
  boundary layer thickness, 238, 240–243, 268, 272, 281–283, 287, 288, 311, 320, 337, 345, 355, 356, 359–361, 364–371, 376, 379, 410, 429, 456, 629, 632, 634
  effect of externally generated turbulence, 329, 330
  in wave flumes, 264, 311, 319, 461, 464
  laminar, 238, 239, 242, 245, 269, 290, 311, 414, 416, 418, 424, 429, 461, 712
  non-uniform, 328, 375, 413, 414, 420, 424, 434
  particle motions, 238, 330
  transition to turbulence, 26, 238, 239, 245, 254, 255, 259, 260, 263, 290, 301–303, 311, 327, 330, 345, 346, 351, 388, 542, 712, 713
  turbulent, 239, 264, 265, 268, 282, 283, 286, 290, 297–299, 311, 319, 326, 373, 420–423, 446, 447, 454, 514, 626, 674, 682, 713
  wall-normal pressure gradient, 245, 247, 276–281
wave drift, 321–324, 404, 405, 461, 463, 482, 492, 493, 636, 637
wave friction coefficient (or factor), xxii, 262, 263, 288–296, 315, 347, 359, 372, 374, 447, 711